Exploring Zynq® MPSoC

With PYNQ and Machine Learning Applications

Exploring Zynq® MPSoC

With PYNQ and Machine Learning Applications

Louise H. Crockett

David Northcote

Craig Ramsay

Fraser D. Robinson

Robert W. Stewart

Department of Electronic & Electrical Engineering

University of Strathclyde

Glasgow, Scotland, UK.

April 2019

This edition first published 2019 by Strathclyde Academic Media.

Cover by Scribble Design. Photography by Kenneth Barlee.

Table of Contents

Foreword

In 2011, Xilinx launched the Zynq product family. This was the first Xilinx SoC family combining an ARM processor system tightly coupled with programmable logic. During that time, we decided to team up with the University of Strathclyde and the team of Professor Bob Stewart and Dr. Louise Crockett to publish *The Zynq Book*, a comprehensive introduction and tutorial to the software and hardware of this innovative architecture.

The overwhelming success of the first publication compelled us to continue the collaboration with the team at the University of Strathclyde, when the second generation of our SoC family, the Zynq MPSoC was introduced.

Today's book, 'Exploring Zynq MPSoC', walks the reader through all the important aspects of the software stack, the multi-processor processing system, and the powerful array of programmable hardware. The MPSoC processor system is equipped with up to four ARM Cortex-A53 processor cores and two ARM Cortex-R5 real time processor cores. In addition one can leverage the FPGA fabric which is extended by powerful arrays of DSP slices and a large provision of distributed memories.

This new architecture makes it especially interesting to free up the processor system from the complex arithmetic and data movement as required in today's emerging AI applications. I really appreciate the special focus in this book on the elaboration of software stack and the programming tools. A special chapter is devoted to the PYNQ programming environment. PYNQ is an open-source project from Xilinx that makes it easy to design embedded systems on the Zynq platform. Using the Python language and libraries, designers can exploit the benefits of programmable logic and microprocessors to build more capable and exciting embedded systems. Part E of the book discusses the implementation of many applications on the Zynq MPSoC platform. This includes a detailed discussion of the FINN-R open-source framework for efficient implementation of neural networks.

This book is an interesting read for a wide spread of embedded system designers, from the software programmers who want an easy software interface and framework for rapid implementation of their machine learning algorithms, to the system designers who want to extract the highest performance by leveraging the programmable hardware capabilities of Zynq MPSoC.

The team at the University of Strathclyde has created an excellent pedagogical book that introduces the reader to the fascinating opportunities of the Zynq MPSoC platform. It is relevant for the first time user as well as the expert who wants to get the most out of the hardware.

I would like to thank and congratulate all the authors who have contributed to this book.

Ivo Bolsens,

Senior Vice President & Chief Technology Officer, Xilinx.

Acknowledgements

The realisation of this book is the product not just of our own efforts, but also of those who have assisted and supported in various ways. We would like to take the opportunity to thank them here.

Firstly, we would like to acknowledge the support of Xilinx in enabling the project, and allowing *Exploring Zynq MPSoC* to become a reality. We would like to acknowledge Patrick Lysaght, Senior Director in Xilinx, San Jose, for his many contributions to this book. Patrick has championed the project since its inception, and has steered many of the strategic decisions relating to content and overall organisation. We also extend a special thanks to Cathal McCabe, Manager of Xilinx University Program in EMEA, who has been our 'book manager' and has been an enormous help throughout — patiently answering questions, giving advice and feedback, and generally helping us out on various fronts. We owe him a great debt of gratitude for all of his inputs and support. Further, we would like to thank Patrick and Cathal for jointly co-authoring Chapter 22 on *PYNQ*.

Also at Xilinx, we appreciate the assistance of those who have given advice and support along the way, or reviewed chapters and given feedback. It has been wonderful to have access to your knowledge and expertise. Our thanks are due to Hugo Andrade, Ivo Bolsens, Sean Fox, Peter Ogden, Parimal Patel, Naveen Purushotham, Yun (Rock) Qu, and Graham Schelle. We also extend special thanks to Giulio Gambardella, Thomas Preusser, Yaman Umuroglu, and Michaela Blott, who contributed Chapter 21 on *Reduced Precision Neural Networks*.

We would also like to thank the team from Craft Prospect Ltd., Phil Karagiannakis, Murray Ireland, and Steve Greenland, for contributing the case study on PYNQ and machine learning for space applications, featured in Chapter 23.

Our colleagues and students at the University of Strathclyde have also been of great help, and we would like to thank them for all of their various support and contributions to the project. In particular, we would like to thank Sarunas (Shawn) Kalade, and Josh Goldsmith, who contributed the *Deep Learning* chapter, and part of the *Software Stacks* chapter, respectively. The following additional colleagues generously gave their time and expertise to review chapters, assist with research, or contribute content: Douglas Allan, Dale Atkinson, Kenny Barlee, Lewis McLaughlin, Andrew MacLellan, Damien Muir, Alan Petrie, Kenneth Stuart, and Rhys Williams. It has been something of a team effort! We would also like to express gratitude to members of our wider research team, who have been generous with their support and encouragement.

Last, and certainly not least, we extend thanks to our various family members and friends, who have supported in their various ways as we worked on this book, with love, encouragement, and patience.

Louise, David, Craig, Fraser and Bob, March 2019.

Exploring Zynq® MPSoC...

Chapter 1

Introduction

Welcome to the book! The pages that follow will provide a comprehensive introduction to the Zynq MPSoC device, an integrated System-on-Chip (SoC) device from Xilinx, which follows on from its predecessor, the Zynq-7000 [7].

The Zynq MPSoC, as it will be known throughout the remainder of the book, is a Multi-Processor System on Chip (MPSoC) [5]. The term MPSoC reflects that it comprises a number of different processing elements, each optimised for particular purposes — for instance, a set of applications processors, real-time processor, and a graphics processor, as well as Field Programmable Gate Array (FPGA) programmable logic. The composition of the device will be expanded upon in great detail in later chapters, but for the time being, we can simply think of the Zynq MPSoC as providing a varied set of the best resources for the job!

Aside from the Zynq MPSoC architecture, design methodologies and software tools are of great importance. Appropriate design methods allow the facilities of the Zynq MPSoC to be leveraged to solve real design problems. With an expanded set of processing elements compared to previous devices, it is important that designers are able to harness the capabilities of the Zynq MPSoC, while achieving desired outcomes in terms of system performance, reliability, cost, power consumption, security, time-to-market... and any other constraints that apply. For that reason, an equally important aspect of this book is to outline options for system development, including design tools and the operating systems that can be deployed on the processing cores. We include a particular feature on the Xilinx SDx tool [6], which enables systems to be described entirely using software code, and then partitioned (under user direction) across the various processing elements available. The SD in SDx stands for *Software Defined* and, as will be discussed further in later chapters, software-based design is becoming an increasingly powerful design method for programmable devices.

Applications of Zynq MPSoC devices are many and varied, and we can anticipate some of the prominent application areas based on prior experience of Zynq-7000, and the expanded set of facilities offered by Zynq MPSoC. These include Advanced Driver Assistance Systems (ADAS), computer vision, 'big data' analytics,

Software Defined Radio (SDR), and high-value monitoring and automation (the Industrial Internet of Things, IIoT).

1.1 Why Should I be Interested?

The market for semiconductor devices traditionally comprises a number of sectors, including logic (fixed and programmable), memory, microprocessors, optical, analogue, discrete components, microcontrollers, sensor systems, and dedicated Digital Signal Processors (DSPs) [4]. Collectively, the global semiconductor market was worth $412.2 billion (US) in 2017, its highest ever level, and an increase of over 20% compared to the previous year [3]. Semiconductor devices enable everything from children's toys, to laptop computers, to control systems for nuclear power generation... to the International Space Station. In short, we can't live without them!

In recent years, system integration has become a strong area of interest. In simple terms, why make the various components needed for a system, and then connect them together afterwards? Surely it would be better to design devices that combine the resources required into a single chip? And thus we arrive at the idea of System-on-Chip (SoC).

Referring to the semiconductor categories identified above, note that SoCs such as Zynq and Zynq MPSoC comprise programmable logic, microprocessors, and memory — the major components normally required in an embedded system. In fact, these devices also incorporate some analogue circuitry, together with arithmetic engines that support Digital Signal Processing (DSP) applications, similar to the functionality provided by a DSP processor. As shown in Figure 1.1, the basic composition of a Zynq MPSoC device is a Processing System (PS), coupled with FPGA Programmable Logic (PL). The two sections are connected via a number of Advanced eXtensible Interface (AXI) interfaces. This high-level structure is very similar to a Zynq-7000 chip.

The main differences, compared to Zynq, are that Zynq MPSoC takes integration a step further, by expanding the selection and number of processors in the PS, expanding the size of the FPGA PL section, and increasing the number and bandwidth of AXI connections between the PS and PL. There are also a number of other enhancements, if we look a little deeper (more on these later).

The need for SoCs is partly driven by the motivation to achieve fast time-to-market in rapidly evolving application areas. Other factors include the reduced engineering effort required to integrate components in the system, minimisation of physical size, and a lowering of power consumption. On the other side of the equation, these relatively complex SoC devices must be tractable to design systems with, and indeed ease of use is constantly improving, due to the development of software design tools and methodologies. Xilinx and partners support a variety of design entry methods and languages, and continue to introduce new features to enable rapid development, including facilities to quickly evaluate different implementation options.

If you are an engineer currently working with programmable logic devices like FPGAs, or on processor-based embedded systems, then the growth in SoCs makes it likely that they will be relevant to you someday very soon. It is worth adding SoC design to your skillset, or at least reading the rest of this book and finding out more about it! If you are a student, then likewise — gaining SoC skills could be very useful for your future career. And finally, if you are a hobbyist, then there are almost infinite possibilities to create cool systems with SoCs! Which, even acknowledging the usual academic and commercial pressures, undoubtedly appeals to students and professionals too.

1.2 The Evolution of Xilinx SoCs — Very Briefly!

Xilinx has traditionally been a *programmable logic* company, specialising in FPGA technology and Complex Programmable Logic Devices (CPLDs, which despite their name can be considered a less complex version of an FPGA). In recent years — since the launch of Zynq-7000 in 2011 — the move towards SoC has seen integration of other building blocks, such that Xilinx now produce devices that are not only composed of programmable logic, but rather programmable logic combined with processors, memory, interfaces, and more.

Since Xilinx invented FPGAs more than three decades ago, and especially in more recent times, there has been an interest in creating flexible embedded systems based on FPGAs. This has been possible due to the availability of 'soft' processors, which can be created using the general purpose programmable logic of an FPGA (rather than by using a dedicated processor chip). Implementing an embedded system in this way is still valid, and enjoys considerable flexibility, but is limited in terms of processor-based performance. In some cases, applications would call for a separate processor chip to be incorporated into a system, with appropriate interfacing to the FPGA.

This led to the development of the Zynq-7000 chip in the early 2010's, which combines the programmable logic of an FPGA with a dedicated, 'hard' processor built in dedicated silicon, and provides fast interconnections between the two parts. The processor in this case is a dual core Arm Cortex-A9 (the 'A' denotes Appli-

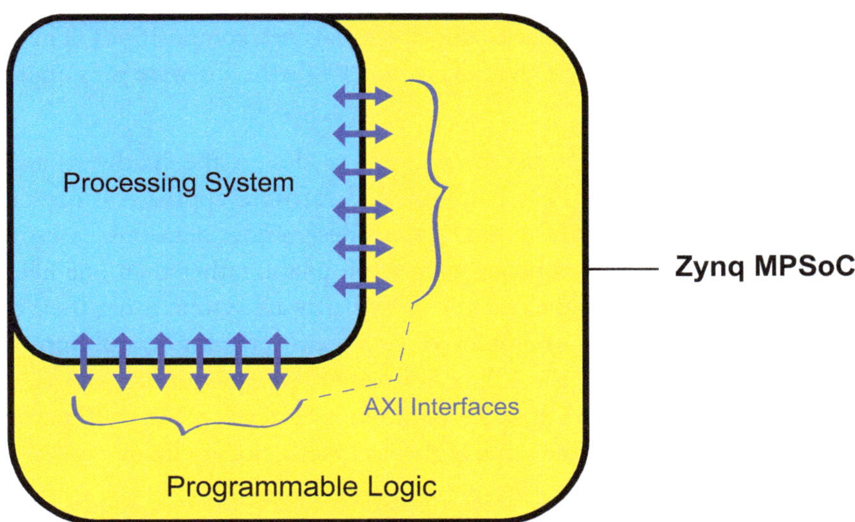

Figure 1.1: A simplified diagram of the Zynq MPSoC architecture

cations processor), which is the same type of processor that can be found in smartphones. This device offers the benefit of enhanced processor capability over the 'soft' processor approach adopted previously, meaning that a complete system could be implemented on a single chip.

Now, in the present day, the Zynq MPSoC takes the SoC concept further, by expanding the set of facilities incorporated into the device. Rather than a dual-core applications processor, there is now the option of a dual- or quad-core, and the processor type is the Arm Cortex-A53, which is 64-bit (as opposed to 32-bit for the A9). There is also a real-time processing system based around two Arm Cortex-R5 cores, a Graphics Processing Unit (GPU), and the option for a Video Processing Unit (VPU) too. The rationale for adding these additional types is that greater performance can be obtained by using processors that are optimised for particular tasks.

It is also worth noting that a new Zynq-7000S has been introduced, with just a single Arm Cortex-A9 core. This means that SoC variations are now available in a selection of different 'sizes' and price points, which makes them suitable for low-complexity, cost-sensitive tasks, right through to very sophisticated systems.

1.3 Design Methods

Now that we have these sophisticated SoC chips, how do we create systems with them? A crucial question!

Actually a number of methods are available, reflecting the composition of SoC systems, the complex evolution of Electronic Design Automation (EDA) tools, and the variety of applications supported. Xilinx and third party software partners are constantly pushing the envelope in terms of making SoC systems as quick, easy, and reliable to design as possible.

We will save a detailed discussion of design methods and flows for Chapter 4, but in the meantime, it is worth outlining simply that the SoC system will comprise (i) a *hardware design*, and (ii) a *software design* (the software design is often referred to as a 'stack', due to its layered composition). The hardware design is mapped to the physical resources available on the SoC device, while the software runs upon one or more of the processors deployed within the system.

Given these fundamental differences, it has been usual to design the hardware and software systems separately, using dedicated tools for each. This 'hardware and software' approach is depicted at the left hand side of Figure 1.2 (the details are abstracted for now). In this design flow, hardware and software development can largely proceed independently, followed by an integration phase, rather than one being dependent upon completion of the other. Engineers generate elements of the hardware system using their choice of tools, with the Xilinx *Vivado* development environment employed for system integration, and implementation onto the target device. Software developers can use the Xilinx Software Development Kit (SDK), or their own choice of development environment if preferred. A possible variation is to develop in a third party tool that leverages Xilinx tools 'under the hood', but even so, the same high-level methodology holds.

More recently, there has been a considerable shift towards software-oriented, hardware/software co-design. Put simply, these tools allow the functionality of the whole system to be described at a high level of abstraction, using software code or a block-based design approach. The functionality is then partitioned across

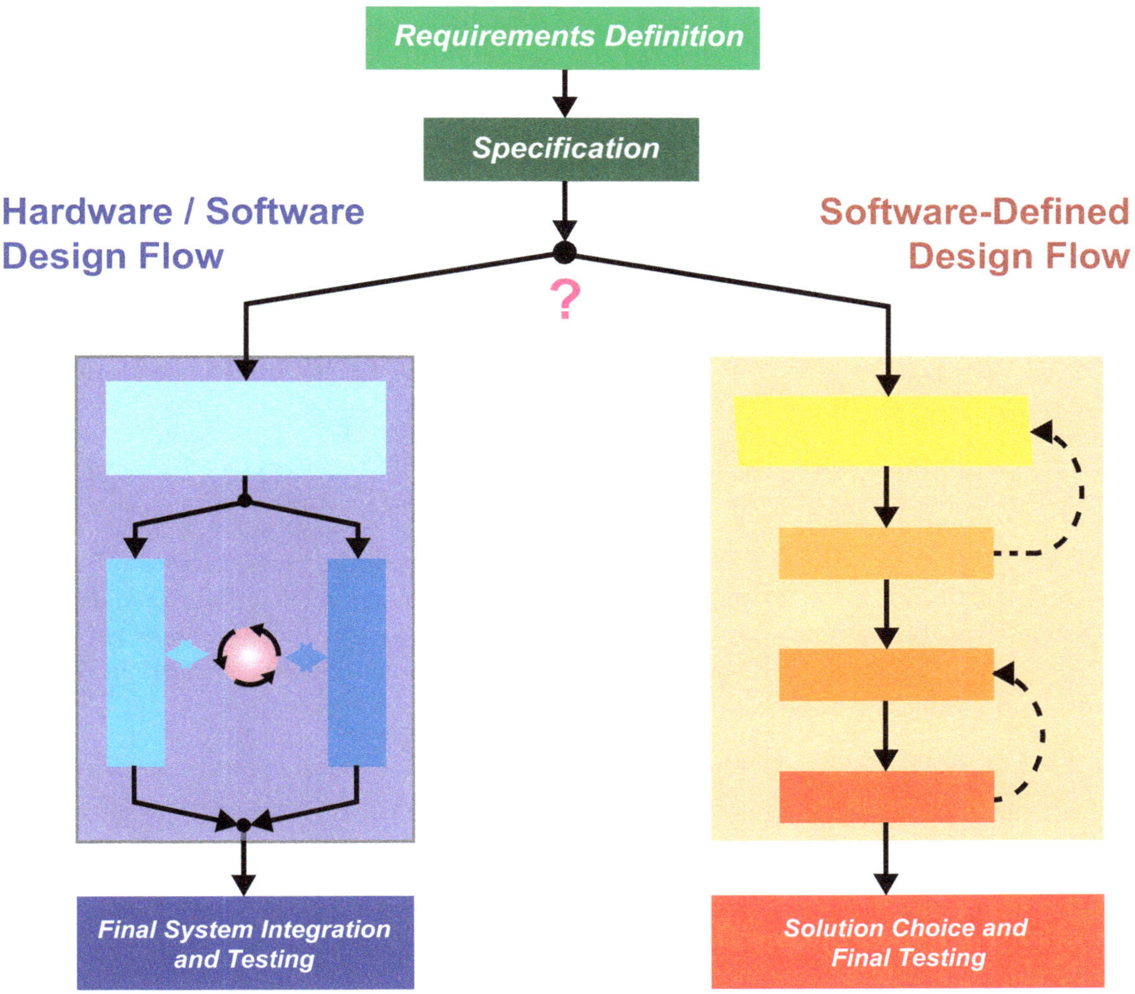

Figure 1.2: Simplified Design Flows for Working with Zynq MPSoC
(left: conventional 'hardware/software' design flow; right: 'software defined' design tool, using SDx)

the hardware and software elements of the SoC, taking account of the capabilities of the resources available, and under the direction of the designer. This method is potentially much faster, because the tools can quickly generate different permutations, depending on how the designer wishes to optimise the implementation, and also as all interfacing between the elements implemented in software and hardware is taken care of automatically. This 'co-design' approach is outlined in the right hand side of Figure 1.2, and reflects the flow of the Xilinx SDx tool, mentioned earlier in this chapter.

We will discuss design methods in more detail in Chapter 4, including filling in the details of the design flows outlined in Figure 1.2. SDx will be the subject of an in-depth review in Part D of the book.

1.4 How to Use this Book

This book is intended as an introduction to the Zynq MPSoC device, and associated design methods, tools, and applications. Hopefully (!) you will find it quite readable and accessible. The book cannot hope to be the answer to all questions however — bear in mind that this device is supported by 1000's of pages of technical literature published by Xilinx, and that is the place to look for the details!

As far as possible, we have tried to address technical issues without assuming too much prior knowledge, although inevitably as the Zynq MPSoC is an advanced, integrated system, there are a number of topics on which it would be beneficial to have some existing background. Useful supporting material is highlighted where appropriate.

1.4.1 Organisation of the Book

The book is organised into five Parts, as outlined below. Each Part consists of a number of chapters on different topics, which collectively represent a theme within the book.

- **Part A: Getting to Know the Zynq MPSoC** — This section of the book provides an introduction to the device, and overview of its architecture, and reviews of design methods and candidate application areas. Part A is positioned at a slightly higher level of abstraction than the rest of the book, and may be particularly suited to technical managers and others who would like to establish an outline understanding of Zynq MPSoC without necessarily delving into the details!

- **Part B: The Zynq MPSoC Architecture in Detail** — Part B expands upon the architecture overview from the Part A with a sequence of dedicated chapters covering different aspects of the device architecture. This includes chapters on the application and real-time processing systems, and facilities for security, safety, and power management, respectively.

- **Part C: Zynq MPSoC Systems Development** — In Part C, the focus is on methodologies and tools for developing systems design on Zynq MPSoC. Here we cover hardware system development, as well as the concept of the software stack and common configurations. Further chapters focus on multiprocessor development, and system booting.

- **Part D: System Design with Xilinx SDx Development Environment** — The fourth part of the book takes an in-depth look at the software-defined design flow, based in the Xilinx SDx development environment.

- **Part E: An Outward Look** — The final, short section of the book takes a wider view, including the SoC 'ecosystem', which represents the spectrum of IP, design tools, hardware development boards, and other available resources that can be leveraged in the design of a Zynq MPSoC system. We also consider some academic case studies, based on previously published work targeting the Zynq-7000 SoC.

1.4.2 Further Sources of Information

At the end of each chapter, you will find a list of references that may be useful for further reading, and web links are provided for convenience (note of course that these are subject to change!). These are drawn from a variety of sources, including a number of reference manuals, tutorials, and other technical literature published online by Xilinx, which are a particularly valuable and authoritative source of information. They provide the deep technical details required to support a design project. You can access the Xilinx support portal at this URL, from which documentation and other useful resources can be obtained:

https://www.xilinx.com/support.html

Note also that particularly important web links are occasionally highlighted within the main flow of the chapter (as above!), rather than via a reference at the end of a chapter.

It is also worth highlighting that this book follows our previous title, *The Zynq Book,* which was concerned with the previously released Xilinx SoC device, the *Zynq-7000* SoC. You may find it useful to refer to *The Zynq Book* for general background, introductory examples, and extended information on certain topics; for instance, the Vivado HLS (high level synthesis) design tool was discussed extensively in *The Zynq Book*, whereas we do not repeat that material in this book. Instead, this book features Xilinx' latest design tool, SDx, which enables hardware/software co-design using a software-based design entry method. Incidentally, SDx leverages the capabilities of Vivado HLS.

To obtain an electronic version of the book, please visit:

http://www.zynqbook.com/

where you can register and download a free PDF copy.

1.4.3 Suggestions for Beginners

If you consider yourself a beginner in terms of SoC design, it would be worth getting hold of the *The Zynq Book* (mentioned in the last section), which includes some introductory material on SoC principles and other relevant background (beginning with *"What is an SoC?"*). Especially given that *The Zynq Book* is available as a free download, we have sought not to repeat the same material again here.

Although Zynq MPSoC systems can be developed using a variety of operating systems, it is anticipated that most designs will feature Linux, and therefore embedded systems development with Linux is a significant topic to be aware of. It is beyond the scope of this book to provide a comprehensive treatment of embedded Linux, but fortunately there are already some very good books available, such as [1].

If you would like to strengthen your background knowledge in computer architectures and operating principles in general, then [2] would also be a useful book to have on your shelf.

1.5 What Next?

Next, we move on to *Part A — Getting to Know Zynq MPSoC.* This section further introduces the MPSoC, including overviews of the device architecture, and the design processes and tools required to create an MPSoC system, and finishes with a discussion of candidate application areas.

1.6 References

Note: All online sources last accessed March 2019.

[1] Christopher Hallinan, *Embedded Linux Primer: A Practical Real-World Approach*, 2nd edition, Prentice Hall, 2011.

[2] David A. Patterson and John L. Hennessy, *Computer Organization and Design: The Hardware Software Interface*, ARM Edition, Morgan Kaufman, April 2016.

[3] Semiconductor Industry Association, "Annual Semiconductor Sales Increase 21.6%, Top $400 Billion for First Time" press release, 5th February 2018.
Available: https://www.semiconductors.org/news/2018/02/05/global_sales_report_2017/annual_semiconductor_sales_increase_21.6_percent_top_400_billion_for_first_time/

[4] Semiconductor Industry Association, *The U.S. Semiconductor Industry: 2017 Factbook*,
Available: http://go.semiconductors.org/2017-sia-factbook-0-0-0

[5] Xilinx, Inc., "Heterogeneous MPSoC" product page.
Available: https://www.xilinx.com/products/silicon-devices/soc/zynq-ultrascale-mpsoc.html

[6] Xilinx, Inc., "SDSoC Development Environment" product page.
Available: https://www.xilinx.com/products/design-tools/software-zone/sdsoc.html

[7] Xilinx, Inc., "Zynq-7000 All Programmable SoC" product page.
Available: https://www.xilinx.com/products/silicon-devices/soc/zynq-7000.html

Part A

Getting to Know Zynq MPSoC

Chapter 2

FPGAs, Zynq, and Zynq MPSoC!

The Zynq MPSoC is an evolution of the Zynq-7000 System on Chip (SoC) — or simply 'Zynq' — the first SoC released by Xilinx. Both of these devices comprise a *Processing System (PS)*, and *Programmable Logic (PL)*, the PL being equivalent to that of a Field Programmable Gate Array (FPGA). As depicted in Figure 2.1, which compares the three device types at a high level, the PS in the Zynq MPSoC is larger and more sophisticated than that in Zynq.

In this chapter, the characteristics of these three device types will be reviewed, and the similarities and differences between them highlighted. We also suggest some reasons for the underlying evolution.

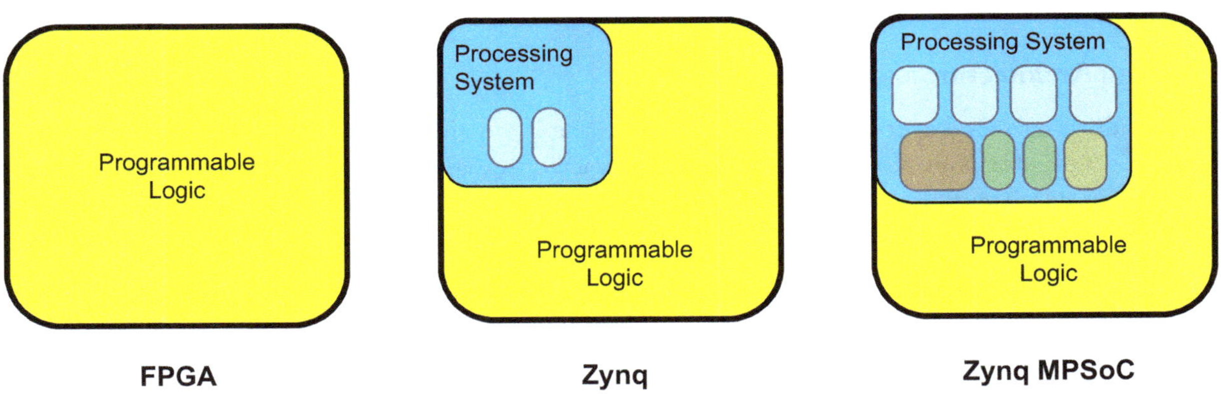

Figure 2.1: High level comparison of FPGAs, Zynq, and Zynq MPSoC

2.1 Technology Timeline

Before proceeding, it is important to mention that Zynq, Zynq MPSoC, and FPGAs are all *current technologies* and complement each other. The 'best' choice depends on the nature of the design being created. All three device classes have their own distinct balance of features, which makes each of them particularly suitable for a range of different tasks and applications.

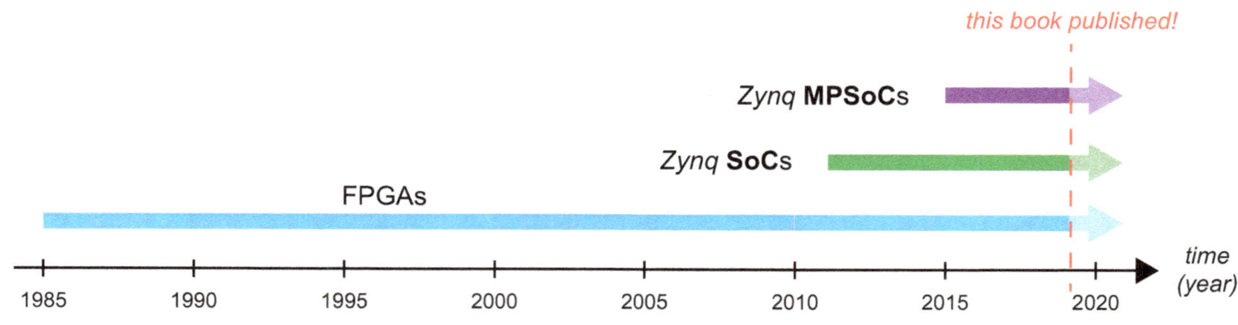

Figure 2.2: Timeline of FPGA, Zynq and Zynq MPSoC introductions

Many readers of this book may be acquainted with the Zynq, and so we begin our review there, which allows comparisons to be drawn with the newer Zynq MPSoC device.

FPGAs, on which the PL section of both SoCs are based, are covered later in the chapter.

2.2 The Zynq-7000 SoC

The Zynq-7000 SoC was the first SoC device released by Xilinx, combining FPGA-based PL with an Arm-based PS. In this section, we will briefly review the architecture and features of Zynq; the reader is directed to our previous book for extended details [3].

2.2.1 Zynq Architecture and Features

To provide a high level overview of the architecture, the Zynq architecture has two parts: the PS and the PL, with a set of interconnections between them, as depicted in Figure 2.3. These interconnections are based on the Advanced eXtensible Interface (AXI) standard, an on-chip communications standard developed by Arm [1].

The rationale for coupling a processing system with PL is to provide a dedicated, optimised resource for running the software-based components of an embedded system, in particular operating systems and software applications, while retaining all of the benefits of FPGA logic (in particular, its great parallelism and reconfigurability.) The AXI interfaces form the connections between the two sections. AXI is a standard that is optimised for SoC applications.

The relative sizes of the PS and PL in Figure 2.3 are not drawn to scale; indeed there are several devices in the Zynq range, each with an identical PS architecture, and different sizes of PL.

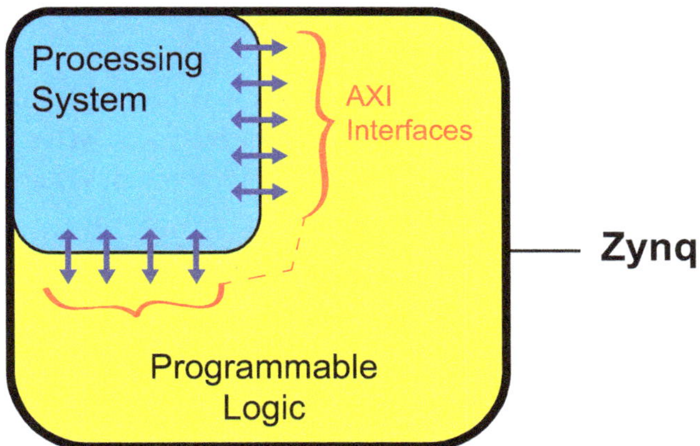

Figure 2.3: High level block diagram of the Zynq SoC device (from 'The Zynq Book' [3])

As will be noted in more detail later, the differences between the Zynq architecture and Zynq MPSoC extend to both parts of the device architecture: PS and PL.

Zynq Processing System

The Zynq PS, indicated by the blue section in Figure 2.3, comprises an application-grade processor along with other components. These are outlined below (see [3] for detailed information).

- **Application Processor Unit (APU)** — This unit includes a dual-core Arm Cortex-A9 processor, along with 256KB of on-chip cache memory.
- **Interconnects and memory interfaces** — These enable communication between the PS and PL, and with external memories, respectively.
- **I/O Peripherals** — A set of integrated peripheral interfaces, covering common standards such as USB, UART, SPI, I2C, etc.

The PS in the Zynq has two processing cores, which gives the designer the option to use both cores for the same operating system, or to use a different operating system on each. Later, we will compare the Zynq PS with the PS of the Zynq MPSoC architecture, and note the expansion of functionality in the new device — in particular, the Zynq MPSoC has up to six processing cores!

More recently, a 'lightweight' version of Zynq — the *Zynq-7000S* — was released, featuring a single Arm Cortex-A9 processor [23].

Zynq Programmable Logic

The PL part of the Zynq is based on either the *Artix-7* FPGA fabric (for the smaller devices in the range) or the *Kintex-7* FPGA fabric (for the larger ones). These two variations represent members of Xilinx' 7-series FPGAs, which also includes the higher performance *Virtex-7*.

Like the equivalent FPGAs, Zynq PL includes DSP48*x* slices (resources for high speed arithmetic), Block RAMs, high speed transceivers, and integrated communications blocks, as well as general purpose logic. Later in this chapter, Section 2.4.3 provides more information about these various elements of modern FPGAs.

Interfaces Between the PS and PL

There are nine interfaces between the two regions of the Zynq. Four of the interconnections are designated as 'general purpose', four as 'high performance', and the remaining interconnection is the 'Accelerator Coherency Port' (ACP), which provides a direct route between the Application Processing Unit (APU, which resides within the PS), and the PL.

2.2.2 Zynq Devices

To provide a brief comparison of the devices in the Zynq-7000 family, consider Table 2.1, which summarises the key features and parameters for each chip (note that the single-core Zynq-7000S family is not included in the table, but equivalent details can be found in [23]).

The PS is identical in all Zynq-7000 chips, the only difference being the maximum supported clock frequency. The PL is similar across the range, with the lower-end devices featuring Artix-7 based logic, and the remainder adopting logic from the Kintex-7. The PL dimensions and numbers of specialist resources such as DSP slices, Block RAM memories, and input / output resources also vary.

2.2.3 Zynq Use Models

Designers may adopt Zynq having previously used FPGAs, or having worked with processors, or both. At the time of its introduction, the selling point of Zynq was to offer a solution for implementing processor-based tasks such as running software and an operating system, *and* FPGA-based processing, on a single device, with high-end performance and high capacity interconnections between the two components.

If previously only an FPGA would have been used, then adding a dedicated hard processor offered increased performance (compared to a 'soft' processor, constructed from general purpose PL resources); while on the other hand, if previously a processor was used in isolation, the presence of FPGA logic enabled certain tasks to be implemented in hardware, providing acceleration and freeing up the processor for other tasks. Systems that previously required both a dedicated processor and FPGA could be reduced from two physical devices to one, with the associated savings in interfacing effort, power consumption, bill of material costs, and so on.

These considerations are discussed further in [3], specifically for the Zynq-7000. However, we find that similar factors motivate the use of the Zynq MPSoC. As described over the coming pages, the Zynq MPSoC

Table 2.1: Comparison of devices in the Zynq-7000 SoC range [23]

	Z-7010	**Z-7015**	**Z-7020**	**Z-7030**	**Z-7035**	**Z-7045**	**Z-7100**
Processor	Dual core Arm Cortex-A9 with NEON and Floating Point Unit (FPU) extensions						
Max. processor clock frequency[a]	866MHz			1GHz			800MHz
Programmable logic	Artix-7			Kintex-7			
No. of Flip-flops	35,200	96,400	106,400	157,200	343,800	437,200	554,800
No. of 6-input LUTs	17,600	46,200	53,200	78,600	171,900	218,600	277,400
No. of logic cells	28K	74K	85K	125K	275K	350K	444K
No. of 36Kb Block RAMs	60	95	140	265	500	545	755
Total amount of Block RAM	2.1Mb	3.3Mb	4.9Mb	9.3Mb	17.6Mb	19.1Mb	26.5Mb
No. of DSP48E1 slices [16]	80	160	220	400	900	900	2,020

a. The maximum supported processor clock frequency depends on speed grade.

provides further integration of processing elements, expanding the PS part of the Zynq to include a real time processing engine, graphics processor, and video codec (on selected devices), along with a more capable application processor.

2.3 The Xilinx Zynq MPSoC

The Zynq MPSoC represents an extension and expansion of the architecture, compared to the Zynq. Although composed of the same high-level elements (a PS and PL, interfaced using AXI interconnects), the PS is more sophisticated, and the PL is updated from Xilinx' *7-Series* to *UltraScale+* FPGA architecture. Across most of the Zynq MPSoC range, the size of the PL is also larger than Zynq series devices.

2.3.1 The Release of Zynq MPSoC

The Zynq MPSoC was first announced in 2015, four years after the initial announcement of Zynq. Over the intervening period, Zynq had seen wide adoption across a variety of application areas, and through this, the need for fully integrated 'FPGA+processor' solutions had been well-established.

The Zynq MPSoC offers an enhanced PS, along with larger PL (a number of different sizes are available). Sub-families have been established within the Zynq MPSoC range to cater for different types of applications and their requirements, particularly in terms of PS resources. These sub-families are distinguished by a two-letter designation: CG, EG, and EV, where the meanings of the two characters are as listed in Table 2.2.

Table 2.2: Zynq MPSoC Sub-Family Designators

Processor System Identifier		**Engine Type**	
C: Dual APU, Dual RPU	**E**: Quad APU, Dual RPU, Single GPU	**G**: General Purpose	**V**: Video

The Zynq MPSoC does not replace the Zynq, but rather it offers an expanded and enhanced solution of similar form. Zynq devices will continue to be the appropriate choice for lower cost and less complex systems.

2.3.2 Zynq MPSoC Architecture and Features

The Zynq MPSoC comprises an expanded set of resources compared to Zynq, with three variations (i.e. the CG, EG, and EV sub-families). The main features of the PS and PL are outlined in Table 2.4. More detail about each of these features will follow in later chapters, when we look at the architecture in detail.

Another important development in the Zynq MPSoC architecture is its power management capabilities. Devices are partitioned into four separate power domains, which can be operated individually. This means that unused parts of the device can be powered down when they are not required, improving overall power efficiency. Security features have also been augmented.

Additionally, it should be highlighted that the Zynq MPSoC benefits from the UltraScale+ FPGA architecture (to be reviewed in Section 2.4.3), which includes certain improvements over the 7-series architecture used for Zynq. Two particularly notable differentiating features are the inclusion of UltraRAM, and the further development of the DSP slice to the DSP48E2 (the DSP48E1 was included in the Zynq).

Devices with several different sizes of PL exist, some which are provided across all the CG, EG, and EV sub-families. The smallest and largest devices are highlighted in Table 2.3. By comparison with Table 2.1 on page 15, note that the Zynq MPSoC range includes devices with considerably larger PL regions than the Zynq series. This makes Zynq MPSoC the better choice when extensive use needs to be made of the PL, for instance for hardware-acceleration of a set of complex algorithms as part of an embedded systems design.

Table 2.3: Basic Resource Comparison (PL of the Smallest and Largest Zynq MPSoCs)

Resource	**Smallest (ZU2*xx*)**	**Largest (ZU19*xx*)**
Flip-flops	94,464	1,045,440
Lookup Tables (LUTs)	47,232	522,720
Maximum number of user Inputs/Outputs[a]	252	668

a. Excluding high speed serial interfaces.

Table 2.4: Overview of Feature Highlights in Zynq MPSoC, by sub-family [24]

Feature	Description	Sub-Family (see Table 2.2)		
		CG	**EG**	**EV**
Application Processor Unit (APU)	Based on the Arm Cortex-A53 architecture, for running OSs and software applications. Either 2 or 4 cores (see right)	2 cores	4 cores	4 cores
Real-Time Processing Unit (RPU)	Arm Cortex-R5 cores for real-time processing tasks	2 cores	2 cores	2 cores
Graphics Processing Unit (GPU)	Dedicated graphics support	-	Included	Included
Video Codec Unit (VCU)	Implemented as a hard IP in the PL, the VCU provides support for H.265 [7] and H.264 [8] video compression standards	-	-	Included
Configuration and Security Unit	Secure boot functionality, Arm TrustZone® support, and voltage and temperature monitoring	Included	Included	Included
Platform Management Unit	For managing power, security, and functional system safety	Included	Included	Included
Memory	256KB on-chip memory, along with interfaces for external memories, which support several memory types	Included	Included	Included
Direct Memory Access (DMA) Controller	Two 8-channel DMA controllers, one each in the low- and full-power domains	Included	Included	Included
High performance interfaces (PS)	PCIe Gen2 USB3.0 SATA 3.1 DisplayPort Gigabit Ethernet	Included (all)	Included (all)	Included (all)
Integrated IP blocks (PL)	PCI Express 150G Interlaken 100G Ethernet MAC	Included[a] - -	Included[a] Included[a] Included[a]	Included[b] - -

a. Included in a subset of devices.
b. Included in all devices.

The various Zynq MPSoC chips also contain varying amounts of other PL resource types (DSP48*x* slices, Block RAMs, etc.) — please refer to the detailed product table for full information [24].

2.4 FPGAs

Of the three device types covered in this chapter, FPGAs are the longest established and the basis of the PL element of the Zynq and Zynq MPSoC devices. Therefore, we will begin our review with a condensed review of FPGA development, followed by some notes on applications, and then a review of the architecture and properties of current FPGA technology.

2.4.1 What is an FPGA?

For readers who are new to the area, it may be useful to provide a brief introduction to FPGAs. FPGA stands for Field Programmable Gate Array. The 'Gate Array' part of the acronym reflects that FPGA devices were originally composed of arrays of logic gates (in fact, the term 'gate array' is no longer strictly true — modern FPGAs are composed not just of simple gates, rather they contain a selection of reconfigurable circuit elements). We will discuss the architecture in more detail a little later. 'Field Programmable' reflects that FPGAs are programmable after manufacture ("in the field") by systems developers and end users, and indeed they can be reprogrammed to implement new hardware functionality as often as required.

FPGA vendors such as Xilinx provide not just the physical devices, but also development tools with which to develop designs for FPGAs, and ultimately program them. To aid productivity, there are also pre-verified cores (referred to as Intellectual Property (IP), reference designs, documentation, and so on. Support also extends, in some cases, to third party software companies and design houses. Zynq and Zynq MPSoC devices share and extend the suite of tools and resources used for FPGA design.

2.4.2 The Development of FPGAs

FPGA architectures have developed in a number of ways since the early days, when an FPGA comprised as few as 64 flip-flops and 3-input Lookup Tables (LUTs, for implementing logic functions). Devices have grown in size, to include far greater numbers of logic elements, and the architecture of these logic elements has evolved. FPGAs are capable of operating at higher frequencies, and of consuming less power. Further, a number of specialist resources have been incorporated, including high speed memory, and support for arithmetic, clocking, and connectivity [15].

Focussing on scale for the moment, consider Figure 2.4. This diagram summarises the expansion of FPGAs in terms of 'logic cells', which are a measure of logic density, slightly abstracted from low level elements to take account of differences between architectures. A snapshot is given of the most modern devices at five-year intervals over the last fifteen years, and these are compared with the first FPGA (from 1985). It is clear that today's FPGAs dwarf those of even 10 years ago, never mind the earliest devices! Notice too that the first FPGA (the XC2064), is represented by the tiny dot on the left hand side of Figure 2.4. In recent years especially, there has been a choice of premium, mid-range, and low-end devices, denoted in the diagram by the red, green, and blue coloured boxes, respectively.

A numerical comparison between the XC2064 and today's largest available device (in terms of logic cells), the Virtex UltraScale+ VU13P device, is presented in Table 2.5. This comparison does not include some of the more advanced features that have been introduced into the architecture in the intervening period.

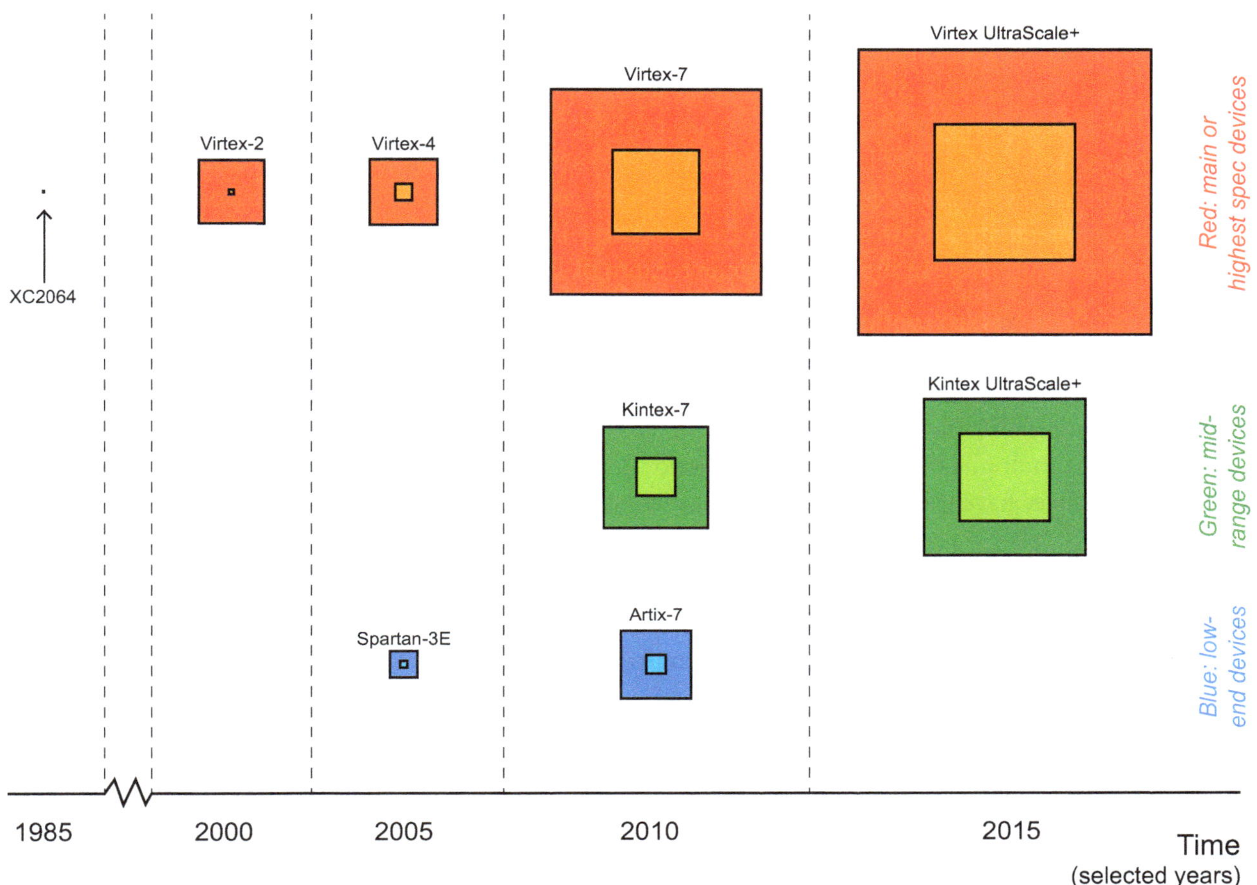

Figure 2.4: Comparison of relative FPGA sizes (selected years and generations), in terms of logic cells (note that inner box represents smallest device, and out box represents largest device, for each generation shown)

Table 2.5: Basic Resource Comparison (Xilinx' first FPGA, versus latest-and-largest FPGA)

Resource	XC2064 (1985)	VU13P (2015)
Flip-flops	64	3,456K
Lookup Tables (LUTs)[a]	64	1,728K
Logic cells[b]	64	3,780K
Maximum number of user Inputs/Outputs	38	832

a. Modern FPGAs have 6-input LUTs; earlier devices had 4-input LUTs.
b. A 'logic cell' is defined as one 4-input LUT and 1 Flip-flop. Reflecting the increased logic density of newer devices, a 6-input LUT and 2 Flip-flops is regarded as ~1.6 logic cells.

The basic architecture of Xilinx FPGAs has remained a two-dimensional array of simple digital logic elements, grouped into CLBs. Each CLB comprises a small number of flip-flops and Lookup Tables (LUTs), where the LUTs are capable of implementing Boolean logic functions, as well as small memories and shift registers. The precise composition of a CLB has evolved over time, with one CLB in modern technology representing a greater amount of logic than in older devices. Retaining the original terminology, CLBs are still connected together with *programmable interconnects* and *switch matrices*, although improvements have been made to this connectivity infrastructure, too. You can see an indication of the array structure of CLBs, switch matrices and programmable interconnects at the bottom of Figure 2.5.

FPGA architecture has continued to evolve, in response to application requirements. Larger memory blocks ('Block RAMs', and more recently, 'Ultra RAMs') provide dense, high speed memory capabilities — allowing, for instance, a significant amount of video data to be stored on the device. Dedicated multipliers were introduced around the year 2000, and these later evolved into integrated arithmetic blocks (DSP48*x* slices) capable of multiplication, addition/subtraction, and other logical functions. Since their introduction in 2004, DSP48*x* slices have themselves developed, with the *x* suffix denoting that several variations exist.

Support for high speed interfacing in the form of 'hardened' IP blocks (i.e. functional blocks physically implemented in silicon on the device) was integrated into select FPGAs, along with high speed serial interface blocks. As FPGA technology has found extensive use in communications infrastructure, data centres, and high performance and cloud computing, these resources are highly valuable.

As is highlighted by Table 2.6, quite aside from their larger size, the latest FPGAs include significant specialist resources and functionality.

Table 2.6: Advanced Resources on the example UltraScale+ FPGA

Resource	VU13P
DSP48*x* arithmetic blocks	12,288
Block RAM memory capacity	94.5Mb
Ultra RAM memory capacity	360Mb
High speed serial interfaces	128
Clock management resources [19]	16 Clock Management Tiles (CMTs)
Integrated 'hard' IP blocks	PCIe, 100G Ethernet, 150G Interlaken [6], System Monitor

In the next section, we will go on to consider the architecture of UltraScale+ FPGAs in a little more detail.

2.4.3 The Modern FPGA Architecture: *UltraScale+*

Modern FPGAs are essentially a two-dimensional array of elements, and in that sense they are similar to the early devices. The composition of the array is, however, much richer in terms of resources; and of course devices are larger, as was highlighted earlier.

Figure 2.5 is an indicative resource layout for an *UltraScale+* FPGA. At a high level of abstraction, the FPGA device layout comprises vertical regions containing different types of resources. The majority of the device is allocated general purpose logic, i.e. CLBs, composed primarily of LUTs and flip-flops. Block RAM and Ultra RAM memory blocks, as well as DSP48*x* arithmetic slices, are arranged in single or double columns on the device, forming thin vertical 'stripes'.

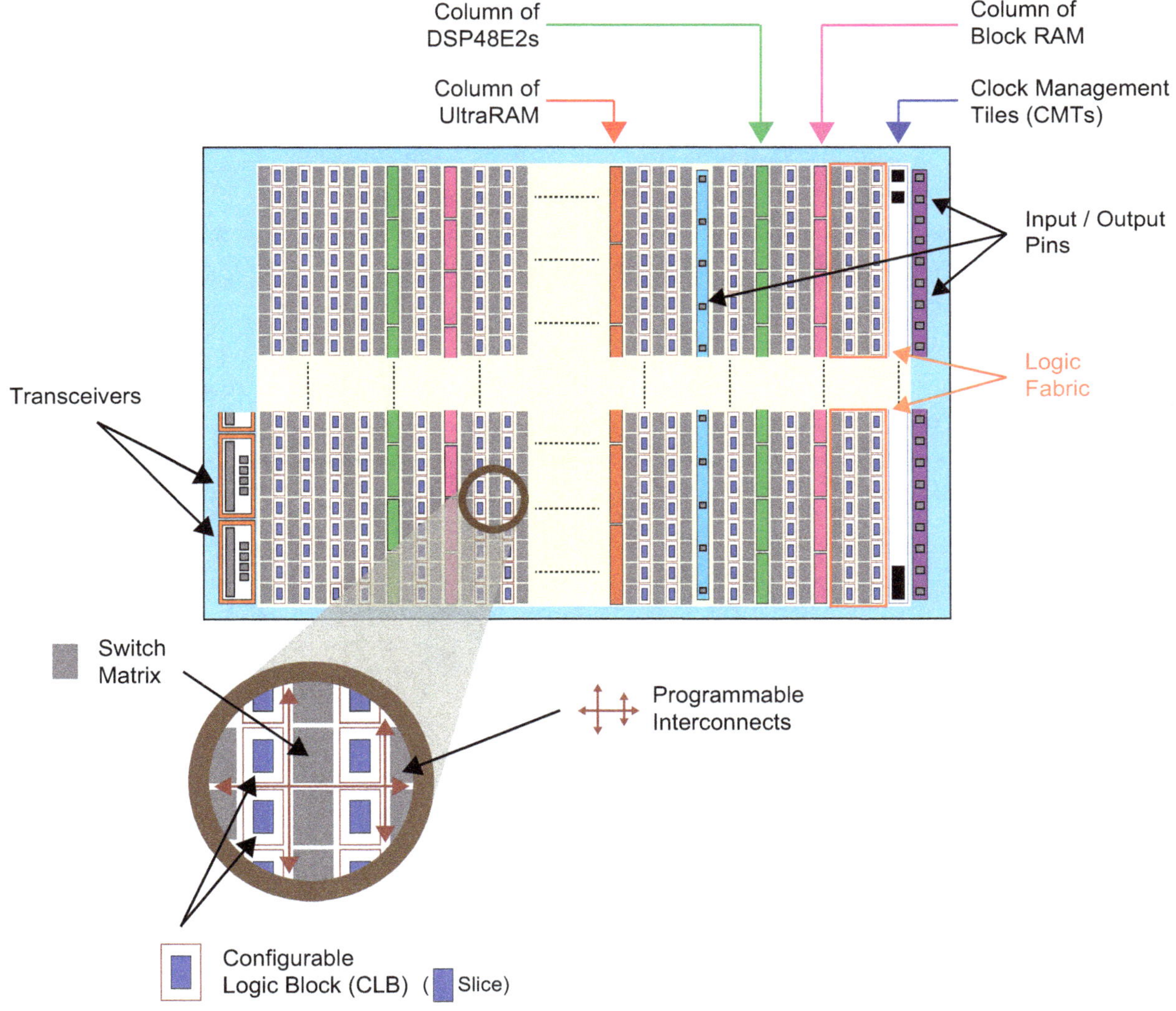

Figure 2.5: Example UltraScale+ layout

In terms of interfacing, Input/Output Blocks are arranged in banks and formed into columns within the main array of resources. IOBs can support a variety of interfacing standards [20]. Additional connectivity is provided in the form of high speed serial transceivers [17], which are normally situated in groups of four at the edges of the FPGA. Close to these, dedicated blocks are provided to support selected communications standards (indicated in Table 2.6). There are also additional resources present in the FPGA structure for configuration, clock management, and system monitoring.

DSP Functionality

Digital Signal Processing (DSP) relies heavily on fixed point multiplication and addition arithmetic. Common DSP tasks such as Finite Impulse Response (FIR) filtering, and the computation of Fast Fourier Transforms (FFTs), are constructed purely from adder/subtractors, multipliers, and sample delays.

Support for these operations is provided via the DSP48*x* slice in Xilinx FPGAs (specifically, the *DSP48E2* variant in UltraScale+ devices). A simplified block diagram of this slice is provided in Figure 2.6, showing the arithmetic operations and wordlengths, but omitting some of the supplementary functionality, such as delay elements, signal paths, and multiplexers. DSP48E2 slices can be cascaded together (without the requirement for fabric resources to be used), in order to create Finite Impulse Response (FIR) filters, or Fast Fourier Transform (FFT) structures, for example. Where longer wordlengths are needed than are available on a single DSP slice — for instance to realise a very wide adder of up to 96 bits — two or more DSP48E2 slices can be combined.

In addition to arithmetic functions, DSP48E2 slices can be used for barrel shifting, pattern detection, and other logical operations. Full information on the DSP48E2, including details of improvements compared to previous DSP48*x* slices, can be found in [21].

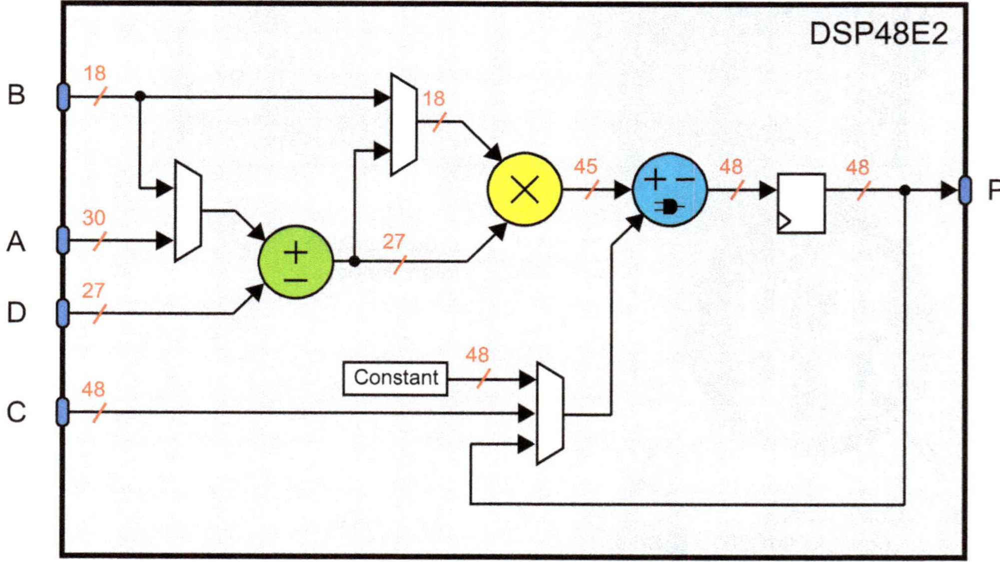

Figure 2.6: Simplified block diagram of the DSP48E2 slice for high speed arithmetic

Memory Support

Memories can be implemented on the FPGA using CLB resources, which is typically the preferred method for storing small amounts of data. For larger memories, there are Block RAMs (capable of storing 36Kb, or acting as two smaller 18Kb sections), and in UltraScale devices, Ultra RAMs with even greater storage capabilities (288Kb each). Larger memories can be created by combining Block RAMs or Ultra RAMs.

Block RAMs and Ultra RAMs are implemented as dedicated physical blocks on the FPGA, as opposed to being constructed from general purpose, low level logic elements. They are capable of high performance operation, running at the maximum clock frequency supported by the device. Ultra RAMs can also be powered down if not used by the current configuration, or put into sleep mode when not needed by the operating design for an extended period. Further details of Ultra RAM can be found in [18].

Taking the alternatives into account, there are four possibilities for data storage, as shown in Figure 2.7.

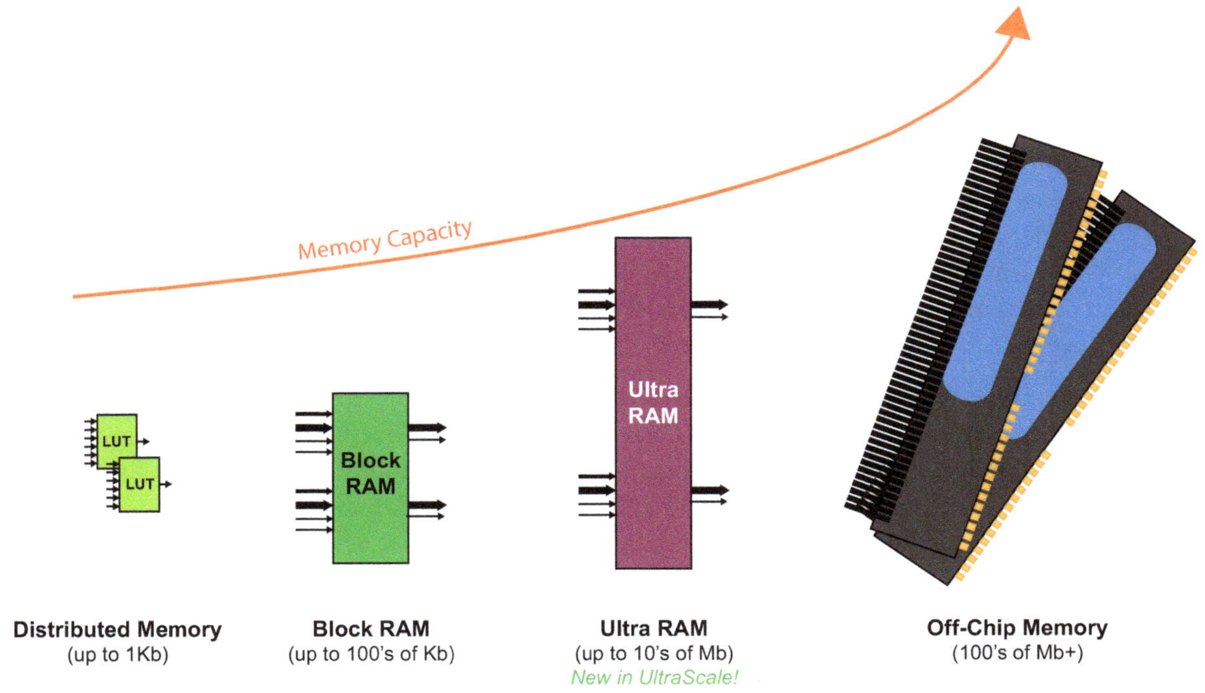

Figure 2.7: Comparison of FPGA memory types

With larger devices, and particularly with the introduction of Ultra RAM, there is the potential to store more data on-chip than in previous generations. This is advantageous because it reduces or eliminates the requirement for off-chip memory, and the implied additional system cost, interfacing effort, power consumption, and latency, along with performance limitations. Each memory type has characteristics that make it the best choice for certain tasks. Depending on the design method adopted, the designer may explicitly target particular memory resources, or allow the synthesis tools to make these decisions.

2.5 Comparison and Discussion: FPGA *v* Zynq *v* Zynq MPSoC

This section will summarise the differences between FPGAs, Zynq, and the Zynq MPSoC, with respect to their architectures, power consumption and performance, and their features for embedded systems implementation. Candidate applications that are particularly well-suited to these devices will also be mentioned. Further discussion of design methods, embedded systems implementation and applications will follow later in the book, focussing on the Zynq MPSoC.

2.5.1 Architectures

In comparing the architectures of FPGAs, Zynq, and Zynq MPSoC devices, we can summarise that three key differences stand out:

- Zynq and Zynq MPSoC both provide a hard processor, whereas FPGAs do not.
- The Zynq MPSoC's PS is larger, more highly specified, and more diverse than Zynq's.
- The largest FPGAs offer more PL than the largest Zynq and Zynq MPSoC devices.

Zynq was Xilinx' first SoC device, combining an application grade processor with FPGA logic. The Zynq MPSoC represents an evolution of Zynq that provides a wider range of processing resources; for instance, Zynq MPSoC offers real-time processors and a graphics processor, as well as an application processor. On the other hand, Zynq has an application processor only. FPGAs do not include a hard processor, although one or more 'soft' processors may be incorporated into an FPGA-based system (i.e. constructed from general purpose logic — see Section 2.5.3 for more discussion on this topic).

All three device types include PL. The architecture of the PL has evolved over the 30+ years since FPGAs were first released, with many different generations having been developed. On the other hand, Zynq and Zynq MPSoC feature regions of PL regions linked to particular generations of FPGA logic. Specifically, the Zynq adopts 7-series logic, while Zynq MPSoC has UltraScale+ logic. A number of differences exist in terms of CLB layout, DSP and memory facilities, connectivity, and clocking and power features.

A quick comparison of resources available is provided in Table 2.7. We can see that the latest FPGAs, while not including a processing system, provide the largest amount of PL — approximately three times more than the largest Zynq MPSoC. It is also notable that Zynq MPSoC devices include a much larger PL section than Zynq, as well as expanded capabilities in the PS.

In FPGAs and the PL section of SoC devices, interfacing resources are comprised of general purpose I/O and high-speed serial transceivers, complemented by hardened IP resources to support standards such as Ethernet and PCI Express. The PS section of Zynq and Zynq MPSoC devices provide further standard interfaces (e.g. CAN, I2C, USB etc.) as hardened resources. The Zynq MPSoC includes some hardened interfaces not present on the Zynq, for instance the Interlaken interface is included as a hardened block on selected Zynq MPSoC chips.

.

Table 2.7: Quick resource comparison between FPGAs, Zynq-7000 SoCs and Zynq MPSoCs

Section	Resource Type	UltraScale+ FPGAs	Zynq-7000 SoCs	Zynq MPSoCs
Processing System	Application processor	n/a	Arm Cortex-A9 (dual core, 32-bit)	Arm Cortex-A53 (dual or quad core, 64-bit)
	Realtime processor	n/a	-	Arm Cortex-R5 (dual core, 32-bit)
	Graphics processor	n/a	-	Arm Mali-400[a]
Programmable Logic	Video processor	n/a	-	H.264 / H.265[b]
	Logic cells (max.)	3,780K	444K	1,143K
	DSP slices (max.)	12,288 (DSP48E2)	2,020 (DSP48E1)	3,528 (DSP48E2)
	Memory types	Distributed BlockRAM, UltraRAM[c]	Distributed, BlockRAM	Distributed, BlockRAM, UltraRAM[c]
	Total BlockRAM & UltraRAM memory	(up to) 454.5Mb	(up to) 26.5Mb	(up to) 70.6Mb

a. In EG and EV devices only
b. In EV devices only
c. In selected devices

2.5.2 Power Consumption and Performance

Considerable progress has been made to reduce power consumption over the years, at the same time as increasing performance (i.e. maximum processing frequency, or 'speed'). As reported in [15], there was more than a 1000-fold reduction in energy consumption (measured per logic cell) between the introduction of Xilinx FPGAs in 1985, and the release of 7-series devices in 2011. Over the same period, performance increased by approximately a factor of 100. It is useful to make a quick comparison for our devices of interest.

Sources of Power Consumption

The power consumption of programmable devices is naturally higher than fixed-function equivalent ICs, due to the underlying architecture required to facilitate programming and hold the device configuration.

In FPGA terms, we usually refer to two elements of power consumption, *static* power and *dynamic* power, which can be defined as follows:

- **Static power** — This comprises power required for the chip to operate, in terms of holding its configuration. Static power occurs due to transistor leakage current (that is, current that passes through the transistor even when it is 'off'), and increases with the size of the device, i.e. as the number of transistors rises. Static power consumption also varies with process technology, voltage, and operating temperature [22].
- **Dynamic power** — The additional power arising from operation of the design on the chip, due to switching activity. It is frequency dependent, e.g. a flip-flop will consume more power if toggling at 200MHz than 100MHz. Dynamic power consumption can vary over time, depending on the activity of circuit elements, and also by the voltage level supplied, and the logic and routing used to implement the design. Engineers can influence dynamic power by optimising their design for low power consumption, for instance by ensuring that circuit elements are not clocked at a higher rate than is required [22].

The PS of Zynq and Zynq MPSoC also have associated power requirements.

Improvements in Power Consumption

Advances in process technology, i.e. the use of ASIC manufacturing processes with progressively smaller feature sizes, accounts in part for the above noted improvements in power consumption and performance. This is largely in line with Moore's Law (that is, the prediction made in 1965 by Gordon Moore, co-founder of Intel, that the number of transistors on a chip would roughly double every two years). It is now widely speculated, however, that the end of Moore's Law may be close [2], [5]. Manufacturing at smaller process geometries is becoming more difficult, and more expensive.

In FPGA terms, performance gains can still be achieved as the process geometry shrinks, but the additional leakage current increases static power consumption [4], [10]. This restricts the potential for further FPGA / SoC power reduction to be achieved purely by making the same style of transistors in smaller dimensions. A shift in the structure of transistors, from the planar style to 'FinFET', has however brought benefits. Xilinx adopted FinFET for the first time in its 16nm devices, which includes UltraScale+ FPGAs and the Zynq MPSoC [13].

In response to manufacturing trends slowing, the focus has shifted towards architectural improvements, and the development of design tools and methodologies that enable systems to be realised more power-efficiently.

In the UltraScale and latterly UltraScale+ FPGA series, new architectural features have been introduced to reduce power consumption, such as improvements to the CLB architecture, DSP48*x* slices and memory structures, as well as a more area-efficient interconnect. There are also device options that can operate at lower core voltages, which allows the designer to trade off power consumption against performance [14]. Techniques such as clock gating (i.e. ensuring that unused sections of the device are not actively working and consuming energy) also play a part in reducing power consumption. An 'intelligent' clock gating feature was introduced with 7-series FPGAs and Zynq, to apply clock gating automatically, without the need for designers to manually incorporate this into their systems [11]. Xilinx design tools are updated regularly to exploit the architectural developments in new generations.

Specifically in the Zynq MPSoC, a Platform Management Unit (PMU) enables control over power domains and the various processing engines that comprise the PS. As most designs will not use all of the features available, power savings can be achieved by turning off those not in active use [12]. The PMU also has other system initialisation functionality and will be discussed further in Chapter 10.

Performance

Computational performance is usually quantified in terms of maximum processing frequency. To compare performance, it is interesting to consider the maximum clock frequencies of PS and PL elements.

Table 2.8: Maximum clock frequencies of UltraScale+ FPGAs, Zynq-7000 and Zynq MPSoC[a]

Component	UltraScale+ FPGAs	Zynq-7000 SoC		Zynq MPSoC
	All Devices	Lower End	Upper End	All Devices
Application Processor	-	max. 866MHz	max. 1GHz	max. 1.5GHz
Real-time Processor	-	-	-	max. 600MHz
Programmable Logic	max. 891MHz	max. 628MHz	max. 741MHz	max. 891MHz

a. Actual figure depends on speed grade and operating voltage.

In summary, the power consumption of FPGA and SoC devices is closely linked, as all SoCs include a PL region that is effectively the same as an FPGA. Comparing Zynq and Zynq MPSoC, power consumption (per logic cell) in the PL of Zynq MPSoC devices is lower, and overall performance is higher, due to the various optimisations in the UltraScale+ FPGA architecture. In terms of the PS, the Zynq MPSoC architecture is more complex, offering greater performance, and also includes additional features for power management.

2.5.3 Embedded Systems Implementation

The desire to implement embedded systems with programmable devices such as FPGAs has motivated the development of Xilinx SoC devices, first the Zynq and more recently the Zynq MPSoC.

An embedded system (typically one or more processors, memory, peripherals, and interconnections, together with connections to external memories or other components) can be created using a single programmable device. With a processor in place, the system can support software applications, usually running on top of an operating system. The programmable nature of the device has the usual benefits of field upgrades and run-time reconfiguration, and the parallel architecture of FPGA logic supports acceleration of suitable tasks.

Zynq and Zynq MPSoC processors have associated memories, and interconnects between selected elements. This Zynq MPSoC has more dedicated processing elements than are incorporated than in Zynq. For instance, a dedicated real time processor has been introduced to complement the application processor.

FPGAs have supported embedded systems design for some time, most notably through the MicroBlaze processor [9], a 'soft' processor configured by the user as an IP core, and constructed from CLB resources on the FPGA. (Here the term 'soft' contrasts with 'hard', which would imply that a dedicated processor was present on the device.)

One of the distinct advantages of soft processors such as the MicroBlaze, is their flexibility. These processors can be customised for the intended application, e.g. floating point support can be omitted if it is not required, which reduces the PL resources necessary to implement the processor. The clear disadvantage is performance — hard processors offer much higher performance than soft processors can achieve. To quantify the difference, MicroBlaze processors can operate at up to approximately 400MHz in UltraScale FPGAs (depending on the configuration of the MicroBlaze core). When compared with the application processors in Zynq and Zynq MPSoC devices (see Table 2.8 on page 27), this is a much lower level of performance.

We can therefore conclude that SoCs are more optimal for embedded systems. They provide dedicated, high-performance processing resources that are capable of operating at significantly higher clock frequencies than FPGA-based soft processors. There is a great diversity of processing resources in the Zynq MPSoC, which enables real-time processing and graphics processing. Meanwhile, the opportunity still exists to use one or more MicroBlaze instances in the PL section, to supplement the primary processor(s) located in the PS.

2.5.4 Applications

FPGAs have become the platform of choice for applications that require computationally complex systems to be implemented, particularly where there is a need for future updates. Due to their reprogrammable nature, FPGAs support field upgrades, and even implement systems with dynamic reconfiguration, i.e. where the device is reprogrammed as part of normal operation. Examples of applications particularly suited to FPGAs include mobile basestation signal processing, video compression and decompression, radar systems, high speed switching and routing infrastructure for data centres, and so on. In each of these cases, the processing requires to be deterministic and high-performance, while providing the potential for new standards, configurations or algorithms to be introduced.

Even when reprogrammability is not a key issue, FPGAs are generally preferred to ASICs where volume is below a certain threshold, due to the high Non-Recurring Engineering (NRE) costs of developing an ASIC. It is generally thought that this threshold is rising [15], meaning that ASICs are now only considered viable for very high volume applications, or when there are special requirements such as a minimal form factor, or a very low power design.

All of the FPGA advantages mentioned so far also apply to Zynq and Zynq MPSoC. By virtue of adding processor capabilities, these devices cater specifically for applications that might otherwise have required a separate processor and FPGA, combining the ability to run software code alongside high-speed parallel processing. In Zynq MPSoC, the dedicated real-time processing capability, GPU and video codec enable systems permits sophisticated systems to be implemented on a single device. Applications include 'big data' analytics, Advanced Driver Assistance Systems (ADAS), broadcast camera equipment, navigation systems and many others. We will go on to explore applications further in Chapter 5.

2.6 Chapter Summary

This chapter has introduced and compared FPGAs, Zynq and Zynq MPSoC. The basic architecture and characteristics of each type of device have been summarised, and their important features have been introduced. We have seen in particular that Zynq and Zynq MPSoC add a processing system to standard FPGA programmable logic, and that the processing system in Zynq MPSoC offers expanded functionality and performance compared to that of the Zynq. It is also clear that the PL provided on SoC devices has expanded considerably with the release of Zynq MPSoC, and we have noted that the power and performance benefits enjoyed by UltraScale+ FPGAs also apply to Zynq MPSoC devices.

2.7 References

Note: All online sources last accessed March 2019.

Some of the Xilinx documentation referred to below has version-specific URLs. If you are working in a newer version of the tools, check for updates on the Xilinx website, or try adjusting the link according to your version.

[1] Arm, Ltd., "AMBA Specifications" webpage.
Available: https://www.arm.com/products/system-ip/amba-specifications

[2] R. Courtland, "Transistors Will Stop Shrinking in 2021, Moore's Law Roadmap Predicts", *IEEE Spectrum* website, 22nd July 2016.
DOI: 10.1109/MSPEC.2016.7551335

[3] L. H. Crockett, R. A. Elliot, M. A. Enderwitz, and R. W. Stewart, *The Zynq Book*, Strathclyde Academic Media, July 2014.
Available: http://www.zynqbook.com.

[4] D. Curd, "Power Consumption in 65 nm FPGAs", Xilinx White Paper, WP246 (v1.2), February 2007.
Available: http://www.xilinx.com/support/documentation/white_papers/wp246.pdf

[5] IEEE Spectrum Magazine, "Special Report: 50 Years of Moore's Law", April 2015 (several articles).

[6] Interlaken Product Alliance, "Interlaken Protocol Definition: A Joint Specification of Cortina Systems and Cisco Systems", Revision 1.2, October 2008.
Available: http://www.interlakenalliance.com/Interlaken_Protocol_Definition_v1.2.pdf

[7] International Telecommunications Union, "H.264, Advanced Video Coding for Generic Audiovisual Services", *Recommendation ITU-T H.264, Series H: Audiovisual and Multimedia Systems*, version 10, February 2016.
Available: http://www.itu.int/ITU-T/recommendations/rec.aspx?rec=12641

[8] International Telecommunications Union, "H.265, High Efficiency Video Coding", *Recommendation ITU-T H.265, Series H: Audiovisual and Multimedia Systems*, version 04, April 2013.
Available: http://www.itu.int/ITU-T/recommendations/rec.aspx?rec=11885

[9] V. Kale, "Using the MicroBlaze Processor to Accelerate Cost-Sensitive Embedded System Development", Xilinx White Paper, WP469 (v1.0.1), June 2016.
Available: http://www.xilinx.com/support/documentation/white_papers/wp469-microblaze-for-cost-sensitive-apps.pdf

[10] M. Klein, "Power Consumption at 45nm", Xilinx White Paper, WP298 (v2.0), August 2016.
Available: http://www.xilinx.com/support/documentation/white_papers/wp298.pdf

[11] F. Rivoallon and J. Balasubramanian, "Reducing Switching Power with Intelligent Clock Gating", Xilinx White Paper, WP370 (v1.4), August 2013.
Available: http://www.xilinx.com/support/documentation/white_papers/wp370_Intelligent_Clock_Gating.pdf

[12] M. Santarini, "Xilinx 16nm UltraScale+ Devices Yield 2-5X Performance/Watt Advantage", *Xcell Journal*, First Quarter 2015.
Available: http://www.xilinx.com/publications/archives/xcell/Xcell90.pdf

[13] M. Santarini, "Xilinx 20-nm Planar and 16-nm FinFET Go UltraScale", *Xcell Journal*, 3rd Quarter 2013, Issue 84.
Available: http://www.xilinx.com/publications/archives/xcell/Xcell84.pdf

[14] K. Subramaniyam, "Proven Power Reduction with Xilinx UltraScale FPGAs", Xilinx White Paper, WP466 (v1.1), October 2015.
Available: http://www.xilinx.com/support/documentation/white_papers/wp466-proven-ultrascale-power-leaders.pdf

[15] S. Trimberger, "Three Ages of FPGAs: A Retrospective on the First Thirty Years of FPGA Technology", *Proceedings of the IEEE*, Vol. 103, No. 3, March 2015, pp. 318 - 331.
DOI: 10.1109/JPROC.2015.2392104

[16] Xilinx, Inc., "7 Series DSP48E1 Slice User Guide", UG479, v1.9, September 2016.
Available: http://www.xilinx.com/support/documentation/user_guides/ug479_7Series_DSP48E1.pdf

[17] Xilinx, Inc., "Leveraging UltraScale Architecture Transceivers for High-Speed Serial I/O Connectivity", WP458 (v2.0), October 2015.
Available: http://www.xilinx.com/support/documentation/white_papers/wp458-ultrascale-xcvrs-serialio.pdf

[18] Xilinx, Inc., "UltraRAM: Breakthrough Embedded Memory Integration on UltraScale+ Devices", WP477 (v1.0), June 2016.
Available: http://www.xilinx.com/support/documentation/white_papers/wp477-ultraram.pdf

[19] Xilinx, Inc., "UltraScale Architecture Clocking Resources: User Guide", UG572 (v1.6), June 2017.
Available: http://www.xilinx.com/support/documentation/user_guides/ug572-ultrascale-clocking.pdf

[20] Xilinx, Inc., "UltraScale Architecture Configuration: User Guide", UG570 (v1.8), December 2017.
Available: http://www.xilinx.com/support/documentation/user_guides/ug570-ultrascale-configuration.pdf

[21] Xilinx, Inc., "UltraScale Architecture DSP Slice: User Guide", UG579 (v1.5), October 2017.
Available: http://www.xilinx.com/support/documentation/user_guides/ug579-ultrascale-dsp.pdf

[22] Xilinx, Inc., "Vivado Design Suite User Guide: Power Analysis and Optimization", UG907 (v2017.4), December 2017.
Available:
http://www.xilinx.com/support/documentation/sw_manuals/xilinx2017_4/ug907-vivado-power-analysis-optimization.pdf

[23] Xilinx, Inc., "Zynq-7000 SoC Data Sheet: Overview", DS190, v1.11, June 2017.
Available: https://www.xilinx.com/support/documentation/data_sheets/ds190-Zynq-7000-Overview.pdf

[24] Xilinx, Inc., "Zynq UltraScale+ MPSoC Product Tables and Product Selection Guide", XMP104 (v.2.3), 2017.
Available: http://www.xilinx.com/support/documentation/selection-guides/zynq-ultrascale-plus-product-selection-guide.pdf

Chapter 3

An Overview of the Zynq MPSoC Architecture

Throughout the following chapter and beyond, we will explore the inner workings of the Zynq MPSoC, focusing on the vast amount of features and resources it provides to the world of embedded applications. We will pay particular attention to the EV device and the similarities this has with other devices in the Zynq MPSoC range. By the end of this chapter, you will have a greater understanding of the elements that combine to form the Zynq MPSoC architecture. You will be further aware of the various processing features available, so that you may begin designing your own embedded systems within the Zynq MPSoC device.

Firstly, we will introduce the range of different Zynq MPSoC devices available. We will then examine the Zynq MPSoC architecture by introducing its Application and Real Time processors, Platform Management Unit, Programmable Logic, and Configuration Security Unit.

3.1 Zynq MPSoC Device Families

All Zynq MPSoC devices contain a Processing System (PS) and FPGA Programmable Logic (PL) within the same chip. The PS is a dedicated and optimised silicon element within the device, containing many different processing units, and platform management and security systems. The PL exists as a programmable region of FPGA fabric, capable of accelerating arbitrary logic functions and is reconfigurable during run-time.

At the time of writing, there are three different Zynq MPSoC device families available and each share similar fundamental components. The device families are known as CG, EG and EV, and are different from one another due to the configuration of resources within the PS and PL. We will begin by first discussing the difference between the PS of each device family and then explore the changes present in the PL.

There are two types of PS available, offering a diverse range of power management and processing capabilities. These are described as follows:

- The first PS contains a dual-core Application Processing Unit (APU) and dual-core Real-Time Processing Unit (RPU). This PS is primarily used by the CG device family.
- The second PS consists of a quad-core APU, dual-core RPU and Arm Mali Graphics Processing Unit (GPU). This PS is used by the EG and EV device families.

Figure 3.1 provides a visual representation of the Zynq MPSoC device families and the PS present in each. For completion, the PL is also included

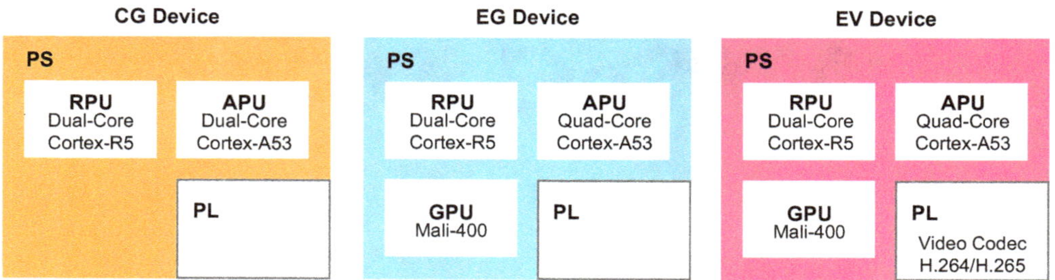

Figure 3.1: Overview of the Zynq MPSoC Families

The resources contained within the PL of each chip varies between family too. Furthermore, the EV device family is the only one to contain a H.264/H.265 Video Codec.

Table 3.1 provides the features of each Zynq MPSoC device family and quantity of FPGA Programmable Logic Cells available. Note the differences in speed between the Application and Real Time Processor of each device family.

Table 3.1: Zynq MPSoC Device Families Overview [1]

	CG Devices	EG Devices	EV Devices
Application Processor	Dual-Core Arm Cotrex-A53 MPCore up to 1.3 GHz	Quad-Core Arm Cortex-A54 MPCore up to 1.5 GHz	Quad-Core Arm Cortex-A54 MPCore up to 1.5 GHz
Real Time Processor	Dual-Core Arm Cotrex-R5 MPCore up to 533 MHz	Dual-Core Arm Cotrex-R5 MPCore up to 600 MHz	Dual-Core Arm Cotrex-R5 MPCore up to 600 MHz
Graphics Processor Unit	-	Arm Mali-400 MP2	Arm Mali-400 MP2
Video Codec	-	-	H.264/H.265
FPGA Programmable Logic Cells	103K-600K	103K-1143K	192K-504K

It is essential to be aware that the FPGA programmable logic operates corresponding to a *Speed Grade*. This is an indication of the timing delays associated with logic elements in the FPGA. At the time of writing, there are five different speed grades for the PL of the Zynq MPSoC. These are 1, L1, 2, L2, and 3, where the higher value corresponds to the fastest FPGA programmable logic. Speed grades that contain the letter 'L' indicate that the FPGA programmable logic is low power. The available speed grades for the PL is dependant on the Zynq MPSoC device family. You can find out which speed grades are available for each device family in [1].

3.1.1 Arm Versus Xilinx Documentation

As you design your own embedded systems with the Zynq MPSoC, you will come into contact with a range of different documentation. Arm and Xilinx both provide reference manuals for their products, which will support you in different ways. We will briefly explain the differences between Arm and Xilinx documentation, and what type of support you should expect from each of them.

Firstly, it is useful for you to understand how Arm license the use of their processor Intellectual Property (IP) to Original Equipment Manufacturers (OEMs), such as Xilinx [2]. The Zynq MPSoC contains both a Cortex-A53 and Cortex-R5 processor which are based on specific Arm architectures (Arm v8 and Arm v7 respectively). Upon selecting these processors, the OEM is given the flexibility to customise particular details of their implementation as required by the end product (in this case, the Zynq MPSoC). The type of customisations that can be applied by the OEM, to the Cortex-A53 and Cortex-R5 processor architectures, is detailed in [5] and [13] respectively. The Zynq MPSoC also contains additional Arm IP that may have also undergone particular customisations. Again, Xilinx will choose implementation details of the IP as set out in the configuration options that Arm provides. Note, the customisations applied by Xilinx to Arm IP in the Zynq MPSoC, are carried out at a silicon level before manufacture. The end user is not capable of making any further changes to the Arm IP.

With the above in mind, we will now discuss what you should expect from Arm and Xilinx documentation. Arm documentation provides technical details for the Arm processor Intellectual Property (IP). This documentation includes the fundamentals of underlying Arm architecture principles and processor core customisation with existing processing elements and optional extensions. Consultation with several Arm documents and manuals will provide a better understanding of the Arm processor IP, however, will not give information on Original Equipment Manufacturer (OEM) configurations. Xilinx documentation will provide information on the specific Zynq MPSoC implementation i.e. OEM changes. In particular, the Zynq MPSoC Technical Reference Manual [4] is the primary reference document for the device.

3.2 Processing System

We now have a better understanding of the Zynq MPSoC device families. Hardware variations are especially notable within the device's PS, e.g. differing speed and quantity of application processing cores. To examine the structure of the Zynq MPSoC processing system, we will study the family containing the most features at the time of writing. This is the EV device family and will serve as the example through the following section. We will mention differences in architecture between device families whenever necessary. Additionally, throughout the lifespan of the Zynq MPSoC, it is possible that other device families will be released. This book may not provide accurate information on the type of resources that will be included in these device families, however, the underlying architecture will be the same.

Figure 3.2 illustrates a simplified overview of the Zynq MPSoC architecture. From the previous section, the APU, RPU, GPU, PL and external memory controller can be quickly identified. Each processing unit is connected to an interconnect known as the Cache Coherent Interconnect (CCI) [3]. The CCI is a helpful addition to the PS architecture and allows processors to share tasks and data between one another dynamically. Primarily, the CCI is used to achieve asymmetric processing, fundamentally providing interconnect support for coherency between processing tasks and ensuring each core operates on the most up-to-date data across the entire PS.

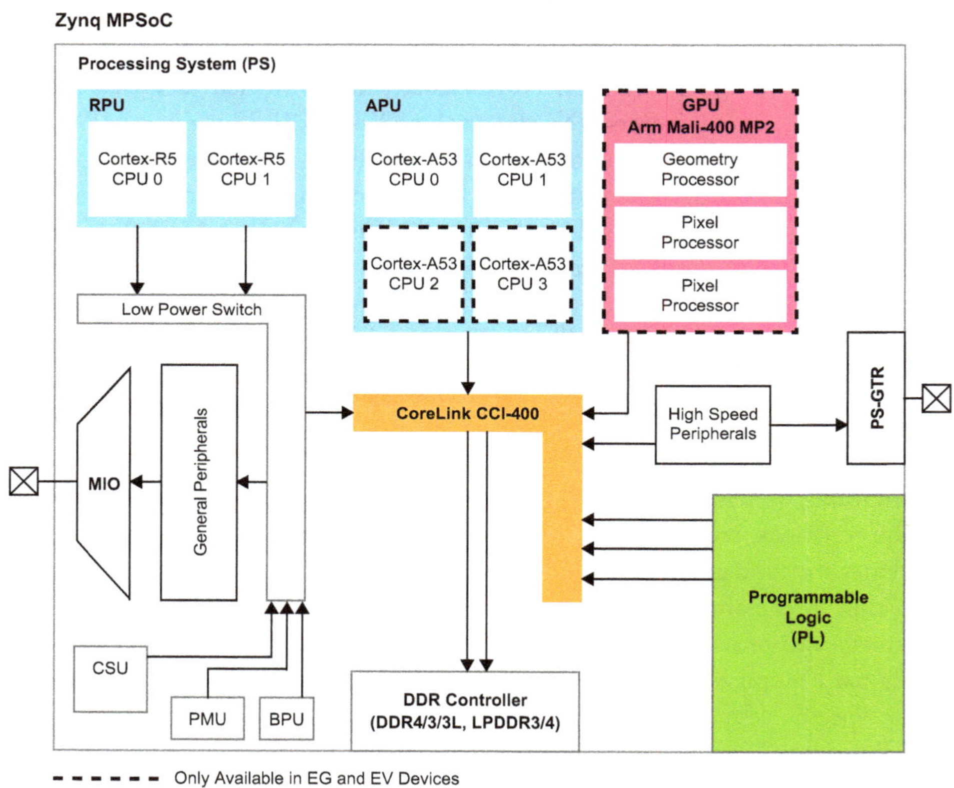

Figure 3.2: Simplified Overview of the Zynq Device Architectures [4]

Zynq MPSoC devices in the EG and EV family contain the Arm Mali-400 MP2 GPU. The GPU is capable of hardware accelerating both 2D and 3D graphics, and includes a geometry processor and two pixel processors. OpenGL [39] and Open VG [15] APIs are supported, allowing the APU to offload graphics processing to the GPU.

General and high-speed connectivity for the PS is also available, allowing the PS to interface and communicate with off-chip components. Connectivity with general purpose peripherals is achieved using the Multiplexed Input/Output (MIO). High-speed connectivity uses a multi-gigabit transmit and receive channel pair, known as a PS-GTR transceiver. General and high-speed peripheral interface blocks, which use the above connectivity, are discussed later in Section 3.2.4.

There are many more resources within the PS than shown in Figure 3.2. The APU and RPU consist of additional interconnects, memories, interfaces, controllers and other associated processing elements. There is also a dedicated Platform Management Unit (PMU), Configuration Security Unit (CSU), and Battery Power Unit (BPU). In addition to the physical processing resources within the PS, power domains and communication paths between processing elements are also of great importance.

3.2.1 Application Processing Unit

The APU houses the Arm Cortex-A53 Multi-Processor Core (MPCore) [5]. Within EG and EV devices, there are four Cortex-A53 MPCores, while CG devices contain two. Figure 3.3, illustrates a simplified diagram of the entire APU.

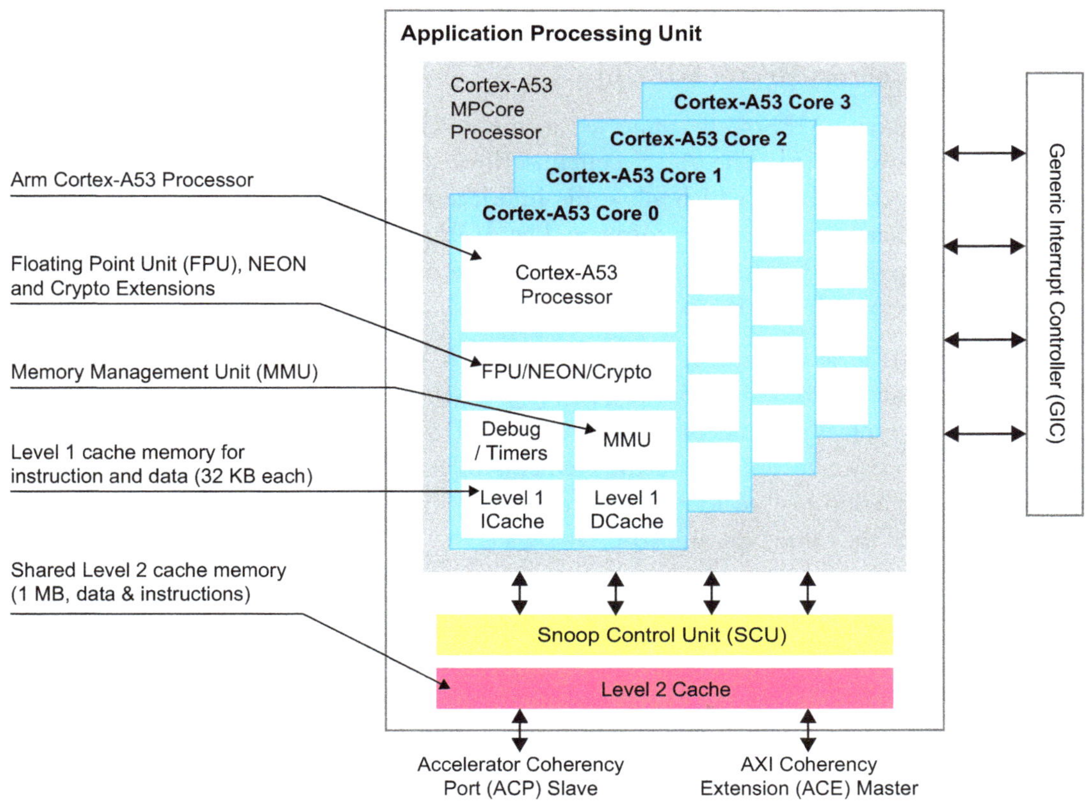

Figure 3.3: Simplified Block Diagram of the Application Processing Unit [4]

Each Cortex-A53 MPCore comprises of the following computational units: Floating Point Unit (FPU), NEON Media Processing Engine (MPE), Cryptography Extension (Crypto), Memory Management Unit (MMU), and dedicated Level 1 cache memory per core (two separate units for instructions and data). The remainder of the APU consists of a Snoop Control Unit (SCU) and Level 2 cache memory.

The Arm Cortex-A53 can operate at up to 1.5 GHz and supports both 32-bit and 64-bit instruction sets. The software can use 64-bit instruction sets and achieve access to a physical memory space beyond 4 GB. All four cores contain their own FPU, NEON and Crypto computational units and separate 32 KB Level 1 cache memories for both data and instructions. Additionally, each core has its own debug and timing resources. The MMU is primarily used to translate between virtual and physical addresses spaces.

Level 1 cache memories allow processing cores to locally store commonly used data and instructions so that they may be retrieved quickly. All four cores share the Level 2 cache memory of 1 MB, for further storage of data and instructions. As all cores share the Level 2 cache memory, it allows them to communicate with one another without the use of external memory.

The SCU governs and controls the access between all cores and the Level 2 cache. The SCU helps maintain the coherency of data between the Level 1 data cache of a processing core and the Level 2 cache memory. It can achieve this without reading and writing to external memory, due to internal buffers allowing a direct cache-to-cache transfer. In addition to this, the SCU also manages transactions between the APU and PL via the Accelerator Coherency Port (ACP) [6].

The CCI within the Zynq MPSoC is connected to the APU's master bus interface using a 128-bit AXI Coherency Extension (ACE) [6]. The ACE contains three additional signals (in contrast to the AXI4 protocol described in Section 3.5.1) which allow the interface to provide cache coherency between components without the use of cache maintenance software. The ACE protocol allows the APU's master interface to access a given address space and retrieve up-to-date data. The data may then be copied locally to Level 1 or Level 2 cache memories to optimise processor performance [4].

The Arm NEON MPE is an extension to Arm series-A processors which provides advanced Single Instruction, Multiple Data (SIMD) facilities [7], [8], [9]. SIMD technology has been designed to accelerate large vector operations typically found in multimedia and DSP style algorithms. SIMD instructions also support the NEON processor to operate on the elements of two input vectors to produce a corresponding output vector. One instruction is used to configure the NEON processor to perform the same operation on multiple data elements of the same type and size.

The Cryptography extension (Crypto) is a dedicated encryption and decryption tool which supports SIMD instruction sets [10]. Standards supported are Advanced Encryption Standard (AES), the Secure-hash algorithm (SHA) functions SHA-1, SHA-224 and SHA-256, and RSA. Crypto also supports finite-field arithmetic, used in algorithms such as the Galois/counter mode and elliptic curve cryptography.

Finally, the APU uses an external Generic Interrupt Controller (GIC) to support system interrupts. This GIC is the Arm CoreLink GIC-400 IP [11], [12], based on the GICv2 specification. Primarily, the GIC provides the APU with registers that manage interrupt sources, behaviour, and routing between processors.

Chapter 6 discusses the APU's architecture and operation in more detail.

3.2.2 Real-Time Processing Unit

The RPU contains a dual-core Arm Cortex-R5 processor [13] for real-time applications. The RPU's architecture provides low latency operation and deterministic performance throughout the entire unit. Figure 3.4 provides a simplified diagram of the RPU.

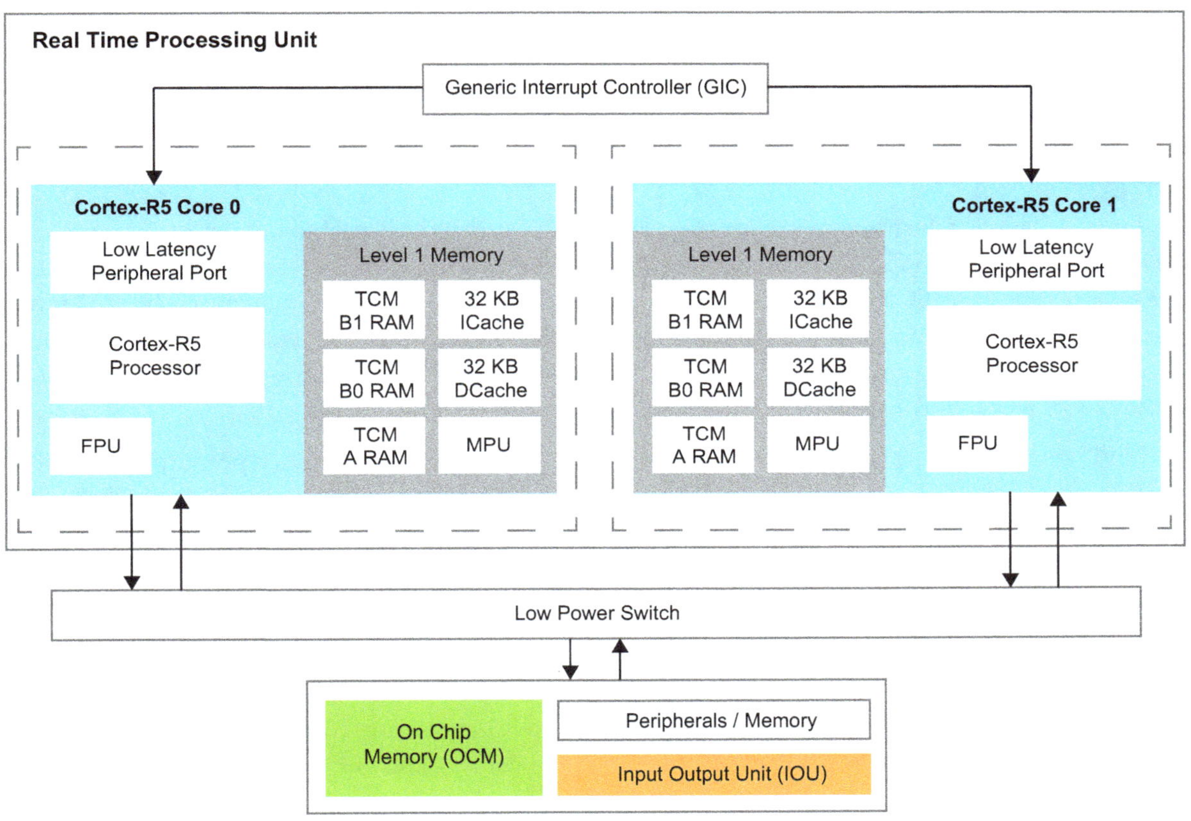

Figure 3.4: Simplified Overview of the Real-Time Processing System [4]

The Cortex-R5 core contains a Floating Point Unit (FPU) for single and double precision arithmetic. The Level 1 memory of each Cortex-R5 core contains three Tightly Coupled Memories (TCMs), two 32 KB caches for data and instructions, and a Memory Protection Unit (MPU).

The Cortex-R5 processing cores within the RPU can operate at up to 600 MHz in EV and EG devices, and 533 MHz in CG devices. Figure 3.4, illustrates the connections between the Arm Cortex-R5 processing cores and the low-power switch. Although the internal details are not shown, the low-power switch is constructed of additional interconnects and processing elements, connected to other peripherals: the Input Output Unit (IOU) and the On Chip Memory (OCM). This connectivity allows the RPU, IOU and OCM to operate together while the Zynq MPSoC device is in low-power mode. As the RPU is part of the low-power domain, it is entirely operational when the Zynq MPSoC is in low-power mode.

There are two primary configurations in which the Cortex-R5 processors may execute software. Split mode allows each processor to work independently, permitting each core to maximise processing performance and latency. Lock-step mode executes the same instructions on both processors. Carrying out instructions in this manner provides redundancy, as if there is a mismatch between the output of the two cores, then an error has occurred. This method of processing is also known as safety mode as it allows the system to determine errors effectively.

Error Checking and Correction (ECC) schemes also provide further error detection and correction support. Within the data set or associated code bits, it is possible to detect up to two errors and correct one error. This scheme is known as a single-error correction, double-error detection (SEC-DED). Refer forward to Chapter 9 for more discussion of safety features and techniques.

The Level 1 data and instruction caches store local data for increased processing performance. To achieve deterministic processing, TCMs are used to provide predictable data loading times to ensure instructions execute on schedule. The TCM RAM divides into two banks labelled 'A' and 'B' which may be accessed by the associated core's Load Store Unit (LSU) concurrently. Concurrent access is possible because each memory bank has dedicated ports connected to the LSU. The TCM blocks labelled 'A' have 64 KB addressable space while the blocks labelled 'B' have 32 KB addressable space.

The MPU manages access requests to the Level 1 memory and external memory, permitting customisation and partitioning of the memory into regions. The regions may then be assigned individual properties. It is also possible to disable the MPU, which will remove memory access management and set areas of memory to their default properties.

The Arm CoreLink PrimeCell GIC PL390 IP [40], based on the GICv1 specification, is used as the GIC within the RPU. This is a dedicated GIC, controlling the outgoing and incoming interrupts from other processing units in the Zynq MPSoC device. The alignment of the GIC's base-address is to the base-address of the Low Latency Peripheral Port (LLPP) of each Cortex-R5 processing core. This base-address alignment is necessary as it ensures that there is low-latency access to incoming and outgoing interrupts.

The architecture and operation of the RPU is discussed further in Chapter 7.

3.2.3 Graphics Processing Unit

The Arm Mali-400 MP2 GPU [4], [14], previously shown in Figure 3.2, consists of a geometry processor (GP) and two pixel processors (PP). The unit uses three integrated MMUs (one per processor) and a Level 2 Cache to store data intermediately.

The GPU can achieve two-dimensional (2D) and three-dimensional (3D) graphic acceleration up to 667 MHz. OpenGL ES 1.1/2.0 [39], and OpenVG 1.0/1.1 [15] Application Programming Interfaces (APIs) allow intensive graphics processing to be offloaded to the GPU. Figure 3.5 provides a simplified block diagram of the GPU, illustrating the geometry and pixel processors and their communication paths.

The Level 2 cache has 64 KB of addressable space and an Advanced Peripheral Bus (APB) [16] slave interface. The APB allows other devices, acting as a master, to control the Level 2 cache.

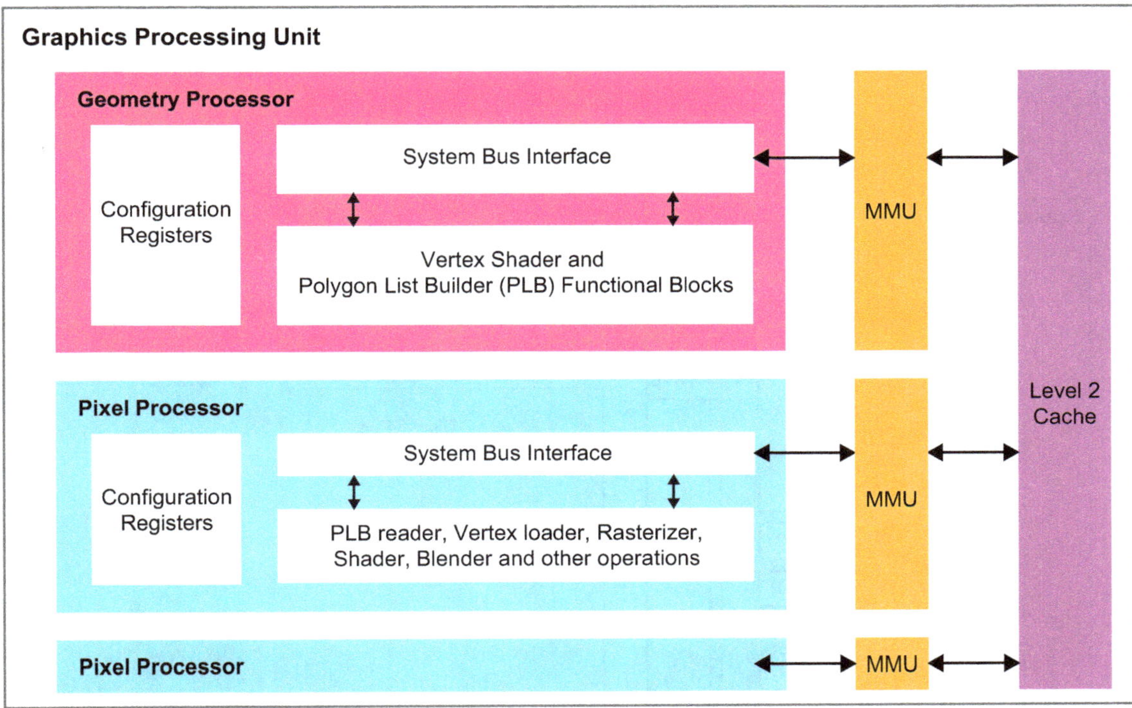

Figure 3.5: Block Diagram of the GPU [4]

Further information about the technical details of the GPU, can be found in the Zynq MPSoC Technical Reference Manual [4] and the Arm Mali GPU Developer Tools Technical Overview [14].

3.2.4 Connectivity

The Zynq MPSoC's PS has many different interfaces for connecting to peripherals. Similar to the Zynq-7000 SoC [17], general peripheral connectivity is primarily achieved using the Multiplexed Input/Output (MIO). The MIO provides a flexible interface that configures the mapping between pins and peripheral interfaces. Seventy-eight (78) of the processing system I/Os are mapped to external peripherals via the MIO.

High-speed serial connectivity to peripherals requiring multi-gigabit communication, can also be achieved using the Serial Input Output Unit (SIOU), found within the Zynq MPSoC's PS. The SIOU is a high-speed serial interface block which supports PCIe, USB 3.0, DisplayPort, SATA and Ethernet interface protocols. There may be only four high-speed serial I/O peripheral interfaces connected in the SIOU at any one time.

Particular connections can also be made accessible to the PL via the Extended MIO (EMIO). The EMIO creates a direct path of communication from the peripheral interfaces in the PS, to the PL resources. Section 3.5.3 covers the EMIO in more detail.

Figure 3.6 provides a diagram illustrating the connectivity of the MIO, EMIO, and SIOU in the PS of the Zynq MPSoC. Note the peripheral interface blocks situated next to the MIO, and the high-speed peripheral blocks contained inside the SIOU.

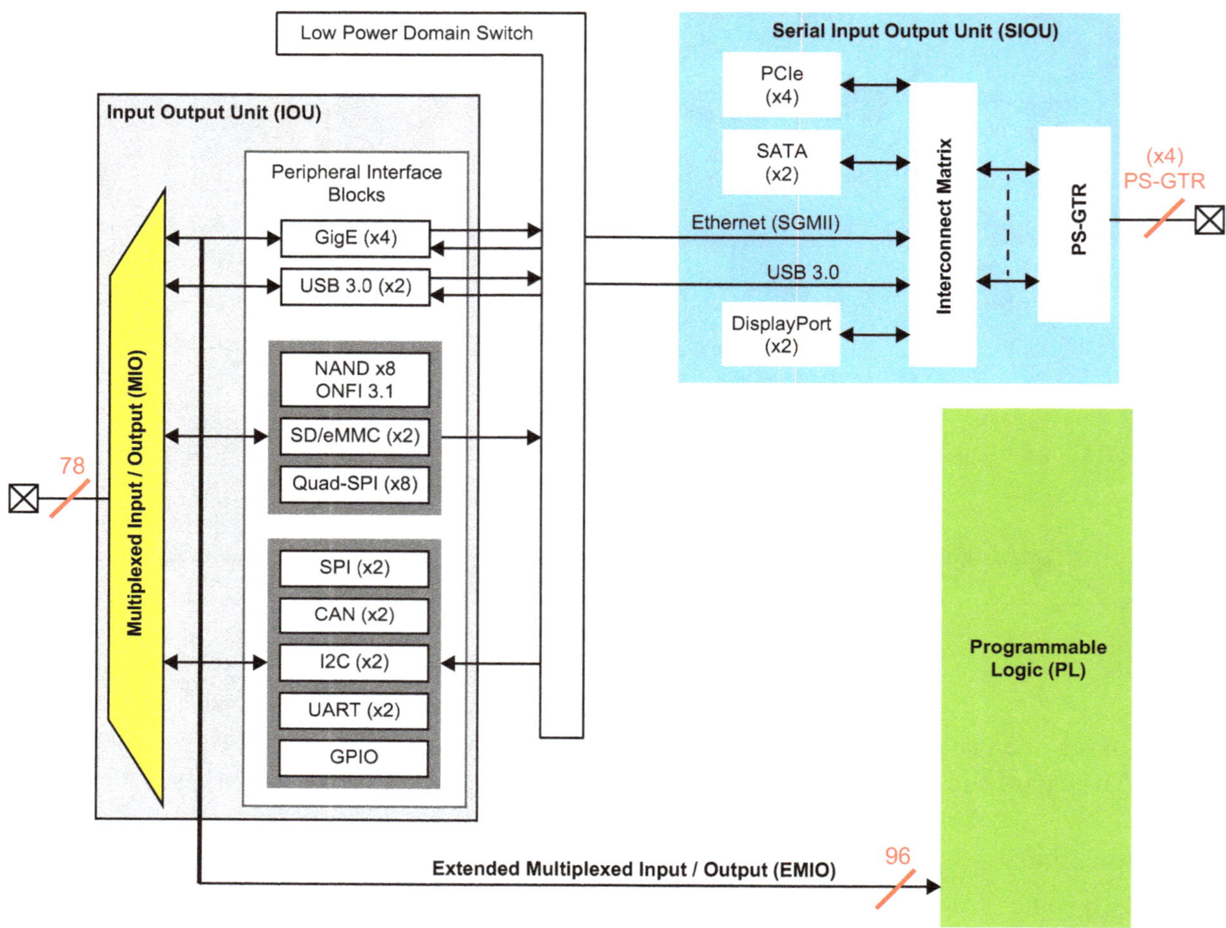

Figure 3.6: Block diagram of the MIO, EMIO, and SIOU connectivity in the Zynq MPSoC [4]

Multiplexed Input/Output

The MIO can be thought of as a very large multiplexer. It is capable of routing several different peripheral interfaces (contained in the peripheral interface blocks in Figure 3.6) to a set of 78 pins. Table 3.2 provides an overview of the peripheral interfaces that are available through the MIO. The MIO shares the USB3.0 and Ethernet peripheral interface blocks with the SIOU.

Note that there are several points to consider when you are assigning pins using the MIO. The Zynq MPSoC Technical Reference Manual [4] provides further information.

Table 3.2: List of General Connectivity I/O Peripheral Interfaces available to the MIO.

I/O Interface	Description
UART (x2)	Universal Asynchronous Receiver Transmitter *Low rate data modem interface for serial communication. Often used for Terminal connections to a host PC.*
SPI (x2)	Serial Peripheral Interface [18] *Standard for serial communications based on a 4-pin interface. Can be used either in master or slave mode.*
CAN (x2)	Controller Area Network *Bus interface controller compliant with ISO 118980-1, CAN 2.0A and 2.0B standards.*
I2C (x2)	I2C Bus [19] *Compliant with the I2C bus specification, version 2. Supports master and slave modes.*
GPIO	General Purpose Input/Output *There are 3 banks of GPIO, each of 26-bits.*
GigE (x4)	Ethernet (RGMII) *Ethernet MAC peripheral, supporting 10 Mbps, 100 Mbps and 1 Gbps modes. The PS MIO only supports the Reduced Gigabit Media Independent Interface (RGMII).*
NAND x8 and ONFI 3.1	NAND Flash Memory Interface *Support for NAND flash memory interface and Open NAND Flash Interface (ONFI) 3.1*
USB 3.0 (x2)	Universal Serial Bus *Compliant with USB 3.0, and can be used as a host, device or flexibly ("on-the-go" or OTG modes, meaning that it can switch between host and device modes).*
SD/eMMC (x2)	Secure Digital and embedded MultiMediaCard Memory Interface *Supports SD 3.0 and eMMC 4.51*
Quad-SPI x8	Quad Serial Peripheral Interface *Serial Communications based on the SPI standard that uses a quad I/O (four-bit data bus). This increases data throughput so that it is faster than a standard SPI interface.*

Serial Input Output Unit

The SIOU contains four high-speed peripheral controllers that each use a single multi-gigabit transmit and receive channel pair, known as a PS-GTR transceiver. The PS-GTR is capable of supporting data rates up to 6.0 Gb/s and can interface to any of the high-speed peripheral blocks using the interconnect matrix shown in Figure 3.6 on page 40. Table 3.3 provides a list of high-speed devices supported by the PS-GTR transceivers. Note that USB 3.0 is available to both the MIO and SIOU.

The interface to the GigE peripheral interface block is different for both the MIO and SIOU, were each use the RGMII and Serial Gigabit Media Independent Interface (SGMII) respectively.

Table 3.3: List of High-Speed Connectivity I/O Peripheral Interfaces available to the SIOU.

I/O Interface	Description
DisplayPort	DisplayPort Interface [20] *A digital display interface primarily used to carry video but can also be used for audio, USB and other forms of data. Up to two channels can be interfaced to the PS-GTR transceivers.*
USB 3.0 (x2)	Universal Serial Bus *Compliant with USB 3.0, and can be used as a host, device or flexibly ("on-the-go" or OTG modes, meaning that it can switch between host and device modes).*
SATA (x2)	Serial Advanced Technology Attachment *SATA uses differential wire pairs for serial communication primarily with storage devices at rates of 1.5, 3.0 and 6.0 Gb/sec depending on the SATA generation (1, 2 or 3 respectively).*
PCIe	Peripheral Component Interconnect Express *Standard for high-speed, serial computer expansion communications.*
GigE (x4)	Ethernet (SGMII) *Ethernet MAC peripheral, supporting 10 Mbps, 100 Mbps and 1 Gbps modes. The SIOU only supports the Serial Gigabit Media Independent Interface (SGMII) to the GigE peripheral interface block. SGMII uses differential pairs for transmit and receive data ports and clocks.*

3.3 Platform Management

Platform management within Zynq MPSoC devices is handled by the Platform Management Unit (PMU) within the PS. The PMU is responsible for the powering-up, resetting and resource monitoring of the entire Zynq MPSoC infrastructure. It also controls the initialisation of the entire system while booting and manages the different power domains throughout the device. The PMU can issue requests and retrieve status information to and from units, independent of other system processors and elements.

In this section, we will first introduce the Zynq MPSoC power modes and power domains. We will briefly discuss the three operational power modes of the Zynq MPSoC and how each of these permit the operation of particular processing elements and power domains within the device. After which, we will summarise the architecture and functionality of the PMU.

3.3.1 Power Modes

The Zynq MPSoC consists of three operational power modes controlled by the PMU. These are the battery powered mode, low-power mode, and full-power mode. Each mode permits the operation of particular processing elements and power domains within the device, and therefore the three modes consume different quantities of power as shown in Figure 3.7.

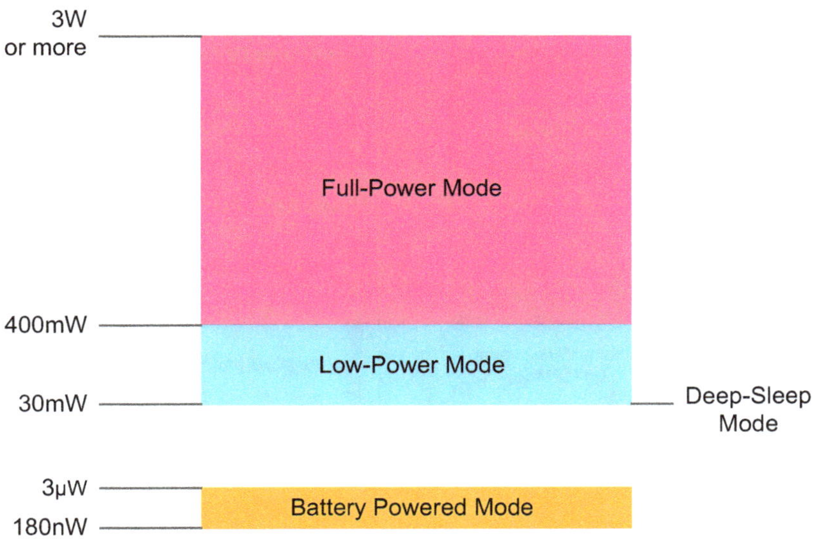

Figure 3.7: Zynq MPSoC Power Modes [21]

Additionally, there is a deep-sleep mode through which the device consumes the lowest possible amount of power while still maintaining the boot and security state of the entire system. During this state, the PMU configures itself to enter a state of sleep or suspension. The PMU will then await a trigger from the MIO, USB or Real-Time Clock (RTC) before waking itself up to perform a pre-specified action. Deep-sleep mode is in contrast to the battery powered mode, where the device has lost its boot and security state; which must then be re-initialised when woken.

The Zynq MPSoC Technical Reference Manual [4] provides further information about each power mode. Zynq MPSoC power modes and domains are also further investigated in Chapter 10.

3.3.2 Power Domains

There are four primary power domains in the Zynq MPSoC: low-power domain (LPD), full-power domain (FPD), PL power domain (PLPD) and battery power domain (BPD). Each is independent of one another and can be isolated to reduce power consumption and to achieve functional isolation (an essential requirement for safety and security applications and tasks). The PMU, which exists within the LPD, facilitates the power-up and power-down of each domain. Figure 3.8 illustrates the processing elements and features accessible by each power domain.

Note that when using the PL, the PLPD will be active and requires the Zynq MPSoC to operate in full-power mode. However, it is possible to shut the PL down when it is not in use and continue to use the APU, and other full-power processing elements, as these exist in separate power domains. Similarly, the PLPD can be operational while the FPD is shut down.

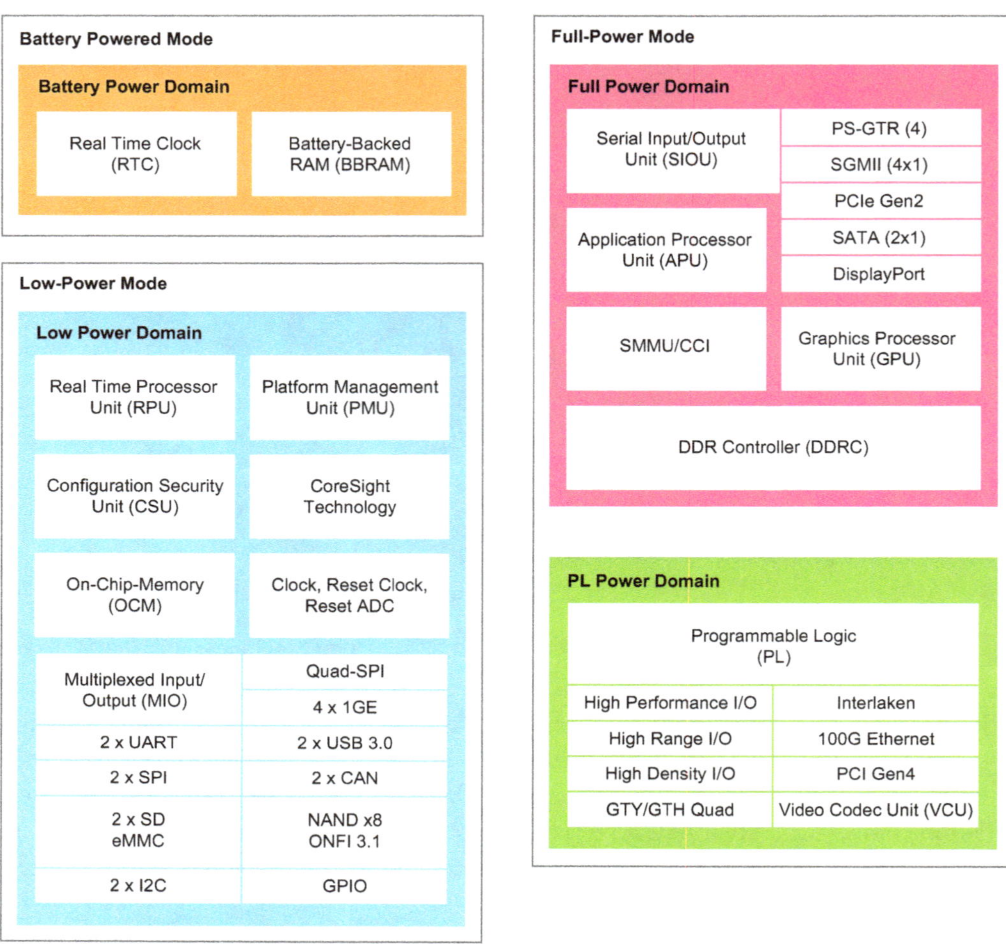

Figure 3.8: Zynq MPSoC Power Domains [21]

3.3.3 Platform Management Unit

The PMU performs several functions that contribute to the overall operation of the Zynq MPSoC. Typical tasks include initiating a power-up and restart of device processing units out of their sleep state upon receiving a wake-up request. Error capture and resolution is also carried out by the PMU.

The PMU contains a triple-redundant processing unit which consists of three hard MicroBlaze [32] processors. These provide increased reliability to the handling of the platform data. The processors are said to be triplicated and form a majority voting system in which the results of two processors will outweigh an error in the third. Voter blocks are used to carry out the logic process by which the voting system between processors takes place. Using redundancy ensures safety-critical systems are at a lower risk of error. Chapter 9 discusses system redundancy in more detail.

There are two primary memories within the PMU: a ROM and a 128 KB RAM. The ROM stores code for the PMU start-up sequence, power control routines and interrupt. The 128 KB RAM stores program data and user/firmware code. Firmware for the PMU, has been provided by Xilinx as discussed in [22]. The typical Zynq MPSoC user will not need to write their own firmware, and can simply use or modify the firmware provided by Xilinx.

The hardware architecture and features of the PMU is described further in Chapter 10. You can also read more about the PMU in the Zynq MPSoC Technical Reference Manual [4].

3.4 Programmable Logic

The PL is a significant component of the Zynq MPSoC, as it provides hardware acceleration for computationally intensive arithmetic. The PL uses the 16nm Kintex *UltraScale+* FPGA fabric [23], that contains all of the features and advantages discussed throughout this section. A diagram highlighting the various elements of the PL, is shown in Figure 3.9.

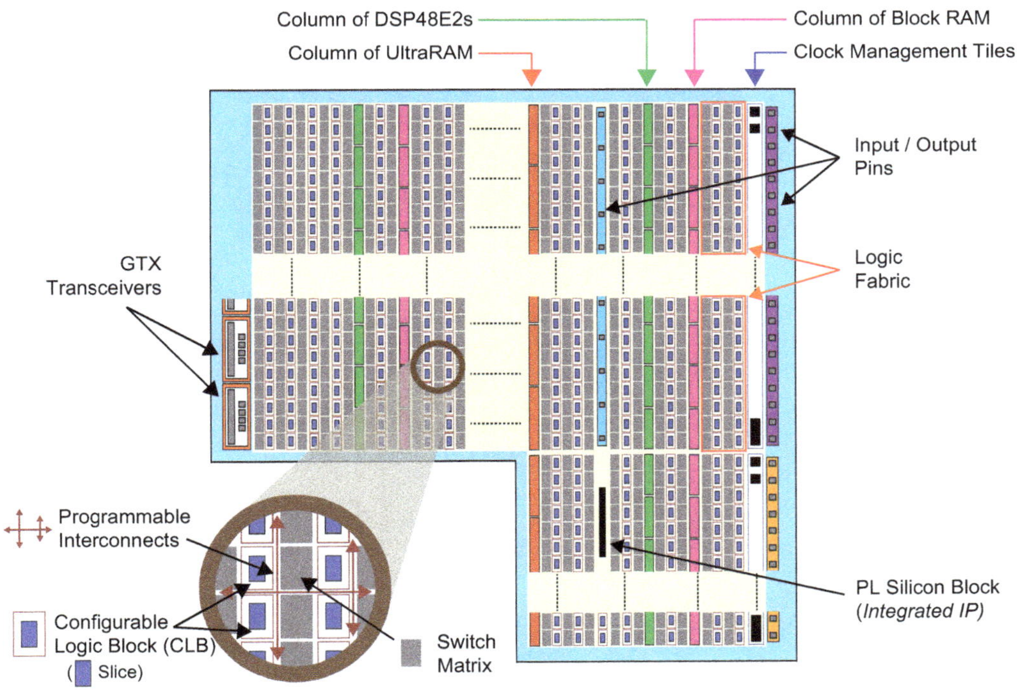

Figure 3.9: Zynq MPSoC Programmable Logic

The following section will provide a brief overview of the Zynq MPSoC logic fabric and dedicated storage and signal processing resources. We will discuss the peripherals contained in the PL, and finally, explore the PL connectivity to the outside world.

3.4.1 The Logic Fabric

The PL is primarily composed of 16nm Kintex *UltraScale+* FPGA fabric, consisting of Slices and Configurable Logic Blocks (CLBs). As previously shown in Figure 3.9, the arrangement of CLBs within the FPGA fabric is in two-dimensional arrays. Each CLB is positioned next to a switch matrix so that it may route data to another PL resource. Additionally, CLBs can connect to other similar resources using programmable interconnects.

A CLB is composed of one slice, which contains the resources necessary for implementing sequential and combinatorial logic circuits. Each slice element within Zynq MPSoC devices is compose of 8 x 6-input Lookup Tables (LUTs), 16 Flip-Flops (FFs), and additional routing logic. Vertically adjacent slices can be chained together to implement large arithmetic circuits through the use of carry logic (noted as Cin and Cout in Figure 3.10). This circuitry consists of multiplexers and logic chain connections which are used to route intermediate signals between adjacent slices.

Figure 3.10 illustrates the composition of a CLB and its constituent components. The UltraScale+ CLB user guide [24] provides further information on CLB resources within the FPGA logic fabric.

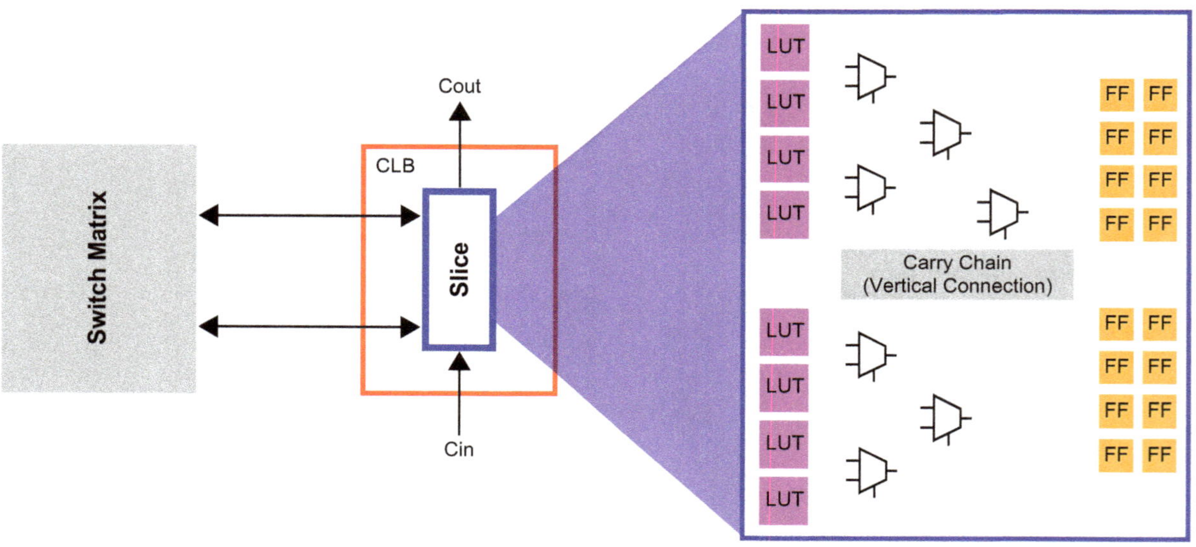

Figure 3.10: CLB connections and internal components [24]

The FPGA logic fabric may be used to construct various arithmetic circuits such as addition, multiplication and division. These circuits may then be combined to carry out arbitrary mathematical functions.

LUTs can be used as small local memories, or connected together to form a larger memory array. This structure is known as Distributed RAM, and due to its resource efficiency and the flexibility of its placement, it is effective when a logic circuit requires storage.

3.4.2 Storage and Signal Processing Resources

Many algorithms and routines accelerated by FPGAs benefit from dedicated storage and arithmetic resources. In addition to the logic fabric, there are three other primary resources for storing and processing data within the Zynq MPSoC PL. These are Block RAMs, UltraRAMs and DSP48E2 slices. The arrangement of each is in columns, situated next to the logic fabric as shown in Figure 3.9 on page 45.

Each of these specialised resources are implemented in dedicated silicon. They are all compact, low power and are capable of operating at the highest clock frequency supported by the PL. Similar to the FPGA logic fabric, each special resource contains a further set of logic elements and features to be considered. A description of each follows:

Block RAM Tiles

Block RAMs in Zynq MPSoC devices can be configured to operate as Random Access Memory (RAM), Read-Only Memory (ROM), and First In First Out (FIFO) buffers while also supporting Error Correction Coding (ECC). These resources are highly optimised memory elements and may be preferred as an alternative to distributed RAM when low latency and high capacity storage is required. Every Block RAM within the PL has a dual port interface (each port is capable of accessing the memory space and stored data while operating under the same, or different clock domains). Additionally, it is possible to cascade Block RAMs with other vertically adjacent Block RAMs, providing a simple method of creating memory arrays.

Each Block RAM can store up to 36 KB of information and may be configured as one 36 KB RAM, or two individual 18 KB RAMs. Block RAMs can be 'reshaped' so that elements of smaller or larger wordlengths can optimise the allocated array of memory. For example, it is possible to reshape a 36 KB Block RAM so that it can contain 4096 elements x 9-bits or 8192 elements x 4-bits. Block RAMs are distributed in columns throughout the FPGA logic fabric so that they are readily available to neighbouring circuit elements, such as those implemented in DSP48E2 slices. Further information about UltraScale+ Block RAM can be found in the UltraScale memory resources guide [25].

UltraRAM Tiles

A single UltraRAM is capable of storing up to 288 Kb. UltraRAMs are only available in selected Zynq MPSoC devices and represent a possible replacement for external memories (such as SRAM) in hardware designs. UltraRAM, have a synchronous, dual port interface and can be cascaded to form larger memory arrays of up to 100 Mb. This is possible without the use of additional logic fabric as UltraRAMs have dedicated routing hardware, allowing them to operate at high clock frequencies. Neighbouring logic circuits can store data within an UltraRAM array and achieve low latency and high capacity storage. For instance, an UltraRAM array is suitable for video processing circuits that require storage of high definition video frames.

It is not possible to reshape UltraRAM blocks. However, they support an addressing scheme of 4096 elements x 72-bits. They exist as a single column in the centre of the UltraScale+ FPGA logic fabric. To reduce routing delay, neighbouring resources which use UltraRAM should be placed nearby. Further information about UltraScale+ UltraRAM can be found in the UltraScale memory resources guide [25].

DSP48E2 Slices

Arithmetic operations necessary for many signal processing applications, benefit from dedicated DSP resources to meet timing constraints. Furthermore, designing large arithmetic circuits using primitive logic elements (LUTs, FFs) consume sizeable areas of logic fabric, placing a demand on FPGA fabric resources.

Dedicated DSP48E2 slices [26] within the PL of all Zynq MPSoC devices, is a hardware multiplier that offers high-speed, efficient signal processing. They are arranged in columns throughout the FPGA fabric, as shown in Figure 3.9 on page 45, to reduce routing delay to neighbouring logic circuits. Additionally, each slice is horizontally aligned with a Block RAM, providing efficient connectivity between the two resource types.

It is possible to cascade multiple DSP48E2 slices if an arithmetic circuit requires greater wordlengths than a single DSP48E2 slice can provide. Cascading slices is possible through the use of cascaded and carry signals throughout the DSP slice and typically doubles the wordlength of most arithmetic functions.

3.4.3 PL Peripherals

Zynq MPSoC devices may include a set of dedicated silicon blocks, also known as Hard IP blocks or Integrated IP, within the PL. There are currently five Hard IP blocks, these are: PCI Express, Interlaken, 100G Ethernet, Xilinx System Monitor (SYSMON) and Video Codec Unit (VCU). The VCU is described in Section 3.4.4.

The PCI Express, Interlaken and 100G Ethernet interfaces all operate within the PL and have many configurable parameters for controlling the connectivity to an external peripheral. The SYSMON block is present in the PL of all Zynq MPSoC devices [27]. This block enables the observation of the PL for issues relating to safety, security, and its physical environment. The SYSMON block uses on-chip power supply monitors and temperature sensors to record measurements, and stores the recorded values in its local registers. These may be accessed using the APB by the Zynq MPSoC processors and PMU.

3.4.4 Video Codec Unit

The VCU, contained only in the PL of EV devices, provides multi-standard video encoding and decoding. It has no direct connection to the PS and must be programmed directly using registers. The unit can support up to eight video streams at the same time and can simultaneously undertake encode (compress) and decode (decompress) functions. The VCU supports High-Efficiency Video Coding (HEVC) H.265 and Advanced Video Coding (AVC) H.264. Additionally, the unit can also process high video resolutions such as 4K x 2K. Further information about the VCU can be found in the Zynq MPSoC Technical Reference Manual [4].

3.4.5 General Purpose Input/Output

The PL of the Zynq MPSoC consists of Input and Output Blocks (*IOBs*) for general purpose communication with the outside world. These are collectively referred to as *SelectIO Resources* [28] and are arranged in banks or tiles of different lengths. Each IOB contains one pad for communicating with external components. The total number of IOBs provided varies depending on the Zynq MPSoC's device and package.

There are three I/O classifications to be aware of when designing systems featuring external interfacing from the PL. These are High-Range (HR), High-Performance (HP) and High-Density (HD) banks. Each has typical use cases summarised in Table 3.4.

Table 3.4: I/O bank classifications

I/O Bank Classification	Operating Voltage (V)	Pins per Bank	Description
High-performance (HP)	1 - 1.8	52	HP I/O banks are suitable for designs requiring high-speed communication with external components, such as memory.
High-range (HR)	1 - 3.3	52	HR I/O banks are capable of supporting a wide range of I/O standards achieving voltages up to 3.3 V.
High-density (HD)	1.2 - 3.3	24	HD I/O banks support slower signalling standards, further described in [28].

3.4.6 High-Speed Connectivity

A further means of communicating between the PL and the outside world is using Gigabit Transceivers (GTX). There are two types of GTX transceiver available within the Zynq MPSoC's PL. GTH transceivers provide low power, high-performance communication at 16.3 Gb/s [29]. GTY transceivers are capable of 32.75 Gb/s, achieving maximum performance for very high-speed applications [30]. Each supports a range of interface standards, including the hard IP blocks available within the PL. The specific device, package, and speed grade of the Zynq MPSoC will determine which transceiver type is available.

GTX Transceivers are arranged in groups of four, also known as a transceiver Quad. Each channel consists of a dedicated Phase Locked Loop (PLL), transmitter and receiver.

3.4.7 JTAG Interface

The Joint Test Action Group (JTAG) [31] ports are responsible for programming and debugging the Zynq MPSoC during the development stage. While debugging, the JTAG interface can be used to manage the flow of data to the device, and communicate errors encountered within the PL and the PS to the user. Upon deploying the device in a commercial or industrial environment, it is usually preferred to have more secure methods of programming the PL with a bitstream, and monitoring the device.

3.5 Processing System and Programmable Logic Interfaces

A significant appeal of the Zynq MPSoC is that it provides dedicated silicon processors in the PS, with the ability to hardware accelerate computationally intensive arithmetic in the PL. In other words, the Zynq MPSoC provides an environment for software and hardware co-processing in a single chip. Communication of instructions and data between the PS and the PL, is achieved using dedicated hardware interfaces.

There are many different types of interfaces; however, the Arm AMBA [33], [34] open standard is the basis of most of the interfaces contained in Zynq MPSoC. This section will begin by examining the Arm AMBA specifications and investigate the use of these interfaces in Zynq MPSoC devices. Subsequently, we will explore and introduce additional interface types.

3.5.1 The Arm AMBA Open Standard

The Arm Advanced Microcontroller Bus Architecture (AMBA) is an open standard responsible for on-chip communications and management. It contains standards for the development of multiprocessor designs that consist of large quantities of controllers and peripherals. Systems using this specification primarily benefit from decreased hardware development time. Many devices, IP cores and blocks produced in industry and by third party manufacturers use some form of AMBA specification.

Arm developed the AMBA standard, released in 1996 for use in microcontrollers. At the time of writing, the standard has undergone numerous revisions and is currently in its fifth version. Typical changes include new and extended interfaces to meet technological demands.

There are six interfaces, extensions and buses within the AMBA open standard, of which five feature in Zynq MPSoC devices. These are the Advanced eXtensibile Interface (AXI), AXI Coherency Extensions (ACE), Advanced High-Performance Bus (AHB), Advanced Peripheral Bus (APB), and Advanced Trace Bus (ATB). Each AMBA interface consists of protocols and interconnects specific to the interface type. We will discuss these as long as they are directly relevant to the operation of the Zynq MPSoC.

AXI - Advanced eXtensible Interface

The AXI bus [6] is used in embedded systems to connect processing elements and IP blocks whilst providing high-bandwidth, low latency communication. It is very suitable for interfacing with memory controllers and can flexibly implement architectures for connectivity over various mediums. The AXI protocol is currently in its fourth revision and is named AXI4. Zynq MPSoC devices use AXI4 within both the PS and PL; it also provides a means of communicating between the two sections of the device.

There are three protocols linked to AXI4, and each provides different advantages and properties which benefit particular tasks. Two of these are 'memory-mapped', which refers to a protocol that issues an address, together with the specified transaction. The data is either stored or read from memory using the given address, which can only be provided by the master.

Memory-mapped protocols support single-beat and burst transfers. A single-beat transfer is when one instance of data is transferred per transaction. During a single-beat transfer, the master must initiate a transaction for each data beat, which results in additional latency.

Maximum performance can be achieved using burst transfers. This method of transferring data is when the slave receives the address from the master and access pattern (a pattern or formula that determines the subsequent address for the data that follows). Not only does this allow multiple data transfers to be carried out with one transaction, but it also reduces the overhead and latency of the transfer.

- **AXI4 [6]** — This protocol is for connections requiring memory-mapped links between processing elements and IP blocks. It is capable of single-beat transfers and can perform burst transfers up to 256 data beats. There are five independent channels for AXI4: read address, read data, write address, write data and write response. Each channel has a dedicated resource for communicating its data. This protocol is suitable for transferring large quantities of information to/from main memory.
- **AXI4-Lite [6]** — A simple, memory-mapped link with reduced handshaking signals, resulting in lower resource allocation than full AXI4. Due to this, single-beat transfers are the only supported method of data transfer (no burst transfers). Therefore this protocol is usually used for low-bandwidth communication with control registers of IP blocks and processing elements.
- **AXI4-Stream [35]** — AXI4-Stream supports point-to-point data streaming. It provides burst transfers of an unrestricted (infinite) size. No address channel is required, as this protocol should be used for a direct flow of data between source and destination within a device. This protocol is particularly useful for signal processing in video, communications and networking applications.

ACE - AXI Coherency Extension

A system can only remain coherent when the local and main memories of its constituent processing elements have a method of sharing and synchronising data. Checks to ensure that a component has shared its most up-to-date information can be carried out in software. However, this would produce additional overhead in a program's execution. The ACE protocol [6] enables the coherency of independent processing elements while ensuring writes to the same memory location are up-to-date and correct (without the need for software).

The ACE protocol is an extension of the AXI interface and consists of five additional channels; three for 'snooping' and two for acknowledgements. When a device is snooping, it is observing data that is being sent or received by one or more devices. Snooping is required in the ACE protocol so that a master can observe read transactions. If the snooping master has the most up-to-date data at the address specified by the read transaction, it may provide the requesting master device with this data.

In addition to this, there also exists an ACE-Lite protocol [6], which is similar to AXI, however, does not contain any of the new Snoop and Acknowledgement channels introduced by ACE. There are instead additional signals on the read address, write address and read data channels. Therefore, ACE-Lite channels can snoop ACE masters, however, cannot be snooped themselves. ACE and ACE-Lite protocols are backwards compatible with AXI, provided that additional channels and signals are disabled.

AHB - Advanced High-Performance Bus

The AHB [36] consists of a single shared channel in which master and slave peripherals communicate information. The AHB is in contrast to AXI, which includes a multi-channel bus with dedicated read, write and response channels. The AHB is for high-bandwidth communication with other processing elements at a reduced cost of channel resources. The protocol is considerably simpler than AXI and can still achieve single beat and burst transfers.

It is possible to configure the data bus within the AHB protocol for 64, 128, 256, 512 and 1024-bit operation. The AHB is available in many areas of the Zynq MPSoC device. The RPU uses the AHB interface to provide fast, dedicated communication to external peripherals and memory, to achieve real-time performance. The SD and SDIO card interface uses the AHB (or AXI) protocol for communication with the host bus target interface. Furthermore, an AHB control port is also used by the SATA host controller interface to provide a facility to manage the peripheral.

APB - Advanced Peripheral Bus

The APB is the preferred protocol for low bandwidth communication between peripherals that do not require high-performance data transfers [16]. This bus type provides low resource utilisation and reduced power consumption as compared to AXI. The interface is very straightforward and easy to use, as there are no handshaking signals between buses. Due to this, burst transfers are not allowed as there is no means of negotiating data transfer parameters. There are two data channels, read and write, which are independent of one another. Since handshaking signals are not available, only one bus may transfer data at any point in time. The APB is present in every processing unit within the Zynq MPSoC's PS, and also in the PL for control of component registers.

ATB - Advanced Trace Bus

Tracing, in software, is the act of using specialised equipment to log data and information about a program or electronic component. The ATB protocol [37] is primarily used to transport format-independent trace information throughout the infrastructure of a device using Arm CoreSight technology [41]. An element using CoreSight is capable of tracing and uses an ATB interface to communicate data throughout the Zynq MPSoC. Xilinx has used Arm CoreSight SoC-400 components to provide debugging functionality to the Zynq MPSoC's PS and PL. All CoreSight components contained in the device support ATB to transport trace profiles throughout the device infrastructure.

3.5.2 PS to PL Interconnects and Interfaces

The primary interface standard used within the PL, for transferring data between IP blocks, is AXI. It is possible to connect multiple AXI ports within the PL from various sources using an AXI interconnect. The AXI interconnect acts a switch within the logic fabric, so that the PL may route traffic from multiple sources to their target destinations. The PS also contains several interconnects, some of which belong to the AXI standard, while others are another form of AMBA standard, depending on the interface.

When communicating between the PL and the PS, AXI provides support for high throughput, low-latency data transfers. The AXI standard implements dedicated ports between the PS and PL. Figure 3.11 illustrates the majority of these connections, only showing resources within the PS that connect directly to the PL. These are the CCI-400, APU (and SCU), system memory management unit (SMMU) [38], central switch and low-power switch. The memory subsystem has also been included to illustrate the various memory resources within the PS that each switch and interconnect can access. The SMMU provides address translation between the PS and the PL, so that the PL can use virtual addresses.

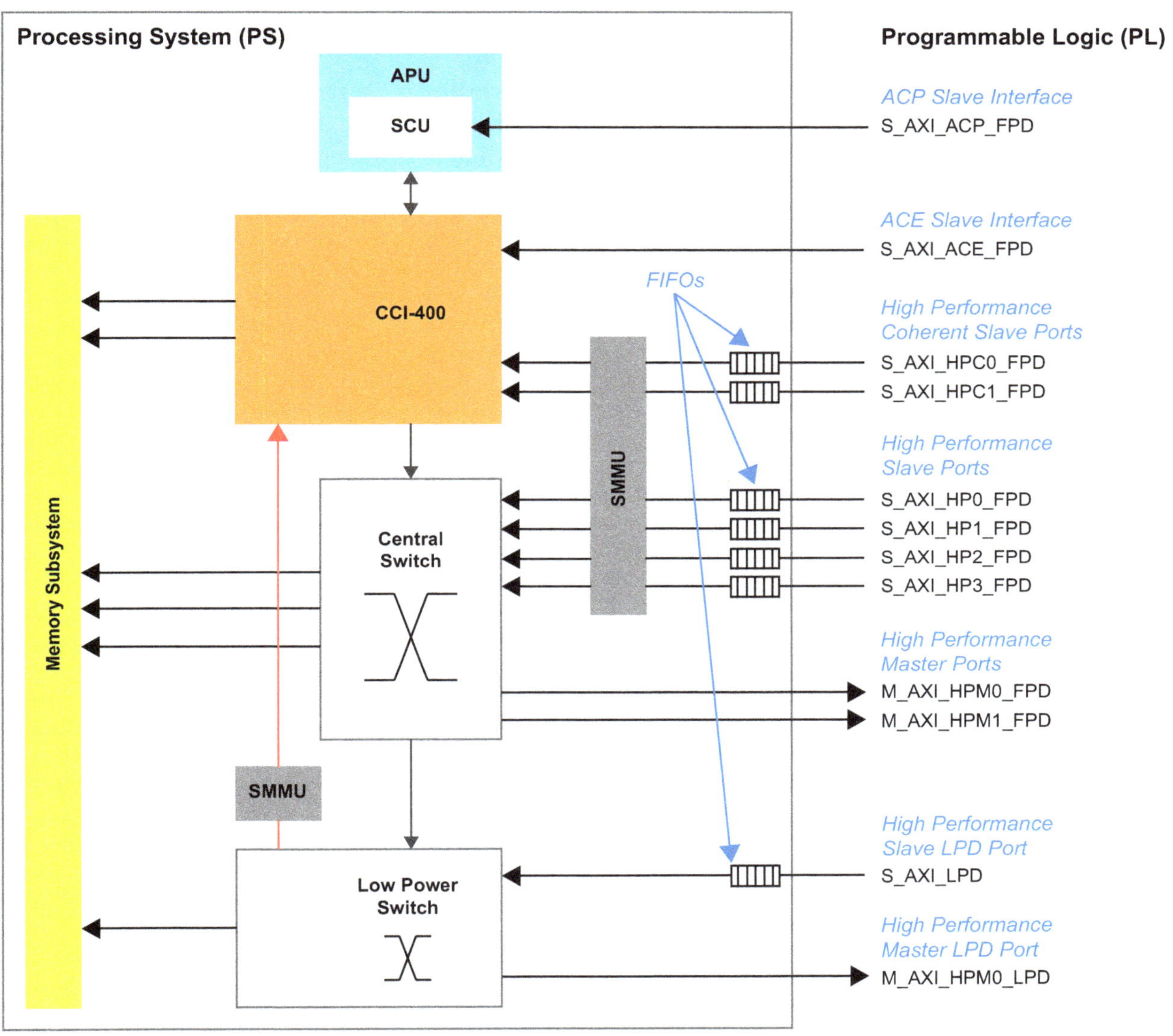

Figure 3.11: AXI interfaces connecting the PL to processing resources within the PS [4]

The interfaces have a naming convention which appropriately represents the role of the master and slave. The first letter of an interface always represents the function of the PS, i.e. an 'S' indicates that the PS is the slave, an 'M' indicates that the PS is the master. Additionally, FPD specifies that the port is part of the full-power domain, while LPD specifies that the port is part of the low-power domain.

There are four interfaces that the PL can use to remain coherent with the PS, three of which are connected directly to the CCI. These are listed as follows:

- ***AXI Coherency Extension*** — The ACE can access the system memory and also the local memory of the APU, via the CCI; sharing up-to-date information.
- ***High-Performance Coherent Ports*** — These high-performance ports are directly connected to the CCI to allow data communication between the PL and PS. Both ports include FIFO buffers to permit burst transfers and high rate communications between the PS and elements in the PL.
- ***Accelerator Coherency Port*** — The ACP uses a subset of the ACE-Lite protocol and provides a single asynchronous connection between the PL and the SCU, within the APU. This connection allows hardware accelerators within the PL to maintain coherency with the Level 1 and Level 2 caches of the APU.

In addition to the coherency ports, there are also low-latency paths between the LPD and PL which could be used by the RPU. Table 3.5 summarises all of the ports shown in Figure 3.11.

Table 3.5: Interfaces between the PS and PL

Interface Name	Address Width	Description	Master	Slave
S_AXI_ACP_FPD	40	Accelerator Coherency Port	PL	PS (FPD)
S_AXI_ACE_FPD	40	AXI Coherency Extension	PL	PS (FPD)
S_AXI_HPC[0:1]_FPD	49	High Performance Coherent Port	PL	PS (FPD)
S_AXI_HP[0:3]_FPD	49	High Performance Port	PL	PS (FPD)
M_AXI_HPM[0:1]_FPD	40	High Performance Master Port	PS (FPD)	PL
S_AXI_LPD	49	High Performance Slave	PL	PS (LPD)
M_AXI_HPM0_LPD	32	High Performance Master Port	PS (LPD)	PL

3.5.3 EMIO Interfaces

As previously mentioned in Section 3.2.4, it is possible to route particular connections from the PS to resources within the PL. By using the Extended MIO (EMIO) to transfer signals between both systems, an Input/Output (I/O) peripheral block within the PS can connect to resources within the PL. Only a subset of I/O peripherals may route to the PL, some with reduced capability. Figure 3.12 shows applicable peripherals and also illustrates the architecture of the EMIO.

The MIO communicates with external peripherals using three, 26-bit banks. Similarly, the EMIO can transfer signals into the PL using three, 32-bit banks, all of which can be routed to a particular I/O peripheral. Once the I/O peripheral and the PL are connected, they may then communicate with an IP core, IOBs or a combination of both.

Each bank is independent of the others and can only be observed and controlled by software. EMIO does not support USB, NAND and QSPI.

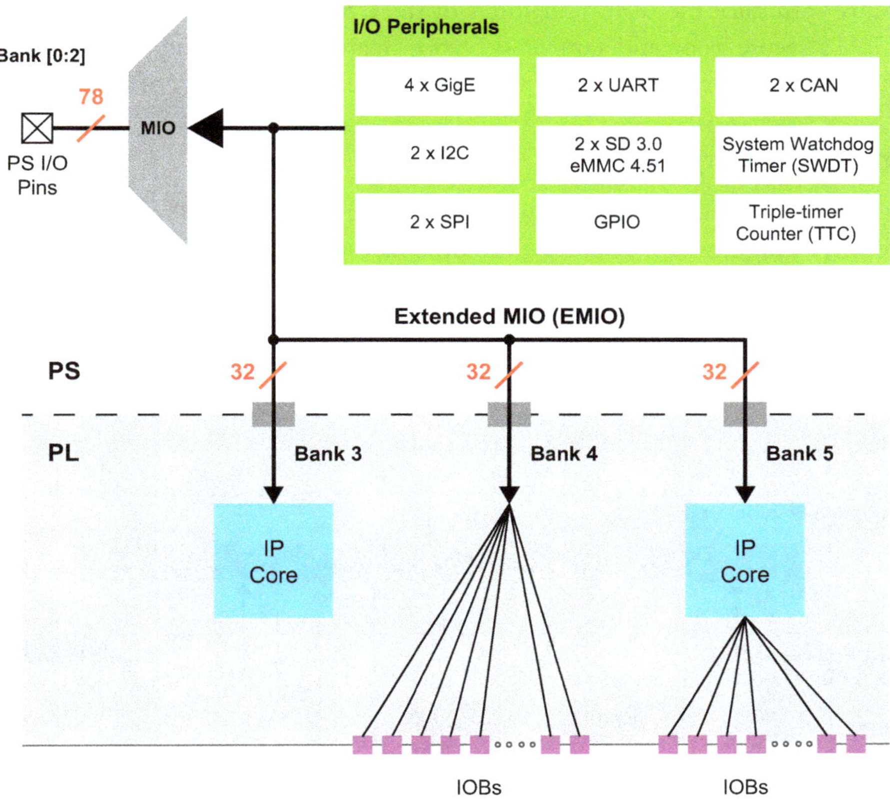

Figure 3.12: Routing I/O peripherals to the PL using the EMIO

3.5.4 Other PS to PL Interfaces

In addition to the high-performance ports, coherency ports and EMIO banks, other dedicated interfaces provide further routing and communication capabilities between the PS and PL. There are interfaces committed to connecting the PL to the Gigabit Ethernet Controller and DisplayPort units contained inside the PS. Other signals include watchdog timers, reset signals, interrupts, Direct Memory Access (DMA) protocol messages and direct GPIO connections to the PMU.

3.6 Security and Configuration

Zynq MPSoC devices contain a broad range of security features and capabilities within their infrastructure. These include a variety of encryption blocks, secure boot facilities, tamper detection and much more. The primary facilitator of security and boot related functions within the device is the Configuration Security Unit (CSU). In addition to this, there are other elements and resources within the chip architecture that contribute to the secure operation of the system.

In this section, we introduce the security features of the CSU and explore its architecture, after which we investigate the CSU's secure boot and tamper detection features. The Zynq MPSoC Technical Reference Manual [4] provides further information on Zynq MPSoC secure boot and configuration.

3.6.1 Configuration Security Unit

The CSU consists of two main blocks known as the Secure Processor Block (SPB) and crypto interface block (CIB) as shown in Figure 3.13. There are also other additional crypto blocks which are for secure applications: AES-GCM [45] [46], SHA-3 [43], and RSA 4096 [42]. The SPB is composed of a triple redundant processor for managing the secure boot. It additionally contains a ROM, local RAM and control/status registers. The CIB provides an interface to the AES-GCM, DMA, SHA-3, RSA and Processor Configuration Access Port (PCAP) [4] blocks.

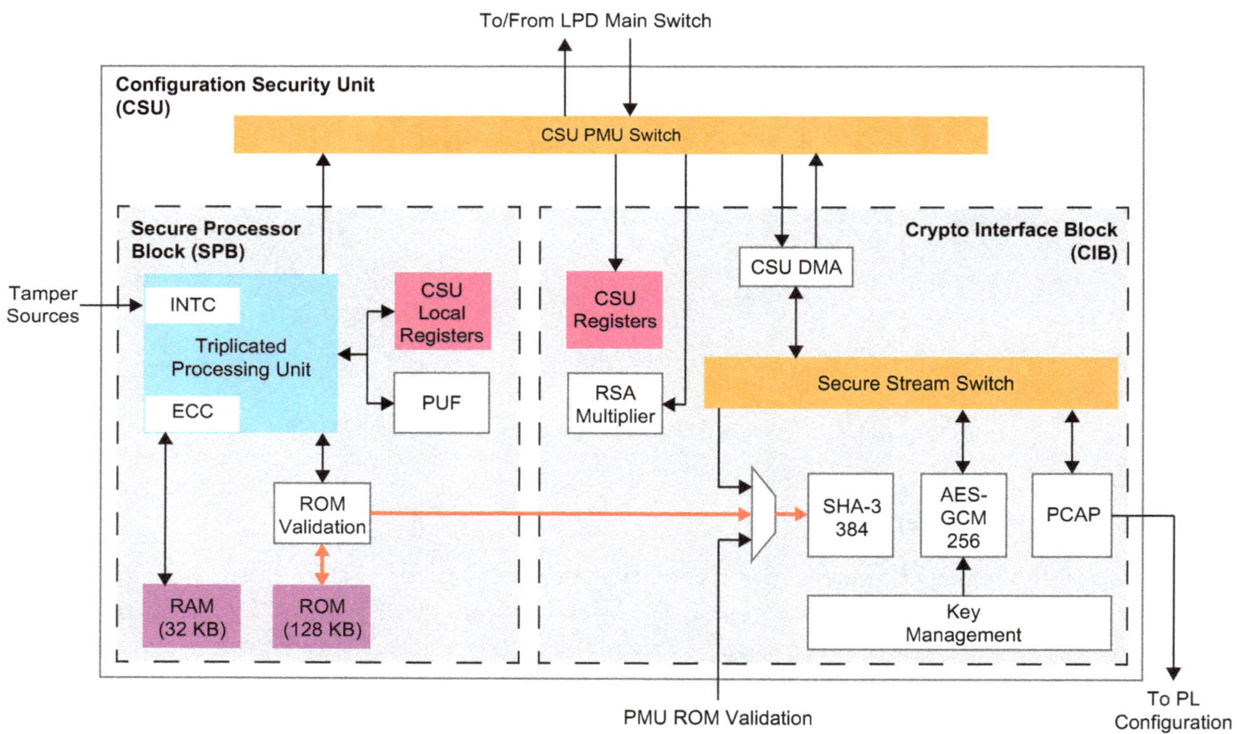

Figure 3.13: Overview of Configuration Security Unit [4]

The SPB's triple-redundant processor manages the secure boot of the Arm processors within the Zynq MPSoC. Each Arm processor will have a boot process specific to its requirements, each of which will be executed by the SPB. The triplicated processing unit also supports tamper detection. There are thirteen tamper response registers which can be enabled while the Zynq MPSoC is operating in secure mode. Upon detecting a tamper event, it is possible to find its exact location and initiate a secure lockdown. The INTC port in Figure 3.13 indicates the tamper source to the processing unit.

The SPB processor uses an ECC memory bus to access the local RAM, preventing errors in transfers and requests. Additionally, all code stored in the processor ROM must pass an integrity check before being executed. A key is generated using the SHA-3 crypto algorithm when the SPB ROM stores software code. Whenever the software code loads from the ROM after this, it is checked using the SHA-3 crypto block. The results of this integrity check will determine whether tampering of the ROM has occurred and whether it is safe to execute the software. Figure 3.13 illustrates the local processor ROM and SHA-3 crypto block connection (highlighted in red) for software code validation.

A Physical Unclonable Function (PUF) [44], is also used by the SPB for the generation of a device-unique encryption key. PUFs are formed during the device manufacturing process, exploiting the randomness present in silicon manufacturing to provide a unique 'fingerprint' for each Zynq MPSoC device.

The right-hand side of Figure 3.13 shows the SHA-3 and AES-GCM crypto algorithm blocks. The entire Zynq MPSoC may use these blocks as necessary by the intended application. The low-power domain switch transfers data between the crypto blocks and other processing elements. Data passes through the CSU PMU switch to the DMA. The DMA is then responsible for transferring data to and from the secure stream switch, otherwise known as the CIB. Data may travel between systems using the CIB without risk of tampering, allowing multiple systems to connect to the same switch. It is here that the crypto algorithm blocks may receive data from other processing elements for encoding and decoding purposes.

RSA provides authentication for the boot image data when the device is in a secure configuration. AES-GCM will then decrypt the image data with a key, which may be loaded using the key management interface from one of the options shown in Table 3.6. Only the ROM may choose the method by which a key is loaded using the Device Key.

Table 3.6: Key management options for secure boot of the Zynq MPSoC

Key Management	Description
BBRAM	The battery-backed RAM (BBRAM) can hold a 256-bit plain text key which may be erased if the battery fails (volatile).
eFUSE	eFUSE is a non-volatile memory, capable of storing a plan text key or an obfuscated key.
Boot Key Register	The boot key register can store a decrypted obfuscated key.
Family Key	The family key is a constant key value which has been hard-coded into the device. This key is used to decrypt the above obfuscated keys whilst the CSU ROM is in operation.
Operational Key	This key is obtained by decrypting the header of the secure boot image using a plain text key obtained from one of the other device key sources.

Chapter 8 provides a detailed investigation about the functionality of the CSU and other security related features of the Zynq MPSoC device.

3.7 Chapter Review

This chapter has introduced the Zynq MPSoC and its two components, the PS and PL. The CG, EG, and EV device families were examined and compared. The general architecture of the PS was explored, showing a cluster of Cortex-A53 application processing cores and Cortex-R5 real-time processing cores, grouped to provide high performance, low latency capabilities for embedded applications. In addition to this, an Arm Mali-400 GPU was found to contain a geometry processor and two pixel processors, and is only implemented in the PS of EG and EV device families.

The PL was found to contain Kintex UltraScale+ FPGA logic fabric. CLBs, slices, Block RAM, UltraRAM, DSP48E2s, integrated peripheral IP blocks and IOBs were amongst some of the UltraScale+ FPGA resources examined.

The interfaces between the PS and PL and the AMBA 5.0 open standard was introduced. Power modes, power domains and the PMU were also investigated. The PMU was shown to manage and oversee the operation of processing components within the Zynq MPSoC device. Finally, the CSU was shown to handle the security and booting procedures of the device, while providing tamper detection during system operation.

You should now have a greater understanding of the processing elements and features that combine to form the Zynq MPSoC architecture. Chapter 4 provides an overview of design tools & methods for developing applications with the Zynq MPSoC.

3.8 References

Note: All online sources last accessed March 2019.

[1] Xilinx, Inc., "Zynq UltraScale+ MPSoC Product Tables and Product Selection Guide", XMP104 (v.2.4), 2018.
Available: https://www.xilinx.com/support/documentation/selection-guides/zynq-ultrascale-plus-product-selection-guide.pdf

[2] Arm, Ltd., "Architectures, Processors and Devices Development Article", Issue A, Version: 1.0, May 2009.
Available: http://infocenter.arm.com/help/topic/com.arm.doc.dht0001a/DHT0001A_architecture_processors_and_devices.pdf

[3] Arm, Ltd., "ARM CoreLink CCI-400 Cache Coherent Interconnect Technical Reference Manual", Issue K, Revision r1p5, December 2015.
Available: http://infocenter.arm.com/help/topic/com.arm.doc.ddi0470k/DDI0470K_cci400_r1p5_trm.pdf

[4] Xilinx, Inc., "Zynq UltraScale+ Device Technical Reference Manual", UG1085, v1.9, January 2019.
Available: http://www.xilinx.com/support/documentation/user_guides/ug1085-zynq-ultrascale-trm.pdf

[5] Arm, Ltd., "Cortex-A53 MPCore Processor Technical Reference Manual", Issue G, Revision r0p4, February 2016.
Available: https://static.docs.arm.com/ddi0500/g/DDI0500.pdf

[6] Arm, Ltd., "AMBA AXI and ACE Protocol Specification", Issue E, February 2013.
Available: http://infocenter.arm.com/help/index.jsp?topic=/com.arm.doc.ihi0022e/index.html

[7] Arm, Ltd., "ARM Cortex-A Series Programmers Guide for ARMv8-A", Issue A, Version 1.0, March 2015.
Available: http://infocenter.arm.com/help/topic/com.arm.doc.den0024a/DEN0024A_v8_architecture_PG.pdf

[8] Arm, Ltd., "Introducing NEON Development Article", Issue A, Version 1.0, June 2009.
Available: http://infocenter.arm.com/help/topic/com.arm.doc.dht0002a/DHT0002A_introducing_neon.pdf

[9] Arm, Ltd., "ARM Cortex-A53 MPCore Processor Advanced SIMD and Floating-point Extension", Issue G, Revision r0p4, January 2016.
Available: http://infocenter.arm.com/help/topic/com.arm.doc.ddi0502g/DDI0502G_cortex_a53_fpu_trm.pdf

[10] Arm, Ltd., "ARM Cortex-A53 MPCore Processor Cryptography Extension", Issue F, Revision r0p4, December 2015.
Available: http://infocenter.arm.com/help/topic/com.arm.doc.ddi0501f/DDI0501F_cortex_a53_cryptography_trm.pdf

[11] Arm, Ltd., "CoreLink GIC-400 Generic Interrupt Controller Technical Reference Manual", Issue B, Revision r0p1, August 2012.
Available: http://infocenter.arm.com/help/topic/com.arm.doc.ddi0471b/DDI0471B_gic400_r0p1_trm.pdf

[12] Arm, Ltd., "GICv3 and GICv4 Software Overview", Issue B, Version 1.0, February 2016.
Available: http://infocenter.arm.com/help/topic/com.arm.doc.dai0492b/GICv3_Software_Overview_Official_Release_B.pdf

[13] Arm, Ltd., "Cortex-R5 Technical Reference Manual", Issue D, Revision r1p2, September 2011.
Available: http://infocenter.arm.com/help/topic/com.arm.doc.ddi0460d/DDI0460D_cortex_r5_r1p2_trm.pdf

[14] Arm, Ltd., "Mali GPU Developer Tools Technical Overview", Issue A, Version 1.0, October 2009.
Available: http://infocenter.arm.com/help/topic/com.arm.doc.dui0501a/DUI0501A_mali_gpu_developer_tools_overview_to.pdf

[15] The Khronos Group. "OpenVG" webpage.
Available: https://www.khronos.org/openvg/

[16] Arm, Ltd., "AMBA APB Protocol Specification", Issue C, Version 2.0, April 2010.
Available: http://infocenter.arm.com/help/index.jsp?topic=/com.arm.doc.ihi0024c/index.html

[17] Xilinx Inc, "Zynq-7000 SoC: Technical Reference Manual", UG585, v1.12.2, July 2018.
Available: https://www.xilinx.com/support/documentation/user_guides/ug585-Zynq-7000-TRM.pdf

[18] NXP, "M68HC11 Microcontrollers Reference Manual (Serial Peripheral Interface, SPI), Revision 6.1, 2007.
Available: http://www.nxp.com/files/microcontrollers/doc/ref_manual/M68HC11RM.pdf

[19] NXP, "I^{2}C-Bus Specification and User Manual", UM10204, Revision 6, April 2014.
Available: http://www.nxp.com/documents/user_manual/UM10204.pdf

[20] *VESA* website.
Available: http://www.vesa.org/

[21] G.Steiner and B.Philofsky, "Managing Power and Performance with the Zynq UltraScale+ MPSoC", Xilinx White Paper, WP482, v1.1, Oct. 2016.
Available: https://www.xilinx.com/support/documentation/white_papers/wp482-zu-pwr-perf.pdf

[22] Wiki.xilinx.com, "Xilinx Wiki — PMU Firmware", January 2019.
Available: http://www.wiki.xilinx.com/PMU+Firmware

[23] Xilinx Inc, "UltraScale Architecture and Product Date Sheet: Overview", DS890, v.3.7, February 2019.
Available: http://www.xilinx.com/support/documentation/data_sheets/ds890-ultrascale-overview.pdf

[24] Xilinx Inc, "UltraScale Architecture Configurable Logic Block User Guide", UG574, v1.5, February 2017.
Available: http://www.xilinx.com/support/documentation/user_guides/ug574-ultrascale-clb.pdf

[25] Xilinx, Inc, "UltraScale Architecture Memory Resources", UG573, v1.10, February 2019.
Available: http://www.xilinx.com/support/documentation/user_guides/ug573-ultrascale-memory-resources.pdf

[26] Xilinx, Inc, "UltraScale Architecture DSP Slice User Guide", UG579, v1.7, June 2018.
Available: http://www.xilinx.com/support/documentation/user_guides/ug579-ultrascale-dsp.pdf

[27] Xilinx, Inc, "UltraScale Architecture System Monitor User Guide", UG580, v1.9.1, February 2019.
Available: http://www.xilinx.com/support/documentation/user_guides/ug580-ultrascale-sysmon.pdf

[28] Xilinx, Inc, "UltraScale Architecture SelectIO Resources", UG571, v1.10, January 2019.
Available: http://www.xilinx.com/support/documentation/user_guides/ug571-ultrascale-selectio.pdf

[29] Xilinx, Inc, "UltraScale Architecture GTH Transceivers", UG576, v1.5.1, August 2018.
Available: http://www.xilinx.com/support/documentation/user_guides/ug576-ultrascale-gth-transceivers.pdf

[30] Xilinx, Inc, "UltraScale Architecture GTY Transceivers", UG578, v1.3, September 2017.
Available: http://www.xilinx.com/support/documentation/user_guides/ug578-ultrascale-gty-transceivers.pdf

[31] IEEE Computer Society, "IEEE Standard Test Access Port and Boundary-Scan Architecture", revision IEEE Std.1149.1-1990 including IEEE Std 1149.1a-1993, February 1990 and June 1993.
DOI: 10.1109/IEEESTD.1990.114395

[32] Xilinx, Inc, "MicroBlaze Processor Reference Guide", UG984, v2016.2, June 2016.
Available: http://www.xilinx.com/support/documentation/sw_manuals/xilinx2016_2/ug984-vivado-microblaze-ref.pdf

[33] Arm, Ltd., "System IP: AMBA Specifications" webpage.
Available: https://www.arm.com/products/system-ip/amba-specifications.php

[34] Arm, Ltd., "ARM Information Centre: AMBA Documentation" webpage.
Available: http://infocenter.arm.com/help/index.jsp?topic=/com.arm.doc.set.amba/index.html

[35] Arm, Ltd., "AMBA 4 AXI4-Stream Protocol Specification", Issue A, Version 1.0, March 2010.
Available: http://infocenter.arm.com/help/index.jsp?topic=/com.arm.doc.ihi0051a/index.html

[36] Arm, Ltd., "ARM AMBA 5 AHB Protocol Specification AHB5, AHB-Lite", Issue B.b, Version 1.0, October 2015.
Available: http://infocenter.arm.com/help/index.jsp?topic=/com.arm.doc.ihi0033/index.html

[37] Arm, Ltd., "AMBA 4 ATB Protocol Specification ATBv1.0 and ATB v1.1", Issue B, Version 1.1, March 2012.
Available: http://infocenter.arm.com/help/index.jsp?topic=/com.arm.doc.ihi0032b/index.html

[38] Arm, Ltd., "ARM System Memory Management Unit Architecture Specification SMMU Architecture Version 2.0", Issue D.c, Version 2.0, June 2016.
Available:
http://infocenter.arm.com/help/topic/com.arm.doc.ihi0062d.c/IHI0062D_c_system_mmu_architecture_specification.pdf

[39] The Khronos Group, "OpenGL Overview" webpage.
Available: https://www.khronos.org/opengl/

[40] Arm, Ltd., "PrimeCell Generic Interrupt Controller (PL390), Technical Reference Manual", Issue B, Revision r0p0, November 2009.
Available: http://infocenter.arm.com/help/topic/com.arm.doc.ddi0416b/DDI0416B_gic_pl390_r0p0_trm.pdf

[41] Arm, Ltd., "CoreSight Technical Introduction: A quickstart for designers", ARM White Paper, August 2013.
Available: http://infocenter.arm.com/help/topic/com.arm.doc.epm039795/coresight_technical_introduction_EPM_039795.pdf

[42] B. Kaliski and J. Staddon, "PKCS #1: RSA Cryptography Standard" (RFC 2437), RSA Laboratories, Version 2.0, September 1998.
DOI: 10.17487/RFC2437

[43] M. Dworkin, "SHA-3 Standard: Permutation-Based Hash and Extendable-Output Functions", National Institute of Standards and Technology, August 2015.
DOI: 10.6028/NIST.FIPS.202

[44] G. E. Suh, S. Devadas, "Physical Unclonable Functions for Device Authentication and Secret Key Generation", *Proceedings of the 44th annual Design Automation Conference (DAC)*, San Diego, USA, June 2007, pp 9 - 14.
DOI: 10.1145/1278480.1278484

[45] NIST, "Announcing the Advanced Encryption Standard (AES)", National Institute of Standards and Technology, November 2001.
DOI: 10.6028/NIST.FIPS.197

[46] M. Dworkin, "Recommendation for Block Cipher Modes of Operation: Galois/Counter Mode (GCM) and GMAC", National Institute of Standards and Technology, November 2007.
DOI: 10.6028/NIST.SP.800-38D

Chapter 4

Design Tools & Methods for the Zynq MPSoC

Before discussing system design with Zynq MPSoC, it is useful to introduce the design tools, development boards, and support materials that support this process. We also provide an overview of the main design flows available. This chapter provides a high level introduction, and more detailed coverage of selected topics will follow during later chapters.

4.1 Anatomy of a Zynq MPSoC Design

At the very highest level of abstraction, a Zynq MPSoC system consists of three elements: the board on which the Zynq MPSoC device sits; the Zynq MPSoC hardware design; and the software layers that run on top of the hardware. There may of course be interfacing to other boards or peripherals, too.

In our discussions, we generally consider the Zynq MPSoC development board to be commercially available (obtained from Xilinx or another vendor). The board enables prototyping of systems on the Zynq MPSoC, with a set of standard facilities and peripheral interfaces. Ultimately, product development will tend to involve the creation of a custom Printed Circuit Board (PCB), but we consider that to be outside the scope of the current book.

The idea of a Zynq MPSoC hardware system is a little abstract, because it refers to a custom design implemented using the reprogrammable hardware of the device, as opposed to the device architecture itself. The hardware system can be altered to any required system design (achieved by reprogramming the chip). On top of the Zynq MPSoC hardware, there is a 'stack' of software elements, which may include one or more operating systems, together with application software and other components. Later chapters will cover the

Zynq MPSoC hardware design and software stack in greater detail; but first we'll begin with an overview of the different elements that comprise a Zynq MPSoC system.

4.1.1 Zynq MPSoC Development Boards

Normally designers begin work using a standard development board, which comprises a Zynq MPSoC device along with a number of other components, including memories and various standard interfaces (USB, Ethernet, DisplayPort, and so on). One such development board is the Xilinx ZCU104, shown in Figure 4.1, which includes a ZCU7EV Zynq MPSoC device. Some of the notable features of this board are summarised in Table 4.1, and a more detailed review can be found in [48].

Table 4.1: Feature summary for ZCU104 development board

Feature Group	Description (see *List of Acronyms* on page 585 !)
Zynq MPSoC device	ZCU7EV, with: Arm Cortex-A53 applications processors (x4) Arm Cortex-R5 real-time processors (x2) UltraScale+ programmable logic (504K logic cells, 1728 DSP slices)
Memory	QSPI flash 2GB DDR4 RAM (attached to PS) DDR4 RAM memory socket (attached to PL)
Connectivity	Ethernet port SATA M.2 port HDMI ports (x2, i.e. separate *in* and *out* ports) DisplayPort port FMC expansion connector USB3 port UART (over USB) MicroSD card slot Pmod ports (x3)
Programming Options	JTAG (over USB) QSPI flash SD card
Debugging and Monitoring	SYSMON (System Monitor) port Arm Trace connector port

Other boards are also available from Xilinx and third party vendors, including Xilinx' ZCU102 [47], and Avnet's Ultra96 [55]. As the selection of boards expands, developers are increasingly able to choose from a selection of different Zynq MPSoC devices with varying combinations of memory and I/O facilities, and at a range of price points.

Development boards are a convenient starting point for most designs, but products are usually developed on custom boards. This enables a free choice from the full selection of Zynq MPSoC chips available, and also

Figure 4.1: ZCU104 development board

allows full customisation for the target application, thus potentially achieving a physically smaller board that includes only the components required, with optimised cost and power consumption. Board design will not be covered in this book, but Xilinx provide extensive support documentation on this topic [42].

4.1.2 The Hardware System

When working with the Zynq MPSoC device, a number of hardware resources are available, as outlined in Chapter 3. This includes the PL, PS, and a selection of peripheral interfaces supporting various standards. It may not be necessary to use all available hardware elements for a particular system, and designers have full flexibility to include only the subset that meets their requirements. The PL in particular provides a 'blank canvas', offering the freedom to implement any system design.

The designer creates their own custom hardware system using Xilinx design tools, optionally augmented by third party design tools. This custom hardware includes the desired elements from the PS, any IP cores[1] or other functionality to be implemented in the PL, I/O connections, memory, clock and interrupt configuration, and interfacing between elements. As an example, the hardware system may be designed to comprise the PS with Ethernet and USB interfaces, and a custom IP core implemented in the PL. The complete hardware system represents the base upon which all software elements of the system are built; software runs on the PS and communicates with the hardware elements of the system. A simple diagram depicting this idea is provided in Figure 4.2.

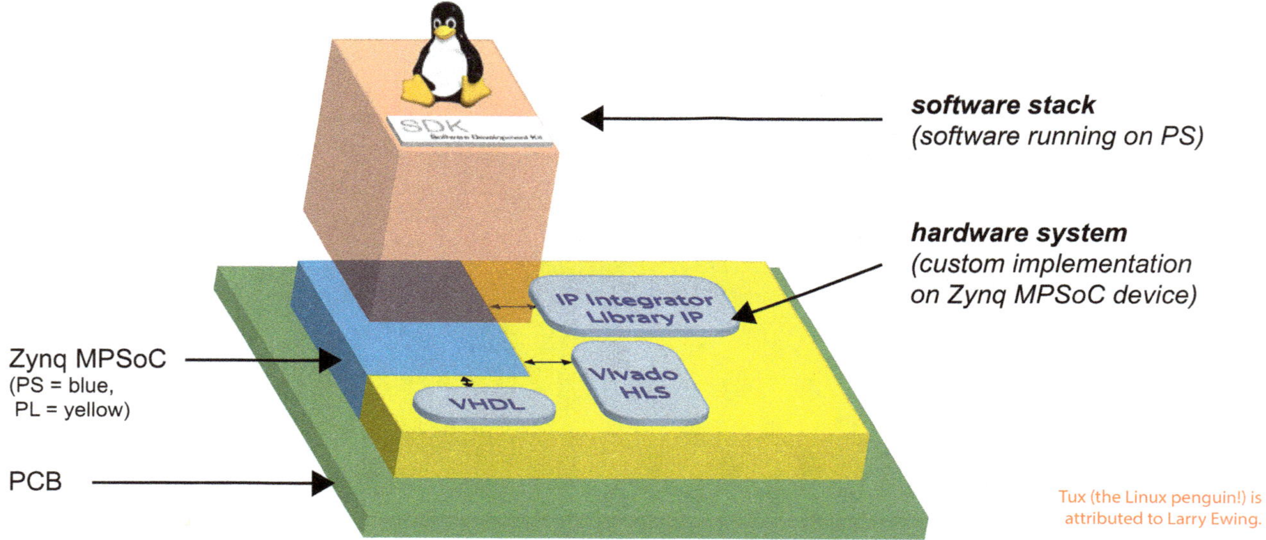

Figure 4.2: Layers of a Zynq MPSoC design

How is this hardware system created? The methodology for designing the hardware system is to use the Vivado IP Integrator tool to create a block diagram for the system. The IP Integrator diagram includes a block representing the PS portion of the Zynq MPSoC device, IP blocks for implementation in the PL, PL-PS interconnections, memory interface blocks, and so on. An example IP Integrator block diagram is shown in Figure 4.3, where the PS block is indicated by the large 'ZYNQ UltraSCALE+' logo. Vivado automatically generates HDL code that represents the 'top-level' of the design, and thereafter, can be used to perform the

1. An Intellectual Property (IP) core is a functional module that is implemented in the PL. The Xilinx IP Integrator tool includes a set of most free IP cores. IP cores can also be used from third party sources (either open source or paid-for), or be custom-designed.

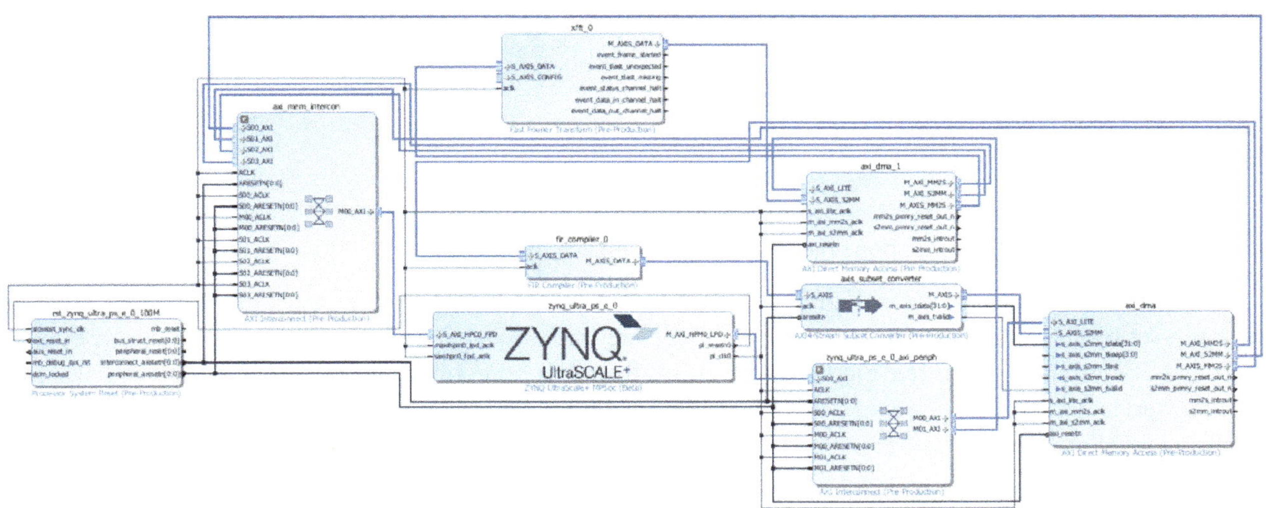

Figure 4.3: Example IP Integrator diagram for a simple Zynq MPSoC design

synthesis, implementation and bitstream generation steps needed to realise the hardware system on the target device, and produce a programming file.

As a side note, all actions performed in Vivado IP Integrator have *Tcl*[1] command equivalents. Therefore, as an alternative to drawing the diagram in IP Integrator, the same hardware system could be created by executing a script. The script-based method may be particularly useful for quickly and reliably replicating a previously created design, for instance. It also provides a means for third party design tools to leverage IP Integrator functionality.

The Xilinx SDx tool can be used for whole-system design (i.e. software, hardware, and the necessary interfacing between all elements), using software-based description as the design method. When designing in the SDx environment, Vivado is called in the background, transparently to the user. SDx is covered in depth later in the book.

Common to both the Vivado and SDx methods is the idea of incorporating IP cores — elements of the design that are implemented within the PL section of the Zynq MPSoC device. Xilinx provides a library of IP cores within IP Integrator that can be parameterised to suit user requirements, and specialist IP cores developed by third parties may be incorporated too, if desired. When using SDx, certain software functions map to existing IP cores, or otherwise, they are designed to be implemented as custom hardware functions.

In more general terms, designers may wish to create their own IP cores to complement existing library IP. A number of possible methods are available for creating IP cores, including HDL development, block-based design, and High Level Synthesis (HLS). In each case, tools are available from both Xilinx and third parties

---

1. Tcl (Tool Command Line) is a scripting language extensively used in Electronic Design Automation (EDA) tools [26].

to support these design methods. IP creation is an important aspect of hardware design, and will be discussed further in Chapter 11, and was also covered in [5] and [6] in the context of the Zynq-7000 device (the same principles and processes apply).

4.1.3 The Software Stack

Software running on the hardware can be considered as a 'stack' of different layers, as indicated in Figure 4.2. We can now start to think about the layers that comprise the stack.

The simplified model shown in Figure 4.4 is a useful starting point. Here, the bottom layer of the software stack is a set of drivers and low level functions. These are implemented either in the form of a Board Support Package (BSP) that sits between the hardware and OS, or as a component part of the OS, and in both cases enable interfacing with the hardware system. Software applications execute on top of the OS, and form the highest layer in the stack. These can include programs such as web browsers, databases, games, and indeed any other type of application that may be desired.

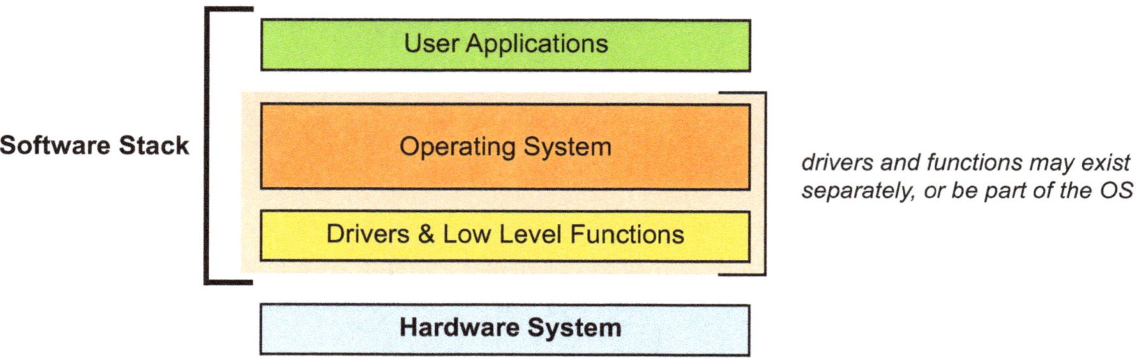

Figure 4.4: Simple model of a software stack

Most notably in Zynq MPSoC, given that it has several processing cores split across the APU and RPU, there could actually be several software stacks — even a different stack for each processing core! There may also be further software stacks in the system, associated with the PMU, video codec, etc., as well as any soft MicroBlaze processor cores implemented in the PL. There are more permutations than in Zynq-7000 (which does not have an RPU, and whose APU has only two application processor cores — or in the Zynq-7000S, only one!).

Where the designed hardware system for Zynq MPSoC includes both APU and RPU, we will generally assume that two different types of operating system are deployed: a desktop / embedded OS on the APU, and a real-time OS on the RPU. Other possibilities also exist, ranging from using the APU or RPU in isolation with a single OS, right through to a more complex scenario where different OSs run on individual cores. We cannot cover all software stack possibilities in this book. Rather, it is more useful to focus on a commonly adopted subset: Linux as the primary operating system, implemented on the A53 processors in the APU; and

FreeRTOS deployed on the R5 cores in the RPU. Extended details about these software stacks, along with the 'Baremetal' application, are provided in Chapter 12.

Xilinx provides tools for developing all layers of the software stacks operating on the PS, as well as on any MicroBlaze soft cores included in the system. Third party tools and OSs can also be leveraged.

In terms of OS development, Xilinx provides PetaLinux, which represents a set of Xilinx-specific Linux development tools and support, while the 'Baremetal' option (a lightweight OS suitable for simple designs) is available via the Xilinx Software Development Kit (SDK). Third party options include other types of Linux, notably the Yocto project [54], which incorporates specific support for Xilinx' Zynq, Zynq MPSoC, and MicroBlaze processors [44], and FreeRTOS for real-time processing [10]. Further possibilities include Windows IoT and Windows Embedded versions, and VxWorks. The Android operating system (commonly associated with smartphones and tablets) is another option, particularly for systems incorporating a touch-screen-type user interface [2]. Guidance on third party OSs can be obtained from this dedicated Wiki page:

http://www.wiki.xilinx.com/3rd+Party+Operating+Systems

Software for Zynq MPSoC can be developed using various different tools. These include the Xilinx SDK provided as part of the Vivado suite for software development. Alternative development environments include third party tools such as Microsoft Visual Studio [19] and Arm DS-5 Development Studio [3].

Further specialist software tools and components are available, including hypervisors and tools for supporting Asymmetric Multi-Processing (AMP), which refers to the scenario where the various cores of a processor host different operating systems. These topics will be covered in Chapter 13.

In summary, a Zynq MPSoC system can be considered to comprise a number of layers. Neglecting the PCB on which the Zynq MPSoC device resides, the lowest level is the *hardware system* developed by the designer, followed by the various layers of *software*, from the low-level drivers all the way up to applications. Chapter 12 will go on to discuss software stacks in more detail, and some of the options available.

4.2 The Design Process

The design process for creating a Zynq MPSoC embedded system can be characterised by a 'design flow', beginning with capturing the requirements for the system to be designed. The designer or design team then considers the project in detail, making key decisions about the approach that will be taken, and putting appropriate plans and schedules in place. In terms of the actual design process, the 'flow' refers to the sequence of stages or processes involved in creating the design — i.e. hardware and software components, and interfacing — and alternatives exist for the design flow. Surrounding issues include profiling, testing, and documentation.

4.2.1 Requirements and Specification

When embarking on a Zynq MPSoC system design, a major initial task is to generate a set of functionality and performance requirements. This will capture the target features of the design, and ultimately it will influence all of the design choices that follow. The captured requirements can be developed into a formal specification for the system.

With the end goals established, this will inform the selection of a target device, estimates of the design effort and resources required, the composition of the design team (hardware, software, system engineering, algorithms specialists, etc.), the choice of design method(s) and development tools, as well as general project planning and scheduling.

4.2.2 Project Decisions and Planning

Undertaking a design based on the Zynq MPSoC is likely to be a complex project comprising a number of different elements, and requiring the input of a design team rather than being completed by a single individual. The system may also involve commercial, security or other particular constraints that must be adhered to. Therefore, the scope in terms of planning and executing a Zynq MPSoC project is wide, and we choose to address this topic simply by posing a few pertinent questions that should be considered.

- **Partitioning** — How will the system functionality be partitioned? In other words, which parts of the system will be implemented in hardware, and which in software? How will they be interfaced?
- **Hardware Requirements** — What are the requirements for the target device? How large a PL is needed? How much on-chip memory does the application require? Are 2 or 4 APU cores needed? Are any special features needed, like the VCU?
- **Software Requirements** — How should the software stack(s) be composed? For each, which is the most appropriate choice for the operating system? Are there any requirements for hypervisors or other special components?
- **Design Tools / Methods** — What general design flow will be adopted? Which tools and methods will be used for developing the hardware and software components of the system?
- **Debugging and Profiling** — Will there be access for all engineers to a physical development board during the design process? Would a virtual platform (QEMU) be useful? What strategies for profiling, testing and debugging will be adopted?
- **Project Scheduling and Resourcing** — What are the skills of the engineers who will work on the project? How long are the various design, interfacing and testing tasks likely to take? And what are the dependencies between these tasks?

Clearly there are a number of factors to bear in mind!

It is also worth noting that the SDx development environment, which is discussed in detail in Part of the book, actually eases some of the above decisions. In particular, SDx provides tools for profiling code and optimising the partitioning of functionality across hardware and software elements.

4.2.3 Design Flow

It is worth revisiting the two design flow models outlined in the Introduction (Chapter 1). These are: (i) the standard method of designing hardware and software separately; and (ii) the newer, 'software-defined' approach based on the SDx tool. These two alternatives, shown in Figure 4.5, represents considerably different methods of designing for Zynq MPSoC, the latter with an increased emphasis on software skills.

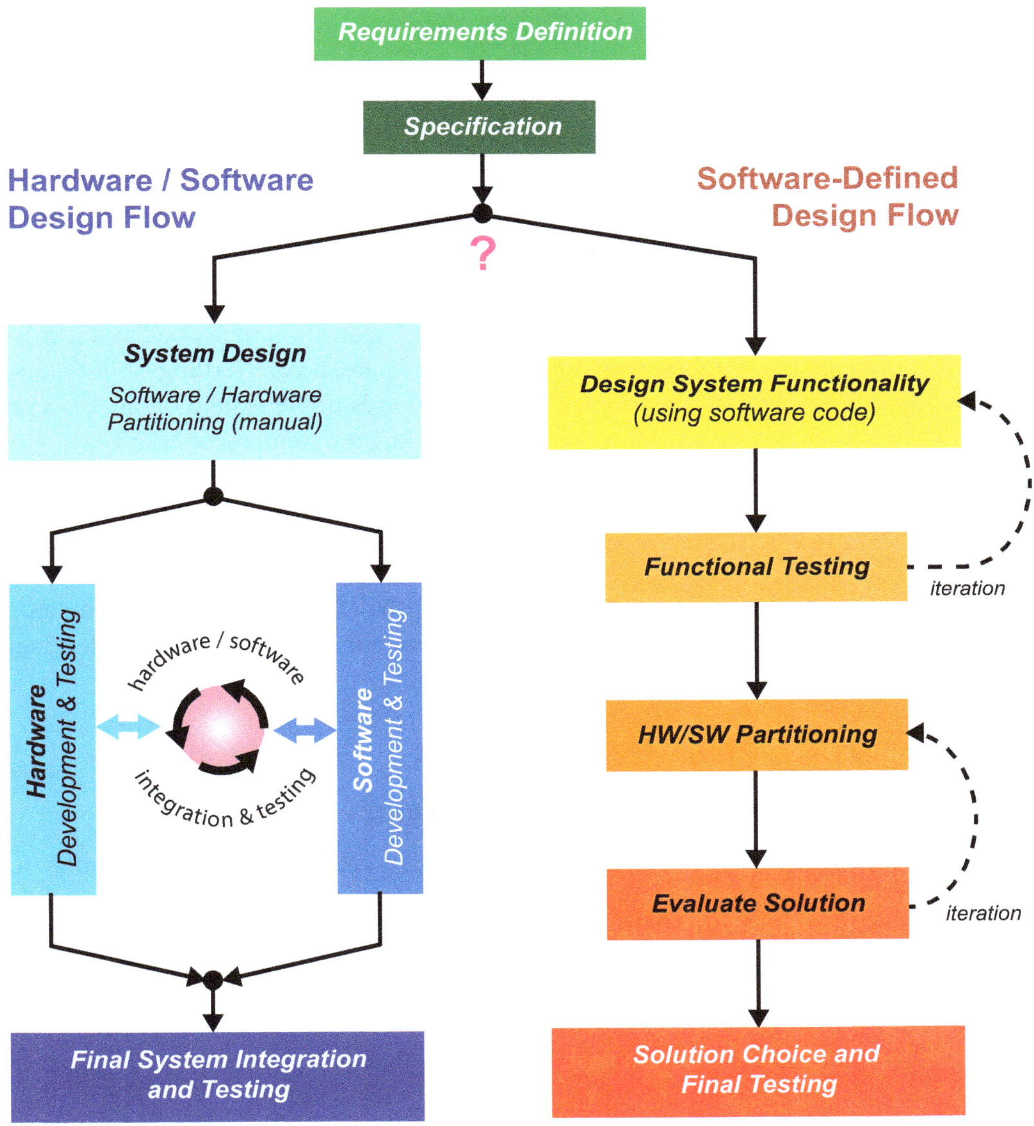

Figure 4.5: Design flow alternatives:
(left) — the 'hardware/software' design flow; (right) — the 'software defined' design flow

Both of these methods will be presented and discussed during the remainder of the book. The standard, hardware/software design flow is covered in Part C of this book (development of the hardware system in Chapter 11, and aspects of software development in Chapters 12 to 15). The software-defined design flow, based on the Xilinx SDx tool, is presented in detail in Part D.

As indicated in Figure 4.5, testing is an integral aspect of both design flows.

4.2.4 Testing

Testing is clearly an important aspect of developing any embedded system. In the 'hardware/software' design flow, software and hardware components can be tested individually (e.g. software in SDK or another development environment, hardware components in an HDL simulator, Simulink, Vivado HLS, etc.) before being integrated, and system testing conducted on a development board. Functional emulation may optionally be performed using a system emulator (QEMU) as a preliminary step, prior to testing on a development board, which allows software to be run as if on the target processor.

In the 'software-defined' design flow, an iterative approach is taken to testing, beginning by verifying the functional correctness of the developed software code. Following on from there, further testing and evaluation takes places while parts of the code are partitioned into hardware, before settling on a final solution. In the software-defined design flow, the hardware part of the solution is implemented on an 'SDx platform', which corresponds to a fundamental hardware system based on the target development board.

4.3 Getting Ready: Design Tools and Development System Setup

Before we can start to develop any designs for the Zynq MPSoC, it is necessary to source the relevant hardware and software components, and to configure the computer system on which development will take place (the 'host' system).

4.3.1 Host System

The first issue to address is the operating system running on the host.

Xilinx tools support various versions of Windows and Linux, and as long as you have one of the supported operating systems installed, you will be in a good position to begin. A list of operating systems supported by Vivado can be obtained from [32]. That said, Linux development for Zynq MPSoC (other than exclusively via SDx — see Chapter 15) will require access to a computer or virtual machine with Linux installed.

In terms of general system requirements, it will be necessary to have at least 50-100GB of free disk space, depending on the combination of tools you wish to install, and allowing some space for the creation of project files.

Operating System Requirements for Linux Development

If you are a regular Linux user, then you should ensure that the Linux distribution you are running is supported by the Xilinx PetaLinux tools. A list of the supported versions is given in [37]. If your installation is one of the supported Linux versions, then you can move straight along to Section 4.3.2.

Unless you already have a supported version of Linux running on the host computer, then there are a few possible options:

1. You may wish to dedicate a separate computer for your Zynq MPSoC development work, and install a supported version Linux on it.

2. You could install a supported version of Linux onto your primary computer, alongside your existing operating system. This would allow you to have a dual-boot configuration with the option to boot into Windows or Linux at start-up. It is beyond the scope of this book to detail how to configure such a setup, but good tutorials are available online. (Just remember to be careful and backup your data before making system changes!)

3. The third and most straightforward option is to install a supported version of Linux in a virtual machine, running on your existing operating system. This option does not require any system changes. One limitation of this approach is that interfacing with the hardware board may be difficult (e.g. programming over JTAG), however in most cases we will simply be copying files to an SD card that is then inserted into a slot on the development board, which can be easily accomplished in this scenario.

The principles and examples covered in this book are largely agnostic to the host OS configuration, and therefore whichever of the three options above you adopt, it should be easy to follow. The only caveat is that we will adopt Ubuntu Linux, which is one of the four operating systems supported for PetaLinux. If you not using Ubuntu but one of the others, then you may observe some slight differences, and if following the tutorials that accompany this book, you may need to adapt the steps a little at times.

4.3.2 Xilinx Software Components

Xilinx provides a set of software tools for systems development. It is useful to introduce these tools and outline their different roles, and also to highlight operating system support for them. First, some general guidance is provided on obtaining the software from Xilinx, and licensing issues.

Sourcing and Licensing Software

All software needed to work with Zynq MPSoC can be downloaded from the Xilinx website, at the URL:

http://www.xilinx.com/support/download.html

In order to download software, users are first required to sign in with a Xilinx account. If you do not already have an account, then there will be an opportunity to register for one first (creating an account is free).

The software components that are covered within this book are:

- Vivado Design Suite - HLx Editions
- Software Development Kit (SDK)
- PetaLinux
- SDx

These can each be located in the download portal, via the link provided earlier in this section. Note that downloads of Vivado and SDx are available for both Windows and Linux operating systems, whereas PetaLinux runs on Linux only (this can be accommodated on a Windows computer by running Linux in a virtual machine).

At the point of installing the Vivado Design Suite, the user will be presented with the choice of tool version to install. Table 4.2 provides a brief summary of options, and confirms that the *System* and *Design* editions are suitable for Zynq MPSoC development, as is WebPACK, provided that your target Zynq MPSoC device is one of the smaller ones. In some cases, software vouchers may be included along with some development kits; normally such vouchers are device-locked to the board supplied in the kit, and in this case, the WebPACK edition will be sufficient.

Table 4.2: Vivado licence options

Version	Features	Purchased Licence Required?	Supports Zynq MPSoC MPSoC?
HL System Edition	As *HL Design Edition*, plus System Generator for DSP	Yes[a]	Yes
HL Design Edition	Synthesis and Place & Route tools; Simulator; *Lab Edition* tools (see below); IP for Debug; Vivado HLS; IP Integrator; Partial Reconfiguration feature.	Yes	Yes
HL WebPACK Edition	As *HL Design Edition*, without the Partial Reconfiguration feature; device limited.	No	Smaller devices only
Lab Edition	Device Programmer; Logic Analyzer; Serial I/O Analyzer.	No	Yes

a. A 30-day evaluation is also available (free of charge).

As part of the software components listed in Table 4.2, it is worth highlighting the *Partial Reconfiguration* feature within Vivado. Partial reconfiguration is a technique that allows one or more portions of an FPGA, or equivalently the PL section of a Zynq or Zynq MPSoC device, to be reconfigured while all other regions of the device continue to operate without being affected [35]. This advanced use of FPGAs and SoCs improves their level of flexibility, by allowing sections of the PL to be time-multiplexed with different functional modules.

If working with the WebPACK edition, it is not necessary to generate a software licence. On the other hand, if you have a licence entitlement or software voucher (e.g. shipped with a kit) for the *Design* or *System* edition, a licence can be created using the Xilinx licensing portal at:

http://www.xilinx.com/getlicense

Licenses can either be generated for individual computers ('node-locked'), or for a license server running on a network ('floating'). Extensive guidance on licence generation and setup is provide in [43]. It is recommended to read the rest of this chapter and the Xilinx licensing documentation, prior to generating a licence.

Vivado Design Suite

Vivado provides an integrated environment for developing the hardware system for Zynq MPSoC (and indeed for all other Xilinx devices). The current version of the tools is denoted *HLx*, reflecting that high-level synthesis capabilities are part of the Vivado Suite.

The Vivado Suite incorporates facilities for various different design methods, including:

- System-level block diagrams (the *IP Integrator* tool, which provides access to a library of parameterisable Xilinx *LogiCORE* IP cores, as well as any user-designed IP cores, and the facility to easily connect them);
- HDL development and simulation, i.e. designing hardware using the VHDL or Verilog languages (supported by the *Vivado Simulator*);
- DSP system development (using *System Generator for DSP*, a block-based tool that runs inside the MathWorks *Simulink* environment);
- High level synthesis from C/C++/SystemC code (the *Vivado HLS* tool);
- Packaging and integration of IP cores created using third party tools, or obtained from third party vendors (introduced into *IP Integrator* using the *IP Packager* feature).

These techniques were previously introduced in *The Zynq Book* [5] and demonstrated in [6] for Zynq-7000 devices, focussing on the Zybo and ZedBoard development boards. The design examples and tutorials featured in this current book will also use some of the same tools and methods.

The development of a hardware system for Zynq MPSoC may combine several of the design methods described above, but in the most straightforward case it will require just one (the IP Integrator tool), where a system design requires only IPs available within the IP Integrator library. Of the other methods, HDL development represents a low level of abstraction, requiring specialist VHDL or Verilog language skills, and arguably involves the highest degree of design effort, leading to the most optimal designs. In constrast, HLS design entry is software-based, and can leverage skills in software coding, which are most prevalent. The development of DSP IPs is closely linked to algorithm development, and the integration of System Generator into Simulink provides access to useful simulation features.

After the design entry stage, Vivado provides an integrated 'flow' comprising the various processes required to take a design from source code and block diagrams, all the way to the bitstream (.bit) file that is used to program the Zynq MPSoC with the hardware system. The main steps involved are:

- **Synthesis** — this stage involves translating the HDL code into a netlist file, which is a representation of the design in terms of low-level gates and connections.

- **Implementation** — comprised of three sub-processes (Translate, Map, and Place & Route), implementation involves converting the netlist representation of the design into a device-specific implementation, where the precise resources, configurations, and signal routes are defined.

- **Bitstream generation** — the last stage is to create a bitstream file from the implemented design. The bitstream file can then be downloaded onto the device to program it with the hardware system.

These three processes can be run interactively using the Vivado GUI, or alternatively can be scripted using Tcl (also known as *project mode* and *non-project mode*, respectively). Various statistics, reports and views are generated, including resource utilisation and timing reports, schematic views and device floorplans. This information can be interpreted by the designer and used to refine the design.

Vivado allows users to set design constraints, or in other words to configure or limit certain aspects of the implementation. Examples include setting I/O pins and standards, defining the positioning of design elements on the PL, and placing limits on the timing of clocks and/or other signals. Constraints are specified in a Xilinx Design Constraints (XDC) file and provided as an input to synthesis and the subsequent process. The XDC file can either be written by hand, edited from a master constraints file (which is normally provided by board vendors), or in some cases the design tools will automatically generate constraints.

A screenshot of the Vivado interface is shown in Figure 4.6, based on a reference design included with the Vivado 2017.1 software release, and with selected sections highlighted. Notice in particular that the panel on the left hand side guides the user through the 'design flow', i.e. steps relating to design entry and simulation, followed by synthesis and implementation, and lastly bitstream generation and device configuration. In the main design window (top right), Vivado shows a block diagram generated using the IP Integrator tool; various other content may be displayed in this area, depending on the task being undertaken.

In addition to the main components of Vivado mentioned so far, there are also tools for programming the target device, and for on-chip signal analysis and debugging. These tools form the basis of the *Lab Tools* version of the Vivado Suite, and are also included in the *Design* and *System* editions.

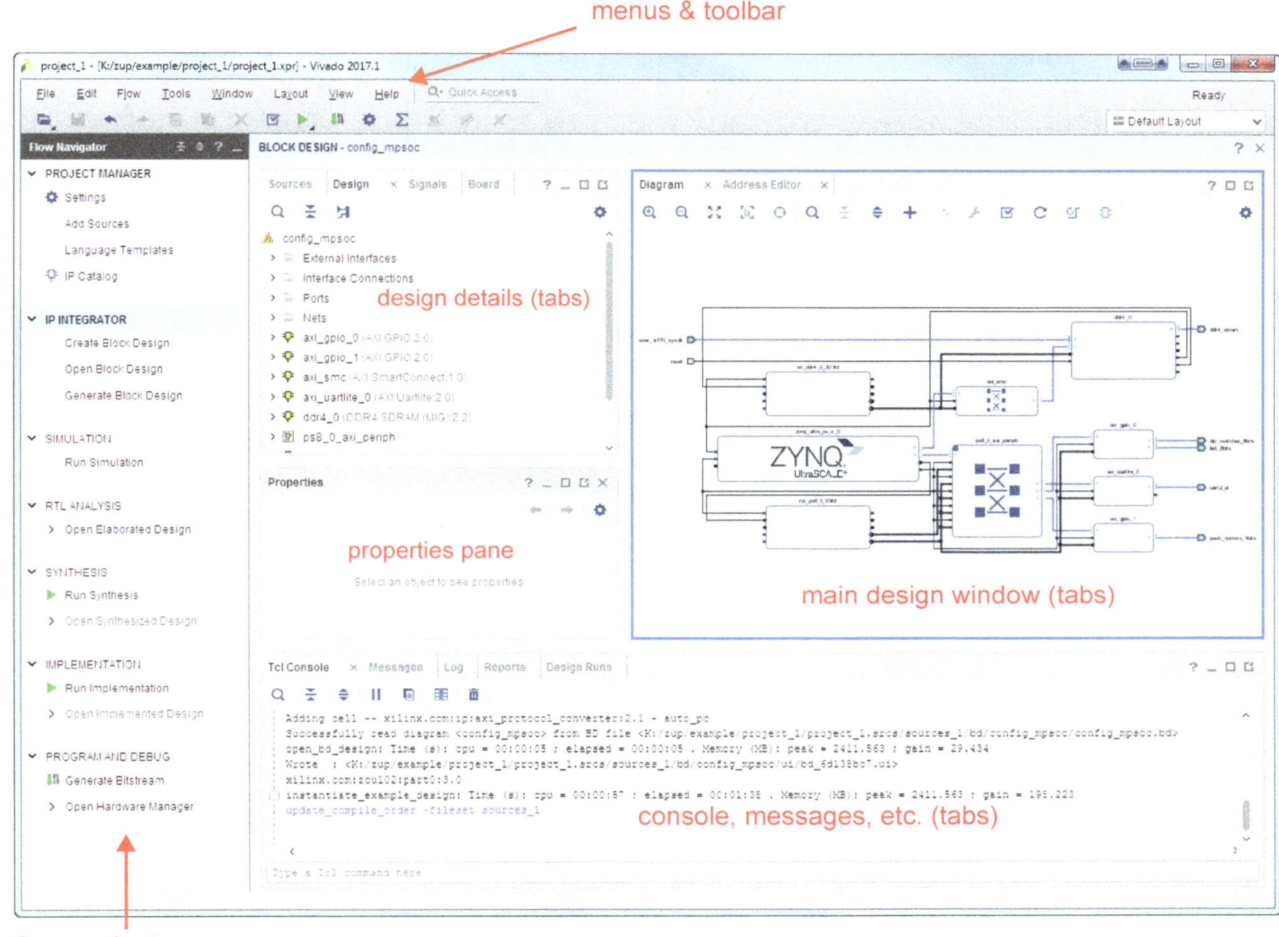

Figure 4.6: Example view of the Vivado interface

Software Development Kit (SDK)

The Xilinx SDK can optionally be installed with Vivado, or alternatively it may be downloaded and installed separately (the latter is convenient for software engineers not involved in design of the hardware system).

SDK is an environment for software development and debugging, based on the open-source Eclipse® platform. It allows applications to be developed for the processor types available within the Zynq MPSoC device, i.e. the Arm Cortex-A53 processors in the APU, Arm Cortex-R5 processors in the RPU, and any MicroBlaze processors used in the design (these may optionally be instantiated in the PL, and also form the basis of the Platform Management Unit (PMU)).

Board Support Packages (BSPs) can also be created in SDK. A BSP is a software interface for a defined set of hardware, including a set of drivers that allows the software and hardware elements of a system to interact. As such, a BSP can be thought of as the lowest level of the software stack, while 'hardware' refers to the custom hardware system that has been designed using Vivado for deployment on the target device.

The operating system options available for generating BSPs in SDK are:

- **Baremetal** — A lightweight OS with basic low-level functions, suited for simple requirements.
- **FreeRTOS** — An OS for real-time applications; needed where deterministic timing is critical.
- **Linux** — A fully fledged OS for rich application development.

Further detail about each of the these OS options is included in Chapter 12.

SDK also includes a variety of other facilities for software development and debugging, including a GNU-based compiler, the facility to program the target device over JTAG, and features for debugging code running on the development system. Further information about SDK, as it relates to Zynq MPSoC development, can be found in [51].

PetaLinux Tools

Linux is one of the most common operating systems for embedded devices, and there are a variety of Linux distributions available, such as *Ubuntu*, *Debian*, *Fedora*, and others. PetaLinux represents a set of tools for Linux development targeted to Xilinx devices, as well as enabling developers to work with other system software components (bootloaders, hypervisors, etc). As PetaLinux is suitable for both soft and hard processors, it can generate an OS for the 'soft' MicroBlaze core, as well as the Arm processors present on the Zynq and Zynq MPSoC. It is also possible to target the Xilinx Quick EMUlator (QEMU) virtual platform in place of a physical processor, which facilitates software development and testing without access to a development board [50].

PetaLinux (or *PetaLinux Tools*) is installed on the host computer and used for development, with some interaction with other Xilinx tools, e.g. Vivado is used to generate the hardware platform on which a Linux system is deployed, while Xilinx SDK can be used to develop software applications. PetaLinux is supplied and supported by Xilinx, and made available via a no-cost license. Using PetaLinux, developers can:

- **Create or Reuse Board Support Packages (BSPs)** — After generating a BSP based on the current hardware system, PetaLinux is able to maintain synchronisation subsequently, i.e. to update drivers when a change is made to the hardware system. This is useful because the hardware system may undergo several iterations during the design process.
- **Customise, Build and Package Linux** — Designers can build the various components that comprise a Linux system, including the kernel, libraries, bootloader, and file system (note: a more detailed introduction to Linux follows in Chapter 15). There is the option to customise any of these components, if

desired. PetaLinux also has a tool for packaging the various parts of Linux into a small set of files that are transferred to the target system, and then booted.

- **Create and Integrate Custom Components** — Templates are available within PetaLinux that can be used to develop customised components (e.g. device drivers, application code, libraries, etc.), and to integrate them into a PetaLinux project.

- **Work with Reference Files** — As part of PetaLinux, users can obtain a reference BSP for their chosen platform from the Xilinx download site; for instance, a BSP is available for the ZCU102 board (note that BSPs are not part of the main PetaLinux download and must be obtained separately). The reference BSP represents a basic 'quickstart' option and can be adapted for custom projects. Additionally, the primary PetaLinux download includes a reference distribution of Linux that includes all main system components, and which has been verified for deployment on Xilinx devices. This set of files can be used as a starting point for user-specific designs.

It is important to note that PetaLinux does not represent a particular distribution of Linux, rather it enables users to work with their desired distribution. Popular distributions for deployment on Xilinx devices are the Xilinx Linux distribution [33], and the Yocto project [44]. Further, we should mention that using PetaLinux is not the only way to develop a Linux system to run on a Xilinx Zynq MPSoC device — a variety of other Linux development tools are available, and ultimately the choice of OS comes down to user requirements and preferences. New Linux users may however find that PetaLinux is a good choice for getting an OS working on the Zynq MPSoC, due to the availability of Xilinx-specific support and documentation [36], [37].

SDx — for the 'Software Defined' Design Flow

SDx is a relatively new tool that builds upon the capabilities of Vivado HLS to generate hardware from high level (C, C++, or SystemC) code [39]. The distinction between these tools is that SDx allows the entirety of a system to be described using software code, whereas Vivado HLS is specifically concerned with generating parts of the hardware system. In SDx, the design starts off as a software program running on the PS. Some sections of the code will be recognise as suitable for offloading to hardware, and these are then generated into IP. SDx manages the interfaces and data transfers between software and hardware automatically. Additionally, the option exists to deploy an OS such as Linux on the target platform.

While this may sound quite simple (!), the Xilinx tools have quite a lot of work to do in the background. They must employ knowledge of the target device (e.g. its performance and available resources), generate IP cores, create the AXI connections between the processor and IP cores, establish all of the necessary low level circuitry, such as clocks, resets, and interrupts, and build the OS. Where Linux is selected, one of the outputs generated is an image that can be loaded onto an SD card, transferred to the development board, and used to boot the system directly. Without SDx, the designer (or design team) is responsible for taking care of all of these aspects manually, and hence the use of SDx can significantly reduce development time.

The designer does still retain high level control over the system implementation generated by SDx. He or she can direct the tool regarding which functions should be partitioned into the PL part of the device, and in doing so, define other aspects of the implementation such as pipelining, parallelism, and interfacing via the use of 'directives'. Integrated profiling tools can be used to report on the relative performance, latency and resource cost of the implementation 'solutions' under consideration, and hence inform design decisions. Extensive further discussion of SDx will follow in Part D of the book.

A Note on Third Party Tools

A number of other development tools are available from third parties, which enable different aspects of design for Zynq MPSoC. These range from HDL simulation environments, to block-based system design, to software and hardware debugging tools; and include both commercial and open-source software. Design houses may adopt a set of preferred tools that suit their requirements.

These third party tools form part of the 'ecosystem' that exists around Xilinx SoC devices, and will be discussed further in Chapter 20.

4.3.3 Hardware Requirements

In addition to a host computer, you should ideally also have a Zynq MPSoC development board, and its associated cables, in order to progress design of a Zynq MPSoC system. An example Zynq MPSoC development board is the ZCU102, which was described in Section 4.1.1 and pictured in Section 4.1 (more on development boards will follow in Section 4.4). On the other hand, if you do not have access to a development board, then it will still be possible to make progress using the Xilinx tools in conjunction with the QEMU emulator. QEMU supports emulation of Zynq MPSoC designs on a functionally equivalent (but not timing equivalent) basis, which makes it suitable for testing APU-targeted software in particular.

Assuming development is undertaken using a board kit, there are some other items of hardware that may be useful. In particular, we need the facility to write files onto an SD card, to be transferred to the development board for programming.

- **An SD card reader/writer** — your computer may have an integrated SD card reader and writer, or if not, then it is possible to purchase a standalone, USB version at low cost ($5-10 US).

- **An SD card** — the medium for storing files, it is important that a Class 10 card is used (lower speed cards may not work). A card of 8GB or 16GB capacity will be large enough for most uses. The cost of such a card is also modest, at around $5 - $10.

- **An Ethernet cable** — an Ethernet cable may be included with your kit, but if not, it would be useful to source one (less than $5 for a short cable). This will be used for data transfer between your computer and the development board. Consider this in conjunction with the next item.

- **USB-to-Ethernet adapter** — If there is no spare Ethernet socket on your computer for making a connection to the development board, then a good alternative is to use a USB-to-Ethernet adapter. This connects to the computer using a USB socket, and provides an Ethernet socket that can be connected to the Ethernet port of the development board. The estimated cost of this item is $15 - $25.

Depending on the system being developed, other cables and external hardware may be needed. For instance, if developing a video processing system with DisplayPort output, then a compatible monitor and cable would also be required.

4.4 Development Boards and Supporting Resources

Design teams often adopt a standard, commercially available development board for prototyping, at least during the initial phases (the eventual outcome may be to develop a custom board). Other projects, particularly student projects and academic studies, may only ever use a standard development board. The advantage of using such a platform is that it is pre-verified by the manufacturer, and there may be reusable components, reference designs and other support available. Additionally, as a large number of other engineers are likely to be working with the same board, help can also be found via support forums and online communities — a good example being

http://www.zedboard.org ,

the website originally launched for the *ZedBoard* (Zynq-7000 board), which has also since expanded to include other products including a Zynq MPSoC System-on-Module (SOM). The ZedBoard website includes forums and an opportunity for community sharing of ideas, problems, and work in progress.

Your development board should be accompanied by various resources (normally via download) that support its use. The important items to look out for are:

- **User guide** — In addition to the user guides and documentation relating to Zynq MPSoC devices, there is also likely to be a guide available for the development board. This will include detailed information on the various components present, as well as guidance on setting up and programming / booting the board.
- **Board support files** — This set of files is copied into the installation directory of the Xilinx development tools, and allows Vivado to provide board-specific options to the user as they develop their designs. For instance, if a system were to include General Purpose Input Output (GPIO), then a GPIO block would be placed into the Vivado IP Integrator design, and connected appropriately. With board support files installed, the GPIO block is pre-populated with available options for the target board, e.g. 8 x LEDs, 4 x push buttons, and so on.
- **Constraints file (*.xdc)** — The constraints file provides a list of all input/output pins on the Zynq MPSoC device, giving their locations (letter-number designators such as 'AE9', 'P7', and so on), and the I/O standards associated with each one. This can be a long list for a board containing a large Zynq MPSoC! The user must edit the file to remove (or comment out) ports that are not used in the current design, and to rename top-level ports if named differently from the defaults given in the constraints file.
- **Targeted Reference Designs (TRDs)** — Xilinx provide TRDs to give designers a headstart in understanding how to use the features of the board, which therefore enables new designs to be progressed

more quickly. Depending on the board in question, there may be a number of TRDs available to illustrate the use of different interfaces or features.

- **Schematics and Bill-of-Materials (BOM) files** — Schematics and BOM files provide a reference useful to those wishing to develop their own custom boards. For instance, a company may wish to develop their own board using a larger Zynq MPSoC device, and/or with a different mix of external memories and interfaces. Understanding how a development board has been designed, and having a list of all of the components used, provides a great starting point!

4.5 Resources and Support

A number of resources are available for Zynq MPSoC design, in terms of the device itself, development boards, and the software tools that can be used to develop systems for Zynq MPSoC. This includes documentation (user guides, reference manuals, etc.), reference designs and tutorials, and other online support. While some reference material is specific to Zynq MPSoC, this is augmented by support for Xilinx SoC and FPGA design generally.

The central point for accessing Xilinx documentation, knowledge base, and forums, is via the link:

http://www.xilinx.com/support.html

Selected types of resources and support are reviewed over the next couple of pages.

4.5.1 Information about Zynq MPSoC Devices

This book is intended to provide an accessible introduction to Zynq MPSoC devices and associated tools and design flows, but it cannot hope to cover Zynq MPSoC in the level of depth possible in Xilinx technical documentation. A number of key documents are published and maintained by Xilinx to provide full detail on various aspects of these devices. It is worth highlighting in particular:

- **Zynq UltraScale+ MPSoC Overview Data Sheet (DS891)** — A compact guide (~40 pages) that includes listings of key features, product tables, and brief, high-level descriptions of the resources present on the device [49].

- **Zynq MPSoC Technical Reference Manual (UG1085)** — A detailed 1,000+ page guide to the Zynq MPSoC architecture! See reference [52].

Several User Guides are common amongst Zynq MPSoC, UltraScale, and UltraScale+ FPGAs, because of the equivalence of the PL architecture. These include guides on CLBs, memories, DSP resources, input/output blocks and high speed transceivers, clocking resources, and so on, and can be easily located via the documentation section of Xilinx support site.

It is useful to highlight that the processing system on the Zynq MPSoC is based around Arm components (i.e. the applications and real-time processors, and graphics processor). Detailed information about these elements of the architecture should be obtained from Arm directly — Xilinx only documents aspects that are Xilinx-specific, whereas the Arm processors are more widely used and have their own dedicated documentation.

4.5.2 Software Tools Support

As outlined in Section 4.3.2, designing for Zynq MPSoC involves becoming proficient with one or more software tools, depending on your role within a development team, or your preferred design methods. There are a few places to look for support with these tools:

- **'Quick Take' videos** — Short videos (most are less than 15 minutes) that introduce design tools, or explain particular features of tools and target devices.
- **Design Hubs and technical documentation** — Design Hubs are web pages that collate all technical documentation and other resources relating to a particular software tool, device, or development board. This includes user guides, data sheets, application notes, white papers, reference designs, and links to other relevant resources. These materials can all be sourced individually. A subset of software documents are specific to the Zynq MPSoC device, e.g. guides for OpenAMP and QEMU for Zynq MPSoC.
- **Training and tutorials** — Formal training courses run by Xilinx or authorised training providers (usually at cost). Xilinx University Program (XUP) additionally runs workshops for the academic community. Self-paced tutorial-style material is also available to download, taking the form of guidance documentation and design files.
- **Knowledge Base** — A catalogue of Answer Records relating to common queries, known issues and other miscellaneous support issues. These are maintained by Xilinx engineers and include detailed version and status information.
- **Community forum** — Xilinx-hosted forums are available for engineers to post questions and engage with other users on topics of interest. Xilinx staff moderate, and may contribute to, forum discussions.
- **Xilinx Wiki** — A community-editable guide for using Xilinx tools and devices. See:

  http://www.wiki.xilinx.com/

- **Xilinx Github** — The portal from which reference code can be obtained from Xilinx. See:

  https://github.com/xilinx

- **Documentation Navigator** — A software application enabling easy access to Xilinx documentation. Available from Xilinx download site, either standalone or as part of Vivado, for installation on the development computer.

Bear in mind that third party sources of information can be useful too — for instance, embedded systems magazines and journals. For those aspects of the development tools and software that are not Xilinx-specific, there will be separate documentation and support. This includes design tools such as MathWorks' MATLAB and Simulink tools, National Instruments' LabView, and also software components that you may wish to deploy on the Zynq MPSoC, including Linux, FreeRTOS, Android, Xen Hypervisor, and so on.

4.5.3 Support for System Design

Xilinx have produced formal guidance for designers to follow in order to achieve favourable results from their system designs, referred to as the *UltraFast Design Methodology*. The stated benefits of this approach, which has been developed based on the experiences of a collection of experienced users, are to "maximise system performance, reduce risk and enable accelerated and predictable design cycles".

There are a small set of UltraFast Design Methodology guides, covering different topics. The main document of interest is the *UltraFast Embedded Design Methodology Guide* [41], with supplementary information collated on this Xilinx webpage [40].

4.6 The Wider Ecosystem

The Zynq-7000 has become well-established over the past five years or so. Its widespread adoption has partly due, in part, to enjoying a rich set of supporting resources, of various types, many of which derive from third parties. This is collectively what we refer to as an 'ecosystem' — the relatively unstructured body of external activities and resources that complement core Xilinx design tools, development boards, and support, and which organically grows and evolves.

The ecosystem includes numerous companies and organisations involved in complementary activities, including the development of standards, design and debugging tools, development boards and add-on cards, and software components of various kinds. It also encompasses training resources, consultancy services, websites and blogs, and the 'community' of engineers and software programmers who work in the area. Having access to these resources makes it quicker and easier to create high quality Zynq-based systems. The presence of an ecosystem also means that developers have more freedom of choice when it comes to the tools, methods and platforms they use to create their system designs.

The ecosystem of Zynq MPSoC, like the device itself, has much in common with Zynq. Many of the resources relevant to Zynq are also relevant for Zynq MPSoC. Tools for multi-processing, for instance, are applicable to Zynq but arguably even more valuable for getting the best out of Zynq MPSoC. With Zynq MPSoC, there are more diverse resources in the PS, and a larger number of processing elements, which extends its design possibilities. The additional processing elements on Zynq MPSoC, such as the graphics and video codec unit, bring new possibilities and therefore expand the scope of the supporting ecosystem.

A conceptual diagram of the ecosystem is provided in Figure 4.7. Many of these bubbles will be discussed in further detail during the remainder of the chapter.

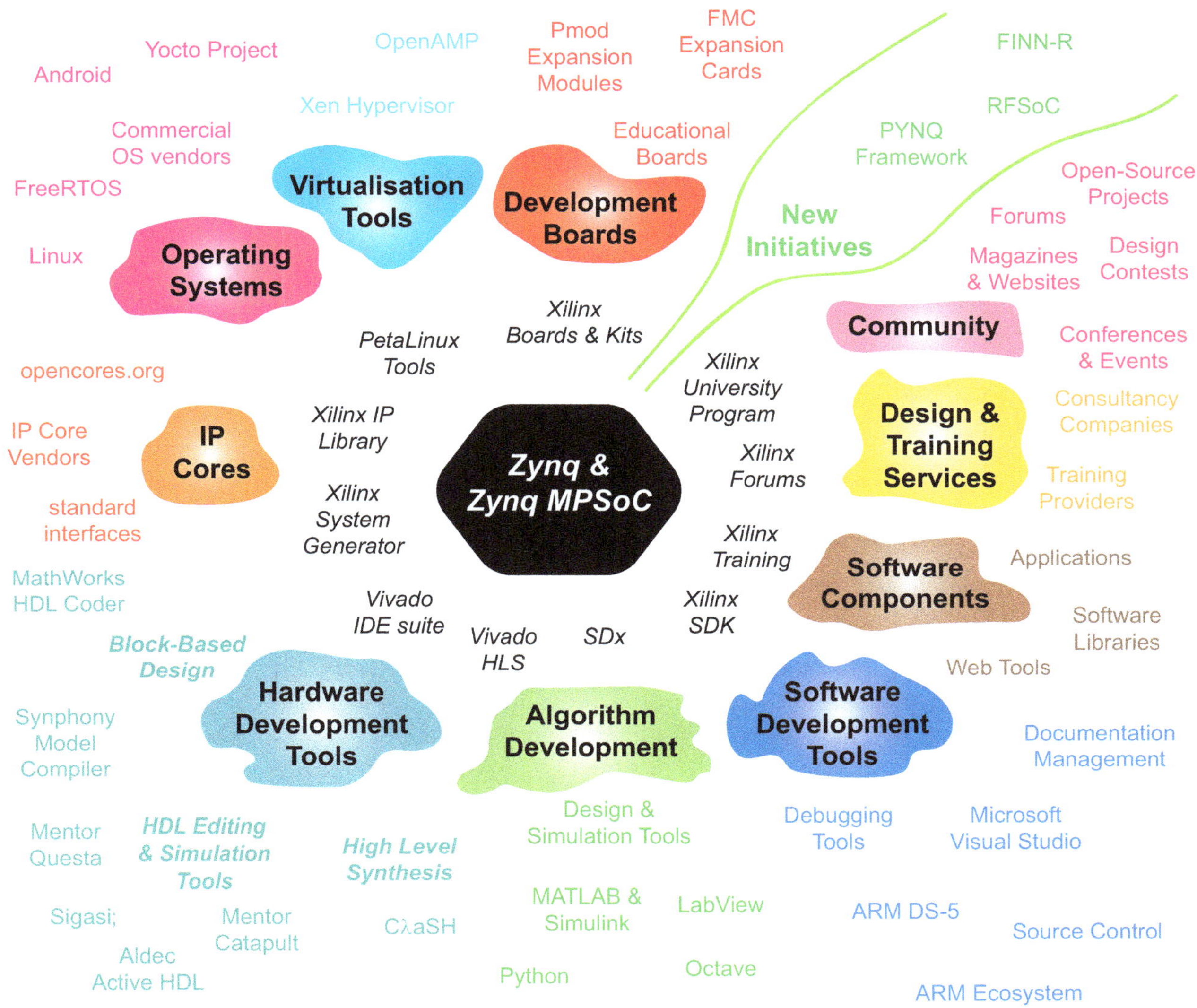

Figure 4.7: High level concept of the Zynq MPSoC 'ecosystem'

4.6.1 Hardware System Design Tools and Components

One of the most fundamental resources for developing an SoC design (of any sort) is the set of design tools available for doing so. These fall into a few different categories, given the multi-faceted nature of the Zynq MPSoC device. We begin by looking at third party hardware system development tools.

IP development is the main aspect of hardware development that can be progressed using third party tools and resources. The designer can opt to generate their own IPs, using one of a variety of different methods, or look for pre-generated IP from a third party (whether free or paid-for). IPs can be packaged into a standard IP-XACT format, compatible with IP Integrator, using Xilinx *IP Packager* tool [45]. It is worth noting that IP development is not Zynq MPSoC-specific; the same principles and processes also apply to Zynq and FPGA systems design.

HDL-based Tools for Creating IP Cores

Vivado includes a code editor and simulator for developing and testing HDL code (VHDL or Verilog), which is a primary method of developing IP. If preferred, however, HDL development can be progressed externally to Vivado, in a development environment such as Mentor Graphics *Questa® Advanced Simulator* [18], or Aldec's *Active-HDL* [1]. Another alternative is the *Sigasi;* tool [25], which offers advanced features for HDL code development and design management, and can be paired with a simulator for debugging and verification.

Block-based Tools for Creating IP Cores

Xilinx provides a block-based design tool for developing IP cores, in the form of System Generator, which runs on top of the MathWorks' *Simulink* tool [14]. Other, third party options also exist.

MathWorks have their own block-based HDL generation tool, *HDL Coder* [12], which facilitates generation of VHDL or Verilog from Simulink designs (using a subset of HDL-synthesisable blocks) as well as a subset of MATLAB code. It is also worth noting that HDL Coder drives Vivado processes from within its own GUI.

IP Core Creation Using Software Languages

As for the previous categories, Xilinx provides its own tools. In this case, for HLS of software-based code into HDL, in the form of *Vivado HLS* and the *SDx* development environment. There are also third party alternatives, such as the Mentor® *Catapult®* tool [15]. HLS methods remain an area of active interest in the research community, one example being the LegUp HLS project [28].

Another category of design tools that has emerged from the open source community uses software languages in a more HDL-like manner. Good examples include *CλaSH* [4], a functional programming approach to hardware design, based on the Haskell language; *MyHDL* [20], which uses Python as the design language, and *Chisel* [27], which adopts Scala.

Third Party IP Cores

Design reuse, utilising cores from third parties, may be adopted when functionality is needed that does not reside in the Xilinx IP library, and when the designer does not wish to create the required IP. This may be for reasons of cost, time-to-market, or in cases where there is a lack of in-house specialist expertise.

There are two main avenues for obtaining third party IP:

- Via open-source channels, where the author of the core has made their IP freely available for reuse. Such cores may be obtained from the website www.opencores.org, for example. Open-sourced cores have the advantage of zero cost, but may not have the reliability, documentation, or support that comes with a commercially obtained IP. Therefore, some verification work may be required to ensure the robustness of open source IPs.
- Purchased as a product from a design house. Some IPs may be extremely complex, representing high value in terms of development cost. This may include, for instance, specialist IPs for video processing, cryptography or communications. Documentation and support for such IPs would normally be available from the vendor, who may also provide reference designs. An example of such IPs are the *logicBRICKS* image and video processing IP cores available from Xylon [53].

Whichever mechanism is used, the ability to source IPs externally has the potential to save significant development time and cost, and is therefore a useful option to have.

4.6.2 Software System Design Tools and Components

Moving on to software aspects of the system, we now consider the various third party software components that may be included in a Zynq MPSoC design, as well as the development tools needed to create them.

Operating Systems

One of the most important decisions in the development of a Zynq MPSoC design will be the choice of OS (or OSs!) to adopt. As there are up to 6 cores on the device (four applications processors and two real-time processors), then there is the potential to use more than one OS.

The predominant choice for the applications processors is likely to be the open-source OS, Linux. Several distributions of Linux are available, such as *Ubuntu* and *Fedora*, each having their own online communities and support mechanisms. One option is to deploy such a distribution to the Zynq MPSoC directly.

Tools are also available to create custom distributions of embedded Linux for target embedded platforms such as the Zynq MPSoC. The Xilinx *PetaLinux* tools perform this function, using a version of Linux that is based on the open-source *Yocto Project* [54]. A *Yocto*-based solution is also available from Mentor on a commercial basis [17].

Other, third party OSs may be ported to the Zynq MPSoC organically over time. For example, *FreeBSD* [9] was previously ported to Zynq by the open source community, and more recently, is was also ported to Zynq MPSoC.

For the RPU, it is likely that a dedicated real-time OS will be required, which will provide significant benefits over a 'baremetal' approach. *FreeRTOS* [10] is free (as the name suggests!) and widely used, and is therefore likely to be a popular choice for the Zynq MPSoC's Arm Cortex-R5 processors. Two closely related alternatives to FreeRTOS are *OpenRTOS* and *SafeRTOS*. Both of these represent commercial software, but offer additional benefits: in the case of OpenRTOS, the ability to modify the core code without obligation to share your changes; while SafeRTOS is a derivative of FreeRTOS that has undergone a higher level of analysis, testing and documentation for compliance with the safety standard IEC 61508.

Other candidate OSs exist for the applications and real-time processors too — see [31] for further details.

Software Development Tools and Resources

Xilinx provides its own Eclipse-based SDK tool, as well as PetaLinux tools, but developers also have the option of using their preference of development environment. Possible options include Microsoft Visual Studio, and Arm DS-5 Development Studio [3]. All of these environments provide facilities for code editing and debugging.

The processor emulation platform, QEMU, can also be a useful tool, especially when physical access to the target platform is not available. In this case, QEMU can be used to emulate PS of the Zynq MPSoC, as well as the MicroBlaze processor of the PMU. QEMU is an open-source project [24], of which Xilinx maintains its own branch particular to MicroBlaze, Zynq, and Zynq MPSoC-based processors [38].

Software Components

A Zynq MPSoC developer may also require other software components to complete their design, such as a hypervisor for virtualisation and AMP. In this regard, the Arm ecosystem may be leveraged too, as a community of products and services have grown up around their processor cores.

Examples of virtualisation software include the open-source Xen Hypervisor project [30], which Xilinx is actively involved in. Xen is supported for deployment on the Zynq MPSoC [46]. Another option is Mentor's *Embedded Hypervisor*, which represents a commercial alternative [16]. OpenAMP (Open Asymmetric Multi Processing) is a further tool for multiprocessing, which Xilinx supports for Zynq MPSoC [34].

Depending on requirements of the software project, it may also be useful to incorporate further libraries and components. A large number of such libraries and components are available (especially if using Linux), including web servers, graphics applications, and software libraries. Examples of software libraries include OpenCV for image and video processing applications [22].

Other Software and Support

There are a few other aspects of software development where tools and resources from the wider community may prove valuable. For instance:

- **Desktop virtual machine** — A virtual machine such as *VirtualBox* is useful when requiring access to a Linux OS for development purposes (for instance when using PetaLinux tools), if a computer with Linux installed is not available.

- **Source control software** — Software development projects, especially when involving multiple engineers, benefit from employing suitable source control. A popular choice currently is *GitHub*, an internet-based service that may be used on a public or private basis for source control and sharing.

- **Documentation tools** — Open source tools such as *Doxygen* can help to create and manage code documentation [8].

4.6.3 Algorithm Development and Interfacing

Slightly separately from our previous section, we ought to mention software-based tools for developing the algorithms for implementation on Zynq MPSoC based systems (including both elements targeted towards the PL, and the PS). There are a multitude of possibilities here, including standard software languages such as C++. A few other possibilities are outlined below, but this is certainly not an exhaustive discussion!

Popular tools for algorithm development include the technical computing software, MATLAB and Simulink [13], [14]. This technical computing suite includes facilities for code- and block-based design, as well as a set of application-specific toolboxes (for instance, supporting communications, image and video systems, machine learning, and fixed point arithmetic). As noted earlier in this chapter, the *HDL Coder* tool provides a direct path to generate HDL code from simulation models. Hardware support packages are also available, and these allow selected development boards to be interfaced with the simulation environment.

LabView is another software tool for algorithm development, simulation, and hardware interfacing [21]. LabView enables algorithms to be developed primarily using graphical methods, and has a strong emphasis on integration with National Instruments hardware for test, measurement, and industrial control systems applications. HDL can be generated from LabView as part of this process.

On the open source side, the Python language [23], supported by libraries like *NumPy* for scientific computing, is a good option for algorithm development. Python has a large user-base and its library support is steadily expanding. Python also forms a primary element of the PYNQ framework, which will be introduced later in the book (see Chapter 22).

Finally, another open-source tool worth being aware of is Octave — an open-source technical computing environment for code-based development and simulation [11].

As indicated earlier, there are a large number of possibilities for algorithm development. An individual choice may depend on programming skills and preferred methods, as well as the availability of specialist libraries, the ability to interface with hardware, and integration with other aspects of the design flow.

4.6.4 Hardware and Peripherals

The development of a Zynq MPSoC-based system requires access to hardware for prototyping, in the form of a development board (often along with expansion cards or modules). We touch on some of the available options in this section.

Development Boards

Xilinx offers a variety of development boards across its portfolio, and these are extensively complemented by third party board vendors. Some boards are general purpose, incorporating a variety of interface types, while others are more application-specific, perhaps targeting high speed connectivity for data-centres, or computer vision for the automotive sector. A subset of kits actually comprise two parts: a 'System-on-Module' and a carrier card (in this case, the Zynq MPSoC chip is hosted on the module, which plugs into a larger carrier card for access to an extensive set of interfaces and debugging facilities).

Vendors of MPSoC-based boards include *Avnet*, *Enclustra*, *Topic Embedded Products* and *Trenz Electronic*. The Xilinx website can be consulted for a full list of boards and vendors.

Expansion Cards and Modules

In the system-on-module approach mentioned in the previous section, the Zynq MPSoC chip resides on the smaller PCB, which is plugged into a larger one with more extensive I/O facilities. Regular development boards include their own peripherals, and usually one or more general purpose ports for adding expansion modules. There are two main standard interface types (Pmod and FMC), which give access to a wide variety of peripheral functionality.

Pmod add-on cards are small, low cost expansion modules, and include functionality ranging from simple GPIO facilities, to communications modules (WiFi, Bluetooth, etc.), lower rate DACs and ADCs, and other sensors and actuators. Pmod ('Peripheral module') is a propreitary interface type developed by *Digilent* [7], which is also supported by other third party vendors, including *Maxim Integrated* and *Analog Devices*. A subset of Pmod expansion modules are shown in Figure 4.8.

FMC expansion cards support faster rates of data transfer (up to 40Gb/s overall), and are typically used for higher value applications. This includes high rate ADCs and DACs, camera modules, serial connectivity, and SDR modules for transmitting and receiving radio signals (an example SDR card is shown in Figure 4.9). FMC vendors include *Abaco*, *Analog Devices*, and *Texas Instruments*.

Figure 4.8: Examples of Pmod expansion modules

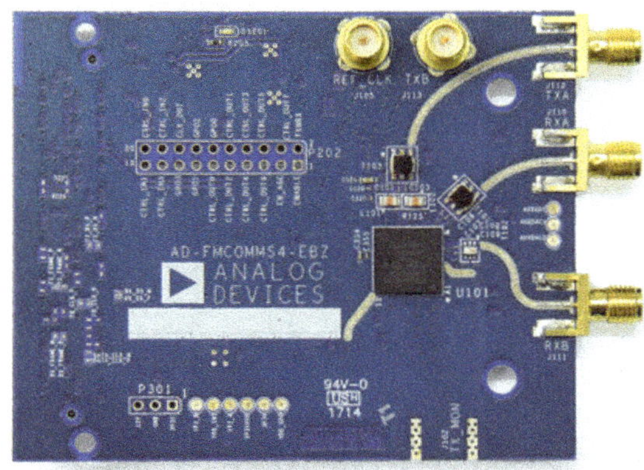

Figure 4.9: Analog Devices FMCOMMS4 SDR card (FMC connection)

4.7 Chapter Summary

This chapter has introduced the various aspects of designing for Zynq MPSoC from a high level perspective. The composition of the Zynq MPSoC, and the roles of hardware and software components have been outlined, with appropriate links made to the device architecture. We have also briefly covered the design flow alternatives and tools available for creating Zynq MPSoC designs, with more detail to follow on these in later chapters. Practical information was included on hardware and software setup requirements, and the available support resources, as was a brief discussion of considerations for Zynq MPSoC system development. Lastly, third party resources from the surrounding 'ecosystem' were also reviewed.

4.8 References

Note: All online sources last accessed March 2019.

Some of the Xilinx documentation referred to below has version-specific URLs. If you are working in a newer version of the tools, check for updates on the Xilinx website, or try adjusting the link according to your version.

[1] Aldec, Inc., "Active-HDL: FPGA Design and Simulation" webpage.
Available: https://www.aldec.com/en/products/fpga_simulation/active-hdl

[2] *Android* website.
Available: https://source.android.com

[3] Arm, Ltd., "DS-5 Development Studio" webpage.
Available: https://developer.arm.com/products/software-development-tools/ds-5-development-studio

[4] *Cλash* website.
Available: https://clash-lang.org/

[5] L. H. Crockett, R. A. Elliot, M. A. Enderwitz, and R. W. Stewart, *The Zynq Book*, Strathclyde Academic Media, 2014.
Available: http://www.zynqbook.com

[6] L. H. Crockett, R. A. Elliot, M. A. Enderwitz, and D. Northcote, *The Zynq Book Tutorials for Zybo and Zedboard*, Strathclyde Academic Media, 2015.
Available: http://www.zynqbook.com

[7] Digilent, Inc., "Pmod Modules" webpage.
Available: https://store.digilentinc.com/pmod-modules/

[8] Doxygen, "Doxygen" webpage.
Available: http://www.stack.nl/~dimitri/doxygen/

[9] *FreeBSD* website.
Available: https://www.freebsd.org/

[10] *FreeRTOS* website.
Available: http://www.freertos.org/

[11] GNU, *Octave* website.
Available: https://www.gnu.org/software/octave/

[12] MathWorks, *HDL Coder* product page.
Available: https://uk.mathworks.com/products/hdl-coder.html

[13] MathWorks, *MATLAB* product page.
Available: https://uk.mathworks.com/products/matlab.html

[14] MathWorks, *Simulink* product page.
Available: https://uk.mathworks.com/products/simulink.html

[15] Mentor, Inc., "Catapult High Level Synthesis" webpage.
Available: https://www.mentor.com/hls-lp/catapult-high-level-synthesis/

[16] Mentor, Inc., "Mentor Embedded Hypervisor" webpage.
Available: https://www.mentor.com/embedded-software/hypervisor/

[17] Mentor, Inc., "Mentor Embedded Linux Development Platform" webpage.
Available: https://www.mentor.com/embedded-software/linux/

[18] Mentor, Inc., "Questa Advanced Simulator" webpage.
Available: https://www.mentor.com/products/fv/questa/

[19] Microsoft, Inc., "Visual Studio" webpage.
Available: https://www.visualstudio.com/

[20] *MyHDL* website.
Available: http://www.myhdl.org/

[21] National Instruments, "What is LabView?" webpage.
Available: http://www.ni.com/en-gb/shop/labview.html

[22] *OpenCV* website.
Available: https://opencv.org/

[23] *Python* website.
Available: https://www.python.org/

[24] QEMU, "QEMU: the FAST! processor emulator" webpage
Available: https://www.qemu.org/

[25] *Sigasi;* website
Available: https://www.sigasi.com/

[26] *Tcl Developer Xchange* website.
Available: https://www.tcl.tk/

[27] University of California, Berkeley, "Chisel" webpage.
Available: https://chisel.eecs.berkeley.edu/

[28] University of Toronto, "LegUp High-Level Synthesis" webpage.
Available: http://legup.eecg.utoronto.ca/

[29] *VirtualBox* website.
Available: https://www.virtualbox.org/

[30] Xen Project website.
Available: https://www.xenproject.org/

[31] Xilinx, Inc., "3rd Party OS Support for Zynq UltraScale+ MPSoC" wiki page.
Available: http://www.wiki.xilinx.com/3rd+Party+Operating+Systems

[32] Xilinx, Inc., "Install - Operating System (OS) support on Vivado Design Tools", Answer Record AR# 54242.
Available: https://www.xilinx.com/support/answers/54242.html

[33] Xilinx, Inc., "Linux" Xilinx wiki page.
Available: http://www.wiki.xilinx.com/Linux

[34] Xilinx, Inc., "OpenAMP" wiki page.
Available: http://www.wiki.xilinx.com/OpenAMP

[35] Xilinx, Inc., "Partial Reconfiguration" webpage.
Available: https://www.xilinx.com/products/design-tools/vivado/implementation/partial-reconfiguration.html

[36] Xilinx, Inc., "PetaLinux Tools Documentation: PetaLinux Command Line Reference", UG1157, v2018.2, June 2018.
Available:
https://www.xilinx.com/support/documentation/sw_manuals/xilinx2018_2/ug1157-petalinux-tools-command-line-guide.pdf

[37] Xilinx, Inc., "PetaLinux Tools Documentation: Reference Guide", UG1144, v2018.2, June 2018.
Available:
https://www.xilinx.com/support/documentation/sw_manuals/xilinx2018_2/ug1144-petalinux-tools-reference-guide.pdf

[38] Xilinx, Inc., "QEMU" wiki page.
Available: http://www.wiki.xilinx.com/QEMU

[39] Xilinx, Inc., "Software Development" web page.
Available: https://www.xilinx.com/products/design-tools/software-zone.html

[40] Xilinx, Inc., "UltraFAST Design Methodology" webpage.
Available: https://www.xilinx.com/products/design-tools/ultrafast.html

[41] Xilinx, Inc., "UltraFast Embedded Design Methodology Guide", UG1046, v2.3, April 2018.
Available: https://www.xilinx.com/support/documentation/sw_manuals/ug1046-ultrafast-design-methodology-guide.pdf

[42] Xilinx, Inc., "UltraScale Architecture PCB Design: User Guide", UG583, v1.14, January 2019.
Available: https://www.xilinx.com/support/documentation/user_guides/ug583-ultrascale-pcb-design.pdf

[43] Xilinx, Inc., "Vivado Design Suite User Guide: Release Notes, Installation, and Licensing", UG973, July 2018.
Available:
https://www.xilinx.com/support/documentation/sw_manuals/xilinx2018_2/ug973-vivado-release-notes-install-license.pdf

[44] Xilinx, Inc., "Yocto Project" Xilinx wiki page.
Available: http://www.wiki.xilinx.com/Yocto

[45] Xilinx, Inc., "Vivado Design Suite User Guide: Creating and Packaging Custom IP", UG1118, v2018.2, June 2018.
Available:
https://www.xilinx.com/support/documentation/sw_manuals/xilinx2018_2/ug1118-vivado-creating-packaging-custom-ip.pdf

[46] Xilinx, Inc., "XEN Hypervisor" wiki page.
Available: http://www.wiki.xilinx.com/XEN+Hypervisor

[47] Xilinx, Inc., "Xilinx Zynq UltraScale+ MPSoC ZCU104 Evaluation Kit" webpage.
Available: https://www.xilinx.com/products/boards-and-kits/zcu104.html

[48] Xilinx, Inc., "ZCU104 Evaluation Board: User Guide", UG1267, v1.1, October 2018.
Available: https://www.xilinx.com/support/documentation/boards_and_kits/zcu104/ug1267-zcu104-eval-bd.pdf

[49] Xilinx, Inc., "Zynq UltraScale+ MPSoC Overview: Advance Product Specification", DS891, v1.7, November 2018.
Available: https://www.xilinx.com/support/documentation/data_sheets/ds891-zynq-ultrascale-plus-overview.pdf

[50] Xilinx, Inc., "Xilinx Quick Emulator User Guide: QEMU", UG1169, v2018.2, June 2018.
Available: https://www.xilinx.com/support/documentation/sw_manuals/xilinx2018_2/ug1169-xilinx-qemu.pdf

[51] Xilinx, Inc., "Zynq UltraScale+ MPSoC Software Developer Guide", UG1137, v8.0, June 2018.
Available: https://www.xilinx.com/support/documentation/user_guides/ug1137-zynq-ultrascale-mpsoc-swdev.pdf

[52] Xilinx, Inc., "Zynq UltraScale+ Device: Technical Reference Manual", UG1085, v1.9, January 2019.
Available: https://www.xilinx.com/support/documentation/user_guides/ug1085-zynq-ultrascale-trm.pdf

[53] Xylon Inc., "logicBRICKS" website.
Available: https://www.logicbricks.com/

[54] *Yocto Project* website,
Available: https://www.yoctoproject.org/

[55] *96 Boards* website, "Ultra96" product page.
Available: https://www.96boards.org/product/ultra96/

Chapter 5

Candidate Application Areas for Zynq MPSoC

As has been noted in the earlier chapters, the Zynq MPSoC is an evolution of the Zynq-7000 SoC, with marked enhancements. It offers a richer, more diverse set of processing elements, as well as improved features for safety, and secure operation. Therefore, the Zynq MPSoC can support more demanding embedded systems applications than the Zynq-7000 SoC. A key factor is the selection of resources the device offers, making it suitable for implementing systems that comprise different types of processing requirements.

In this chapter, we begin by outlining candidate sets of requirements that map well to the Zynq MPSoC device, and which would make it a clear choice as an implementation platform. Following on, a small set of application examples are chosen, and we explore how these could be implemented using the Zynq MPSoC. We also introduce the Zynq RFSoC, a device which includes PL and PS components very similar to those of the Zynq MPSoC family, and adds additional features including front end RF Analogue-to-Digital Converters (RF-ADCs), and Digital-to-Analogue Converters (RF-DACs). With its capability of sampling at GHz rates, the Zynq RFSoC therefore brings a number of mobile/wireless *Software Defined Radio* (SDR) applications to the fore.

5.1 What Makes my System a 'Zynq MPSoC System'?

As noted above, the Zynq MPSoC is distinct from the Zynq-7000 and other SoCs, because of the set of resources it comprises. Any particular system design may use all of these, or a subset, but the motivation for adopting Zynq MPSoC is clearest when full advantage is taken of its various processing elements.

A 'typical' (if there is such a thing!) Zynq MPSoC system is likely to feature demanding OS and application software, beyond that which could be supported by the dual core Arm Cortex-A9 processor on the Zynq-7000 SoC. It might also need to support Asymmetric Multi-Processing (AMP), where different OSs are run across the four available cores. Further, the application may require real-time software operation, which is enabled on the Zynq MPSoC by the Arm Cortex-R5 processors. Certain applications may involve a rich Graphical User Interface (GUI), which would be better supported by the Mali GPU that is integrated into the Zynq MPSoC. The application may have specialist requirements in terms of interfacing, security, memory density, video encoding, etc. that map well to the Zynq MPSoC. In common with the Zynq-7000, the presence of PL provides the ability to hardware-accelerate suitable tasks; the distinction being that the Zynq MPSoC range of devices afford a greater amount of PL.

Applications with comparatively modest multi-processing requirements, and which can be realised within the PL dimensions available, would be more cost-effectively implemented using a Zynq-7000 device.

Away from the multi-processing theme, applications with a dominant requirement for deterministic 'dataflow' processing (as opposed to software-based processing) would map better to a high-end FPGA such as a Virtex UltraScale+. If processors are required, such devices offer the ability to include one or more soft processors in the PL alongside the classic FPGA dataflow-style computational architectures. At the other end of the complexity scale, lower cost FPGAs are good implementation options for less demanding applications that have a lesser emphasis on multiprocessing (fewer concurrent things to do in software!).

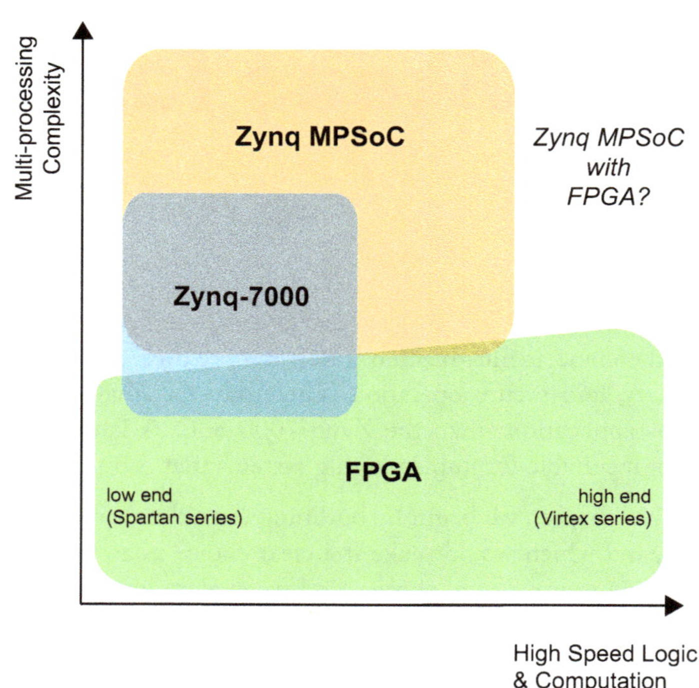

Figure 5.1: Indication of device suitability with balance of requirements

These general categories are indicated in Figure 5.1, and suggest how devices may be selected for increasing logic-based computation, against multiprocessing requirements and complexity. The majority of application requirements can be covered by choosing a single chip; extremely demanding applications may need a combination.

Next, we choose a small set of application examples and demonstrate why Zynq MPSoC (in particular) would be an appropriate platform on which to implement them. These are intended to provide a few indicative candidate application examples, only — the possibilities are not limited to these areas.

5.2 Application Areas for the Zynq MPSoC (and RFSoC)

The three main families of Zynq MPSoC target different general applications areas, albeit with obvious overlap, leaving the final device selection to the designer. The EV devices are typically adopted for demanding applications in the image and video processing domain. The EG devices are aimed at applications that are computationally demanding, across a range of applications, including for instance navigation systems, and cloud computing services. CG devices are targeted at applications with lower computational requirements, such as ultrasonics, or sensor-based systems running gateways and local processing.

Table 5.1 presents a few example applications that the CG, EG and EV MPSoC devices might typically be used for. The fourth column of the table lists example applications for a device that is closely related to the Zynq MPSoC: the Zynq RFSoC. Section 5.5 will introduce more background and detail on the Zynq RFSoC architecture [29], compare it to the Zynq MPSoC families, and review some applications that it is suited for. The RFSoC is a close relative (cousin or closer!) of the CG and EG Zynq MPSoC device families, but also includes some very specific analogue RF transceiver circuitry, making it a quite revolutionary device.

Table 5.1: Some applications for the Zynq MPSoC device families, and including the Zynq RFSoC

Zynq MPSoC Families			**RFSoC Devices**
CG Devices	**EG Devices**	**EV Devices**	
IoT gateway	Flight navigation	Situational awareness	Software defined radio
Motor control	Air traffic control	Surveillance	5G New Radio
Smart grids	Aerospace control	Smart vision	Massive MIMO
Ultrasound imaging	Secure solutions	Image manipulation	Beamforming
Traffic engineering	Networking	Graphic rendering	Beamsteering
Drone control	Cloud computing security	Human machine interface	DOCSIS implementation
Instrumentation	Data centre applications	Automotive driver assist	Phased array radar
Process control	Machine vision	Video processing	Satellite communications
Optical systems control	Mobile edge computing	Interactive display	RF test & measurement
Sensor processing & fusion	Robotics	Image / scene stitching	GHz carrier IF systems

In the following sections, we choose a small set of application examples, and demonstrate why Zynq MPSoC and/or Zynq RFSoC would be a particularly appropriate platform on which to implement them. These are intended to provide a few indicative candidate application examples, only — the possibilities are not limited to these areas — and there is a myriad of applications suitable for Zynq MPSoC in particular.

5.3 Drones

Drones (also known as Unmanned Aerial Vehicles, or UAVs) have become increasingly popular in recent years, transitioning from an expensive resource for the military and other specialist applications, to mainstream use. Today, drones are available that span the range from inexpensive toys ($50 - $250), to professional equipment for uses such as aerial videos and photography ($100's to $1000's), to specialist ($10,000+), for military and scientific applications.

A number of other uses of drones are emerging. One high-profile application is for logistics, as first proposed by Amazon for delivering parcels (initially to some level of public disbelief!), in 2013 [1]. Another logistics application was launched in Switzerland in 2017, for delivering even more sensitive cargo including blood and medical samples [16].

The proliferation of drone technology presents a number of challenges, particularly in the case of autonomous drones, which have no immediate user control. Ensuring that they fly safely, complying with airspace restrictions and avoiding collisions even in imperfect weather conditions, is an important issue. Concerns about privacy and security exist, requiring counter-measures to protect against interception, tampering, and so on.

Since its release, the Zynq-7000 has been adopted by a number of drone manufacturers, and the fast pace of development continues, with new and emerging demands for features. For instance, the Zynq chip hosts machine-learning algorithms in the ZeroTech *Dobby AI* drone [14], while DJI's *Inspire 2* drone achieves cinema-quality filming and advanced positional agility of the drone, using Zynq [13]. Another drone manufacturer, Aerotenna, have released a Zynq-based drone development platform that has been successfully adopted for agricultural uses (controlled spraying of fields) [12].

The Zynq MPSoC device is not likely to find its way into the lower end of the drone spectrum, but in the case of more advanced drones, and especially with the rapid evolution in this area, Zynq MPSoC may be a very good fit. The Zynq MPSoC possesses the range of processing resources, safety and security features, and flexibility, required for next-generation drone technology. Its low power consumption is also significant, given that the maximum achievable flight time of a drone is an important factor. Next, we explore this further by considering the various processing requirements involved in drones.

5.3.1 Flying

Drones come in a variety of forms, including fixed-wing, aeroplane-like vehicles, to multi-rotor 'copters. Of these, the quadcopter is the most widely encountered, with its four rotors equally spaced around a central point. Some have 6 rotors ('hexacopters'), or even 8 rotors ('octocopters').

Basing our example on the quadcopter type, for the drone to fly in a stable and controlled manner, it must have independent control of the speed of all four rotors. Driving the rotors is the only means of controlling the drone's flight — it does not have any other steering aids. Each type of motion requires the rotors to be controlled in a different way, as shown in Figure 5.2 [3]. It is also possible to combine these effects, such that a drone might rotate anti-clockwise while ascending and moving rightwards, for instance.

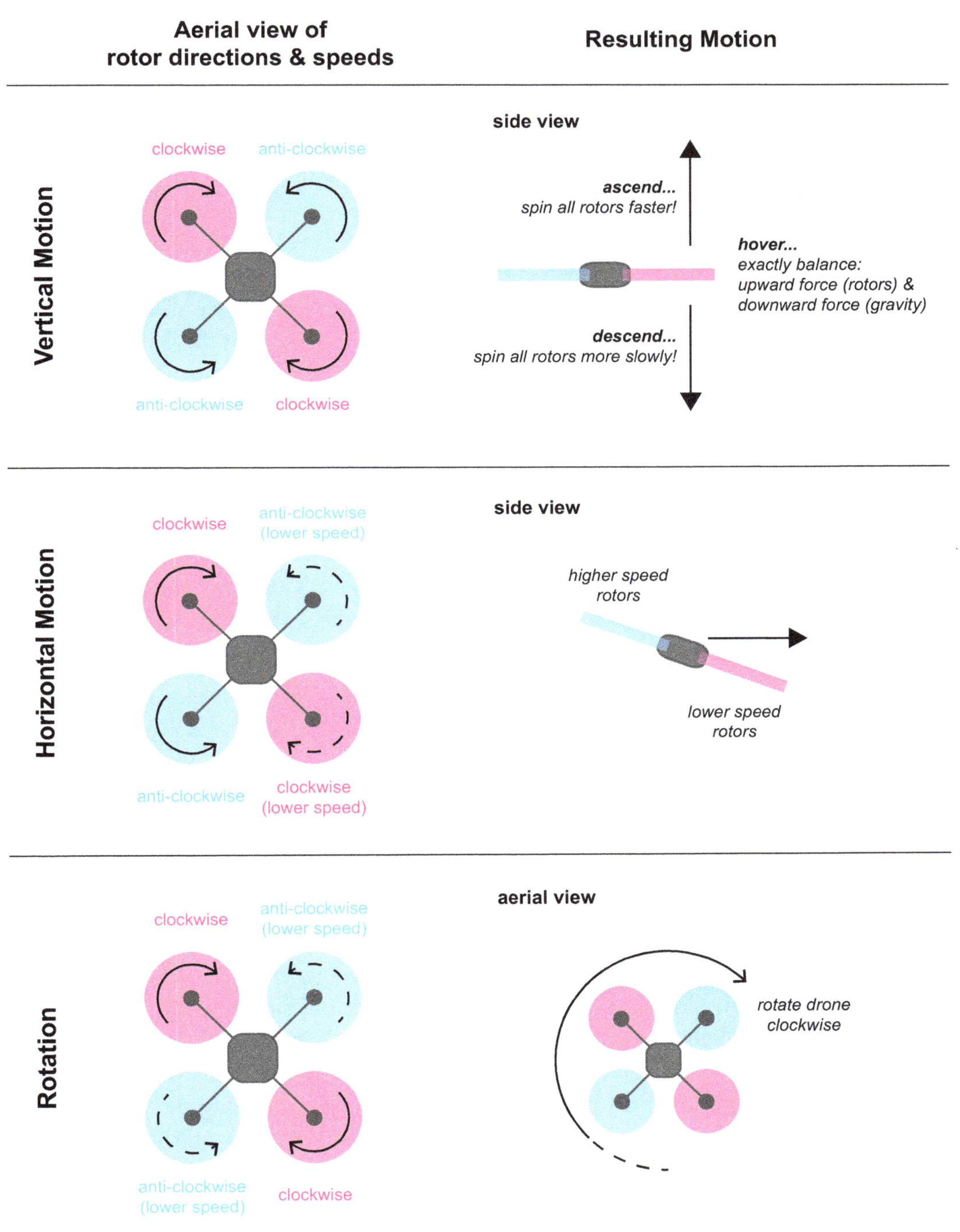

Figure 5.2: Drone movements induced by varying rotor speeds

The pilot flies the drone using a joystick or similar controller, the inputs from which are then translated into control signals to the rotors. For stable and precise control of the drone, motor control must be executed in a tight feedback loop, integrating readings from on-board accelerators and gyroscopes (which measure the motion of the drone), and in some cases a barometer (which measures altitude) and a compass (which measures orientation), to make real-time adjustments to rotor speeds. This processing is undertaken in the Inertial Measurement Unit (IMU) of the drone. Active stabilisation of this form is especially important for outdoor use, where the effects of wind can be disruptive.

Without going into the precise details of drone flight control implementation (the reader may wish to other articles for details [3], [8]), it is clear that there are requirements for multiple sensor inputs (perhaps with pre-processing), the generation of multiple motor control outputs, and low-latency parallel processing of the flight algorithms. The system may also integrate redundancy to enhance the safety of the drone (see Chapter 9 for a discussion of safety principles). These requirements are well-served by the architecture of the Zynq MPSoC: particularly the parallel processing enabled by the PL, real-time software processing on the RPU, and low-latency interfacing between these resources. Crucially, Zynq MPSoC can implement the fast feedback loop needed for fine control of the drone's motion.

5.3.2 Video Processing

One popular use of drones is aerial photography (still images and video), with end applications ranging from professional (and amateur) film-making, to news and sports broadcasting, real estate (property videos, etc.), surveillance, safety inspection, and precision-agriculture.

To maximise the quality of the images and/or videos produced, the stability of the on-board camera must be addressed. The camera should, as far as possible, be isolated from external forces acting on the drone, such as gusts of wind, and the vibrations produced by the rotors. The orientation of the camera should also be independent of the orientation of the drone (as far as possible). These needs can be met mechanically, using a gimbal, with further improvements achieved through the use of Digital Image Stabilisation (DIS) software, which can be run in real-time (the Zynq MPSoC's RPU would be suitable here) or as a post-processing stage.

Professional quality recording may require some combination of 4K (or higher) resolution, fast frame rates, and High Dynamic Range (HDR) video, resulting in vast quantities of data to be stored, usually to an SD card (a lightweight medium) which exploits the Zynq MPSoC's integrated support for SD interfacing. Live-broadcasting of drone footage is also desirable in the case of news and sports programmes. Streaming video from drones presents its own challenges, and areas of recent research include compression methods suitable for cameras with high mobility [20], and networks of drone-based cameras to cover sports and other live events [21]. Further, advanced imaging features include the ability to track a target object in the video. All of these video processing tasks require sophisticated processing. Software running on the Zynq MPSoC's APU, ideally with the benefit of hardware-acceleration in the PL, would be an ideal implementation.

5.3.3 Sensing

Aside from the capture of video and photographs, drones can carry a variety of different sensor types. This may include ultrasonic sensors (for distance measurements), various types of advanced imaging data (e.g.

thermal, hyperspectral, Radio Frequency (RF) sensors, RADAR (Radio Detection and Ranging), LiDAR (Light Detection and Ranging), and so on. Making these sensors sufficiently compact and lightweight for small drones is a challenge, but miniaturisation advancements are being made, for instance in RADAR [6].

The capture of visual images should also be considered a sensory input. Visual imaging supports 'computer vision' functionality such as object detection, collision avoidance, and gesture recognition, all of which enable advanced drone flight. For instance, researchers at Massachusetts Institute of Technology (MIT) have demonstrated an obstacle detection and avoidance system based on stereo-vision cameras [7]. Computer vision techniques can also be used for more conventional purposes, for instance people-counting, traffic monitoring, and remote inspection. On-board processing has the potential to reduce datasets, and hence storage and/or communication overhead, and could be suitably enabled by the Zynq MPSoC. Data could also be uploaded to 'the cloud' for further processing.

Some applications use 'sensor-fusion' to combine measurements from multiple sensors to provide a more complete or accurate dataset. This can include, for example, different types of image data (visual, hyper-spectral, thermal, etc.), and ultrasonic, LiDAR, or RADAR sensors. The 'fusion' aspect may involve pre-processing of individual sensor data, as well as the task of combining the various sources, and is analogous to the idea of sensor-fusion in vehicles (to be discussed in Section 5.4).

5.3.4 Communications and Navigation

Various aspects of wireless communications are associated with drones, perhaps the most obvious being the link between a user controller, and the drone, for directing flight. In addition, there may also be a return path for reporting of diagnostic information, and transfer of data captured by the drone (e.g. video or sensor data).

As drone applications become more demanding, the requirement for high-bandwidth, low latency communications will increase. This may be driven by the desire for higher quality video, or an increased payload of sensors, for instance. A more exotic possibility is for drones to carry mini cellular basestations, to provide pop-up coverage in the case of emergencies, or to serve rural or hard-to-reach areas [18]. Drones are also used to identify areas of poor mobile coverage, for instance in sports stadiums, meaning that their 'sensing' task is actually to measure and record the strength of wireless communications signals at a desired set of locations [5]. This is actually an excellent application for the Zynq RFSoC, discussed in Section 5.5.

Drone positioning is another important aspect, and some drones are equipped with GPS sensors which allow them to determine their geographical location. GPS readings are combined with measurements from the drone's IMU, to precisely determine position (location and orientation). Functionality based on positional awareness includes the option to 'return home', i.e. to direct the drone back to a designated location at the conclusion of its flight. At this point, a drone equipped with automatic landing functionality would gently lower itself to the ground, based on awareness of its altitude. Altitude can be determined based on GPS signals, or other sensor measurements such a from a barometer or ultrasonic sensor.

Communications functionality is demanding, requiring extensive interfacing and processing, as well as real-time adjustments based on location and orientation. This can make use of several Zynq MPSoC resources.

Some of the features discussed here, and in earlier parts of this section, are illustrated in Figure 5.3.

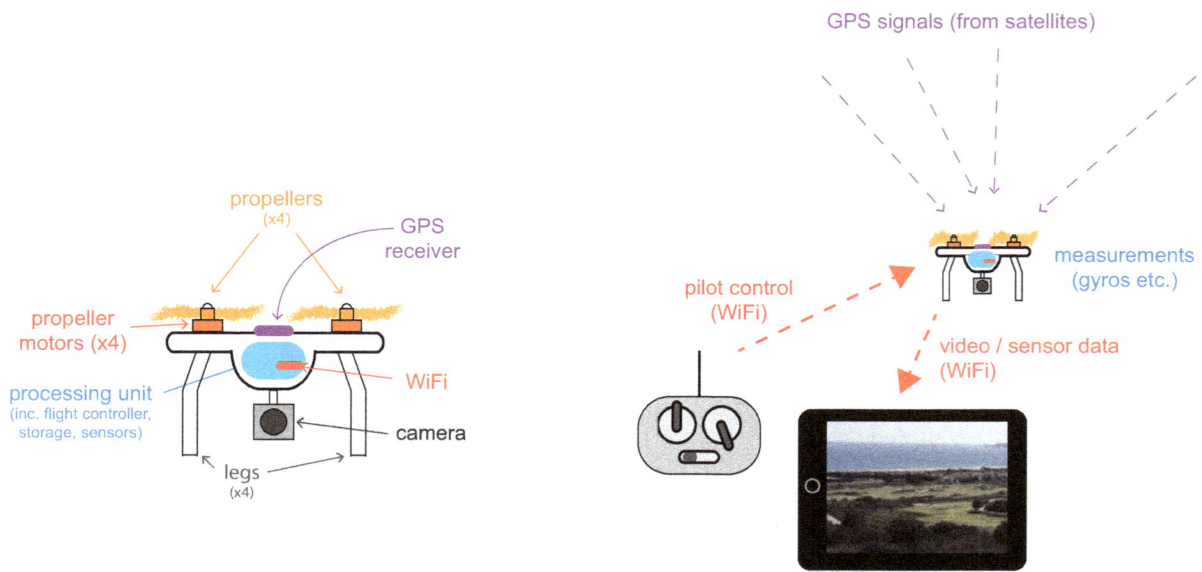

Figure 5.3: Drone features and communications

5.3.5 Advanced Drones

Autonomous flying is an important area for further development. Being 'autonomous' infers the ability to self-navigate from source to destination, detect and react to obstacles and hazards, comply with airspace restrictions, and even interact with other drones or ground-stations. Autonomous flying therefore implies a high degree of situational awareness, which may be derived from fusing data from imaging cameras, GPS, and other sensors, and aspects of decision making (for instance, if an object is determined ahead, how should the drone divert its path?). In 2018, drones of this level of sophistication are starting to become commercially available. The Zynq MPSoC's resources map well to these requirements.

The ability to fly a single drone autonomously leads to the possibility of flying a set of drones in formation (sometimes referred to as a 'swarm'). Maintaining the formation requires significant awareness, communication, and coordination. A swarm could enable distributed sensing and processing tasks that cannot be undertaken by a single drone, while providing collective decision-making and the ability to self-organise and recover from failures. Drone swarms are already at various stages of development, and have been demonstrated for both military and civilian applications [11],[19].

Autonomous and swarm-based flying adds significant complexity to the drone's on-board system. Therefore, autonomous flying drones may be well-suited to deployment on Zynq MPSoC devices, especially for demanding sensing and processing applications. There are also further challenges in terms of ensuring security and safety [10], which the Zynq MPSoC is well-placed to address.

5.4 Smart and Autonomous Vehicles

Recent years have seen an increase in the amount of technology available in cars for driver assistance. The term Advanced Driver Assistance Systems (ADAS) is in widespread use, with most new cars now providing the option of at least one ADAS function (such as a lane departure warning system, or blind-spot detection). Meanwhile, one of the longer-term themes in the automotive sector is not just to assist the driver, but essentially to take over their role in driving the car. Several companies, including traditional auto manufacturers like Renault and Ford, as well as technology companies such as Google and Uber, are developing driverless cars.

Perhaps less obviously, there is actually a plethora of automation levels, and cars that offer fully automated driving in all situations have yet to be realised. A scale of driver assistance has been defined by Society of Automotive Engineers, and adopted by the United Nations, and the US Department of Transportation [4]. As shown in Table 5.2, the scale ranges from no driver assistance (Level 0), to fully automated driving (Level 5). At the time of writing, commercially available vehicles are limited to Levels 1 and 2, with the first Level 3 car announced in late 2017. It is expected that Level 3 cars will be available from the early 2020s.

Table 5.2: Levels of Automation in Road Vehicles [4]

Category	Level	Description
Driver performs part or all of the dynamic driving task	Level 0	No driving automation
	Level 1	Driver assistance *(adaptive cruise control OR lane centring, with driver supervision)*
	Level 2	Partial driving automation *(adaptive cruise control AND lane centring, with driver supervision)*
Automated Driving System (ADS) performs all of the dynamic driving task	Level 3	Conditional automated driving *(in dense freeway traffic at low speeds)*
	Level 4	High level of automated driving *(automated driving within a city centre or geo-fenced location)*
	Level 5	Full automated driving *(automated driving everywhere!)*

In all cases, there must be a safe backup if the Automated Driving System (ADS) fails. Up to and including Level 3, this relies on the driver resuming control. For Levels 4 and 5, the ADS must itself bring the vehicle to a stop safely.

In the remainder of this section, we will outline some of the features and functionalities required for ADAS and autonomous vehicles, and draw parallels with the resources available on the Zynq MPSoC, demonstrating why the device is a good fit for these applications.

5.4.1 What Does 'ADAS' Cover?

The term ADAS encompasses a variety of driver assistance functions, many of which are well-known. These include:

- **Navigation** — provision of directions for the driver, based on GPS readings and mapping data.
- **Lane departure warning** — alerting the driver when the car drifts out of its lane, which might indicate loss of concentration by the driver.
- **Blind spot detection** — the ability to detect a vehicle in the driver's 'blind spot', and provide an alert (often in the form of a warning light integrated into the wing mirror).
- **Adaptive cruise control** — the vehicle automatically reduces its speed when it detects a hazard ahead.
- **Speed sign detection** — automatic reading of speed signs, with information relayed to the driver via the dashboard.
- **Automatic parking** — the car manoeuvres itself into a parking space.

All of these systems require positional awareness, whether at the macro-level for navigation, reasonably long distances (e.g. detecting a slowdown ahead on the motorway) or down at centimetre precision for manoeuvring into parking spaces. The need to understand the vehicle's environment (lane markings, traffic signs, pedestrians, other vehicles) is another vital component. ADAS systems therefore need to incorporate multiple sensors to interpret the vehicle's surroundings, and to feed these various sets of data into a decision making process.

5.4.2 Sensing Requirements and Implementation

ADAS systems normally include multiple sensors to capture data about the scene around the car, including the proximity of other vehicles and objects. Generating this information requires that a number of sensors are integrated around the vehicle, providing different types of sensor data. This includes visual data from cameras (which enables detection of road signs and markings), as well as ultrasonic and radar sensors to measure the distance from other vehicles and objects. Radar can also measure the Doppler shift, and hence velocity, of other vehicles.

LiDAR (Light Detection And Ranging) operates using laser light with similar principles to radar, and is quickly emerging as a compelling technology for autonomous vehicles in particular. Although LiDAR systems are prohibitively large and expensive at the present moment, there is considerable interest in developing lower cost, more compact solutions suitable for mass market vehicles [2].

Figure 5.4 provides an indication of these various sensors placed around a car. Note that the ultrasonic sensors operate for short ranges (up to a few metres), to assist the driver with parking. Long distance radar detection facilitates *adaptive cruise control*, where the car's speed is adjusted in response to slower moving traffic detected ahead. Radar is also used for medium range functions, such as *blind spot detection*, and *cross traffic alert*, i.e. to warn the driver of traffic crossing behind the car when reversing. Camera systems at the front, rear, and sides can be used to extract visual elements from the scene, such as road markings and traffic signs. LiDAR is significant because it can provide better distance resolution than radar at medium distances (10's to 100's of metres), with full 360 degree coverage — which is perfect for autonomous vehicles.

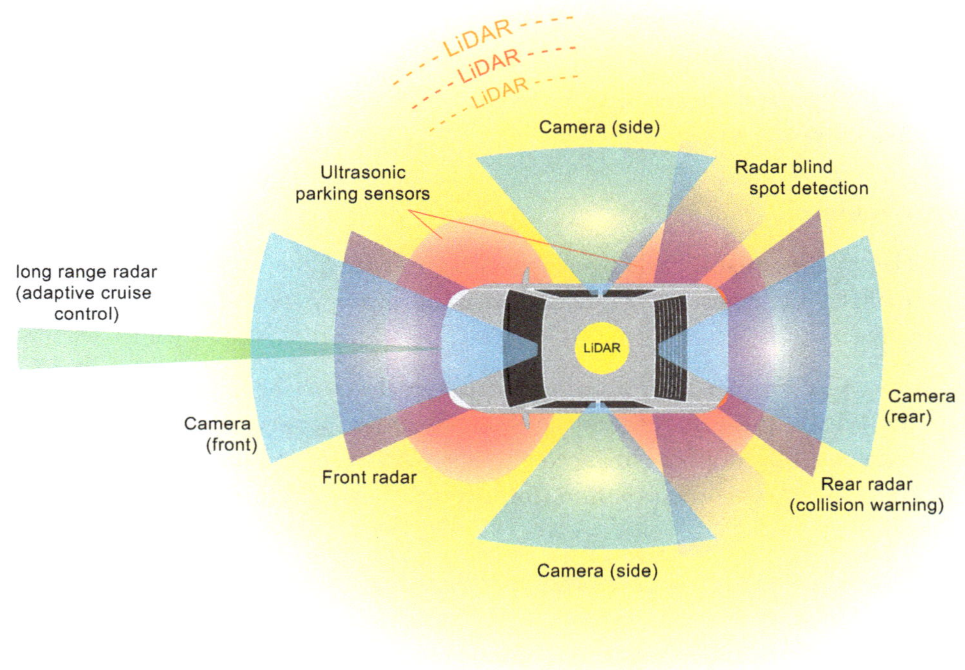

Figure 5.4: Aerial view of a car equipped with sensors for ADAS / autonomous driving

Why so many different types of sensors? The reason is because they each have different capabilities, in terms of detection range, resolution, and robustness in different environment conditions (such as snow, fog, bright sunlight etc.). The combination of data from different types of sensors enables greater robustness, which is important in such a safety critical application.

The large amount of data generated by the vehicle's sets of sensors must then be pre-processed, combined ('sensor fusion'), and analysed to interpret the vehicle's surroundings. This implies extensive signal processing on individual sensor inputs, as well as precise synchronisation and further processing to combine the data, interpret hazards and take necessary action [17]. All of this processing (encompassing both PL-, and PS-based elements) must take place with as little delay as possible, given the high speed at which vehicles may be travelling, and the response times needed to avoid accidents.

The Zynq MPSoC represents a well-suited implementation platform for ADAS systems, given that it has multiple processors for executing parallel software tasks, including those with real-time capability, as well as high-speed interfaces for handling the large volume of sensor inputs. The PL can be leveraged to hardware-accelerate the pre-processing of sensor inputs, and the low-latency connections between the PL and PS provide a distinct advantage in this time and safety critical application. Furthermore, safety features such as lock-stepping of the Arm Cortex-R5 processors, and processor / PL redundancy can be incorporated into the design to improve its robustness.

Finally, it is worth noting that automotive grade Zynq MPSoC chips are available. These can operate over an extended temperature range, and are certified for the automotive safety standard, ISO26262 ASIL 3. This and other safety features are reviewed in Chapter 9.

5.4.3 Autonomous Vehicles

Autonomous vehicles present many challenges. For full autonomy, they must be capable of fully interpreting their environment in all situations, and reliably making appropriate and safe decisions — even when something unexpected occurs.

The amount of data collection needed for autonomous vehicles may be expanded compared to ADAS systems, given that the vehicle must understand its environment to a very high degree, in order to make human-like driving decisions. This implies enhanced computer vision techniques to identify different types of objects in view, and even to anticipate their motion (for instance, having detected a pedestrian, where is he or she likely to move next?).

Machine learning techniques will play an important part in being able to classify objects within the vehicle's field of view, e.g. other vehicles, pedestrians, cyclists, street furniture, etc. [22]. The challenge is to ensure safe behaviour at all times, so the vision system must also be able to understand unusual objects and occurrences, as well as routine ones, and potentially to learn from its own experience [24]. For instance, if the car was to encounter a tractor, pram, mobility scooter, pedestrian-with-an-umbrella, skip (dumpster in the USA!), or object falling from a vehicle ahead, it must be able to react appropriately. Creating robust intelligence for autonomous vehicles will be a major element of reaching Levels 4 and 5 on the scale of autonomy (Table 5.2).

The fusion of the processed sensor information with geographical data can be used to build up a picture of the environment for future journeys. Data from vehicles passing through a particular location can even be shared, to help improve interpretation of the surroundings, as well as equipping them with an awareness of other vehicles in proximity. The overall traffic situation can be managed via connected road infrastructure, such as traffic signs and road signs, thereby helping to reduce congestion and improve journey times for road users. Therefore, vehicle-to-vehicle communications (V2V), as well as vehicle-to-infrastructure (V2I) communications, will play an important part in realising fully autonomous vehicles.

5.4.4 Connectivity and Security

The integration of ADAS systems, and in particular, features for autonomous vehicles, requires advanced connectivity. Some of the processing tasks discussed in the previous section, in particular for machine learning and data aggregation between vehicles, cannot take place on the vehicle itself, but will instead require access to a network, and to Cloud or Mobile Edge Computing (MEC) facilities [23]. There will also be wireless links to connect with other vehicles, and the traffic infrastructure in proximity. This means that highly robust wireless communication is needed. Further, low latency is crucial given the speeds at which decisions need to be made in automotive applications (for example, to recognise a cow in the middle of the road, and apply the brakes!). 5G communications, as reviewed in Section 5.5.5, is designed to provide the reliability and low latency required by ADAS and autonomous vehicles.

There are a number of public concerns about autonomous vehicles [25], as evidenced by interest in accidents involving them. Traffic safety is the most prominent concern, according to a recent survey, and security is also a prominent issue. In this regard, people's concerns range from privacy issues (for instance, personal details and route information being compromised) all the way to a hacker taking over the vehicle they are riding in, and the threat of terrorism [26]. Cybersecurity is therefore a top priority for the automotive and technology companies involved, as well as for government regulators. There is currently a move towards cybersecurity standardisation for autonomous vehicles, given the importance of achieving robust security practices, and maintaining public trust [27].

Making autonomous vehicles secure involves both physical security (such that its electronic systems cannot be tampered with), and security of communications. The security features of the Zynq MPSoC device can be leveraged in ADAS and autonomous vehicles, as discussed in Chapter 8, while its processing capabilities and tools are well placed to support the cryptography required in 5G communications systems (and beyond).

Zynq MPSoC may also form part of the communications network, MEC facilities, and road management infrastructure that underpins the deployment of autonomous vehicles.

5.5 Interfacing to the RF Analogue World — the Evolution to Zynq RFSoC

Since the first DSP-enabled Virtex II FPGAs in the early 2000s (which featured on-chip multipliers), FPGAs have been used extensively for many radio and mobile/wireless applications including basestation implementation, switching, security and networks, with solutions typically featuring the FPGA as a workhorse for the computational work. A suitable analogue chip with ADC and DAC facilities was required to connect the digital (FPGA) world with the analogue (or RF) world.

The Zynq-7000 family of devices has been used extensively over recent years for applications in mobile and wireless communications, and also radar systems. For example, one of the best known and widely used Software Defined Radios (SDRs) of the last 10 years are the ubiquitous USRP devices, such as the Ettus USRP E310 [34], which uses the Zynq Z-7020 along with the Analog Devices AD9361 [32] chip. The AD9361 chip has a two channel transmitter (Tx) and a two channel receiver (Rx), or equivalently, can be described as a 2 x 2 RF transceiver. It can tune over a range from 300 MHz to 6 GHz, with signal bandwidths in the range 200kHz to 56MHz. Other available off-the-shelf integrated SDR solutions include the Analog Devices Active Learning Module (ADALM) Pluto SDR, which is based on the Zynq 7010 and AD9363 RF transceiver. In recent years, there was also the PicoZed SDR [35], featuring the Zynq 7035 device, and the AD9361 chip (shown on the FMC module from Figure 4.9 on page 91).

In these Zynq-7000 SDR solutions, the DSP and radio management is performed by the Zynq chip, and the ADC and DAC functions are performed by an externally connected analogue/RF chip. Figure 5.5(a) illustrates the point of interface between the Zynq device and such a chip. Software platforms such as MATLAB and Simulink provide support packages to allow easy use and programming of these Zynq-7000 Series SDRs [42]. We can of course evolve from the Zynq 7000 series to the Zynq MPSoC family and implement SDR with suitable radio front transceiver front-ends, again such as the Analog Devices SDR transceiver family.

Similar to the Ettus USRP E310 Zynq-7000 SDRs (Figure 5.5(a)), fully integrated and higher performance Zynq MPSoC SDR development kits are available, such as the Raptor SDR featuring a Xilinx Zynq MPSoC (the XCZU9EG with 2520 DSP slices and 600k system logic cells, quad-core Arm Cortex-A53 and dual core Arm Cortex-R5). The main components of the Raptor Zynq MPSoC are as shown in Figure 5.5(b), comprising the Zynq MPSoC and the RF-DACs and RF-ADCs. (There are of course other interfacing devices, memories, chips and peripherals on the board that are not shown in the figure.)

Now, if there was a solution to the technology challenge of integrating the RF transceiver (i.e. the RF-DACs and RF-ADCs) on to the Zynq MPSoC device/substrate, then a true (and very powerful) single chip SDR solution will have been created. And this is exactly what has been achieved with the creation of the new member of the MPSoC family (or indeed a very close cousin) — the Zynq RFSoC device.

Building on the pioneering work at Xilinx on RF-DACs and RF-ADCs [40], the RFSoC is now the ultimate SDR device, and the first to actually realise the ideas of Joe Mitola in his seminal paper of 1995, "*The Software Radio Architecture*" [41]. Looking back to the list of examples of possible Zynq RFSoC applications given in the fourth column of Table 5.1 on page 99, the applications are distinguished by the fact that they all require very high sampling rates to both acquire signals from RF radio spectrum and to create signals for the RF radio spectrum, exactly as provided by the Zynq RFSoC.

5.5.1 The Zynq RFSoC Device Family Compared to MPSoC

The Zynq RFSoC, announced in 2017, is a close relation to the Zynq MPSoC. It is targeted towards the implementation of high value, integrated radio systems and incorporates the required RF analogue circuitry on-chip to offer a single chip SDR solution for many applications. Key features of the Zynq RFSoC that differentiate it from the Zynq MPSoC families are the addition of:

- *High rate data converters* — 8 or 16 RF-ADCs and RF-DACs, each operating at up to 4Gsps (sample rate, f_s = 4GHz) and 6.4Gsps (sample rate, f_s = 6.4 GHz), respectively, and with 12 or 14 bit resolution.
- *Integrated digital up-converters (DUC) and digital down-converters (DDC)* — Capable of processing real or complex (I/Q) data, and used for front-end processing in RF spectrum based systems.
- *High PL DSP Slice Density* — Zynq RFSoC offers up to 4272 DSP slices (more than any MPSoC device) supporting DSP computational processing for implementation of expensive algorithms.
- *Hardened resources for high performance Soft Decision Forward Error Correction* — SD-FEC coding and decoding, supporting Turbo codes, and Low-Density Parity Check (LDPC) codes.

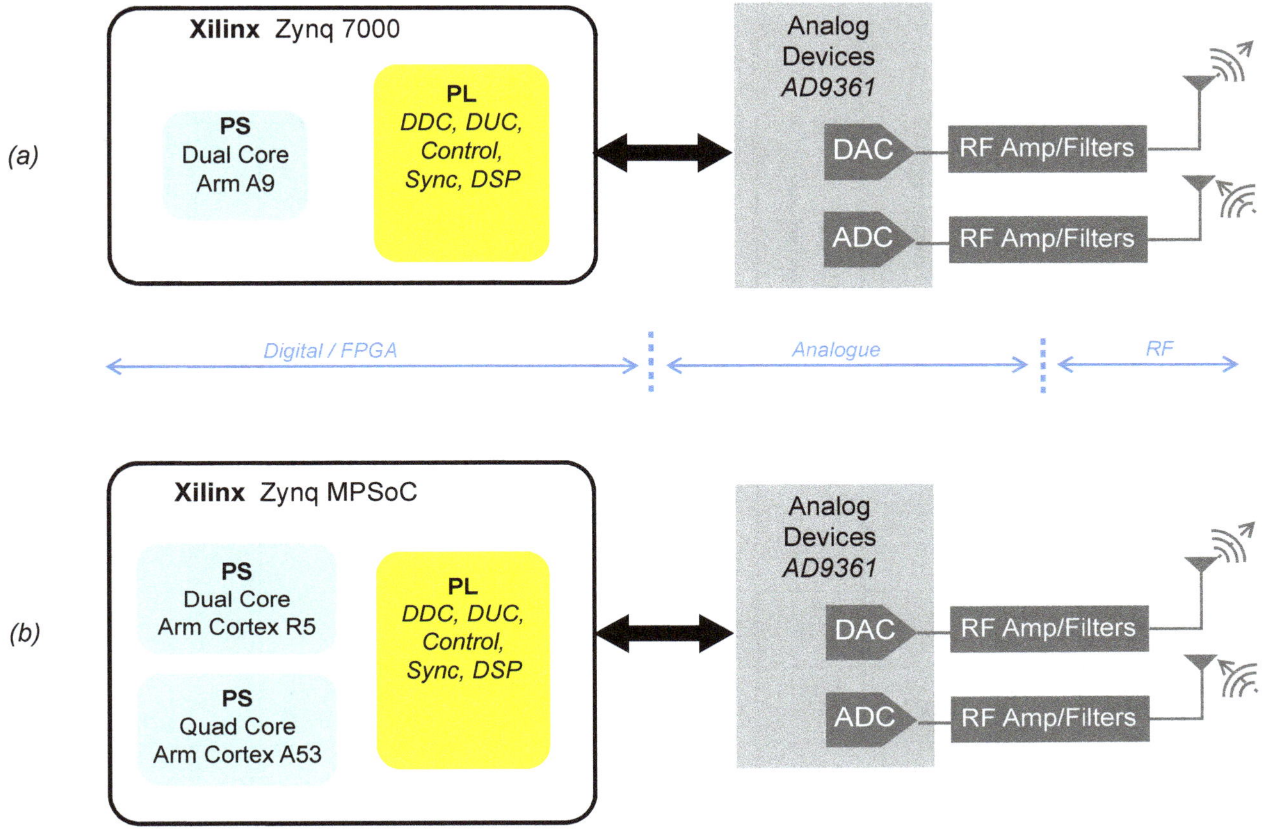

Figure 5.5: General block diagram of (a) USRP SDR architecture using Zynq-7000 Series [34] and (b) the Raptor SDR Development Kit architecture using Zynq MPSoC [33]

As with other Zynq families, a range of Zynq RFSoC devices are available, offering different features and options (for more information see the latest selection guides [30]).

Table 5.3 compares the main device architecture features of the Zynq MPSoC [31] with those of the Zynq RFSoC. The Zynq RFSoC product family shares the same architecture for the PS as the Zynq MPSoC, and features a quad core Arm Cortex-A53 application processor, and a dual core Arm Cortex-R5 real-time processor. A notable difference is that graphics and video processors are omitted from the PS of the RFSoC, and indeed, graphics processing is less likely to be a targeted application in an RF wireless communications system. Given the overall similarities between the two device architectures, their associated design flows and software components etc., then they substantially share the same ecosystem.

As might be expected for a device aimed at DSP intensive applications, the Zynq RFSoC's PL contains higher numbers of DSP slices than Zynq MPSoC devices, and has a comparable number of system logic cells

Table 5.3: Comparing the Zynq MPSoC and Zynq RFSoC PL and PS Architectures

Family	**PS**			**PL**					**RF-DAC, RF-ADC, Comms**			
	APU Arm Cortex-A53 cores	**RPU Arm Cortex-R5 cores**	**Graphics Processing Mali-400**	**DSP Slices (smallest to largest)**	**System Logic Cells (smallest to largest)**	**BlockRAM (Mb)**	**UltraRAM (Mb)**	**Video Codec Unit (in PL, H.264/.265)**	**RF-DAC 14 bit**	**RF-ADC 12 - 14 bit**	**DUC and DDC rates**	**SD-FEC: Soft Decision Forward Error Correction**
CG	Dual	Dual	-	240 to 2520	103k to 600k	4.5 to 32.1	13.5 to 27.0	-	-	-	-	-
EG	Quad	Dual	1	240 to 2530	103k to 1143k	4.5 to 34.6	13.5 to 36.0	-	-	-	-	-
EV	Quad	Dual	1	728 to 1728	192k to 504k	4.5 to 11.0	13.5 to 27.0	1	-	-	-	-
RFSoC	Quad	Dual	-	3145 to 4272	678k to 930k	27.8 to 38.0	13.5 to 22.5	-	8 or 16	8 or 16	1x to 40x	8

to the largest Zynq MPSoC devices. The Zynq RFSoC also ensures a large number of Block RAMs, with the available range being 27.8 to 38.0Mb, compared to the much wider variation in the Zynq MPSoC (Block RAM in the range 4.5 to 34.6Mb). High speed interfaces are also included, similar to the Zynq MPSoC, and in this case they are targeted at supporting the demanding requirements of mobile fronthaul and backhaul networks, including implementation of the Common Packet Radio Interface (CPRI) standard [38]. A simplified diagram of the RFSoC architecture is shown in Figure 5.6.

As listed in the example applications in Table 5.1, the RFSoC is targeted towards current and emerging wireless standards such as 5G NR (New Radio), and features of massive Multiple-Input Multiple-Output (MIMO) radio, as well as applications in phased array radar. With the ability to sample and process up to 16 transmit and 16 receive channels on a single chip, reaching rates that permit directly sampling the radio frequency spectrum of signals well into the GHz range, the RFSoC provides great potential for developing highly flexible next-generation radio equipment. More detailed information about the RFSoC can be obtained from Xilinx webpages [30]; for specific background on the RF-sampling technology, please refer to [37].

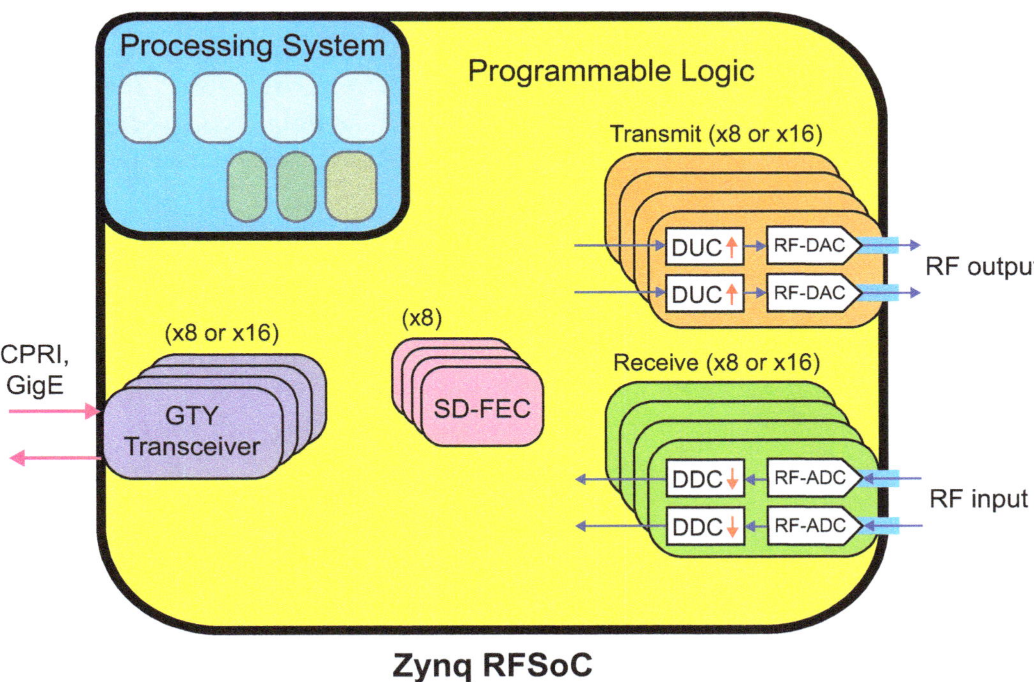

Figure 5.6: Simplified architecture of the RFSoC, including both SD-FEC, and high rate RF-DACs and RF-ADCs [28]

5.5.2 RF Sampling: RF-ADCs and RF-DACs, and Single Chip SDR

The Zynq RFSoC brings all of the core components together to facilitate SDR, including direct RF sampling data converters, and interfaces such as Common Public Radio Interface (CPRI) and Gigabit Ethernet-to-RF. The data converters take the form of RF Analogue-to-Digital Converters (RF-ADCs) and RF Digital-to-Analogue-converters (RF-DACs). They are integrated with the device's PL to allow the pre- and post-processing stages of decimation, interpolation, channelisation, and so on, to be efficiently and effectively implemented.

The RF-ADC can support sample rates of more than 4Gsps (giga samples per second), or f_s > 4 GHz. To help illustrate just how fast this is, we might also express these large numbers as 4000 *million* samples per second, or 4000 Msps.) The RF-DAC output stage can sample at more then 6.4Gsps, or f_s > 6.4 GHz, and generate output carrier frequency bands above 3.2GHz (i.e. above f_s /2) using strategies that exploit the Nyquist band images above half of the sample rate [28], or the so called 1st and 2nd Nyquist bands. The Zynq RFSoC data converters also include efficient implementation of digital down-converters (DDCs) and digital up-converters (DUCs) with programmable numerically controlled oscillators (NCOs), enabling the implementation of complex mixers.

Using the Zynq RFSoC, the Zynq MPSoC + ADC/DAC architecture of Figure 5.5(b) is now implemented via the single Zynq RFSoC device as shown in Figure 5.7. This device fully integrates the analogue RF components on to the monolithically integrated Zynq RFSoC, i.e. the software defined radio is now on a single chip. Zynq RFSoC family devices have different variants in terms of the number of input and output channels (8 or 16), SD-FEC implementation (not on all devices), and ADC sample rates (either 2Gsps or 4Gsps); see [29] for more information.

The integrated RF data converters of the Zynq RFSoC provide an SDR platform that is a fully programmable, direct RF sampling SDR. Therefore, RF signal processing has essentially moved into the digital domain, given the ability to digitise very large sections of the RF spectrum that are used for mobile/wireless communications. Generally, power consumption is reduced, as interfacing devices between the MPSoC / FPGA and RF analogue chip are no longer required, and similarly there is a reduction in system footprint due to the single chip design. Also, using a set of integrated tools and working with fewer components, enabled through Vivado, then design cycle times should be shorter.

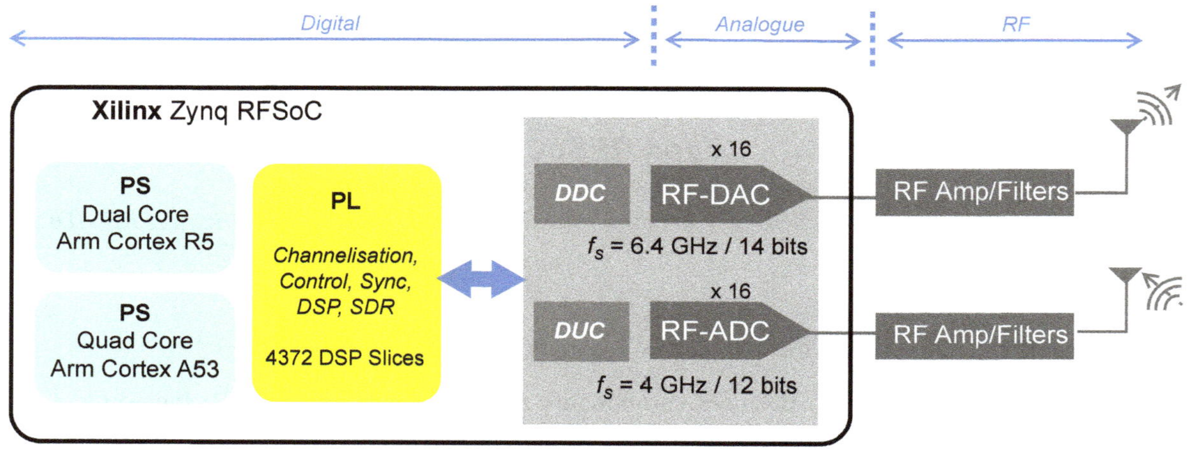

Figure 5.7: The single chip Xilinx Zynq RFSoC featuring on-chip RF-DACs and RF-ADCs. (Compare to the SDR implementations in Figure 5.5 featuring a separate (off-chip) RF Transceiver for RF-DAC and RF-ADC).

5.5.3 Direct Sampling Zynq RFSoC Data Converter Subsystem

The key DSP and SDR features of the Zynq RFSoC data converters are summarised in Table 5.4. The RF-ADCs and RF-DACs are structured as tiles on the device, with each tile containing either 2 or 4 RF-ADCs, or four RF- DACs. Each tile also has Phase Locked Loops (PLLs) and the required clock handling logic for routing of analogue and digital logic paths [28]. Depending on the Zynq RFSoC family generation, the number of channels, wordlength resolution and sample rates vary. For first generation devices, the RF-ADC wordlengths are either 12 or 14 bits resolution, with sample rates up to 4.096 GHz, and the RF-DAC wordlengths are 14 bits with sample rates up to 6.554 GHz. Newer, third generation devices raise RF-DAC rates to 10Gsps, and RF-ADC rates to 5Gsps. More precise device performance figures for the RF-ADCs and RF-DACs, such as Spurious Free Dynamic Range (SFDR), Noise Spectral Density (NSD), Effective Number of Bits (ENOB) etc. are available in [39].

Table 5.4: Key features of the Zynq RFSoC RF-DACs and RF-ADCs

RF-ADC	RF-DAC
4 x RF-ADCs 12 or 14 bit resolution Up to 4GHz sample rates	4 x RF-DACs 14 bit resolution 4GHz bandwidth output, and up to 5GHz RF output
DDCs for decimation 80% passband, 89dB stopband attenuation	DUCs for interpolation 80% passband, 89dB stopband attenuation
Complex spectrum mixing facility 48 bit NCO per RF-ADC fs/4 and fs/2 modes Real or complex signal operation	Complex spectrum mixing facility 48 bit NCO per RF-DAC fs/4 and fs/2 modes Real or complex signal operation 1st and 2nd Nyquist zone RF-DAC operation

With very high sampling rate RF-DACs and RF-ADCs, Zynq RFSoC enables direct sampling of the radio frequency spectrum. Consider Figure 5.8, which shows some of the main bands of radio frequencies for communications. The classic Nyquist theorem states that signals must be sampled at greater than twice the signal bandwidth in order to retain all information; this is often restated in terms of baseband signals (i.e. signals starting from 0Hz), such that we might state the need to sample at greater than twice the *maximum frequency*. So, for Figure 5.8, noting the baseband for modern RF communications starts low at MHz ranges, if we were to 'sample' at say, 6GHz, then all spectrum components in the range 0 to 3GHz would be captured

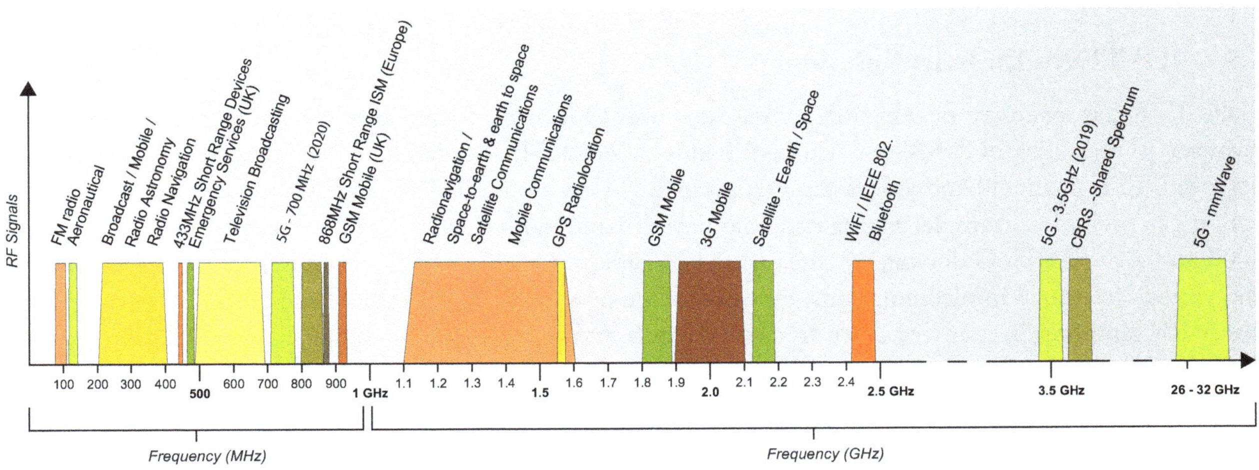

Figure 5.8: General review of RF Communications spectrum from 80MHz through 1GHz to 26 - 32GHz

and available in the digital domain. This allows us to consider the Zynq RFSoC as a true digital radio, or SDR by virtue of this direct RF sampling. Above 3GHz, we can still access the spectrum either via 1st or 2nd order Nyquist strategies to create signals. For frequencies in, say, the 20GHz carrier range, we can use front end analogue RF oscillators, and modulate/demodulate to intermediate frequencies that are within the baseband region of the RF-DACs and RF-ADCs.

Figure 5.9 illustrates again the simple architecture of a full SDR with high speed RF-DACs and RF-ADCs at the front end, where all other modulators, demodulators, receivers etc. are implemented using DDCs, DUCs, and DSP implemented on the Zynq RFSoC's PS and PL.

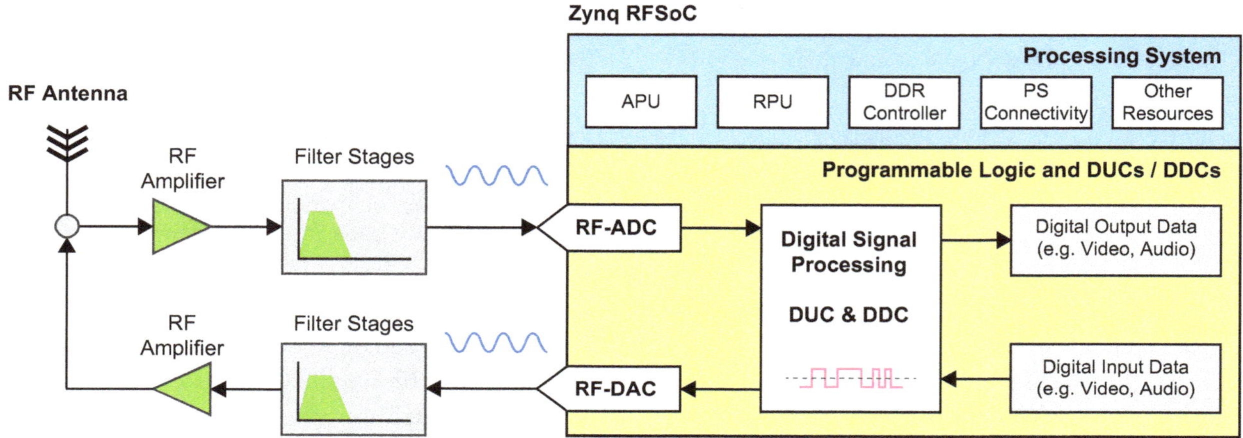

Figure 5.9: The ultimate SDR, with ADCs and DACs sampling at rates of a few GHz

5.5.4 4G LTE Multicarrier Solutions

4G LTE has extensive penetration across the globe. Operating frequencies vary from country to country, however in the likes of USA and Europe, bands at 600MHz, 800MHz, 1800MHz, 2.1GHz are widely deployed. LTE channel bandwidths can vary from 1.4MHz up to 20MHz, and techniques like carrier aggregation can yield, for example, aggregated channels of bandwidth 5 x 20 = 100MHz. Solutions using Zynq-7000 and Zynq MPSoC devices are already widely deployed in 4G LTE basestations, cloudRANs, backhaul and various fronthaul implementations. However, if we now have available the Zynq RFSoC, we can perform direct RF sampling in many of these frequency bands and design multi-channel single SDR chip solutions, implementing algorithms for the 4G LTE PHY layer implementation and appropriate MAC layer controls, virtually across the entire 600MHz to 3GHz frequency bands.

The Zynq RFSoC's PS can be used to control the PHY and MAC layers, with the PL running the channelisation and filtering DSP. As required, Digital Pre-Distortion (DPD) can also be performed on the PS and PL.

5.5.5 5G Mobile and Wireless Implementations

5th Generation radio (so-called 5G) is all about new radio standards, and in many ways, new thinking. As well as technology driving faster data speeds, lower latencies, massive deployment of radios (densification), and proliferation of devices in the Internet of Things (IoT), 5G will also bring more agile and frequency-tunable radios, and allow the development of end-to-end use cases. This will lead to many more new applications, and new business models for mobile and wireless operators. The Zynq RFSoC will be a key enabling device for rollout of 5G deployments in the next few years [39].

The sampling rates afforded by the Zynq RFSoC means that the signals for transmission in new frequency bands being designated for 5G at 700MHz and 3.5GHz can be sampled digitally — and therefore a complete SDR solution can be developed and implemented on the RFSoC (with of course suitable front end RF amplifier stages and pre-filters). The multiple channels available on a single Zynq RFSoC device (up to 16 channels transmit, and 16 channels receive), also means that multi-antenna techniques such as beamforming, beamsteering and MIMO implementations can be implemented. The PS can run the control, baseband and MAC style processing, and the matrix type DSP computation of beamforming, beamsteering, digital pre-distortion etc. can be optimised to run using the PL, leveraging the DSP slices in particular.

One prospect for 5G and/or next generation mobile and wireless implementations is the more widespread availability of shared spectrum bands. These shared spectrum bands will allow new operators such as communities, industrial premises, or campuses to run their own private LTE/5G networks using lower cost disaggregated radio access networks. Recent years has seen TV White Space (TVWS) shared spectrum, most notably in the UK [44] and USA [45], allow Dynamic Spectrum Access (DSA) for the UHF bands (470 to 790MHz) in regions of the country where certain bands are not being used. Similarly, CBRS (Citizen Broadband Radio Service) [43] is now being deployed in the USA at 3.5GHz (see Figure 5.8). Both run databases for the management of the dynamic allocation of these spectrum bands. Within these shared bands, a multitude of different standards/waveforms (e.g. LTE, 5G NR, 802.11xx) may be used, requiring SDR designs with frequency agility, and the flexibility to select dynamically from different frequency bands, and adopt different channel bandwidths.

The Zynq RFSoC will therefore be an excellent design platform for DSA. For example, using TVWS, the RFSoC could digitize the entire UHF band from 400MHz to 800MHz. Individual allocated channel bands in TVWS are 8MHz in the UK and 6MHz in the USA, and some rules of operation allow more than one channel to be allocated, or in some cases adjacent bands to be aggregated. Therefore, by digitising the spectrum with the Zynq RFSoC's RF-ADCs, filtering stages can be designed to extract the bands of interest, whether just one channel of 8MHz (or 6MHz), or the aggregation of multiple channels that are either contiguous (neighbouring) or non-contiguous (for example, two 8MHz channels separated by 64MHz).

5.5.6 5G mmWave Implementations using IF Architectures

5G is expected to bring more densification of devices by using mmWaves, i.e. carrier frequencies at 20GHz and higher. Using mmWave frequencies in the 5G bands at 26GHz to 32GHz ranges (see Figure 5.8), and at higher frequencies of 60GHz, will bring new spectrum bands into play, and allow much wider channel bandwidths and higher data rates (by simple virtue of the much wider channel bandwidths). For example, at a

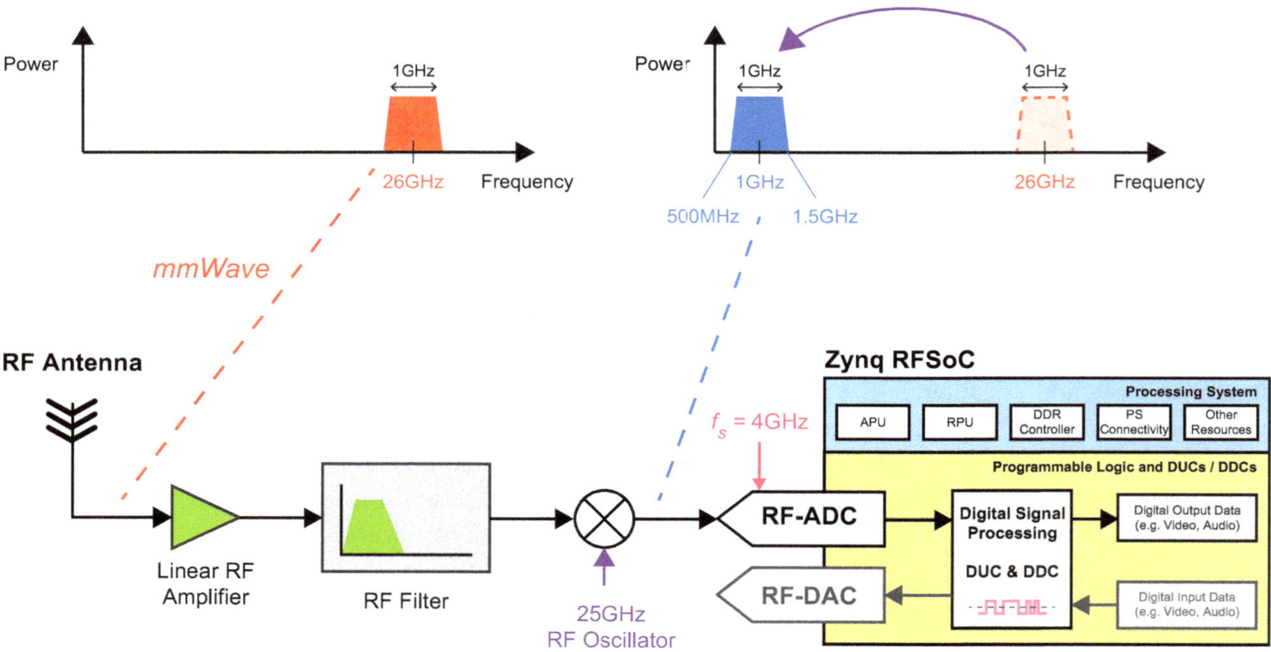

Figure 5.10: Using the Zynq RF as part of an IF Receiver, demodulating from a 26GHz signal to baseband

carrier frequency of 700MHz, a 100MHz frequency band represents has a 14% bandwidth-to-carrier ratio. At 26GHz, a 100Mz channel bandwidth is less than 0.4% bandwidth-to-carrier ratio. Clearly, therefore, there is considerably more spectrum available at these higher frequencies for future communications, and 5G will roll-out in 26/28GHz around the globe.

In this type of mmWave application, the carriers frequencies are well outside of the direct RF sampling bandwidths (which for current Zynq RFSoC devices is often described as direct RF sampling for sub-6GHz signals). However, the Zynq RFSoC can be configured in an Intermediate Frequency (IF) type architecture, whereby front end 26GHz RF oscillators can demodulate the signal of interest down to an IF within the range of the Zynq RFSoC's ADCs. For instance, to create a 26GHz system with, say, 1GHz channels, we can use a first RF oscillator and demodulating/filtering stage to mix a 1GHz bandwidth signal from the 26GHz carrier, down to the band 500 - 1500MHz, and then sample at 12 bits resolution with a sampling rate of f_s = 4GHz. This is illustrated in Figure 5.10. We can expect to see the versatile Zynq RFSoC enabling 5G mmWave basestations as the technology begins to roll out from 2020 onwards.

5.6 Chapter Summary

In this chapter, we first established the system general characteristics that might make Zynq MPSoC the best implementation platform, in favour of a Zynq or FPGA device. Three diverse application areas were then selected as case studies, namely drones, smart vehicles (extending from ADAS systems into autonomous vehicles), and 5G communications. In particular for the last of these applications, we introduced the Zynq RFSoC device, and noted its strong applicability for 5G. As well as providing a brief overview of each area, we considered potential system requirements for each area, and highlighted features of the Zynq MPSoC (or Zynq RFSoC) that can be used to achieve them.

5.7 References

Note: All online sources last accessed March 2019.

[1] 60 Minutes Overtime, "Amazon Unveils Futuristic Plan: Delivery by Drone", *CBS News website*, 1st December 2013.
Available: https://www.cbsnews.com/news/amazon-unveils-futuristic-plan-delivery-by-drone/

[2] E. Ackerman, "LIDAR That Will Make Self-Driving Cars Affordable", *IEEE Spectrum*, Vol. 53. Issue 10, September 2016, pp. 14.
DOI: 10.1109/MSPEC.2016.7572525

[3] R. Allain, "How do Drones Fly? Physics, of Course!", *Wired magazine*, May 2017.
Available: https://www.wired.com/2017/05/the-physics-of-drones/

[4] Auto Alliance, "Levels of Automation".
Available: https://autoalliance.org/wp-content/uploads/2017/07/Automated-Vehicles-Levels-of-Automation.pdf

[5] A. Blake, "AT&T Deploys Drones to Find and Fix 'Dead Zones' in Congested Sports Stadiums", *The Washington Times*, 6th October 2016.
Available: http://www.washingtontimes.com/news/2016/oct/6/att-deploys-drones-to-find-and-fix-dead-zones-in-c/

[6] D. Coldewey, "Echodyne's pocket-sized radar may be the next must-have tech for drones (and drone hunters)", *TechCrunch.com website*, 12th May 2017.
Available: https://techcrunch.com/2017/05/12/echodynes-pocket-sized-radar-may-be-the-next-must-have-tech-for-drones-and-drone-hunters/

[7] A. Conner-Simons, "Self-Flying Drone Dips, Darts and Dives Through Trees at 30 mph", *Computer Science and Artificial Intelligence Laboratory (CSAIL) webpages of Massachusetts Institute of Technology (MIT)*, 26th October 2015.
Available: http://www.csail.mit.edu/drone_flies_through_forest_at_30_mph

[8] F. Corrigan, "Drone Gyro Stabilization, IMU and Flight Controllers Explained", *DroneZon website*, June 2017.
Available: https://www.dronezon.com/learn-about-drones-quadcopters/three-and-six-axis-gyro-stabilized-drones/

[9] Federal Communications Commission, "United States Frequency Allocations: The Radio Spectrum", *U.S. Department of Commerce*, October 2003.
Available: https://www.ntia.doc.gov/files/ntia/publications/2003-allochrt.pdf

[10] R. J. Hall, "An Internet of Drones", *IEEE Internet Computing*, vol. 20, issue 3, May 2016, pp 68 - 73.
DOI: 10.1109/MIC.2016.59

[11] D. Hambling, "The next era of drones will be defined by 'swarms'", *BBC website*, 27th April 2017.
Available: http://www.bbc.com/future/story/20170425-were-entering-the-next-era-of-drones

[12] S. Leibson, "Ag Tech Precision Spray Autopilot uses Aerotenna's new, Zynq-based Smart Drone Dev Platform", *Xilinx Xcell Daily Blog*, 11th September 2017.
Available: https://forums.xilinx.com/t5/Xcell-Daily-Blog/Ag-Tech-Precision-Spray-Autopilot-uses-Aerotenna-s-new-Zynq/ba-p/793030

[13] S. Leibson, "Lights! Action! Drone! The DJI Inspire 2 drone operates the 5K camera: a Zynq SoC its "internal engine"", *Xilinx Xcell Daily Blog*, 12th December 2016.
Available: https://forums.xilinx.com/t5/Xcell-Daily-Blog/Lights-Action-Drone-The-DJI-Inspire-2-drone-operates-the-5K/ba-p/738127

[14] S. Leibson, "Zerotech's palm-sized Dobby AI drone uses DeePhi machine-learning algorithms running on 3W Xilinx Zynq Z-7020 SoC", *Xilinx Xcell Daily Blog*, 27th June 2017.
Available: https://forums.xilinx.com/t5/Xcell-Daily-Blog/Zerotech-s-palm-sized-Dobby-AI-drone-uses-DeePhi-machine/ba-p/774764

[15] Ofcom, "UK Frequency Allocation Chart (UKFAT)", online interactive tool.
Available: http://static.ofcom.org.uk/static/spectrum/fat.html

[16] T. Ong, "The first autonomous drone delivery network will fly above Switzerland starting next month", *The Verge website*, 20th September 2017.
Available: https://www.theverge.com/2017/9/20/16325084/matternet-autonomous-drone-network-switzerland

[17] S. Patole, M. Torlak, D. Wang, and M. Ali, "Automotive Radars: A Review of Signal Processing Techniques", *IEEE Signal Processing Magazine*, Volume 34, Issue 2, March 2017, pp. 22 - 35.
DOI: 10.1109/MSP.2016.2628914

[18] M. Russon, "Nokia and EE Trial Mobile Base Stations Floating on Drones to Revolutionise Rural 4G Coverage", *International Business Times*, 15th August 2016.
Available: http://www.ibtimes.co.uk/nokia-ee-trial-mobile-base-stations-floating-drones-revolutionise-rural-4g-coverage-1575795

[19] M. Schuler, "Watch: Over 100 Micro-Drones Swarm in Formation", *gCaptain website*, 13th January 2017.
Available: http://gcaptain.com/watch-over-100-micro-drones-swarm-in-formation/

[20] X. Wang, A. Chowdhery, and M. Chiang, "SkyEyes: Adaptive Video Streaming from UAVs", *Proceedings of the 3rd Workshop on Hot Topics in Wireless*, New York, USA, October 2016, pp. 2 - 6.
DOI: 10.1145/2980115.2980119

[21] X. Wang, A. Chowdhery, and M. Chiang, "Networked Drone Cameras for Sports Streaming", *Proceedings of the 37th International Conference on Distributed Computing Systems (ICDCS)*, Atlanta, USA, July 2017, pp. 308 - 318.
DOI: 10.1109/ICDCS.2017.200

[22] A. Davies, "The Wired Guide to Self-Driving Cars", *Wired* magazine, 13th December 2018.
Available: https://www.wired.com/story/guide-self-driving-cars/

[23] M. Chen, Y. Tian, G. Fortino, J. Zhang, and I. Humar, "Cognitive Internet of Vehicles", *Computer Communications*, Vol. 120, May 2018, pp. 58 - 70.

DOI: 10.1016/j.comcom.2018.02.006

[24] J. Janai, F. Guney, A. Behl, and A. Geiger, "Computer Vision for Autonomous Vehicles: Problems, Datasets and State-of-the-Art", *arXiv Computing Research Repository*, April 2017.

Available: https://arxiv.org/abs/1704.05519

[25] L. M. Hulse, H. Xie and E. R. Galea, "Perceptions of autonomous vehicles: Relationships with road users, risk, gender, and age", *Safety Science*, Vol. 102, February 2018, pp. 1 - 13.

DOI: 10.1016/j.ssci.2017.10.001

[26] T. Liljamo, H. Liimatainen, and M. Pöllänen, "Attitudes and concerns on automated vehicles", *Transportation Research Part F*, Vol. 59, Part A, November 2018, pp. 24 - 44.

DOI: 10.1016/j.trf.2018.08.010

[27] T. Bechor, "Cybersecurity for Autonomous Vehicles Must Be a Top Concern for Automakers", *the institute* IEEE news source blog, 23rd January 2019.

Available: http://theinstitute.ieee.org/ieee-roundup/blogs/blog/cybersecurity-for-autonomous-vehicles-must-be-a-top-concern-for-automakers

[28] Xilinx, Inc., "Zynq UltraScale+ RFSoC RF Data Converter 2.1", PG269 v2.1, December 2018

Available: https://www.xilinx.com/support/documentation/ip_documentation/usp_rf_data_converter/v2_1/pg269-rf-data-converter.pdf

[29] Xilinx, Inc., "Zynq UltraScale+ RFSoC Data Sheet: Overview", DS889, v1.7, February 2019.

Available: https://www.xilinx.com/support/documentation/data_sheets/ds889-zynq-usp-rfsoc-overview.pdf

[30] Xilinx, Inc., "Zynq UltraScale+ RFSoC Product Tables and Product Selection Guide", February 2019.

Available: https://www.xilinx.com/support/documentation/selection-guides/zynq-usp-rfsoc-product-selection-guide.pdf

[31] Xilinx, Inc., "Zynq UltraScale+ MPSoC Product Tables and Product Selection Guide", February 2019

Available: https://www.xilinx.com/support/documentation/selection-guides/zynq-ultrascale-plus-product-selection-guide.pdf

[32] Analog Devices, "AD9361 Analog Devices RF Agile Transceiver" product page.

Available: https://www.analog.com/en/products/ad9361.html

[33] Xilinx, Inc., "Raptor SDR Development Kit" webpage.

Available: https://www.xilinx.com/products/boards-and-kits/1-scfgkc.html

[34] Ettus Research, "USRP E310" product page.

Available: https://www.ettus.com/all-products/e310-kit-1/

[35] ZedBoard.org, "PicoZed (SDR) Development Kit" webpage.

Available: http://zedboard.org/product/picozed-sdr-development-kit

[36] Analog Devices, "Active Learning Module SDR - Pluto featuring Zynq Z-7010" webpage.

Available: https://www.analog.com/en/design-center/evaluation-hardware-and-software/evaluation-boards-kits/adalm-pluto.html

[37] A. Collins, "All-Programmable RF-Sampling Solutions", *Xilinx White Paper*, WP489, v1.0.1, April 2017.

Available: https://www.xilinx.com/support/documentation/white_papers/wp489-rfsampling-solutions.pdf

[38] *Common Public Radio Interface* website.
Available: http://www.cpri.info

[39] Xilinx, Inc., "Understanding Key Parameters for RF-Sampling Data Converters", *Xilinx White Paper*, WP509, v1.0, February 2019.
Available: https://www.xilinx.com/support/documentation/white_papers/wp509-rfsampling-data-converters.pdf.

[40] C. Erdmann, E. Cullen, D. Brouard, R. Pelliconi, B. Verbruggen, J. Mcgrath, D. Collins, M. De La Torre, P. Gay, P. Lynch, P. Lim, A. Collins, and B. Farley, "A 330mW 14b 6.8GS/s dual-mode RF DAC in 16nm FinFET achieving -70.8dBc ACPR in a 20MHz channel at 5.2GHz", *Proceedings of the IEEE International Solid-State Circuits Conference (ISSCC)*, San Francisco, USA, February 2017, pp. 280 - 281.
DOI: 10.1109/ISSCC.2017.7870370

[41] Joe Mitola. "The Software Radio Architecture", *IEEE Communications Magazine*, Volume 33, Issue 5, May 1995, pp. 26 - 38.
DOI: 10.1109/35.393001

[42] MathWorks, "Design and prototype SDR Systems with MATLAB and Simulink" webpage.
Available: https://www.mathworks.com/discovery/sdr.html

[43] *CBRS Alliance* website.
Available: https://www.cbrsalliance.org/

[44] Ofcom, "TV White Space Databases" webpage.
Available: https://www.ofcom.org.uk/spectrum/spectrum-management/TV-white-space-databases

[45] Federal Communications Commission (FCC), "White Space Database Administrators Guide" webpage.
Available: https://www.fcc.gov/general/white-space-database-administrators-guide

Part B

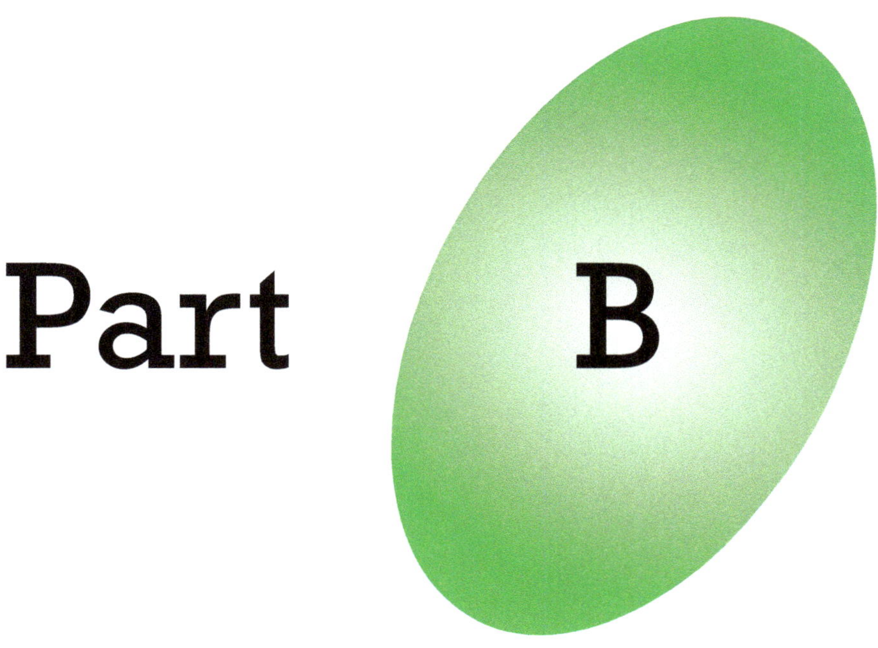

The Zynq MPSoC Architecture in Detail

Chapter 6

The Application Processing Unit

An application processor is a type of microprocessor unit, primarily used for processing the main application code within an embedded system. It is capable of controlling and interfacing to peripherals and specialised processors.

In this chapter, we will investigate the role of the Cortex-A53 MPCore processor within the Zynq MPSoC device. We describe in detail the Armv8-A architecture and explore the signals and interfaces, memories, interrupts, processor extensions and peripherals that are part of the Application Processing Unit (APU). Finally, we introduce application processor virtualisation.

6.1 The Cortex-A53 MPCore Processor

The Zynq MPSoC has an Arm Cortex-A53 application processor, suitable for tasks requiring high performance and power efficiency. Figure 6.1 shows the Cortex-A53 MPCore processor and highlights its processor cores, management resources, control units, memories and I/O interfaces. Each of these components form three distinct groups: the Cortex-A53 MPCore, the Governor, and the Level 2 Memory System [10].

The Cortex-A53 MPCore processor cluster contains four Cortex-A53 cores in EG and EV devices, while CG devices contain two. Each Cortex-A53 core consists of its own FPU, NEON and Crypto computational units, together with separate 32 KB Level 1 cache memories for both data and instructions.

The Governors provide their respective cores with clock, debug, and tracing resources. The Level 2 Memory System provides the entire Cortex-A53 MPCore processor with communication to devices outside the APU.

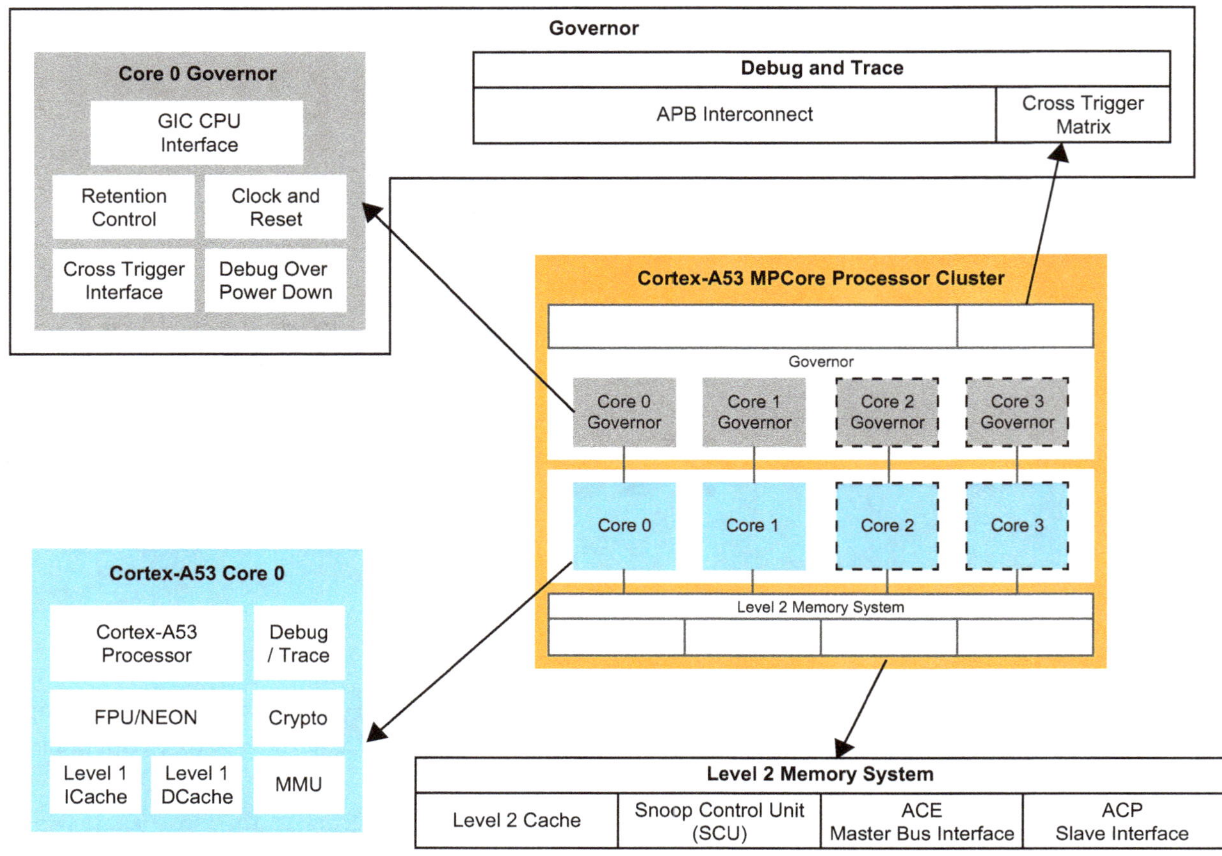

Figure 6.1: The Cortex-A53 MPCore showing the Cortex-A53 cores, Governor and Level 2 Memory System [12]

6.2 Fundamentals of Armv8-A

The Cortex-A53 processor uses the Armv8-A processor architecture described in [9]. We will now investigate several concepts and terminology that you may encounter when using the Cortex-A53 MPCore processor. Initially, we will explore the Armv8-A execution states and study several key aspects associated with the operation of the processor core. This investigation will include the A32 and T32 instruction sets, processor modes, and privilege levels, that were previously associated with the Armv7-A architecture. These are still relevant as they are used in Armv8-A architectures.

New concepts and definitions relating only to the Armv8-A architecture will also be introduced, including the A64 instruction set, processor exception levels, and the security model [9]. This section aims to provide you with a fundamental understanding of the Armv8-A architecture of the Cortex-A53 MPCore processor.

6.2.1 64-Bit/32-Bit Execution States

Before the release of Armv3, Arm has used 32-bits for most aspects of its processor cores (prior to Armv3, 26-bits was used for the address bus) [4]. The use of 32-bits can be seen on the data bus, address bus and processing width of relevant Arm processors. It is only now, as the Arm architecture has approached its 8th version, in which a 64-bit core is available. The Cortex-A53 processor adopts a 64-bit core.

So that Armv8-A architectures can continue to support operation in a 32-bit execution state, the processor's execution environment must be able to change between 32-bit and 64-bit operation.

An execution state characterises a processor's execution environment [9] and defines each of the following.

- The processor's supported register widths.
- The processor's supported instruction sets.
- Significant aspects of the processor's execution model, Virtual Memory System Architecture, and programmer's model.

The Cortex-A53 MPCore processor may use two possible execution states. These are the AArch64 and AArch32 execution states [9].

AArch32

AArch32 is a 32-bit execution state compatible with the Armv8-A architecture. It was supported by previous Armv7-A architectures and definitions that contain the TrustZone Security [7] and Virtualisation Extensions [8]. Features of the AArch32 execution state [9] are listed as follows.

- Previous Thumb 2 and Arm 32-bit instruction sets (further described in Section 6.2.2) are supported and are fully backwards compatible with previous Armv7-A architectures.
- The execution state features 13 general-purpose 32-bit registers, and a 32-bit Program Counter (PC), Stack Pointer (SP), and Exception Link Registers (ELRs).
- The Armv8 exception model (discussed in Section 6.2.3) can be mapped onto the previous Armv7 (AArch32) Privilege Level (PL) system.
- The AArch32 exception model, supports the Armv7 exception model's use of processor modes (discussed in Section 6.2.4).
- AArch32 supports 32-bit Virtual Addresses. The Virtual Memory System Architecture maps these to a Physical Address that can support up to 40-bits.
- The entire processor state is available in the Current Program State Register (CPSR).

AArch64

The AArch64 is a 64-bit execution state [9] and allows Armv8-A architectures to take advantage of increased address space, general purpose registers and a range of other features. A description of these follows:

- In the 64-bit execution state, the Armv8-A processor uses a new Arm 64-bit instruction set named A64. The processor can now perform operations on 64-bit wide registers. Instructions are still supported using 32-bits.
- The core can support 31 general-purpose 64-bit registers, and a 64-bit PC, SP and ELRs.
- The use of the Armv8 exception model, which consists of four Exception Levels (EL0 - EL3). The exception levels provide a hierarchy of execution privilege (further discussed in Section 6.2.3).
- AArch64 supports 64-bit Virtual Addresses. The Virtual Memory System Architecture maps these to 40-bit Physical Address maps.
- Several PSTATE elements hold the processor's current state, which can be operated on directly using specific instructions in the A64 instruction set.
- Each system register now has a suffix indicating the minimum Exception Level in which to access it.

Rules for Changing Execution State

Changing between different Execution States (AArch32 or AArch64) requires the Exception Level to be increased or decreased. That is, taking a processor exception to a higher Exception Level or returning from an exception to a lower Exception Level. This change is known as *interprocessing* [9].

When an exception rises to a higher Exception Level, the execution state can either remain the same or increase from the AArch32 state to the AArch64 state. It may not decrease from the AArch64 state to the AArch32 state.

When returning from an exception to a lower Exception Level, the execution state can either remain the same or decrease from the AArch64 state to the AArch32 state. It may not increase from the AArch32 state to the AArch64 state.

Section 6.2.3 provides further discussion on Exception Levels and Execution State.

6.2.2 Instruction Sets and Programming Languages

Cortex-A series processors support Arm and Thumb instruction sets [9]. Instruction sets are a group of commands that can be issued directly to a processor with very little or no abstraction between the software and underlying architecture. For example, two fundamental commands that may be issued to a processor are a *store* operation (storing data in memory) and a *load* operation (loading data from memory).

There are three primary instruction sets that are used by the Cortex-A53 MPCore processor within the Zynq MPSoC. These are the A32, A64 (Arm) and T32 (Thumb) instruction sets. When a processor is using a particular set of instructions, it is said to be in an *instruction set state* relative to the chosen set. For example, if the processor uses the A32 instruction set, it is in the A32 state.

It is not possible for a processor to execute an instruction set when it is not in the corresponding *instruction set state*. For example, a processor operating in the A32 state cannot execute T32 instructions. Similarly, a processor operating in the T32 state cannot execute A32 instructions.

The AArch32 Execution State may only use the A32 and T32 instruction sets and the AArch64 Execution State may only use the A64 instruction set.

The A32 and T32 Instruction Sets

The A32 instruction set, also known as the original Arm instruction set, provides 32-bit instructions covering a vast range of features and operations. The Thumb (T16) instruction set was first used by Armv4T architectures and was capable of 16-bit instructions. Thumb-2 (T32), used by Armv6 architectures and beyond is composed mostly of 16-bit instructions, but also includes 32-bit instructions. Currently, the AArch32 processor Execution State can use the A32 and T32 instruction sets.

When the processor fetches a 16-bit T32 instruction from memory, it interprets the word into a 32-bit instruction. This conversion allows the T32 subset to achieve performance similar to the A32 instruction set, while optimising code density better than the original Thumb instruction set.

The Arm Architecture Reference Manual, Armv8 [9], provides further information about assembly language for each subset.

The A64 Instruction Set

The A64 instruction set provides a series of 32-bit instructions similar to the A32 and T32 subsets, and currently may be used in the AArch64 processor Execution State. There are new encodings, assembly language and other features available to the programmer in the A64 instruction set that are not available in the A32 instruction set. These include access to a larger virtual address space, 64-bit addressable physical memory space, enhanced support for advanced SIMD (NEON), and various register improvements such as 64-bit registers.

The Arm Architecture Reference Manual, Armv8 [9] provides further information.

6.2.3 Exception Levels

An exception is taken by the processor core whenever a halt to the normal execution of a program occurs. This exception will place the processor in its associated mode of operation, depending on the given exception. The result will cause the exception level to increase between EL0-EL3 to correctly handle the exception.

Figure 6.2 illustrates where each processor mode has been allocated in the Exception Model and to which exception level.

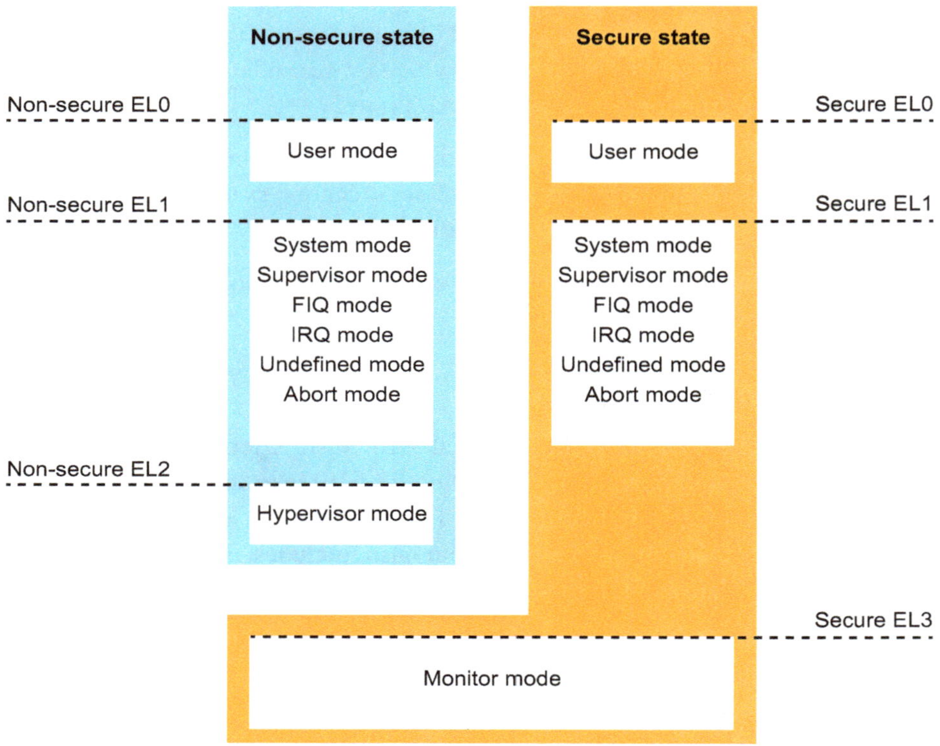

Figure 6.2: The Secure and Non-Secure state with processor modes and privilege levels [4]

Each exception level in the Armv8 Exception Model have different purposes. A description of these are as follows:

- **EL0** is known as unprivileged software execution and has the lowest software execution privilege.
- **EL1** provides support for increased software execution privilege. Restricts unprivileged user access to protected resources.
- **EL2** provides support for processor virtualisation and the implementation of hypervisors.
- **EL3** supports a Secure state (further discussed in Section 6.2.5).

If a fast interrupt exception occurs in user mode, the processor will increase the exception level to EL1 to enter Fast Interrupt reQuest (FIQ) mode. Before handling an exception, the processor preserves its state, and critical elements of the original program, so that it may recommence once the handler routine has finished. This preservation is known as context switching [4].

6.2.4 Processor Modes

A processor, operating in the AArch32 or AArch64 processor Execution State (Section 6.2.1), requires a method of securely catching interrupts, managing exceptions and accessing protected resources. Processor modes have been provided to handle the above issues. These are known as the processor's *modes* of operation [4]. There are nine modes as shown in Table 6.1.

Changing the *mode* of operation can be carried out in software. Additionally, an external or internal exception may also cause the mode to change (Section 6.2.3 discusses this further).

Table 6.1: Modes of Operation for Armv8-A Architectures [4]

Mode	TAG	Exception Level (EL)	Security State
User	USR	EL0	Both
Fast Interrupt	FIQ	EL1	Both
Interrupt	IRQ	EL1	Both
Supervisor	SVC	EL1	Both
Monitor	MON	EL3	Only Secure State
Abort	ABT	EL1	Non-Secure State
		EL3	Secure State
System	SYS	EL1	Both
Undefined	UND	EL1	Non-Secure State
		EL3	Secure State
Hypervisor	HYP	EL2	Only Non-Secure State

The first is the User (USR) mode which is the normal operating mode for the execution of Arm or Thumb programs and applications. Modes other than the standard user mode are collectively known as privileged modes. Each privilege mode can map directly to the Armv8 exception model's use of Exception Levels, discussed previously in Section 6.2.3.

User

User mode is suitable for an software applications running on an operating system so that protected system resources and specific instructions may be restricted. Programs that execute in this mode operate under Exception Level 0 (EL0). EL0 is known as the unprivileged mode, indicating that it cannot access protected system resources.

Fast Interrupt (FIQ) and Interrupt (IRQ)

The processor enters the *Fast Interrupt* mode when it has encountered a fast interrupt (FIQ) exception. The processor enters the *Interrupt* mode when it has detected a standard interrupt (IRQ) exception. An FIQ exception has priority over an IRQ exception, such that IRQ handlers are disabled until the issued FIQ service routine has completed.

Supervisor

The Supervisor Call (SVC) exception is the operating system's protected mode that the processor enters when taking a supervisor call. An SVC requests a protected operating system function for a user. The Supervisor mode has increased Exception Level (EL1) and can grant requests to protected functions. The processor will also enter Supervisor mode on reset.

Monitor

Monitor mode is a secure mode which is always in the Secure state, regardless of other processor status registers. The processor enters this mode only when a Secure Monitor Call (SMC) exception occurs. Software operating in Monitor mode has access to copies of both the Secure and Non-secure system registers. These registers provide the processor with a method of changing between the Secure and Non-secure states. Arm security and further detail of the Monitor mode is described later in Section 6.2.5.

Abort

The processor enters this mode after a data or instruction abort exception has occurred.

System

A privileged mode for executing operating system tasks that need access to protected system resources. It cannot be entered by any exception and has the same registers available as User mode. This restriction ensures that the state of the task is not corrupted by the additional registers associated with the other exception modes.

Undefined

The processor enter this mode when an undefined instruction exception occurs.

Hypervisor

A hypervisor is a virtual machine monitor that creates and runs virtual machines, discussed further in Chapter 13.

Hypervisor mode is a Non-secure mode only available in the Non-secure state. It is implemented as part of the virtualisation extensions in Armv7-A processors, and is available in all Armv8-A processors. This mode provides all of the functionality for virtualisation with the target processor and executes at EL2.

Hypervisor mode can only be accessed by taking an exception from a Non-secure EL1 or EL0 mode. It is also possible by returning from an exception in Secure Monitor mode. By default, entering the hypervisor mode using a Hypervisor Call (HVC) is initially undefined until enabled by secure software in the Secure state. Enabling the HVC exposes a unique set of APIs to a Guest OS (described further in Section 6.9).

6.2.5 Armv8 Security Model

It is important to consider why it is necessary to have security measures within an embedded device such as the Cortex-A53 MPCore processor. The main purpose of security within a system is to ensure that essential resources cannot be written to, read from, damaged or made unavailable to genuine users. It is possible for devices to be attacked, putting vital resources at high levels of risk. Arm has detailed several ways in which a device may be attacked by what are known as *attack vectors*. Further information can be found in [7].

We will explore Arm security technology, which has been designed to combat malicious software threats and attacks. Previously, the Armv7-A architecture provided the OEM with the option of implementing a processor security extension known as TrustZone Security, for particular Cortex-A processors [6]. The Armv8-A profile already contains similar functionality due to EL3, which was described in Section 6.2.3.

The Secure and Non-Secure State

System users may intentionally or unintentionally put the system at risk by malicious software attacks. Therefore, separating critical system resources from other non-critical resources is beneficial.

Within the Armv8-A profile, there are two security states: a *Secure state* and a *Non-secure state* [7]. The Non-secure state is also known as the Normal World. The Normal World is used to describe not only the processor's execution state, but all of the additional features such as memory and peripherals that are available within that state. The Secure state is similarly known as the Secure World. Just like the Normal World, this is used to describe the processor's execution state and all available features accessible within that state.

Both the Normal World and Secure World can operate simultaneously on the same physical core through virtualisation. Combining both worlds on a single physical core is possible because the Armv8-A architecture is capable of providing two virtual cores [7]. In addition to this, each Virtual Central Processing Unit (VCPU) has separate system registers and a memory address space. These allow the execution of instructions in each state to be isolated from one another. An operating system can then run in parallel with a trusted operating system. Figure 6.3 provides an example.

Swapping Between States

Each virtual processor shares the physical processor by executing instructions in a time-sliced fashion. Swapping between security states is achieved by entering a processor mode known as *Monitor Mode.*

The manner in which the Normal World enters this mode is firmly controlled, so as to remain secure. Switching from a Non-secure state to a Secure state may be triggered through a dedicated instruction known as a Secure Monitor Call (SMC), or by a subset of hardware exceptions [7].

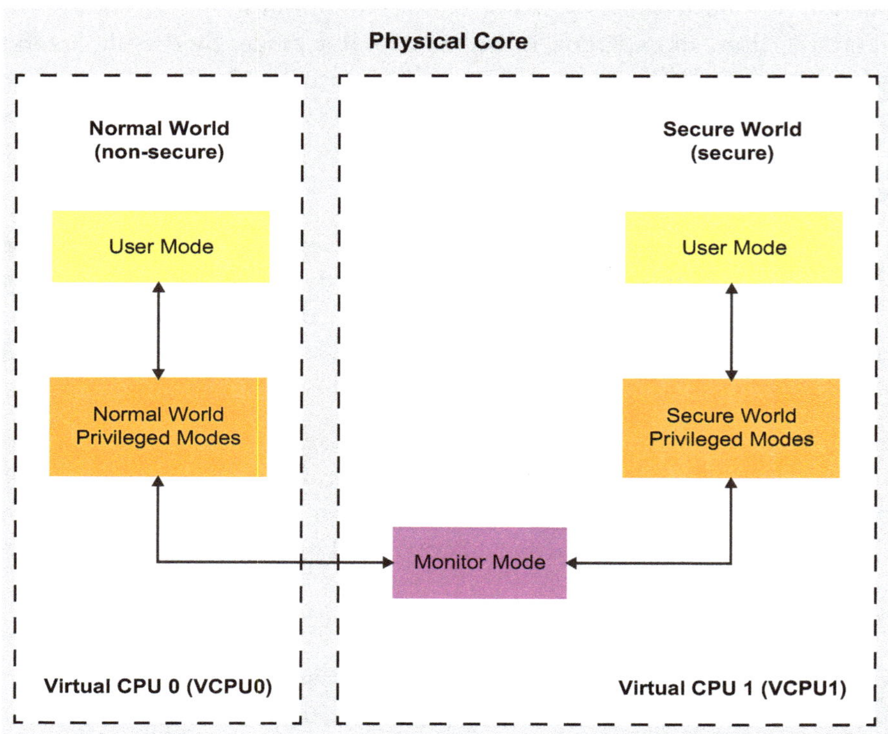

Figure 6.3: Two virtual cores, using one physical core, implementing the secure and non-secure state [7]

Entry to Monitor Mode via the Secure World is readily available and can be achieved by directly writing to the Current Program Status Register (CPSR), or using any of the methods available to the Normal World.

The Monitor Mode uses the software to robustly maintain and manage a switch between Secure and Non-secure states. When the processor is operating in Monitor Mode, it is always executing instructions in the Secure World. The software ensures that it saves the state the processor is leaving, and correctly restores the state the processor has switched to. In this way, the monitor may *context switch* resources required by both worlds. This method will help critical system resources to remain protected from attack vectors.

Each Cortex-A53 core has a Normal World (Non-secure state) and a Secure world (Secure state). Each physical core may transition between the Secure World and Normal World at any point in time, and may do this independently of other cores in the cluster.

Security Model when EL3 is using AArch64

EL3 may operate in the AArch64 state. This allows exception levels below EL3 to either remain in the same execution state or decrease to AArch32. Figure 6.4 contains a diagram illustrating the AArch64 security model.

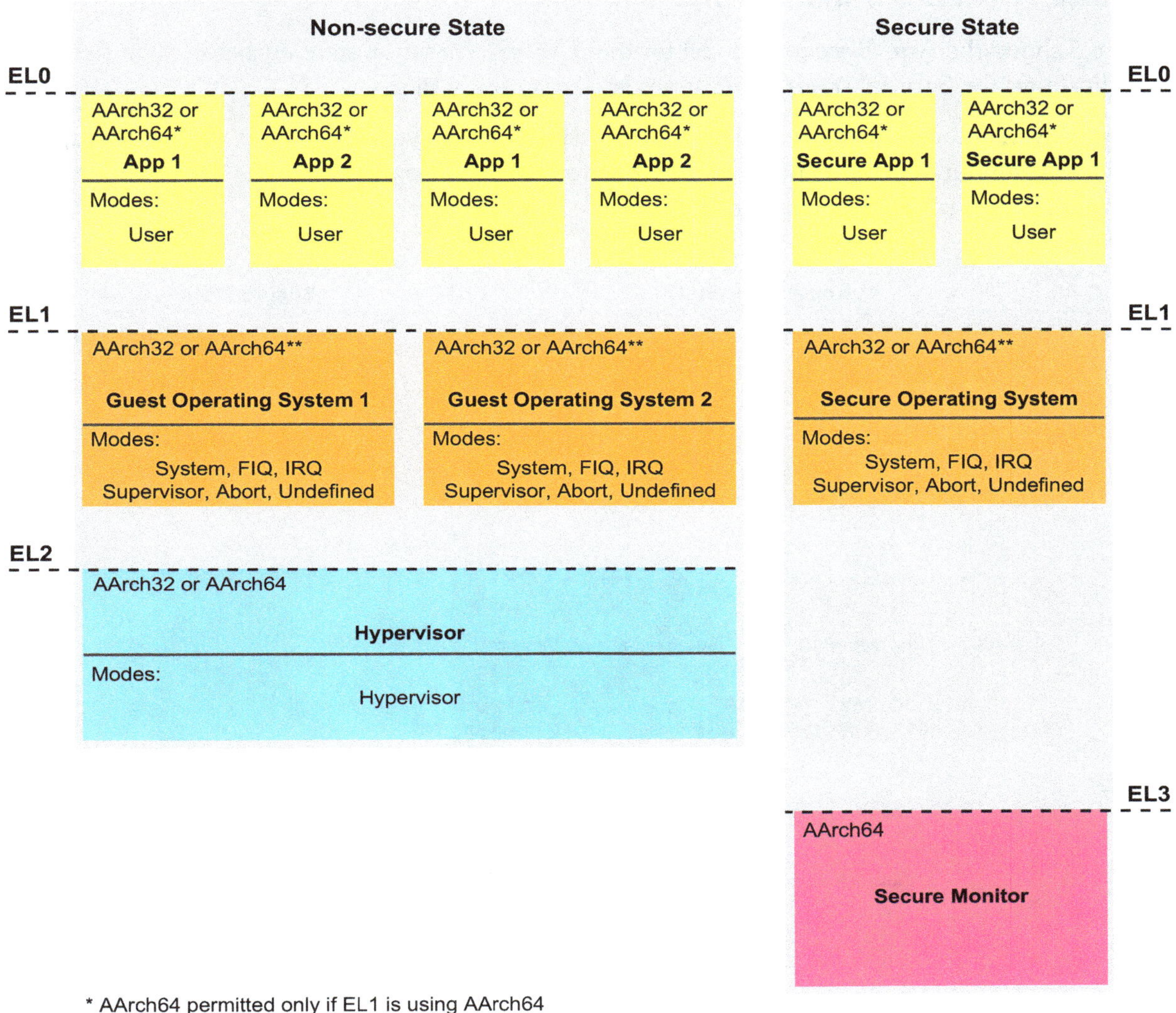

Figure 6.4: The AArch64 security model (when EL3 is using AArch64) [9]

It is easy to identify the Secure state and Non-secure state in Figure 6.4. Within each state are the respective processor modes, applications and operating systems. The hypervisor mode is also shown in EL2 and is only accessible in the Non-secure state. Similarly, the Secure Monitor shown in EL3 is only present in the Secure state.

As previously mentioned (Section 6.2.1) it is not possible for the execution state to increase from AArch32 to AArch64 when returning from an exception. Therefore AArch64 is only permitted in EL1 if EL2 is using AArch64. Similarly, AArch64 is only allowed in EL0, if EL1 is also using AArch64.

Security Model when EL3 is using AArch32

Figure 6.5 shows the Armv8 security model for the AArch32 execution state. As before with AArch64, the diagram illustrates the Security states, applications and operating systems.

As previously mentioned (Section 6.2.1) it is not possible for the execution state to increase from AArch32 to AArch64 when returning from an exception. Therefore all exception levels (in the Secure and Non-secure states) may only use the AArch32 execution state.

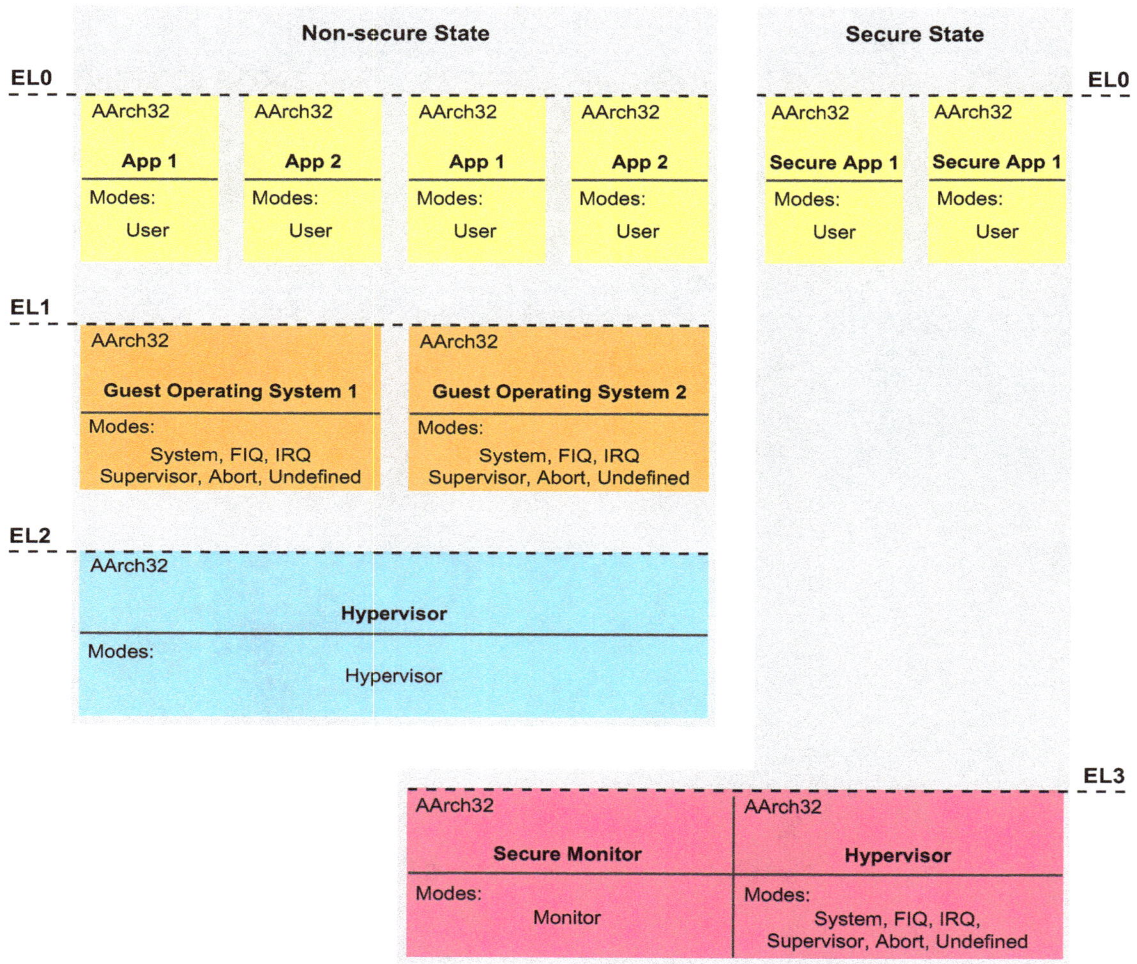

Figure 6.5: The AArch32 security model (when EL3 is using AArch32) [9]

In contrast to the AArch64 security model, EL3 now contains the secure operating system within the Secure state. As detailed in [9], this is to provide software compatibility with the Virtual Memory System Architecture Version 7 (VMSAv7) implementations associated with the Armv7 architecture.

6.3 Signals and Interfaces

The Cortex-A53 MPCore processor has many input and output ports. Several of these use an interface from the AMBA open standard, described previously in Section 3.5.1. There are two primary interfaces for communicating between the APU and other processing elements. These are the Master Memory Interface [10] and the Accelerator Coherency Port (ACP) [10], [12]. Other interfaces include tracing ports, interrupt sources, cross trigger interfaces, resets, and clocks.

Figure 7.7 illustrates the signals and interfaces of the Cortex-A53 application processor and their connections in the Zynq MPSoC. The remainder of this section will provide an overview of each of the signals and interfaces available on the Cortex-A53 MPCore processor. We will discuss functionality and specific usage requirements where appropriate.

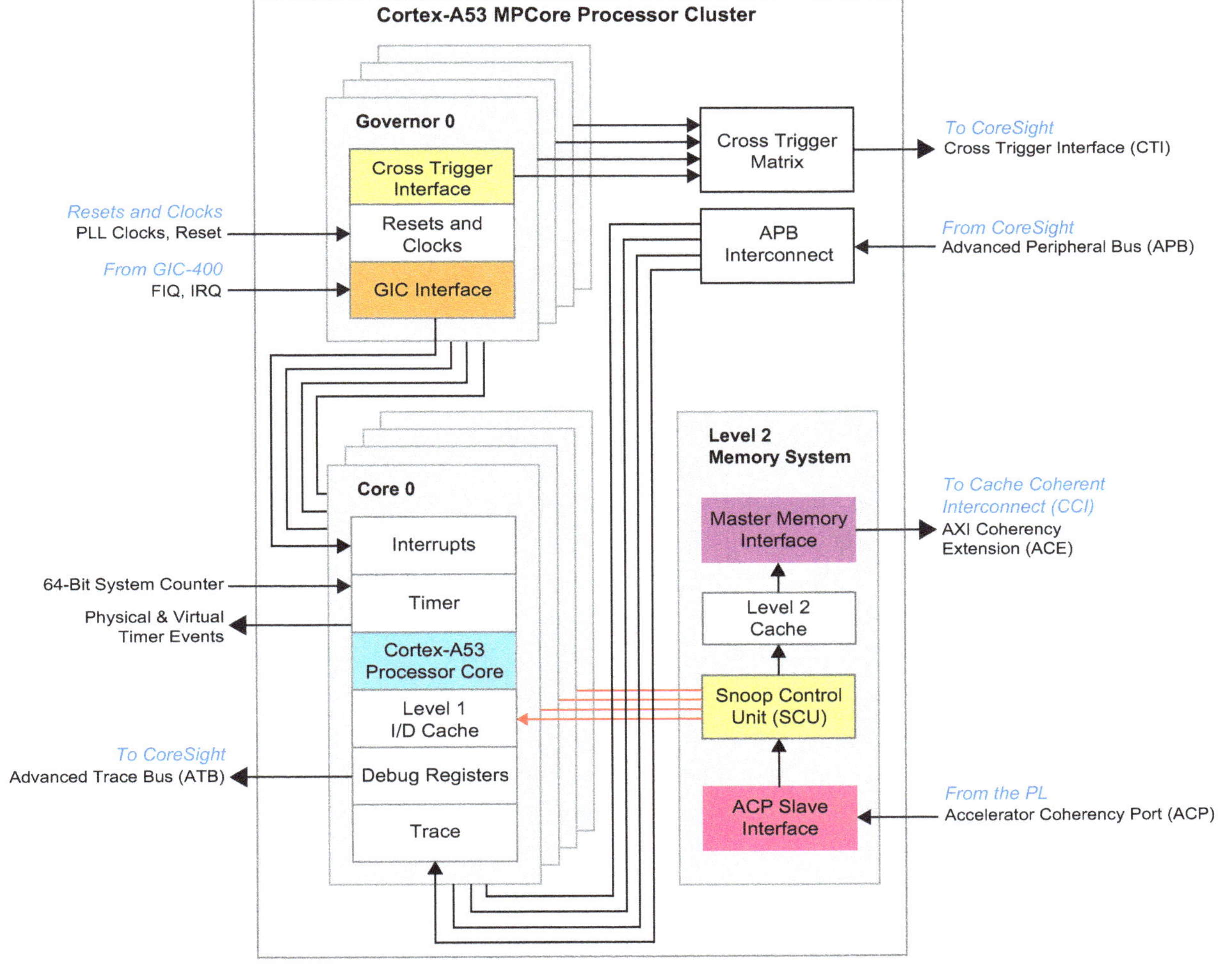

Figure 6.6: Signals and interfaces of the Cortex-A53 MPCore processor [11]

6.3.1 The Master Memory Interface

The master memory interface uses the AXI Coherency Extension (ACE) protocol to communicate with the Cache Coherent Interconnect (CCI). Previously, the Cortex-A9 application processor in the Zynq SoC used AXI to communicate between processing elements throughout the device [14]. The Cortex-A53 application processor in the Zynq MPSoC uses ACE, which is part of the AMBA open standard (described in Section 3.5.1). The ACE protocol is similar to AXI, however, supports hardware coherency. This allows for system-level communication of up-to-date and information across the entire Zynq MPSoC.

ACE Operation

A concept known as *barrier transactions* is used in the ACE protocol to manage the order of transactions efficiently. A barrier transaction is where the ACE protocol stalls the processing flow of data until a particular condition has been met [1]. There are two types of barrier transaction. The first is a memory barrier, which is used to ensure that transactions issued before the barrier are fully completed before transactions that are issued after the barrier. The memory barrier could be used to make sure that data has been successfully written to a memory location before another master initiates a read operation on the same memory space. The other type of barrier transaction is a synchronisation barrier, which prevents all master devices in a system from issuing a transaction until all previous transactions have successfully resolved.

Barrier transactions are possible using ACE since there are additional control signals added to the existing read address, read data, write address and write data channels of the AXI protocol. These allow for barrier signalling between components and indicate which masters should manage the barrier transactions. There are also additional signals for snooping which indicate the type of snoop (read or write) to be carried out [1].

ACE Snooping

Snooping is a major factor in achieving coherency between the APU and the remaining Zynq MPSoC components. Coherency relies on all master devices listening to every transaction issued by other masters. Upon a read operation being carried out, the master device with the most current data will provide it to the requesting master. Similarly, when a write operation is detected, the local copy held by the master 'listening in' will be invalidated [2]. Snooping makes it is possible to achieve coherency and ensures write operations to the same memory location are discernible by two or more master devices.

The ACE snooping cache coherency protocol requires three additional snoop channels alongside the existing AXI standard [1]. The first is the snoop address channel, which is used to provide a cached master device with address and control information of the ongoing snoop transaction. Another is the snoop data channel that allows a cached master device to pass data to a device performing the snoop. The last is the snoop response channel which indicates the type of response to a snoop transaction. Every snoop has a response associated with the transaction, which indicates whether the cached master device is going to pass data onto the snoop data channel.

The ACE interface is connected directly to the Level 2 cache of the Cortex-A53 MPCore. This interface arbitrates memory requests from the Level 2 cache and the rest of the Zynq MPSoC. The ACE interface continuously snoops on other masters via the CCI. Snooping allows the APU to remain coherent with the remainder of the Zynq MPSoC.

6.3.2 ACP Slave Interface

The Accelerator Coherency Port (ACP) is a cache coherent slave interface connected to the SCU within the APU's Level 2 Memory System. The PL is connected to this interface, whose purpose is to allow FPGA hardware accelerators access to the Level 2 cache of the APU. Due to using a hardware interface to handle coherency, as opposed to using software code, the ACP may increase performance and improve power consumption.

It is important to note that the APU is unable to access memories that are local to the PL. The coherency port only provides one-way communication for hardware accelerators, which are tightly-coupled to the APU.

ACP Granularity of Instructions

The ACP is a low-latency path that can be used to access any coherent areas of memory in the APU's Level 1 and Level 2 cache, and the DDR memory. Initially, ACP read transactions are processed by the SCU to check if the required data is contained within the APU's Level 1 data cache. If the required data is retrieved, then it is returned to the hardware accelerator operating in the PL. However, if the data cannot be found (also known as a cache miss) the read transaction will check the Level 2 cache. If another cache miss occurs, the read transaction is forwarded to the DDR memory, which impacts system performance and power consumption.

The APU's Level 2 cache memory is shared with many system processing elements. As a result, PL communication through the ACP is non-deterministic. It is important to consider how transactions between a hardware accelerator in the PL and the ACP should be issued to minimise cache misses and prevent cache thrashing (eviction of old data that could be important).

The ACP is not suitable for transactions where the granularity of instructions is fine-grained. Fine-grained granularity is when there are many small transactions that all accumulate additional latency overheads. Too many transactions, using the ACP, can cause cache thrashing or many transactions reaching their destination late. Furthermore, the ACP is not suitable for accessing single address locations in the APU's cached memory. If a cache miss occurs, then the transaction may be forwarded to the DDR memory, decreasing performance. Access to single DDR memory locations is faster through the Zynq MPSoC's high-performance ports instead (discussed further in Section 11.3).

Large data transactions are also not suitable using the ACP, as the APU's cores and the ACP compete with one another for access to system components outside the APU. As a result, the APU's processor cores could be starved access to external system resources if large amounts of data is transferred.

A typical application may include macroblock processing of video data, which consists of a small array of video samples (e.g. 32x32 samples). Instructions are issued to transfer a small array of video samples from the PL to the APU's Level 2 cache. Since the transaction has low overhead and the amount of data to be transferred is low, this type of transaction may be optimal using the ACP.

There are many factors that should be considered when designing systems that use the ACP, including transaction size and the cache-line length of the ACP. Further information can be found in [12].

ACP Limitations

As we now have a better understanding of the type of instructions to issue to the ACP, we will now investigate restrictions imposed on its operation.

The ACP slave interface is a subset of the ACE-Lite protocol [1]. The ACE-Lite protocol can be considered as an AXI slave interface with restrictions on several ports, preventing changes in the type of transaction issued from the PL to the Level 2 cache of the APU. Amongst some of these limitations (further described in [1]) are the read data and write data bus widths, which must be 128-bit.

Each ACP transaction must communicate from the PL to the APU Level 2 cache using the incrementing burst type. An incrementing burst transaction stores each transferred byte is in the adjacent memory location to the previous byte. In contrast, a fixed burst will transfer data to the same address location, without incrementing to the next adjacent memory space.

There are further limitations imposed on the ACP. It may only accept particular transactions as detailed in [1].

ACP Performance Bottlenecks

The ACP and Cortex-A53 cores are all connected to the Level 2 cache. Contention between each interface could cause a drop in overall performance of the PL masters or the Cortex-A53 cores, as they both compete for access to resources outside of the APU block [12].

The ACP only supports up to four outstanding transactions [12]. These can be any combination of reads and writes, however, there may only be one outstanding transaction per AXI ID. The AXI ID signal (AXI4 protocol) is used for ordering transactions when they become out-of-order. If there are two outstanding transactions issued by a PL master using the same AXI ID, the ACP interface will stall (preventing other masters from access) until the first outstanding transaction is complete.

Lastly, ACP performance can increase significantly if write transactions to the Level 2 cache contain full cache lines of data.

ACP Usage

As previously discussed, the ACP provides a low latency path between the accelerators implemented in the PL, and the APU's Level 2 cache. There are a set of steps involved in communicating between the PS and a PL accelerator using the ACP. These are detailed in [12] and are listed as follows:

1. The Central Processing Unit (CPU) prepares input data for the accelerator within its local cache space.
2. The CPU sends a message to the accelerator using one of the HPM AXI master interfaces to the PL.

3. The accelerator fetches the data through the ACP, processes the data, and returns the result through the ACP.

4. The accelerator sets a flag by writing to a known location to indicate that the data processing is complete. The status of this flag can be polled by the processor or can generate an interrupt.

6.3.3 Cross Trigger Interface (CTI)

Each Cortex-A53 core is connected to a Cross Trigger Matrix (CTM) via a Cross Trigger Interface (CTI), as shown in Figure 6.7.

The CTI and CTM support debug logic within each core to communicate events to other CoreSight components inside and outside of the APU [10]. For example, it is possible for a CTI within a Cortex-A53 core to generate an interrupt to other connected CoreSight components when a trace trigger event has occurred. The generated signal is known as cross triggering and may be used to communicate trace trigger events to other CoreSight components across the Zynq MPSoC.

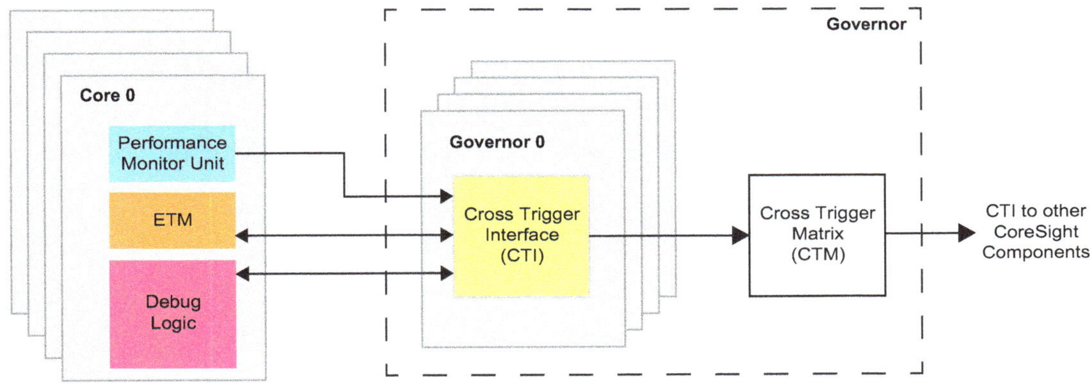

Figure 6.7: Cross Trigger Interface (CTI) allowing the APU to cross trigger other CoreSight components

In addition to setting trigger events between the debug logic in the Cortex-A53 and other CoreSight components, the CTI also enables the Embedded Trace Macrocells (ETM) and Performance Monitor Units to interact with each other and other CoreSight components [10]. The ETM is a CoreSight module that is capable of capturing traces of real-time instructions from the Cortex-A53 processor. The Performance Monitor Unit is a statistics and data gathering module which has been included in the Cortex-A53 processor as a useful tool for behaviour monitoring, profiling and debugging.

The Zynq MPSoC Technical Reference Manual [12] provides further information about CoreSight components and their operation in the Zynq MPSoC.

6.3.4 Advanced Peripheral Bus (APB) Debug

There is one interface which can access the debug registers of the Cortex-A53 processors. This interface is the AMBA 3 Advanced Peripheral Bus (APB) which is connected to the APU via a CoreSight debug access port (DAP) [10]. The DAP is routed to external package pins using a Joint Test Action Group (JTAG) interface. Figure 6.8 contains a diagram illustrating the APB and its connection to the JTAG interface.

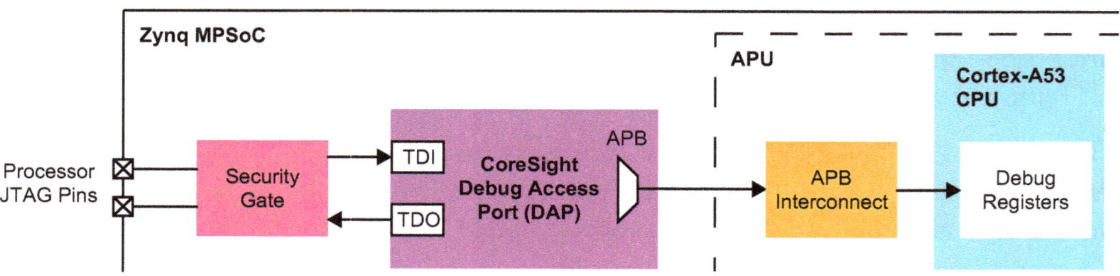

Figure 6.8: APU debug path, showing the Arm DAP and APB interconnect [12]

6.3.5 GIC Interface

The GIC CPU interface, within the APU, is connected to a GIC-400 to support interrupts. The GIC CPU interface supports the management of interrupts from various sources in the Zynq MPSoC.

Registers are available for managing interrupt sources, behaviour, and routing to one or more cores. Section 6.7 further discusses interrupt sources and signals, and Section 6.9 provides further information on APU interrupt virtualisation.

6.3.6 Trace Interface

Each Cortex-A53 core has a dedicated AMBA 4 Advanced Trace Bus (ATB) interface for capturing trace information [10]. The ATB allows for non-invasive inspection and debugging of processor registers.

The source of a trace will usually compress and format the trace information into packets and subsequently send it onto an ATB. The trace information will pass through several Arm CoreSight components before finally emerging at its destination. Trace information created by the APU could proceed to a Trace Memory Controller (TMC) or a Trace-Port Interface Unit (TPIU). Figure 6.9 provides a diagram illustrating the ATB path as seen by the APU.

The TPIU allows trace data to be routed out to pads or used by the PL to monitor PS activity. A TMC provides on-chip memory for storing and buffering trace data. The TMC has two possible configurations; as an Embedded Trace Router (ETR) or Embedded Trace FIFO (ETF).

An ETR is capable of routing trace data into the PS interconnect (using an AXI bus) so that memory, such as the external DDR or internal OCM, may store the trace data. An ETF stores trace information using the attached SRAM as the FIFO memory. The ETF is capable of buffering trace data in bursts.

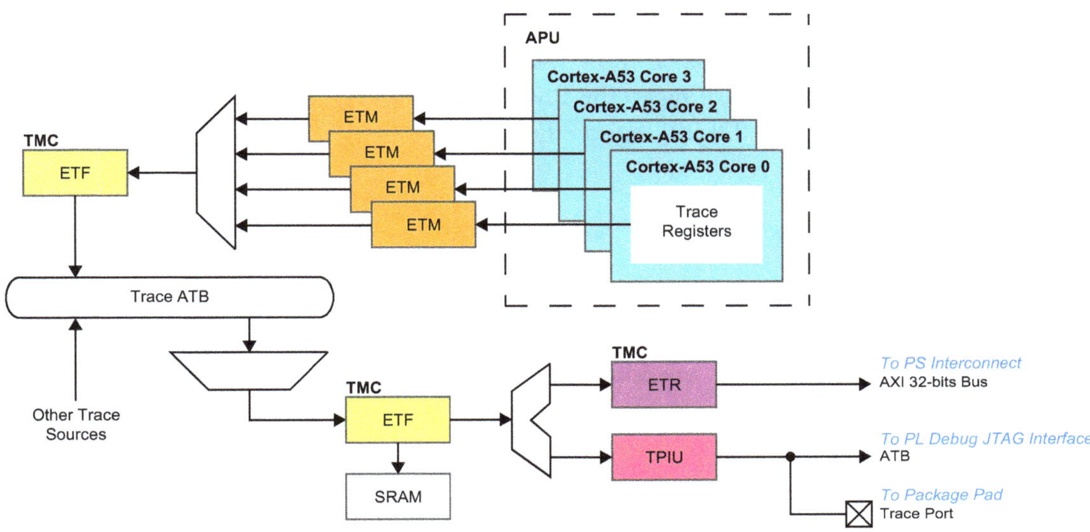

Figure 6.9: APU debug path, showing the Arm DAP and APB interconnect [12]

6.3.7 Other Signals

The remaining signals interfaced to the Cortex-A53 processors are timers, reset, and (not shown in Figure 6.6) power control. Additionally there is a configuration port so that the Cortex-A53 MPCore can be programmed. A brief description of these interfaces follows.

Reset

Each individual Cortex-A53 processor core may be independently reset. A reset may either be triggered by the FPD System WatchDog Timer (SWDT) or a software register write. The APU reset should be primarily used for software debug. The following steps detailed in [12], can reset each Cortex-A53 core.

1. Write a 1 to the corresponding bit of the REQ_SWRST_INT_EN register to enable the interrupt to the performance monitor for this process.

2. By writing a 1 to bits [0:3] of the REQ_SWRST_TRIG register, you can request the reset of the four APU cores.

Timers and Counters

The APU is connected to its own SWDT (separate from the RPU), which can reset the Cortex-A53 MPCore and reset the entire FPD. There are also several generic timers. This includes the system counter which is 64-bits wide. It is possible for software to access the CNTFRQ register so that the frequency of the counter can be read or modified. All Cortex-A53 cores are connected to a counter input, allowing them to capture each increment of the system counter.

Power Control

The APU power control signals are managed by the PMU in the LPD. APU wake-up requests for each Cortex-A53 core can be initiated by writing to the PMU's GPI1[3:0] register. Power-down requests can be initiated by writing to the PMU's GPI2[3:0] register. Further information can be found in [12].

6.4 Memory Management Unit

A Memory Management Unit (MMU) is used in the Cortex-A53 MPCore to translate a *virtual address* into a *physical address*. A *physical address* is required by the underlying hardware when storing data in main memory. The *virtual address* space is a set of addresses that are used by the program executing in software. The compiler and linker also use these when placing data in memory. A virtual address map can be implemented contiguously even if the physical memory is fragmented.

Within large systems, such as the Zynq MPSoC, many processing elements often connect to the same main memory through an address and data bus (typically of AMBA standard in the Zynq MPSoC). Several of these processing elements may load and execute multiple tasks that require their own private virtual memory space. These tasks do not require knowledge of other programs that may be running at the same time, or of the physical memory of the system in which they are contained. Additionally, without dedicated memory protection for tasks and processes, it is possible for each processing element to access the physical memory space of another and either attack or unintentionally corrupt its contents.

Many processors and devices connected to the same memory system may contain different virtual and physical address maps. To provide address translation between the virtual addresses used by a software program and the physical addresses used by main memory, the operating system (OS) executing on the Cortex-A53 processor must program the MMU. After configuration of the MMU, address translation is automatically carried out in hardware. The software executing in the Cortex-A53 processor is unaware of the address translation, and simply uses the virtual addresses to store data and instructions. It is possible for several programs operating at the same time to use the same virtual address space.

Further information about the operation of the MMU with the Cortex-A53 MPCore can be found in [10].

6.5 Memory Systems

There are three memory systems, as seen by the APU, within the Zynq MPSoC device. Figure 6.10 illustrates and highlights each of these.

The instruction and data caches of each Cortex-A53 core are part of the Level 1 memory system, while the shared Level 2 cache is part of the Level 2 memory system. The Level 3 memory system contains all other memories, such as the off-chip DDR memory (main memory), on-chip memory (OCM), and also includes PL memories.

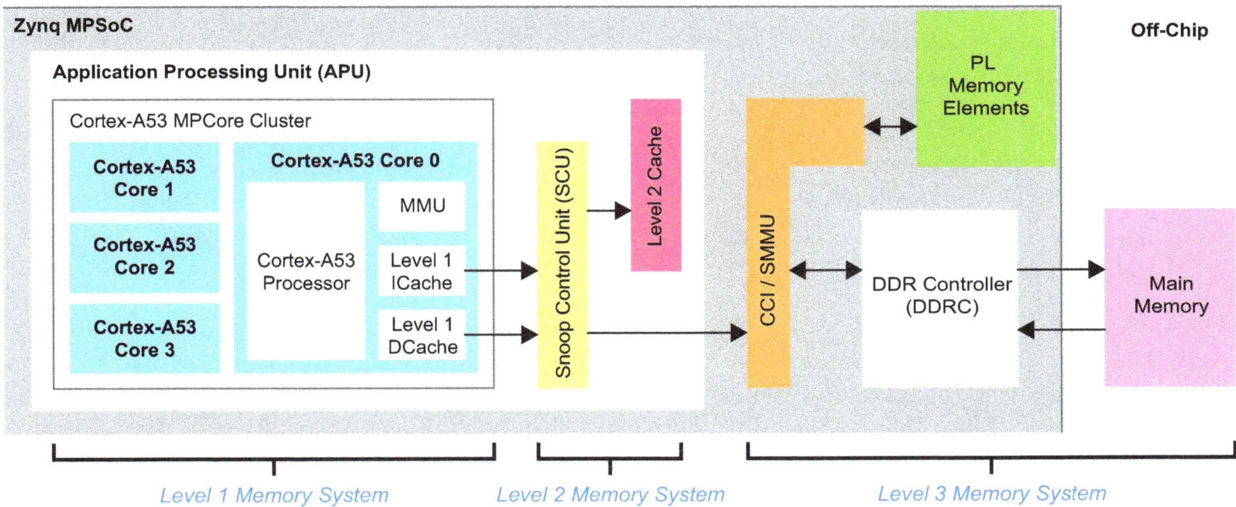

Figure 6.10: APU Memory Systems

6.5.1 Level 1 Memory System

Each Cortex-A53 core contains two caches for instructions and data in their Level 1 memory systems. In addition to this, there are also two levels of Translation Lookaside Buffers (TLBs). A TLB stores the most recent translations between virtual and physical addresses, which enables them to be retrieved with better efficiency later. Two micro TLBs in each core provide both separate instruction and data caches. The main TLB handles any TLB misses from the micro TLBs.

The Cortex-A53 core contains an Instruction Fetch Unit (IFU) responsible for retrieving instructions from memory and predicting the outcome of branches in the instruction stream. Instructions can either originate from external memory or the instruction cache within a Cortex-A53 core. The cache is accessed using the instruction cache controller and associated linefill buffer within the IFU. Once the instructions are processed, they are passed to the Data-Processing Unit (DPU) for further processing.

A DPU is present in each Cortex-A53 core and is responsible for storing processor states that are visible to the software executing on the processor. Instructions are received from the IFU and are decoded and executed.

The DPU also executes instructions provided by the IFU that require a transfer of data between the Level 1 memory system and other system memories. Communicating with the Data-Cache Unit (DCU), responsible for all load and store operations, will transfer data accordingly.

Further information about the Cortex-A53 MPCore's Level 1 memory system can be found in [10].

6.5.2 Level 2 Memory System

The Level 2 memory system contains a 1 MB cache and an SCU, essential for maintaining cache coherency between all Cortex-A53 cores and arbitrating Level 2 cache requests.

The SCU provided inside this subsystem contains a dedicated connection to each Cortex-A53 core within the APU. A connection also exists between the SCU and the master memory interface, Level 2 cache, and the ACP [10]. The SCU contains buffers which implement direct cache-to-cache transfers between processor Level 1 memories. These buffers allow processors to move data without the need to read or write to an external memory system.

For the SCU to provide coherency support, it must store duplicate copies of the Level 1 data-cache tags. It uses these to snoop on other processors in the cluster. The retrieval of Level 2 cache tags is in parallel with the SCU duplicate tags. If both sets of tags request the same data, a read action is performed on the Level 2 cache before snooping one of the other processors.

As an additional note, the Cortex-A53 MPCore processor uses the MOESI protocol to maintain data coherency between multiple cores. More on this can be found in [10].

6.6 Processor Extensions

Each Cortex-A53 core within the APU contains its own set of processor extensions. These are the Cryptography, NEON MPE, and VFP extensions, which are discussed further in this section.

6.6.1 Cryptography (Crypto)

The AArch32 and AArch64 Execution States support the Cryptography (Crypto) extension. The Crypto extension carries out the Advanced Encryption Standard (AES) for encryption and decryption, and also implements SHA1/256 and RSA. These techniques execute using A64, A32 and T32 advanced SIMD instructions. Each Cortex-A53 core contains its own individual Crypto extension and should not be confused with the Crypto Interface Blocks (CIBs) contained inside the Configuration Security Unit (CSU).

Further information about the Cryptography extension can be found in [9]. The extension is also provided with a programmers model presented in [11].

6.6.2 NEON Media Processing Engine

The Arm NEON Media Processing Engine (MPE) is an extension to Arm Cortex-A processors which provides advanced Single Instruction, Multiple Data (SIMD) processing facilities [5]. NEON has been designed to accelerate large vector operations, such as those typically found in multimedia and DSP style algorithms. The SIMD architecture allows the NEON engine to operate on the elements of two input vectors to produce a corresponding output vector containing results. One instruction is used to configure the engine to perform the same operation on multiple data elements of the same type and size. Not only does this boost the performance of Digital Signal Processing (DSP) algorithms, but it also reduces power consumption as unused resources are switched off. The Cortex-A53 MPCore contains a NEON extension in each of its processor cores, so that they each have dedicated access to their own NEON engine and are able to take advantage of its parallel vector processing capabilities.

NEON MPE Architecture

Figure 6.11 contains an illustration of the NEON MPE architecture. There are two sets of input registers, Q2 and Q1, each which hold a vector of length *N*. Both input vectors will distribute their elements between the processing lanes. It is possible for there to be vectors of different sizes, which will affect the number of processing lanes used by the engine. Each element of the vector must have the same data type and must perform the same operation in each lane. An output vector, indicated by register Q0, stores the results. Since one instruction is used to configure the operation in each lane (which all operate in parallel), the NEON MPE is considered to have a SIMD architecture.

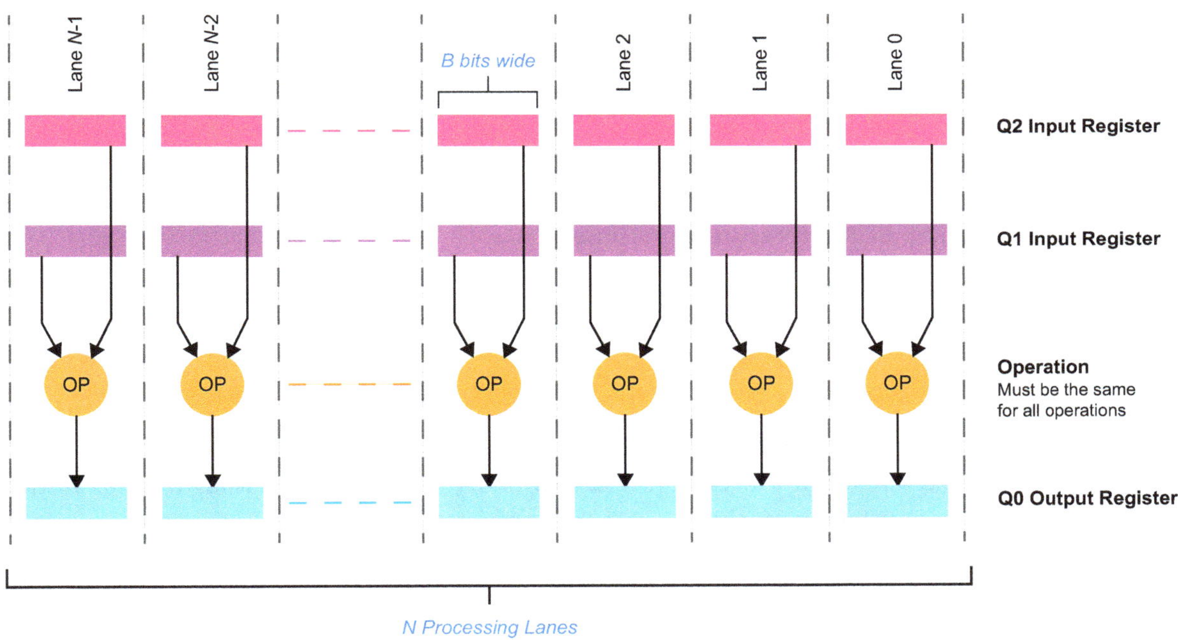

Figure 6.11: The NEON MPE architecture, showing SIMD capabilities

The NEON's input registers are 128-bits wide. NEON does not support 128-bit operations; therefore registers are either subdivided into two 64-bit lanes, four 32-bit lanes, eight 16-bit lanes or sixteen 8-bit lanes. It is not possible for carry or overflow logic to transfer from one lane to another.

The Cortex-A53 MPCore processor supports the A32 and A64 instruction sets. Each contains Advanced SIMD instructions for issuing tasks to the NEON engine. The operation of the NEON MPE will depend on the instruction set used by the Cortex-A53 MPCore processor. The A64 advanced SIMD NEON instructions are similar to those in the A32 instruction set.

A32 instructions support 8-bit, 16-bit, 32-bit and 64-bit signed and unsigned integers with the NEON engine. Additionally, NEON also supports the 32-bit single precision floating-point data type. When using the A64 instruction set, NEON supports all of the previous single precision, integer and floating point data types, with the addition of 64-bit double-precision floating point [3].

The Execution State of the Cortex-A53 processor, which has different register configurations for the NEON engine, also affects its behaviour. In the AArch32 Execution State, there are sixteen 128-bit vector registers for processing data elements. When using the AArch64 Execution State, there are thirty-two 128-bit vector registers. Further information about NEON operation with the Cortex-A53 processor can be found in [3], [4] and [5].

The NEON MPE shares its input and output registers with the Vector Floating Point (VFP) extension. The NEON engine and VFP extensions are independently configured; therefore software does not have to switch between their register configurations explicitly. As a final note, the NEON MPE does not support double precision floating point operations using the A32 instruction set, but these can alternatively be performed by the VFP.

6.6.3 Vector Floating Point Unit

The Arm Floating Point Unit (FPU) extension, also known as the Vector Floating Point (VFP) unit, applies arithmetic operations to vectors of data. The VFP applies the same operation to each vector element in sequence, due to its lack of parallel processing capabilities.

The VFP's primary purpose is to accelerate the processing of floating point operations in parallel with the Cortex-A53 processor. The VFP supports single-precision and double-precision floating point data types.

Similar to the NEON MPE, it is important to understand that the operation of the VFP extension relies on the instruction set in current use by the Cortex-A53 processor. If the processor is using the A32 or T32 instruction sets, the VFP extension will operate on a sequence of data elements from the shared NEON register bank and only contain sixteen 128-bit registers. However, if the processor is using the A64 instruction set, the shared register bank may now use thirty-two 128-bit registers.

The VFP is expected only to be used to process data elements of single and double precision. The NEON MPE should operate on all other data types as it contains pipelining. As an additional note, the VFP is also fully IEEE-754 compliant, unlike the NEON MPE. Further information about the floating-point hardware can be found in [3].

6.7 Interrupts

The APU uses an interrupt controller to support and manage interrupts. This is a GICv2 controller and it is backwards compatible with GICv1 controllers, such as that employed by the RPU. Interrupts can originate from many sources throughout the Zynq MPSoC, including interrupts from the PL or several processors and IP blocks contained inside the PS. The GIC CPU interface supports the management of interrupts in a system where many may originate from various sources.

The GIC-400 [13] used by the APU within the Zynq MPSoC has been configured to provide a variety of different features, including support for the Arm architecture's security and virtualisation extensions. Additionally, there are accessible registers for managing interrupt sources, behaviour and routing between processors.

This section will investigate different types of interrupts and their priorities and states. We will then provide a brief overview of the GIC-400 controller.

6.7.1 GICv2 Interrupt Types

The GICv2 architecture supports four types of interrupts. These are peripheral interrupts, Software-Generated Interrupts (SGIs), virtual interrupts and maintenance interrupts for virtual machines.

Peripheral Interrupts

There are two types of peripheral interrupts defined by the GICv2 architecture. Each peripheral interrupt is either edge-triggered or level-sensitive.

- Private Peripheral Interrupts (PPIs) are unique to a single processor, and are not accessible to any other processor or combination of processors.
- Shared Peripheral Interrupts (SPIs) can be routed by the distributor to any specified combination of processors, and originate from various sources within the Zynq MPSoC. These are wired interrupts which are physically connected to the GIC.

Software Generated Interrupts

Inter-processor communication (between the APU's processor cores) can be achieved using SGIs. An SGI is a section of developed software which writes to the GICD_SGIR register within the GIC. An SGI operates using edge-triggering within the associated software.

In addition to normal operation, the GIC also contains virtualisation extensions. When a virtual SGI occurs, management registers contained within the GIC virtualisation extension allow the target guest OS to receive the identity of the processor that generated the interrupt. Upon writing to the management registers, a hypervisor (a virtual machine monitor that creates and runs virtual machines, discussed later in Chapter 13) is then capable of generating a virtual interrupt. A virtual interrupt appears to a virtual machine, operating within the processor, as an SGI.

Virtual Interrupt

A virtual interrupt is a means of communicating interrupts within a processor that is using a virtual machine. For example, hypervisors will receive all interrupts in the non-secure state and distribute relevant interrupts to each of the virtual machines in the form of a virtual interrupt. Virtual interrupts are only available if the relevant processor has a hypervisor to perform the translation from physical to virtual interrupt.

Maintenance Interrupt

The maintenance interrupt is a level-sensitive interrupt that is used to communicate between a processor's hypervisor and the GIC's virtual extensions. Typically, this is used to signal whether a group of interrupts are becoming enabled or disabled and may require the attention of the hypervisor.

6.7.2 GICv2 Interrupt Priority and States

SGI and SPI interrupt requests are all assigned a unique ID for them to be identified correctly by the interrupt controller. Pending interrupts, for each CPU, are held by the interrupt distributor. The distributor selects the interrupt with the highest priority so that it can issue it to its target CPU interface before all others.

Interrupts that have a higher priority are assigned lower values. For example, the priority field value 0 will always indicate the highest priority. When interrupts have equal priority, the distributor resolves the matter by selecting the interrupt with the lowest ID first.

The interrupt distributor is responsible for holding the entire list of connected interrupts and processors and is also responsible for sending SGIs to their target CPUs. Upon sending the interrupt with the highest priority to its target CPU, the distributor will then receive an acknowledgement back. The distributor then updates the status of the interrupt to *active*. It is then the responsibility of the acknowledging CPU to end that interrupt.

There are four interrupt states that apply to an interrupt interface between the GIC and connected CPUs. These are listed as follows.

- **Inactive** — An interrupt in the *inactive* state is not *active* or *pending*. This is the inactive state when the interrupt is not being used by the GIC or a connected processor.
- **Pending** — An interrupt in the *pending* state has been asserted in hardware or generated by software. It has been received by the GIC and waiting to be serviced by the target processor.
- **Active** — Interrupts that are *active* have been sent to and acknowledged by the target processor. They are in the process of being serviced and are yet to be completed. The target processor must place the interrupt back into the *inactive* state after completion.
- **Active and pending** — An interrupt with an *active and pending* state is one which is currently being serviced by a processor and also contains a *pending* interrupt from the same source. This is typical of a source which has issued more than one interrupt consecutively.

6.7.3 GIC-400 Overview

The GIC-400 contains the virtualisation extension and is therefore able to support virtual interrupts for a processor containing a hypervisor and one or more virtual machines. The GIC-400, with the virtualisation extension, is composed of five functional blocks. Namely a CPU interface, a virtual CPU interface, a distributor, a virtual distributor (with the aid of a hypervisor) and an AXI interface [13].

The GIC-400 can support the Cortex-A53 processor in the following two ways:

- The GIC-400 supports interrupt distribution when the Cortex-A53 processor contains an operating system that executes on native hardware. This is the simplest configuration of the GIC-400 as it does not need to distribute virtual interrupts as the Cortex-A53 processor contains no hypervisor or virtual machine.
- The GIC-400 supports interrupt distribution when the Cortex-A53 processor has been configured as a host machine (when the processor contains a hypervisor and one or more virtual machines). This configuration supports the distribution of virtual interrupts.

This section investigates the GIC-400's operation for each of the above cases.

GIC-400 Operation without Virtual Interrupts

Figure 6.13 provides a diagram illustrating the connections between the GIC-400 and the Cortex-A53 processor. The Cortex-A53 processor contains an operating system that executes on native hardware. Note that the virtual distributor and virtual CPU interface blocks are not in use as the Cortex-A53 processor does not need virtual interrupt support.

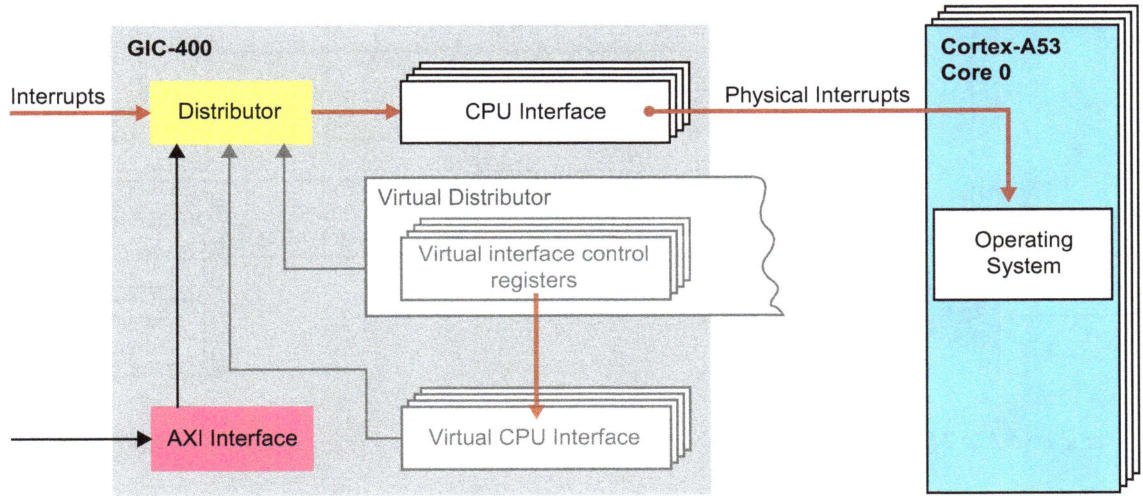

Figure 6.12: The GIC-400 as used by the Cortex-A53 MPCore processor [13]

The distributor block is responsible for consolidating all interrupt sources. Initially, the distributor determines the priority of requesting interrupts before distributing the interrupt to the relevant CPU interface block for further processing. In software, the distributor provides the processor with the state of each interrupt and a method to set or clear the pending state of a peripheral interrupt.

The AXI interface provides control and configuration of the GIC-400 to the user. In particular, this allows you to program features of the Distributor, CPU interfaces and virtual CPU interfaces.

Upon an interrupt entering the pending state, the CPU interface block checks whether it has sufficient priority to be signalled to its target processor. This process is carried out by comparing the priority value of the pending interrupt with a threshold value held in the GICC_PMR register of the target Cortex-A53 processor.

The CPU interface is also capable of masking an active interrupt of lower priority, to signal a pending interrupt of higher priority to the target processor. Further information can be found in [13].

GIC-400 Operation with Virtual Interrupts

Figure 6.13 provides a diagram illustrating the connections between the GIC-400 when the Cortex-A53 processor is host to a hypervisor and one or more virtual machines. Note that GIC virtualisation support will not be covered extensively in this section; Section 6.9.2 provides further discussion about interrupt virtualisation and grouping.

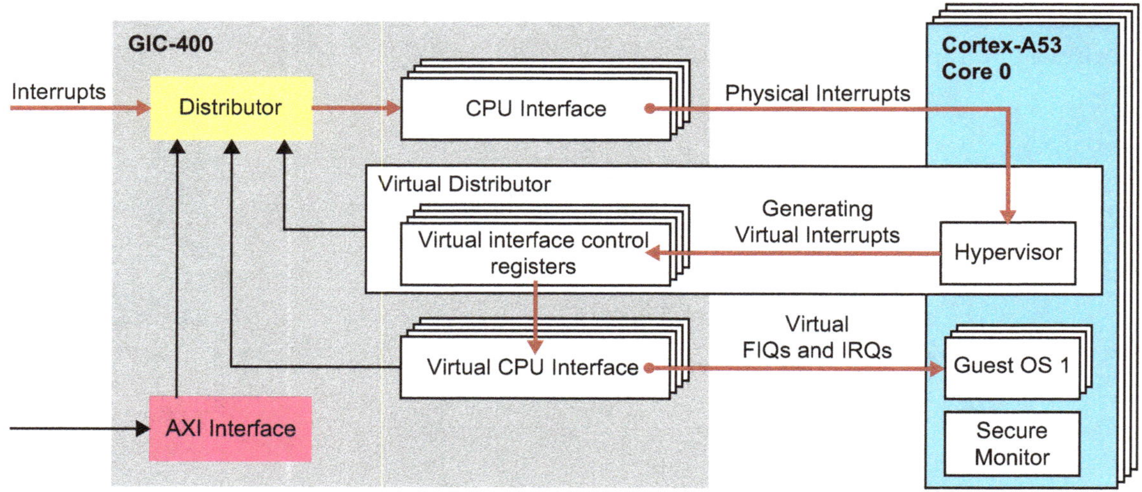

Figure 6.13: The GIC-400 as used by the Cortex-A53 MPCore when configured as a host machine [13]

Both the virtual and physical CPU interfaces provide similar control registers. These registers are provided so that the virtual machine cannot distinguish between them. The hypervisor is responsible for accepting pending interrupts and then signalling them to the appropriate virtual CPU interface so that the target virtual machine may receive them.

Further information about GIC-400 interrupts grouping and virtualisation are discussed in Section 6.9.2, Interrupt Virtualisation.

6.8 Power Management

The Zynq MPSoC has many mechanisms to support the management of power throughout the chip. It is possible to shut down entire processors and domains to save energy. Similarly, the APU also provides methods that allow users to control both its dynamic and static power dissipation.

This section aims to explore the Cortex-A53 processor power management techniques. We will introduce the APU's power islands and power modes, and investigate MPCore shutdown methods.

6.8.1 Power Domains

Power domains allow a system, containing many components, to isolate sections so that they may operate under different power conditions or modes. The APU has three power domains which can be controlled independently from one another [10].

Note that the APU's power domains are exclusive to the Zynq MPSoC's full-power domain (FPD). In the Zynq MPSoC Technical Reference Manual [12], the APU's power domains are referred to as power islands as not to confuse them with the Zynq MPSoC's power domains. In the following discussion, we will refer to them as power domains.

Table 6.2 describes the power domains and their associated components.

Table 6.2: Power Domains supported in the APU

Power Domain	Components	Further Description
PDCORTEXA53	Snoop Control Unit (SCU) Level 2 Cache Controller	This domain is a part of the PS full-power domain (FPD). Debug Registers in the debug domain are also affected.
PDL2	Level 2 Data and Tag RAM Level 2 Victim RAM SCU Duplicate TAG RAM	
PDCPU[4]	Cortex-A53 Core 0, 1, 2 & 3	This includes the advanced SIMD and floating-point extensions, Level 1 TLB, Level 1 processor RAMs and debug registers within each core [12].

The acronym PD (representing the words *power domain*) is used to prefix the names of the power domains shown in Table 6.2. For example, PDCPU[4] represents the *power domain* for each of the Cortex-A53 processor cores.

It is possible for the PDCORTEXA53 and PDL2 power domains to remain active while all of the cores are powered down, allowing the level 2 cache to continue to be snooped by other connected devices. Figure 6.14 provides a diagram illustrating the Cortex-A53 MPCore processor power domains.

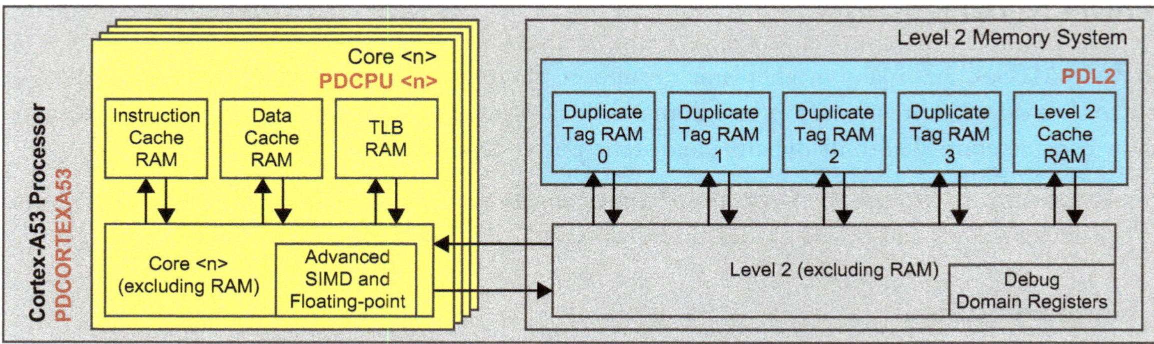

Figure 6.14: Cortex-A53 MPCore processor power domains [10]

6.8.2 Power Modes

There are several supported power modes for the APU's power domains. These either place the Cortex-A53 cores in a standby state, shut them down, or allow them to operate as normal.

Normal Mode

The Cortex-A53 MPCore is considered to be in its normal mode of operation when full processor functionality is available to the user. The design of the Cortex-A53 processor is such that gating can be used to disable the clocks and I/Os of unused functional blocks. In this state, dynamic power is only consumed by logic performing an operation.

Standby Mode

The Cortex-A53 MPCore processor is capable of entering into a standby state using two methods. These are the MPCore Wait for Interrupt (WFI) and Wait for Event (WFE) features [10]. They are available to all cores within the APU and typically follow the same operation.

The WFI and WFE features disable most of the MPCore's clocks, so that it may enter a low-power state. In this state, the MPCore's power draw is significantly reduced and includes static leakage current. Logic is used to wake up the MPCore, and this causes it to draw a small amount of current that contributes to dynamic power overhead. Entering the WFI or WFE low-power state is achieved by executing the associated instruction in software. Each processor core is capable of entering these states independent from others.

Upon executing the WFI or WFE instruction, the associated processor core will wait for all instructions to complete before entering the low-power state. The processor core will also wait for any explicit memory accesses that have occurred before executing the instruction. These include instructions that require a response or data from the level 2 memory system and also any instructions that updates the cache or previously issued to the SCU. These are listed as follows:

- Load instructions and store exclusive instructions.

- Cache and TLB maintenance operations.

If all the processor cores are in the WFI low-power state, the shared level 2 memory system will also enter a WFI low-power state. This state is known as L2 Wait for Interrupt. Further information can be found in [10].

MPCore Shutdown Modes

It is possible to initiate individual Shutdown Modes for each of the processor cores in the APU, by shutting down the PDCPU power domain of each core. Doing so will reset the configuration of all states in the PDCPU domain. Steps to power down the MPCore can be found in [10], [12].

The Cortex-A53 MPCore processor can also be placed in a Cluster Shutdown Mode. This mode is where the PDCORTEXA53, PDL2 and PDCPU power domains are all shut down, resetting the configuration of all states in the affected power domains. Cluster Shutdown Mode may be carried out with or without a system driven level 2 flush (when the level 2 memory is emptied/cleared). Further information on using the Cluster Shutdown Mode can be found in [10].

The Platform Management Unit (PMU) is responsible for power gating any of the unused blocks on the Cortex-A53 MPCore processor. Power gating is handled by the software operating on the PMU further discussed in Chapter 10.

6.9 System Virtualisation

Virtualisation, in the context of computing, refers to the action of creating a computing resource, such as an operating system or storage device, in software. The computing resource may not physically exist, however, it appears to do so through the use of virtualisation software. A design that requires multiple operating systems may execute on the same CPU, through virtualisation.

Operating systems that run, in a virtual environment, are classified as Guest OSs. They run inside a Virtual Machine (VM). It is possible to have multiple VMs per CPU; however, it is only possible to create and manage VMs through the use of a hypervisor (virtualisation software). The hypervisor provides a layer of abstraction between the VMs and the native hardware resources.

Figure 6.15 provides a diagram which illustrates the use of the Cortex-A53 CPU as a traditional computing system alongside one that uses virtualisation. The traditional system has several applications which are executed using an OS running on the native hardware. The virtual processing system is more complex and draws on knowledge of the Armv8 security model, previously described in Section 6.2.5.

As shown in Figure 6.15 (right), the physical processor has been separated into two VCPUs, as defined in the Armv8 security model. In the Non-secure World (VCPU 0), a hypervisor has been used to generate two VMs. Each VM is host to a Guest OS which can execute user applications. In the Secure World (VCPU 1), a Secure OS runs secure software to maintain the operation and security of the Arm processor. The secure monitor is used to manage switching between the secure and non-secure states.

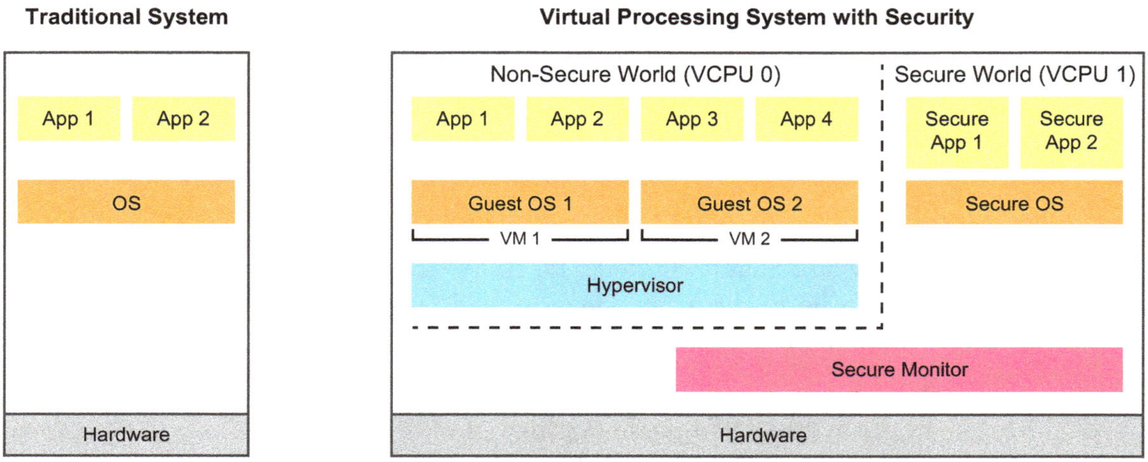

Traditional System

App 1	App 2
OS	
Hardware	

Virtual Processing System with Security

Non-Secure World (VCPU 0) | Secure World (VCPU 1)

App 1, App 2, App 3, App 4 | Secure App 1, Secure App 2

Guest OS 1 (VM 1), Guest OS 2 (VM 2) | Secure OS

Hypervisor

Secure Monitor

Hardware

Figure 6.15: Traditional system (left), virtual processing system with security (right) [7]

Operating systems are usually designed to run on native hardware. They expect to be executing at the highest privilege level and to have control over the entire processing system. In a virtual environment, the Guest OS uses the lowest privilege level. Due to this, the Guest OS is incapable of issuing privileged instructions, and instructions which are necessary for configuring the native hardware. Issuing these type of instructions can only be managed by the VM, which operates at a higher privilege level than the Guest OS. Since there can be many VMs and Guest OSs, it is necessary to consider the arbitration of system resources such as memories and interrupts.

The level of abstraction required to arbitrate system resources is dependent on the underlying hardware. However, the main approaches to achieving this can be categorised into two groups:

- **Full Virtualisation** — This is when the Guest OS is not aware that it is virtualised. The VM is responsible for managing all privileged instructions. The Guest OS has not been modified in any way and can still issue de-privileged (user mode) instructions on the native hardware.

- **Paravirtualisation** — The Guest OS has been amended so that it is aware that it is virtualised. The modification provides the Guest OS with direct communication with the VM. A particular API is provided by the VM to the Guest OS so that it may issue privileged instructions. These are known as hyper-calls or hypervisor calls [10]. The Guest OS can still issue de-privileged instructions on the native hardware.

The remainder of this section will investigate the arbitration of system resources in a virtual environment. We will cover two important matters of virtualisation within the APU. These are address translation in a virtual environment, and interrupt virtualisation.

6.9.1 Address Translation in a Virtual Environment

Running several operating systems in a virtual environment requires a method of translating virtual and physical addresses. In the Non-secure state, a process exists which correctly executes the mapping of addresses from EL2 (the hypervisor exception level). To achieve this mapping, the process requires an Intermediate Physical Address (IPA) between translation stages. The translation process is performed in two steps as follows [10], [12].

1. The Guest OS in EL1 manages the translation of a Virtual Address (VA) to an IPA. At this stage, the Guest OS considers the IPA to be the Physical Address (PA) used by hardware.

2. The IPA is subsequently mapped to a PA by the hypervisor. The Guest OS is not aware of this stage. Hardware may then use the PA. The hypervisor is primarily responsible for creating the stage 2 translation table and arbitrating the use of the shared physical memory.

6.9.2 Interrupt Virtualisation

The complexity of handling interrupts in a virtual environment has been reduced due to the GIC-400's virtualisation extension. This extension provides a virtual CPU interface for each processor in the system and a list of active and pending interrupts, accessible by the hypervisor.

The virtual CPU interface is used to provide physical connections to a connected processor, allowing virtual interrupts to signal the processor such that they are indistinguishable from physical interrupts. This process is carried out similarly to a standard CPU interface. The VM operating within the target processor may access the control and status registers for the virtual interrupts using the virtual CPU interface.

The GIC-400 provides support for managing a list of active and pending interrupts for the VMs running on the connected processor. The control registers are accessed and managed by the hypervisor, which is also operating on the connected processor. Together, the hypervisor and GIC-400 form a virtual distributor. This appears to the VMs operating in the connected processor as a physical GIC distributor.

GICv2 Interrupt Grouping

The GICv2 architecture classifies issued interrupts into groups. Interrupt group assignment varies depending on whether the issued interrupt is destined to the Secure or Non-secure world [13].

- A Secure interrupt is assigned to group 0 and signalled to the target processor as an FIQ. The priority of an FIQ is higher than an IRQ. As as a result, IRQ handlers are disabled until the servicing of the issued FIQ is complete.

- A Non-secure interrupt is assigned to group 1 and signalled to the target processor as an IRQ.

The virtual CPU interface also maps virtual interrupts using groups. Group 0 and 1 issue interrupts to the target VM as vFIQs and vIRQs respectively. The processor views the issued virtual interrupts as regular physical interrupts.

GIC-400 Virtualisation Support

The entire system responsible for handling interrupts in a virtual environment can be seen in Figure 6.16. Interrupts intended for the VCPU operating in the secure state are issued using a FIQ. Interrupts destined for the VCPU operating in the non-secure state are issued using IRQs.

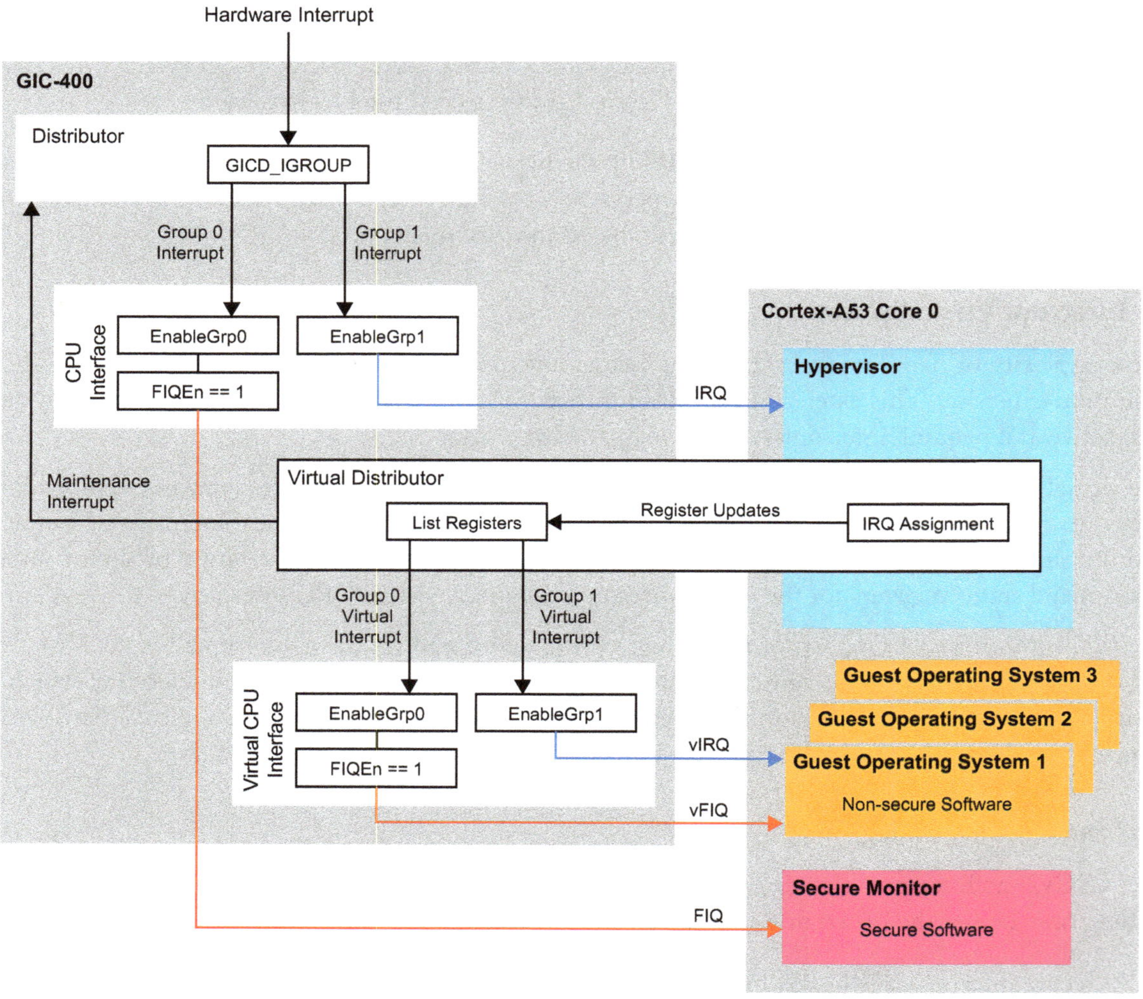

Figure 6.16: The GIC-400 as used by the APU with virtualisation support [13]

The hypervisor receives physical IRQs from the CPU interface and initially determines whether the interrupt is for itself, or destined for a Guest OS [13]. If the interrupt is for a Guest OS, the hypervisor must determine the following:

- The Guest OS which was responsible for handling the interrupt.
- Has the target Guest OS configured the issued interrupt as an FIQ or an IRQ?
- The interrupt priority, as set by the Guest OS's interrupt priority configuration.

The hypervisor will then update the list registers contained inside the virtual distributor. This update will add the issued interrupt to a list of pending interrupts for the virtual machine hosting the target Guest OS. If the interrupt is an FIQ or IRQ, it will be sent using a vFIQ or vIRQ respectively.

6.10 Chapter Review

This chapter has provided an overview of the Arm Cortex-A53 MPCore processor and investigated the fundamentals of the Armv8-A architecture. In particular, the associated security model and exception levels were explored. The Cortex-A53 MPCore interfaces were then examined, highlighting that there is a master memory interface, an ACP connected between the PL and SCU, and several additional ports for other features.

The APU extensions were then investigated, which showed that a NEON MPE, VFP and Crypto extension was available within each Cortex-A53 core. The Crypto extension was found to be capable of implementing AES for encryption and decryption, SHA1/256, and RSA. The NEON MPE was seen to support vector processing using the advanced SIMD instruction set. It was also established that the VFP is fully IEEE-754 compliant, unlike the NEON MPE.

The APU's power management features were explored, including power domains and modes. Interrupt support using the GIC-400 was investigated, including GICv2 interrupt types, prioritisation and states. Finally, virtualisation with respect to the APU was discussed. In particular, interrupt virtualisation and address translation in a virtual environment were considered.

6.11 References

Note: All online sources last accessed March 2019.

[1] Arm, Ltd., "AMBA AXI and ACE Protocol Specification", Issue E, February 2013.
Available: http://infocenter.arm.com/help/index.jsp?topic=/com.arm.doc.ihi0022e/index.html

[2] A. Stevens, "Introduction to AMBA 4 ACE and big.LITTLE Processing Technology", Arm White Paper, June 2011 (revised July 2013).
Available: https://www.arm.com/files/pdf/CacheCoherencyWhitepaper_6June2011.pdf

[3] Arm, Ltd., "ARM Compiler Version 6.5 armasm User Guide", Issue F, June 2016.
Available: https://static.docs.arm.com/dui0801/f/DUI0801F_armasm_user_guide.pdf

[4] Arm, Ltd., "ARM Cortex-A Series Programmers Guide for ARMv8-A", Issue A, Version: 1.0, March 2015.
Available: http://infocenter.arm.com/help/topic/com.arm.doc.den0024a/DEN0024A_v8_architecture_PG.pdf

[5] Arm, Ltd., "ARM Cortex-A53 MPCore Processor Advanced SIMD and Floating-point Extension", Issue G, Revision r0p4, January 2016.
Available: http://infocenter.arm.com/help/topic/com.arm.doc.ddi0502g/DDI0502G_cortex_a53_fpu_trm.pdf

[6] Arm, Ltd., "ARM Architecture Reference Manual ARMv7-A and ARMv7-R edition", Issue C.c, May 2014.
Available: http://infocenter.arm.com/help/index.jsp?topic=/com.arm.doc.ddi0406c/index.html

[7] Arm, Ltd., "ARM Security Technology", Issue C, Version 1.0, April 2009.
Available: http://infocenter.arm.com/help/topic/com.arm.doc.prd29-genc-009492c/PRD29-GENC-009492C_trustzone_security_whitepaper.pdf

[8] Arm, Ltd., "A-Profile Architectures", webpage.
Available: https://developer.arm.com/products/architecture/cpu-architecture/a-profile

[9] Arm, Ltd., "ARM Architecture Reference Manual, ARMv8, for ARMv8-A architecture profile", Issue A.k, September 2016.
Available: http://infocenter.arm.com/help/index.jsp?topic=/com.arm.doc.ddi0487a.k_10775/index.html

[10] Arm, Ltd., "ARM Cortex-A53 MPCore Processor, Technical Reference Manual", Issue J, Revision r0p4, June 2018.
Available: http://infocenter.arm.com/help/topic/com.arm.doc.ddi0500j/DDI0500J_cortex_a53_trm.pdf

[11] Arm, Ltd., "ARM Cortex-A53 MPCore Processor Cryptography Extension, Technical Reference Manual", Issue F, Revision r0p4, December 2015.
Available: http://infocenter.arm.com/help/topic/com.arm.doc.ddi0501f/DDI0501F_cortex_a53_cryptography_trm.pdf

[12] Xilinx, Inc., "Zynq UltraScale+ Device Technical Reference Manual", UG1085 (v.1.9), January 2019.
Available: http://www.xilinx.com/support/documentation/user_guides/ug1085-zynq-ultrascale-trm.pdf

[13] Arm, Ltd., "CoreLink GIC-400 Generic Interrupt Controller, Technical Reference Manual", Issue B, Revision r0p1, August 2012.
Available: http://infocenter.arm.com/help/topic/com.arm.doc.ddi0471b/DDI0471B_gic400_r0p1_trm.pdf

[14] Xilinx Inc, "Zynq-7000 SoC, Technical Reference Manual", UG585, (v1.12.2) July 2018.
Available: https://www.xilinx.com/support/documentation/user_guides/ug585-Zynq-7000-TRM.pdf

Chapter 7

The Real-Time Processing Unit

The fundamental goal of real-time systems is to guarantee a bounded response-time for time sensitive tasks. While software must be carefully implemented to provide these guarantees, suitable hardware is a prerequisite! *Determinism, responsiveness,* and *dependability* are key attributes of all real-time systems. These attributes provide the motivation for the design differences between the Real-time Processing Unit (RPU) and other General Purpose Processors (GPPs), including the APU's processors.

In this chapter, we will look at the architecture of the RPU and how this can be used by a developer to realise real-time aspects of their system.

7.1 Introduction

7.1.1 What is Real-Time Processing?

When talking about processing in general, we would typically judge correctness by a system's output. This verifies logical correctness of the system's function. The fundamental difference for real-time processing is that its correctness is judged not only by function, but also by the time in which it is performed. Time constraints (a.k.a. *deadlines*) are often defined as the time between a (physical) event occurring and the required system response. For example, a car's anti-lock braking system must always respond to a driver's braking action before a wheel has the opportunity to 'lock up' — a number of milliseconds determined through research and regulation. The requirement of any real-time system is to be logically correct and guarantee that temporal deadlines are met. Note that it should be a *guarantee*, based on worst-case scenarios and not the common-case.

Because of these strict guarantees, determinism and predictability are an important aspect of designing real-time systems.

So, what type of applications actually demand real-time processing? There are many different examples, which are commonly grouped into three categories, based on the deadline requirement:

- **Hard deadlines**: A single missed deadline can have catastrophic consequences. This is typically seen in mission critical systems such as anti-lock braking or pacemakers. Dependability of hard real-time systems is a vital aspect as failure can often cause damage to equipment or even loss of life.

- **Firm deadlines**: Like hard deadlines, as soon as a deadline expires, the output of the corresponding task is no longer useful. However, infrequently missing a deadline is not so serious — it just impacts quality of service. A tangible example of firm deadlines is a digital TV set-top box. Decoding of sequential frames can be thought of as a periodic task. While missing the deadline for decoding a frame has no catastrophic consequences, missing many deadlines would cause visible jitter. Note also that a late frame has no value — displaying it would only cause more disruption than simply discarding it.

- **Soft deadlines**: Unlike hard or firm deadlines, an output can still be somewhat useful soon after a soft deadline expires. Missing deadlines only impacts quality of service (with no other serious consequences). For example, an air-conditioning unit could be classified as a soft real-time system. Temperature measurements can be used as feedback to control a cooling system. This is classified as soft because acting on a temperature measurement late is still better than no action at all.

There are 3 major design considerations for all of these real-time systems, as defined in [1]. These guide our later discussion in Section 7.3 and Section 7.4.

1. **"Time is the most precious resource"**. When talking about time for real-time systems, determinism is king. Determinism is what allows us *guarantee* that we will meet deadlines. Section 7.3 discusses how the RPU's features can provide more deterministic execution, as well as responsiveness, granting the developer as many clock cycles as possible before a deadline passes.

2. **"Reliability is crucial"**: For hard real-time systems, reliability can mean the difference between correct operation and the loss of life. While distinct from real-time processing, reliability is also a big part of designing mission critical systems. Section 7.4 discusses the safety features in the RPU which can be employed to ensure reliability. See Chapter 9 for more information on device-wide safety features.

3. **"The environment under which a computer operates is an active component"**: The wider environment should always be considered when a designing the real-time processing system. For example, one would not deign an ADAS controller without also considering the vehicle itself! This environment is what should dictate the deadlines that are designed against. We do not cover this heavily here as it is application-dependent but, suffice it to say, analysis of the wider system should inform the specifications of the real-time controller.

7.1.2 Why a Different Architecture for Real-Time Processing?

It is common for GPPs to favour throughput over responsiveness and added dependability/safety features. Designing for both is difficult as features for responsiveness (such as the low interrupt latency seen later) usually result in a trade-off with throughput.

Determinism can also be difficult to provide with common architectures. In particular, memory access is one of the biggest causes of variable execution time. Reading and writing from main memory can take hundreds of clock cycles, so processors usually include caches — small but very fast local memories. Caches help speed up memory accesses by storing blocks of instructions/data, prompted by any non-cached access within that block. When new data is to be cached, other entries in the cache must be evicted to make space. This makes the system performance difficult to guarantee because the memory performance depends on the current state of the caches. This is especially difficult to reason about when code can be initiated by asynchronous interrupts with unpredictable timing. The RPU must be able to make guarantees about worst case response time, so the typical GPP memory hierarchy should also be modified.

Many hard real-time applications are mission critical, such as aircraft control or anti-lock braking. These applications demand some extra dependability features, usually realised through redundancy. There are overheads associated with these features in terms of silicon, and often even performance. This level of dependability is not needed for most general purpose applications (e.g. web browsing on a laptop) so the features are simply not considered in most GPP architectures.

7.2 Overview

This overview runs through the main features present in the Zynq MPSoC RPU. The most important features from a developer perspective are put in context in Section 7.3 and Section 7.4. At the heart of the RPU there is a cluster of two Arm Cortex-R5 cores. These are processors optimised for real-time applications, sitting in the middle of the Cortex-R range. Zynq MPSoC CG devices can clock the RPU at up to 533 MHz, while EG and EV devices run up to 600 MHz — all with a performance of 1.67 DMIPS/MHz [2]. They implement the 32-bit Armv7-R instruction set as well as Thumb-2 technology for improved code density [3]. Note that this is not compatible with the 64-bit Armv8 instruction set of the Cortex-A53s, found in the APU. A simplified view of the system is shown in Figure 7.1.

Each core has a Floating-Point Unit (FPU) that supports single and double precision floating-point arithmetic using the VFPv3 instruction set [3]. The Cortex-R5s can be configured to be in one of two modes, which are collectively known as "split/lock":

- *Split mode* for performance. Each processor runs independently.
- *Lock-step mode* for safety. Each processor runs the same instructions, giving the redundancy required to detect errors in operation.

There are 32 KB Harvard caches for each core. The "Harvard" architecture dictates that there are distinct caches for storing instructions and data. While these caches are always available, they may not be suitable for all real-time applications and can be disabled. Alongside the caches, there are 128 KB Tightly-Coupled Memories (TCM) which provide deterministic access with a latency comparable to the caches. Also for added safety, all caches and TCMs have Error Checking and Correction (ECC). The scheme used can detect up to 2 bits of error and correct 1 erroneous bit within each 32 bit word. The memory protection unit can optionally be used to set access attributes for different memory regions — detecting and preventing accidental or malicious access to restricted areas of the address space. There is even a Built-In Self-Test (BIST) which can be used to detect random hard errors which usually signal permanent failure.

Dedicated, low-latency 32-bit master AXI interfaces exist between each core and the interrupt controller. Another 64-bit master AXI interface allows for access to main memory and shared peripherals. Finally, a 64-bit slave AXI interface provides a route to the TCMs for the system's Direct Memory Access (DMA) controller. This facilitates efficient offloading of data transfer tasks to and from the TCMs.

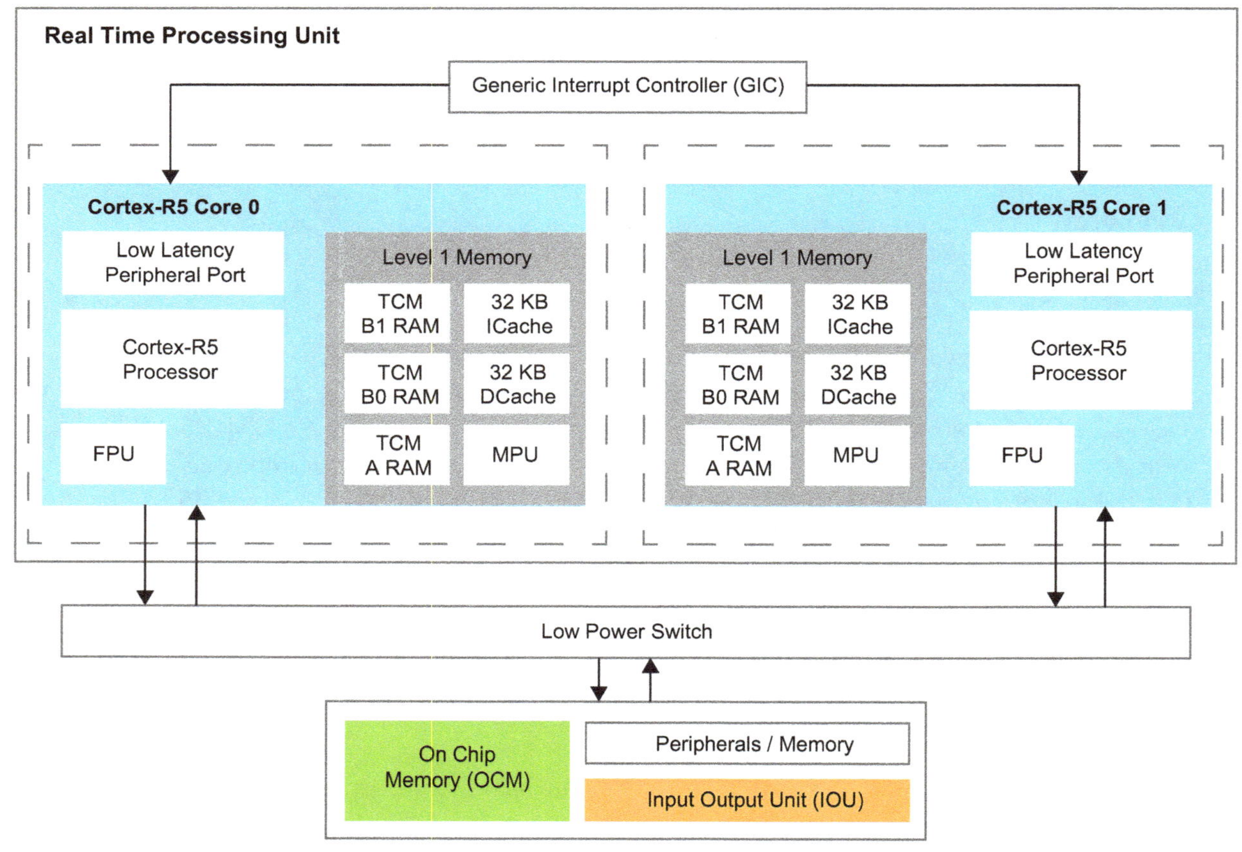

Figure 7.1: Simplified RPU Architecture

There are also some features to facilitate system analysis and debug. There is a debug Advanced Peripheral Bus (APB) interface to a CoreSight Debug Access Port (DAP), which is accessible to the developer via JTAG. This exposes both instruction and data traces, including information such as changes in program flow, current processor instruction state, and so on [5]. There are also registers which can be configured to monitor the system performance. These can be used to capture counts for up to 3 events types. Event types can be clock cycles, cache misses, instructions issued, execptions, stalls, interrupts, and ECC errors, to name a few.

7.3 Determinism and Responsiveness

Determinism and responsiveness are two key elements of a real-time system, so how does the RPU help achieve them?

7.3.1 Tightly Coupled Memories

One of the most important features a developer can make use of here is the memory hierarchy. This is usually designed as a trade off between cost and performance, using caching at various levels. Caching helps improve average performance by exploiting the temporal and spatial locality of data/instructions in memory, requiring no extra effort of the programmer. While this is useful, some elements of a real-time system need to make guarantees on strict response times. Although caching can hugely improve performance, determinism is usually more important when designing real-time systems. The TCM can be used to improve *worst-case* response time, making execution more deterministic.

Figure 7.2 shows a basic memory hierarchy for the RPU. Note that the TCM is considered to be a level 1 memory — alongside the Harvard caches. TCMs always act as non-cacheable, non-shareable memories [4]. .

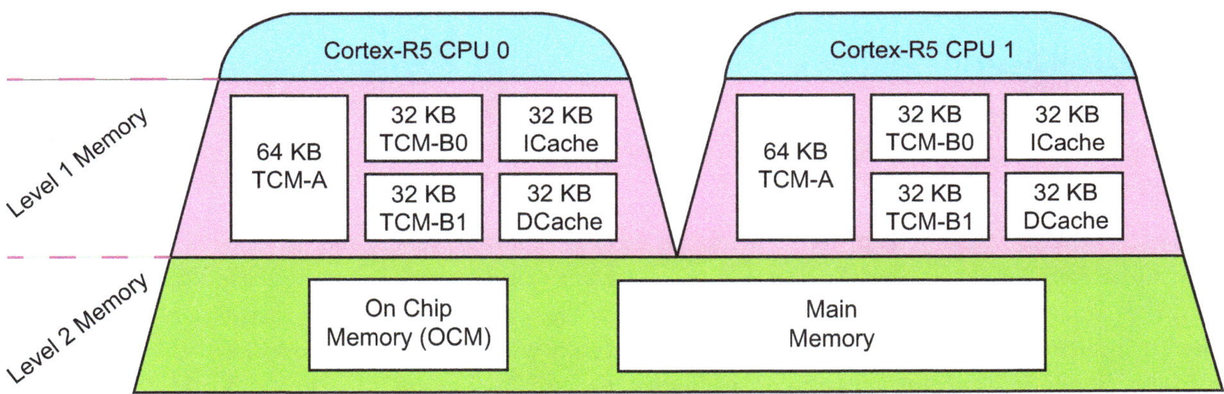

Figure 7.2: RPU Memory Hierarchy

Each core has access to two 64 bit wide, 64 KB TCM banks, TCM-A and TCM-B. Although usually visible to the programmer as a single memory, the TCM-B banks are actually comprised of two interleaved 32

KB memories. This interleaving further improves performance. Each TCM bank has a dedicated port and can be accessed concurrently with other TCMs. Typically the banks are used as follows:

- TCM-A to store *instructions* for interrupts for exceptions which cannot afford the delay associated with a cache miss.
- TCM-B to store blocks of *data* to be processed which must complete deterministically.

Although typically accessed from one of the Cortex-R5 cores, the TCMs can be accessed from elsewhere in the system such as the APU and DMA controller. This is facilitated by the AXI slave interface. Even within the Cortex-R5 cores there are multiple units which may contend for access to the TCM. These include the Load/Store Unit (LSU) as well as speculative access from the PreFetch Unit (PFU). Some arbitration is needed to control which device has access to the TCM in a given scenario. Explicit accesses from the LSU are given the highest priority, followed by speculative accesses from the PFU, and the lowest priority is assigned to the AXI slave interface. This can be an important point to remember when using the system DMA controller to move data to/from the TCM memories!

The TCMs are mapped into the main address space so the programmer only needs to use a particular address range in order to make use of the TCM. The addresses are mapped both globally, for access over the AXI slave interface, and locally for each Cortex-R5 core. The mapping also depends on the mode of CPU operation (split mode/lock-step mode) which is discussed further in Section 7.4. The mapping is shown in Table 7.1.

As an example, a system on the RPU may process video frames captured from a camera which are routed through the FPGA. The designer may choose to store the interrupt handler for retrieving a new frame in TCM-A (from 0x0000_0000) and store the frame data in TCM-B (from 0x0002_0000). This helps decouple the response time of processing a frame from any lower priority background tasks running on the RPU.

7.3.2 Interrupt System

An important element of responsiveness is interrupt handling. The latency between an event occurring and the corresponding Interrupt Service Routine (ISR) being dispatched can be a major component in the overall responsiveness in the system.

Before considering the interrupt controller at all, there are a few aspects of the Cortex-R5's Armv7-R architecture to be familiar with. These can speed up the worst-case interrupt response time quite significantly. Firstly, the architecture features *restartable* instructions. The currently executing instruction must terminate before an interrupt can be handled by the processor. Even with optimistic assumptions, this can introduce significant latency in interrupt response. Consider the example in Figure 7.3, where a "load multiple" (LDM) is issued just before an interrupt arrives at the simplified processor.

The interrupt latency for this example can be defined as the time between the interrupt arrival and the first interrupt handling instruction being executed. The majority of this delay is caused by the LDM instruction using the execute stages of the pipeline. Here it is assumed that the data to be loaded is found in cache. If this

Table 7.1: RPU L1 Address Map

	R5_0 View (Start Address)	**R5_1 View (Start Address)**	**Global View (Start Address)**
Operating in split mode			
R5_0 TCM-A (64 KB)	0x0000_0000	N/A	0xFFE0_0000
R5_0 TCM-B (64 KB)	0x0002_0000	N/A	0xFFE2_0000
R5_0 I-Cache (32 KB)	I-Cache	N/A	0xFFE4_0000
R5_0 D-Cache (32 KB)	D-Cache	N/A	0xFFE5_0000
R5_1 TCM-A (64 KB)	N/A	0x0000_0000	0xFFE9_0000
R5_1 TCM-B (64 KB)	N/A	0x0002_0000	0xFFEB_0000
R5_1 I-Cache (32 KB)	N/A	I-Cache	0xFFEC_0000
R5_1 D-Cache (32 KB)	N/A	D-Cache	0xFFED_0000
Operating in lock-step mode			
R5_0 TCM-A (128 KB)	0x0000_0000	N/A	0xFFE0_0000
R5_0 TCM-B (128 KB)	0x0002_0000	N/A	0xFFE2_0000
R5_0 I-Cache (32 KB)	I-Cache	N/A	0xFFE4_0000
R5_0 D-Cache (32 KB)	D-Cache	N/A	0xFFE5_0000
R5_1 slave port is not available for lock-step mode			

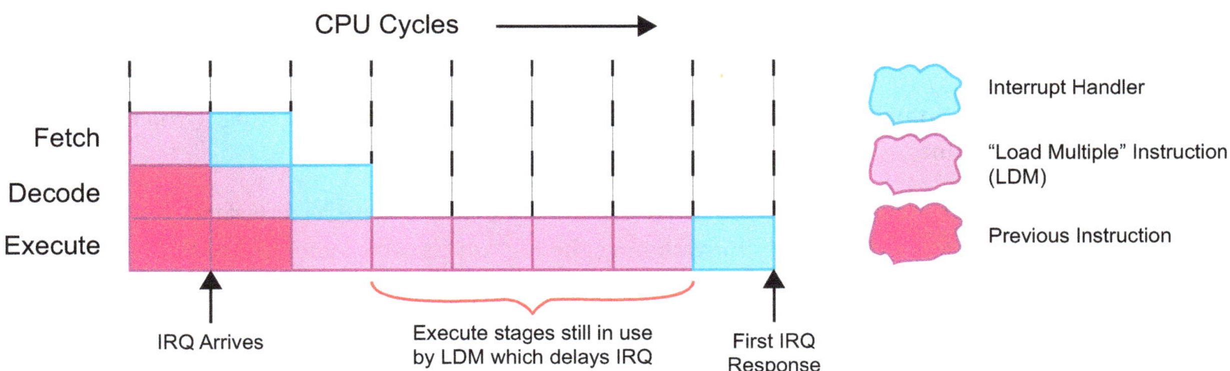

Figure 7.3: Simplified Pipeline Showing Interrupt Latency with Non-restartable Instructions

is not the case, the delay can become orders of magnitude larger. Importantly, it is not the common case which limits the guarantees of real-time systems, but the worst case.

The worst case latency here can be can be reduced using the restartable instructions present in the Cortex-R5s. With these, the interrupt handler instructions will pre-empt the current instruction and gain faster access to the execute stages. This helps to decouple the interrupt latency from the characteristics of the previous instruction, hence reducing the jitter. A few caveats with the use of restartable instructions are:

- A processing overhead is incurred as the interrupted instruction must be *restarted*. Intermediate stages may be repeated.

- It is not possible to restart every type of instruction. For example, it is difficult to ensure correct restart behaviour when accessing memory marked as a device type or strongly-ordered (peripheral registers, etc.) as the number and order of accesses will often affect the peripheral behaviour.

While the scenario in Figure 7.3 helps illustrate the concept, keep in mind that the Cortex-R5s have multiple pipeline stages within each of the fetch, decode, and execute stages depicted here.

Another aspect of the Armv7-R instruction set pertinent to interrupt handling is the software overhead when context switching to an interrupt service routine. There are three instructions included to help minimise overhead when entering and exiting interrupt handlers: SRS (save return state), RFE (return from exception), and CPS (change processor state) [3].

Now some of the processor-side elements of interrupt performance have been addressed, we now consider the interrupt controller logic. The interrupt system for both Cortex-R5 cores is controlled by a shared Generic Interrupt Controller (GIC). The fact that the GIC serves both cores is important when operating in lock-step mode. It can be configured to only dispatch interrupt requests to one core — these interrupts can then be duplicated, keeping both processors in sync during lock-step. The GIC used within the RPU is the PrimeCell Generic Interrupt Controller (PL390) and the reader is referred to [6] for full details. The main features include:

- Support for several interrupt sources including Software Generated Interrupts (SGI), Shared Peripheral Interrupts (SPI), and Private Peripheral Interrupts (PPI).

- Programmability of interrupt sensitivity (i.e. level or edge sensitive), targeting (which CPUs are informed), and handling.

It consists of a distributor unit and two CPU interfaces, along with two AXI interfaces for control and configuration. The distributor is responsible for assigning the highest priority interrupts to the targeted CPU interfaces. The CPU interfaces forward interrupts to their respective Cortex-R5 cores and can mask lower priority interrupts from the distributor while the current interrupt is being handled by the CPU. It is integrated with the Cortex-R5s as shown in Figure 7.4.

There are two different improvements to interrupt latency which are caused by the way the GIC has been integrated with the Cortex-R5s. First of all, there is a dedicated AXI interface (via the low-latency peripheral

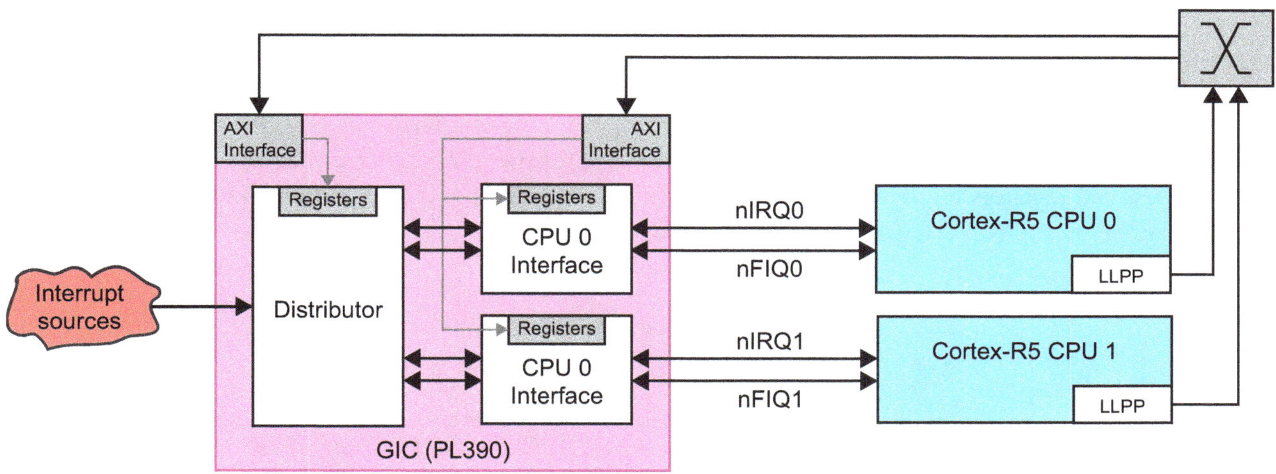

Figure 7.4: Block Diagram of RPU Interrupt System

port) between the CPUs and the GIC. This means there is no extra delay due to contention for a shared AXI interface. This interface allows access to the registers in the GIC distributor and each CPU interface. With this, each core can configure interrupt sensitivity and targeting, as well as per core interrupt masking while servicing an interrupt.

Another point to note is the distinction between the two interrupt lines going to each Cortex-R5 core. These are the IRQ (interrupt request) and FIQ (fast interrupt request). The IRQ signals an interrupt of standard importance while the FIQ signals a more important (usually single) non-maskable interrupt. A safety critical or otherwise latency-sensitive interrupt (e.g. a block, non-DMA data transfer) are good candidates for being routed as an FIQ signal. The reduction in response time for FIQ mainly comes from:

- The FIQ has a higher priority than all IRQs. This means than an FIQ can pre-empt and interrupt an IRQ. Conversely, all IRQs will be masked while handling the FIQ.
- The FIQ entry is at the end of the exception vector table. This means that the overhead normally incurred due to the branch instruction to the main ISR can be avoided. As there are no table entries after the FIQ entry, there is no need to branch. Just place the rest of the handler code directly after the exception vector table, as shown in Table 7.2.
- The FIQ execution mode has a set of "banked" registers. A register is said to be banked when the same identifier between two modes points to different physical registers. The banked registers (r8 through r14 for FIQ mode) are only accessible through a single mode. This reduces the need to save and restore register state upon entering and leaving the handler as it is the only code to run in the FIQ mode. Furthermore, the handler can rely on values from the previous call being retained in the banked registers.

If multiple FIQ sources were used, most of the performance benefits would be diminished. The non-maskable behaviour, the exception vector table layout, and the banked registers only boost performance when one FIQ source is used. So, systems are usually designed with a maximum of one source for each CPU FIQ.

Architectural features such as the dedicated LLPP (low-latency peripheral port) and the restartable instructions do not require any extra effort from the programmer. However, making use of the FIQ system does. Using the banked registers effectively has traditionally required hand-coded assembly. While this can still be valuable, compilers such as GCC (GNU Compiler Collection) now have some optimisations for Arm FIQ handlers [7].

Table 7.2: Example Exception Vector Table

Exception	Address	Example Instruction
Reset	0x0000_0000	LDR PC, [PC, #RESET_handlr_offset]
Undefined Instruction	0x0000_0004	LDR PC, [PC, #UNDEF_handlr_offset]
Software Interrupt	0x0000_0008	LDR PC, [PC, #SWI_handlr_offset]
Abort (prefetch)	0x0000_000C	LDR PC, [PC, #PABT_handlr_offset]
Abort (data)	0x0000_0010	LDR PC, [PC, #DABT_handlr_offset]
Reserved	0x0000_0014	N/A
IRQ	0x0000_0018	LDR PC, [PC, #IRQ_handlr_offset]
FIQ	0x0000_001C	*First FIQ handler instruction (no branch required!)*
	0x0000_0020	*Remainder of FIQ handler*
	...	

7.4 Safety within the RPU

Redundancy can be enabled in a few ways to help protect the RPU against errors. This is a vital consideration for safety critical systems, such as most automotive applications. We touch on safety aspects here as it applies to the RPU — see Chapter 9 for a more complete view of functional safety. With the features explained in this section, errors in memories and even the internal logic of the processors can be mitigated. The source of error can be anomalies due to radiation, other extreme environments such as temperature, or even maliciously injected faults using EM or lasers!

7.4.1 Memory Protection

The protection of level 1 memories within the RPU is enabled by two systems:

- Memory Protection Unit (MPU): controls access to the L1 and external memory. Regions can be programmed with configurable access permissions and memory types. This helps catch erroneous or malicious software which accesses restricted regions in the address space.
- Error Checking and Correction (ECC): can be applied to L1 memories to be able to correct 1 bit of error and detect 2 bits of error by storing a number of redundant bits. This helps to defend against memory corruption without the power penalty incurred by duplicating the memories entirely.

First, we discuss the MPU in more detail. The MPU is included in lieu of a more traditional MMU to better fit a real-time environment. One job of a MMU is to perform translation from a virtual address space to a physical one. This leads to a few different non-deterministic behaviours, including page faults and misses in the TLB. This is undesirable for real-time environments, so the MPU is designed without these features — instead focusing only on memory protection.

The MPU can be configured with up to 16 regions. Upon an error (permission fault, alignment fault, or an access to an area outside all regions), the MPU raises an abort exception. This interrupts the Cortex-R5 core and the appropriate abort handler is dispatched. Note that errors are not only caused by explicit load/stores (a data abort) but also by speculative prefetches performed by the processor (a prefetch abort). These have different handlers, as shown back in Table 7.2.

Each region has a configurable base address, size, attributes and access permissions. The attributes define how memory accesses are performed. This includes the following groups of options:

Table 7.3: MPU Region Attributes

Attribute Group	Attribute	Description
Memory type	Strongly ordered	*All accesses* will be performed in instruction order and size.
	Device	All *device accesses* will be in instruction order (normal accesses can still be moved around these) and size.
	Normal	Accesses may be reordered, altered (e.g. a byte load is run as word load to memory, merging of two half-word stores...), and caching is possible.
Sharing	Shared	Address range holds data that several processors share. This has implications on the use of L1 caching.
	Non-shared	Address range holds data that only a single processor uses.

Table 7.3: MPU Region Attributes

Attribute Group	Attribute	Description
Caching	Non-cacheable	No caching is used.
	Write-through cacheable	Caches are used and data is written synchronously back to L2 memory.
	Write-back cacheable	Caches are used and data is only written back to L2 memory when the block is evicted from cache.
Allocation	Read allocation	Data is brought into the cache only when it is read (writes do not trigger the caching of a new block).
	Write allocation	Data is brought into the cache upon a read or a write.

There are also access permissions which exist alongside the memory attributes. These allow the choice of read-only, read/write, or no access. Access permissions are distinct for the Privileged mode and other modes. Regions may also be marked as "execute never" to prevent instructions from being fetched.

The MPU supports overlapping regions and subregions. This lets a developer construct complex layouts with the 16 available regions. The higher the region number, the higher the priority of the attributes and permissions. For example, when region 4 and 15 overlap, the permissions of the overlapping addresses are inherited from region 15. Each region can also be split into 8 equal sized subregions. A subregion can be disabled such that the attributes and permissions of an overlapping lower priority region apply instead. These techniques can be applied to produce exceptions upon events such as stack overflows quite easily. This is much preferable to silently overflowing a stack and encountering unexpected behaviour later.

With the MPU configuration, there are a couple of situations in which the region settings are overridden:

- Addresses mapped to TCM are always treated as normal, non-shared memory as each processor has its own TCM.
- Addresses mapped to peripheral ports (such as the LLPP to the GIC) are always treated as non-cacheable and execute never.

Now consider the error checking and correction for the L1 caches and TCM. For each 32 bit chunk, the L1 caches and TCM store a number of redundant bits which can be used for Single-Error-Correction, Double-Error-Detection (SEC-DED). The operation is largely passive from the programmer's perspective, once it has been enabled. The checking and correction is performed in hardware as shown in Figure 7.5. Upon each load, the corresponding ECC bits are used to check for errors in the fetched data. Corrections can be committed back to memory or passed directly on to the next stage..

There can be performance implications when using the ECC scheme, even with error-free operation. For example, consider writing *one* byte to L1 memory. In order to write the data, a new set of ECC bits must be

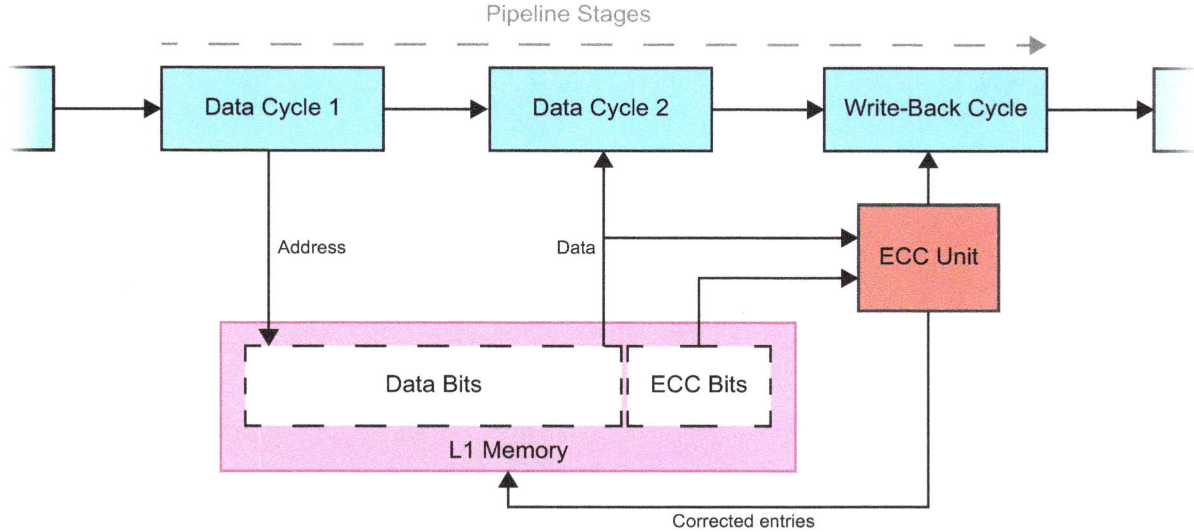

Figure 7.5: Block diagram of ECC scheme in pipeline [8]

calculated. However, the data chunk which the ECC is applied to is larger than a byte. The remainder of the chunk must first be read, joined with the new byte, then new ECC bits must be calculated before finally performing the write! This process is called read-modify-write. In this scenario, a single write instruction infers a read, calculation, and the original write.

When a single error in a word is detected, it can be automatically corrected. There are two mechanisms for this:

- *Correct inline*: uses the ECC bits to correct the data read from memory and this new data is used directly. The corrected data is not written back to memory at time of correction.
- *Correct-and-retry*: uses the ECC bits to correct the data read from memory and this new data is written back to memory. The read instruction is re-executed and the corrected data is read, assuming no more errors have occurred.

Correct-and-retry takes more cycles (at least 9) to complete, however this does not affect the average performance significantly as memory errors should be very infrequent. Correct inline is used for errors detected upon a read from the AXI-slave interface (from outside the RPU) and correct-and-retry for errors detected upon an instruction/data read from one of the Cortex-R5 cores.

So, correctable errors are handled using correct inline or correct-and-retry. How does the ECC scheme handle uncorrectable errors such as 2 or more bits in error? The RPU can be configured to generate an abort exception upon detected errors or ignore them. Correctable errors can also be configured to generate abort exceptions. It is worth noting that the ECC scheme cannot entirely guarantee error free memory operation.

Problems arise if more than two bits are in error. The ECC scheme may incorrectly interpret this as a valid chunk or a single error and attempt to correct it, still producing an incorrect result. "Hard errors" can also affect the reliability. These are physical faults such that a bit in memory cannot be accessed reliably. The ECC scheme will not always be able to recover from these.

7.4.2 Split/Lock Modes

The Cortex-R5s support two configuration modes. The combination of these two modes is termed "split/lock". The two modes are:

- Split mode: each CPU operates independently. Also known as performance mode.
- Lock-step mode: both CPUs execute the same instructions for safety. Also known as safety mode.

The mode configuration is static, such that switching between modes is only permitted while the processor group is held in reset. The SLCLAMP and SLSPLIT signals can be used to select the lock-step and split modes respectively.

Generally, when safety is not a main concern, the performance benefits of split mode are worthwhile. The processor group operates like a typical dual-core system. The main caveat to note with split mode is coherency between the two cores. There is no built-in coherency between the two sets of L1 cache. When sharing data between the cores the MPU can be used to mark an area of memory as non-cacheable/shareable. This avoids issues of coherency in the L1 caches (by not using them!).

Lock-step mode uses the second CPU as redundant logic, executing the same instructions as the first. There is extra logic which compares the outputs of both CPUs in an effort to detect errors from the internal logic. There is an opportunity for errors in the logic due to radiation to affect both processors in the same way. This would not be flagged as an error, so avoiding errors common to both cores is a design goal. To assist with this, the two processors are separated physically, and in time — i.e. one processor will run 2 clock cycles behind the other. In fact, the term "lock-step" comes from a military term for marching very close in single file. This is analogous to one processor following the actions of the other, close behind. The lock-step configuration is as shown in Figure 7.6.

Note the delays required to have the CPUs execute an instruction at different instances *and* resynchronise before the checking logic. The checking logic uses the CPU output pins and the interface to L1 memory. Monitoring the memory interface enables erroneous writes to be caught early enough in the pipeline to be stopped before committing anything to memory — even with the 2 cycle delay!

The L1 memory system also changes slightly in lock-step mode. The TCM and caches do not need to be duplicated for safety because there is already an ECC scheme for this. Now that both CPUs are performing the same instructions, the TCMs associated with each core can be combined into a single, larger TCM in lock-step mode. This was introduced back in Table 7.1.

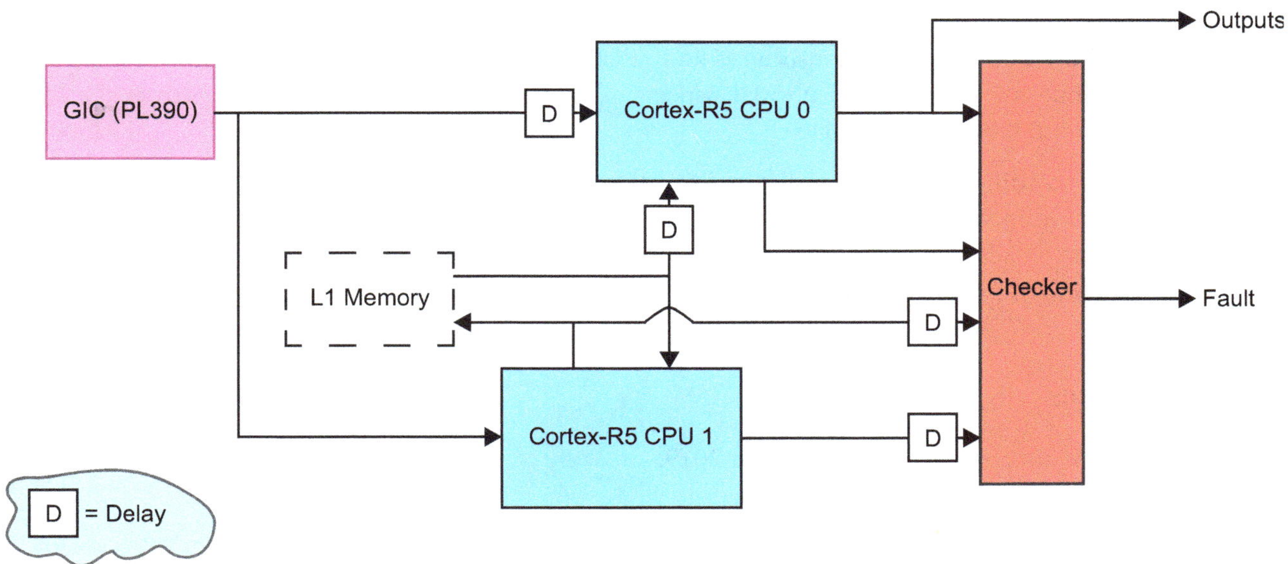

Figure 7.6: Lock-step configuration of Cortex-R5 group [10]

The fault signals (DCCMOUT[7:0] and DCCMOUT2[7:0]) can be raised upon a discrepancy between the outputs of the two cores. This is connected to the PMU error handling system and exposed via the JTAG error register. The PMU can be configured to behave in the following ways upon an RPU lock-step error:

- Assert PS_ERROR_OUT signal
- Generate an interrupt handled by the PMU
- Generate a system reset (SRST)
- Generate a power-on-reset (POR)

Lock-step mode can be entered while the RPU is held in reset. There should be code in the reset handler which ensures interrupts are only dispatched to CPU 0. The full sequence used to enter lock-step mode is listed in [9].

7.5 Chapter Review

This chapter has provided a detailed look at the real-time processing unit. The main features have been explored from the viewpoints of responsiveness and reliability — the defining characteristics of many real-time systems. While this is far from a tutorial, aspects of configuring important features (such as the MPU and split/lock modes) have been introduced.

For any further reading, the reader is directed to [9] and [4] initially. Note that both the Xilinx and Arm documents are recommended as they complement each other. Arm discusses the details of the architecture while generalising details like cache sizes and number of cores while Xilinx describe the implementation choices within the Zynq MPSoC devices.

7.6 References

Note: All online sources last accessed March 2019.

[1] K. G. Shin and P. Ramanathan, "Real-time computing: a new discipline of computer science and engineering", *Proceedings of the IEEE*, vol. 82, no. 1, pp. 6-24, Jan 1994.
DOI: 10.1109/5.259423

[2] Arm, Ltd., "Cortex-R5 Overview".
Available: https://developer.arm.com/products/processors/cortex-r/cortex-r5

[3] Arm, Ltd., "ARM Architecture Reference Manual: ARMv7-A and ARMv7-R edition", vC.b, July 2012.
Available: https://silver.arm.com/download/download.tm?pv=1603196

[4] Arm, Ltd., "Cortex-R5 Technical Reference Manual", vr1p2, September 2011.
Available: http://infocenter.arm.com/help/topic/com.arm.doc.ddi0460d/DDI0460D_cortex_r5_r1p2_trm.pdf

[5] Arm, Ltd., "ARM CoreSight ETM-R5 Technical Reference Manual", vr0p0, July 2013.
Available: http://infocenter.arm.com/help/topic/com.arm.doc.ddi0469b/DDI0469B_etmr5_r0p0_trm.pdf

[6] Arm, Ltd., "PrimeCell Generic Interrupt Controller: Technical Reference Manual", vr0p0, November 2009.
Available: http://infocenter.arm.com/help/topic/com.arm.doc.ddi0416b/DDI0416B_gic_pl390_r0p0_trm.pdf

[7] GNU, "GCC ARM Function Attributes".
Available: https://gcc.gnu.org/onlinedocs/gcc/ARM-Function-Attributes.html

[8] N. Werdmuller, "Addressing functional safety applications with ARM Cortex-R5", v1, January 2015.
Available:
https://community.arm.com/groups/embedded/blog/2015/01/22/addressing-functional-safety-applications-with-arm-cortex-r5

[9] Xilinx, Inc., "UG1085 — Zynq UltraScale+ Device Technical Reference Manual", v1.9, January 2019.
Available: http://www.xilinx.com/support/documentation/user_guides/ug1085-zynq-ultrascale-trm.pdf

[10] Chris Turner, "Safety Features and Standards for Cortex-R Processors in Embedded Systems", ARM TechCon, October 2012.
Available: http://www.armtechforum.com.cn/2012/16_Safety_Features_and_Standards_for_Cortex-R_Re-Ordered_slide_BDL.pdf

[11] J. Yiu, "A Beginner's Guide on Interrupt Latency - and Interrupt Latency of the ARM® Cortex®-M processors", v4, April 2016.
Available: https://community.arm.com/docs/DOC-2607

[12] International Electrotechnical Commission, "Functional Safety and IEC 61508 Homepage".
Available: http://www.iec.ch/functionalsafety/

[13] International Organization for Standardization, "ISO 26262-1:2011: Road vehicles - Functional Safety Abstract", November 2011.
Available: http://www.iso.org/iso/home/store/catalogue_tc/catalogue_detail.htm?csnumber=43464

Chapter 8

Security in Zynq MPSoC

This chapter focuses on the security features in the Zynq MPSoC — including cryptography services, reacting to tamper attempts, and other run-time protections. We explain the *what*, *how*, and most importantly, *why* of these features. The aim is to provide an appreciation of why you really should use the available security features in your designs, and an idea of how to go about doing so.

For readers who have worked with FPGA devices before, you may remember a time when security nearly boiled down to a single question: "Should I encrypt my bitstream?"[1]. As programmable devices have become ever more complex, and common attacks have become ever more sophisticated, this simple question has exploded into many different considerations — some of which are distinct, while others are subtly interwoven. In the following sections we run through the numerous security features of the Zynq MPSoC and why they should be important to a security-conscious designer. We approach system security as three more digestible sub-topics:

1. **Information Assurance:** for keeping system configuration and data secret and unmodified using cryptographic techniques. This forms part of the chip's preventative measures.

2. **Anti-tampering:** for protecting our system's hardware from hands-on attackers with physical access to the device. This is details the chip's detection and response measures.

3. **Isolation:** as a precaution to stop any spread of attacks or faults between different parts of the system — again, returning to some final preventative measures.

1. Incidentally, whenever security is a concern, the answer to this is almost always "Yes, of course!".

Keep in mind that all three of these topics are usually employed in tandem to prevent sophisticated attacks — there is rarely a 1:1 relationship between attacks and countermeasures. To whet the reader's appetite, however, Table 8.1 lists some example attacks that we will address throughout this chapter.

Table 8.1: Example types of attacks

Attack	Description
IP theft	Reverse engineering or cloning our IP, etc.
Side-channel attacks	Recovering secrets by 'listening in' on our device as it performs cryptography.
Glitching	Inducing anomalies in our device by injecting pulses into clock signals, EM blasts, or using lasers!
Focused Ion Beam Probing	Using a focused Ion Beam to modify our device at a semiconductor level.
Environmental	Exposing our device to extreme temperatures and voltages to exploit anomalous behaviour.

So, without any further ado, let's dive into to the security features of the Zynq MPSoC!

8.1 Information Assurance for Configuration Security

For most embedded systems, especially once deployed in the Big Bad World™, there are elements we want to keep secure. For Zynq MPSoC systems, we often want to protect two different sets of information — system configuration (bitstreams and software images) and any user data. For system configuration, we often want to defend against threats such as IP theft (someone simply cloning the device to undercut our hard work, or even reverse engineering parts of the system) and someone running unauthorised code. These map to the properties of confidentiality and authentication respectively.

When talking about information assurance, 'security' often to refers to system *integrity, authentication* and *confidentiality*. These properties are vital to securing our software and bitstream:

- **Integrity** ensures that our system configuration has not been modified.
- **Authentication** extends the property of integrity, proving that the configuration really is from a trusted developer, and not an attacker.
- **Confidentiality** ensures that parts of our configuration cannot be understood by an unauthorised person. This is typically what people think of when talking about encryption.

These are distinct properties but they are commonly used together. Consider each property in turn for a moment. An encrypted or 'confidential' bitstream can still be modified even if the real contents are hidden from you ([1] discusses a real example of this). A bitstream with good integrity checking/authentication can be plainly seen, but any attempt to modify the content will prove difficult. However, using integrity *and* confidentiality results in a bitstream which cannot be easily read or modified.

The Zynq MPSoC has hardware support for cryptography, providing confidentiality and authentication for both system configuration during boot-time and user data at run-time. We will go on to explain the architecture of these blocks, why/when each block should be used, and how to use them.

8.1.1 Configuration Security Unit Introduction

The Configuration Security Unit (CSU) is the heart of the Zynq MPSoC's crypto capabilities and ensures their proper use during secure boot. As seen in Figure 8.1, the CSU can be thought of as two blocks:

- The Crypto Interface Block (CIB) encapsulates all of the hardware crypto units and related key management. The crypto units can be used both during a secure boot and from the application at run-time. The main functions include SHA-3/384 hashing, AES-GCM-256 encryption/decryption, and RSA acceleration.
- The Security Processor Block (SPB) is the trusted arbiter for (secure) booting. This is a small, self-contained processing system which can make use of the CIB to authenticate and decrypt secure configurations. The SPB is not user-accessible, so there is no opportunity for malicious software to influence its security functions.

These two blocks are interfaced with the rest of the Zynq MPSoC exclusively via the CSU PMU Switch.

Secure Processor Block

The SPB is a small processing system specifically designed to handle system configuration at boot-time. It also monitors for any tampering attempts at run-time, including voltage/temperature alarms and JTAG activity. The block is largely isolated from the rest of the Zynq MPSoC and is not user accessible. This helps prevent any malicious or erroneous software elsewhere in the system from disturbing the SPB's proper handling of critical tasks. Defences are also implemented against hardware-level faults or tampering, as discussed next.

First of all, the system ensures that no modified code will be executed. The SPB will only execute code which is stored in its on-chip 128 KB metal-masked ROM. This code is developed by Xilinx, and is implemented as part of the mask used during fabrication. Having this code on-chip already introduces some difficulty for an attacker. Even if a modification is successful, there is a SHA-3 based integrity check performed on the ROM automatically at boot to detect such faults. The SPB is also protected while executing code by redundancy — both by having triple redundant MicroBlaze cores and ECC for the dedicated internal RAM.

While redundancy features are traditionally considered as a protection for safety against radiation randomly flipping bits, they have also become a topic in security research. Many published attacks work by deliberately exposing a device to radiation (EM pulses, lasers, temporary unexpected voltages) to induce controlled faults. Without these redundancies, induced faults can defeat secure boot by skipping instructions [3], or performing

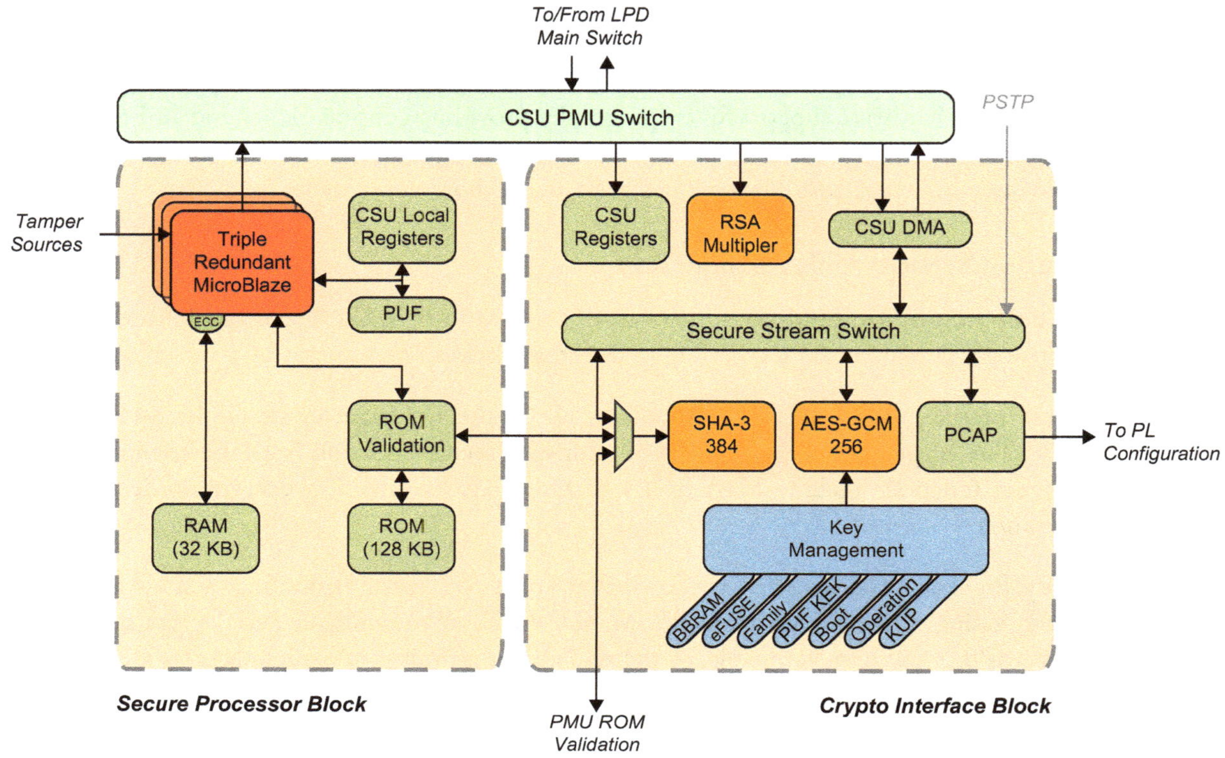

Figure 8.1: Overview of the Configuration Security Unit

some more complex analysis against cryptography, e.g. [2]. More about how these redundancy features are implemented can be read in Chapter 9. Similarly there are attacks which force instructions, or part of instructions, to be skipped (causing corruption) by glitching an external clock signal [4]. The SPB helps prevent this by running from an internal, uninterruptible clock source.

The SPB also controls a Physical Unclonable Function (PUF). PUFs exploit the inherent, uncontrollable randomness present in silicon manufacturing. Normal digital devices are designed with tolerances to account for this randomness, but PUFs actively exploit it to provide a unique fingerprint for each device. Here, the uniqueness is guaranteed by the physical process — not by a manufacturing employee. This fingerprint can then be used to generate a per-device key for secure boot. See "Storing Keys in Encrypted (Black) Form" on page 195 for more on the use of the PUF.

Crypto Interface Block

The CIB contains hardware crypto services which can be employed by the SPB during boot and a user application at run-time. An overview of its structure was shown in Figure 8.1 on page 182. We can think of the CIB as a set of crypto cores (coloured orange in Figure 8.1), a set of key stores (blue), and supporting logic for data transfer (green). The main features include:

- **Key management:** provides long-term storage for secret keys (AES keys and RSA key hashes) and temporary registers for temporary/intermediate keys.
- **Crypto blocks:** provides hardware-accelerated AES-GCM (256 bit), SHA-3 (384 bit), and RSA (4096 bit). These help a designer to implement data confidentiality (*attacker can't decipher it*), integrity (*we know if an attacker makes a modification*), and authentication (*we know the data originates from who we expect*).
- **Secure Stream Switch (SSS)**: facilitates data transfer between the AES-GCM block, SHA-3 block, PCAP interface, JTAG interface, and the CSU DMA (which can be configured from a user application).
- **PCAP:** provides an interface to configure the FPGA fabric via the secure stream switch — with data from either the CSU DMA, JTAG interface, or the AES-GCM output directly.
- **CSU DMA:** provides dedicated data transfer logic for large blocks between the SSS and a user application.

Unlike the SPB, elements of the CIB are user accessible so a user application can make use of its crypto services at run-time. The crypto blocks and PCAP are accessible via the SSS. User requested data transfer to/from these blocks can be performed by the CSU DMA.

Before continuing, there are two aspects of the CIB's structure (Figure 8.1) worth clarifying. Firstly, most blocks that use the SSS are connected directly, but why is the SHA-3 block different? The SHA-3 block is also used for integrity checking the CSU and PMU ROMs on boot (whereas the other SSS-connected blocks are not). This boot-time ROM integrity checking then merits a direct path to the SHA-3 block which leads to the extra multiplexer for the SHA-3 input in the overview diagram. Secondly, why is the RSA block not driven by the CSU DMA, like the AES-GCM and SHA-3 blocks? The main use of the RSA block is to authenticate SHA-3 digests. These digests are only 384 bits long! Because the input and output data is typically so small, the overhead incurred by configuring the DMA is likely not worth it.

8.1.2 Crypto Blocks

The main crypto accelerators exist as three units, each with complementary purposes. These are AES, SHA, and RSA units. We will discuss each of them in turn, giving some background of how the algorithms work and why we may choose to use them. Just a little appreciation for the algorithms can go a long way towards demystifying its proper use. This is nothing new for readers with a background in crypto, but we will also touch on how to program these blocks.

One note regarding all of these blocks — they run from the internal oscillator during boot, but can be configured to run from a faster PLL clock after boot. This helps prevent common clock glitching attacks during the secure boot cryptography.

AES-GCM

In terse, technical language, this is a hardware AES implementation which supports 256 bit keys with the Galois/Counter mode (GCM). We are going to take a step back and try to give an appreciation of what this really does for us.

First, we look at the AES encryption algorithm — ignoring the GCM aspect for now. The algorithm was selected by the National Institute of Standards and Technology (NIST) after a transparent and open process based on input from interested parties and is now used worldwide. Simply put, its job is to take some data and scramble it up using a key which is kept secret. The output is then unintelligible to anyone without the secret key. So, we can store an encrypted file off-chip and be confident that an attacker cannot recover the file contents as long as the key remains secret. (Section 8.1.3 describes how keys can be kept "secret" using on-chip key storage.) The 'scrambling' process is reversible so decryption is possible.

To appreciate how the algorithm works, we visualise the 4 fundamental steps in Figure 8.2. The data is represented as a 4x4 grid of bytes. These 4 steps are combined to form a single 'round' — and 14 of these rounds make the full AES-256 algorithm. Each step has a unique purpose:

- **SubBytes** provides *non-linearity*. This is done by substituting the value of each byte with another value (usually implemented as a table look-up).
- **ShiftRows** provides *diffusion* by ensuring that there is some cross-pollination between each of the columns. This helps avoid each column being encrypted in isolation. Each row is cyclically shifted by a unique number of bytes.
- **MixColumns** provides *diffusion* by mixing together all of the values in a given column.
- **AddRoundKey** provides *key-dependence*. This is achieved by XORing the state with a round key (derived from the main key — there is one made for each of the 14 rounds).

AES encryption is deterministic. If we encrypt the same block of data twice (with the same key) the output will be the same. When a file has been encrypted simply block by block, an attacker could potentially see patterns of repeated data and extract some useful information. To combat this, different modes of operation are employed which provide unique output for any repeated blocks. These modes typically make use of an incrementing counter value or feedback from previous outputs. We use the GCM mode [5]. It is often chosen for its good performance with reasonable resources. The input data is never passed through the AES algorithm itself, but rather a counter value is "encrypted" and then combined with the input data. The initial counter value is user-controlled by an Initialisation Vector (IV). Encrypting counters allows blocks to be processed with some level of concurrency as the output of one block is not needed at the input stage of the next. Also, the output of each block (plus some extra data) is passed through a Galois multiplication, then

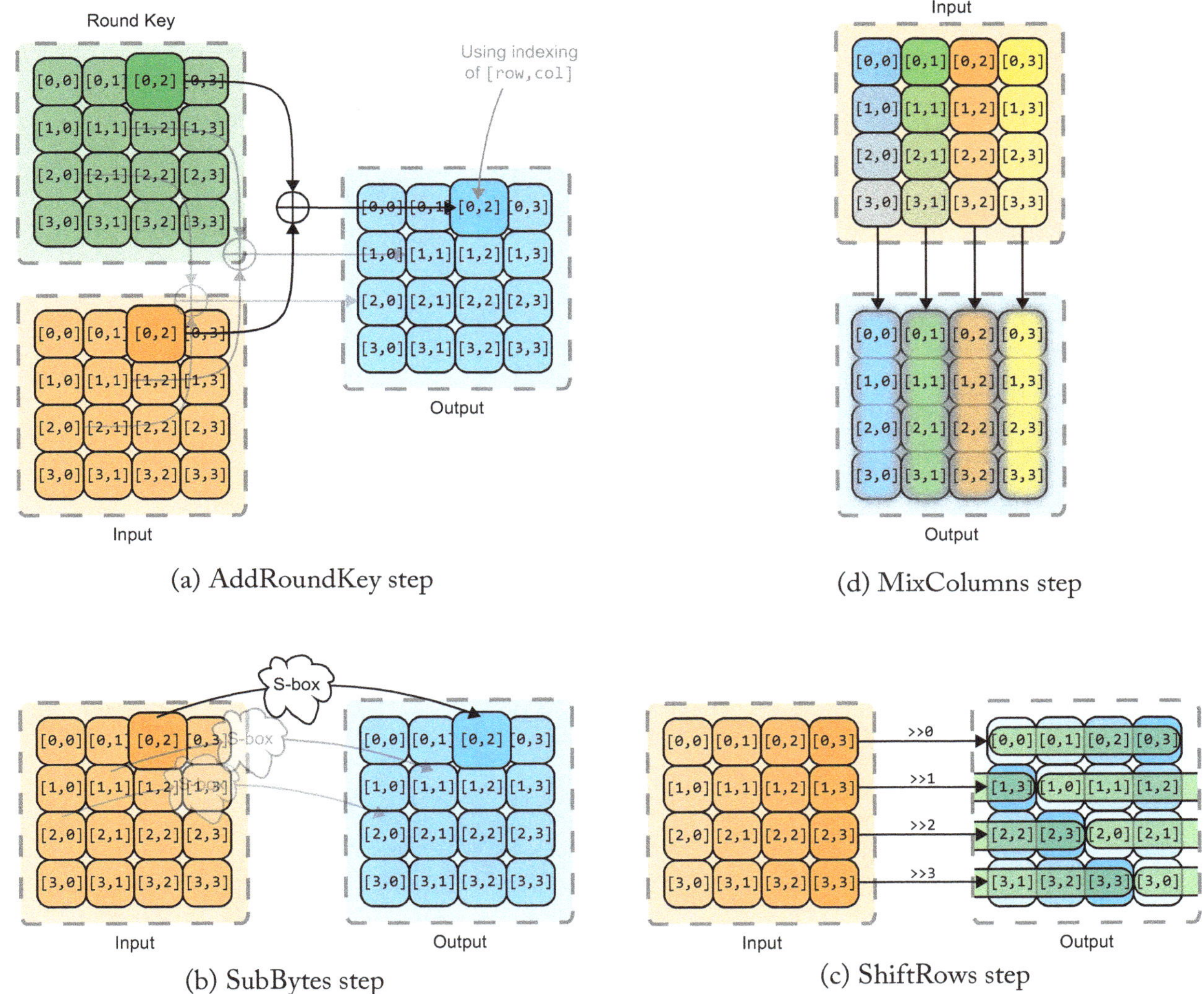

(a) AddRoundKey step

(d) MixColumns step

(b) SubBytes step

(c) ShiftRows step

Figure 8.2: Overview of steps in an AES round

combined and encrypted to provide a Message Integrity Code (MIC). The use of the MIC can preclude attacks such the bit-flip attack mentioned earlier in [1]. The full encryption process is visualised in Figure 8.3:

The IV should be randomly generated for every file to make the most of AES-GCM. The IV is contained within the boot header, if used. The IV vector also serves a second purpose in the Zynq MPSoC implementation. Only the most significant 96 bits of the IV are used by the AES-GCM — the least significant 32 bits are replaced by an incrementing counter value. This means we are free to encode some extra options in the least significant 32-bits of the IV. These are used for a key rolling technique which is discussed next.

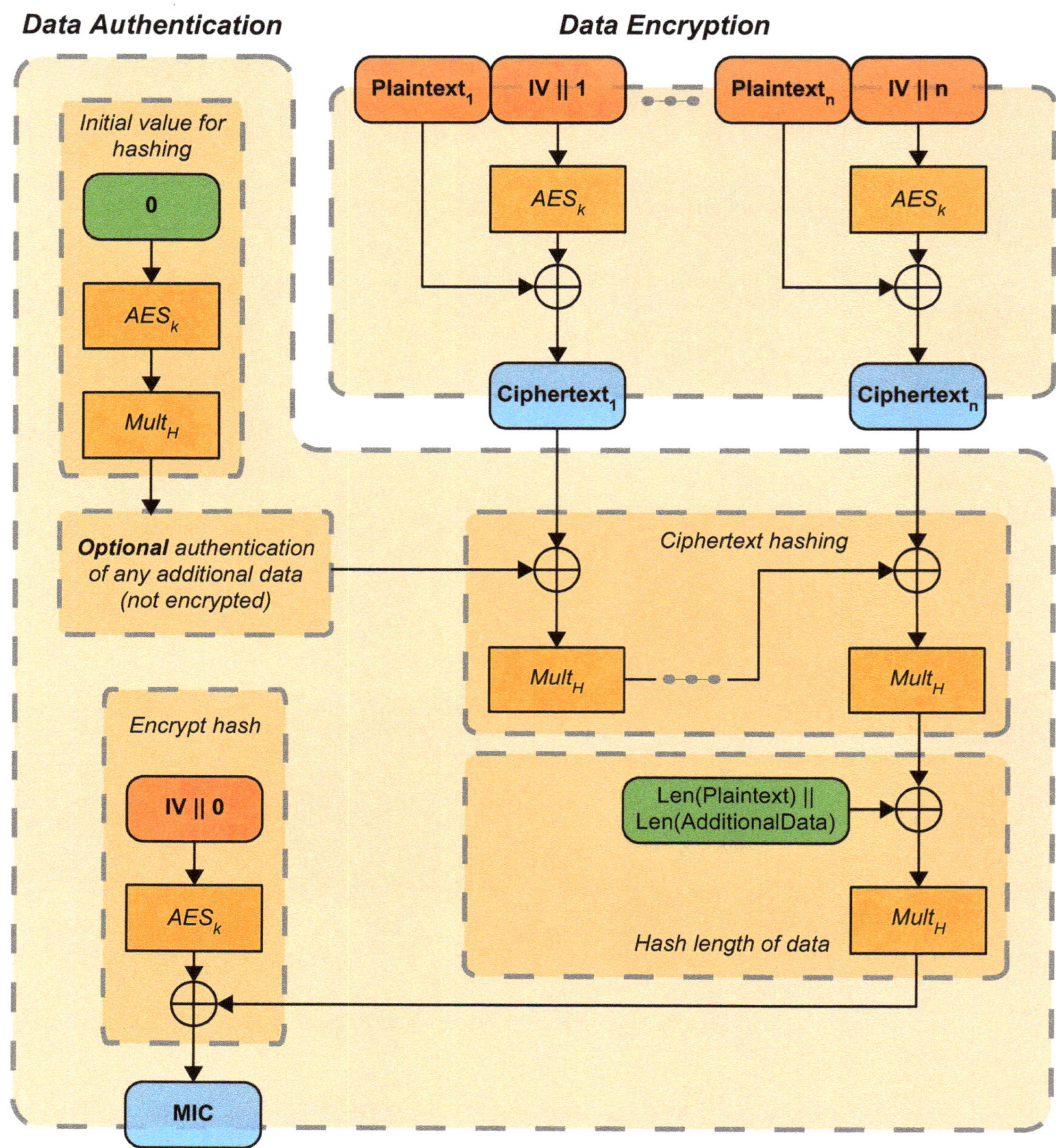

Figure 8.3: Flow diagram for AES-GCM encryption

The security considerations with Zynq MPSoC now include countermeasures against Side-Channel Attacks (SCA), as such attacks are becoming ever more feasible. These attacks attempt to extract secret, on-chip keys through analysis of power or electromagnetic measurements taken as a device decrypts many blocks of data. One SCA countermeasure available is the use of key rolling. This is designed to reduce the use of a single key, making it impossible for an attacker to gather enough measurements to recover the key under normal operation. This is achieved by splitting the file to be encrypted (the FSBL image, for example) into small blocks that are encrypted separately. Each block is encrypted with a unique key and IV. The initial key is stored as usual (on-chip or in the boot header) but all following keys and IVs are encrypted and stored at the end of the previous block. This minimises use of a single key and, importantly, limits exposure of the new keys to an attacker via side-channels. The least significant 32 bits of the IV are used to set the block size (i.e. how much data is encrypted with a single key). The key rolling process is demonstrated in Figure 8.4.

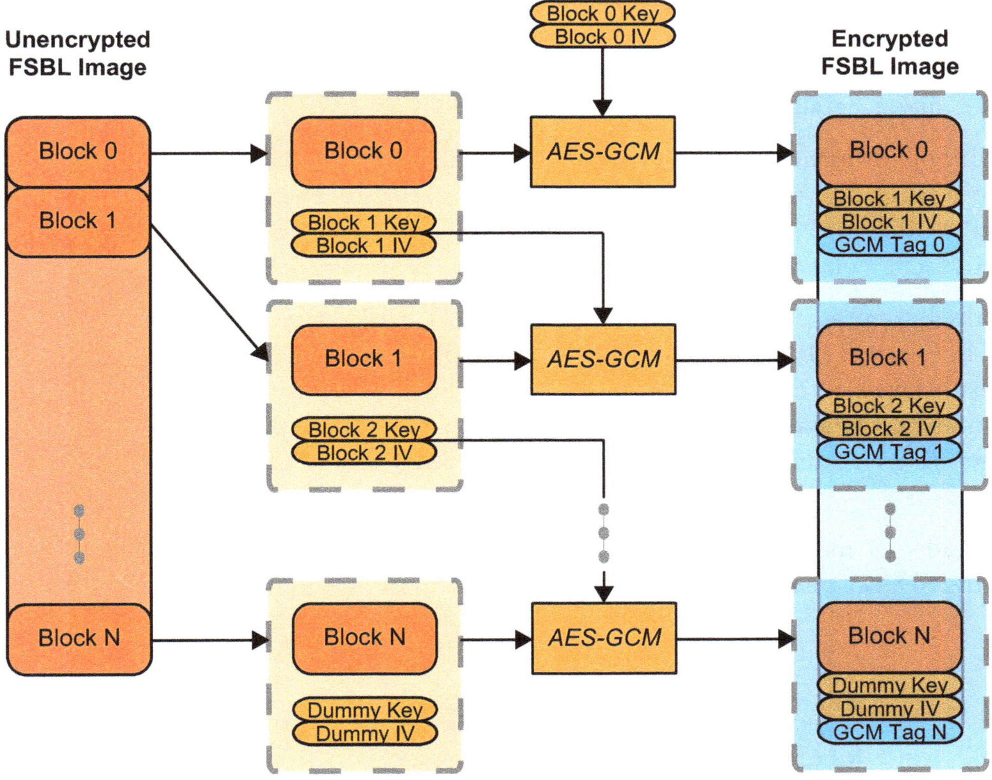

Figure 8.4: Example of key rolling with a FSBL image

The key rolling technique is implemented in the CSU ROM, not the hardware unit directly. This means it is available for secure boot, but not at run-time from a user application — although the same technique can easily be reproduced. For more on its use during secure boot, see See "System Booting" on page 353.

In summary, the AES core implements AES-256 for ensuring data confidentiality, using the Galois/ Counter Mode for added authentication and protection against correlation of ciphertext, as well as a key rolling technique available via the CSU ROM for combating SCA.

See Chapter 12 of the Technical Reference Manual [6] for details of programming the AES core from your own application at run-time.

SHA-3

The SHA-3/384 unit is a hardware accelerator for the Secure Hash Algorithm 3, providing a 384-bit output (or 'digest'). Much like AES, this algorithm was selected by NIST [7], and they actually have a common co-designer, Joan Daemen! The purpose of SHA-3 is to provide integrity. The goal of cryptographic hashing is to provide a one-way function which is not feasibly reversible, outputting a fixed length digest. It is the one-way property which is exploited for integrity checking. When a SHA-3 digest is distributed alongside a (possibly corrupted or modified) file, a device can locally calculate the SHA-3 digest of the file and compare it to the trusted SHA-3 digest. It is extremely difficult for an attacker to generate a different file that produces the same SHA-3 digest, because the function is one-way.

For Zynq MPSoC in particular, the SHA-3 unit is used for integrity checks of both the CSU and PMU ROMs at boot-time. A user application may also use it for integrity checking. Above and beyond integrity checking, it is also often used in conjunction with the RSA unit to generate signatures — providing authentication. The RSA section on page 189 discusses these signatures in more detail.

SHA-3 works by repeatedly passing a set of bytes, called the state, through a permutation function (denoted as '*f*' in Figure 8.5). Before each permutation, a block of input data is mixed with a subset of the state. The digest is taken from a subset of the final state. The majority of the mathematical finesse is hidden within the permutation function. It importantly provides diffusion such that a single bit-flip of the input is amplified to generate an entirely different digest. An overview of the SHA-3/384 algorithm is shown in Figure 8.5.

Again, the reader is directed to Chapter 12 of the Technical Reference Manual [6] for details of programming the SHA-3 core. Note that the input's length must be a multiple of 104 bytes — add padding when required.

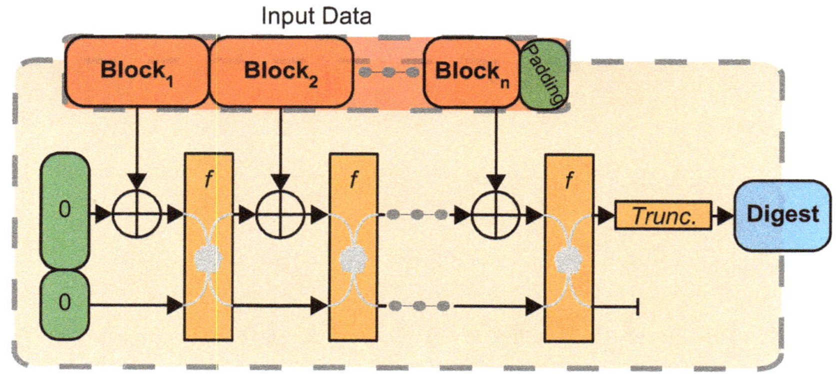

Figure 8.5: SHA-3/384 algorithm overview

RSA

The RSA unit provides hardware acceleration for the fundamental building block of RSA — an *asymmetric* encryption algorithm. As an asymmetric algorithm, RSA uses two different keys for encryption and decryption (one public key and one private key) instead of the shared secret key used for AES. This alleviates the challenge of securely sharing a secret key between two devices. RSA is quite computationally expensive, so it is commonly used to communicate encrypted keys for AES (or similar), before switching to that algorithm instead. In the context of Zynq MPSoC, we use RSA more often for checking signatures of files to ensure integrity and authenticate their origin. The signatures are SHA-3 hashes which are then passed through RSA with the author's private key. Anyone with the author's public key can not only verify the integrity of the file, but also be sure that it has not been modified by anyone other than the original author (as only they should know the author's private key!). Signature checking is used extensively in the Zynq MPSoC secure boot process to detect unauthorised firmware.

The asymmetry of the RSA algorithm relies completely on the idea that hardware can multiply two large prime numbers together, but it is *very* difficult to factor the result back into the two original primes. This assumption[1] is used when generating keys for RSA — where a lot of RSA's mathematical details lurk [9]. For the purposes of the present discussion, it is the encryption and decryption process that are more interesting.

For both encryption and decryption, RSA follows a similar mathematical structure:

$$y = x^e \pmod{n}$$

given an input x, a key exponent e, a modulus n, and an output y. This is called modular exponentiation. While it is simply described in mathematical notation, modular exponentiation can be very slow when implemented in software. Of course exponentiation can be implemented by repeated multiplication (and division for the modulo), but keep in mind that these words can theoretically be thousands of bits long!

The RSA core accelerates this modular exponentiation with support for:

- Modular exponentiation using a **Montgomery multiplier** [10][11]. This avoids the need for costly divisions by n and instead uses additions and division by powers of 2 (i.e. bit shifts).
- Optional calculation of prerequisites for Montgomery multiplication ($R^2 \bmod n$, where R is a constant based on n and the system word length). This is already pre-calculated by the BootGen software for signed firmware, so the developer does not need to think about it!
- Configurable 4096-bit and 2048-bit key support.

The RSA core is supported by a local RAM with 132 words, each 192-bits wide. The user has access to this RAM via a set of supporting registers. A 'done' signal can be exposed via a CSU interrupt or by simply polling the *RSA_CORE*'s *STATUS* register.

---

1. This assumption is being challenged as quantum computing peers over the horizon [8]

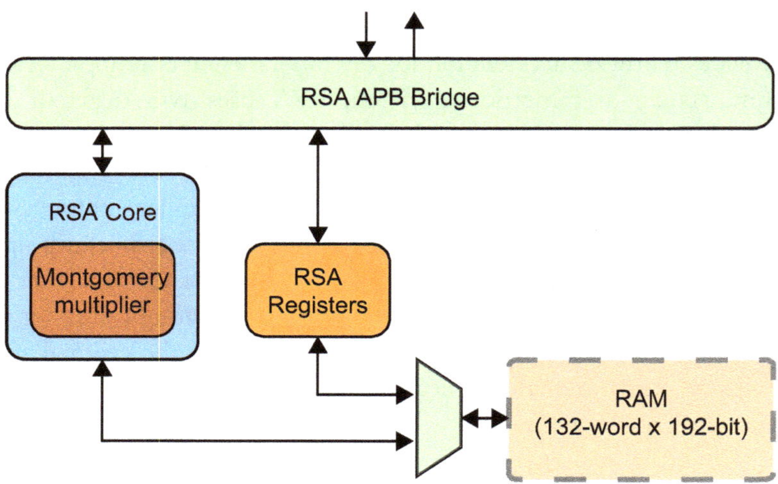

Figure 8.6: RSA Core Architecture overview

Further programming details can be found in Chapter 12 of the Technical Reference Manual [6] and by referring directly to the standalone XilSecure library [12].

8.1.3 Key Management

Storing keys internally is a vital aspect of cryptography for embedded designs. If an attacker has access to our secret keys, they can decrypt our sensitive files because the algorithms are publicly available (and should be!). We get a long way towards protecting these keys by storing them in a small on-chip memory. These memories can only be affordable by being very restrictive in size (otherwise we would just store the whole application on-chip!). We will cover options for long-term key storage, and options for using intermediate keys held in dedicated registers.

Boot Image Keys

The key management must be able to store keys which persist between power cycles, unlike the rest of the chip, which loses its configuration at power off. This is one of the key elements (pun intended) that makes secure boot possible.

The available options are similar to that of Zynq-7000, but with one new addition. These are:

- **eFUSE:** facilitates a permanent, one-time-programmable key located inside the chip. These can be configured independently for each chip by electronically 'blowing' bits of the eFUSE array.
- **Battery-Backed RAM (BBRAM):** the on-chip BBRAM can hold a 256 bit key for AES. Complementary to the eFUSE's permanence, the rewritable BBRAM provides the opportunity for key erasure (in reaction to tamper detection) and agility (update at run-time). It is supported by an external battery.

- **Boot Image:** new to the Zynq MPSoC, a key can actually be stored in the boot image itself. This is achieved securely by only storing a version of the key which has been obfuscated in some way — never the real key itself. This obfuscation can be achieved by encrypting the real key using AES with another key already on-chip, e.g. a key generated from the PUF, or the 'family key'[1].

For each of the three storage options available, there are different programming methods and states the key can be stored in. A key may be stored in plain text, or it can be further protected by first obfuscating it with the family key, or fully encrypting it with a PUF key-encryption key. These three states (plain text, obfuscated, and encrypted) are often referred to using the colours red, gray, and black, respectively. This stems from the 'RED/BLACK' concept used in security regulations to talk about how devices handling national security information (red) must be separated from devices that handle already encrypted data (black) [13]. We use 'gray' to show that an obfuscated key is somewhere in between red and black — not plain text, but not completely cryptographically sound either. A summary of the key storage possibilities are listed in Table 8.2.

Table 8.2: Boot Image Key Option

Features	BBRAM	eFUSE	Boot Image
Programming Method	Internal via software	Internal via software	Bootgen
	External via JTAG	External via JTAG	Bootgen + PUF registration software
		PUF registration software	
Programming Verification	CRC32	CRC32	N/A
Key State During Storage	Red	Red, black (or gray)	Black (or gray)

So now that we have an idea of the different available options for storing a boot image key, why would one of these be chosen over the others? The different options available provide unique properties which are useful for some applications and undesirable for others.

BBRAM can be useful for high security applications, where strict regulations require keys to be updated regularly, and keys must be zeroised if any tampering is detected. Of course, if we automatically erase the key upon detecting a tamper attempt (or just lose battery power), the device will "brick" until it is reprogrammed (unless backup boot images are available). Quite conversely, the eFUSE array provides a truly permanent storage which requires no external circuitry to preserve. The downside here is that we cannot update the key (it is one-time-programmable!). If a motivated attacker obtains this key, they will be able to clone your application, and maybe reverse engineer it. Storing a black (or gray) key in the boot image itself has some nice

1. The family key is a hard-coded AES key which is the **same for all Zynq MPSoC devices**. Its use for key obfuscation is intended only as an added barrier against IP theft by contract manufacturers — hence the term 'obfuscated' rather than 'encrypted'. See "Storing Keys in Obfuscated (Gray) Form" on page 194.

advantages too. This really limits the use of the on-chip key (in this case, the key from the PUF). We only use the on-chip key to decipher the black key, then switch to using the deciphered black key. This is useful as a further countermeasure against SCA which can recover keys through analysis after "listening in" on a device as it uses the same key many, many times [14].

Now, we take a closer look at how these key stores are implemented.

Battery-Backed RAM

In terms of hardware, the BBRAM is essentially just a 288-bit dedicated SRAM array, much like any other on-chip RAM. When available, the BBRAM is powered from the auxiliary supply (VCCAUX), but as soon as power is lost, this automatically switches over to the external battery (VCCBAT).

The BBRAM can be written to via the APB, as shown in Figure 8.7. This is accessible via software running in the RPU and APU, or over the PJTAG interface. There is no read-back path to the user at all — unlike previous Zynq devices. With previous Zynq devices, the read-back path (which can be disabled) is used to verify that the key has been correctly written, but in doing so, is one more potential risk that the developer needs to actively navigate when trying to properly deploy a secure design. Instead, in the Zynq MPSoC, the user writes the key to the BBRAM and also supplies a CRC32 value for that key. The BBRAM controller then verifies that this CRC value matches that of the written key, without ever passing the key itself back to the user. This default behaviour in the Zynq MPSoC means there is no need for a developer to actively disable a read-back feature (which is very easy to forget!).

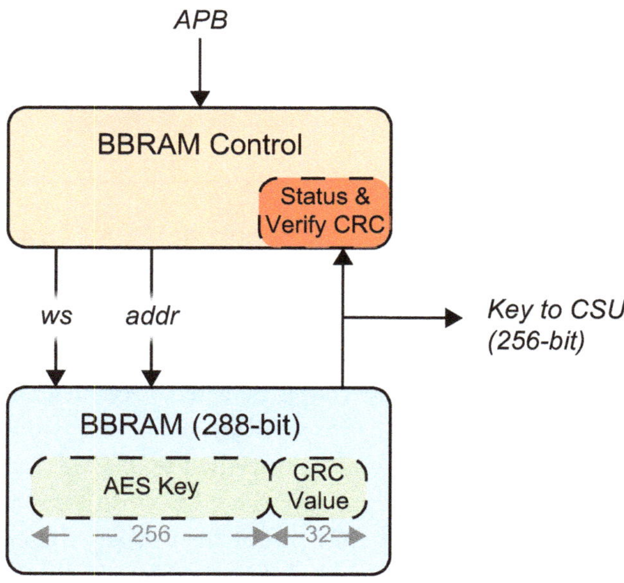

Figure 8.7: BBRAM Control Structure

By writing to the BBRAM, the user can change keys on-the-fly from internal software and also zeroise a previous key by writing all '0's.

In practice it is recommended to use Xilinx's standalone XilSKey library (packaged with the SDK and also available at [12]) to program the BBRAM. For more practical guidance, the user is referred to [17]. As always, a reference for the full set of registers can be accessed online at [15]. Note that there are useful flag bits in BBRAM_STATUS, such as a flag bit to check if the key has already been zeroised.

eFUSEs

Another key storage option is the array of eFUSEs. This is a collection of one-time-programmable electronically-controlled fuses. Each bit of this array consists of:

1. A fuse link — small metal wiring which can be programmed, or 'blown'.
2. A programming circuit which can temporarily short a fuse link to blow it.
3. Sensing circuitry to provide a logical '1' or '0' given the physical state of the fuse.

An example of fuse links in both the intact and blown states are shown in Figure 8.8. More about the physical eFUSE technology can be found at [16].

Figure 8.8: A visualisation of an intact and blown eFUSE link, based on [16]

The Zynq MPSoC allows for eFUSE storage of an AES key as well as hashes of 2 RSA public keys, 256-bits for user defined data, read/write disable flags, and flags to enforce certain behaviour at boot. An exhaustive list is found in the table entitled "Zynq UltraScale+ MPSoC Security eFUSEs" in [6].

As with the BBRAM, there is now no read-back path for the AES key. Verification of a successful AES key write is performed within the eFUSE controller using a user-supplied CRC value. Read-back is still available for the non-AES key eFUSEs.

Like zeroising the BBRAM, in theory the eFUSE AES key *could* be set to all ones in an attempt to destroy the key in response to a tamper event. This behaviour is not always good practice however. Although this would preclude an attacker's attempts to read the key using imaging similar to that shown in Figure 8.8, the act of blowing all of the intact fuses can give away information about which bits in the key are zeros! Such a 'simple power analysis' attack is particularly realistic because each bit of the key is programmed sequentially — not simultaneously.

The eFUSEs used for storing data (AES key, user data, and RSA public key hashes) have extra flag eFUSE bits for write disables. This prevents the data from being maliciously altered by blowing extra fuses (e.g. setting a trusted key to all '1's). The AES key also has a read disable flag which disables the CRC verification, while the RSA public key hashes have an invalidate flag.

Again, this is only useful if we know how to program it! Working with the eFUSE array is a little more complex than the BBRAM, so use of the Xilinx library is highly recommended. Again, more practical guidance can be found in [17].

Storing Keys in Obfuscated (Gray) Form

Storing keys in gray form is a technique introduced to try to help protect IP in commercial production situations (i.e. minimising the trust required of a contract manufacturer) [18]. Usually to program devices on a large scale, the encrypted design and a plain (red) key are both given to the manufacturer. There is nothing to prevent an employee decrypting the design with the supplied key to reverse engineer it, or use it for other purposes — such as IP theft!

Storing keys in a gray form allows a developer to pass on their encrypted design along with an *obfuscated* key. Ideally, this would mean that the manufacturer could not decrypt the design at all, alleviating concerns about IP theft at the factory. There are some limitations to this technique which we will return to after discussing how it works.

The obfuscation is performed by a trusted PC (not the device itself) using AES-GCM. The red key is used as the input data, and the 'family' key is used as the key. The family key is an AES key, known only by Xilinx, that is hard coded into each device, and is **the same for all Zynq MPSoC devices**. This provides some level of obfuscation which, importantly, can be deployed identically to each chip. (If it was unique for each manufactured device, the obfuscation would need to be performed by the manufacturer, giving them full trust regardless!). The process of generating a gray key is shown in Figure 8.9. Note that the user image is still encrypted with the red key and not the gray or family keys. As the gray key is obfuscated to some extent, it can be stored off-chip (in the boot image header) or on-chip (with eFUSES).

The secrecy of the family key is the weak point of this technique. Once an attacker knows the family key, it becomes trivial to 'unobfuscate' any gray key. The key is hard-coded into every Zynq MPSoC device — albeit not in a user-accessible way. It is this "break once, attack everywhere" potential that means we do not encourage using this technique alone when the gray key is stored in the boot image header. It does, however, provide a valuable, additional layer of protection when used in conjunction with the other techniques in this chapter — particularly if outsourcing to contract manufacturers. For example, an encrypted design can be passed off to a manufacturer with an obfuscated key, and upon first boot the device then configures the eFUSEs or BBRAM with a device-specific key for subsequent boots.

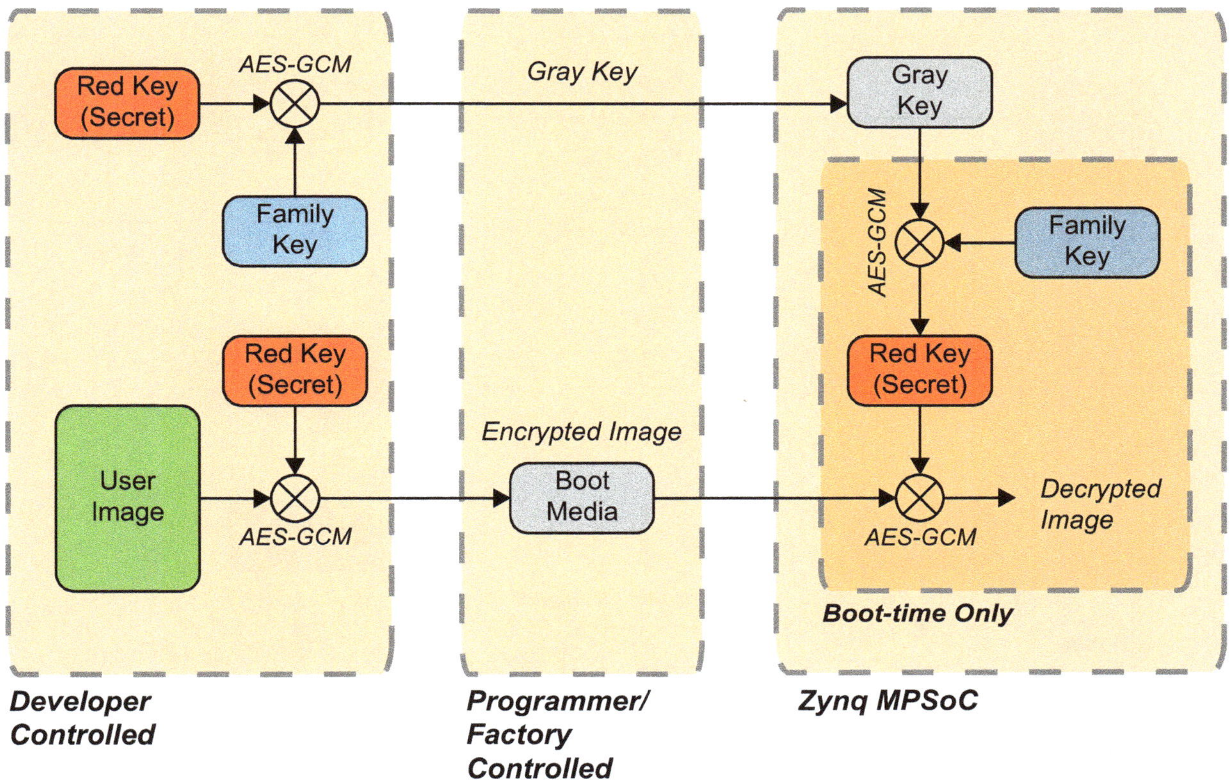

Figure 8.9: Device programming with gray key

Storing Keys in Encrypted (Black) Form

Storing keys in an encrypted form appears to be very similar to the obfuscation used for gray keys, but its advantages are a little different:

1. To facilitate safe key storage in an authenticated boot header
2. Or, to make hardware attacks more difficult through diversity of logic technology (the ring oscillator based PUF's combined with SRAM or eFUSE)

This is aimed at a very different environment than the gray keys (where the person programming a device is untrusted).

The core difference between generating gray and black keys is the use of a device-specific key during obfuscation. The family key has a fixed value for all Zynq MPSoC devices whereas a PUF can be used instead to generate a unique key for every single device off the production line. This key generated by the PUF is called the PUF Key-Encryption Key (PUF KEK). Introducing the PUF KEK to the programming process results in Figure 8.10. Note that the generation of the black key is now performed *inside* each device, as opposed to

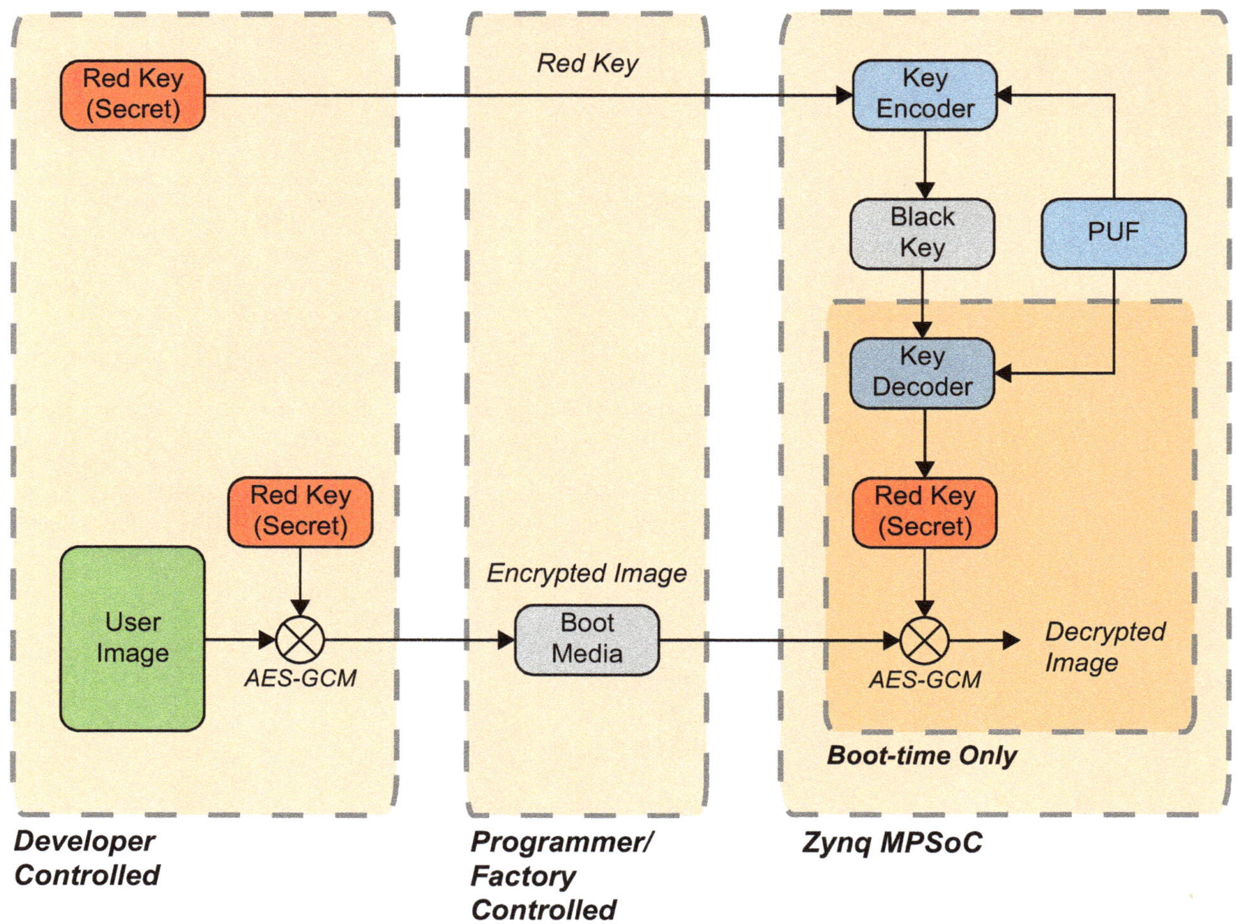

Figure 8.10: Device programming with black key

generation on a separate PC for gray keys. This is required because the PUF KEK is not only unique for each device, but also cannot be directly read out by a user. As with the gray key, the black key may be stored in eFUSE or boot image headers.

Until this point, it has been assumed that the PUF KEK is unpredictable (an attacker cannot easily guess it) and is reliable (the device can always regenerate the same PUF KEK). Now we go on to discuss how this is actually implemented.

As mentioned earlier, there is always some degree of randomness present in silicon manufacturing techniques. The PUF exploits this to provide an unpredictable, but reproducible, fingerprint for each device. In other words, the physical process of silicon manufacturing is our root-of-trust. So, how are the manufacturing variations exploited to provide a key?

The PUF in Zynq MPSoC is based on sets of ring oscillators. A ring oscillator is a very simple circuit (essentially a loop of logical NOT gates) which oscillate at a certain frequency. The oscillation frequency will change slightly with the variations from manufacturing. The PUF works simply by comparing the frequencies of pairs of ring oscillators — if the first is faster than the second, it contributes a '1' bit to the key, otherwise it uses a '0' bit. The exact pairing of oscillator circuits is defined by a user input. The basic structure is shown in Figure 8.11. Note that, instead of directly comparing the oscillation frequency, counters are employed to measure over many oscillations instead. This helps to amplify any variation between oscillators.

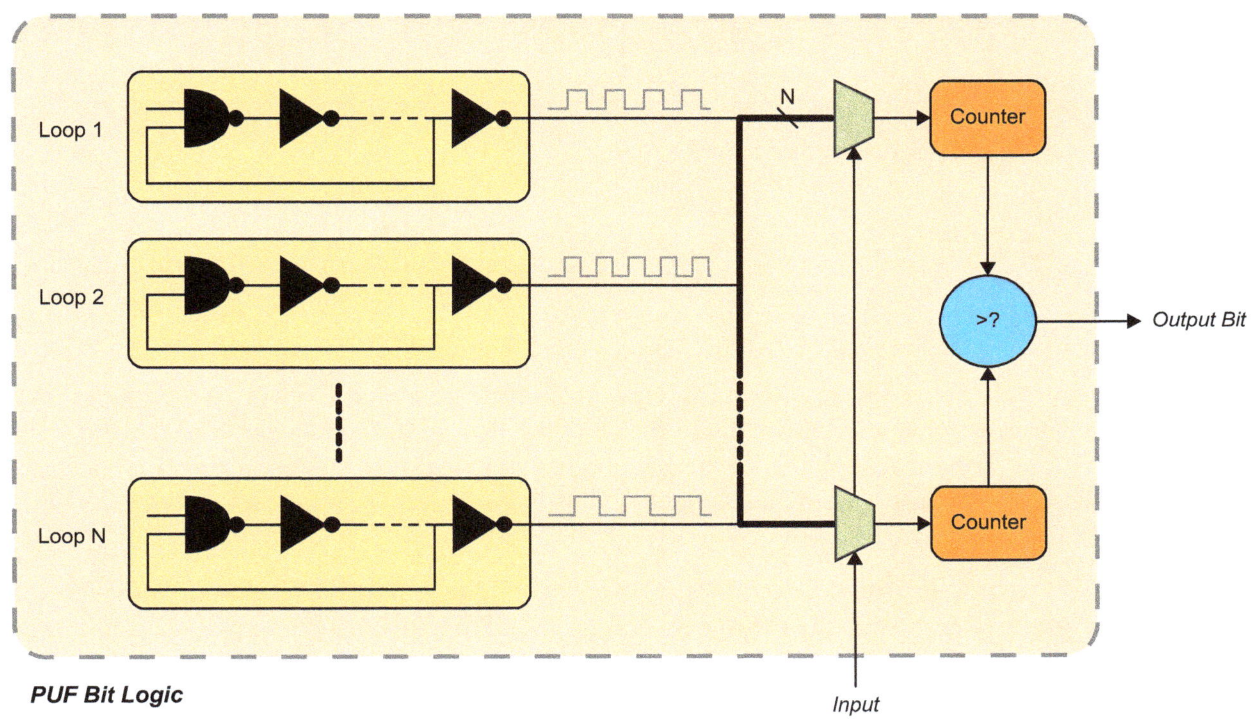

Figure 8.11: Structure of 1 bit of the ring oscillator PUF[19]

Beyond the logic shown in Figure 8.11, there is also some extra redundancy to ensure that the key is generated reliably. Many factors can affect each ring oscillator's frequency, including environmental aspects such as temperature and voltage. 4 Kb of helper data is stored to ensure the PUF KEK is recreated correctly over the entire guaranteed operating temperature/voltage for the whole life of the device. The helper data can be kept in either eFUSEs or the boot image header. This helper data is defined in Table 8.3.

Table 8.3: PUF helper data definition

Field	Size (Bits)	Description
Syndrome	4096	Bits to assist in error correction of the PUF signature
Aux	24	A Hamming code that allows the PUF to perform some error correction
Chash	32	A hash of the original PUF signature to verify correctness of regenerated signatures

The CSU exposes only 3 user commands for the PUF:

1. Registration
2. Re-registration
3. Reuse

The PUF KEK is first required when creating a black key from a given red key. At the time of this key loading, the **registration** command initialises the PUF for the first time, creating the KEK and the corresponding helper data. The resulting black key and helper data may either be stored in the eFUSE array or exported for use in a boot image header. The black key and helper data must be stored together (either both in eFUSEs or both in the boot image).

Re-registration is used for subsequent boots. The re-registration command takes previously generated helper data and uses it to regenerate the original KEK from the PUF. The CSU does this for any boot which demands use of a black key. The location of the helper data (eFUSE/boot image) is defined in the boot image header.

The **reuse** command is used for further encryption/decryption with a previously registered KEK (only valid for eFUSE helper data). This allows a user application to use the PUF KEK at run-time if the helper data is stored in eFUSEs. Device-specific keys are often desirable for use within the user application. For example, it can be used to encrypt other user-level keys for off-chip storage.

From a practical perspective, most of this interfacing will be performed by the XilSKey library. This provides helpers and examples for PUF KEK registration and use [12]. Xilinx tutorials also demonstrate the registration and use of the PUF KEK for working with black keys during boot.

Key Registers

In conjunction with the non-volatile key storage, there are a set of registers to aid key management. These include:

- Boot key: holds a decrypted black or gray key while it is in use.
- Operational key: holds a key which was set within an encrypted boot image header, decrypted using another device key. This (optional!) feature limits the use of the on-chip key — limiting the exposure of the key which is needed for side-channel attacks under normal operating conditions.
- Key update register: is a write-only register. This allows us to use the AES block as an accelerator during at run-time, without exposing boot-time keys to the application. This also facilitates the key rolling technique, discussed on page 187.

Key Selection

Throughout the key management section so far, we have discussed seven unique on-chip key sources: BBRAM, eFUSE, Family key, PUF KEK, boot key register, operational key register, and the key update register. So, what dictates which key is actually passed to the AES block?

From a user's perspective, the *aes_key_src* register is used to select either the key update register (for crypto from a user application at run-time) or a 'device key'. The selection of a particular device key is controlled strictly by the CSU ROM which parses the boot image header. Figure 8.12 shows this selection process and surrounding gating.

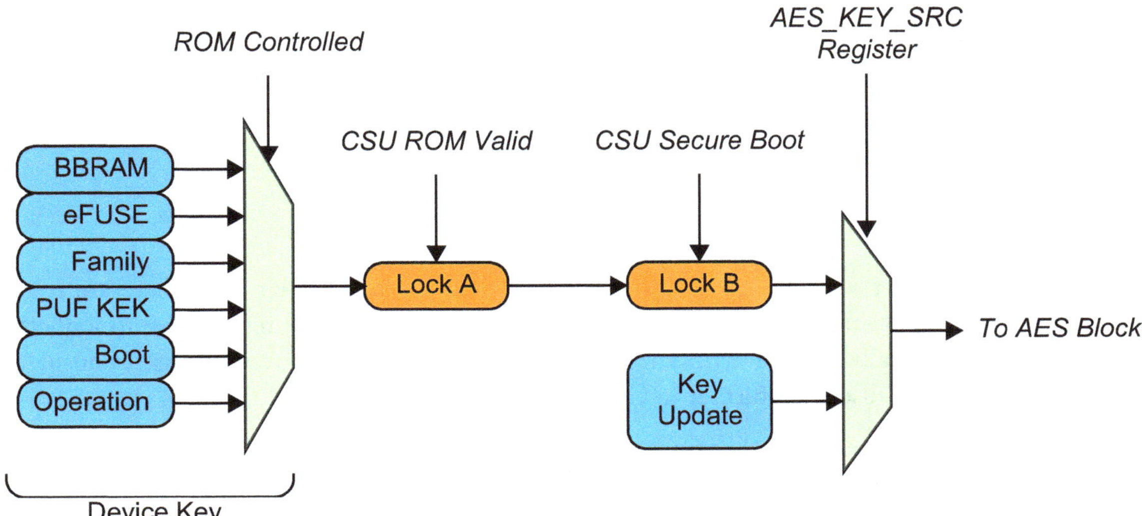

Figure 8.12: Key selection sequence

We see that the device key is carefully arbitrated by the CSU. The CSU ROM controls the key source selection (by parsing the boot image header), and it also controls the device key's exposure through two locks. The device key can only ever be used if the CSU ROM is valid (still passes the integrity check) and the boot header actually specifies that secure boot should be used. It is important to note that the device key can be used by the user application at run-time, but only if it is also used during boot. The user application cannot change the source for the device key — this is set solely by the CSU ROM. The user application may choose to use a new key by writing it to the key update register and setting the *aes_key_src* register accordingly (0x0 to use the key update register or 0x1 for the device key).

An Aside on RSA Key Management

The previous sections have largely concerned the storage of AES keys only. While RSA keys are also an important part of information assurance with Zynq MPSoC, the protection of configuration data is done with **public** RSA keys. The private counterpart should never leave the developer's computer or company. Because the device only holds public RSA key material, we do not need to take such intricate measures to protect it. We simply use eFUSE bits to store this in a non-volatile, on-chip fashion. The main caveat is that RSA keys can (and should!) be quite long compared to their symmetric counterparts and dedicating thousands of eFUSE bits to a single public RSA key is not practical. Instead, a shorter hash of the public RSA key is stored — meaning that fewer bits are required for storage on-chip, but that an attacker still cannot reasonably alter the key without detection.

Another point of note is the distinction between primary and secondary RSA keys (referred to more in Chapter 14). This separation is really only a policy to help with key management, such as key revocation [17]. Primary private keys are to be kept by the developer under stringent protections and as such will rarely be revoked — so the Zynq MPSoC only provides storage for two primary keys. These primary keys are only ever used to sign secondary keys. Secondary keys are then handed out to employees more freely and are used to sign the actual boot images. Revocation of secondary keys is then a much more likely requirement, so the Zynq MPSoC can work with up to 32 different secondary keys throughout its lifetime.

8.2 Anti-Tampering

While cryptography is a vital part of securing embedded systems and their data, there are other aspects which are also very important to consider. One such aspect is detecting and responding to any attempts an attacker makes to physically tamper with the device. Note that this also overlaps with considerations of functional safety, which is discussed further in Chapter 8.

First of all, we discuss the options Zynq MPSoC provides for detecting different 'tamper events'. With these events detected, we will cover some possibilities for reacting to tamper events. We conclude with some additional, passive precautions which can be employed in Zynq MPSoC designs to help preclude some of the common tampering scenarios.

8.2.1 Monitoring

During run-time, the CSU's main task is monitoring and responding to tamper events. It can be configured to monitor 13 unique events, listed fully in Table 8.4. These can be broadly grouped into monitoring of system voltage rails, internal temperatures, PL configuration corruption, activity on external pins, and an internal trigger register. Keep in mind that these events can be caused by either the chip's natural environment, or by a deliberate attack.

Table 8.4: Tamper monitoring sources

Register	Event Description	
csu_tamper_12	Analog Mixed Signal (AMS) voltage alarm for...	Gigabit serial Transceivers (GT)
csu_tamper_11		Processing system's GPIO (PS I/O) bank 3
csu_tamper_10		Processing system's GPIO (PS I/O) banks 0, 1, and 2
csu_tamper_9		Physical-layer DDR memory interface (VCC_PSINTF-P_DDR)
csu_tamper_8		Processing system's auxiliary supply voltage (VCCPAUX)
csu_tamper_7		Low power domain's internal logic supply voltage (VCCPINT_LPD)
csu_tamper_6		Full power domain's internal logic supply voltage (VCCPINT_FPD)
csu_tamper_5	AMS temperature alarm for...	APU/Full power domain
csu_tamper_4		RPU/Low power domain
csu_tamper_3	Corruption (SEU) detected in PL configuration — supported by the Soft Error Mitigation IP core [20][18], which can also perform correction	
csu_tamper_2	External JTAG activity. This detection is active unless the user explicitly disables and of the JTAG security gates in the CSU's *jtag_sec* register.	
csu_tamper_1	Event on an external pin, routed through MIO. Used for events external to the chip — such as monitoring a tamper evident case.	
csu_tamper_0	A dummy event triggered by writing to the *csu_tamper_trig* regitser. Often used while testing tamper response or for multi-stage responses	

The internal trigger caused by writing to the *csu_tamper_trig* register merits particular attention. This can be used in a few different ways:

1. **Software driven event**: Inducing a tamper response from software-based monitoring. For example, this could be used with an intrusion detection system running on Linux in the APU — such as Tripwire [21].

2. **Emulating a tamper event**: Thoroughly testing your tamper response functions is essential, however, it can be very difficult to safely induce tamper events relating to temperature and voltage. Using the *csu_tamper_trig* register, response testing can be carried out by just writing to a register rather than pushing a chip outside of its normal operating environment.

3. **Multi-staged response**: Allowing a response to a tamper event (i.e. a 'tamper penalty') to be built from both custom code and a hard-wired system response. See in "Multi-Stage Response" on page 202 for more on how this is achieved.

8.2.2 Response

The Zynq MPSoC system will continuously monitor for all tamper events, but it is up to the developer to decide exactly how the CSU should respond to each event. This is done primarily by setting flag bits within the corresponding *csu_tamper_X* registers (as listed in Table 8.4 on page 201). Each register has the following structure (don't worry, we will explain the new terms afterwards!):

Table 8.5: Tamper response bits for csu_tamper_X registers

Bit	Response	
4	Erase the AES key in BBRAM — as well as any other option listed below	
3	Secure lockdown and 3-state all I/O	*Mutually exclusive options!*
2	Secure lockdown	
1	System reset	
0	System interrupt (IRQ)	

Note that only one of the 4 responses in the lower part of the table may be taken, as each is mutually exclusive (while erasing BBRAM is not). If multiple bits are set, the most significant set bit is used. Also, note that these bits are set-only, i.e. they can not be cleared until a power-on reset is issued. This is done such that any errant software can only make the tamper penalty more severe, and never lessened.

Elaborating on the different options a little, a 'secure lockdown' [22] means that the on-chip RAM, all caches and the PL are cleared, and the processing systems enter a lockdown mode which is only left after a power-on reset. The option to '3-state all I/O' puts all I/O pins in a high-impedance state — effectively disconnecting them from the board. This forms the strictest tamper penalty, removing all external status signals.

Multi-Stage Response

A multi-stage response technique offers the ability to impose other tamper penalties, above and beyond the *csu_tamper_X* register bits we have already seen. This is achieved with a combination of the *csu_tamper_trig* register and the system interrupt tamper penalty.

We demonstrate how this can be implemented in Figure 8.13 using the example of responding to a FPD temperature alarm by clearing user data *and then* entering a secure lockdown.

Here we see that the *csu_tamper_trig* register is used to trigger the system tamper penalty, while the original tamper event only triggers a system interrupt — handled by user code. In our example, the user code performs tamper event logging or clearing of user data, but it could alternatively perform many different operations. We discuss some potential responses in the following section.

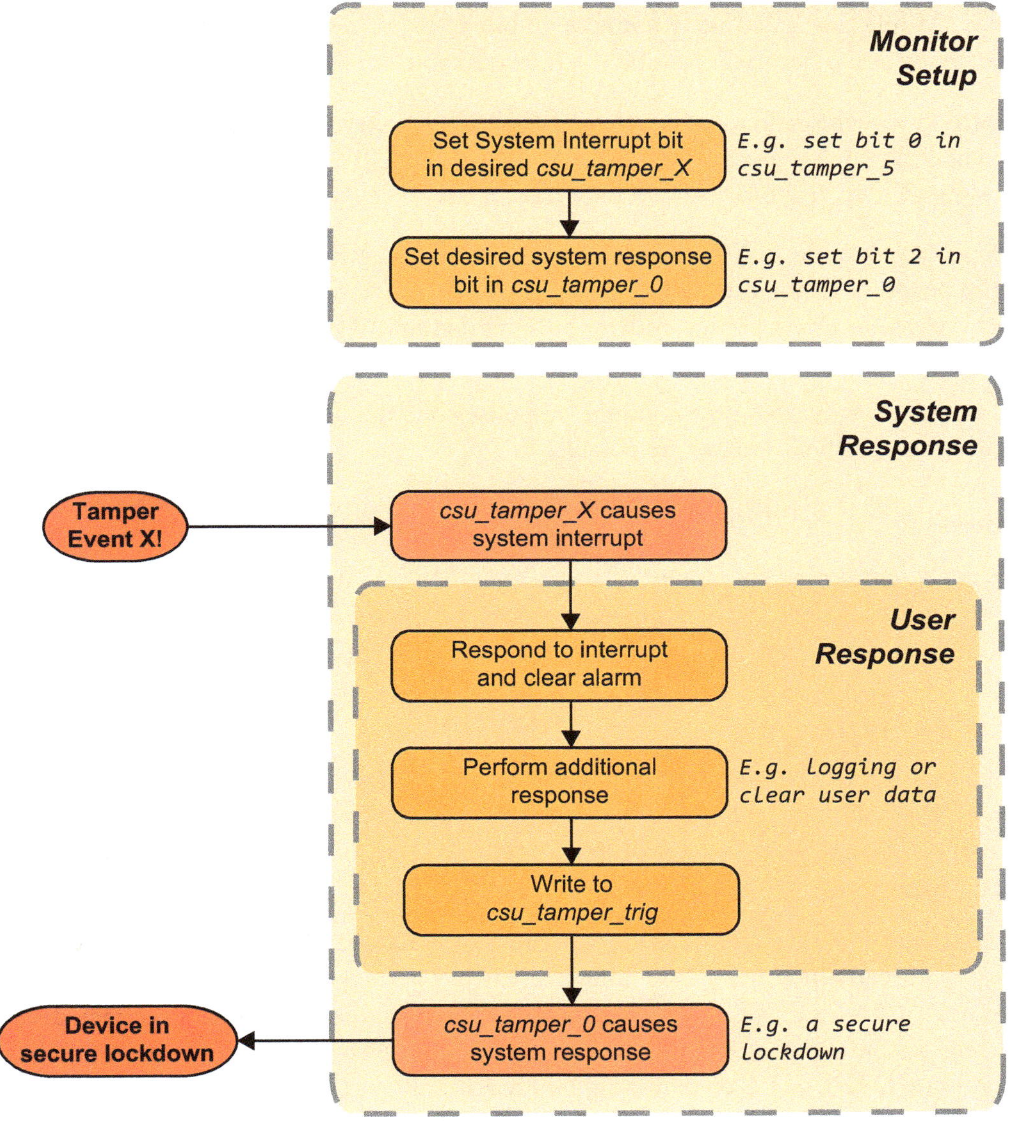

Figure 8.13: Flow of setting up and executing a multi-stage tamper response

Potential User Responses

User-defined tamper responses can vary with application, but we touch on a few common responses here.

- **Clearing sensitive user data**: For applications which process or store private user data, a sensible tamper response may be to clear any unencrypted copies from memory, and/or clear user keys.
- **Non-volatile logging**: when a device is intended to be tamper-evident, a good response technique can be logging of detected tamper events. To make this log persistent, events could be logged in the 256-bit user eFUSE. While this is definitely not enough bits to log something like a stack trace, it can be used to log that an event has occurred, possibly also storing more detail elsewhere.
- **RSA public key revocation**: wiping the BBRAM AES key is a standard system response, but revocation of RSA keys is just as important. While these eFUSE based keys cannot be wiped, there are corresponding eFUSE flag bits which revoke these keys.

Stringent tamper penalties do make things difficult for attackers, but a designer may also wish to consider the effects of false positives on the user. For example, when erasing or revoking keys as a tamper penalty, it is possible to brick a device — either permanently or until reprogramming. This might not be allowable in some safety-critical applications or other high-availability services.

We have presented a few of the most common responses but this list is by no means exhaustive! Many, many more application-specific responses are possible.

8.2.3 Precautions

Aside from actively monitoring for tamper events, a Zynq MPSoC design can adopt passive precautions to preclude some tampering efforts. These precautions generally require the developer to take some proactive action to set up, but it is not at all arduous!

Protecting the JTAG Interface

The JTAG interface is an early port-of-call for any hardware attacker. Remember, you likely used the JTAG interface for device programming, debugging, and analysis during development — and that can be done just as easily by an attacker! Disabling the interface ensures that no unwanted debug or programming features are exposed in a final design.

The JTAG interface is automatically disabled under some secure boot scenarios. However, for other scenarios (or just pure certainty) the JTAG interface can be permanently and explicitly disabled. This is governed by the *SEC_CTRL* eFUSE register's *JTAG_DIS* bit. The disabled JTAG interface does not allow any access to debug or configuration features such as the Zynq MPSoC device test access port (TAP), Arm core debug access port (DAP), or PL configuration. Only access to part type identification codes (IDCODE) remains, and all other commands perform nothing (a BYPASS).

Unique Identifiers

Unique Identifiers (UIs) can be used by a user application as an anti-cloning measure. Suppose that an attacker captures your design's plain, unencrypted bitstream from a single device. The attacker could potentially clone your design by simply putting the captured bitstream on their own Zynq MPSoC device, likely with the intent to sell on many cloned devices. However, if the design actively tries to verify a device's UI, it can detect when it is running on a cloned device (with an unexpected UI) and refuse to function.

Now, the secure boot flow described in Chapter 14 already precludes cloning attempts (as long as the key is kept secret) but the *additional* use of UIs provide an extra layer of security. An example use case of how to verify UIs — in a manner which scales for large production lines — is discussed in [18]. UIs can also be used to gate certain features per customer, depending on price or other business factors.

Now that we have discussed why UIs are useful, what hardware support for UIs does Zynq MPSoC provide? The two main features are:

1. **Device DNA**: is a device-specific 96-bit serial number set by Xilinx during manufacture. It is particularly suited for use in anti-cloning measures as it is unique to each device and also preprogrammed (i.e. easy!). Device DNA is implemented as a set of eFUSE bits.

2. **User eFUSE**: Some or all of the 256-bit user eFUSE bits can also be used for UI purposes. The difference from device DNA is that the user eFUSE value can be decided by the user — it is not prescribed by Xilinx. This allows the for 'unique' identifiers to be given to groups/classes of devices for business reasons, rather than having a truly unique ID. User eFUSE can also be used for high-assurance application which may not fully trust the uniqueness of prescribed device DNA values.

Disabling Status Signals

High-assurance applications may want to hide sensitive status signals from outside the device. For example, such signals may leak information which is useful to attackers during fault injection. Built-in status signals (not generated by user code) are of particular concern here. These include *PS_ERROR_OUT* and *PS_ERROR_STATUS*, which signal the presence of an error and a more specific system status, respectively. These signals can be disabled by user code (i.e. only after a successful boot) using the PMU's global registers — namely with *PMU_GLOBAL's ERROR_SIG_X* registers. Note again that this does not apply during boot time, as it is enforced by user code!

8.3 Security Through Isolation

Isolation can be an effective measure against common hardware attacks and erroneous or malicious software. The goal is to prevent any master within the system from accessing peripherals and memory regions which are not necessary for their operation. These masters can be physical (i.e. an IP core in the PL) or more conceptual (i.e. the separation of 'secure' and 'non-secure' software). This isolation helps to prevent attacks through software, Trojans in 3rd party IP cores, and common hardware-level tampering such as DMA attacks. Also, from a functional safety perspective, isolation can be employed to separate safety critical software and peripherals from the rest of the design — preventing the spread of faults due to erroneous software/IP cores. In this section we will touch on how the Armv8 architecture provides a foundation for isolation techniques, and then we go further, showing how these techniques can be extended to the whole Zynq MPSoC system — including IP cores in the PL and hardware peripherals such as UART controllers, GEMs, and internal memories.

8.3.1 Isolating Software with Virtualisation and Armv8

As software becomes an ever larger and more complex piece within the programmable SoC puzzle, it begins to make more sense to split software into multiple levels of trust. In a Zynq MPSoC system there may be many layers, ranging from a small, trusted world handling cryptography all the way to untrusted, third-party applications running on top of Linux. There are clear security benefits in limiting the access of the least trusted software, while allowing more privileges to fully trusted software.

Virtualisation is a powerful isolation technique for software running in the APU. As discussed further in Chapter 13, virtualisation allows multiple virtual 'guest' operating systems to share the same physical resources. The guests are orchestrated by a layer of trusted software called the hypervisor. The hypervisor acts as an arbiter, dividing resources (including memory, CPU time, and peripherals) between guest OSs as seen in Figure 8.14.

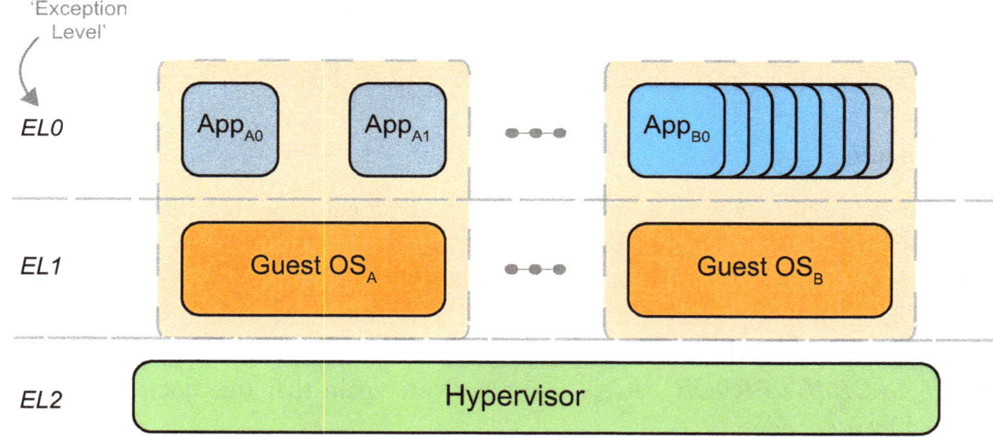

Figure 8.14: Example software stack using virtualisation

Virtualisation can be employed as a security mechanism to segment and isolate software worlds which should have different access restrictions. For example, there may be one guest for crypto operations which is allowed access to stored secret keys and reserves access to a UART, while two other guests are denied access to the keys and UART as they run user applications such as a web server or GUI.

Now we have an idea of why virtualisation should be used for security, let's touch on how this isolation is implemented. The APU's Armv8-A architecture provides hardware support for running many virtualised operating systems at near-native performance. These hardware features are namely:

1. Introduction of a new exception level for hypervisors (EL2) and supporting protections of the existing exception levels for guest operating systems (EL1) and their applications (EL0)

2. The System Memory Management Unit's (SMMU) ability to perform *two-stage* address translation

The use of exception levels (introduced in "Processor Modes" on page 131) allows guest operating systems in EL1 to run as if they have full control while disallowed behaviour (writing to certain registers, memory-management operations, etc.) is caught and 'trapped' at a hardware level. A trapped operation then signals an exception to EL2, returning execution to the hypervisor to decide how best to continue. These hardware-enforced traps go a long way towards isolating all guests from the hypervisor, but next we consider memory management techniques which focus more on isolating the guests from each other.

The SMMU (or more specifically, the Arm CoreLink MMU-500 [23]) provides two important features. The first is two-stage address translation, which we cover here, and the second is extra system protections, covered later in "PS Address Space Protection with SMMU" on page 208. The two-stage translation allows guest operating systems to make use of virtual memory with hardware-accelerated translation (stage 1), while the hypervisor retains complete control of how each guest's address space is mapped to the physical device (stage 2). This makes it simple for the hypervisor to isolate the address spaces of different guests, and control which guests have access to each memory-mapped peripheral. The two-stage address translation is shown in Figure 8.15.

In summary, the use of virtualisation on the APU can help separate user software into isolated guests. The hypervisor governs the execution of each guest and their address maps with the assistance of the Cortex-A53's Armv8 architecture and the SMMU. The reader is referred back to "System Virtualisation" on page 155 for details of the hardware implementation of these features.

While virtualisation can definitely be a useful layer of isolation in a Zynq MPSoC system, it is not the full picture. **The hypervisor only enforces isolation within the APU itself**. Different guests running on the Cortex-A53s are stopped from tampering with each other and can be prevented from accessing other peripherals in the address space. However, there are many other bus masters within a typical Zynq MPSoC system which still have free reign of the same address space! Even if untrusted software is isolated, another bus master could dump sensitive areas in memory or completely corrupt our software! These bus masters include other processors, such as the Cortex-R5s in the RPU, IP cores programmed in the PL, DMA-capable FPD peripherals such the GPU, SATA, DP, and PCIe interfaces, as well as some of the Input Output Peripherals (IOP) such as SPI and GEM interfaces. To fully implement isolation, we begin to consider the complete Zynq MPSoC system (not just software!) in the next section.

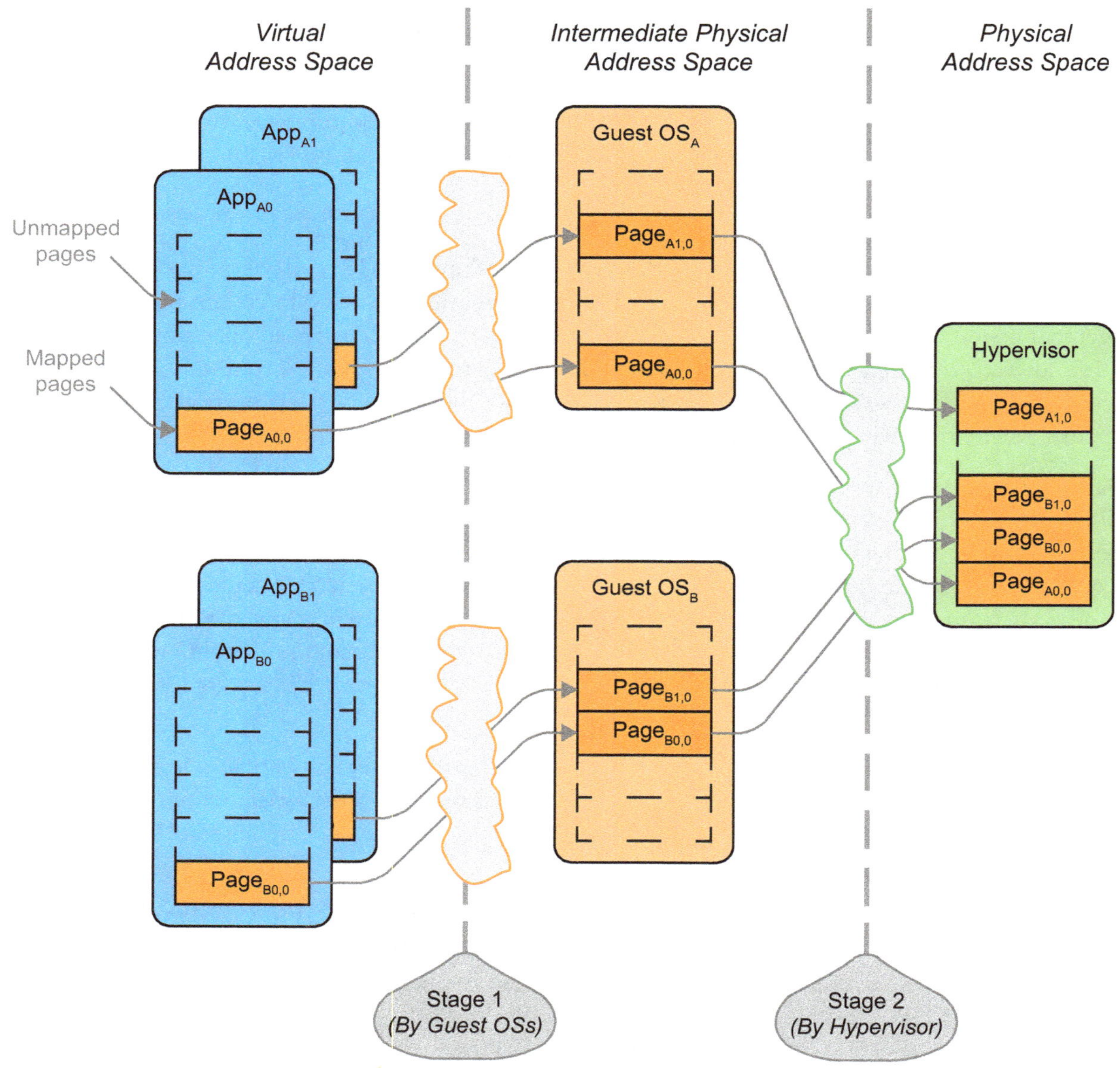

Figure 8.15: Use of two-stage address translation in virtualisation

PS Address Space Protection with SMMU

As well as the SMMU's role in supporting virtualisation using its two-stage address translation, it also has a role in system-wide isolation. The simplified Zynq MPSoC diagram provided in Figure 8.16 shows that all bus masters which connect to the CCI first come through the SMMU. This allows the SMMU to also provide some protection against other masters in the system corrupting system memory, even without the dedicated protection units discussed later. By restricting DMA-capable I/O devices to a predefined physical address space, common DMA attacks against the PS address space are prevented without the advanced configuration required of other protection units.

8.3.2 Extending Isolation to a Complete System

For isolation techniques to be effective when applied to a complex SoC system, the software, PL, and hardware peripherals must share a common scheme to separate the 'secure' elements from 'non-secure' elements. This can be achieved with Arm's TrustZone technology [24] which is supported and extended by custom Xilinx units such as the XMPU and XPPU in the Zynq MPSoC. Remember that the whole system is orchestrated through AXI buses. For an isolated system, a TrustZone status flag is propagated through the AXI buses and units such as the SMMU, XMPU, and XPPU use this flag to properly guard the endpoints.

We anticipate that manual isolation techniques will be implemented largely by experts as it is a quite advanced feature. Because of this, we cover only what these techniques can achieve and defer to the Technical Reference Manual [6] for full practical details.

Arm TrustZone

TrustZone technology is the foundation used to split the entire Zynq MPSoC system into two security states: 'secure' and 'non-secure'. This is a software-controlled but hardware-enforced system. It affects the whole system because a secure/non-secure flag is attached to AXI transactions — i.e. all communication between masters and slaves on the device. Note that these security states (secure or non-secure) are distinct from the exception levels used for virtualisation. In fact, they can work in tandem!

Looking at TrustZone from the perspective of the Cortex-A53s, all code is executed in either the secure state (the 'secure world' which is entered on boot) or the non-secure state (the 'normal world'). Any AXI transaction generated on behalf of normal world software will have the non-secure flag set, while only secure world software can choose to generate either secure (or non-secure) transactions. The transition between the two worlds is tightly controlled by hardware, as introduced in "Armv8 Security Model" on page 133. Each world has a unique set of system registers and can have unique address spaces. TrustZone is often used to create a Trusted Execution Environment (TEE) where a small, authenticated Secure OS is run in the secure world, and a more extensive Rich OS is run in the normal world which can request Secure OS actions via an exposed API. Example Arm code is available online, as well as commercial TEE solutions [25].

Now looking beyond the APU, TrustZone states are carried system-wide by using a non-secure flag which is mapped to the AxPROT [1] attribute of the AXI buses. We have already looked at how the security state is handled by the TrustZone-aware Cortex-A53s, but how is this scheme extended to other masters and slaves that are not TrustZone-aware? The security state of system memories and peripherals (slaves) are generally configured by the XMPU and XPPU blocks described on page 210. The simplified placement of these blocks is shown in Figure 8.16. There are, however, some system registers which are only accessible from a secure state — including the system-level configuration registers (SLCRs), debugging interfaces, and PMU/CSU registers. The security state set for other bus masters deserves a little more discussion.

Firstly, the RPU's Arm Cortex-R5s do not implement TrustZone technology, so we use bits in an SLCR (more precisely the *slcr_rpu* register within *LPD_SLCR_SECURE*) to set the security state for each Cortex-R5 core. This register is only accessible by another secure master and can be further protected by a write protection lock, disabling all write access until the next power-on reset. This "always on" or "always off" approach to security states precludes the idea of dividing software worlds, as is possible in the APU, but this is less of an issue for the smaller, real-time applications that are generally suited to the RPU. The TrustZone configuration for the Cortex-R5s is an example of a programmable master — where a register selects between secure and non-secure. This technique is used for most other masters too, including the LPD and FPD DMAs, as well as USB, SATA, PCIe and DisplayPort interfaces [15]. The main exception to this is the handling of PL-to-PS transactions. It is assumed that an encrypted bitstream will be designed as part of the secure world. Because of this, the PL is in control of the security state of PL-to-PS AXI interfaces. If this assumption sounds unrealistic (if you are concerned about IP Trojans, etc.) then note that the following sections covering XMPU and XPPU allow protection of slaves on a per-master basis, allowing individual access control for each AXI-connected IP core in the PL.

In short, TrustZone technology allows a complete Zynq MPSoC design to be split into exactly two different security worlds. We have seen how this applies to software running in the APU, how the security separation is carried over into the rest of the design via AXI buses, and how non-TrustZone-aware slaves/masters are integrated properly.

Memory Protection with XMPU

The Xilinx memory protection unit (XMPU) is a TrustZone-enforcing, region-based memory protection unit. Its purpose is to isolate regions of memory peripherals not only by TrustZone security states, but also on a per-master basis. A total of 8 XMPU blocks protect the DDR controller (using 6 blocks), the OCM, and FPD peripherals (including GPU, SATA, and PCIe). The placement of these blocks can be seen in Figure 8.16. Each block consists of:

- A slave AXI port to receive transactions
- A master AXI port to forward transactions with a redirect (or 'poison') attribute for disallowed transactions
- A slave APB port for accessing control registers (including a lock register)
- Interrupts to signal access violations back to the rest of the system

A XMPU block controls 16 regions, each defined by a start and end address. These regions are aligned to 1 MB boundaries (a.k.a 'apertures') for the DDRAM controller and 4 KB for the OCM and FPD peripherals. These regions may overlap and the region with the highest ID number will take highest priority. If a request matches none of the enabled regions, the XMPU will continue with a default action which is configurable through the control registers (either 'allow' or 'poison').

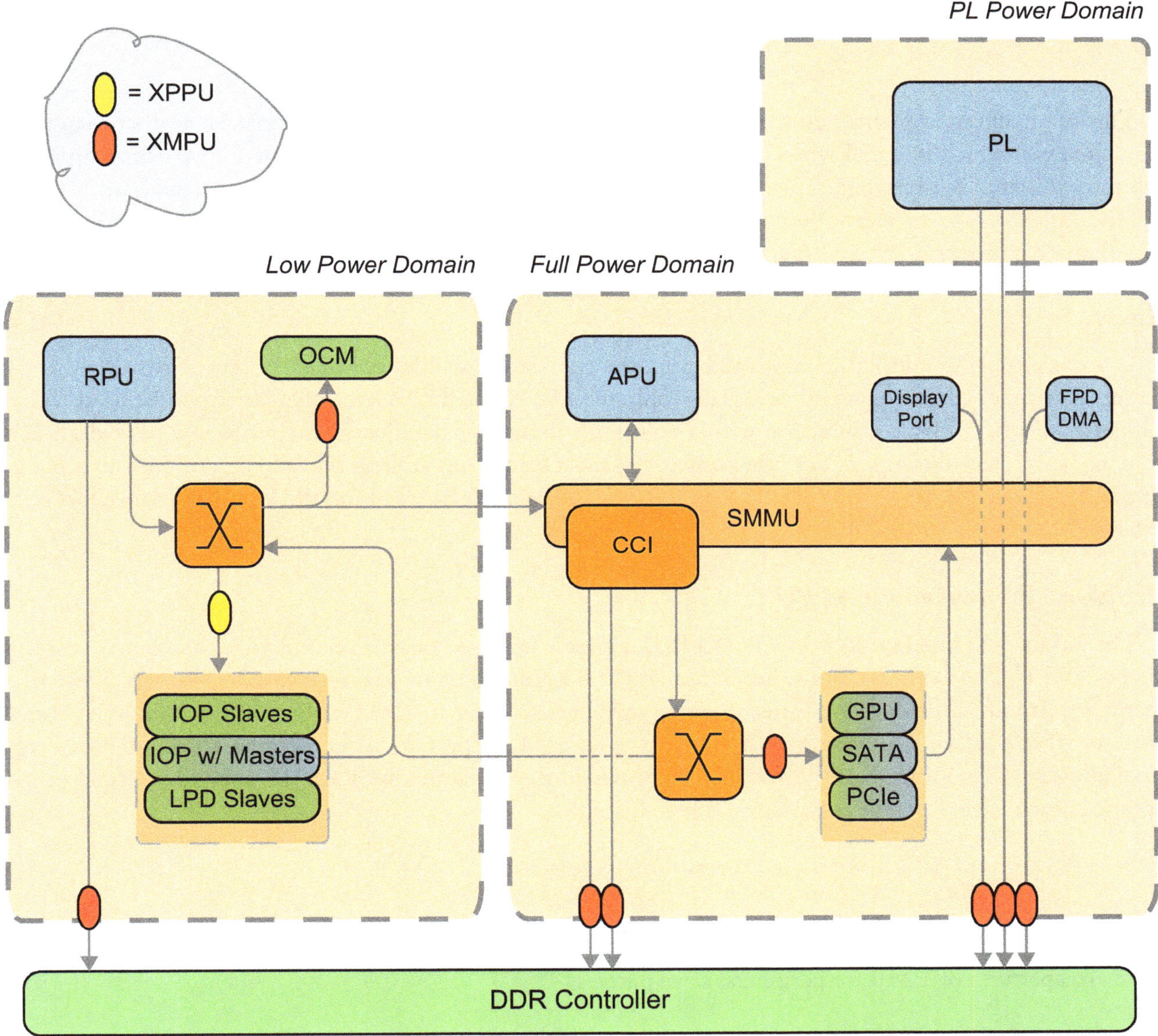

Figure 8.16: Simplified placement of memory protection units

In order to validate an incoming transaction, two things are checked: is the master of this transaction allowed in this region, and is the transaction in the correct TrustZone security state? Validation of the incoming master ID is performed with two values which are programmed into the XMPU control registers — a bit mask (*[MASK]*) and a resultant value (*[OK_ID]*). Using a bit mask and value allows us to select which bits in the master ID are important, and also define what these bits should be. The XMPU then performs the following check for a given incoming transaction's master ID (*IN_ID*):

$$[OK\_ID] \,\&\, [MASK] == IN\_ID \,\&\, [MASK] \quad (8.1)$$

This often allows us to craft bit masks and values to match a subset of available masters (rather than just a single master) which is useful when sharing resources. Note that this does not allow for matching upon any arbitrary subset of masters, but in practice this is not a big restriction. For example, there are six XMPU blocks for the DDR controller where only a few unique masters can reach each block. Therefore we only have to differentiate between a few masters at once and not the full set. TrustZone security states are also enforced through a region's control registers — i.e. non-secure transactions are disallowed from accessing secure regions.

If a transaction should fail to pass validation, the transaction will be poisoned in one of two ways. If the target slave supports poison attribute signals (only the OCM and DDR controller) they will be used. When supported slaves receive a transaction with a poison attribute, it is handled with a write-ignore/read-all-zero response to protect the region. For other slaves, the most significant address bits will be poisoned to a predefined value, in effect redirecting the transaction to a poison sink device. This results in either a data abort or an interrupt to the processor.

Peripheral Protection with XPPU

The Xilinx peripheral protection unit (XPPU) is a look-up based protection unit with the same motivation as the XMPU. The main difference lies in how it has been designed to accommodate peripherals, rather than memories. It must protect many, varied on-chip peripherals, rather than the fairly uniform memory apertures used by the XMPU. The single XPPU instance is used to protect LPD peripherals, control registers (including itself!), and message buffers used for inter-processor communication. This requires a total of 400 apertures each with different permissions. These include:

- 128 apertures of 32 B for inter-processor message buffers
- 256 apertures of 64 KB for peripheral slave ports
- 16 apertures of 1 MB for peripheral slave ports
- 1 aperture of 512 MB for the QSPI memory controller

The XPPU's master ID verification works with two structures:

1. A central list of 20 master ID mask/value pairs, similar to those used in the XMPU. The first 8 are predefined and the remaining 12 are programmable.
2. A 20-bit register for every aperture which allows or disallows each master specified in the central list independently.

Unlike the XMPU, this allows for letting any combination of masters share access to a single aperture — which is essential for slaves such as inter-processor buffers. If a transaction is invalid, the address bits can be poisoned, which redirects the transaction to a predefined dummy sink device.

8.3.3 Isolation Summary

Isolation can be an extremely useful precautionary technique for building secure systems. If one element in the design is attacked or behaves erroneously (be it a hardware block or a software task) the effects can be contained by isolation. This not only improves system-wide security by preventing an attacker from gaining further control of a device, but it can also be useful from a functional safety perspective to prevent the spread of errors and faults.

At first we introduced virtualisation on the APU to provide isolation between many different software stacks. This allows for isolation of an arbitrary number of software tasks but it is only enforced within the APU itself, leaving other bus masters open to tamper with the processing system. The protection features of the SMMU can help here by restricting which physical addresses other masters can access. We continued by looking at a system-wide isolation system — although virtualisation remains a *very* important tool for isolation in Zynq MPSoC.

Arm TrustZone is the foundation of this system-wide isolation and, unlike virtualisation, it takes a binary approach. Everything falls into either a 'secure' or 'non-secure' domain. We walked through how this affects software running on the APU and then how the TrustZone security states are propagated through the rest of the system using AXI bus attributes. We then rounded off with a brief discussion of two protection units and how they relate to implementing isolation on bus slaves. The XMPU and XPPU provide TrustZone-awareness to system slaves (memories and peripherals respectively). They also provide finer grain permissions, allowing for per-master access control.

This whistle-stop tour of isolation features has covered what can be achieved with the Zynq MPSoC in terms of using isolation to preclude certain attacks and prevent the spread of faults. Practical implementation has been left outside of the scope of this chapter since, as it stands, properly and fully implementing these techniques demands a thorough understanding of the interconnects and architecture of the full Zynq MPSoC system. For those seeking to implement a system using these features, [25] is a very good place to start.

8.4 Chapter Summary

Security within modern SoC designs has undoubtedly become an important and nuanced topic. This chapter has offered a guide through some of the most important security features that can be employed on the Zynq MPSoC from 3 perspectives:

1. Trying to keep our system configuration and user data safely away from prying eyes and malicious modification (i.e. information assurance)

2. Trying to keep our system performing safely and its secrets hidden even when an attacker is physically tampering with the hardware (i.e. anti-tampering)

3. Trying to mitigate any eventual attack or fault by containing it in a single part of the system using isolation techniques.

We appreciate that this can be a daunting landscape to navigate but it really can be an important one to consider, especially in commercial and high assurance contexts. The real need for these features has been imparted here, through our inclusion of background information and reference to common, real attacks. This fundamental appreciation should provide the motivation needed for the reader to identify any security measures relevant to them and know where to look for further guidance!

8.5 References

Note: All online sources last accessed March 2019.

Some of the Xilinx documentation referred to below has version-specific URLs. If you are working in a newer version of the tools, check for updates on the Xilinx website, or try adjusting the link according to your version.

[1] S. M. Trimberger, and J. J. Moore, "FPGA Security: Motivations, Features, and Applications", *Proceedings of the IEEE*, Vol. 102, No. 8, August 2014.
DOI: 10.1109/JPROC.2014.2331672

[2] M. Joye, A .K. Lenstra, J. Quisquater, "Chinese Remaindering Based Cryptosystems in the Presence of Faults", *Journal of Cryptology* 12(4):241–245, 1999.
DOI: 10.1007/s001459900055

[3] N. Timmers, A. Spruyt, and M. Witteman, "Controlling PC on ARM using Fault Injection", Riscure, 2016.
Available: https://www.riscure.com/documents/controlling_pc_on_arm_using_fi_fdtc_2016.pdf?1470724913

[4] H. Bar-El, H. Choukri, D. Naccache, M. Tunstall, and C. Whelan, "The Sorcerer's Apprentice Guide to Fault Attacks", 2004.
Available: https://eprint.iacr.org/2004/100.pdf

[5] M. Dworkin, "Recommendation for Block Cipher Modes of Operation: Galois/Counter Mode (GCM) and GMAC", National Institute of Standards and Technology, November 2007.
DOI: 10.6028/NIST.SP.800-38D

[6] Xilinx, Inc., "Zynq UltraScale+ Device Technical Reference Manual", UG1085, v1.9, January 2019.
Available: http://www.xilinx.com/support/documentation/user_guides/ug1085-zynq-ultrascale-trm.pdf

[7] M. Dworkin, "SHA-3 Standard: Permutation-Based Hash and Extendable-Output Functions", National Institute of Standards and Technology, August 2015.
Available: 10.6028/NIST.FIPS.202

[8] P. W. Shor, "Polynomial-Time Algorithms for Prime Factorization and Discrete Logarithms on a Quantum Computer", SIAM Journal on Computing, 1997, Vol. 26, No. 5, pp. 1484-1509.
DOI: 10.1137/S0097539795293172

[9] B. Kaliski and J. Staddon, "PKCS #1: RSA Cryptography Standard" (RFC 2437), RSA Laboratories, Version 2.0, September 1998.
DOI: 10.17487/RFC2437

[10] P. L. Montgomery.: "Modular Multiplication without Trial Division", Math. Computation, Vol. 44, pp. 519-521, 1985.
Available: http://www.ams.org/journals/mcom/1985-44-170/S0025-5718-1985-0777282-X/S0025-5718-1985-0777282-X.pdf

[11] Xilinx, Inc., "Zynq UltraScale+ MPSoC Software Developer Guide", UG1137, v8.0, June 2018.
Available: http://www.xilinx.com/support/documentation/user_guides/ug1137-zynq-ultrascale-mpsoc-swdev.pdf

[12] Xilinx Inc., "GitHub Repo for Xilinx embeddedsw".
Available: https://github.com/Xilinx/embeddedsw

[13] Committee on National Security Systems, "CNSS 4009 — Glossary", April 2015.
Available: https://www.cnss.gov/CNSS/issuances/Instructions.cfm

[14] A. Moradi, and T. Schneider, "Improved Side-Channel Analysis Attacks on Xilinx Bitstream Encryption of 5, 6, and 7 Series", Ruhr University Bochum, 2016.
Available: https://eprint.iacr.org/2016/249.pdf

[15] Xilinx Inc., "Zynq UltraScale+ MPSoC Register Reference", UG1087, v1.7, February 2019.
Available: https://www.xilinx.com/html_docs/registers/ug1087/ug1087-zynq-ultrascale-registers.html

[16] N. Robson et al., "Electrically Programmable Fuse (eFUSE): From Memory Redundancy to Autonomic Chips", *IEEE Custom Integrated Circuits Conference*, San Jose, CA, 2007, pp. 799-804.
DOI: 10.1109/CICC.2007.4405850

[17] Xilinx Inc., "Programming BBRAM and eFUSEs", XAPP1319, v1.0, July 2017.
Available: https://www.xilinx.com/support/documentation/application_notes/xapp1319-zynq-usp-prog-nvm.pdf

[18] Xilinx Inc., "Developing Tamper-Resistant Designs with Zynq UltraScale+ Devices", XAPP1323, v1.1, August, 2018.
Available: https://www.xilinx.com/support/documentation/application_notes/xapp1323-zynq-usp-tamper-resistant-designs.pdf

[19] G. E. Suh, S. Devadas, "Physical Unclonable Functions for Device Authentication and Secret Key Generation", *Proceedings of the 44th annual Design Automation Conference (DAC)*, San Diego, USA, June 2007, pp 9 - 14.
DOI: 10.1145/1278480.1278484

[20] Xilinx Inc., "Soft Error Mitigation Controller v4.1: LogiCORE IP Product Guide", PG036, April 2017.
Available: https://www.xilinx.com/support/documentation/ip_documentation/sem/v4_1/pg036_sem.pdf

[21] Tripwire, Inc., "Tripwire Homepage".
Available: https://www.tripwire.com/

[22] Xilinx Inc., "Zynq-7000 SoC: Technical Reference Manual", UG585, v1.12.2, July 2018.
Available: https://www.xilinx.com/support/documentation/user_guides/ug585-Zynq-7000-TRM.pdf

[23] Arm, Ltd., "ARM CoreLink MMU-500 System Memory Management Unit Technical Reference Manual", vr0p0, 2013.
Available: http://infocenter.arm.com/help/topic/com.arm.doc.ddi0517a/DDI0517A_corelink_mmu_500_r0p0_trm.pdf

[24] Arm, Ltd., "SoC and CPU System-Wide Approach to Security".
Available: https://www.arm.com/products/security-on-arm/trustzone

[25] Arm, Ltd., "Development of TEE and Secure Monitor Code".
Available: https://www.arm.com/products/security-on-arm/trustzone/tee-and-smc

[26] Xilinx Inc., "Secure Solutions Overview", 2011.
Available: https://www.xilinx.com/publications/prod_mktg/secure-solutions-overview.pdf

[27] Xilinx Inc., "Zynq UltraScale+ MPSoC: Embedded Design Tutorial", UG1209, v2017.1, July 2017.
Available: https://www.xilinx.com/support/documentation/sw_manuals/xilinx2017_1/ug1209-embedded-design-tutorial.pdf

[28] Xilinx Inc., "Zynq UltraScale+ MPSoC Processing System v3.0: LogiCORE IP Product Guide", PG201, April 2017.
Available: https://www.xilinx.com/support/documentation/ip_documentation/zynq_ultra_ps_e/v3_0/pg201-zynq-ultrascale-plus-processing-system.pdf

Chapter 9

Safety Features & Techniques

Safety is an important consideration in a wide variety of embedded systems — from consumer electronics to spacecraft, and everything in between. But what do we mean by safety exactly? And what features and design methods are available for Zynq MPSoC to help ensure safe systems? We address these topics (and more) within this chapter.

9.1 An Introduction to Safety

It is useful to begin by defining what is meant by the term 'safety' in the context of embedded systems. A good general definition of safety is:

"... keeping people and the environment safe from harm".

For instance, the controller for a boiler system would not allow continued operation if a fault condition had been detected, and there was a risk of harmful gases escaping. A wind turbine will angle its blades when experiencing high winds, to prevent the risk of damage to the turbine itself, and potentially hazards to the surrounding environment.

In this section, we also take the opportunity to position the concept of safety in embedded systems, within other aspects of system dependability, and to provide some examples of embedded systems where safety considerations are particularly important.

9.1.1 Safety and Dependability

The taxonomy developed by Avizienis et al [1] identifies a number of attributes that contribute to system *dependability*. They define the term 'dependability' as "the ability to avoid service failures that are more frequent and more severe than is acceptable".

Safety is one attribute of *Dependability*, alongside the attributes of Availability, Reliability, Integrity, and Maintainability (formally defined in [1]). For instance, the system must be ready to be operated when needed (availability), and it must do so correctly (reliability). It must operate without causing harm to people or the environment (safety), and without unauthorised changes (integrity). Finally, the system must have the ability to undergo modification as required (maintainability). Meanwhile, the concept of *Security* is said to comprise a subset of the above attributes, together with one additional attribute — Confidentiality — as illustrated in Figure 9.1.

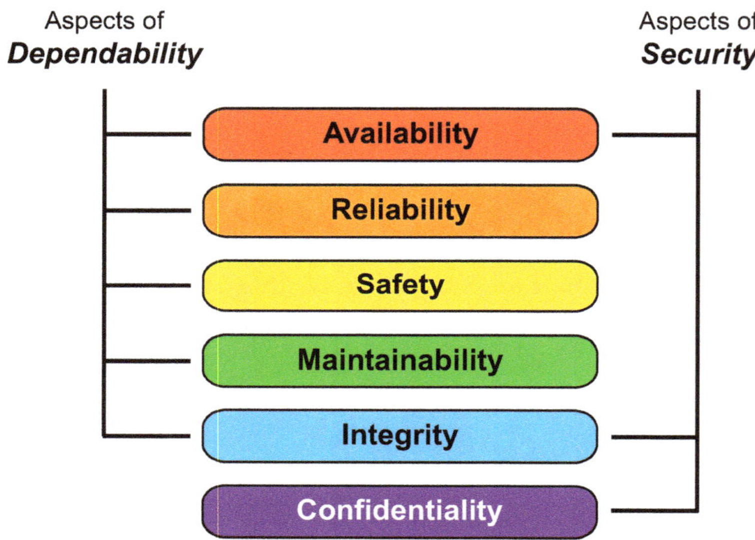

Figure 9.1: Taxonomy and attributes of Dependability and Security (adapted from [1])

The salient point here is that ensuring Safety is a distinct and important aspect of creating a dependable system. It is also worth noting that Security, while related, is a separate issue that demands specific attention.

In the taxonomy introduced in Section 9.1.1, it was established that safety and security are important but separate aspects of embedded systems. However, it is also possible to draw links between the two — and in particular to note that a breach of security may lead to safety also being compromised. In other words, an attacker who successfully defeats security counter-measures may be able to maliciously alter the operation of a system, such that it behaves in an unsafe or hazardous manner. This could be a particular concern in the power industry, or in manufacturing automation applications, for instance.

Meanwhile, consider a sensor network for collecting various types of data in a smart city scenario. In this case, it is possible to conceive of a security breach without safety implications, for instance if the attacker was purely interested in 'stealing' the collected data.

For these reasons, we consider safety on its own merits within this chapter, with a nod to security considerations where appropriate. Security in Zynq MPSoC is discussed separately, in Chapter 8.

9.1.2 Safety Application Examples

It is fairly easy to identify applications for which safety is an important attribute. These include applications in transportation, such as aircraft landing systems, automatic braking in cars, and even simply traffic lights at a junction (imagine all approaches simultaneously set to green!). Indeed, the demands for safety in transportation area are only likely to increase, given the movement towards Advanced Driver Assistance Systems (ADAS), and the emergence of autonomous vehicles.

Aircraft and spacecraft operate in potentially hazardous conditions due to radiation (in addition to other hazards such as weather, turbulence, space debris, and so on). Radiation effects have the ability to cause errors internally within the digital hardware of an embedded system, such as single event upsets wherein a register value is 'flipped' from a 0 to a 1 (or vice versa). The malfunction of an aircraft or spacecraft carries considerable risk, and therefore particular steps must be taken to mitigate the effects of radiation. More on this topic later.

The need for safety in industrial applications is clear too, and might include such examples as control systems for rotating machinery, robotic welding systems, and automated assembly lines. The effect of uncontrolled, or incorrectly controlled, motion could cause considerable danger to workers. Industries involving hazardous materials, nuclear systems, high voltages etc. will present their own particular demands.

Safety is also of key importance in a range of other application areas, ranging from medical systems, to defence electronics, right through to consumer electronics like kitchen appliances!

9.2 Functional Safety

9.2.1 What is Functional Safety?

When considering the safety of embedded systems, there is a strong tendency to focus on *functional safety*, i.e. the ability of the system to behave in a safe manner at all times [6]. Thus, other aspects of safety are put to the side, for instance the safety of the materials used in the system, and physical properties such as the effects of ageing. These factors are of course still important, but can be effectively separated from functional aspects, and subject to a separate design and review process. Put another way, we separate the active safety measures (e.g. detecting fault conditions and responding accordingly) from the passive ones (e.g. using a particular fire-safe material in the manufacture of the system). In the context of Zynq MPSoC, passive safety measures include the packaging of chips — the casing is designed to protect from damage, and also the effects of radiation (which can cause bits to 'flip' on the device, introducing errors in its operation).

The rest of this chapter therefore discusses *functional safety*. This means ensuring that the embedded system always behaves in a safe way, even when (for instance) unexpected input sequences are applied, or undesirable external events occur. Achieving functional safety could include ceasing operation in a safe manner, as in the term 'fail-safe'.

9.2.2 Errors, Faults and Failures!

Before going further, it is useful to define a subset of key terms used in the functional safety domain[1]. The terms *error*, *fault*, and *failure* are similar but have distinct meanings:

- **Error** — When there is a discrepancy between the measured, observed or calculated signal or state, and that which is expected (e.g. a deviation from a theoretically correct result). This can occur due to a *fault*, or due to unexpected operating conditions, and may lead to *failure*.
- **Fault** — An abnormal condition that may cause the failure of system or functional component.
- **Failure** — The termination of the correct behaviour of a system or functional component.

Faults can occur due to a variety of reasons. We can broadly classify them into categories, as shown in Table 9.1 with some SoC-based examples of each type.

Table 9.1: Examples of Fault Types

Category	Example
Random faults (occurring in hardware)	Single Event Upsets (SEUs) and Single Event Latch-ups (SELs) that alter the values of individual bits held in registers and memories, arising due to the effects of radiation.
	Random faults attributed to the physical reliability properties of the circuit (e.g. device manufacturing defects, ageing, shielding offered by packaging).
Common cause failures (failure of more than one component, due to the same fault).	Power supply issues (e.g. voltage / current outside permitted range of operation; power supply interruption or excessive noise; reset/power-on sequencing issues).
	Clock issues (e.g. interruption to clock source; frequency drift; jitter).
Systematic faults	Flaws in the design and realisation of hardware and/or software components of the system (e.g. software and hardware design bugs, timing violations, inadequate or incomplete testing).
	Flaws in the specification of the system (e.g. incomplete definition of possible input patterns; inadequate of numerical ranges and/or precision in signal paths).

1. Additional terminology is also defined within published safety standards — see Section 11.2.3

As a note on terminology, some Xilinx literature refers to Single Event Effects (SEEs, with a similar meaning to SELs / SEUs), with those SEEs that result in errors that do not damage the device being referred to as 'soft errors'.

Steps can be taken to mitigate each of the above types of faults and failures, by following robust, safety-oriented processes for all stages from initial specification to final verification. As there are several aspects contributing to the safety of an SoC-based embedded system (everything from device manufacture and packaging, to radiation-related random errors, to design and specification issues for hardware and software elements), then this is a complex matter! It is therefore helpful that safety standards exist to guide the process and ensure that a comprehensive approach is taken. Achieving certification according to one of these standards generates confidence that the system will operate safely.

9.2.3 Handling of Faults, Errors and Failures

Again drawing on [1], we can summarise the different ways in which a system may react to faults and their possible consequences, based on the definitions of the terms *fault*, *error* and *failure* given in Section 9.2.2.

Faults and Errors

If fault handing is fully embedded within a design, then *faults* can be detected as they occur, and appropriate action taken to compensate for them. Hence, the component in which the fault occurs can be immediately restored to a state of correct operation, and in this way, the fault is prevented from leading to the failure of the component. This is important because it means that the fault cannot propagate to other components and cause a wider scale problem (if not handled properly, a single fault could ultimately result in *failure* of the system as a whole).

We also defined that an *error* may occur, either due to an *internal fault* (which we assume has not been detected and corrected), or due to unexpected operating conditions (which might be termed an *external fault*). If the component includes logic to detect and correct errors, then this also has the effect of restoring the component to its correctly operational state, and thus preventing any adverse knock-on effects to other parts of the system.

These possible actions in response to faults and errors within a system component are depicted in Figure 9.2. Systems that can compensate for faults and errors in this way are normally referred to as '**fault-tolerant**'.

Component and System Failures

As mentioned above, a *fault* or *error* within a component can lead to its *failure*. For instance, consider the example of a component that divides an input signal, *N*, by a value held in memory, *D*. If a random bit error (e.g. due to radiation) caused the value of *D* to change from a significant value to zero, then it would cause a divide-by-zero error. This results in the component failing, because it does not operate as required.

The failure of one component can lead to another — for instance if Component A does not provide the required input to Component B (in our example case, the result of the division operation), this can cause the failure of Component B, and even of the system as a whole. Further, system failure can lead to a safety hazard if not appropriately contained, and to harmful consequences for the operator or the surrounding environment.

On the other hand, if the *failure* of a component can be detected, then its effect can be mitigated to ensure overall system safety. This can be achieved in one of two ways:

- If the detected failure cannot be compensated, then the system can react to the problem by moving to a known safe state, and ceasing normal operation (the nature of this safe state would depend on the type of system involved — it might mean that it shuts down and displays an information message, or falls back to a basic level of functionality). The operator would then need to intervene to address the problem before the system could be restarted as normal. This approach is normally known as **'fail-safe'**.

- If the detected component failure can be compensated, then the effects of the failure can be cleared, and correct system operation can resume. This requires more sophisticated design but has the benefit that the system can continue operating safely without disruption, and indeed the operator may be unaware that the failure of a component within the system occurred. This is known as **'fail-operational'**.

Possible responses to a component failure are shown in Figure 9.3. If the component failure is not detected, and no action is taken, the failure propagates to the next component in the system (a simplification — there may be more than one), causing it to also fail. The consequence is overall system failure, potentially leading to a safety hazard. In summary, the design is unsafe because it does not take steps to ensure safety if a component failure occurs. This is shown by the set of orange and red states in the diagram.

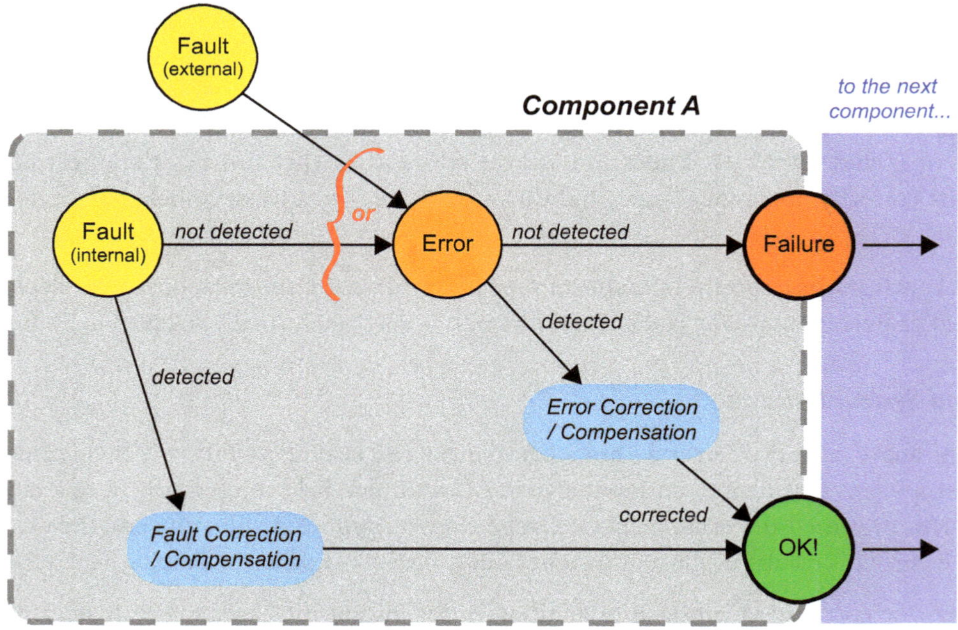

Figure 9.2: Possible Consequences of Fault and Errors in a System Component [4]

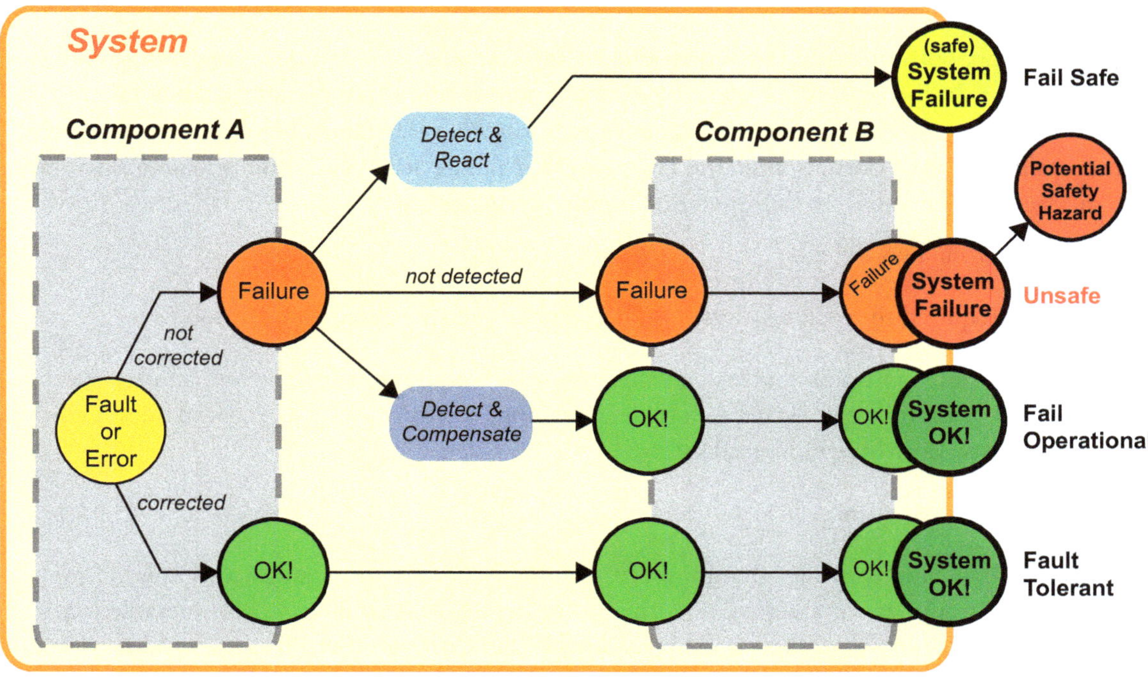

Figure 9.3: Possible Consequences of the Failure of a System Component [4]

If the system is able to detect the failure of a component, but not to compensate for it, the system can still react by going to a safe state. This means that the system will also fail, but it will do so in a safe manner. This sequence is shown at the top of Figure 9.3.

A better alternative would be that the system can compensate for a component failure, if it detects one. In our example, this means that the input supplied to Component B has been corrected, and it can then operate as normal. The system is said to be fail-operational, because it can continue to operate safely even if a failure occurs to Component A.

The final sequence shown is the correction of an internal fault or error, within Component A. This means that Component A does not fail, and that subsequent operation can continue as expected. This is referred to as **'fault tolerant'**, reflecting that low-level faults can be overcome.

It is also worth noting that a 'fault-tolerant' system may also be 'fail-safe' or 'fail-operational'. Fault tolerance corresponds to the ability of the system to tolerate low-level faults or errors, whereas the two failure modes refer to how the system behaves in the presence of a component failure. A preferred scenario would involve handling both low-level faults/errors, while also including the functionality to detect and react or compensate for component failures.

Faults, errors, component failures etc. can all be recorded for diagnostic analysis, thus enabling future safety improvements.

Fail-Safe vs Fail-Operational

To quickly recap, both of the above terms refer to the response of a system in response to a single failure occurring within it. In the case of *fail-operational*, the system continues to operate safely, while *fail-safe* means that the system ceases normal operation, and goes to a known, safe state. The overriding similarity, and the most important aspect of both, is that the response to a failure leads to a safe outcome. However, certain systems may require the ability to fail-operational rather than fail-safe.

Some of the factors involved are:

1. The time required to reach a safe state, and potential hazards during this period.

- Example: In transportation systems, a vehicle may be travelling at speed when a failure occurs. During the time to reach a safe state, safety hazards may continue to be encountered. Fail-operational functional would provide enhanced safety.

2. Tolerance of down-time.

- Example: If a failure in an aircraft safety system required maintenance, this might imply that the aircraft was grounded for a period, and unable to operate. This could have considerable impact in terms of inconvenience to passengers, a potential knock-on effect to other services, and lost revenue for the airline.

3. Cost factors.

- Example: The incorporation of redundancy and highly specified components into certain cost-sensitive products would not be deemed justified. For instance, buyers tend to value a lower purchase price for a coffee machine, over a very high degree of reliability. Fail-safe is therefore likely to be more appropriate here.

In all cases, the approach to functional safety should ensure that risk is reduced to an acceptably small level for the application. However, it is not possible to eliminate risk altogether. Taking care to anticipate sources of risk, to constrain faults, errors and failures such that they occur in detectable and classifiable ways, and to design systems that can successfully handle these occurrences, are all steps that can be taken to minimise the risk of safety hazards occurring.

A Note on Common Cause Failures

Common Cause Failures (CCFs) are failures which can occur concurrently, due to the same root cause. For instance, multiple functional components within an embedded system could simultaneously fail, due to a common input clock drifting away from its specified frequency.

The examples discussed so far have dealt with single points of failure, but the simultaneous failure of multiple components would present a more difficult (perhaps impossible!) scenario for the system to recover from. Therefore, it is better to take a preventative approach to CCFs, by anticipating possible causes, and

Figure 9.4: Selected functional safety standards

providing a level of diversity (for instance in the above example, supplying different clock sources to redundant components). Mitigating measures for CCFs in Zynq MPSoC are discussed further in Section 9.3.5.

9.2.4 Functional Safety Standards

Functional safety standards have been developed for use in various different domains, including for electronic and electrical systems in industrial applications, for the automotive section, aviation, and railways. Perhaps the most notable standard, in the sense that several others derive from it, is International Electrotechnical Commission (IEC) standard 61508: *Functional safety of electrical/electronic/programmable electronic safety-related systems*, otherwise known as IEC 61508 [6]. This standard is generally applicable to a variety of systems, whereas the derivative standards shown in Figure 9.4 define particular safety measures tailored to their respective domains.

One of the most demanding areas, in terms of functional safety, is the automotive industry. Automotive safety (cars and light goods vehicles) is covered by the International Standards Organisation (ISO) standard number 26262 (commonly referred to as ISO 26262). We will now look at this standard in a little more detail, as a case study.

Case Study: Automotive Safety Standard (ISO 26262)

ISO 26262 has been the agreed safety standard for electronic and electrical equipment in road vehicles (cars and light utility vehicles) since its publication in 2011 [7]. The standard was developed from IEC 61508 to cater more specifically for the requirements of the automotive industry.

The ISO 26262 standard defines a set of four Automotive Safety Integrity Levels (ASILs), which each specify a different level of reliability and fault detection. This reflects that cars contain systems which are safety-critical to different extents; for instance, the correct operation of the braking system is more important than the correct display of information on the dashboard. Functions therefore fall into different categories for the required level of safety, with individual functions categorised as ASIL-A / B / C / D (with increasingly stringent safety requirements).

Categorisation is based on an assessment of three factors:

- The likelihood of exposure to a hazard (very low, through to high probability of occurrence);
- Controllability by the driver, if the hazard does occur (easy, through to difficult/impossible to control);
- The severity of the hazard (no injuries / slight or moderate injuries, through to life-threatening or fatal injuries).

In our example, the braking system would be categorised as ASIL-C or D because, although the likelihood of a hazard occurring would be low, it could be difficult to control, and potentially life-threatening. Having identified the level of safety required, the braking system can then be designed to reduce the level of residual risk to an acceptably low level. This might include fault prevention, as well as including mechanisms to effectively handle faults when they do occur. There may also be an element of fault forecasting, which implies that the system can be pro-active in anticipating problems, which will help to prevent hazardous situations from arising.

The ASILs defined in ISO 26262 specify requirements in terms of three metrics (Table 9.2):

- **Single Point Fault Metric (SPFM)** — undetected faults which cause an immediate impact (in terms of violating a safety target);
- **Latent Fault Metric (LFM)** — faults which may be present and undetected for a period before causing an impact, e.g. a bit error occurring in memory that is not accessed immediately;
- **Failures in Time (FIT)** — the number of failures that can be expected to occur in every 1 billion device-hours of operation.

Table 9.2: Outline of ASIL safety integrity categories for ISO 26262 [7]

Safety Level	Single Point Fault Metric (SPFM)	Latent Fault Metric (LFM)	Failure in Time (FIT)
ASIL-D (most critical)	> 99% of faults detected	> 90%	< 10 failures in 10^9 hours
ASIL-C	> 97% of faults detected	> 80%	< 100 failures in 10^9 hours
ASIL-B	> 90% of faults detected	> 60%	< 100 failures in 10^9 hours
ASIL-A (least critical)	> 60% of faults detected	-	-
Quality Managed (QM)	-	-	-

It is also worth highlighting the 'Quality Managed' category in Table 9.2, which sits below the ASIL safety levels, and covers functions which are either unrelated to safety, or have a very low probability of impacting on overall system safety (for instance, entertainment systems). Automotive developers and manufacturers can set their own targets for this category.

ISO 26262 certification at a given ASIL is achieved by satisfying a number of requirements in terms of design processes (including software elements), tools, components, documentation, etc. Car manufacturers rely on incorporating components from a large number of third party suppliers, and therefore certification of these components is an important aspect. Embedded systems based on Zynq MPSoC devices are eligible for ISO 26262 certification (as well as IEC 61508), provided that the design and testing processes used, and accompanying documentation, also meet the requirements of the standard. A summary of factors relevant to the certification of Xilinx based designs is provided in [18].

A new version of ISO 26262 is due for publication in 2018, which will expand the scope of the standard to cover additional vehicle types (e.g. buses and heavy goods vehicles). It will also extend guidance on semiconductor components, and address current and emerging challenges including ADAS systems, cybersecurity issues, and autonomous vehicles.

9.3 Design Principles and Architectural Support

There are several strategies that may be employed in the implementation of safety critical systems [8]. This section goes on to summarise some of the common design principles for safety systems, and where appropriate, highlights the architectural support present in Zynq MPSoC, and safety related IPs available in Vivado.

9.3.1 Redundancy

One popular approach for designing safety critical systems is to introduce redundancy. This usually means implementing more than one instance of a design unit, and determining the correct output based on a 'voting' mechanism — i.e. redundancy in the physical sense. As this implies extra resources, a possible alternative for cost-sensitive applications is to repeat the same task multiple times ('temporal redundancy'), although he additional latency involved may not be suitable for all types of application, and the single instance reduces failure-resistance. For that reason, the rest of our discussion focuses on physical redundancy.

How Does Redundancy Work?

Redundancy involves implementing multiple, functionally identical copies of a design unit, and comparing the results produced. The result that occurs most often (i.e. which has the most 'votes') is then selected, on the basis that random errors (e.g. SEUs) are highly unlikely to affect all instances simultaneously in the same way. Thus, redundancy provides the ability to detect faults, and to identify the correct result. Another benefit of redundancy is that, if one of the redundant implementations were to fail, the system could continue to operate (albeit with a reduced level of resistance to any subsequent faults or failures).

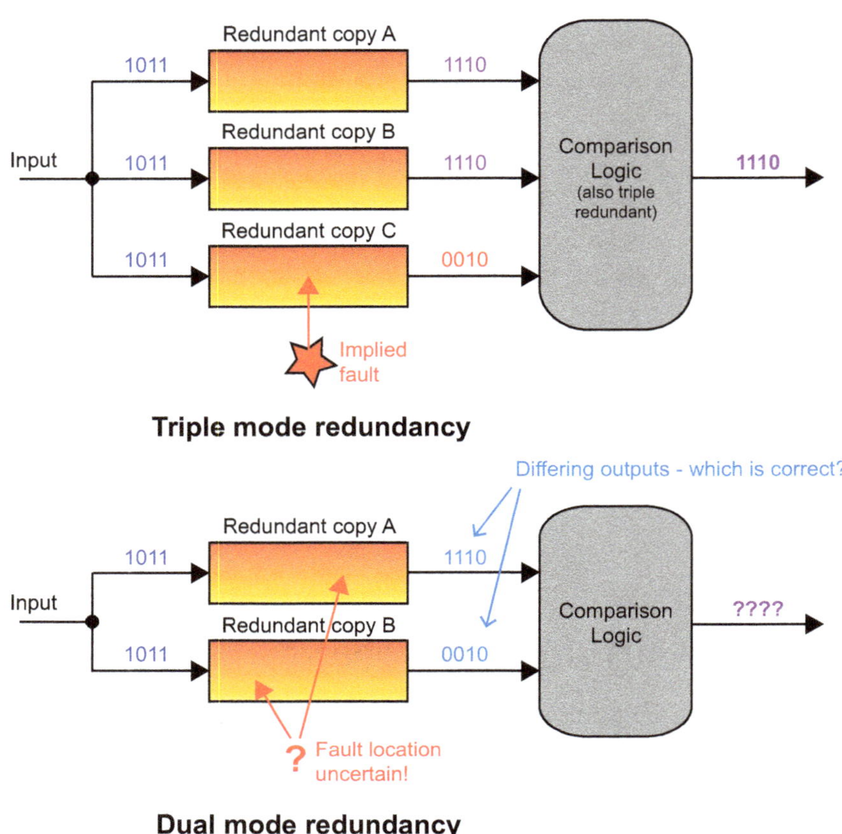

Figure 9.5: Sketches of dual and triple mode redundancy schemes

Redundancy with three instances is often used, referred to as Triple Mode Redundancy (TMR). With only two outputs to compare, and if they differ, there would be no way of telling which is correct; but with three, there should always be at least a 2-1 vote in favour of the correct result. These options are depicted in Figure 9.5. Dual mode redundancy does still have the benefit of identifying that a fault has occurred, which allows the system to take appropriate mitigating action (such as resetting a component, or repeating a sequence of operation). The realtime processors on the Zynq MPSoC device can be operated with dual redundancy, and this feature will be discussed further in Section 9.3.4.

A potential weakness in a TMR implementation is the checking logic which counts the votes, and which could itself be subject to a random fault. To address this issue, the checking logic can be replicated such that it is also redundant. The required level of correctness can be strict, i.e. the results must be bit-by-bit identical, or specified based on a threshold [8].

Higher levels of redundancy are occasionally deemed necessary — using four or five instances provides a level of fail-operational capability (adopting four instances allows for one failure, while still maintaining at least the three-instance minimum that is required to detect faults; five instances would allow two failures, and so on). This approach was used in the design of the Space Shuttle, which provided via four redundant primary processors and a backup processor [9].

Redundancy can be implemented at various hierarchical levels within a system, from low-complexity logic functions and software components, all the way up to processor level, and safety critical systems may employ multiple layers of redundancy. More complex systems often consider redundancy in conjunction with diversity (coming up in Section 9.3.2).

Advantages and Disadvantages

The key benefit of redundancy is that it provides fault / failure detection capability, and protection against failures, and thus improves the safety of the system. This also benefits dependability more generally.

In the context of SoC-based systems, implementing hardware components in triplicate implies greater resource cost and power consumption, even to the extent that a larger device is required. Redundancy within software elements increases the load put the processor(s), which may have performance implications. Further, the enhanced complexity of implementing a system with redundancy schemes requires greater design effort. These factors should all be considered in conjunction with safety requirements, when deciding on an appropriate level of redundancy.

Implementing Redundancy in Zynq MPSoC

In Zynq MPSoC, there are various mechanisms for incorporating redundancy. We can broadly categorise them as:

- **Architectural features that are inherently redundant** — Notably, the Zynq MPSoC Platform Management Unit (PMU) and Configuration Security Unit (CSU) are both implemented with TMR [19]. Each of these units is fundamental to the overall safe and secure operation of the device, so TMR is employed to provide highly dependable correct operation.

- **Architectural features with multiple instances (which facilitates redundancy)** — Zynq MPSoC features multiple cores, with the exact number depending on the device. The designer has the option to use these as part of a redundancy scheme, if desired, but could also opt to deploy the processors independently. In particular, the pair of RPUs can be operated in a special redundant 'lock-step' mode, which will be explained further in Section 9.3.4.

- **Components that are purposely implemented with redundancy by the designer** — This last category refers to redundant software and hardware components that the designer may wish to include as part of their safety scheme. Given the presence of multiple processors and PL on the device, there is vast scope for developing custom implementations with redundancy incorporated at one or more levels of hierarchy. Support is available in IP Integrator for implementing TMR via a set of pre-verified IP blocks (as discussed in [16] in the context of an example TMR MicroBlaze processor design).

In each case, our use of the term 'redundancy' implies that all of the copies of the component are equivalent, e.g. three instances of a design unit implemented on the PL that specify the same functionality in exactly the same way, and which are implemented using a physically separate but otherwise identical set of hardware resources. Another, related approach is to introduce an aspect of *diversity*.

9.3.2 Diversity

Diversity is similar to redundancy, in the sense that it entails multiple copies of a design unit. The distinction is that, when diversity is employed, the design units are implemented and/or operated in a different way from each other.

How Does Diversity Work?

Diversity is similar to redundancy, but with differentiation between the design units, and/or the operation of the design units. Aspects of dissimilarity are purposely introduced to reduce the possibility of systemic errors (e.g. software and hardware design bugs) and possible CCFs that could affect multiple instances in the same manner.

Physical diversity entails that multiple instances (e.g. three, in the case of triplication) are created, and that these instances have identical functionality, but differ from one another in terms of implementation. Aside from the core functional aspects, memories, connecting buses, etc. that could possibly be shared, are often avoided. Instead, each diverse instance has as little in common with the others as possible, which reduces the potential for a single fault to affect all instances.

Temporal diversity can also be implemented, meaning that the operation of the instances is purposely distributed in time (for example, execution of the modules may be separated by a time delay, Δt, with re-alignment prior to comparison and voting, as in the TMR scheme). Temporal differentiation provides a level of protection against transient faults that could otherwise result in a common fault.

An approach combining temporal and physical diversity is shown in Figure 9.6. Here, the colours of the module blocks represent functionality (top) and implementation (bottom) — hence the three copies shown have the same functionality but differing implementations. The execution of the three modules is staggered over time, with appropriate delays inserted to re-synchronise the outputs, prior to the comparison stage.

Implementing Diversity in Zynq MPSoC

Zynq MPSoC provides considerable scope for incorporating diversity into designs. This stems partly from its inherent multi-processor capability, which offers clear opportunities for implementing multiple instances of system components using physically different resources, even to the extent of running alternative hardware and software versions of a particular algorithm in parallel. The hardware/software approach would require both very different implementations (reducing the potential for common design bugs) while also using physical resources that are physically separate, with different power sources, clocks, and IO facilities, thus reducing the potential for CCFs.

The Zynq MPSoC device comprises four power domains:

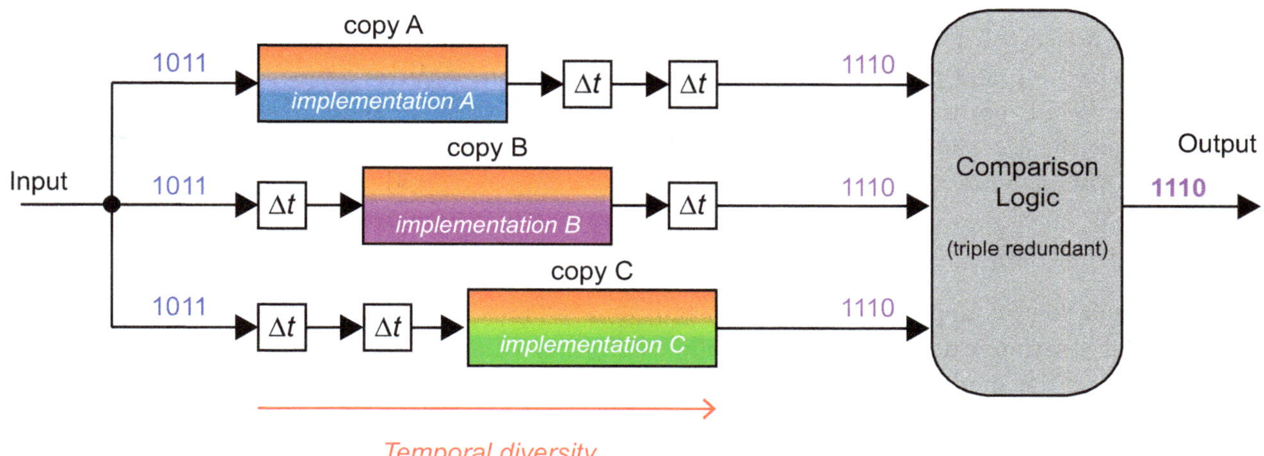

Figure 9.6: An example combined temporal and physical diversity scheme (temporal / physical diversity could each be applied alone)

- Battery Power Domain (BPD)
- Low Power Domain (LPD; contains R-5 real-time processors)
- Full Power Domain (FPD; contains A-53 application processors)
- PL Power Domain (PLPD)

These domains are physically separated from each other on the chip. In some cases there are also power-gating options for individual processor cores [13]. With the exception of the battery power domain, which is reserved for core encryption and time-keeping tasks, key design components could be diversified across two or more of these domains.

Within the PL power domain, it is possible to create functionally equivalent copies of a design unit, each with a different physical implementation. These would be synthesised to generate alternative netlists that would require different sets of resources and routing, and thus produce different layouts on the PL. Taking this a step further, it can be ensured that modules are completely independent of each other by adopting Xilinx' Isolation Design Flow (IDF), which will be introduced in Section 9.3.3.

The idea of diversity is also linked to the 'dual-lockstep' mode previously mentioned in conjunction with the pair of Arm Cortex-R5 real-time processors; this mode is further outlined in Chapter 7 on The Real-Time Processing System.

9.3.3 Isolation Design Flow

The power domains listed in Section 9.3.2 (BPD, LPD, FPD, and PLPD) are physically separated from each other on the chip, and can be isolated from each other using the PMU. A somewhat related approach can be applied to the PL portion of the system, by introducing a physical separation between modules. This is facilitated using the *Isolation Design Flow* (IDF), which is a Xilinx design methodology for safety and security applications. Although not specific to Zynq MPSoC, it can be used to good effect in Zynq MPSoC systems.

Why Isolate?

In safety and security applications in particular, it may be desirable to enforce the physical separation of certain modules within a design, and to carefully control any communication between them. This has the effect of making the modules more independent, and therefore from a safety perspective, there is better potential to implement redundancy without the risk of CCFs. An indicative IDF floorplan is shown in Figure 9.7, while real examples of IDF floorplans produced for a functional safety design can be found in [2].

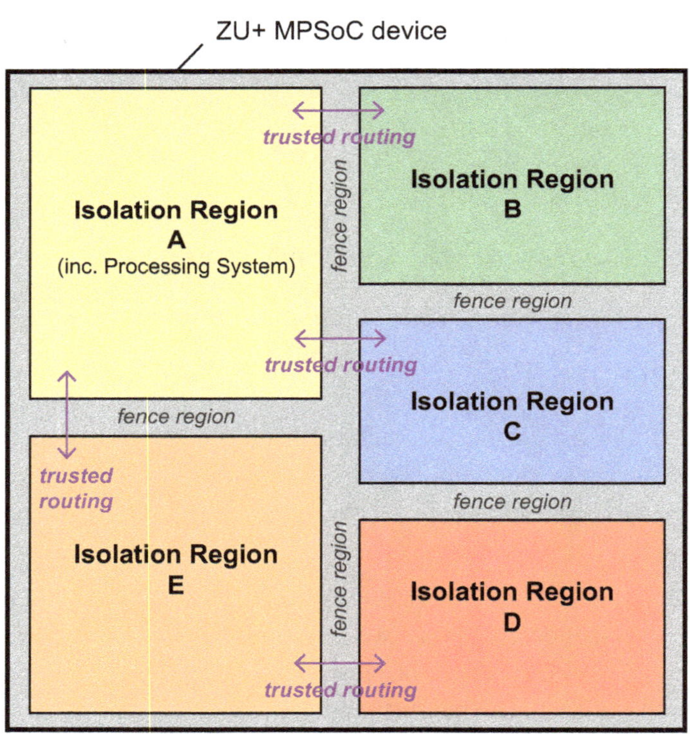

Figure 9.7: An indicative floorplan for a design employing IDF

Using the Isolation Design Flow

The IDF is followed in Vivado via the adoption of some additional steps incorporated into the standard hardware design flow, based on IP Integrator. In particular, as the physical separation of modules is sought, then floorplanning plays a more significant role than usual. The allocation of functional modules to defined regions of the chip is required at an early stage in the design process, and these isolated modules are separated by 'fences', i.e. strips of otherwise unused logic and routing resources, such that a physical gap exists between them. Communication is allowed only via dedicated 'trusted routing' paths. These features are highlighted in Figure 9.7.

The IDF also incorporates a tool for checking the successful implementation of isolation, Vivado Isolation Verifier (VIV). This tool is effectively a set of Design Rule Checks (DRCs) which establish that the various criteria required for isolation are met, e.g. that fencing between modules is correctly implemented, and that pins from different isolated modules are not placed adjacent to each other (which, from a safety perspective, would introduce a risk of CCFs).

Support for the IDF principally takes the form of Xilinx documentation and reference designs, available from [12]. Some of the examples featured relate to security-focused designs, however the flow and principles presented are equally applicable to safety.

9.3.4 Real-Time Processors: Dual-Lockstep Mode

One of the most important safety-related features provided on the Zynq MPSoC is the ability of the two Arm Cortex-R5 processors in the RPU to operate in a self-checking manner.

The RPU processor cores can operate in two different modes: either as a normal dual-core processor, or in a safety mode. In the latter, known as 'dual-lockstep' mode, the two processor cores undertake the same operations, with a small time offset, using a checking mechanism to ensure correctness. A conceptual diagram showing the operation of the RPU in dual-lockstep mode is provided in Figure 9.8, while extended details may be found in Chapter 7, on the Real Time Processing System.

Dual-lockstep mode employs diversity, such that the two processor cores undertake the same processing operations, but slightly offset in time (by 2 clock cycles). This mitigates against a radiation event simultaneously affecting both sets of operations in the same manner. Synchronisation and checking logic is required to ensure that both cores produce the same outputs — if not, a fault is flagged, the handling of which is defined as part of the system design (for instance, the RPU can generate an interrupt to be handled by the PMU according to user-defined code).

Although dual lockstep mode reduces the potential workload of the RPU by 50% compared to the alternative, 'performance mode', wherein the two cores may be used for separate sets of processing tasks, it does represent an important resource for implementing safety-critical systems.

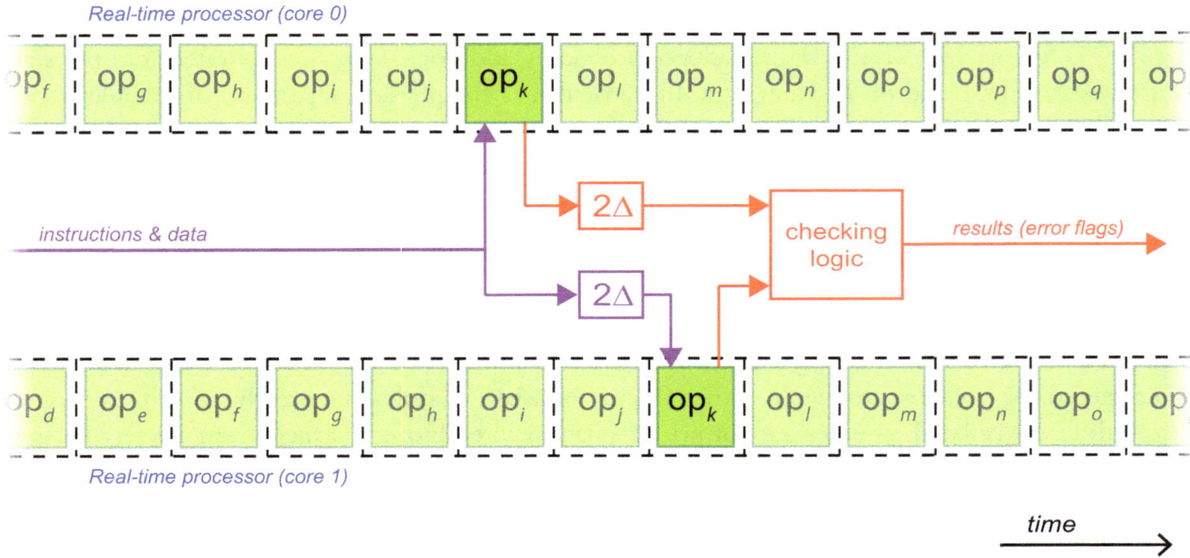

Figure 9.8: Intuitive diagram of lockstep operation (operating both Arm Cortex-R5 cores with a time offset and checking mechanism)

9.3.5 CCFs and System Monitoring

As discussed earlier in this chapter, there are potential sources of CCFs in embedded systems, for instance due to clocks, power sources, and IO resources that are shared by multiple elements within the system. Environmental conditions like temperature can also be a cause of CCFs, e.g. if the temperature were to exceed the specified operating range, this may simultaneously affect more than one subsystem in the same way. One of the most important elements to be protected by system monitoring is the RPU — operation of the Arm Cortex R-5 processor cores in lock-step mode is itself a safety measure, but the two cores could fail simultaneously due to a CCF.

As part of a proactive strategy to mitigate possible failures, monitoring of certain aspects of the system can be undertaken. This is supported in Zynq MPSoC by dedicated System Monitor ('SysMon') components in each of the PS and PL sections of the device.

How Does System Monitoring Work?

System monitoring is undertaken by measuring the values of internal and supply voltages, on-chip temperatures, external analogue signal voltages etc. to ensure that they remain within their specified operating ranges. The SysMon component also includes a 10-bit ADC that allows external signals to be measured.

The PS and PL SysMon instances are very similar in operation, catering for the two separate regions of the Zynq MPSoC device. The few key differences between them include the sampling frequency of the ADC

(1MHz for the PS and 200kHz for the PL SysMon), the options for programming interfaces, and the availability of auxiliary analogue inputs in the PL but not PS SysMon component, as detailed in [19].

Implementing System Monitoring in Zynq MPSoC

The designer has the option to use either or both of the two SysMon components on the Zynq MPSoC device, i.e. the PS SysMon, and the PL SysMon. The latter needs to be instantiated within the hardware design (Vivado block diagram) in order to allow access its full functionality.

The PS and PL SysMon components are configured and interrogated via a set of dedicated registers. This allows the designer to, for instance, set their own custom thresholds for voltage conditions on different signals, and to specify the conditions for raising alarms (the PL and PS SysMon components each provide a set of 16 different alarm outputs). These act as interrupts to the PMU processor, where software routines determine what actions to take in response to the reported issues. For instance, the PMU may respond by performing a reset of one of the power domains. There is also the option to have the system perform an automatic shutdown if the die temperature exceeds a critical upper limit (by default, 125°C), in order to prevent damage to the device.

Further, detailed information about the use of the PL and PS SysMon components is available in [17] and [19], respectively.

9.3.6 Error-Correcting Code (ECC) Memory

While random faults in logic circuits can be mitigated using TMR and diversity techniques, special treatment is appropriate for memories in safety applications, due to the potentially large amounts of data held, of different types, and which is susceptible to corruption by SEUs.

How Does ECC Memory Work?

ECC memory works by adding extra bits to the data held in memory, such that parity checks can be conducted to detect and potentially correct for errors. The degree of error protection increases with the number of parity bits added, and although this implies an overhead in terms of storage capacity and access times, ECC is generally considered worthwhile in safety systems due to the data integrity it provides.

ECC is based on Hamming codes [5], which are well-covered in standard textbooks on computing and information theory. The most basic parity check provides the ability to detect the presence of an odd number of bit errors within a data word (a single error being the most likely), although not to determine its position. This means that the error cannot be corrected. Slightly increasing the parity component allows information about the position of the error to be generated, such that one bit error within the data word can be corrected, and a second error detected but not corrected. Higher levels of error protection (correction of more than one error) can be achieved by using more parity bits.

The disadvantages of ECC are the additional storage implied, and the processing overhead of performing the parity generation and correction processing, which means a slightly increased time for writing to memory, and reading from memory, respectively.

Implementing ECC Memory in Zynq MPSoC

The Zynq MPSoC device contains a number of memories and caches within its various processing elements. Critical memories included in the CSU and PMU components are pre-equipped with ECC, which functions transparently. ECC support is also available in other elements, and can be incorporated as desired, e.g. by specifying options of the RPU and APU subsystems, and of the Double Data Rate (DDR) external memory controller.

An ECC core is available for inclusion in IP Integrator designs [11], such that memories implemented in the PL, and transmissions of data to/from external memory, can optionally include error protection. As a further measure, interleaving may be applied to ECC-protected memories, which reduces the potential for clustered SEUs to cause uncorrectable errors.

Extensive further information about ECC features and options in Zynq MPSoC are available in [19].

9.3.7 Fault Injection and Testing

An important practical issue to consider is how to verify a safety-critical system through testing. After all, such designs are developed to be robust to errors that occur randomly (and hopefully very rarely) as a result of environmental conditions such as radiation. The ability to synthesise such errors in a controlled and efficient manner is necessary for testing purposes.

In doing so, it is also worth bearing in mind the different types of memory within the Zynq MPSoC device that can be affected by errors. This includes configuration memory (where the contents of bitstream files are loaded), as well as Block RAMs, distributed RAMs and flip-flops on the PL, and the various memories within the PS.

One of the features supporting fault injection and testing is the Soft Error Mitigation (SEM) Controller, which is available as a block (IP core) in IP Integrator [15], and compatible with UltraScale and UltraScale+ devices (FPGAs as well as SoCs). The primary function of the SEM Controller it to mitigate errors in configuration memory. Used in this way, the core may be integrated into production designs; it also offers error injection features for use in the development phase.

Further fault injection testing mechanisms can be incorporated as part of the design verification process. This may require more custom solutions, given that the nature of the implemented functionality will differ from system to system. For instance, a module could be incorporated to randomly flip bits in a particular data path, synthetically introducing the SEU type errors ordinarily produced due to radiation effects, and thus providing a way to exercise and test error mitigation measures such as redundant logic implementations [3]. Similarly, software can be designed to incorporate fault injection for testing purposes, with the testing process aided by an emulation platform such as the Quick EMUlator (QEMU) [14]. The topic of fault injection testing is an area of active interest in the research community.

9.4 Chapter Summary

In this chapter, we have considered the theme of safety as it applies to SoC based designs. The role of safety as part of creating 'dependable' systems was explored, according to the taxonomy proposed in [1] (and illustrated in Figure 9.1 on page 218), and the concept of functional safety was introduced. We defined various items of terminology used in functional safety, including different types of faults and how they arise (including SEUs due to radiation effects, and CCFs that might occur from an out-of-range supply voltage, for instance). Examples of detecting and correcting faults and errors (...or not!) were also provided.

Safety standards were briefly discussed, and a short case study on ISO 26262 was included, representing an example in automotive safety. With the move towards autonomous vehicles in particular, this area is likely to generate highly challenging functional safety requirements in the years ahead. An indication of safety standards relevant to selected other application areas was also included (referring to Figure 9.4 on page 225), spanning medical to railway to nuclear industries. It was mentioned that Zynq MPSoC devices and their associated Xilinx design tools and methodologies provide a basis for certification according to such standards.

The remainder of the chapter was devoted to explaining various techniques relevant to safe design, including redundancy, diversity and isolation, and the features incorporated in Zynq MPSoC devices and design tools to support safe design. In particular, we explored the dual-lockstep mode of the RPU, ECC memory, the IDF for creating hardware designs that include physical isolation of modules, and the SysMon component for helping to identify potential causes of CCFs. A combination of these features and techniques can be used to develop Zynq MPSoC based system designs with a high degree of functionally safety.

9.5 References

Note: All online sources last accessed March 2019.

[1] A. Avizienis, J. Laprie, B. Randell and C. Landwehr, "Basic Concepts and Taxonomy of Dependable and Secure Computing", *IEEE Transactions on Dependable and Secure Computing*, Vol. 1, No. 1, January - March 2004, pp. 1 - 23.
DOI: 10.1109/TDSC.2004.2

[2] G. Corradi, R. Girardey and J. Becker, "Xilinx Tools Facilitate Development of FPGA Applications for IEC61508", *Proceedings of the 2012 NASA/ESA Conference on Adaptive Hardware and Systems*, pp. 54 - 61.
DOI: 10.1109/AHS.2012.6268669

[3] S. Di Carlo, G. Gambardella, P. Prinetto, F. Reichenbach, T. Lokstad, and G. Rafiq, "On enhancing fault injection's capabilities and performances for safety critical systems", *Proceedings of the 17th Euromicro Conference on Digital System Design*, Verona, Italy, 2014, pp. 583 - 590.
DOI: 10.1109/DSD.2014.12

[4] R. Grave, "Autonomous Driving — From Fail-Safe to Fail-Operational Systems", Elektrobit company, December 2015.
Available: https://d23rjziej2pu9i.cloudfront.net/wp-content/uploads/2015/12/09163552/Autonomous-Driving-From-Fail-Safe-to-Fail-Operational-Systems_TechDay_December2015.pdf

[5] R. W. Hamming, "Error detecting and error correcting codes", in *The Bell System Technical Journal*, vol. 29, no. 2, pp. 147-160, April 1950.
DOI: 10.1002/j.1538-7305.1950.tb00463.x

[6] International Electrotechnical Commission, *Functional Safety: Essential to Overall Safety*, 2015.
Available: http://www.iec.ch/about/brochures/pdf/technology/functional_safety.pdf

[7] International Organization for Standardization, ISO standard 26262, 2011.
Available: https://www.iso.org/news/2012/01/Ref1499.html (document may be downloaded from this page at cost).

[8] J. H. Lala and R. E. Harper, "Architectural Principles for Safety-Critical Real-Time Applications", *Proceedings of the IEEE*, Vol. 82, No. 1, January 1994, pp. 25 - 40.
DOI: 10.1109/5.259424

[9] J. R. Sklaroff, "Redundancy Management Technique for Space Shuttle Computers", *IBM Journal of Research and Development*, Vol. 20, No. 1, 1976, pp. 20 - 28.
DOI: 10.1147/rd.201.0020

[10] J. Yoshida, "Auto SoCs: Race to ASIL D", *EE Times* [online], published 9th March 2017.
Available: http://www.eetimes.com/document.asp?doc_id=1331459

[11] Xilinx, Inc., *ECC v2.0 LogiCORE IP Product Guide*, PG092, June 2017.
Available: https://www.xilinx.com/support/documentation/ip_documentation/ecc/v2_0/pg092-ecc.pdf

[12] Xilinx, Inc., *Isolation Design Flow* webpage.
Available: https://www.xilinx.com/applications/isolation-design-flow.html

[13] Xilinx, Inc., "Managing Power and Performance with the Zynq UltraScale+ MPSoC", WP482, v1.1, October 2016.
Available: https://www.xilinx.com/support/documentation/white_papers/wp482-zu-pwr-perf.pdf

[14] Xilinx, Inc., QEMU wiki page.
Available: http://www.wiki.xilinx.com/QEMU

[15] Xilinx, Inc., "Soft Error Mitigation Controller v4.1: LogiCORE IP Product Guide", PG036, April 2018.
Available: https://www.xilinx.com/support/documentation/ip_documentation/sem/v4_1/pg036_sem.pdf

[16] Xilinx, Inc., "Triple Modular Redundancy (TMR) Subsystem v1.0 product guide", PG268, November 2018.
Available: https://www.xilinx.com/support/documentation/ip_documentation/tmr/v1_0/pg268-tmr.pdf

[17] Xilinx, Inc., "UltraScale Architecture System Monitor User Guide", UG580, v1.9.1, February 2019.
Available: https://www.xilinx.com/support/documentation/user_guides/ug580-ultrascale-sysmon.pdf

[18] Xilinx, Inc., "Xilinx Reduces Risk and Efficiency for IEC61508 and ISO26262 Certified Safety Applications", WP461 (v1.0), April 2015.
Available: https://www.xilinx.com/support/documentation/white_papers/wp461-functional-safety.pdf

[19] Xilinx, Inc., "Zynq UltraScale+ Device Technical Reference Manual", UG1085, v1.9, January 2019.
Available: https://www.xilinx.com/support/documentation/user_guides/ug1085-zynq-ultrascale-trm.pdf

Chapter 10

Platform Management Features

The Zynq MPSoC contains several different types of processing units that have their power states and errors managed from a single on-chip component. This component is known as the Platform Management Unit (PMU). The PMU is responsible for coordinating platform management requests from the APU and RPU. It is also capable of responding to errors that occur throughout the system. Additionally, the PMU is an essential element in booting the Zynq MPSoC.

In this chapter, we will cover three topics involving the Zynq MPSoC's platform management features. These are as follows:

1. **Power modes and domains** — The Zynq MPSoC's power modes and domains will be explored. This investigation will include the Zynq MPSoC's three operational power modes and four primary power domains. Additionally, we identify the other remaining power domains in the Zynq MPSoC.

2. **The PMU's architecture** — The PMU is the primary topic of discussion in this chapter. Its architecture is described, including the triplicated processing unit, memories, peripherals, and interrupt system. Furthermore, error management using the PMU is also discussed.

3. **The PMU's firmware** — The PMU's associated firmware that provides routines and functions for responding to system interrupts and events is reviewed. The firmware's architecture, execution flow, and modules are discussed. Note that Xilinx has provided a template for the PMU's firmware, as discussed in Section 10.3.1.

10.1 Power Modes

The PMU is responsible for managing the Zynq MPSoC's power mode. There are three operational power modes in which the device can be configured: battery powered mode, low-power mode, and full-power mode. Each mode permits the operation of particular processing elements and power domains within the device.

Figure 10.1 provides a diagram illustrating the power consumption of each operational mode. Note that, low-power mode consumes up to 150mW and full-power mode consumes approximately 3W. Additionally, another power mode known as deep-sleep mode is also shown. Deep-sleep mode consumes the lowest amount of power while still maintaining the boot and security state of the Zynq MPSoC.

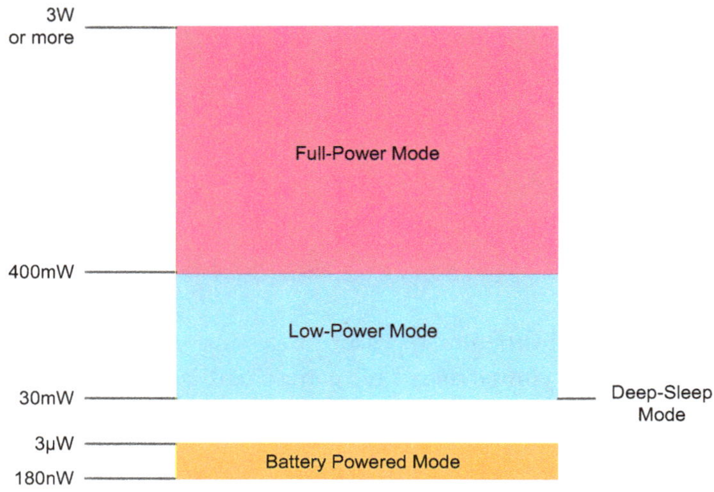

Figure 10.1: Zynq MPSoC Power Modes [2]

10.1.1 Battery Powered Mode

Zynq MPSoC devices can be programmed to switch off their system elements to conserve power. One method of conserving power is to switch the device to battery powered mode. Two hardware blocks, known as the battery-backed RAM (BBRAM) and the Real-Time Clock (RTC), are used to support battery powered mode. Each of these hardware blocks receive power from a battery provided off-chip. The battery supplies power to the battery power domain, discussed further in Section 12.3.2.

The Zynq MPSoC will lose its security and boot state upon entering battery powered mode. When powering up, the Zynq MPSoC must reinitialize the boot and security state to enter low-power or full-power mode. Alternatively, the deep-sleep mode can be used to improve the time taken to boot the Zynq MPSoC. However, this will consume more power. Deep-sleep mode is discussed in Section 10.1.4.

10.1.2 Low-Power Mode

Low-power mode provides power to hardware blocks and processors that are present in the low-power domain. Blocks that are powered up in low-power operation include the RPU, PMU, CSU and IOP. However, this does not include the APU or PL which are only accessible using full-power mode. The low-power mode also includes all available on-chip peripherals apart from the DisplayPort and SATA interface blocks.

The number of powered system elements, and their operating frequencies, is a major factor in determining the total power dissipation of the Zynq MPSoC when operating in low-power mode.

10.1.3 Full-Power Mode

Full-power mode allows power to be driven to all of the Zynq MPSoC power domains. There are several domains, and each contain separate power rails for power distribution. Section 10.2 further discusses the Zynq MPSoC power domains.

Similar to low-power mode, the total Zynq MPSoC power dissipation depends on the number of powered system elements and their operating frequencies.

10.1.4 Deep-Sleep Mode

The Zynq MPSoC can enter a deep-sleep mode that consumes the lowest amount of power while still maintaining the boot and security state of the entire system. During deep-sleep mode, several processing elements in the PS are powered down. These include the Cortex-R5 processors, the system monitor, the APU's debug block, the phase-locked-loops (PLLs) in the PS, and any component in the full-power domain. Additionally, the PL's internal power is switched off.

There are two sets of memories in the PS that can continue to hold their data in deep-sleep mode. These are the OCM and TCM, however, only one of these memories can be maintained at a time. The only operational elements in the PS are the RTC and BBRAM. During deep-sleep mode these hardware blocks use the VCC_PSAUX supply described in Section 10.2.1.

Upon entering deep-sleep mode, some system elements are not powered down and are instead placed in suspended operation. A system element in suspended operation, enters a state of no-activity to save power. The PMU, eFUSE, CSU, and several peripherals in the LPD, are placed in suspended operation during deep-sleep mode. In particular, the PMU is placed in a suspended state, as it needs to wait for an interrupt or wake event to power-up the Zynq MPSoC into an operational mode. Interrupts or wake events used to power up the Zynq MPSoC from deep-sleep mode, can be issued by the MIO, Ethernet, RTC or USB.

There are two main advantages of using deep-sleep mode over battery powered mode. Deep-sleep mode will retain the security and boot state of the Zynq MPSoC, and less time is spent booting or restarting the system. The battery powered mode should be selected if low power consumption is of particular importance, or the Zynq MPSoC is going to be powered down for an extended period of time.

10.2 Power Domains

Power optimisation is a key design specification of the Zynq MPSoC. Particular sections of the Zynq MPSoC may operate at lower voltages, or be shut down completely through the use of multiple voltage levels and power gating logic. Processing elements that have power supply characteristics in common are grouped together to form sections. These are formally known as power domains, of which the Zynq MPSoC has several. Power domains provide the option of switching off unused resources to save power. Furthermore, power domains make it possible for multiple voltage sources to provide power to resources.

Our focus in this section will be on the Zynq MPSoC's four primary power domains, as each of these are host to several critical components and processing elements. These are the battery power domain (BPD), low-power domain (LPD), full-power domain (FPD), and PL power domain (PLPD).

The power domains should not be confused with the Zynq MPSoC power modes. The power domains refer to a subset of resources on the chip that share power supply characteristics. The power modes permit the operation of particular processing elements and power domains within the Zynq MPSoC.

Each power domain is independent of the others and can be isolated to reduce power consumption and to achieve functional isolation (an essential requirement for safety and security applications and tasks). Figure 10.2 illustrates the processing elements and features accessible by each power domain.

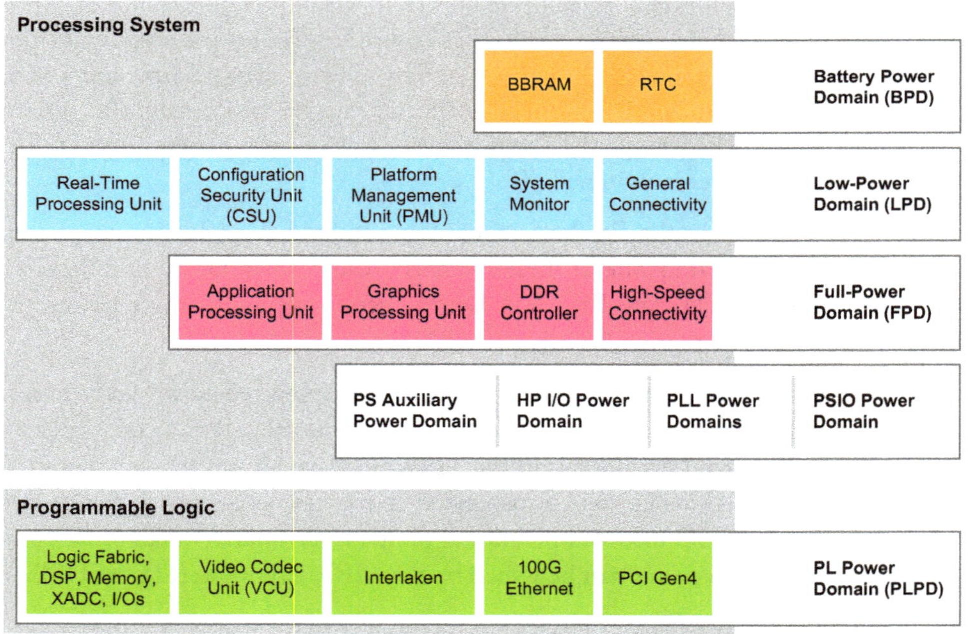

Figure 10.2: The processing elements and features accessible in each Zynq MPSoC power domain [2]

The Zynq MPSoC contains other power domains primarily associated with peripheral and memory interfaces not shown in Figure 10.2. These are listed in Section 10.2.5.

10.2.1 Battery Power Domain

The BPD is host to a BBRAM and RTC module as shown in Figure 12.3. The BPD is of particular importance when the Zynq MPSoC is in the battery powered mode, as it contains all the necessary components (BBRAM and RTC) to wake the Zynq MPSoC, so that it may begin the boot process. The RTC may also serve further purposes, such as providing an alarm to wake the Zynq MPSoC out of deep-sleep mode.

The supply of power to each hardware block in the BPD can be from one of two power supplies. These are the Zynq MPSoC's auxiliary supply (VCC_PSAUX) and an off-chip battery supply (VCC_PSBATT).

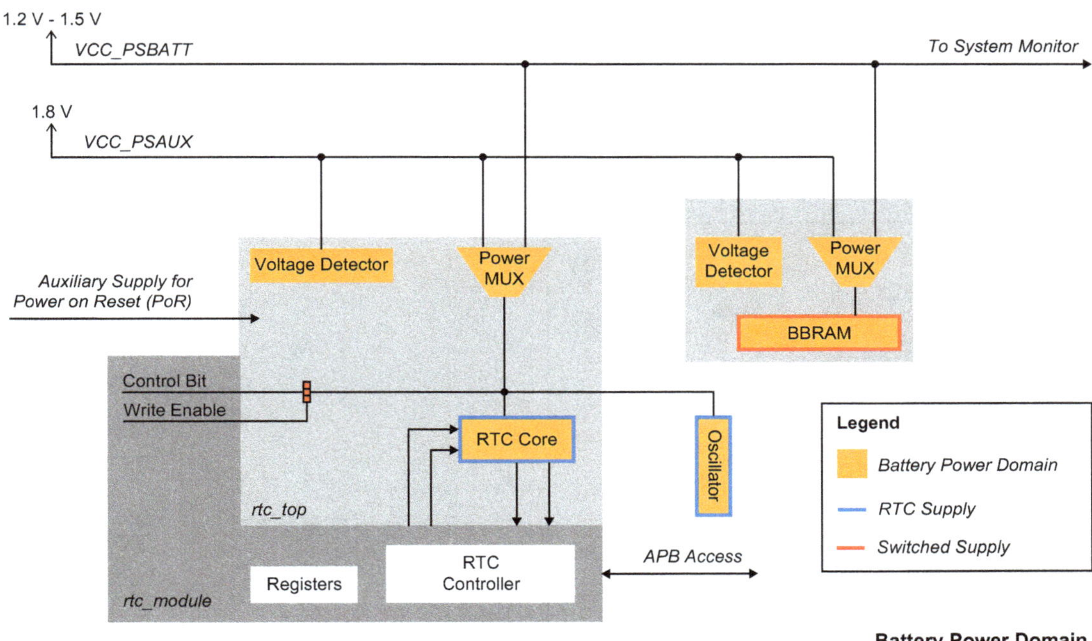

Figure 10.3: Hardware blocks connected in the BPD [1]

The BBRAM is a static RAM array responsible for storing the device AES key. The BBRAM normally receives power from the Zynq MPSoC's auxiliary supply VCC_PSAUX, that exists in the PS Auxiliary power domain. However, if VCC_PSAUX is switched off, the BBRAM will receive power from the battery supply VCC_PSBATT. Swapping between power supplies is handled by the power multiplexer (MUX) and voltage detector.

The RTC provides the real-world time and date, even if the Zynq MPSoC has been switched off. It can be programmed to interrupt all processors at a pre-determined time and date. Similarly to the BBRAM, the RTC will use the VCC_PSAUX power supply. However, if this is not available, the RTC will use power supplied from VCC_PSBATT.

The RTC controller and user interface do not receive power from the BPD. Instead, they are connected to the LPD power rail and will therefore only be operational when the supply for the LPD is on.

10.2.2 Low-Power Domain

Figure 10.4 illustrates the PS hardware blocks and processors present in the LPD. These include the following significant system elements:

- RPU
- PMU
- CSU

In addition to the above, the LPD also includes the Input Output Peripheral (IOP) unit, on-chip memory (OCM), and Low-Power Direct Memory Access (LP-DMA). Several power supply sources are shown to the left of the LPD diagram in Figure 10.4, with LPD resources shown to the right.

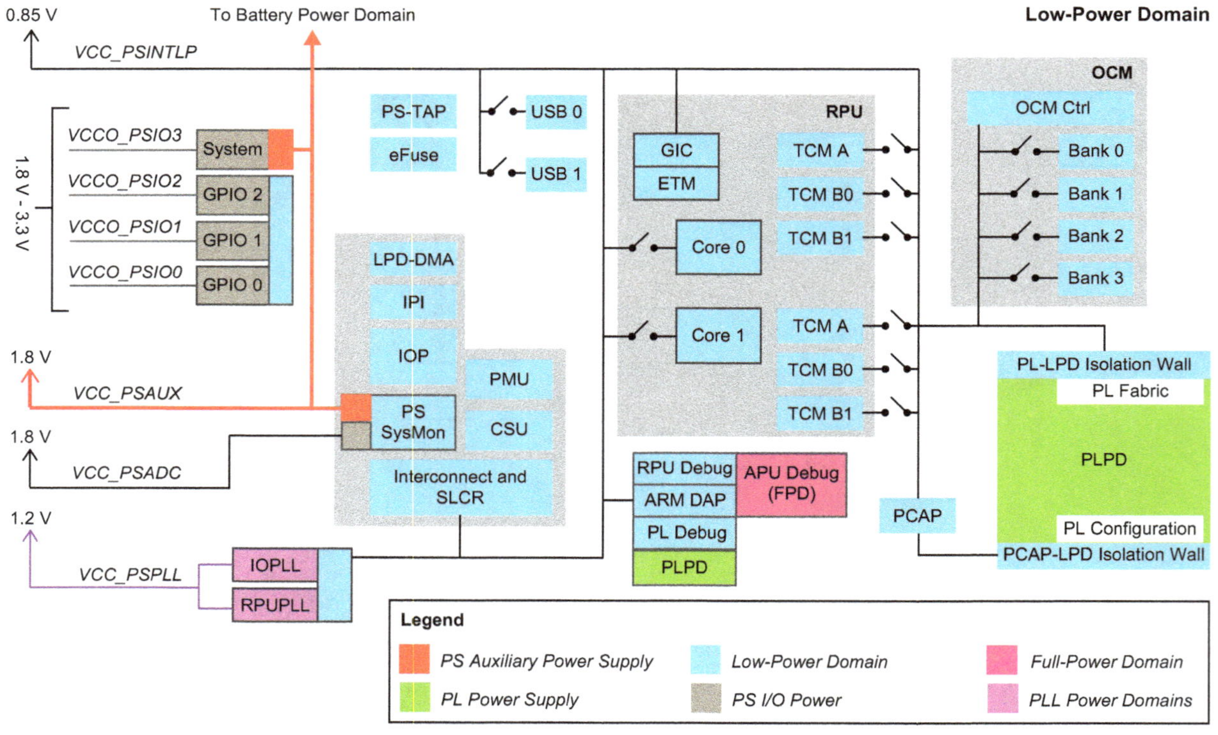

Figure 10.4: Hardware blocks and processors connected in the LPD [1]

The LPD contains many switches which indicate that a system element can be isolated from its power source. Isolating a system element in this manner is known as power gating. Power gating can be applied to the RPU's processor cores and Tightly-Coupled Memories (TCMs). Additionally, it is possible to independently power gate the USB interfaces and turn off unused banks in the OCM to reduce power consumption. The PMU is responsible for power gating system elements in the LPD.

10.2.3 Full-Power Domain

Figure 10.5 provides a diagram illustrating the architecture of the FPD. The FPD is host to the following significant system elements.

- APU — *There is a quad-core APU in EV and EG devices and a dual-core APU in CG devices.*
- Full-Power Direct Memory Access (FP-DMA).
- GPU — *The GPU is only available in EV and EG devices.*
- DDR Controller (DDRC).
- High-speed I/O peripherals.

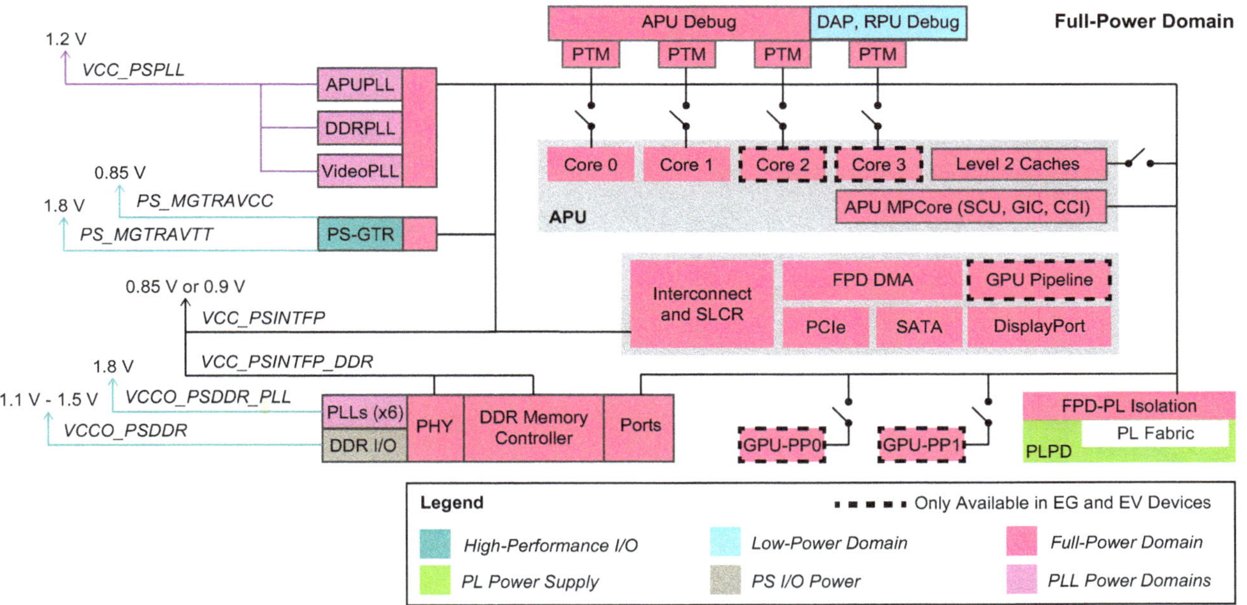

Figure 10.5: Hardware blocks and processors connected in the FPD [1]

To the left of the diagram are several voltage supplies contributing to the operation of system elements within the FPD. There are two primary power supplies; VCC_PSINTFP_DDR and VCC_PSINTFP. The former provides power to the DDR's physical interface block and memory controller. The latter powers the remaining system elements in the FPD.

As shown by the switches in Figure 10.5, it is possible to individually power gate each Cortex-A53 processor core, the Level 2 Caches, and the GPU's pixel processors (GPU-PPx). The PMU is responsible for power gating system elements in the FPD, providing flexible control of power consumption.

The FPD has several interfaces to the PL fabric described below. Each of these interfaces are investigated further in Chapter 11: Hardware System Development on page 263.

- There are six high-performance PL bus master interfaces. These interfaces provide master hardware blocks in the PL, access to PS slave devices. The typical use for these interfaces is to support PL access to the external DDR memory.
- There are two high-performance FPD bus master interfaces. These interfaces provide master devices in the PS, access to the PL. The APU and FP-DMA can use these interfaces when communicating with hardware blocks in the PL.

The FPD is connected to multi-gigabit transmit and receive channel pairs, known as PS-GTR transceivers. These transceivers exist in the FPD and allow the PS to communicate with high-speed peripherals.

To use the FPD, it must be powered during the initial booting of the Zynq MPSoC.

10.2.4 PL Power Domain

Figure 10.6 provides a simplified diagram of the PLPD. The PL resources are shown in each of the white blocks. Any resource may be switched off as required to reduce power consumption.

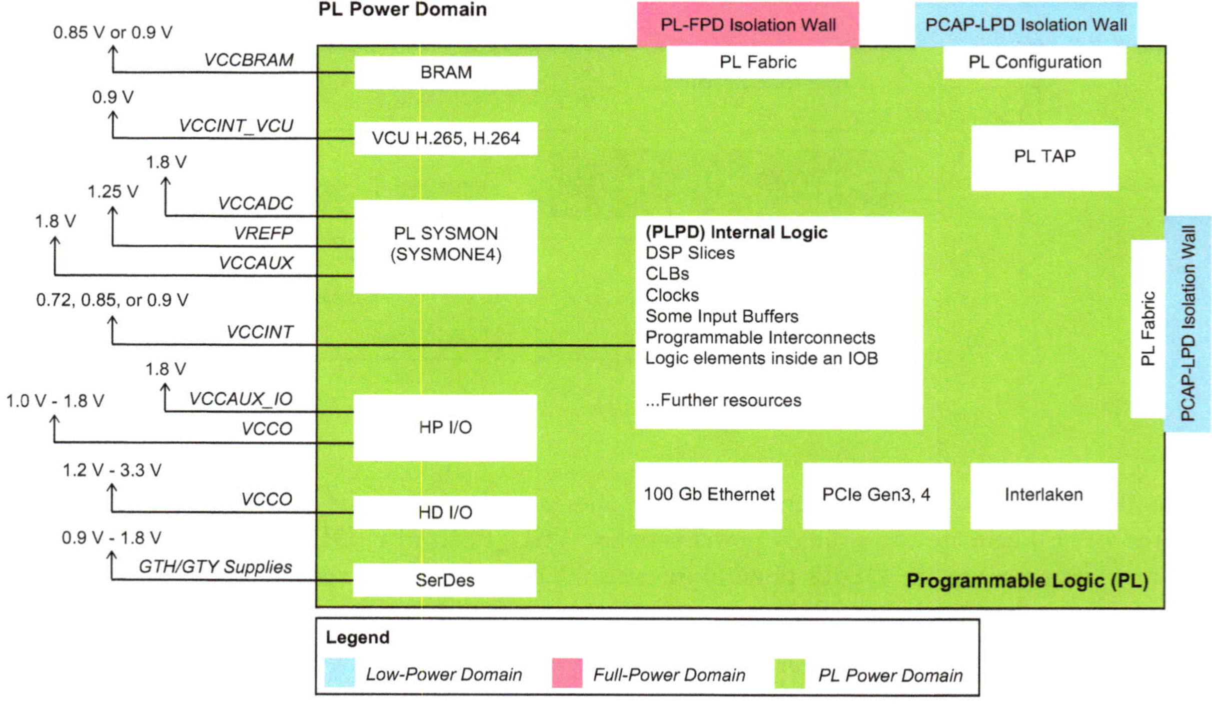

Figure 10.6: Hardware blocks and processors connected in the PL power domain [1]

There are several different supplies that provide power to resources in the PL. Each supply has its own purpose. Further information about the PL power supplies can be found in [4].

Note that not all Zynq MPSoC devices will contain every PL resource shown in Figure 10.6. For a complete list of resources contained in each Zynq MPSoC device, refer to [3].

10.2.5 Other Power Domains

There are more power domains in the Zynq MPSoC for isolating peripherals, interfaces and other system resources. Each domain can be shut down to decrease power consumption. The additional power domains are as follows.

- **PS Input Output (PSIO) Power Domain** — The PSIO domain hosts the dedicated GPIO pins for connecting to the MIO in the PS LPD.
- **Phase Locked Loop (PLL) Power Domains** — The PLL power domains host the PLLs distributed between the LPD and FPD.
- **High-Performance Input/Output Power Domain** — This domain is associated with the PS-GTR transceivers in the PS of the Zynq MPSoC.
- **PS Auxiliary Power Domain** — The PS Auxiliary power domain is used in the LPD and BPD. This power domain was described previously in Section 10.2.1 on page 243.

The Zynq MPSoC Technical Reference Manual [1] provides further information about each of these power domains.

10.3 Platform Management Unit

The PMU is significant element in the Zynq MPSoC, as it controls many system-critical operations and maintains the device's power state at all times. It is important to consider the PMU's many responsibilities in the Zynq MPSoC.

The PMU is responsible for the following:

- Executing power maintenance tasks for power masters in the PS, i.e. the Cortex-A53 and Cortex-R5 processors. Tasks include initiating a power-up and restart of the target processor after a wake-up request.
- Maintaining the Zynq MPSoC while it is in sleep mode and, upon receiving a request through a trigger mechanism, waking-up the resources responsible for handling the request.
- Managing errors that occur in the Zynq MPSoC and responding accordingly.

- Minimising power consumption and maximise battery life (if a battery is used as the main source of power).
- Executing a sequence of functions after a Power On Reset (POR) has been asserted (before the CSU reset signal is released). Chapter 14 discusses these functions in more detail.
- Executing procedures for the following: power-up, power-down, reset, Memory Built-In Self Repair (MBISR), Memory Built-In Self Test (MBIST) and the scan clear function. The scan clear function, also known as scan zeroisation, writes zeros into all storage elements within a resource.

Figure 10.7 provides a detailed diagram illustrating the entire PMU, including its interfaces, internal interrupt controller, and clock and reset signals.

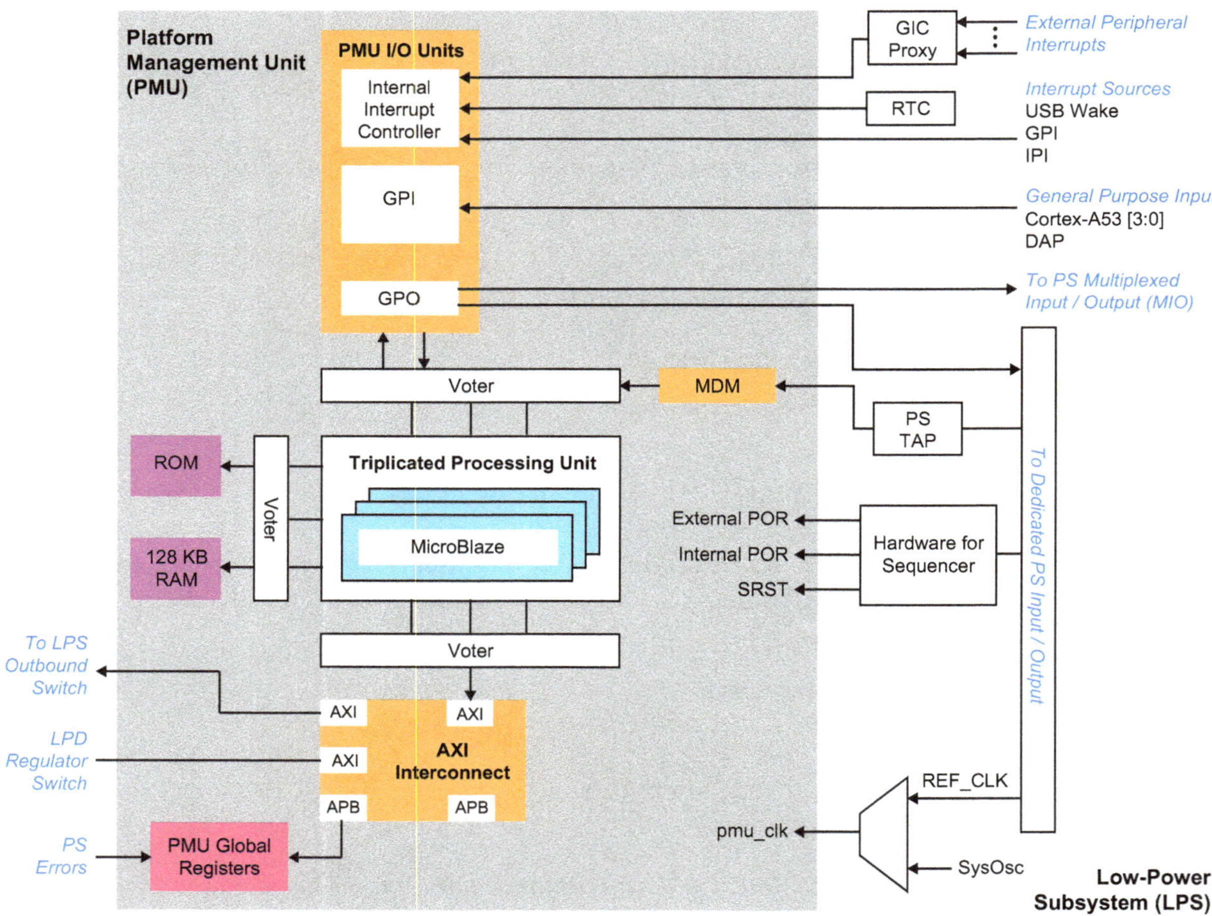

Figure 10.7: Overview of Platform Management Unit with external connections [1]

In the remainder of this section we will consider the PMU's underlying architecture, and signals and interfaces to other system elements. These principles are essential to understand the PMU's connectivity to other processing units in the Zynq MPSoC device.

10.3.1 PMU Processor

The PMU is host to a triple-redundant processing unit, consisting of three hardened MicroBlaze [6] processors. The triplicated processing unit is surrounded by voting logic that implements a majority voting system. The results from each of the three processors are combined to produce a single output. If one of the three processors fails, the remaining two processors are capable of maintaining the correct operation of the processing unit. Safety-critical systems are at a lower risk of error when using triple processor redundancy. Redundancy was described previously in Chapter 9.

The PMU's triplicated processor does not contain any caches, but does have a small local RAM. This RAM uses error checking and correction (ECC). ECC corrects single errors and when multiple errors have occurred, a system error is generated. ECC improves the reliability of the overall processing system.

The ROM, as illustrated in the PMU overview diagram in Figure 10.7 on page 248, contains important information for the PMU. The contents of the ROM include the PMU's boot code, interrupt vectors, and the primary service routines in which the PMU may execute. It has also been preconfigured to store scripts for pre-boot and post-boot tasks. These are described in the Zynq MPSoC Technical Reference Manual [1].

The ROM contains a fixed set of default functions that cannot be modified. If it is required to change the basic functionality of the PMU, the contents of the ROM must be overridden. The PMU 128 KB RAM is available for this purpose, capable of adding, extending and overriding the instructions and routines within the ROM.

PMU Firmware

The PMU's ROM provides code for basic platform management functionality. However, an application may require custom routines, functions, or modifications to the PMU's default operation. The PMU's RAM provides memory space for user firmware that can override the PMU's default operation, which is provided in the ROM (as the ROM cannot be modified).

To load the firmware into the RAM, an external master must access the RAM through the APB interface, while the PMU is in a sleep state. Loading the PMU firmware into RAM is described further in [1], [7] and [11].

Xilinx provides model firmware for the PMU that Zynq MPSoC developers can edit, as shown in [7] and [8]. The firmware's software architecture, execution flow, and modules are further described in Section 10.5.

It is also possible to use the Zynq MPSoC QEMU to emulate the operation of the PMU in software. Further information is provided in [9] and [10].

10.3.2 Interfaces and AXI Interconnect

The PMU uses several interfaces to maintain communication with other processing units and resources in the Zynq MPSoC. These interfaces can be seen in Figure 10.7 on page 248, and include AXI connections, clock and reset signals, power and control interfaces, and interrupt signals and sources. They are described further in this section.

AXI Connections and Interconnect

An AXI Interconnect enables AXI connections to be established within the PMU. The interconnect supports access to the PMU's global registers for the PMU's triplicated processor and other on-chip processors. It also supports access to the PMU's RAM, via the triplicated processor and voting logic.

Clock and Reset Signals

The clock and reset signals are integral to the operation of the PMU. The clock input is provided by the System Oscillator (SysOsc) contained inside the PS System Monitor [1]. The PMU controls gating of the SysOsc clock. When the PMU processor encounters an error which cannot be corrected, the SysOsc clock is disabled.

There are two methods in which the PMU can be reset. These are the POR and system reset (SRST). A POR will clear all the logic, RAMs and registers within the system and prepare the Zynq MPSoC for booting. A SRST will only reset the PMU processor subsystem and the PMU interconnect. It will also reset some of the local and global registers, leaving most of them in the same states as before the reset was applied.

You can learn more about each of these resets and the SysOsc clock in [1].

General Purpose I/Os

The PMU has several General Purpose Inputs (GPIs) and General Purpose Outputs (GPOs) that are used for a variety of purposes. These include wake from sleep signals, error reporting, peripheral configurations and signals that are used for handshaking. There are four GPI banks (GPI0 - GPI3) that are only accessible to the PMU processor. There are also four GPO banks (GPO0 - GPO3) that are connected to various elements in the Zynq MPSoC.

Extensive information about the PMU's general purpose I/O registers is available in [1]. Bit level information about each of the registers can be found in [5].

Dedicated PS I/Os

The PMU is connected to dedicated input/output pins in the PS, via the MIO. Many of these connections can be used to issue wake-up requests from the MIO to the PMU. Others are used to issue power supply control signals from the PMU to the MIO. Two connections in particular are used to power up the FPD and PLPD. MIO[32] is used to initiate a power-up of VCC_PSINTFP (one of the FPDs primary power supplies) and MIO[33] is used to initiate a power-up of VCCINT (the power supply for the PL's logic fabric).

JTAG Interface

In addition to the above general purpose I/O connections, the PMU also contains a MicroBlaze Debug Module (MDM) connected to a JTAG interface from the PS Test Access Port (TAP) controller. This connection allows the PMU to retrieve information from a data or instruction register in the JTAG boundary scan path. The purpose of this connection is to provide the PMU with the ability to communicate to external systems when it encounters an error.

10.3.3 Local and Global Registers

There are global and local registers within the PMU, described as follows:

- The global registers are non-critical to the safety and security of the Zynq MPSoC. These include power, isolation, error capture, system power state and reset request registers. Other bus masters connected to the PMU can access the global registers.
- The local registers are responsible for the safe and secure operation of the Zynq MPSoC. These may only be accessed and maintained by the PMU to prevent other processing elements within the system from causing errors; which in turn may damage the device or compromise its operation.

Detailed information about the contents of the global and local registers, can be found in [5]. The global registers are denoted PMU_GLOBAL and the local registers are denoted PMU_LOCAL.

10.3.4 Programmable Interval Timers

Programmable Interval Timers (PITs) are counters that are initially programmed with an upper limit (maximum count value). The PIT will count to this value and output an interrupt signal to the PMU's interrupt controller when the upper limit has been reached. PITs are useful for event scheduling, or as a delay counter in embedded designs.

The PMU contains four 32-bit PITs known as PIT0, PIT1, PIT2 and PIT3. Each of their clocks are driven by the fixed system oscillator (SysOsc). PIT1 has a special purpose, as it is used by the PMU's scheduler contained in its firmware. The scheduler uses PIT1 to maintain interval timing for tasks, described further in Section 10.5.2.

Further configuration and information about the PITs can be obtained by writing and reading to the registers in the PMU I/O Module. These registers are described in [1] and [5].

10.3.5 Interrupts

If you refer back to Figure 10.7 on page 248, the PMU has an internal interrupt controller and an external (with respect to the PMU) GIC proxy block to support interrupts connected to the APU and RPU. There are several interrupts connected directly to the interrupt controller from various sources, including the RTC.

The purpose of the interrupt controller is to provide elements of the PS and the PL, with the means of communicating to the PMU, power-up, power-down, reset, and other system critical requests. In this section, we aim to provide further details surrounding the PMU's interrupt system by exploring the method in which the PMU receives an interrupt and investigating the GIC proxy block and inter processor interrupt channels.

Receiving an Interrupt

It is important to be aware of what happens when the PMU's interrupt controller receives an interrupt. The process is detailed as follows:

1. The interrupt controller receives an interrupt. The software code in the ROM consults with the *pending interrupt register* (IRQ_PENDING) within the PMU's I/O unit. The IRQ_PENDING register will determine which interrupt is pending and requires maintenance.

2. The interrupt service routine associated with the pending interrupt is sourced from the ROM or RAM (the RAM may contain an interrupt service routine in its firmware). If several interrupts are pending, then the interrupt priority detailed in [1], determines which interrupt should be issued first.

3. The RAM may contain firmware that overrides the routines configured in the ROM. The priority of the interrupt may also be overridden by the firmware. If the RAM contains no firmware, the interrupt priorities and service routines are maintained by the ROM.

The GIC Proxy

The GIC proxy block is used to collate the interrupts that are connected to the APU and RPU GICs. The proxy makes it possible for the PMU to detect an interrupt that has been issued to the APU or RPU, when they are powered down. The GIC proxy block is capable of supporting the PMU in low-power mode.

Upon detecting an interrupt via the GIC proxy, the PMU may then take the relevant action of waking up the processing unit responsible for handling the interrupt, or carrying out a pre-determined action in accordance with the firmware contained inside its RAM or ROM.

The GIC proxy interrupts are controlled by twenty-five 32-bit registers, as set out in the low power domain system-level control register bank (LPD_SLCR). These are detailed in the Zynq MPSoC Register Reference [5].

The PMU's Inter-Processor Interrupt Channel

The PMU's interrupt controller is also connected to some of the inter-processor interrupt (IPI) channels. The purpose of the IPI in the Zynq MPSoC is to provide a processor with the ability to interrupt another. This is useful when a processor requires another to perform an action. In the Zynq MPSoC, there are eleven IPI channels. Four channels are assigned to the PMU interrupt controller. These channels may not be reprogrammed as they are fixed in hardware i.e. only connected to the PMU interrupt controller. The remaining channels are connected to other processing units in the Zynq MPSoC.

The IPI channels are capable of sending 32-bytes of information between connected processors. The channels use memory buffers to help support the inter-processor interrupt structure. The message buffers are capable of storing request and response messages between processors using the IPI channels. There are only 8 sets of memory buffers: seven sets are assignable to any processor using an IPI channel, and one set of buffers are dedicated to the PMU.

The PMU has a set of special rules regarding the IPI. Primarily, the PMU is assigned four IPI channels (fixed in hardware) and a dedicated memory buffer. In addition to this, channel 0 of the IPI (IPI0) has a dedicated purpose. IPI0 instructs the PMU to enter sleep mode. Sleep mode is useful if the PMU RAM must be accessed by an external master, via the APB interface (IPI0 is required when loading the PMU's firmware into RAM).

Only information about the IPI that is relevant to the PMU has been discussed in this section. More details on this subject are available in [1].

10.4 Error Management

The PMU is responsible for error capture throughout the Zynq MPSoC. Error signals are routed to the PMU, where they are recorded in the global registers block. On the event that an error has occurred, the PMU can take appropriate action depending on the type of error.

There are registers and error input signals made available by the PMU to capture software and system-level hardware errors respectively. Software errors are those that occur during the execution of software code in the PMU's ROM and firmware, and the CSU's ROM. System-level hardware errors are the result of an error occurring in one of the Zynq MPSoC's processing units or system elements. This type of error will propagate an error signal to the PMU. Note that when the PMU is executing its pre-boot routine, software errors are not recorded by the PMU. However, all other errors are recorded.

When a software or system-level hardware error occurs, it is recorded in the PMU using error status registers. All of these errors are capable of triggering an interrupt to the PMU's processor. The interrupt can be masked (ignored by the PMU's processor) depending on the error that occurs. This is possible using the interrupt error registers contained inside the PMU's global registers block.

In this section, we discuss the error status and enable registers, the registers involved in triggering an error interrupt to the PMU, and the actions that can be taken to handle an error that has occurred.

10.4.1 Error Status and Enable Registers

The PMU provides error capture and resolution for the Zynq MPSoC. The *error status* of system elements is captured in the global registers block in the PMU. These registers are denoted as ERROR_STATUS_1 and ERROR_STATUS_2 respectively. Their bit level descriptions are provided in [5]. If an error has occurred, the status of the error will remain in the *error status register* until it is cleared, even if there is a SRST or an internal POR.

Errors that affect the *error status registers*, can be enabled or disabled by configuring the *error enable registers*. There are two error enable registers denoted as ERROR_EN_1 and ERROR_EN_2 in the global register block. Writing a bit '0' to a location in these registers will prevent the associated error from affecting the *error status registers* (or the interrupt error registers, described in Section 10.4.2). Writing a bit '1' to a location in the *error enable registers* will allow the error to affect the error status (and interrupt error) registers.

10.4.2 Interrupt Error Registers

Interrupt error registers help with the management of errors. In particular, they are used to determine whether an interrupt should propagate to the PMU processor when an error has occurred in the Zynq MPSoC. The PMU processor will then perform an action as set out in its firmware to respond to the error. Similar to the error status and enable registers, the *interrupt error registers* are contained inside the global registers block.

The interrupt mask register (ERROR_INT_MASK_x) is used to determine whether an error interrupt will propagate to the PMU processor. If an interrupt is enabled for an error, the bit location of that error in the interrupt mask will be equal to a bit '0'. If an interrupt is disabled for an error, the bit location of that error in the interrupt mask will be equal to a bit '1'.

If an error interrupts the PMU processor, there must be firmware available to handle the error/interrupt. If there is no firmware available, a *no-firmware* error will occur.

10.4.3 Error Handling

When an error occurs, an action can be carried out. We have already covered one of these methods in which the PMU can be informed of an error by using an interrupt (discussed in Section 10.4.2). There are three other actions that can be executed as the result of an error. These are as follows:

- Generation of a SRST.
- Generation of a POR.
- Assertion of the PS_ERROR_OUT signal on the Zynq MPSoC device. This signal should be asserted for the purpose of communicating an error with the outside world. Disabling the PS_ERROR_OUT signal may be beneficial for secure designs, as this prevents the outside world from knowing that the device is in an error state.

To perform one, or several of the above actions, the relevant mask register(s) must be configured. This process is similar to that described previously for error interrupts (Section 10.4.2). The mask registers for the SRST, POR, and the PS_ERROR_OUT signal are contained in the global register block. The mask registers are configured by writing to the associated enable and disable registers.

For a list of error sources thoughout the Zynq MPSoC device, and the state of their associated masks, the reader is directed to [1].

10.5 PMU Firmware

The PMU's triplicated MicroBlaze processor loads and executes software code from two available memories. These are the 32KB ROM (which cannot be modified) and the 128KB RAM (which can be modified). The code contained within each memory allows the PMU to carry out power-up, reset, and control functions and routines for the Zynq MPSoC.

The PMU's ROM is preloaded with various functions known as the PMU Boot ROM (PBR) functions. The PBR is responsible for carrying out tasks before and after booting of the Zynq MPSoC. To carry out more complex platform management techniques, the PMU will need to use custom PMU firmware, also known as the PMUFW. The custom PMUFW is written to and stored in the 128 KB RAM, so that it can be loaded and executed by the processor.

This section aims to provide the fundamentals of the PMUFW by exploring its features and increasing its accessibility for those not yet familiar with its execution. We will investigate the PMUFW's composition and execution. Additionally, the firmware's modular functionality will also be reviewed, where adding and customising software modules can be easily achieved. Finally, loading a custom PMUFW into the RAM is also discussed, as this routine must be carried out in a particular way.

Note, Xilinx has provided a template for the PMUFW; developers do not have to write their own firmware from scratch. The firmware template provided by Xilinx can be accessed using the Xilinx SDK, where the source code has been made accessible. The template may be modified to suit the developer's needs and application constraints, as described in [8] and [11].

10.5.1 Composition of the PMU Firmware

The PMUFW is composed of two components, namely its *Base Firmware* and *Modules*. The Base Firmware provides rudimentary Application Programming Interfaces (APIs) for calling ROM and reset services, and utilities [8], [11]. These are part of the PMUFW's general APIs.

The Base Firmware also provides three handlers for modules (described further below). These handlers are for initialising a software module, handling an event, and handling an IPI. A handler is a software routine for processing and responding to a system event or interrupt. The handlers are part of the PMUFW's core APIs, discussed in [11].

Modules provide the actual functionality of the PMUFW. The *Modules* component is named as such because it is entirely modular. Additional functionality can be added to the PMUFW in the form of software modules. Modules help by simplifying the addition of new functionality to the PMUFW and they support easy customisation of existing functionality.

The Base Firmware and Modules work together to provide the overall functionality of the PMUFW. The Modules call on the APIs provided in the Base Firmware to accomplish their tasks and routines. A block diagram illustrating the composition of the PMUFW can be seen in Figure 10.8.

Note that there are four default modules in the PMUFW. These are Error Management, Power Management, the Scheduler and the Safety Test Library. Each are illustrated on the left side of the modules block in Figure 10.8. There is also a fifth module that represents a user created, custom module. There may be up to 32 modules at one time in the firmware (additional modules can be created by using existing modules as templates). Furthermore, unused modules can be disabled to save memory (disabling modules is further discussed in [11]).

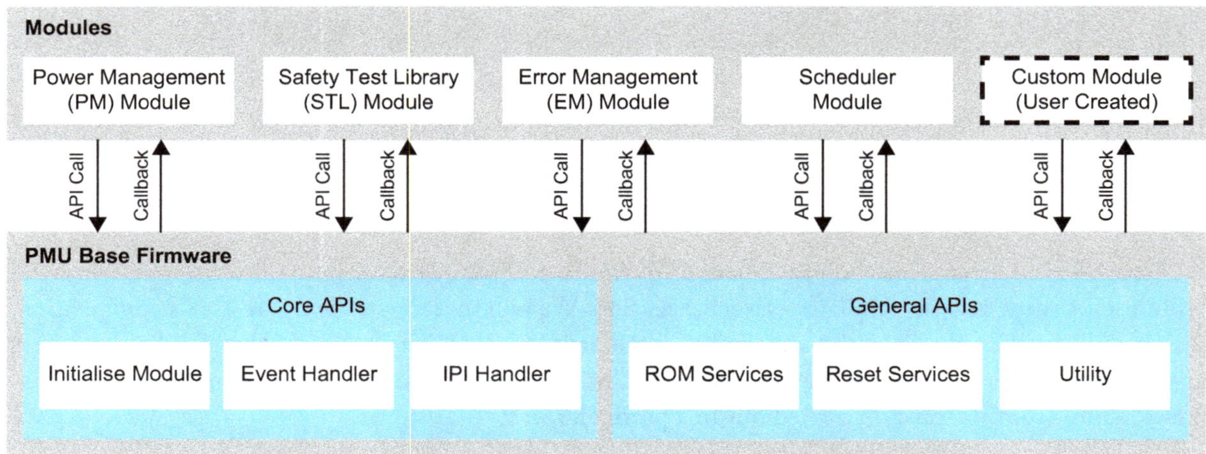

Figure 10.8: The composition of the PMU's firmware [11]

10.5.2 Default Modules

As shown in Figure 10.8, there are four default modules. Each are detailed as follows.

- **Error Management (EM)** — Errors that are generated throughout the Zynq MPSoC are handled by the EM module in the PMUFW. The error handlers within the EM module, may be customised to decide what action should be carried out when an error occurs. The type of actions that can be performed in response to an error were discussed previously in Section 10.4.3.
- **Power Management (PM)** — The PM module is responsible for handling interrupts that are triggered by power management events throughout the Zynq MPSoC. The module contains a set of handlers that determine which action should be taken to respond to a power management event. The PM module can also access the PMU ROM to obtain handlers for control of particular hardware resources.
- **Scheduler** — The scheduler module is responsible for maintaining the time and status of a set of tasks required by the PMUFW. The scheduler uses PIT1, to maintain interval timing for tasks. There may be up to 10 tasks that are prioritised using a number. '0' is the highest priority and '9' is the lowest. Further information about the scheduler's data structure and operation can be found in [11].
- **Safety Test Library (STL)** — The STL is a repository of software safety procedures and routines. These software safety mechanisms are used together with hardware safety features for detecting random hardware faults.

10.5.3 Execution of the PMU Firmware

The PMUFW execution flow consists of three phases. These are initialisation, post-initialisation, and wake-up. A diagram illustrating the transitions between each phase is shown in Figure 10.9.

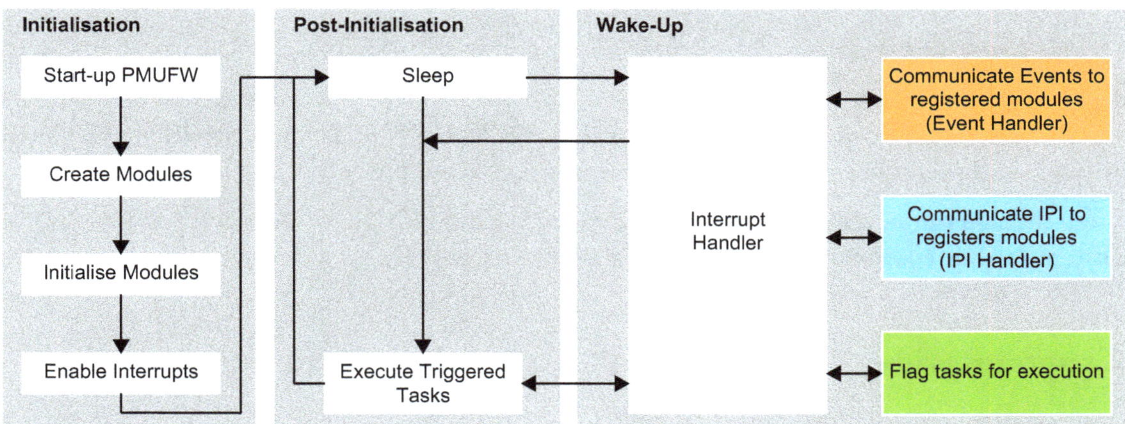

Figure 10.9: An overview of the execution flow of the PMU's firmware [11]

As shown in Figure 10.9, the firmware enters the initialisation phase first, then enters the post-initialisation phase. The firmware will then proceed to move back and forth between the post-initialisation and the wake-up phase. The operations carried out in each phase are described below.

Initialisation Phase

On start-up, the firmware enters the initialisation phase. During this state of operation, the firmware will execute several functions and routines. These include built in self-tests (BISTs), hardware initialisations, and module creation and initialisation (custom modules are briefly discussed in Section 10.5.4).

During this period, interrupts to the PMU are disabled and will not be enabled until the initialisation phase is complete. This is to stop interrupts from propagating to the PMU (causing interruptions in code execution) and to prevent system resources from being used before they have been successfully initialised.

Post-Initialisation Phase

After initialisation, the firmware will enter the post-initialisation state and fall into sleep mode. The firmware will continue to sleep until an interrupt to the PMU is triggered. The firmware will then move to the wake-up phase by entering into the interrupt handler state.

Wake-Up

In this phase, the firmware will now service the received interrupt. This involves determining which module the interrupt is for and either calling the module's Event Handler (if the interrupt corresponds to a registered event) or the module's IPI handler (if the interrupt has a corresponding module ID). The firmware will then flag any tasks that need to be executed with respect to the handler called. The firmware will then return to the post-initialisation phase and execute the triggered tasks. Once complete, the firmware will fall back into sleep mode until another interrupt occurs.

10.5.4 Custom Modules

Custom modules can be added to the PMUFW to extend the PMU's functionality in the Zynq MPSoC. Additional modules should be entirely self-contained in the PMUFW and may use additional API's to call in firmware functions and routines from the *Base Firmware* (described previously in Section 10.5.1).

A custom module can be simply created by adding code to the PMUFW template provided in the Xilinx SDK. Custom module code and examples, are detailed further in [11].

10.6 Chapter Review

This chapter has introduced the Zynq MPSoC's power domains, modes of operation, and PMU.

In the early part of this chapter, the Zynq MPSoC was seen to use three operational power modes: battery powered mode, low-power mode, and full-power mode. The Zynq MPSoC can also be placed into deep-sleep mode that consumes the lowest amount of power while still maintaining the boot and security state of the entire system. Subsequently, the Zynq MPSoC's primary power domains were explored. These included the battery power domain, low-power domain, full-power domain and PL power domain.

The PMU's architecture was introduced, including its triplicated processing unit, memories, peripherals, and interrupt system. In particular, it was noted that the ROM of the PMU contains a fixed (read-only) set of functions and routines for the Zynq MPSoC platform. In contrast, the RAM is capable of being written to by a master processor, so that new functions and routines may be added (or to override existing functions in the ROM).

The software composition of the PMU's firmware was also explored. The firmware was found to have a modular architecture allowing new functionality to be easily added. Additionally, it was noted that the default firmware contains four modules. These are error management, power management, scheduling and the safety test library. Xilinx provide a firmware template, that can be modified by the developer to add custom modules if desired.

10.7 References

Note: All online sources last accessed March 2019.

Some of the Xilinx documentation referred to below has version-specific URLs. If you are working in a newer version of the tools, check for updates on the Xilinx website, or try adjusting the link according to your version.

[1] Xilinx, Inc., "Zynq UltraScale+ Device Technical Reference Manual", UG1085, v1.9, January 2019.
Available: http://www.xilinx.com/support/documentation/user_guides/ug1085-zynq-ultrascale-trm.pdf

[2] G.Steiner and B.Philofsky, "Managing Power and Performance with the Zynq UltraScale+ MPSoC", Xilinx White Paper, WP482, v1.1, October 2016.
Available: https://www.xilinx.com/support/documentation/white_papers/wp482-zu-pwr-perf.pdf

[3] Xilinx, Inc., "Zynq UltraScale+ MPSoC Product Tables and Product Selection Guide", XMP104, v.2.4, 2018.
Available: https://www.xilinx.com/support/documentation/selection-guides/zynq-ultrascale-plus-product-selection-guide.pdf

[4] Xilinx, Inc., "Zynq UltraScale+ Device Packaging and Pinouts", UG1075, v.1.7, January 2019.
Available: https://www.xilinx.com/support/documentation/user_guides/ug1075-zynq-ultrascale-pkg-pinout.pdf

[5] Xilinx, Inc., "Zynq UltraScale+ MPSoC Register Reference", UG1087, v1.7, February 2019.
Available: https://www.xilinx.com/html_docs/registers/ug1087/ug1087-zynq-ultrascale-registers.html

[6] Xilinx, Inc, "MicroBlaze Processor Reference Guide", UG984, v2018.2), June 2018.
Available: https://www.xilinx.com/support/documentation/sw_manuals/xilinx2018_2/ug984-vivado-microblaze-ref.pdf

[7] Wiki.xilinx.com, "Xilinx Wiki — Getting Started: Overview of the Xilinx Zynq UltraScale+ MPSoC & Zynq-7000 AP SoC Design Flow", February 2019.
Available: http://www.wiki.xilinx.com/Getting+Started

[8] Wiki.xilinx.com, "Xilinx Wiki — PMU Firmware", January 2019.
Available: http://www.wiki.xilinx.com/PMU+Firmware

[9] Xilinx, Inc., "Zynq UltraScale+ MPSoC Quick Emulator User Guide, QEMU", UG1169, v2016.2, June 2016
Available: https://www.xilinx.com/support/documentation/sw_manuals/xilinx2016_2/ug1169-zynqmp-qemu.pdf

[10] Wiki.xilinx.com, "Xilinx Wiki — MPSoC Power Management", September 2018.
Available: http://www.wiki.xilinx.com/MPSoC+Power+Management

[11] Xilinx, Inc. "Zynq UltraScale+ MPSoC: Software Developer Guide", UG1137, v8.0, June 2018.
Available: https://www.xilinx.com/support/documentation/user_guides/ug1137-zynq-ultrascale-mpsoc-swdev.pdf

Part C

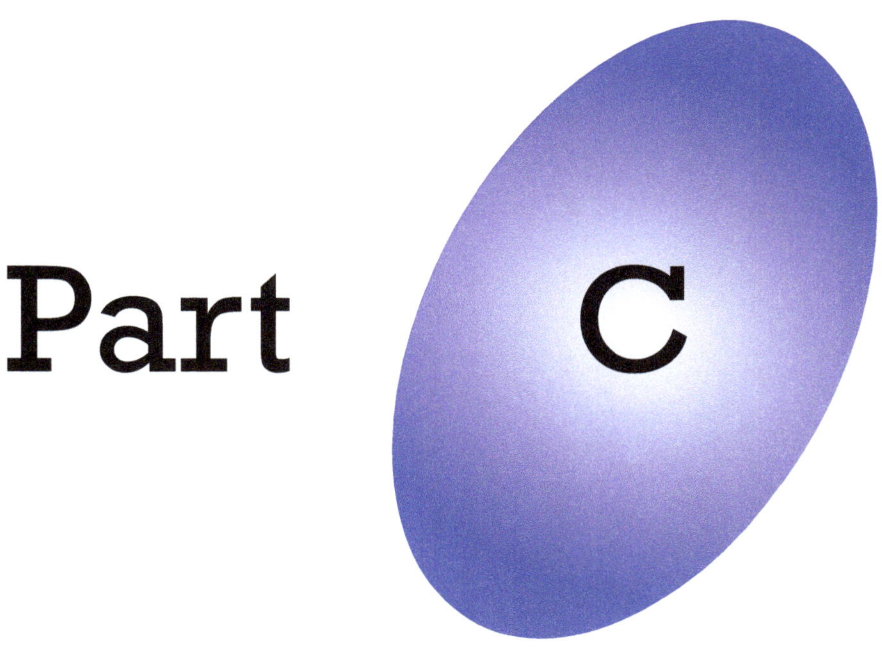

Zynq MPSoC Systems Development

Chapter 11

Hardware System Development

Up until now, this book has detailed the various processing units, safety, security, and platform management features of the Zynq MPSoC device. In this part of the book, system development with the Zynq MPSoC will be explored and will call upon knowledge previously learned in other chapters. This chapter in particular will dive into the Zynq MPSoC's hardware infrastructure and system-level communication between each of its processing units. A particular emphasis will be placed on the communication interfaces between the PS and PL, as these are essential for communicating with hardware accelerators in the FPGA logic fabric.

The primary motivation for using a Zynq MPSoC device is to leverage the FPGA for accelerating functions and tasks. The discussion in this chapter will provide you with a better understanding of the PL, and how you can use it in your own system design. The chapter is outlined as follows:

- Hardware/software co-design is introduced and the roles of the Zynq MPSoC's processing units are described. In particular, the role of the PL in the Zynq MPSoC is discussed.
- An overview of the Zynq MPSoC hardware system is provided, detailing its interfaces and signals, interconnects, and memories. Subsequently, the purpose of each PS-PL AXI port is explored.
- The Zynq MPSoC interrupt system and device memories are also detailed. Finally, the fundamentals of transferring data throughout the Zynq MPSoC device is described.

As there are differences between Zynq MPSoC device families, discussions and illustrations in this chapter will be primarily based on EV devices (the device family with the most features at the time of writing). Throughout this chapter, we will explicitly mention any differences between device families when necessary.

11.1 Heterogeneous Computing with Zynq MPSoC

Zynq MPSoC devices have several processing units and features that each have their own special processing capabilities. A significant challenge of designing embedded systems with the Zynq MPSoC is to use each of it's constituent processing units to their best effect and, when required, have each operate in parallel with one another. This is not an easy task, however, there are methodologies and tools that can help.

Hardware/software co-design is an important task during the development of an embedded system, as it can have a significant effect on the overall system performance, power and resource consumption. The underlying process involves partitioning an embedded design into hardware and software components. The motivation for doing so is to leverage the capabilities of each execution environment. Hardware components, such as those found in FPGAs, can speed up most aspects of embedded systems, due to their inherent parallel processing capabilities. In contrast, software components lend themselves better to sequential processing, typically implemented in a microprocessor or microcontroller.

The Zynq MPSoC device can enable hardware/software co-design effectively during its application development stage. There is a general methodology outlined in [2] that will help the system designer determine the best course of action to take, when assigning a task to a processing unit. The general principles of this methodology is shown in Figure 11.1.

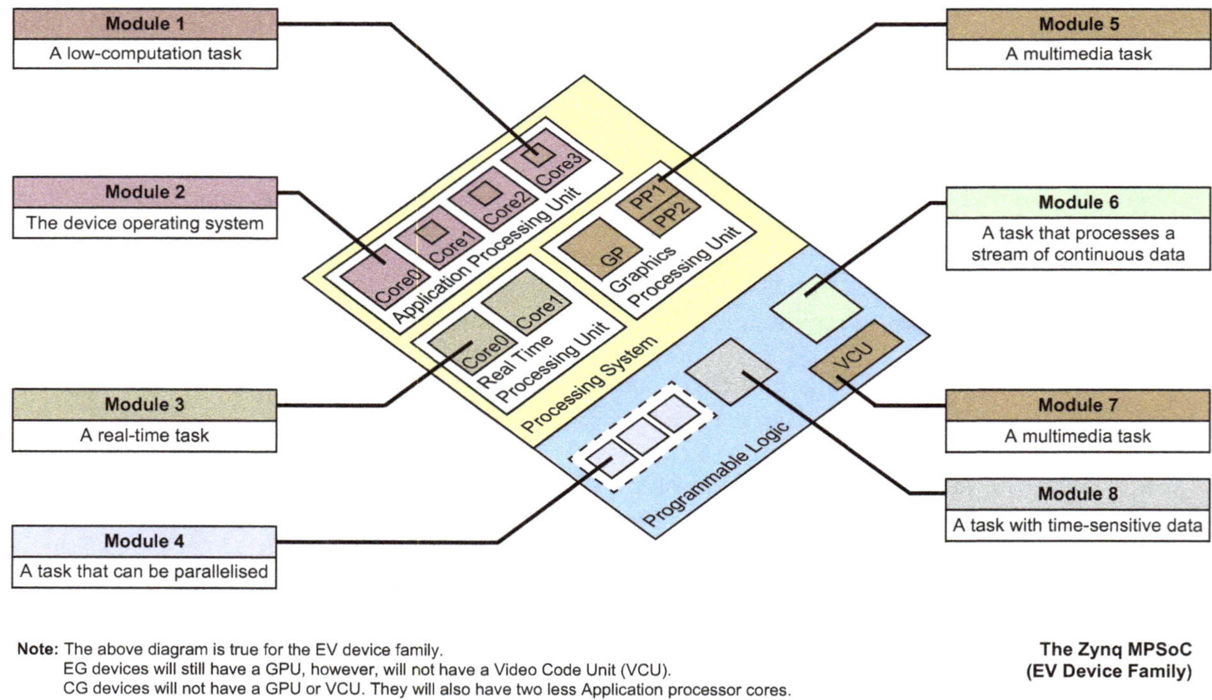

Figure 11.1: General principles of hardware/software partitioning using the Zynq MPSoC [2]

Each processing unit provides their own unique functionalities for hosting embedded applications and increase the opportunities for hardware/software partitioning. For example, if an application needs to continuously process data, then the PL could be the most suitable for this task. The PL will provide a reasonable trade-off between processing time and resource consumption.

In Figure 11.1, the Cortex-A53 application processors host the device's operating system. It provides drivers for peripheral interfaces and can participate in the execution flow of an application. Furthermore, it is capable of controlling hardware accelerators in the PL using PS-PL interfaces.

As shown, the PL is capable of efficiently implementing functions and routines that can be separated into multiple smaller tasks. The speed-up in computation is realised when the smaller tasks are executed simultaneously. The parallel execution native to FPGAs allow for concurrent execution of algorithmic functions and tasks. Additionally, it is also suitable for time-sensitive data, as it has deterministic performance. If a task processes streaming data, it should be executed in the PL if it can be effectively implemented using the FPGA fabric [2].

The challenges involved in hardware/software co-design, lie in identifying the routines and functions that should be executed in hardware or software. This crucial step in embedded system design can be approached using a new Xilinx development environment named SDx [4].

SDx (detailed in Chapter 11) provides the developer with an entirely software based environment for hardware/software co-design. Individual software functions, written as part of a complete embedded system, can be selected to execute in the PL by calling High-Level-Synthesis (HLS) [5]. HLS reduces the complexity and time taken to manually create hardware designs, by converting software functions directly into RTL. This type of development is particularly useful for Zynq MPSoC devices, as software can be developed on the APU and RPU, while targeting software functions for acceleration in the PL.

A significant challenge when designing a system with the Zynq MPSoC is to determine how each processing unit can communicate information effectively with one another. This problem exists for each processing unit in the Zynq MPSoC that needs to share data. For example, what physical medium exists for the APU to communicate with the PL?

Communication between processing units is described at a software level using Linux, openAMP, and the Xen hypervisor in Chapter 13. However, this chapter aims to explore the underlying hardware infrastructure that we can use to control the data movement and communication between processing units and elements. These are the Zynq MPSoC's interfaces and signals, memories, and interconnects throughout the device. Each component plays a vital role in the transfer of data in our embedded system design.

Before diving into the world of SDx, Linux, openAMP, and Xen, it is recommended that you become familiar with the Zynq MPSoC's hardware infrastructure. The remainder of this chapter will explore the Zynq MPSoC's hardware system, PS-PL communication paths, interrupts, and memories. Section 11.6 details how data is moved throughout the Zynq MPSoC using Direct Memory Access (DMA) controllers.

11.2 Hardware System Overview

The Zynq MPSoC is a heterogeneous multiprocessor platform consisting of several processing units. Each processing unit is capable of executing a set of instructions as set out by their underlying architecture. To leverage the Zynq MPSoC to the best of its ability, all processing units must be used based on their capabilities to effectively reduce power consumption and increase processing performance. Therefore, it is necessary for data to be shared effectively. Data sharing is an important challenge when developing the hardware system in the Zynq MPSoC device. It is essential that processors are able to share or retrieve data using the underlying hardware infrastructure. Additionally, an application may also require communication to the outside world (off-chip).

The challenge associated with sharing data between processing units, must first be considered at the system-level of the Zynq MPSoC (its physical infrastructure). Other than the Zynq MPSoC's processing units, the majority of it's resources and processing elements can be broken down into three categories: interfaces and signals, interconnects, and memories.

- *Interfaces and signals* are provided by processing units and system resources. These are gateways that are connected to *The Interconnect*, or other hardware resources, so that data may propagate from one system element to another.
- Interconnects (sometimes collectively referred to as *The Interconnect* in Zynq MPSoC devices) provide a connection for each of the processing units and elements in the hardware system. *The Interconnect* makes it possible for data to travel from one point on the chip to another.
- *Memories* are essential for storage, allowing data and instructions to be saved until later required by a processing unit.

This section will introduce the basic hardware infrastructure of Zynq MPSoC devices. Discussion will include the fundamentals of the Zynq MPSoC's interfaces and signals, interconnects, and memories. Each of these topics will increase your knowledge of the hardware infrastructure and support the detailed discussion of these topics later in this chapter.

11.2.1 Interfaces and Signals

Zynq MPSoC devices are packed full of features that provide an abundance of solutions to complex embedded system designs. One of the main challenges during system development is deciding which signals and interfaces to use when communicating with many of the Zynq MPSoC's processing resources and elements. This is particularly evident when communicating between the PS and PL.

An illustration of the Zynq MPSoC's interfaces and signals can be seen in Figure 11.2. Note that the illustration provided is not complete. The reader is directed to [1] for more a complete diagram. Previously, the Multiplexed Input/Output (MIO), Serial Input Output Unit (SIOU), and PL signals were introduced in Chapter 3. These topics will not be discussed in this section. Instead, the interfaces and signals between the PS and PL will be detailed (illustrated in Figure 11.2 with a dashed red box).

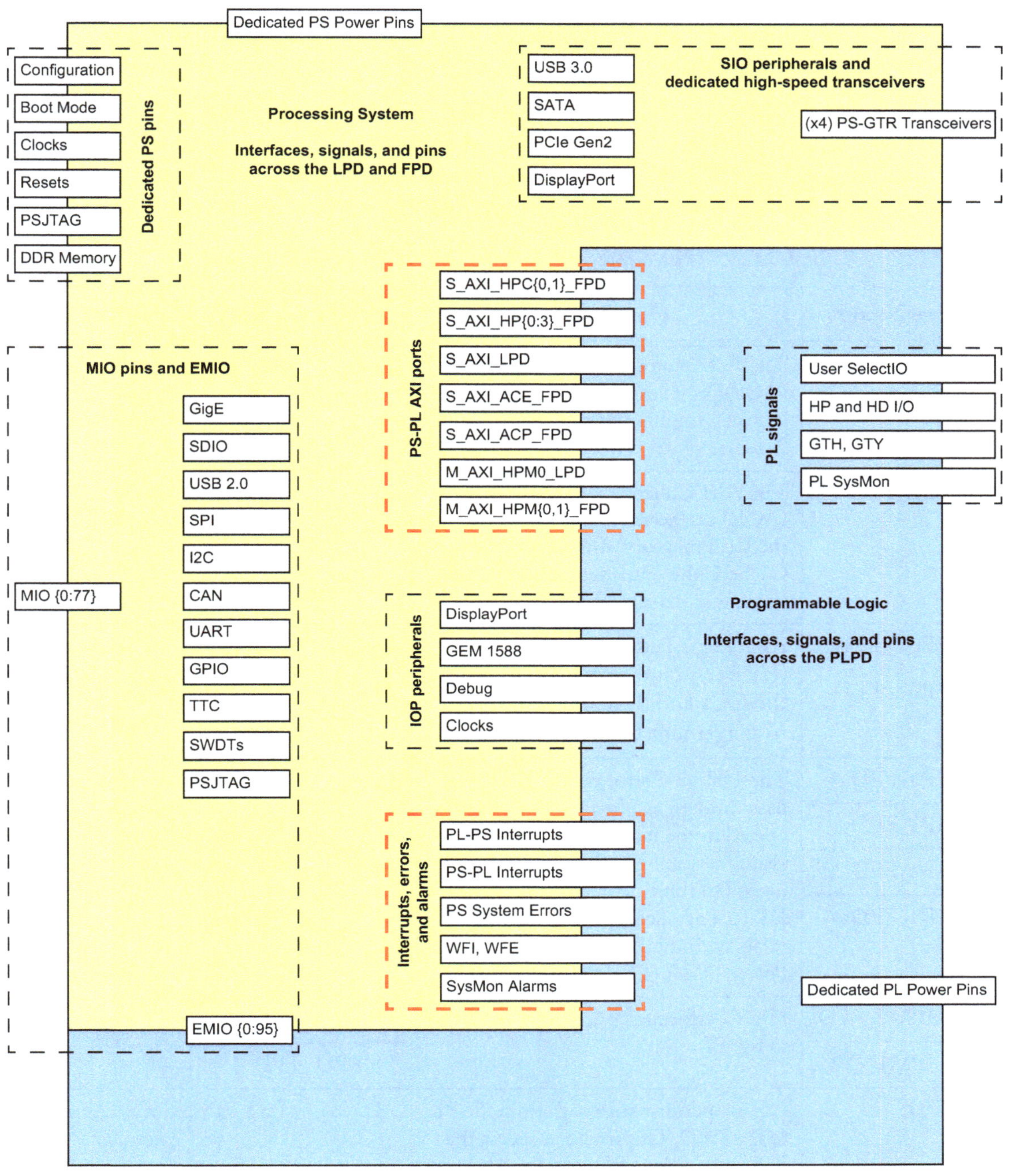

Figure 11.2: A simplified diagram of the Zynq MPSoC's interfaces, signals, and pins [1]

PS-PL AXI Ports

The PS can interface to hardware accelerators in the PL using the PS-PL AXI ports provided. These are high bandwidth, low latency connections.

There are 12 PS-PL AXI ports in Zynq MPSoC devices, as described in Table 11.1. Note that there is a naming convention for these interfaces. The first letter of an interface always represents the function of the PS, i.e. an 'S' indicates that the PS is the slave, an 'M' indicates that the PS is the master. FPD specifies that the port is part of the full-power domain, while LPD specifies that the port is part of the low-power domain.

Table 11.1: AXI ports that connect the PS and PL for transferring data [1]

Interface Name	Description	Master	Slave	Further Reading
S_AXI_ACP_FPD	The Accelerator Coherency Port (ACP) provides a coherent path between the PL and the APU's Level 2 cache.	PL	PS FPD	The ACP was discussed previously in Section 6.3.2 on page 139.
S_AXI_ACE_FPD	The AXI Coherency Extension (ACE) can access system memory and the local memory of the APU, via the Cache Coherent Interconnect (CCI); sharing up-to-date information.	PL	PS FPD	See Section 11.3.2 on page 274.
S_AXI_HPC0_FPD	Each High-Performance Coherent (HPC) port is directly connected to the CCI and System Memory Management Unit (SMMU).	PL	PS FPD	See Section 11.3.4 on page 277.
S_AXI_HPC1_FPD		PL	PS FPD	
S_AXI_HP0_FPD	These High-Performance (HP) ports pass through the SMMU and are connected to the interconnect's central switch in the FPD. They are connected to three dedicated ports on the DDR controller. However, sharing exists with the DisplayPort and Full-Power DMA (FP-DMA).	PL	PS FPD	See Section 11.3.4 on page 277.
S_AXI_HP1_FPD		PL	PS FPD	
S_AXI_HP2_FPD		PL	PS FPD	
S_AXI_HP3_FPD		PL	PS FPD	
M_AXI_HPM0_FPD	High-Performance port from the FPD to the PL.	PS FPD	PL	See Section 11.3.6 on page 278.
M_AXI_HPM1_FPD		PS FPD	PL	
S_AXI_LPD	A high-performance path from the PL to the LPD. This port can access the RPU when the FPD is powered down.	PL	PS LPD	See Section 11.3.5 on page 277.
M_AXI_HPM0_LPD	Low-latency high-performance port that interfaces the LPD to the PL.	PS LPD	PL	See Section 11.3.6 on page 278.

As indicated in the last column of Table 11.1, each of the ports are discussed in more detail later in this chapter (apart from the ACP that was discussed earlier in Chapter 6).

PS-PL Interrupts

The PL is capable of triggering 16 interrupts to the PS, each of which are assigned a priority level (for prioritised interrupt handling). The PL interrupts propagate through the Zynq MPSoC, via interrupt controllers, to arrive at their target processor. The interrupts can be in the form of an *Interrupt ReQuest (IRQ)* or *Fast Interrupt reQuest (FIQ)*.

There are also four inter-processor interrupt (IPI) channels connected to the PL. These are useful for providing direct communication with other connected processing units in the PS. There are also connections from the PL to the APU's legacy FIQs and IRQs, and interrupt connections from the PL to the legacy nFIQ and nIRQ of both RPU cores (lower-case n represents active-low signals).

In Section 11.4, the entire Zynq MPSoC interrupt system is discussed in more detail.

11.2.2 The Interconnect

The Zynq MPSoC's interconnect is particularly complex. Compared to the Zynq SoC [3], there are many more features, switches, and blocks that control the flow and transfer of data in the Zynq MPSoC device. It is not necessary to learn every aspect of the Zynq MPSoC's interconnect architecture. Unless you are an advanced developer, you will only need to understand the fundamentals of the interconnect.

The interconnect is located in the PS of the Zynq MPSoC. It facilitates the communication between master and slave elements using AXI point-to-point channels. The interconnect is comprised of many high-performance switches that support the AXI protocol for read, write, and response transactions between system elements [1].

A simplified diagram of the Zynq MPSoC's interconnect is shown in Figure 11.3. Note that there are many more resources, processing elements, and connections than those shown. The reader is directed to [1] for a detailed diagram of the Zynq MPSoC's interconnect architecture.

There are several AXI connections in the Zynq MPSoC's interconnect. Many connections use high-performance switches that support several AXI bus protocols in the Arm AMBA family. The Arm AMBA bus connections are used by the interconnect to support many of its features and capabilities, including Quality-of-Service (QoS) support for improving the prioritisation of AXI transactions, and debug and test monitoring for Zynq MPSoC system elements. The reader is directed to [1] for more information about AXI QoS.

The interconnect in Figure 11.3 has been separated into three categories. These are the LPD, FPD, and PL power domain (PLPD). Note that 5 out of 6 DDR controller ports (shown in red) are to the FPD. Only one is connected in the LPD.

The remainder of this section will highlight the features available in the Zynq MPSoC interconnect, and also provide a summary of the interconnect switches. Additionally, we will mention some of the AXI functional blocks in the interconnect (not shown in Figure 11.3).

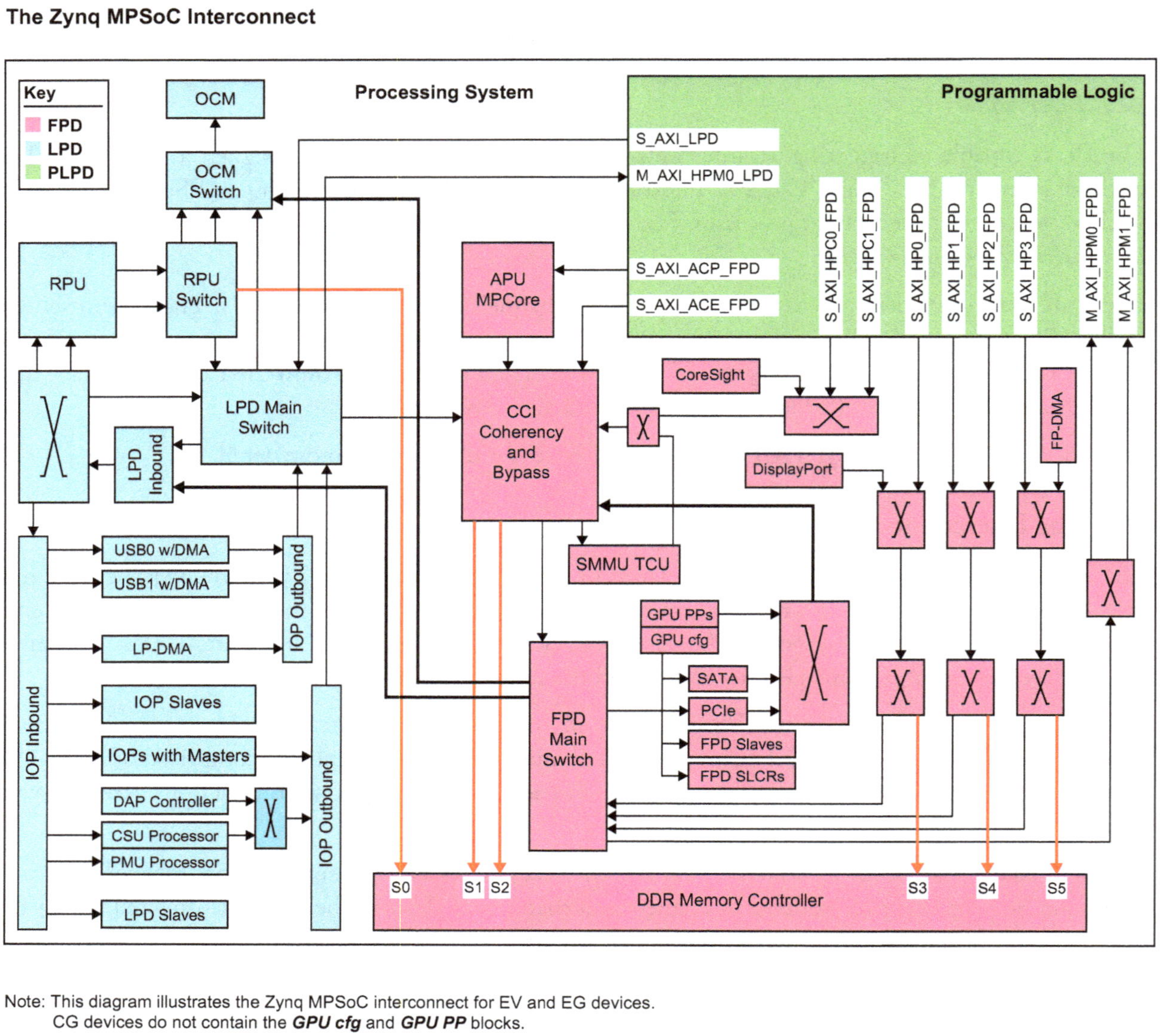

The Zynq MPSoC Interconnect

Note: This diagram illustrates the Zynq MPSoC interconnect for EV and EG devices.
CG devices do not contain the ***GPU cfg*** and ***GPU PP*** blocks.

Figure 11.3: A simplified diagram of the Zynq MPSoC's interconnect in the PS [1]

System Coherency using the CCI

When a processor caches data, it is storing that data in a local memory that is easily accessible and typically has low-latency. This improves the performance of that processor core, because it can retrieve data quickly as opposed to the higher latency incurred from accessing main system memory. The processing units in the Zynq MPSoC have their own local memories (caches) for temporarily storing data.

One of the challenges associated with a multiprocessor environment, such as the Zynq MPSoC, is ensuring all processing units are up-to-date with data that is produced throughout the system. A problem can occur when two or more processing units are storing shared data in their own local caches. If one modifies this data, the system may become cache incoherent, as the others have not been informed (and consequently will use the wrong data in their functions and routines). The objective is to ensure that all processing units can remain cache coherent with one another so that they all have the same view of the shared data. There are two methods of achieving cache coherency while still maintaining good system performance, as described below:

- **Software Coherency** — In a software solution, the processing unit responsible for producing new data must clean/flush their caches before allowing another to read the shared data from memory. If the processing unit is consuming the shared data, the process is similar. However, they must invalidate their cache instead before reading the new data. This is because the data they have initially stored is no longer up-to-date. Unfortunately, this method requires a lot of processor intervention.

- **Hardware Coherency** — A hardware solution can improve system performance, as this simplifies the software involved. A processing unit's physical interface that is used for requests, snoops on the requests generated by other processing units, as to provide them with the most up-to-date data if available in their cache (snooping is described in Section 11.3.1). The improvement in performance is significant, as there is no need to perform cache flushing/invalidation normally required by a software solution.

The CCI is a helpful addition to the PS architecture and allows processors to share tasks and data between one another using hardware coherency. Primarily, the CCI is used to achieve asymmetric processing, fundamentally providing interconnect support for coherency between processing tasks and ensuring each processor core operates on the most up-to-date data across the entire PS.

There are two ways in which the CCI can be leveraged to support coherency in the Zynq MPSoC. The first method is full (two-way) coherency between master devices in the system. Masters that are fully coherent can provide each other up-to-date data between caches. This is also known as snooping, further described in Section 11.3.1. Coherency can also be carried-out using I/O coherency (one-way). An interface using I/O coherency can retrieve data, but cannot share data with other connected masters (snooping is one-way).

The CCI uses the Arm CoreLink CCI-400 IP that is detailed further in [6]. Coherent PL interfaces are discussed further in Section 11.3.

System Memory Management Unit

The SMMU is responsible for address translation of connected system masters and a series of protection services for slave clients. The SMMU uses the Arm CoreLink System MMU-500 IP, as described in [7]. The SMMU is discussed further in Section 8.3.1.

Interconnect Switches

The switches in the Zynq MPSoC's interconnect are based on the Arm CoreLink NIC-400 [8] network interconnect IP. Some of the significant switches are summarised below:

- **FPD Main Switch** — This is one of the Zynq MPSoC's primary switches, responsible for connecting masters in the FPD, to the LPD slaves. This switch has a direct path to the On-Chip Memory (OCM) (via the OCM switch) that bypasses the LPD main switch entirely, as to minimise the latency between the FPD and OCM. Additionally, this switch allows masters in the FPD, access to the LPD's peripheral registers (via the LPD inbound switch).

- **LPD Main Switch** — A lot of data movement in the LPD is controlled by this switch. In particular, it facilitates access to the LPD inbound switch, OCM switch, and PL (via the M_AXI_HPM0_LPD interface). Importantly, it provides a connection to the CCI in the FPD, capable of providing a path for the RPU and Low-Power DMA (LP-DMA) for cache coherent operation with other processing units (I/O coherency only). Additionally, this switch also controls incoming traffic from the Input/Output Peripherals (IOPs) and hardware blocks in the PL (S_AXI_LPD).

- **OCM Switch** — The OCM switch facilities the movement of traffic between the OCM, FPD main switch, LPD main switch, and RPU switch.

- **RPU Switch** — This switch is responsible for connecting the RPU to port 1 of the DDR memory controller. It also facilitates transactions between the RPU, OCM switch, and LPD inbound switch.

- **LPD Inbound Switch** — Processing units in the Zynq MPSoC, such as the RPU and APU, can access the Platform Management Unit's (PMU's) global registers and RAM using the LPD inbound switch. Additionally, communication with the RPU, Configuration Security Unit (CSU), LP-DMA and IOPs can also be carried out via this switch.

Additional AXI Blocks

Lastly, there are two other AXI blocks that you may find in the Zynq MPSoC interconnect (not shown in Figure 11.3). These are the AXI Timeout block, and AXI/APB isolation block. The AXI Timeout block prevents the interconnect from hanging due to a non responding slave. The AXI/APB isolation block functionally isolates an AXI/APB master from its associated slave when either has to be powered-down. Further information on each of these can be found in [1].

11.2.3 Memories

Apart from processor caches and PL memories, there are two other significant system memories worth exploring. These are the OCM and Double Data Rate (DDR) Memory.

The OCM has a 128-bit AXI slave port and contains 256 KB of RAM. It operates up to a clock frequency of 600 MHz and supports low latency access for the RPU MPCores. It also uses Error Correcting Code (ECC). The DDR memory controller communicates with the Zynq MPSoC through six AXI data ports and an AXI control interface. The controller communicates with the system's main memory off-chip and is capable of supporting DDR3, DDR3L, LPDDR3, DDR4 and LPDDR4.

For further information about each memory, turn to Section 11.5.

11.3 PL Interfacing

Sooner or later during system development, it will be necessary to communicate with a hardware accelerator or function in the PL. The interface required will depend entirely on the application's constraints and the requirements of the hardware block operating in the PL. A summary of the PS-PL AXI interfaces can be found in Section 11.2.1. In this section, we aim to explore each of these interfaces in detail and point out their differences.

Figure 11.4 provides a diagram illustrating the PS-PL AXI ports.

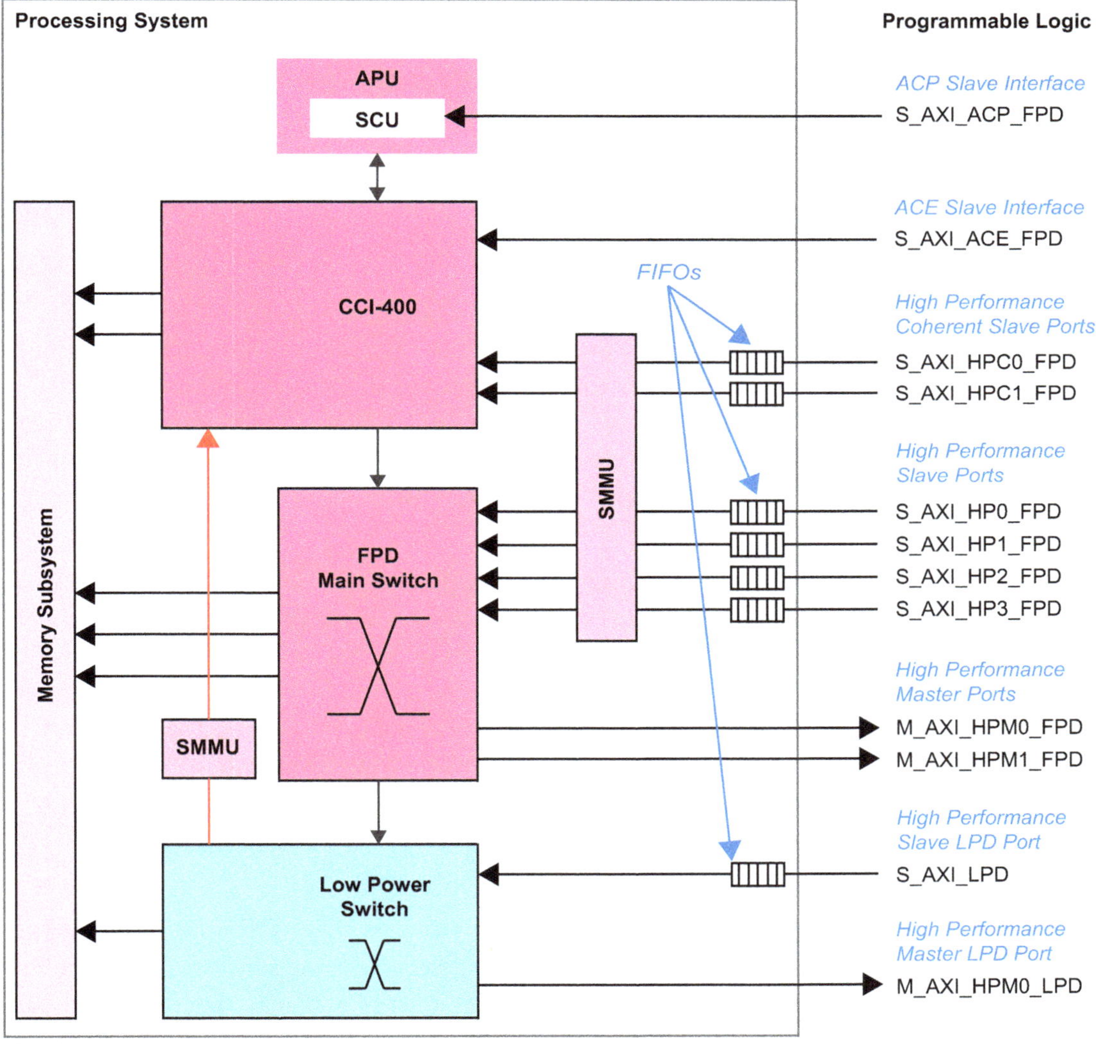

Figure 11.4: AXI interfaces connecting the PL to processing resources within the PS [1]

11.3.1 Snooping

Before we begin investigating the PS-PL AXI ports, it is necessary to understand the concept of 'snooping' and why it is an essential technique for coherency between two or more processing units.

In a multiprocessor system, such as the Zynq MPSoC, there will be several caches in each of its constituent processing units, including caches in the PL. Cache coherence is when each cache in a system has a method of synchronising and sharing information, so that they are each capable of obtaining the most up-to-date data. In a software environment, checks to ensure each cache is up-to-date would produce additional execution overhead. Instead, snooping is used on the physical interface of each cache (via the CCI-400 in the Zynq MPSoC) so that cache coherency can be performed in hardware.

An interface that is snooping, is observing data that is being sent or received by one or more connected interfaces. Snooping is required so that a master device can observe read transactions by other devices. If the snooping master has the most up-to-date data at the address specified by a read transaction, it may provide the requesting master device with this data. This process allows a system to be coherent between each of its constituent caches.

11.3.2 AXI Coherency Extension (ACE) Interface

The S_AXI_ACE_FPD port uses the ACE protocol as described in the AMBA AXI and ACE Protocol Specification [9]. The S_AXI_ACE_FPD is a PL-master interface connected to the CCI-400 in the PS. It is capable of supporting full coherency (two-way) between masters in the PS and hardware blocks in the PL. The full ACE protocol uses five additional channels in comparison to the AXI interface. Three channels are used for snooping and two are for acknowledgements. The ACE protocol enables the coherency of independent processing elements while ensuring writes to the same memory location are up-to-date and correct (without the need for software).

It is possible to connect a fully coherent ACE interface to a PL master hardware block. This connection will allow the PL master to store its cache in the FPGA's dedicated memory resources or logic fabric. The ACE interface can then be used to remain coherent with other coherent masters in the PS, and coherent masters in the PL. In particular the APU can snoop into PL cached masters (this is not possible using the S_AXI_ACP_FPD interface). Since the S_AXI_ACE_FPD port is directly connected to the CCI-400, it does not pass through the SMMU. Therefore it is unable to take advantage of physical and virtual address mapping in the PS. Therefore, it has no virtualisation support in the PS.

ACE-Lite Interface

The S_AXI_ACE_FPD port can also use the ACE-Lite protocol [9]. ACE-Lite is similar to AXI4, however, does not contain any of the new snoop and acknowledgement channels introduced by the full ACE. There are instead additional signals on the read address, write address and read data channels. Therefore, ACE-Lite channels can snoop ACE masters, however, cannot be snooped themselves. ACE and ACE-Lite protocols are backwards compatible with AXI4, provided that additional channels and signals are disabled. However, using the AXI4 protocol for the S_AXI_ACE_FPD port is not recommended by Xilinx [1].

A PL master may use the ACE-Lite protocol when they have no cache, as a mechanism for snooping the coherent caches of other masters. This is an example of one-way coherency (also known as I/O coherency). The PL master is capable of issuing transactions that can be stored in the caches of other coherent masters. There are several points to consider before using ACE-Lite in a hardware system, as detailed in [1].

The Limitations of the ACE

There are a few limitations of the full ACE and ACE-Lite protocols. These are summarised below:

- For full, two-way coherency using ACE, the transactions must be limited to a burst length of 64-bytes when the CCI snoops a PL master.
- If using ACE-Lite, it is possible that long bursts from the PL to the PS may cause the APU MPCore to hang. This is due to the path in which the transaction is propagating (between the CCI and DDR memory). If this path is used for a long period of time, others will not be able to access the memory until the transaction is complete. Burst lengths should be limited (no greater than 16, as suggested in [1]) to allow other systems access on their respective DDR controller ports.

An ACE System Cache in the PL

Implementing a system cache in the PL is a simple example of using the S_AXI_ACE_FPD port for a real application. Usually an AXI-ACE interconnect is implemented in the PL to interface one or more coherent hardware accelerators to the DDR memory, via the ACE interface. Instead, all ACE transactions are issued to a system cache in the PL (constructed from Block RAMs) using point-to-point interfaces between the system cache and hardware accelerators. This removes the need for an AXI-ACE interconnect and improves APU performance, as reading from the system cache is faster than reading from DDR memory. Additionally the bandwidth of the DDR memory/controller is minimised, as access to these systems is reduced.

A diagram illustrating this application is shown below in Figure 11.5.

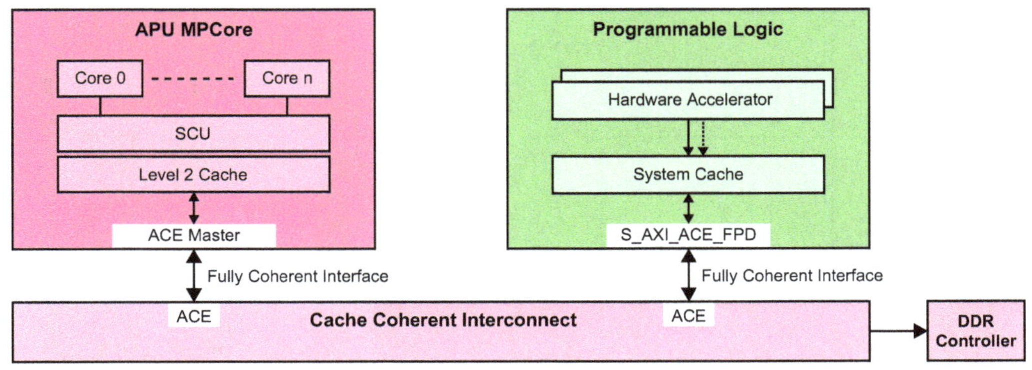

Figure 11.5: A system cache implemented in the PL that uses the ACE interface [1]

11.3.3 The AXI FIFO Interface

Before investigating the high-performance PS-PL AXI ports, we will briefly summarise the AXI FIFO interface. Masters in the PL and the PS DDR memory controller, are connected through high-bandwidth, low-latency data paths, via the high-performance PS-PL AXI ports. The AXI FIFO interface, also known as an AFI, is included in each of the slave high-performance ports between the PS and PL. The AFI provides high-throughput communication as summarised below:

- As the DDR memory controller can be preoccupied with other transactions, there can be a large variation in access latency from hardware accelerators in the PL. The AFI helps reduce the variation in access latency and allows data to stream continuously between the DDR memory and hardware accelerators in the PL [1].
- The clock rate at which data is written into a high-performance interface from the PL, could be different from the clock rate in the PS. The AFI helps to adjust the rate of data when transferring it between the PS and PL clock domains.
- Finally, the AFI converts between the AXI4 standard (in the PL) and the AXI3 standard (in the PS).

The AFI has two AXI ports for connecting to the PL and the interconnect in the PS. A block diagram of the AFI is shown in Figure 11.6.

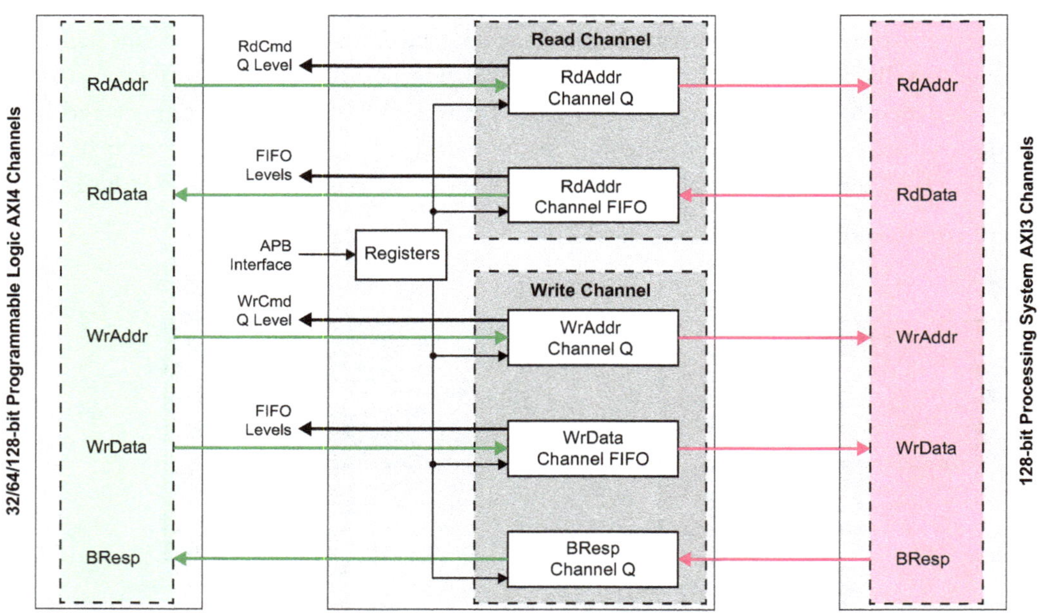

Figure 11.6: The AXI FIFO interface implemented in the high-performance PS-PL AXI ports [1]

The reader is directed to [1] for further information about the AFI.

11.3.4 PL-FPD AXI Masters

There are six PL High-Performance (HP) AXI master interfaces to the FPD that can access all slave devices in the PS. Primarily these ports are designed to access the DDR memory as they are high-bandwidth communication paths. There are four HP AXI masters denoted as S_AXI_HPn_FPD and two High-Performance Coherent (HPC) AXI masters denoted as S_AXI_HPCn_FPD (described below).

Each of the S_AXI_HPn_FPD interfaces are connected to the DDR memory controller via several AMBA switches. Their connections are outlined as follows:

- **S_AXI_HP0_FPD** — This interface shares its port on the DDR memory controller with the DisplayPort master in the PL. These interfaces are connected to port 3 of the DDR memory controller.
- **S_AXI_HP{1,2}_FPD** — Both ports share exclusive access to port 4 of the DDR memory controller. This exclusivity provides high-throughput, low-latency communication with DDR memory.
- **S_AXI_HP3_FPD** — This interface shares its port on the DDR memory controller with the FP-DMA controller.

All six of the HP AXI master interfaces pass through the SMMU in the PS. The SMMU is capable of performing physical and virtual address translation. This connection allows each interface to support virtualisation with the APU (discussed previously in Section 6.9).

I/O Coherent AXI Masters

Of the six high-performance AXI4 master ports between the PL and PS, there are two that are connected to the CCI-400. These are the S_AXI_HPC0_FPD and S_AXI_HPC1_FPD ports, and they are capable of I/O coherency using the ACE-Lite interface. They may be used to provide I/O coherency between a hardware accelerator in the PL and the level 1 and level 2 caches of the APU.

These ports can access the DDR memory controller by passing through the CCI. This consequently has a longer latency to the DDR memory in comparison to the data paths inherent to the S_AXI_HPn_FPD ports.

11.3.5 PL-LPD AXI Master

A high-performance AXI master interface is connected between the PL and LPD (S_AXI_LPD). It is capable of low-latency access to the OCM and Tightly Coupled Memories (TCMs). This interface is particularly useful when the FPD has been powered down, as it is still capable of providing the PL with high-performance access to the LPD. Unfortunately, due to the topology of the interconnect, this port has a long latency to the DDR controller.

This port has two modes of operation; physical and virtual mode. In physical mode, the transactions are not routed through the SMMU block (shown in Figure 11.4 on page 273). In virtual mode, the transactions are routed through the SMMU block so that the physical address can be translated into a virtual address. The route the transactions take in virtual mode is as follows: PL, LPD, FPD (SMMU/CCI) and then to the LPD.

11.3.6 PL-PS AXI Slaves

There are two high-performance interfaces that communicate between the FPD and the PL, and one high-performance interface that communicates between the LPD and the PL. These are as follows:

- In the FPD, there are two interfaces to the PL; M_AXI_HPM0_FPD and M_AXI_HPM1_FPD. These interfaces are suited for providing FPD masters in the PS access to the memories in the PL, so that they can transfer large amounts of data.
- There is one high-performance slave interface from the LPD to the PL (M_AXI_HM0_LPD). Similarly, this interface is suited for providing LPD masters (such as the LP-DMA) in the PS access to the memories in the PL, so that they can transfer large amounts of data. This interface is low-latency and can be accessed when the FPD is powered down. Note that the APU does not have access to this interface due to an ID converter in the path [1].

11.3.7 Selecting a PL Interface

In this section, we summarise each PS-PL AXI interface and provide a method of selecting one for a hardware system design. The PS-PL AXI interfaces are detailed in terms of data width and coherency in Table 11.2. Note that interfaces with multiple data widths can be programmed to use the widths shown.

Table 11.2: Summary of the PS-PL interfaces [1]

PL Interface	Type	Master	Data Width	Coherency	Description
S_AXI_HP{0:3}_FPD	AXI4	PL	128/64/32	Non-Coherent	Connected from the PL to the FPD main switch.
S_AXI_LPD	AXI4	PL	128/64/32	Non-Coherent	Connected from the PL to the IOP in the LPD.
S_AXI_ACE_FPD	ACE	PL	128	Fully Coherent	Connected between the PL memory and CCI.
S_AXI_ACP_FPD	AXI4	PL	128	I/O Coherent	I/O coherent with APU level 2 cache and CCI.
S_AXI_HPC{0,1}_FPD	AXI4	PL	128	I/O Coherent	I/O coherent with CCI.
M_AXI_HPM{0,1}_FPD	AXI4	PS	128/64/32	Non-Coherent	Connected between FPD masters and PL slaves.
M_AXI_HPM0_LPD	AXI4	PS	128/64/32	Non-Coherent	Connected between LPD masters and PL slaves.

Figure 11.7 provides a flow-chart that details a method of selecting a PS-PL AXI interface for a hardware system design using the Zynq MPSoC. Included are some of the benefits of using each interface and issues that need to be considered if that interface was selected.

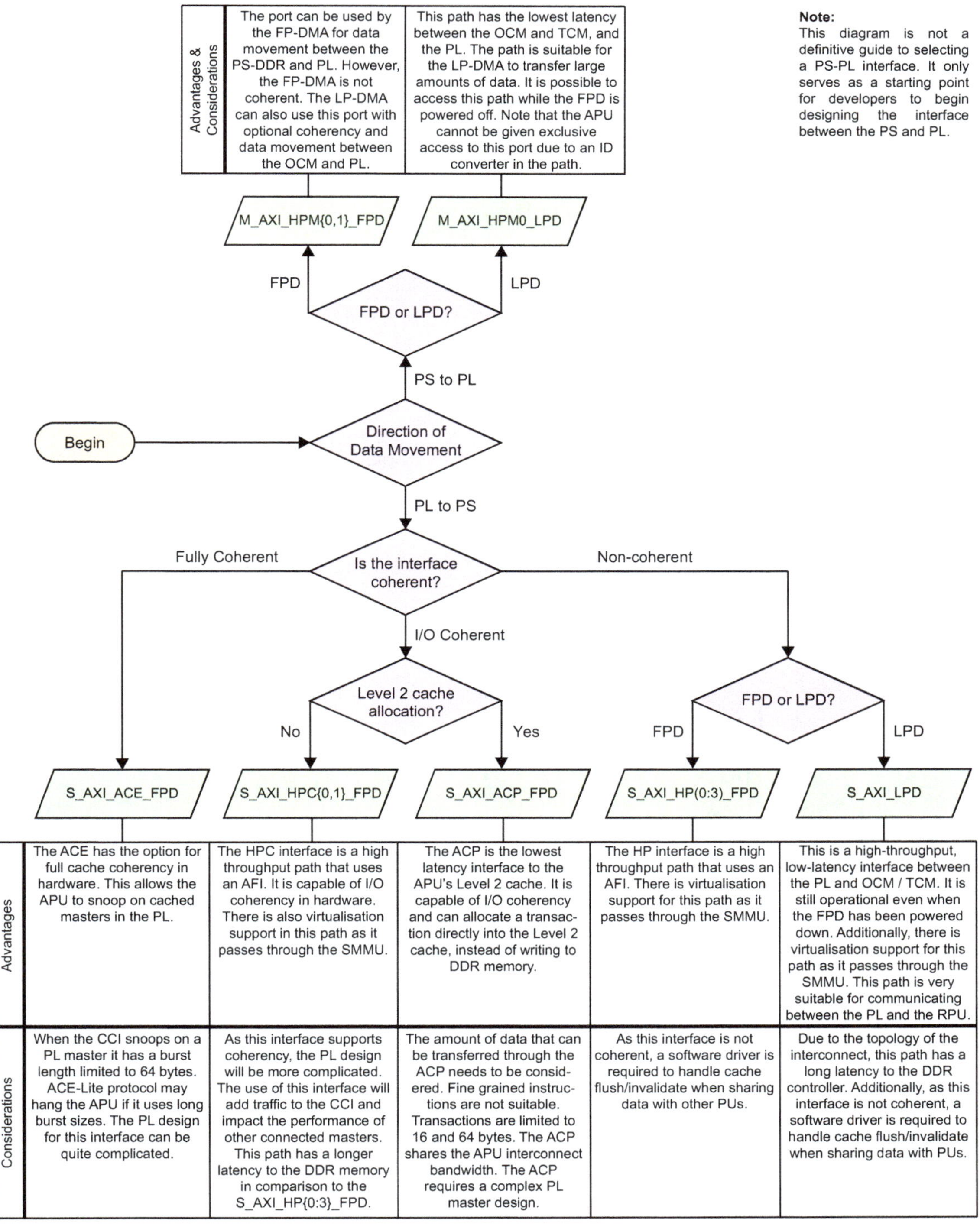

Advantages & Considerations
The port can be used by the FP-DMA for data movement between the PS-DDR and PL. However, the FP-DMA is not coherent. The LP-DMA can also use this port with optional coherency and data movement between the OCM and PL.
This path has the lowest latency between the OCM and TCM, and the PL. The path is suitable for the LP-DMA to transfer large amounts of data. It is possible to access this path while the FPD is powered off. Note that the APU cannot be given exclusive access to this port due to an ID converter in the path.
Note:
This diagram is not a definitive guide to selecting a PS-PL interface. It only serves as a starting point for developers to begin designing the interface between the PS and PL.
M_AXI_HPM{0,1}_FPD
M_AXI_HPM0_LPD
FPD
LPD
FPD or LPD?
PS to PL
Begin
Direction of Data Movement
PL to PS
Fully Coherent
Is the interface coherent?
Non-coherent
I/O Coherent
Level 2 cache allocation?
No
Yes
FPD or LPD?
FPD
LPD
S_AXI_ACE_FPD
S_AXI_HPC{0,1}_FPD
S_AXI_ACP_FPD
S_AXI_HP(0:3)_FPD
S_AXI_LPD
Advantages
The ACE has the option for full cache coherency in hardware. This allows the APU to snoop on cached masters in the PL.
The HPC interface is a high throughput path that uses an AFI. It is capable of I/O coherency in hardware. There is also virtualisation support in this path as it passes through the SMMU.
The ACP is the lowest latency interface to the APU's Level 2 cache. It is capable of I/O coherency and can allocate a transaction directly into the Level 2 cache, instead of writing to DDR memory.
The HP interface is a high throughput path that uses an AFI. There is virtualisation support for this path as it passes through the SMMU.
This is a high-throughput, low-latency interface between the PL and OCM / TCM. It is still operational even when the FPD has been powered down. Additionally, there is virtualisation support for this path as it passes through the SMMU. This path is very suitable for communicating between the PL and the RPU.
Considerations
When the CCI snoops on a PL master it has a burst length limited to 64 bytes. ACE-Lite protocol may hang the APU if it uses long burst sizes. The PL design for this interface can be quite complicated.
As this interface supports coherency, the PL design will be more complicated. The use of this interface will add traffic to the CCI and impact the performance of other connected masters. This path has a longer latency to the DDR memory in comparison to the S_AXI_HP{0:3}_FPD.
The amount of data that can be transferred through the ACP needs to be considered. Fine grained instructions are not suitable. Transactions are limited to 16 and 64 bytes. The ACP shares the APU interconnect bandwidth. The ACP requires a complex PL master design.
As this interface is not coherent, a software driver is required to handle cache flush/invalidate when sharing data with other PUs.
Due to the topology of the interconnect, this path has a long latency to the DDR controller. Additionally, as this interface is not coherent, a software driver is required to handle cache flush/invalidate when sharing data with PUs.

Figure 11.7: A flow-chart for selecting a PS-PL interface [1]

11.4 The Interrupt System

The Zynq MPSoC device is a diverse multiprocessor environment where interrupts are an essential form of communication between its processing units. Previously, the topic of interrupts has been explored for the:

- APU's GIC-400 (Section 6.7 on page 148),
- RPU's GIC PL390 (Section 7.3.2 on page 166),
- PMU's GIC proxy and local interrupt controller (Section 10.3.5 on page 251).

In this section, we will further explore the Zynq MPSoC's interrupt system architecture, including the interprocessor interrupts (IPIs) and the types of interrupt used within the system.

11.4.1 Interrupt System Overview

A system diagram illustrating an overview of the Zynq MPSoC interrupts can be seen in Figure 11.8. This diagram will help you visualise the interrupt system, and the type of interrupts you can expect to find.

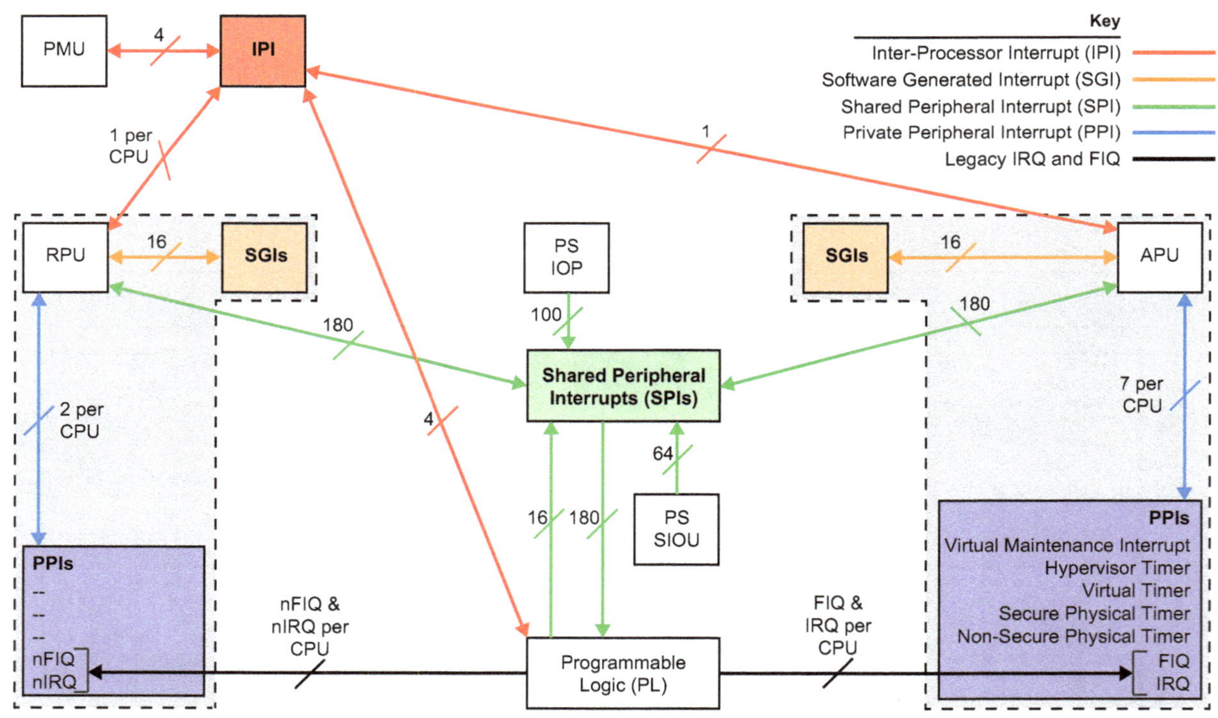

Figure 11.8: System-view of the Zynq MPSoC interrupts

The details of each GIC has not been shown to make the diagram in Figure 11.8 clearer. Also, the PMU's entire interrupt system has not been shown as it is quite large. The PMU is only shown connected to the IPI. Further information about the PMU's interrupts can be found in Section 10.3.5 or [1].

Before investigating interrupt types, we will briefly discuss the physical mechanism that allows an interrupt to propagate through a system. This is known as an interrupt's *sensitivity type*, of which there are two methods in Zynq MPSoC devices; edge-triggering and level-sensitive interrupt propagation.

- **Edge-triggering** indicates that the interrupt will be asserted when a rising or falling edge is detected on the interrupt signal. The interrupt remains pending/active until cleared by the receiving processor [10].
- **Level-sensitive** interrupts are asserted when the interrupt signal level is active. In other words, if the interrupt is active-low or active-high, the interrupt signal will have to be driven low or high respectively to be asserted. The pending/active state of a level-sensitive interrupt, is determined by the peripheral that asserts the interrupt signal. If the peripheral stops asserting the interrupt, it will be removed from the pending/active state into an inactive state.

11.4.2 Types of Interrupts

Primarily there are four types of interrupt used in the system shown in Figure 11.8 (the APU also supports virtual and maintenance interrupts, described previously in Section 6.7). These are Private Peripheral Interrupts (PPIs), Shared Peripheral Interrupts (SPIs), Software Generated Interrupts (SGIs), and Inter-Processor Interrupts (IPIs). IPIs are detailed in Section 11.4.4. The others are described as follows.

Private Peripheral Interrupts

The processors in both the APU and RPU are connected to their own set of private interrupts. These are known as PPIs. The PPIs for the APU are summarised in Table 11.3.

Table 11.3: The APU's Private Peripheral Interrupts (PPIs)

Int. ID	Name	Description
31	Legacy IRQ Signal	An IRQ signal for each processor core that is connected to the PL.
30	Non-secure physical timer	A physical timer event that occurs in the Non-secure state.
29	Secure physical timer	A physical timer event that occurs in the Secure state.
28	Legacy FIQ Signal	An FIQ signal for each processor core that is connected to the PL.
27	Virtual timer	Virtual timer generated event.
26	Hypervisor timer	A physical timer event in hypervisor mode.
25	Virtual maintenance interrupt	Unlike the other PPIs that are external, this particular PPI is internal to the GIC-400. This PPI is generated by the virtual CPU interface to indicate when a hypervisor action is required (discussed in Section 6.7.3).

The sensitivity type for the PPIs are fixed. The method in which they are asserted (edge-triggering or level-sensitive) cannot be changed. As the sensitivity type is fixed, the GIC responsible must be programmed to accommodate the required sensitivity type. SDK device drivers are used to program the GIC [1].

All PPIs are active-low level-sensitive except the legacy IRQ and FIQ signals between the RPU and PL [1]. These are instead inverted as they become active-high at the PS-PL interface and active-low in the associated interrupt sensitivity register (ICDICFR1).

Shared Peripheral Interrupts

As shown in Figure 11.8, the Zynq MPSoC's network of SPIs is densely populated. An SPI, can be routed by the distributor to any specified combination of processors, and originate from various sources within the Zynq MPSoC. These are wired interrupts which are physically connected to both of the APU and RPU GICs, and the PL.

The SPIs originate from the IOP and SIOU blocks in the PS of the Zynq MPSoC. Additionally, the PL also provides 16 SPIs. In total there are 180 SPIs (in EV devices) accessible to the APU, RPU, and PL.

Each SPI has a fixed sensitivity type and cannot be changed (except SPIs from the PL). To accommodate this, the SDK device drivers must be used to program the GIC accordingly [1]. Additionally, the SPIs have a rule regarding each of their sensitivity types. If the interrupt is level-sensitive, then after acknowledgement, the handler must clear the interrupt accordingly. If the interrupt is edge-triggered, the source that generates the interrupt must ensure that the edge pulse is long enough. This requirement is further described in [1].

A list of the SPIs in the Zynq MPSoC can be found in [1] and their corresponding registers in [11].

Software Generated Interrupts

SGIs can be routed to any specified combination of processors, acting as a mechanism for one processor to interrupt another (or itself) in the APU or RPU. Note that SGIs cannot be used for interrupting across the Zynq MPSoC's processing system i.e. a Cortex-R5 processor could not interrupt a Cortex-A53 processor using an SGI. The inter-processor interrupt is used for this purpose instead. The respective GIC for each MPCore does not have a set of inputs for an SGI, as they are generated inside the GIC. The APU and RPU each have their own set of 16 SGIs that are generated following a procedure.

An SGI is generated in the RPU by writing the SGI interrupt number to the *PL390.enable_sgi_control* register (ICDSGIR). When writing to this register, the target processor(s) should be specified. An SGI is cleared by writing a '1' to the associated bits in the interrupt clear-pending *PL390.enable_sqi_pending* (ICDICPR) register, or by reading the interrupt acknowledge *PL390.control_n_int_ack_n* (ICCIAR) register.

The APU uses a similar approach. An SGI can be generated by writing to the GICD_SGIR register. As before, the SGI interrupt number should be issued along with the target processor(s). An SGI may also be set to pending by writing a '1' to the GICD_SPENDSGIRn register. The SGI can be cleared by writing to the associated interrupt clear-pending register GICD_CPENDSGIRn [10].

All SGIs are edge-triggered. This indicates that the interrupt will be asserted when a rising-edge has been detected on the interrupt signal. The interrupt remains asserted until cleared using the methods above.

11.4.3 Interrupt Prioritisation, States, and Handling

We will briefly detail the interrupt priority system, states, and how they are handled in Zynq MPSoC devices. This topic was also previously covered in Section 6.7, for interrupt handling in the APU.

Interrupt Priority System

All GICs in the Zynq MPSoC use an interrupt priority system to determine which interrupts should be issued before all others. The process is quite simple, all interrupt requests (PPIs, SPIs, and SGIs) are assigned a unique ID so that they may be identified correctly by the interrupt controller. Inside each GIC, there is an interrupt distributor that holds a list of pending interrupts. The distributor will select interrupts that have the highest priority first and issue them to their target processor interfaces. Lower priority interrupts then follow.

An interrupt with a high priority, will be assigned a lower value. For example, the priority field value 0 will have a higher priority than an interrupt with a priority field value 4. If interrupts have equal priority, the distributor may resolve the matter by selecting the interrupt with the lowest ID first.

SPIs have special circumstances regarding their handling. As an SPI can be targeted to any number of processors, it is possible that the interrupt may be taken by more than one processor at the same time. As described in [1], logic ensures that only one processor takes the interrupt, and another will receive a spurious interrupt ID (1023 or 1022), or end up with the next pending interrupt due to timing.

Interrupt States and Handling

The state of an interrupt is held by the corresponding GIC of the target processor. There are four states that determine the assertion level of an interrupt. These are inactive, pending, active, and active and pending.

Initially, an interrupt may be in an *inactive* state. This indicates that the interrupt is not being used by the GIC, or a connected processing resource. The interrupt will enter a *pending* state if an interrupt controller receives an interrupt request. The *pending* state indicates that the interrupt has been asserted somewhere in hardware, or generated by software, and it has been received by an interrupt controller and waiting to be serviced by the target processor.

Once the target processor has acknowledged the interrupt, the interrupt status **could** change from *pending*, to *active and pending*. An interrupt with an *active and pending* state is one that is currently being serviced by a processor, but also has another *pending* interrupt from the exact same source. This will happen if the source has issued more than one interrupt consecutively.

If there are no consecutive interrupts, a *pending* interrupt will be set to *active* upon the interrupt controller receiving a processor acknowledgement. An *active* interrupt is in the process of being serviced and is yet to complete. Once the target processor has handled the interrupt, it will place it back into an *inactive* state.

Further information about handling interrupts can be found in [1].

11.4.4 The Inter-Processor Interrupt

The purpose of the IPI in the Zynq MPSoC, is to provide a processor with the ability to interrupt another. This is useful when a processor requires another processor to perform an action. In the Zynq MPSoC, there are eleven IPI channels that are assigned as follows:

- Seven channels are reprogrammable, however, default to the APU MPCore {0}, RPU0 {1}, RPU1 {2}, and the PL{7:10} (the PL is assigned 4 channels by default). Reprogramming the channels is possible as the IPI is already distributed between each of these processor's interrupt controllers.

- The remaining four channels are assigned to the PMU interrupt controller. These channels are known as IPI{3:6}. They may not be reprogrammed as they are fixed in hardware (i.e. only connected to the PMU interrupt controller). The PMU has special rules regarding the IPI, described in Section 11.4.4.

The IPI channels are capable of sending 32-bytes of information between connected processors. The channels use memory buffers to help support the IPI structure. The message buffers are capable of storing request and response messages between processors using the IPI channels. There are only 8 sets of memory buffers: seven sets are assignable to any processor using an IPI channel, and one set of buffers are dedicated to the PMU. Note, the message buffers have eight request buffers and eight response buffers resulting in 16 buffers per set. Overall, there are 128 total message buffers.

The IPI channel architecture can be seen in Figure 11.9. All connected processors can communicate to each other. It is also possible for a processor to send an interrupt to itself.

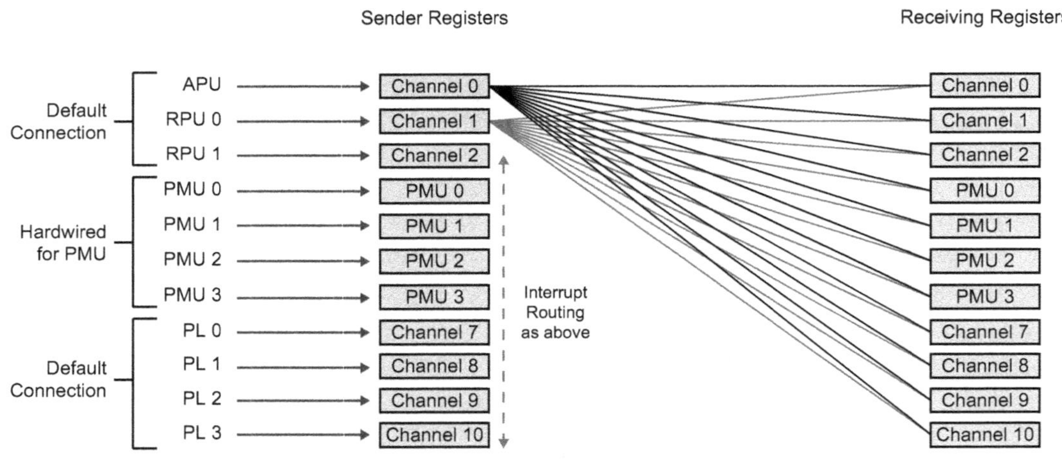

Figure 11.9: The IPI channel architecture [1]

Further information about the IPI channels, message buffers, and architecture can be found in [1].

11.5 Memories

The Zynq MPSoC has several memory devices and interfaces. Aside from the memories that exist in the PL and processor caches, the Zynq MPSoC has two significant built-in memories worth exploring. These are the OCM and TCM attached to the RPU. Additionally, there is also the DDR memory controller and interface for external (off-chip) memory. As the TCMs were discussed previously in Section 7.3.1, the OCM and DDR memory will be discussed in this section.

11.5.1 The Global Address Space

First off, we will explore an interesting subject regarding the Zynq MPSoC's global system memory map. As shown in Figure 11.10, the address space stretches over 1TB and has been implemented so that it is suitable for 32-bit processors and 64-bit processors, such as the RPU and APU MPCores respectively (the APU may also operate in 32-bit mode). Additionally, a list of 64-bit masters have been provided.

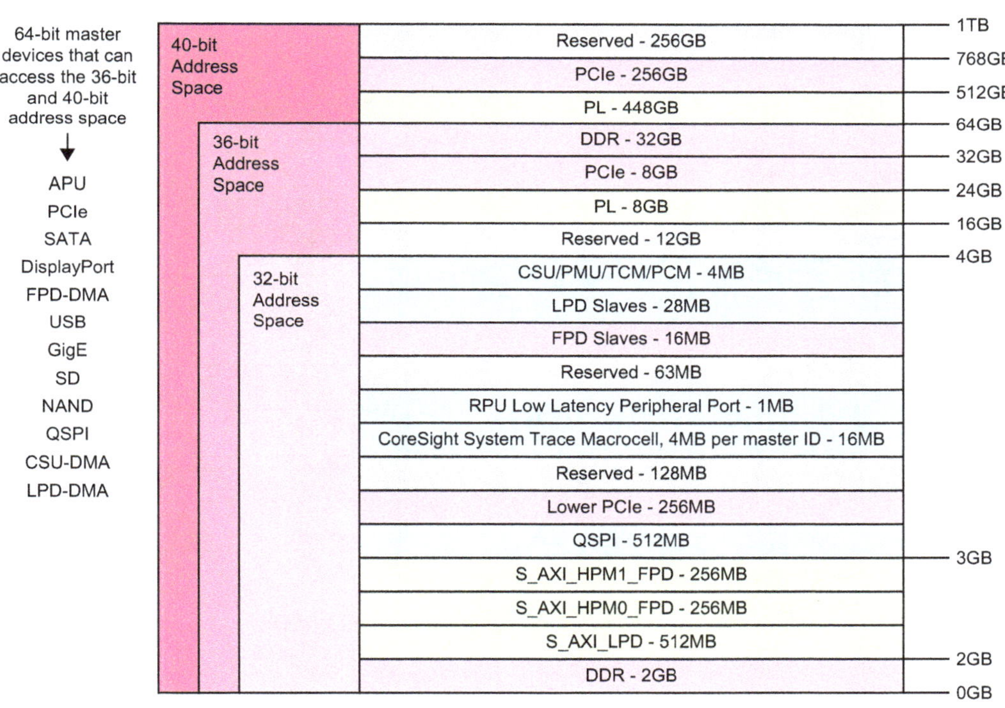

Figure 11.10: The Zynq MPSoC's global address space [2]

Due to the way in which the address space has been configured, 32-bit processors are capable of communicating with the majority of the Zynq MPSoC's memories, on-chip peripherals, and other processing elements. When increasing the address space to 36-bits, 64-bit masters can optimise their access to commonly used resources such as the DDR memory and PL. Lastly, the 40-bit address space allows significant access to the Zynq MPSoC's processing resources. Further information can be found in [1] and [2].

11.5.2 On-Chip Memory

The OCM contains 256 KB of RAM that is accessed by other system elements using its 128-bit AXI slave port. It operates up to a clock frequency of 600 MHz and supports low latency access for the RPU MPCores. It also uses ECC to detect multi-bit errors and recover from a single-bit fault in memory.

When reading and writing to the OCM, note that it uses a double-width memory of 256-bits. Maximum bandwidth can be achieved when the read and write operations are a multiple of 256-bits, in addition to using 256-bit aligned addresses. When the requested address is not aligned, or the writes to memory are not 256-bits, the OCM will perform a read-modify-write operation on the memory instead [1].

The OCM unit is shown in Figure 11.11. All transactions between the OCM and the rest of the Zynq MPSoC pass through the OCM's Xilinx Memory Protection Unit (XMPU). The XMPU enforces the isolation of memory regions on a per-master basis. In the OCM, the regions (also known as aperatures) are aligned to 4KB. Therefore the OCM is divided into 64 blocks of 4KB, each having their own security attributes in accordance with the associated master, and also Arm TrustZone security states. Memory protection with XMPU was detailed previously in Section 8.3.2.

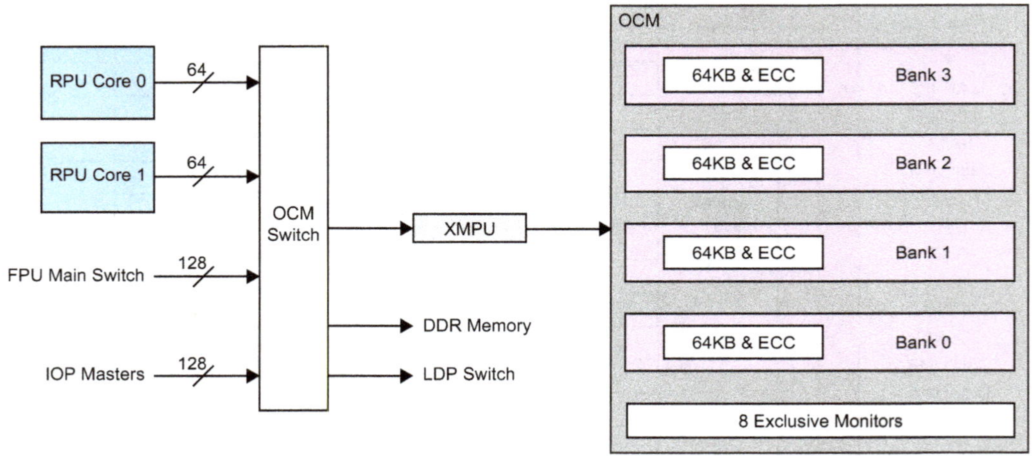

Figure 11.11: The architecture of the On-Chip Memory [1]

The OCM is connected to the RPU's cores, FPD main switch, IOP masters, DDR memory, and LPD switch via the OCM switch.

As illustrated in Figure 11.11, the OCM is also separated into 4 banks. Each bank has a separate power island that can be powered-up and powered-down by the PMU. It is possible for the OCM to be placed in retention when in a powered-down state, as discussed previously for deep-sleep mode in Section 10.1.4. A table indicating the address range of each memory bank can be seen in Table 11.4.

Table 11.4: The OCM memory map [1]

Memory Bank	Size	Address Range
0	64 KB	FFFC_0000 — FFFC_FFFF
1	64 KB	FFFD_0000 — FFFD_FFFF
2	64 KB	FFFE_0000 — FFFE_FFFF
3	64 KB	FFFF_0000 — FFFF_FFFF

Further information regarding the configuration, architecture, and operation of the OCM unit, can be found in [1].

11.5.3 DDR Memory Interface

As there is a lot of information regarding the DDR memory controller and interface, we will summarise its basic architecture. Shown in Figure 11.12, is the multi-protocol DDR system and its constituent functional blocks. These are the DDR PHY (physical layer), DDR Controller, AXI to APB bridge, AXI Performance Monitor, and six instances of the XMPU block.

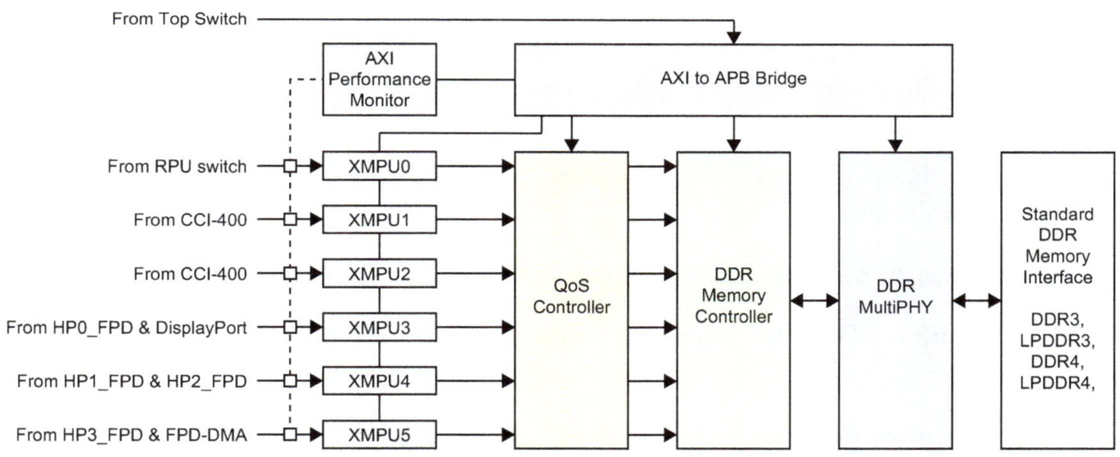

Figure 11.12: The functional blocks in the DDR Memory Controller [1]

The DDR system issues read and write requests from each of its six ports using AXI interfaces. The memory controller is responsible for issuing these commands on the DDR PHY interface that is physically connected to pins on the DDR memory device off-chip. Each port has a path through an XMPU for memory protection (discussed previously in Section 8.3.2). There is also an APB control interface for issuing configuration commands. Finally, the DDR supports ECC memory in both 32-bit and 64-bit data width modes.

As shown, the DDR memory system is connected to the Zynq MPSoC's interconnect using 6 AXI interfaces. Each originate from different locations in the system and are summarised in Table 11.5. Note that port 3, 4, and 5 are shared using NIC-400 switches (described earlier in Section 11.2.2).

Table 11.5: DDR port connections to the rest of the Zynq MPSoC

DDR Port	Interface Width	Power Domains	Connected Resources & Interfaces
0	64-bit	LPD	RPU Switch
1	128-bit	FPD	Cache Coherent Interconnect
2	128-bit	FPD	Cache Coherent Interconnect
3	128-bit	PLPD	PS-PL interface: S_AXI_HP0_FPD
		FPD	DisplayPort interface
4	128-bit	PLPD	PS-PL interface: S_AXI_HP1_FPD
		PLPD	PS-PL interface: S_AXI_HP2_FPD
5	128-bit	PLPD	PS-PL interface: S_AXI_HP3_FPD
		FPD	FP-DMA

The DDR memory system is capable of interfacing with DDR3, DDR3L, LPDDR3, DDR4, and LPDDR4 memories. The reader is directed to [1] for a complete list of example memory configurations for the DDR memory system.

The DDR memory has a list of conditions that must be met to connect to the list of compatible memory types above. These are detailed as follows:

- **Maximum memory density** — Maximum total memory density must be no larger than 34 GB.
- **Total Data width** — The total data width compatible with the DDR system can either be 32-bits or 64-bits.
- **Component memory density** — The connected memory density must be either 0.5, 1, 2, 4, 6, 8, 12, 16, or 32GB. In LPDDR4 6, 12 and 24 GB is not supported.
- **Number of ranks** — The maximum number of chips that can be connected is 2.
- **Number of row and bank address bits** — The number of row address bits is limited by the controller at 16. The number of bank address bits is 3.

Most of the information regarding the DDR memory system has been summarised in this section. Extensive information about its configuration, operation, and architecture can be found in [1].

11.6 Data Movement Fundamentals

Up until this point, we have explored a lot of the underlying hardware infrastructure of the Zynq MPSoC. This included the interrupts, memories, and PS-PL interfaces present in its architecture. We will now investigate a very important aspect of embedded system design. As we have explored much of what the Zynq MPSoC has to offer in its hardware system, we can now discuss **data movement** between its constituent processing elements and memories.

Our discussion will begin by detailing the fundamental principles of moving data in an embedded system. The transfer of data is generally required between processors, peripherals, and memories. With regards to SoC and MPSoC devices, data may be moved between its internal processing elements, or off-chip to another system resource. The primary objective of transferring data remains the same for all applications; data should be moved as efficiently as possible. The method in which data is transferred depends entirely on the application constraints, the type of data involved, and the native hardware of the system. Deciding how data should be moved is a key challenge encountered by an embedded system designer.

For the remainder of this section, we will explore data movement within the Zynq MPSoC device. Initially, programmable I/O and Direct Memory Access (DMA) is described. Then, we investigate the movement of data from the PS to PL using the FP-DMA and LP-DMA controllers. Subsequently, we will explore the transfer of data from the PL to the PS using AXI DMAs.

11.6.1 Direct Memory Access

Transferring data can be achieved by executing instructions in software using a processor. This method of accessing memory is known as programmable I/O and is useful if the associated processor only needs to transfer a small amount of data. Large data transfers using this method is inefficient, as the processor will be busy accessing memory and unable to perform any other tasks. DMA is essential for embedded systems that are focused on performance. The general concept is quite simple; a hardware subsystem capable of DMA, can access the main system memory with minimal processor support. These hardware subsystems are known as DMA controllers, and they relieve the processor of the workload associated with accessing main system memory for large data transfers.

Figure 11.13 provides an example system that will require the movement of data in the Zynq MPSoC device. The basic principle of this diagram is to read data from the off-chip memory and write to a hardware accelerator operating in the PL. Note that some system elements have been removed for clarity.

As shown, the PL contains a DMA controller, a hardware accelerator, and two AXI interconnects. The purpose of the DMA controller is to communicate with the DDR memory, via the AXI interconnect and high-performance interface. The DMA controller minimises the amount of support required by the APU MPCore by speaking directly to the DDR memory for the hardware accelerator. In order for the DMA controller to begin operation, the APU must configure it with a set of parameters that will determine how data is transferred. Once the APU has configured the DMA controller, it may operate autonomously without any APU involvement. If required, the DMA controller can interrupt the APU when it has finished its allocated task, or if an event has occurred.

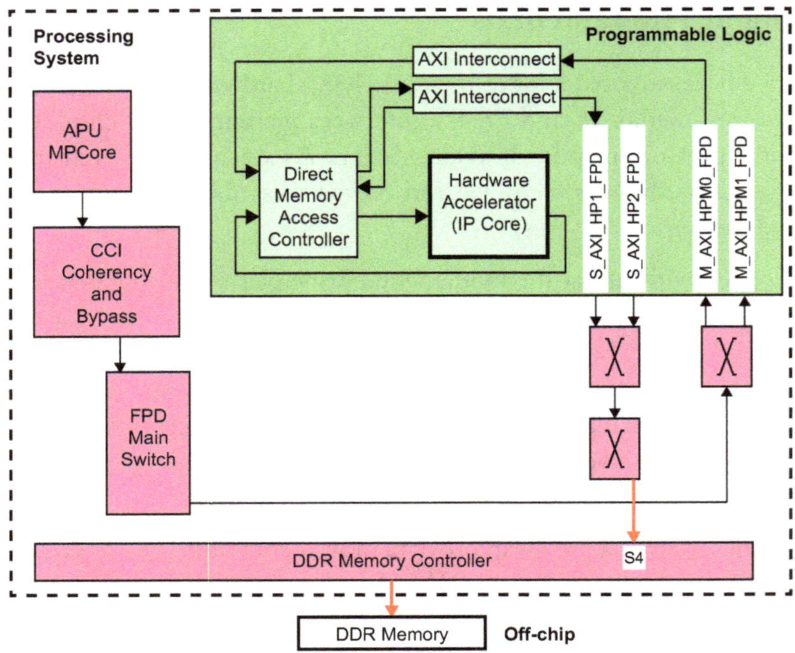

Figure 11.13: An example system transferring data between a hardware accelerator in the PL and main memory [1]

The diagram shown in Figure 11.13, is one of many possible system configurations. For example, the RPU may also configure DMA controllers in the PL, if the APU is powered down. There are also several other DMA controllers in the Zynq MPSoC device. The LPD and FPD have their very own DMA controllers for transferring data between processing units and system memories with minimal processor support. Peripherals in the MIO, including USB and Ethernet, have their own DMA controllers for transferring data directly into memory. The PL can have any number of DMA controllers, so long as there is enough FPGA logic resources and bandwidth on the PS-PL AXI interfaces and DDR memory.

The FP-DMA and LP-DMA controllers, and AXI DMA controllers each support two modes of operation. These are simple DMA mode, and scatter/gather mode. Each are described as follows.

Simple DMA Mode

Simple DMA mode is the easiest way of using a DMA controller (if present, the scatter/gather interface is not used in this mode). The processor core in charge will initially configure the DMA controller using its control interface. A transaction is created by writing directly to the DMA's registers. The parameters are the source and destination addresses, and access pattern (these are known as a buffer descriptor). The DMA controller will begin moving data when the processor core in charge writes to its enable register. Once the transfer is complete, the DMA controller asserts an interrupt that if enabled, will generate an interrupt out. This interrupt is useful as it can be connected to a processor, informing it that the transaction has completed.

Scatter/Gather Mode

As mentioned earlier, the DMA controller has an interface to support scatter/gather mode. This interface allows the DMA controller to fetch buffer descriptors that have been preloaded in system memory. Buffer descriptors are the set of parameters that are normally used to configure the DMA controller, as in simple DMA mode. However, the DMA controller can fetch these descriptors from system memory without the help of the processor core in charge. As the DMA controller is managing its own configuration, a processor does not need to intervene with DMA management. This maximises system performance.

Further information about scatter/gather mode can be found in [14].

11.6.2 The AXI Interconnect

Previously shown in Figure 11.13, the AXI interconnect, provided by Xilinx, allows the connection of multiple master and slave AXI4 interfaces. Normally, the AXI4 interface will only connect one master and one slave together. The AXI interconnect can be used when more than one connection is required.

The AXI interconnect is very useful for connecting multiple AXI enabled systems to the PS-PL AXI interfaces, as there are a limited number of physical ports available. Further information about the AXI interconnect IP provided by Xilinx, can be found in [13].

11.6.3 DMA Controllers in the PS

We will now explore the operation of the FP-DMA and LP-DMA controllers that exist in their respective power domains in the PS. These are the most efficient method of transferring large amounts of data across the Zynq MPSoC device (while removing processor involvement). Each DMA controller in the PS is almost the same. Both have 8 independent channels for communicating with slave clients (one device per channel), and have support for unaligned transfers, interrupts, simple DMA mode, and scatter/gather mode.

There are some differences between the FP-DMA and LP-DMA. The FP-DMA is connected to a 128-bit AXI bus and is not coherent. This is because the FP-DMA transfers are sent directly to the DDR memory controller, and do not pass through the CCI. The LP-DMA is connected to a 64-bit AXI bus and is optionally I/O coherent, as it can pass through the CCI. If coherency is required using the FP-DMA, software support will be required.

A significant difference between the LP-DMA and FP-DMA is the size of the common buffer. The FP-DMA has a 4KB buffer and the LP-DMA has a 2KB buffer.

A diagram illustrating the internal architecture of the PS DMAs can be seen in Figure 11.14. Note that there are 8 DMA channels, a set of read and write interfaces, and a control interface using the APB. Also shown is a MUX and a Common Buffer at the DMA controller's input and output respectively. The purpose of the MUX is to allow data to be read by each of the DMA channels. Similarly, the common buffer allows data that has been received in a transaction, to be stored until it can be written to the slave client.

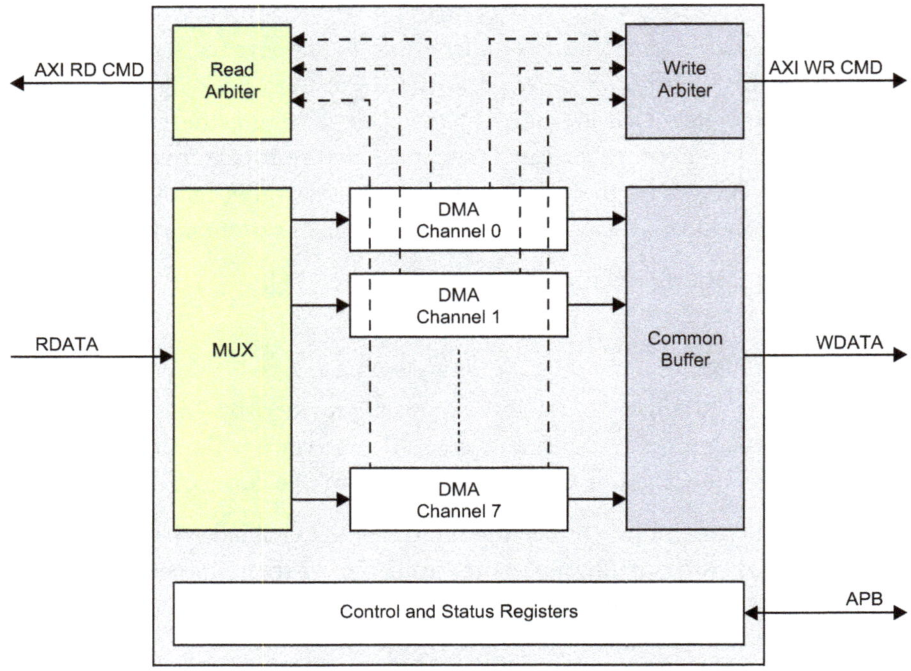

Figure 11.14: The architecture of the FPD and LPD DMAs in the PS [1]

Primarily, the FP-DMA and LP-DMA are used to transfer large amounts of data into the PL using one of the following PS-PL AXI interfaces:

- **M_AXI_HPM(0,1)_FPD** — These interfaces are suitable for the FP-DMA controller to access.
- **M_AXI_HPM0_LPD** — This interface is suitable for the LP-DMA controller to access.

Transferring data using one of the PS DMAs will use the AXI Read and Write Arbiters, as shown in Figure 11.14.

The AXI Read Arbiter uses its control interface (AXI RD CMD) to read buffer descriptors from memory (to indicate the next read address and access pattern). Once a buffer descriptor is sourced, the RDATA interface may be used for reading data from a memory resource in the PS, or the main system memory. Once data has been read, it will be stored in the common buffer.

The AXI Write Arbiter uses its control interface (AXI WR CMD) to obtain a buffer descriptor from memory. The descriptor is used to write the data from the common buffer into the PL, via one of the above PS-PL AXI interfaces.

For each of the above transactions to work, the correct access patterns and system addresses must be used. Further information about the PS DMAs can be found in [1].

11.6.4 Simple AXI Communication

Before discussing the operation of DMA controllers in the PL, we will first describe the most basic form of data movement between the PS and PL.

AXI4-Lite (discussed previously in Section 3.5.1) is an interface standard capable of reading and writing to registers in the PL. To allow this type of communication, the hardware accelerator (also known as an Intellectual Property Core or IP Core) in the PL must have an AXI4-Lite interface [12]. The IP Core will contain a set of addressable registers that can be read and written to using the AXI4-Lite interface. Data is written to the register address location (using programmable I/O) to update its value in the PL.

AXI4-Lite should generally be used to communicate control and status information to an IP Core in the PL. The reason for this, is that it is only capable of single-beat transfers (for every read/write instruction, there is one element of data). Therefore, this protocol should only be used for low-bandwidth communication with registers of IP blocks. As you will learn in Section 11.6.5, AXI4-Lite is actually how a processor operating in the PS, configures a DMA controller IP Core in the PL.

Figure 11.15 provides an illustration of how AXI4-Lite may be used to communicate between the PS and PL of the Zynq MPSoC device. As shown, the example illustrates a connection between the APU MPCore and an IP Core in the PL using AXI4-Lite. The connection could also be implemented using the other M_AXI_HPM1_FPD port or the M_AXI_HPM0_LPD port instead. However, note that the M_AXI_HPM0_LPD port is not exclusively accessible to the APU, due to the interconnect topology described in Section 11.3.6.

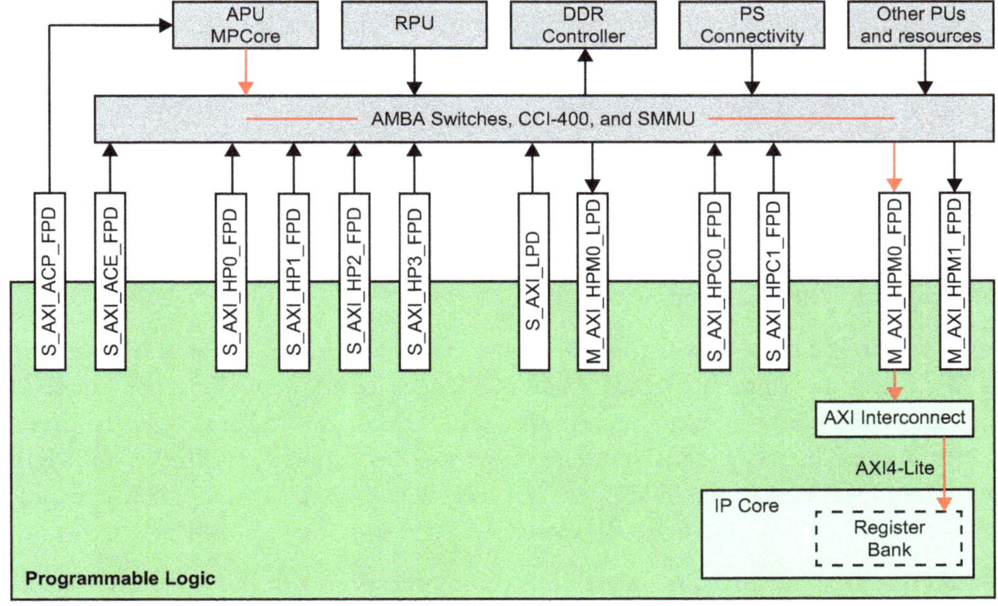

Figure 11.15: An example of AXI4-Lite connection between an IP Core in the PL and the APU [12]

11.6.5 The AXI DMA

Xilinx provide an AXI DMA IP Core [14] so that hardware accelerators in the PL, can communicate with main system memory. The AXI DMA provides high-bandwidth communication using the AXI Memory-Mapped and AXI stream interfaces. A diagram illustrating the primary input and output ports of the AXI DMA IP, is shown in Figure 11.16.

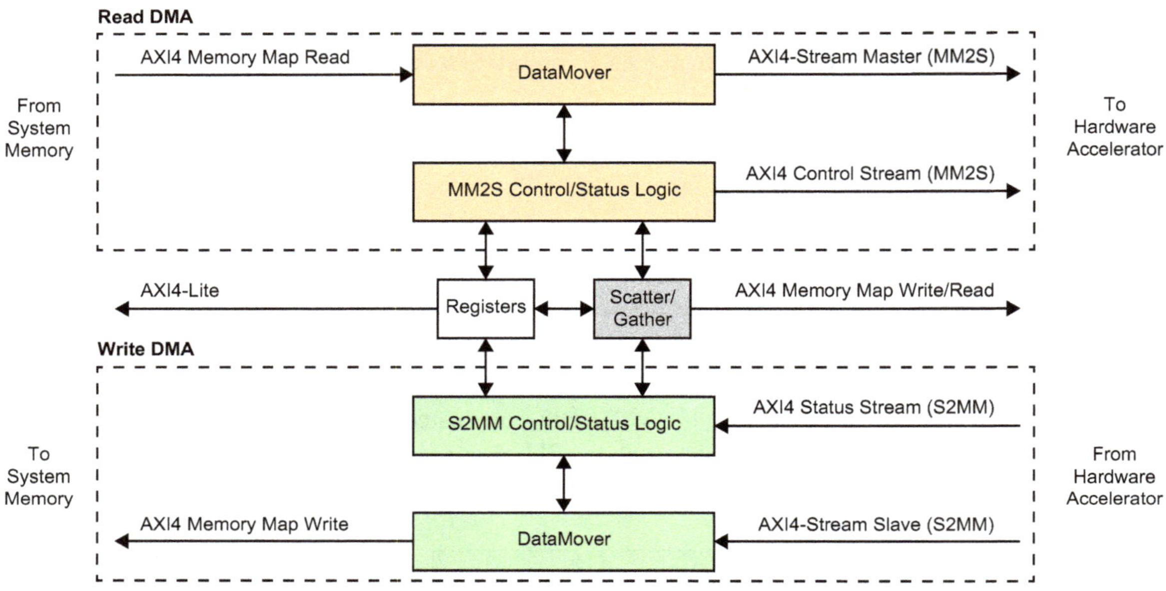

Figure 11.16: The AXI interfaces on the AXI DMA Controller [14]

As shown, the AXI DMA IP Core can have two data movers inside its architecture. One data mover is used to read from system memory (shown by the orange blocks). The other is used to write data to system memory (shown by the green blocks). Each channel operates independently from one another and can be disabled/enabled during hardware system development.

Reading from system memory uses the *AXI4-Stream Master* interface, denoted by Memory-Mapped to Stream (MM2S). The other interface *AXI4 Control Stream (MM2S)* provides the target IP Core with additional application and control data. Similarly, the write DMA uses the *AXI4-Stream Slave* interface for writing data into system memory. This interface may also be denoted as Stream to Memory-Mapped (S2MM). The write DMA also has an additional interface *AXI4 Status Stream (S2MM)* for receiving status updates and application data from the target IP Core.

The AXI4-Lite interface provides low-bandwidth communication with the PS, as described previously in Section 11.6.4. The scatter/gather interface is optional, and allows the DMA to fetch descriptors that have been preloaded in system memory (with minimal help from a processor core). The DMA can then configure itself for its target address, length of transaction, and other control parameters.

AXI DMA Connection in the Zynq MPSoC

An example of the AXI DMA connected in the PL is shown in Figure 11.17. This example is one of many possible system configurations using the Zynq MPSoC. The AXI DMA uses the AXI4 Memory-Mapped interface when communicating with the DDR Controller. The AXI4 Memory-Mapped protocol supports burst-transfers. These type of data transfers use an address and an access pattern provided by the master (a pattern or formula that determines the subsequent address for the data that follows). Multiple data transfers can then be carried out with one transaction, as the access pattern is used to determine the next address to access. This method reduces the overhead and latency of the data transfer.

When passing data to the target IP Core, the AXI DMA uses the AXI4-Stream interface, which allows burst transfers of an unrestricted (infinite) size. No address channel is required, as this protocol should be used for a direct flow of data between source and destination within the device.

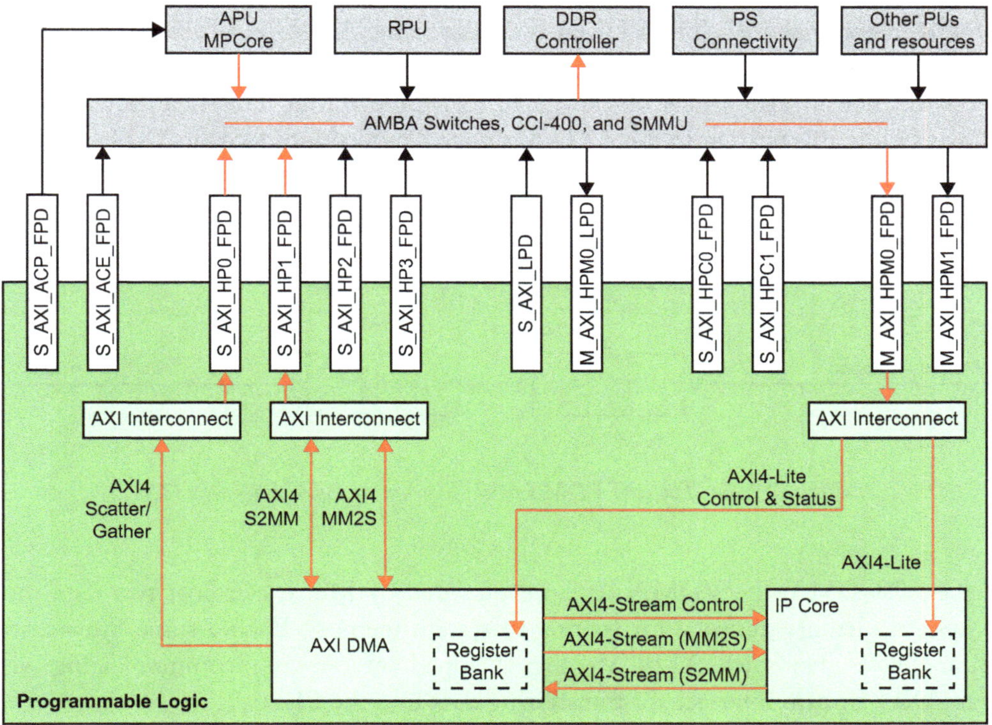

Figure 11.17: An example of the AXI DMA connected in the PL of the Zynq MPSoC [12]

There are several different connections in the example shown in Figure 11.17. The first of these is the connection between the DMA and S_AXI_HP1_FPD port, via the AXI interconnect. This is the DMA's primary data path to read and write to the main system memory. We have connected it to the PL's high-performance ports, such that it has a high-throughput path to the DDR controller. In particular, the first high-performance port (HP1_FPD) is used, as it has an exclusive connection to port 4 of the DDR controller.

The optional scatter/gather port has been connected to the S_AXI_HP0_FPD, as a means of fetching buffer descriptors from main memory. The AXI4-Lite control and status interface is connected to the M_AXI_HPM0_FPD for communicating with the Arm processors in the PS. This connection allows the Arm processors to configure the AXI DMA and receive status information.

When reading from system memory, the DMA uses the AXI4 MM2S channel, also known as Memory-Mapped to Stream. The retrieved data is transferred to the AXI4-Stream (MM2S) channel to be sent to the IP Core. The IP Core will send any data back to the DMA controller using the AXI4-Stream (S2MM) channel, also known as Stream to Memory-Mapped. When writing to system memory, the DMA controller will use the AXI4 S2MM channel.

11.6.6 The AXI Video DMA

The AXI Video DMA (VDMA) IP Core [15] can provide high-performance transfer of video frame data between the DDR memory and PL. The VDMA is similar to the AXI DMA (described in Section 11.6.5 on page 294). It has control and status logic, a data mover block, and AXI4-Lite registers as shown in Figure 11.18. Additionally, there is also a new block known as a Line Buffer. This is an asynchronous buffer for holding pixel data before it is written to the AXI4 Memory-Mapped interface, or AXI4-Stream interface.

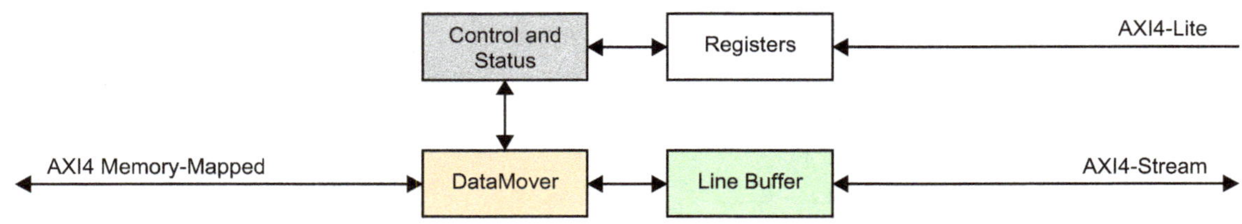

Figure 11.18: The AXI VDMA block diagram and AXI interfaces [15]

Similar to the AXI DMA, the VDMA IP Core provided by Xilinx, can host two data movers; one for reading from system memory and one for writing to system memory. Each assume the structure shown in Figure 11.18, and have their own AXI4 Memory-Mapped interface for communicating with the DDR memory and an AXI4-Stream interface for transferring data into the PL.

The reason for using the VDMA IP instead of the AXI DMA, is that it has been optimised for transferring video data between system memory and the PL [15]. The VDMA is capable of efficiently performing DMA operations on video frame data and can asynchronously transfer video frames on read and write channels. This is particularly suited when the PL needs to buffer video data for different clock domains, or to wait for another task to finish.

The VDMA can support up to 32 frame buffers over a 64-bit address space. There is a Data Realignment Engine (DRE) for unaligned access to memory. The DRE allows frame buffers to begin at any address in memory.

A frame buffer example in the Zynq MPSoC device, is shown in Figure 11.19. The general premise of this example is to buffer the video frames of an incoming High-Definition Multimedia Interface (HDMI) signal, using the AXI VDMA IP Core. Once the video frames are buffered, they are read from system memory and written onto the AXI-Stream (MM2S) channel.

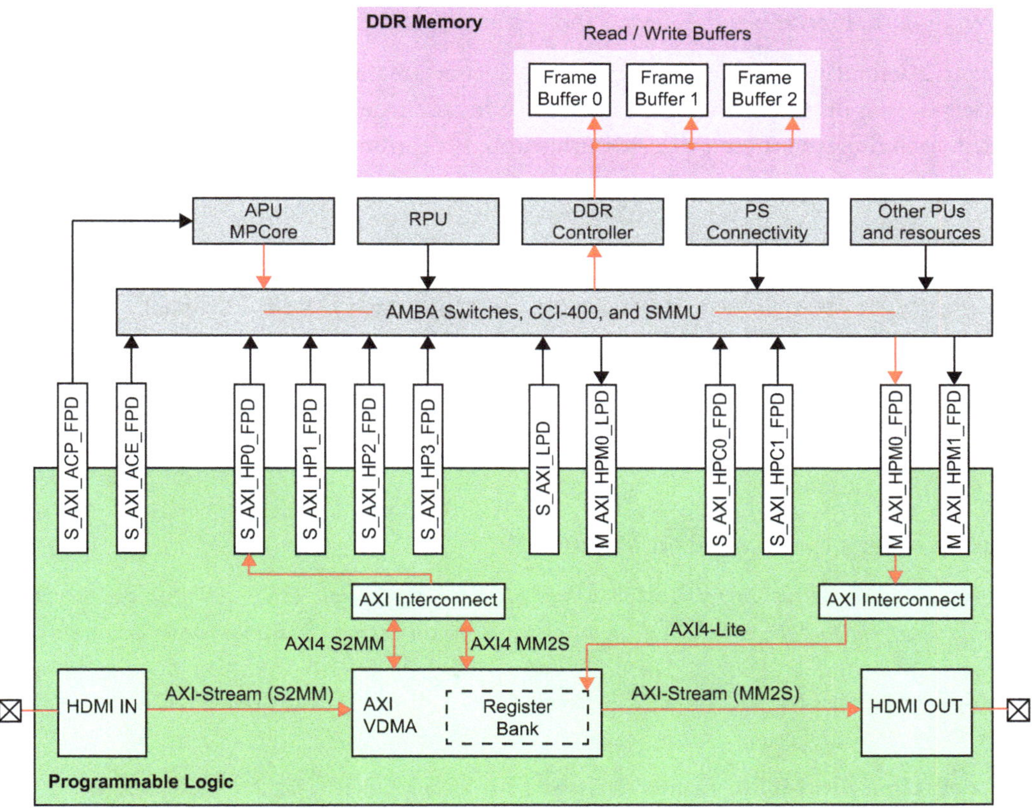

Figure 11.19: The AXI VDMA video frame buffer example

Note that the above example only uses three frame buffers. The AXI VDMA provides a method of synchronising the reading and writing of video frames through Genlock synchronisation. Frame synchronisation is necessary as video frame data can be produced faster or slower than it is consumed. The Genlock synchronisation prevents channels from accessing a frame at the same time. Once each channel has finished operating on their frame buffer, they may swap over. Each channel of the VDMA may operate as the master or slave in this context, leading or following the other channel respectively.

All together there are four Genlock and Dynamic Genlock synchronisation modes that can be used to configure each channel of the VDMA. Each channel under the above modes have different affects on frame synchronisation. Further information about the VDMA and its Genlock synchronisation modes can be found in [15].

11.7 Chapter Review

We have now come to the end of this chapter and have thoroughly explored the Zynq MPSoC's hardware infrastructure and data movement techniques. This included a detailed analysis of the Zynq MPSoC's interfaces and signals, interconnect, interrupts, and memories. Each of these system elements were found to be essential to moving data throughout the device and communicating between processing units and resources.

This chapter in particular explored the PS-PL AXI interfaces and provided a simple flow-chart to help system developers decide on the most suitable interface for their design. The Zynq MPSoC interrupt system was also detailed, including the inter-processor interrupts for communication between processing units in the device.

Finally, data movement throughout the Zynq MPSoC device was investigated. Direct Memory Access controllers can move large quantities of data without involving an Arm processor; allowing the associated Arm processor to carry-on with its own tasks until the data is transferred. The PS DMAs, PL AXI DMA, and PL AXI VDMA were all described. The PL DMA controllers in particular were presented with example connections in the FPGA logic fabric.

11.8 References

Note: All online sources last accessed on March 2019.

Some of the Xilinx documentation referred to below has version-specific URLs. If you are working in a newer version of the tools, check for updates on the Xilinx website, or try adjusting the link according to your version.

[1] Xilinx, Inc., "Zynq UltraScale+ MPSoC Technical Reference Manual", UG1085 (v.1.7), Dec. 2017.
Available: http://www.xilinx.com/support/documentation/user_guides/ug1085-zynq-ultrascale-trm.pdf

[2] Xilinx, Inc., "Zynq UltraScale+ MPSoC Embedded Design Methodology Guide", UG1228 (v1.0), Mar. 2017.
Available:
https://www.xilinx.com/support/documentation/sw_manuals/ug1228-ultrafast-embedded-design-methodology-guide.pdf

[3] Xilinx Inc., "Zynq-7000SoC, Technical Reference Manual", UG585, (v1.12.1) Dec. 2017.
Available: https://www.xilinx.com/support/documentation/user_guides/ug585-Zynq-7000-TRM.pdf

[4] Xilinx Inc., "SDSoC Environment User Guide", UG1027 (v2017.4), Jan. 2018.
Available: https://www.xilinx.com/support/documentation/sw_manuals/xilinx2017_4/ug1027-sdsoc-user-guide.pdf

[5] Xilinx Inc., "Vivado Design Suite User Guide", UG902 (v2017.1), Apr. 2017.
Available: https://www.xilinx.com/support/documentation/sw_manuals/xilinx2017_1/ug902-vivado-high-level-synthesis.pdf

[6] Arm, Ltd., "ARM CoreLink CCI-400 Cache Coherent Interconnect Technical Reference Manual", Issue K, Revision r1p5, Dec. 2015.
Available: http://infocenter.arm.com/help/topic/com.arm.doc.ddi0470k/DDI0470K_cci400_r1p5_trm.pdf

[7] Arm, Ltd., "ARM CoreLink MMU-500 System Memory Management Unit, Technical Reference Manual", Issue D, Revision r2p1, Jun. 2014.
Available: https://static.docs.arm.com/ddi0517/d/DDI0517D_corelink_mmu_500_r2p1_trm.pdf

[8] Arm, Ltd., "ARM CoreLink NIC-400 Network Interconnect, Technical Reference Manual", Issue C, Revision r0p2, Dec. 2013.
Available:
http://infocenter.arm.com/help/topic/com.arm.doc.ddi0475c/DDI0475C_corelink_nic400_network_interconnect_r0p2_trm.pdf

[9] Arm, Ltd., "AMBA AXI and ACE Protocol Specification", Issue E, Feb. 2013.
Available: http://infocenter.arm.com/help/index.jsp?topic=/com.arm.doc.ihi0022e/index.html

[10] Arm, Ltd., "ARM Generic Interrupt Controller Architecture Version 2.0, Architecture Specification", Issue B.b (v2.0), July 2013.
Available: http://infocenter.arm.com/help/index.jsp?topic=/com.arm.doc.ihi0048b/index.html

[11] Xilinx, Inc., "Zynq UltraScale+ MPSoC Register Reference", UG1087 (v1.5), December 2017.
Available: https://www.xilinx.com/html_docs/registers/ug1087/ug1087-zynq-ultrascale-registers.html

[12] S. Erusalagandi, "Leveraging Data-Mover IPs for Data Movement in Zynq-7000 AP SoC Systems", Xilinx White Paper, WP459 (v1.0), Jan. 2015.
Available: https://www.xilinx.com/support/documentation/white_papers/wp459-data-mover-IP-zynq.pdf

[13] Xilinx, Inc., "AXI Interconnect v2.1, LogiCORE IP Product Guide", PG059, Dec. 2017.
Available:
https://www.xilinx.com/support/documentation/ip_documentation/axi_interconnect/v2_1/pg059-axi-interconnect.pdf

[14] Xilinx, Inc., "AXI DMA v7.1, LogiCORE IP Product Guide", PG021, Oct. 2017.
Available: https://www.xilinx.com/support/documentation/ip_documentation/axi_dma/v7_1/pg021_axi_dma.pdf

[15] Xilinx, Inc., "AXI Video Direct Memory Access v6.2, LogiCORE IP Product Guide", PG020, Nov. 2016.
Available: https://www.xilinx.com/support/documentation/ip_documentation/axi_vdma/v6_2/pg020_axi_vdma.pdf

Chapter 12

Software Stacks

with Josh Goldsmith, University of Strathclyde

This chapter introduces some of the main software stacks available for the Zynq MPSoC. The term "software stack" refers to a set of base software which a developer can use to support their own application. Even if you are unfamiliar with the term, you have likely used software stacks before, such as the Xilinx bare-metal stack or even the Arduino platform for hobbyist projects.

Consider programming an entire embedded system from first principles. Before touching the application, there is a need for boot code (likely in assembly), drivers for peripherals such as SD cards, and then support for a file system. The aim of a software stack is to implement the common base functionality and abstract it away from the application developer. The advantage of using a pre-existing software stack should be clear; there's no need for every developer to create their own bootloaders, drivers and operating systems — the value that most developers add is in the application that sits atop the stack. After all, if we have programmed further, it is by standing on the shoulders of giants.

The rest of this chapter will give an overview of some of the Xilinx supported software stacks for the Zynq MPSoC, along with guidance on the advantages of each. The aim here is to nudge developers towards a path of least resistance, despite the lack of a clear "one size fits all" solution.

12.1 Bare-Metal Software Stack

A bare-metal software stack is provided directly by Xilinx and is distributed as part of the Xilinx SDK tool. It provides a set of low level drivers and libraries for basic functionality including I/O and access to hardware features [1]. Simplicity is the main factor which sets the bare-metal stack apart from the alternatives. It is a very simple, single-threaded environment (i.e. no direct way to use multiple cores). This makes it the most simple stack to get up and running, especially when using the Xilinx SDK. If the distinctive features of other stacks are not valuable for the application, bare-metal will usually offer the easiest configuration and lowest overhead. While this was more feasibly an option for Zynq-7000 devices, it is a less likely choice for a complete Zynq MPSoC design as it would be difficult to take full advantage of the hardware, due to the complexity of coordinating cores within (or between) the APU and RPU. However, it can be a very useful platform for making smaller, independent applications that run on a single core of a wider Zynq MPSoC system.

12.1.1 What does Bare-Metal give us?

The bare-metal software stack provides a simple, single-threaded environment with some convenient support for the Zynq MPSoC hardware. The support includes all peripherals in the processing system and a subset of available programmable logic IP.

The composition of the bare-metal stack is shown in Figure 12.1. There is a clear layered structure which leads to the term software *stack*. As with all of the stacks presented in this chapter, the bottom layer represents the Zynq MPSoC hardware. Each subsequent layer provides some added level of abstraction — e.g. the drivers abstract the hardware implementation, while the libraries generally abstract away driver implementations.

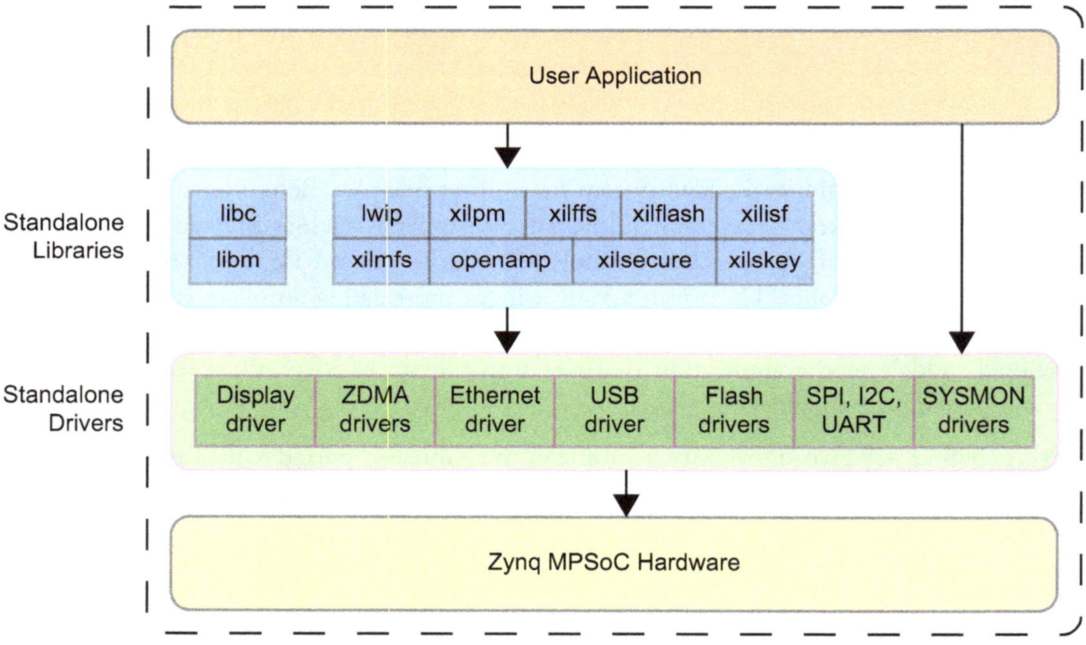

Figure 12.1: Bare-metal software stack

More formally, each layer is composed as follows:

- **Standalone drivers**: support for all PS peripherals, optional PL peripherals, and routines for using the processors within the APU, RPU, and PMU. Processor-centric functions include configuring interrupts and caches. "Standalone" implies use without an additional operating system.

- **Libraries***:* provide higher level interfaces for networking, file systems, encryption, as well as C standard libraries (`libc` and `libm`).

- **User application***:* Either the user's custom application or one of the supplied examples (including "hello world", first stage bootloader, and tests)

Generally, the purpose of the stack is to provide abstraction from one layer to the next. In that case, the developer only needs to understand the layer directly beneath their application. Note that Figure 12.1 shows the application also directly communicating to the standalone drivers, instead of going through the library layer. This is because not every driver has a corresponding library — some drivers are not complex enough to merit another layer of abstraction. The following sections overview the available libraries and drivers. For full details, the reader is referred to [3] for libraries and [4] for standalone drivers.

12.1.2 C Standard Libraries

The "C standard library" defines some core C routines which most developers will rely on without a second thought. These include routines such as `malloc()` and `fopen()`. The functionality is defined by the ANSI C standard but the implementation will be dependent on the target platform.

The C standard library for the Zynq MPSoC is based on the Newlib library. This is an implementation which targets embedded systems and can be found in the wild anywhere from commercial tools to hobbyist software for game consoles. The two main sets of modules are:

- **libc** which provides standard C such as routines from stdio and stdlib.

- **libm** which provides standard math routines. This is enhanced beyond the Newlib implementation.

All header files which compose these libraries are supplied along with the Xilinx SDK (and online at [2]), but it can still be difficult to identify exactly which one you need by file name alone. As a reference, Table 12.1 provides a summary of some of the `libc` header files.

Table 12.1: libc header descriptions[1]

Header file	Description
alloca.h	Allocate space in the stack
assert.h	Diagnostics
ctype.h	Character handling
errno.h	Error handling
inttypes.h	Integer type conversion
math.h	Mathematics
setjmp.h	Non-local goto (bypasses normal function call behaviour)

Table 12.1: libc header descriptions[1]

Header file	Description
stdint.h	Width-based integer types (e.g. int16_t)
stdio.h	I/O facilities
stdlib.h	General utilities including string conversion, random numbers, memory management, searching, and sorting
time.h	Time facilities

Similarly, Table 12.2 summarises the functions provided by `libm`, grouped by type of function. All entries are defined in math.h.

Table 12.2: libm function listing[1]

Function type	Supported functions
Algebraic	cbrt, hypot, sqrt
Elementary transcendental	asin, acos, atan, atan2, asinh, acosh, atanh, exp, expm1, pow, log, log1p, log10, sin, cos, tan, sinh, cosh, tanh
Higher transcendental	j0, j1, jn, y0, y1, yn, erf, erfc, gamma, lgamma, and gamma_ramma_r
Integral rounding	ceil, floor, rint
IEEE standard recommended	copysign, fmod, ilogb, nextafter, remainder, scalbn, fabs
IEEE classification	isnan
Floating point	logb, scalb, significand
User defined error handling	matherr

12.1.3 Standalone Libraries

Alongside the C standard libraries, the bare-metal stack provides a set of middleware libraries to facilitate networking, file systems, cryptography, and platform management. These sit on top of the standalone drivers and provide a higher level of abstraction. Table 12.3 gives an overview of the available libraries.

Table 12.3: Descriptions of standalone libraries

Library	Description
lwip141	Lightweight TCP/IP stack (version 1.41)
openamp	OpenAMP for asymmetric multiprocessing (more detail in Chapter 13)
xilffs	FAT file system implementation
xilflash	Xilinx flash library for Intel/AMD CFI compliant *parallel* flash
xilisf	Xilinx In-system and *serial* flash library
xilmfs	Xilinx memory file system
xilpm	Power management API library
xilsecure	API to CSU for SHA-3 hash functions, AES for symmetric cryptography, and RSA for authentication
xilskey	Xilinx "secure key" library for programming cryptographic keys to the eFUSE bits and the battery-backed RAM

These middleware libraries are supplied with the Xilinx SDK, but can also be accessed directly from [5].

12.1.4 Standalone Drivers

The standalone drivers sit in the layer just above the Zynq MPSoC hardware. The support covers all peripherals embedded in the processing system, as well as a selection of optional IP within the programmable logic. Again, full details are best found in an installation of the Xilinx SDK. Without those tools, the same files can be inspected via the git repository at [5] (`embeddedsw/XilinxProcessorIPLib/drivers/`) — but ensure the driver versions match any installed tools when browsing the git repository directly.

12.1.5 How can we use it?

The easiest way to get started with the bare-metal stack is using the Xilinx SDK. The drivers and libraries can be selected through the settings in the board support package (BSP). Documentation for each driver can be viewed within the SDK and example code can be imported. As mentioned previously, all of the source code is available online at [5], and could be used from an alternative environment with a suitable cross-compiler. The SDK is the recommended flow, however.

The design flow for using the bare-metal software stack is shown in Figure 12.2. A description of the hardware can be imported, via a custom Vivado project, or a pre-defined example. The SDK can then generate a BSP based on the hardware description. For bare-metal, this will include the standalone drivers and libraries. The user code exists as a separate application project, which references the BSP. Debugging and profiling can be performed iteratively, before the SDK is used to generate a bootable image for standalone use on the board.

The SDK offers integrated access to example applications and Doxygen documentation for each Xilinx library and driver in an effort to make getting started with the bare-metal stack as painless as possible.

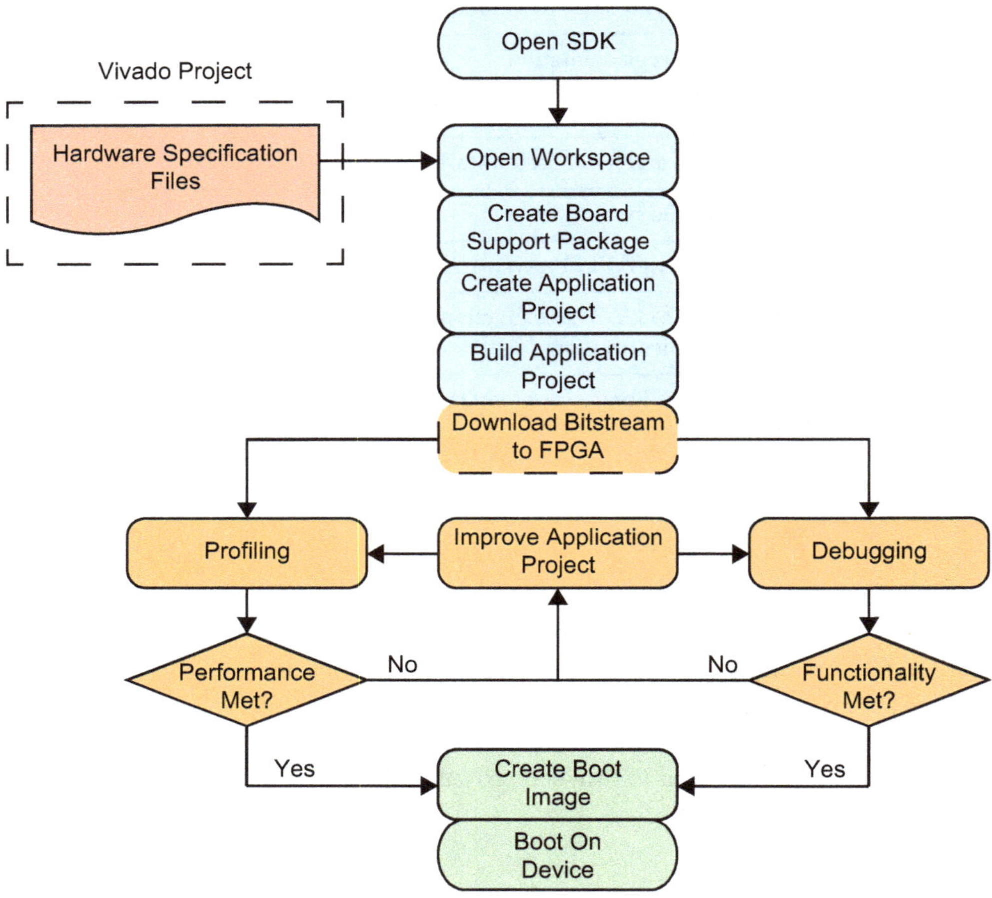

Figure 12.2: Design flow using the bare-metal stack

12.2 FreeRTOS Software Stack

FreeRTOS is a popular, open source, real-time operating system (RTOS). When writing code to handle *a single* real-time task, using the bare-metal stack would be a sensible option. The bare-metal stack provides a base set of drivers within a *single-threaded* environment. However, real systems usually have more complicated requirements than a single, simple function. We often want to run multiple conceptual tasks which can interact and synchronise in order to build a larger, more useful system (multi-tasking). This is where FreeRTOS comes in.

12.2.1 What does FreeRTOS give us?

FreeRTOS provides an environment where multiple tasks (or "threads") can cooperate, sharing the processor and other resources. It is designed to minimise overhead in terms of both execution and complexity. This is an important goal as real-time systems demand quite careful analysis to ensure safety critical code will meet its deadline. The bulk of the environment is provided by just a few core FreeRTOS features. These focus on prioritising, scheduling, running tasks (a primary concern of all other OSs too!), as well as synchronisation between tasks. The base FreeRTOS kernel is comprised of:

- **task.c and task.h**: The task handling code.
- **queue.c and queue.h**: A simple queue system for passing messages between tasks. This is also used for synchronisation between tasks.
- **list.c and list.h**: A doubly-linked list implementation which is used by both `task.c` and `queue.c`.
- **port.c, portASM.S, and portmacro.h**: Provides a processor agnostic interface to the rest of the FreeRTOS kernel. These compose the hardware dependent part of FreeRTOS, specific to each supported processor architecture.

There is one major aspect of typical OS functionality which is missing here — device drivers.

The only hardware dependent part of FreeRTOS is the processor specific code in `portmacro.h` and friends. We are free to use whatever set of standalone drivers we see fit. The set of standalone drivers from the bare-metal stack (see Chapter 12.1) is usually used here. They are convenient and well integrated with the Xilinx toolchain. As we are supplying our own set of drivers, many would consider FreeRTOS more of a "threading library" than a full blown operating system. Let's not minimise it though — FreeRTOS is more than enough to get a rather complex system up and running.

The FreeRTOS stack can be thought of as shown in Figure 12.3. It sits adjacent to the standalone Xilinx drivers. The FreeRTOS kernel concerns itself with the execution of tasks, while drivers know how to communicate out to the larger system. Although we can still use the Xilinx supplied standalone drivers, there are some restrictions. These drivers are not aware of the FreeRTOS kernel, and do not use its synchronisation or mutual exclusion functions. So, each device must only be accessed by a single thread to avoid synchronisation issues.

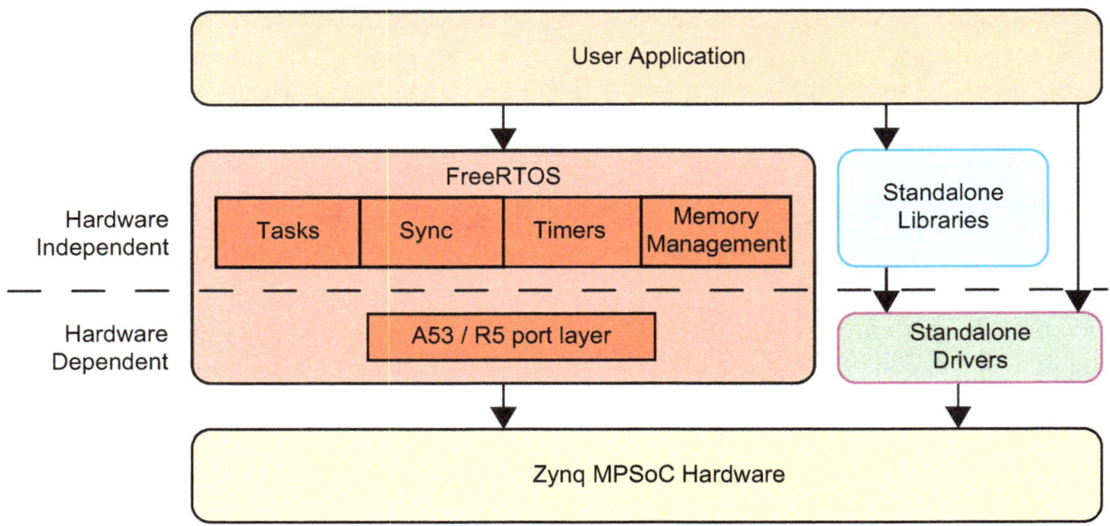

Figure 12.3: FreeRTOS stack with standalone drivers

12.2.2 Tasks

We have used the term "tasks" to describe the different responsibilities of a larger system. Through FreeRTOS's eyes, a task is a single C function. The current state of the function execution can be represented by a task control block (TCB). Part of the hardware dependent code allows FreeRTOS to save the state of the current task (register values, stack information, etc.) to a TCB and load another task's state from another TCB. This interruption of tasks lets the FreeRTOS scheduler swap between tasks to give the appearance of near simultaneous execution.

FreeRTOS will also assign a status to each task. One of 4 states can be assigned to a task at a time:

- **Running**: This is the currently executing task. There can only be as many running tasks as there are processing cores available to FreeRTOS.
- **Ready**: Ready tasks include all tasks which are ready to execute, but are not currently executing. Upon a scheduler "tick", the highest priority ready task is chosen to become running.
- **Blocked**: A task can be "blocked" when waiting for an event, such as receiving a new message on a queue or an external hardware event. The scheduler knows that blocked tasks have no useful work to do, and will not bother giving any blocked tasks CPU time.
- **Suspended**: Suspended tasks are similar to blocked tasks except they have been explicitly excluded from the ready state using the `vTaskSuspend()` function. No event will return a suspended task to the ready state — an explicit call to `vTaskResume()` is needed.

The life-cycle of a task can be thought of as a state machine, traversing though these different states upon certain events/function calls.

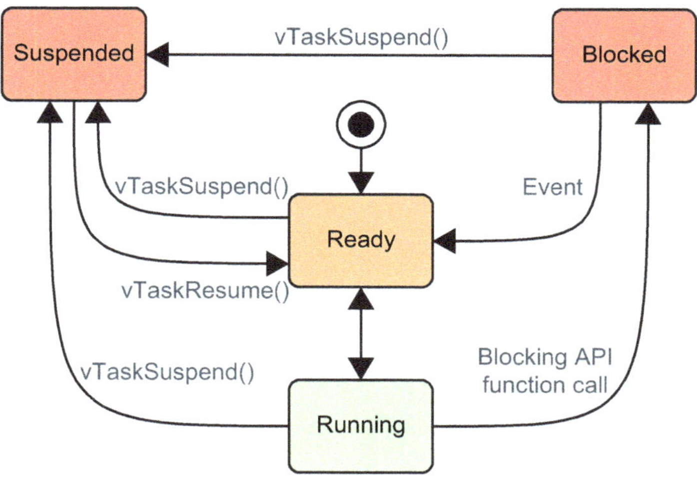

Figure 12.4: Life-cycle of a FreeRTOS task

The scheduler can switch between tasks to share CPU time. This is guided by the user defined task priorities. However, it is rare that all the tasks operate in isolation from each other. This leads us on to the synchronisation features within the FreeRTOS kernel.

12.2.3 Inter-task Synchronisation

FreeRTOS supports a range of inter-task synchronisation and message passing features. These are necessary to properly implement any type of dependence between tasks — including notification of events, passing data/messages, and ensuring exclusive access to a shared resource. Traditionally, these features were all derived from the message queue functionality, so we are going to review FreeRTOS queues first.

Generally, a FreeRTOS queue is configured to hold a finite number of a particular data type. The messages are added to the end of the queue and removed from the front, in a "first in first out" (FIFO) fashion. Messages are always *copied* (not by reference) into the queue. The memory management is handled by FreeRTOS — there is no need to manually create buffers. This API is simple, perhaps to the point of appearing restrictive. What if we want to send variable sized messages? What if we need a queue for multiple different types of message? Or what if we cannot afford to have FreeRTOS copy a large message?

To avoid the "by copy" approach to adding messages to the queue, we could simply configure the queue to hold pointers to the real data. Now, the pointer is copied but the full data is not. Similarly, many different types of message can be accepted by a queue by passing a structure with two fields. One field can identify the type of message, and the other can hold (or point to) the real data. Both of these techniques are nicely demonstrated in the FreeRTOS+UDP IP stack[6]. In particular, look for the `xNetworkEventQueue`, visualised in Figure 12.5.

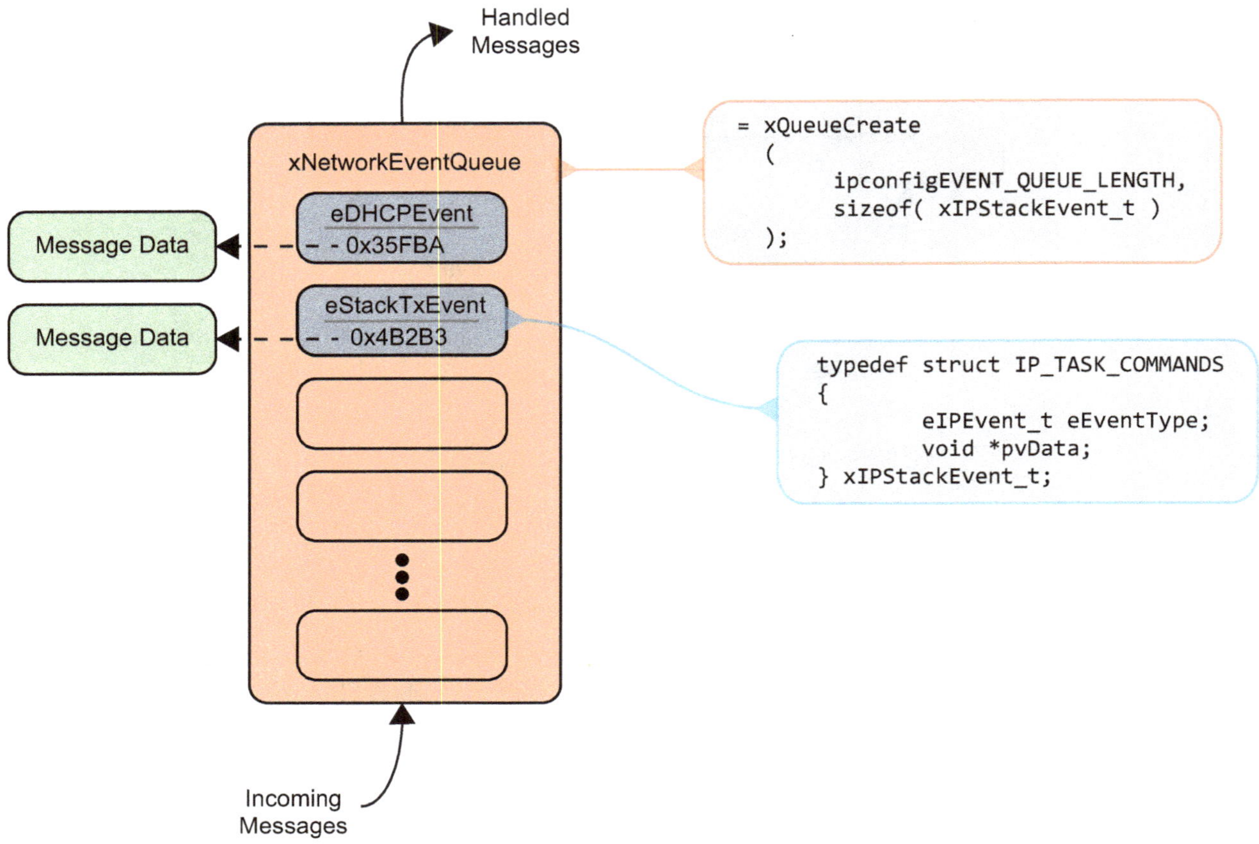
Handled
Messages
xNetworkEventQueue
eDHCPEvent
0x35FBA
eStackTxEvent
0x4B2B3
Message Data
Message Data
Incoming
Messages
= xQueueCreate
(
ipconfigEVENT_QUEUE_LENGTH,
sizeof(xIPStackEvent_t)
);
typedef struct IP_TASK_COMMANDS
{
eIPEvent_t eEventType;
void *pvData;
} xIPStackEvent_t;

Figure 12.5: Flexibility in queues found in FreeRTOS+UDP stack

The `xNetworkEventQueue` queue is instantiated with space for a fixed number of `xIPStackEvent_t` structures. This structure defines one field to identify the type of event and another as a pointer to the event data. Note that passing a pointer to the data lets us avoid the overhead of copying the full set of data (although now it is up to us to ensure the data is held in a valid state!) *and* works around the need for all messages to be of the same size. Variable sizes are quite important here as the example DHCP event will likely have quite a small payload, while the "`StackTx`" event (a user request to send data) could be much larger.

Message passing is important, but it is not the only feature that queues provide. They also provide *synchronisation*. Think about how you would write code for two tasks: one which writes messages to the queue, and another which consumes messages. This actually forms the basis of a classic computer science problem called the "producer–consumer problem". There are two exceptional circumstances we need to handle:

- A consumer reading from the queue when it is already empty
- A producer writing to the queue when it is already full

A simple approach would be to have either task "busy wait" when the queue is empty or full. A busy wait is essentially a loop which does no useful work except checking the state of the queue. This wastes lots of CPU time! The implementation in FreeRTOS works more efficiently by making use of the ability to block and resume tasks — allowing us to avoid busy waiting. If a producer tries to write to a full queue, or a consumer tries to read from an empty queue, the task is blocked by the kernel and will use no further CPU time. The task will then be resumed by FreeRTOS when the queue becomes "available" (non-empty for a consumer, and non-full for a producer). As an example, this means that a task which processes samples of some sort could run until it exhausts all available samples, and then uses no unnecessary CPU time until the next sample arrives. It is also important to note that the FreeRTOS implementation does not restrict the number of unique tasks which can act as consumers or producers for a given queue. This makes the implementation a little more complex, but we don't have to worry about that as an application developer!

The data synchronisation implemented for queues is also extended to provide process synchronisation where no data is passed — namely semaphores and mutexes. These can ensure that multiple tasks can catch-up with each other or can safely share a shared resource. For further reading on these synchronisation techniques, see [7] and [8]. As well as the queue-based features, a new lightweight "task notification" was introduced in 2015 with FreeRTOS v8.2, providing a more restrictive means of synchronisation which can be up to 45% faster[9].

Synchronisation features are really difficult to implement properly from first principles. Race conditions between tasks are easy to overlook, and would often go undetected until disaster strikes. This is one of the big reasons to use a mature project such as FreeRTOS. As Linus' law states: "given enough eyeballs, all bugs are shallow"[11].

12.2.4 Other Utilities

FreeRTOS does give us some utilities beyond the core goal of multi-tasking. These include software timers and memory management. The timer mechanism lets us call a function either at a single specified time in the future (one-shot), or periodically (auto-reload). Execution is still within the timer service task, so it should never attempt to block. For example, calling `vTaskDelay()` or using non-zero block times for accessing queues will cause problems. Full details can be found on the FreeRTOS website[10], and within the distributed source code.

From FreeRTOS v9.0, objects such as tasks, queues, semaphores, mutexes, and so on may be created with user supplied memory addresses. This is a form of *static* memory allocation. This approach gives the most control to the programmer. The exact location of data is known at compile time, the RAM footprint is also known at compile time, and there is no need to handle the event of failure with dynamic allocation. However, the typical approach is to use the FreeRTOS' own, custom dynamic heap allocation. The motivation for a special implementation (not the standard libc `malloc()` and `free()` functions) is that we need determinism for real-time applications. Calls to `malloc()` can take different amounts of time from call to call, which will make meeting guaranteed deadlines difficult.

To allow trade off between performance and feature set, FreeRTOS provides 5 different memory management modules:

- **heap_1**: The most simple scheme. Memory can be allocated but never freed.
- **heap_2**: Allows memory to be freed but will not rejoin adjacent free regions.
- **heap_3**: Just wraps the standard `malloc()` and `free()` to ensure they work safely in the FreeRTOS multi-tasking environment
- **heap_4**: Allows joining of adjacent freed regions as well as placement at absolute addresses.
- **heap_5**: Allows allocation which spans across multiple non-contiguous regions.

All of these features are demonstrated within the demos supplied with the FreeRTOS source code. Browse to the folder FreeRTOS/Demo/Common/Minimal and have a look at how tasks, queues, semaphores, mutexes, timers, and more can be used in practice.

12.2.5 How can we use it?

The easiest way to get up and running with FreeRTOS is through the Xilinx SDK. When creating a new board support package, we can choose FreeRTOS — as opposed to bare-metal. This supports both the Cortex-A53 and Cortex-R5 processors. Alternatively, the full source code can be downloaded from [12]. This comes with demo applications for both sets of processors as well as separate demo files showing how the hardware independent FreeRTOS features can be used.

In terms of design flow, the broad steps are the same as with the bare-metal stack (seen back in Figure 12.2). The only difference is how we go about designing and implementing the user application itself. The task-based approach that FreeRTOS offers, lets us take a more logical, top-down approach to writing the software. We can identify the logically distinct parts of the system and then implement them in near isolation. Often this can be directly related to a high-level system diagram. Using tasks does introduce some new challenges as well. For example, there must be some careful consideration of the priorities assigned to each task. When done poorly, some important tasks could be completely starved of a resource and even bring the rest of the system down along with it. We shouldn't consider function alone when partitioning the system in to tasks, but also priorities, and overzealous inter-task synchronisation. Issues such as these can make partitioning the system into tasks a challenge — but let's not forget the inherent benefits!

There are a few points specific to the Zynq MPSoC to keep in mind if using FreeRTOS:

- The processing system's triple-timer counter 0 (TTC0) must be enabled. This drives the FreeRTOS scheduler "tick" — essentially the heartbeat for all multiprocessing decisions.
- Any custom interrupt handlers must be configured without overwriting (or being overwritten by) the FreeRTOS configuration. Custom handlers must be set in FreeRTOS' existing structure *after* the

scheduler is started. This is demonstrated in `IntQueueTimer.c` from the Zynq MPSoC example code[12].

While looking at these FreeRTOS examples, the distinctive function naming conventions will likely stand out. These function and variable names follow the FreeRTOS style guide[13]. Reading this style guide is highly recommended as the names convey a lot of information, including what each function returns, where it is defined, and what its scope is! For example the `vTaskStartScheduler()` function should be read from left to right as: "This function returns **void**, is defined in **`task.c`**, and its job is to **start** the **scheduler**.

From a legal perspective, FreeRTOS can be quite attractive for commercial applications — it has been recently released under the permissive MIT license[14]. Along with this, Amazon Web Services (AWS) has become the steward of the project, indicating that it is not likely to disappear any time soon.

12.3 Linux Software Stack

Linux is a full-blown operating system which is in widespread use. It supports many different architectures, ranging from slow cookers to super computers... with Zynq MPSoC somewhere in between. Linux will likely be the platform of choice for most developers using the APU, unless the application has a specific need for a different stack.

Before we go any further with Linux specifically, consider why using a "full" operating system on the Zynq MPSoC could be useful. One major advantage for application developers is the high level of *abstraction* provided — the application developer programs to a set of OS service APIs and can mostly ignore implementation details. These services include file systems, networking, inter-process communication, to name a few. The abstraction provided by the operating system also promotes *modularity* within our system. For example, if we program to the networking API, the code is entirely decoupled from the implementation of the network, which could be Ethernet, WiFi, or even some custom radio interface. Another advantage of widely-used operating systems is code reuse. If we pick an operating system which is popular enough, such as Linux, we can benefit from a vast wealth of existing projects! This holds particularly true for Linux as it spans a chunk of the embedded market, many desktop users, and even servers. In a true community effort, any of these Linux users can contribute back to the software ecosystem. If someone has written a good web server application for that operating system (not necessarily with embedded applications in mind), this can save us development time. If someone has written a good graphics library for that operating system, this can save development time, and so on.

The downside to using a larger operating system is the flip-side of the abstraction provided by the operating system. As the overall system is more complicated, it is more difficult (but not impossible) to make strict guarantees about the performance. Each part of the system can have subtle effects on the performance of others (CPU time available, even down to effects on caching). This does not have to be a deal-breaker for Zynq MPSoC devices however. As discussed in Chapter 13, we can run multiple software stacks, each addressing very different requirements. A common configuration is Linux running on all Cortex-A53 cores and more sensitive real-time code running on each of the Cortex-R5 cores.

To summarise, the motivation for using a larger operating system boils down to productivity — we can then focus on the product's added value, rather than reinventing wheels. But the question remains: Why Linux in particular, which variant is best, and why not commercial alternatives? This can be a difficult choice which combines both design and commercial considerations. However, some initial points to consider are:

- Linux is *free*, both in terms of "beer" (cost) and "speech" (liberty). Other open source alternatives are available. Commercial alternatives are not free in either senses.

- Linux is quite popular on the desktop for many developers. This carries over onto embedded platforms through personal preference and experience, as many of the skills are directly transferable between domains.

- The software community around Linux embraces open source licenses. This gives a developer the right to take existing software projects and adapt it for their own needs (often with stipulations about sharing any modifications). Open source can also be advantageous when working with non-x86 architectures, such as the Arm cores in the Zynq MPSoC. We can recompile almost any open source software for our Arm processors when needed. With closed source software, we could only pick from software distributed for our particular architecture.

12.3.1 What does Linux give us?

Like all operating systems, Linux gives us an environment to run our tasks (known as *processes* in Linux). Linux does not only share CPU time between tasks (like FreeRTOS for Zynq MPSoC), it can also schedule tasks between multiple cores of the Cortex-A53. This is known as symmetric-multiprocessing (SMP). Linux offers many services such as networking and file systems to the user application. Productivity is improved even further by allowing the a design to include untrusted, third party applications which may be executed with minimal permissions. The structure of a Linux system is a little different from FreeRTOS. A simplified representation is shown in Figure 12.6.

The device drivers are now part of the operating system's core — the "kernel". This is one of the differences in structure from the FreeRTOS stack. Note that there is also now a strict distinction between the execution of user applications and kernel code. These two realms are called "user space" and "kernel space" respectively. User applications can request work to be done by the kernel using the system call interface.

The kernel provides abstraction of hardware dependent features, such as networking and backing storage. Network devices are exposed to user space as implementation agnostic network interfaces. Backing storage is exposed as a block device, completely decoupled from the choice of file system and hardware (SPI Flash, SD card, SATA hard disk drive, etc.).

Naturally there must be some code responsible for managing and scheduling the different processes on the system. This is accompanied by a memory management scheme which makes use of virtual memory. This allows each process to live within a unique virtual address space, so the kernel can ensure that there are no accidental or malicious memory accesses from one user process to another (or into the kernel space). Process management, as well as memory management, require some architecture dependent code. This code is

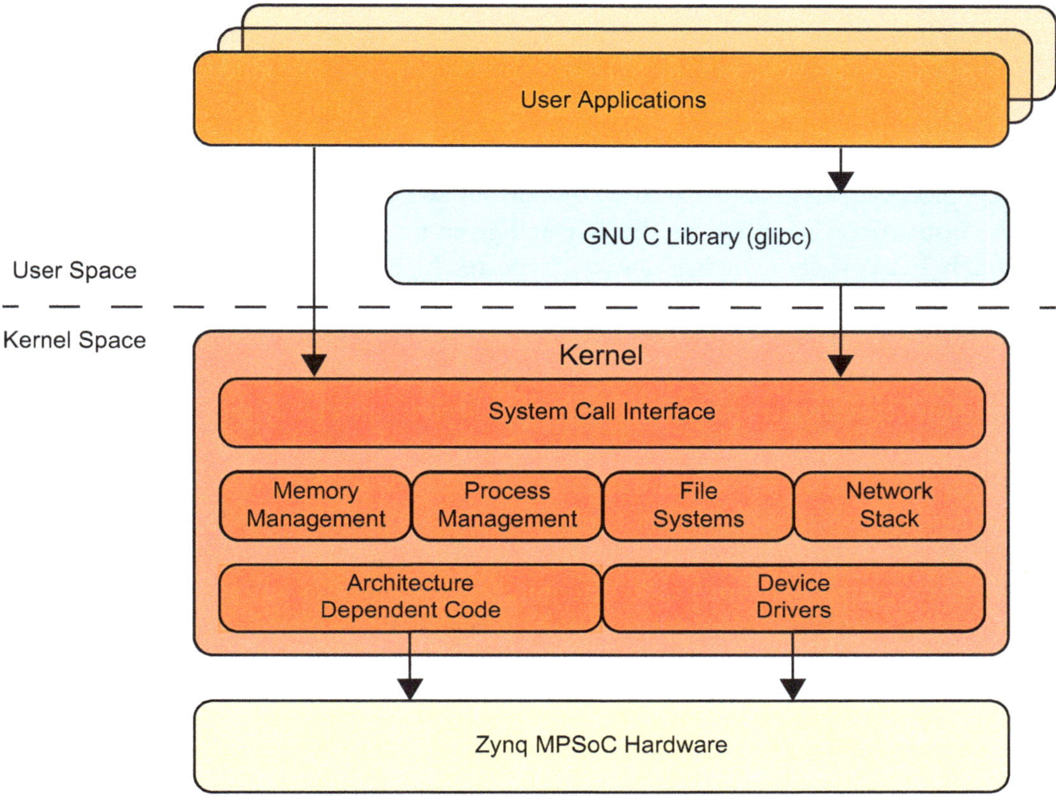

Figure 12.6: Linux software stack

responsible for tasks such as saving/restoring process state and communicating with the hardware memory management unit (MMU). This is the only section of code written specifically for the APU processor architecture.

In the following sections, we will look at the main features of the kernel in a little more detail. This is done mainly with an emphasis on what is useful for the application developer (the *what*), rather than the details of the kernel implementation (the *how*).

12.3.2 The Kernel Space and User Space Divide

As already touched on, there is a strict difference between the "user space" and "kernel space". This is a key point for ensuring user applications play nicely with others. With virtual memory, we can prevent a user process from reading or writing to the memory of another process, or even worse, the kernel. By forcing all calls to privileged functions through the system call interface, we can prevent user processes from tampering with OS structures and devices. These features help provide the possibility to incorporate untrusted third party software.

Oversimplifying things slightly: user code runs in user space and kernel code runs in kernel space. The interesting point here is how this is enforced. What is preventing user code jumping to an address within the kernel space and executing code? We are going to look at this quickly as an introduction to the exception levels in the Cortex-A53. This has become more important with the possibility of using hypervisors on Zynq MPSoC, as discussed in Chapter 13.

The Cortex-A53 processors have some hardware support for *exception levels*. This is a concept related to the permissions of the instructions being run on the core at a given time. There are 4 different exception levels, named EL0 through EL3, where a higher number indicates higher privilege. The gist of the separation between user space and kernel space execution comes down to these exception levels. User space code runs in EL0 (the least privileged level) and kernel code generally runs in EL1, which has more privileges. The other two levels are not really important here — they are for virtualisation and secure states. The general sequence taken when a transition between user space and kernel space is performed (a "system call") is shown in Figure 12.7. This depicts a user program which requests some kernel-level activity via a system call. This is very much like an interrupt or function call which transcends exception levels. The user program prepares for the call, followed by a `SVC` instruction (SuperVisor Call). This jumps from EL0 mode to EL1 — the mode in which the kernel code runs. The kernel responds by running from a predefined address, which is not dictated by the untrusted user program! The kernel code determines which function was requested, similar to identifying the proper routine for a given interrupt source. The requested function is performed in kernel space. Execution is then returned from EL1 to EL0 using the `ERET` instruction (Exception RETurn).

This is how user applications *should* operate, but now we consider an application which does not play nicely! What is actually stopping them from executing kernel level code? EL0's access to certain features is prohibited at a hardware level by the Arm core. This includes access to special registers and regions of memory, with which the user program could suppress fault signals, freely change exceptions levels, and be nefarious in innumerable distinct ways. The other side of this is that system boots into a privileged level. Before every single return to user space, execution is returned to EL0, ensuring there is no opportunity for user programs to perform foul play. This is a practical security measure not provided by bare-metal and FreeRTOS stacks. This allows us some freedom to run code written by others without the need to worry about hardware level side effects on other processes.

Figure 12.7 is deliberately vague about loading arguments to registers. The details of this for each call are defined by the system call interface, and application binary interface (ABI) below that. This may sound a little cryptic, and that is completely OK! The truth in practice is that all of these system calls tend to be performed through a `libc` library — commonly "`glibc`" from the GNU team. If we need to issue a "file open" call to the operating system, we just call the `libc` function "`fopen`". Libc will handle all of the complexity of the system call on our behalf. We *could* do this manually, but it is not often required.

Making all system calls through this one interface means that there is a surprising amount of independence between the user and kernel space. In fact, we can often swap out a Linux kernel or the user space programs without any real repercussions. This works as long as the user space applications have a version of `libc` to talk to that is appropriate for the kernel version. This means a carefully crafted kernel will work well with many different sets of user space applications, and vice versa. While this may not seem like a big deal, it will be very useful when testing different root file systems after creating the perfect drivers and Linux kernel for your system!

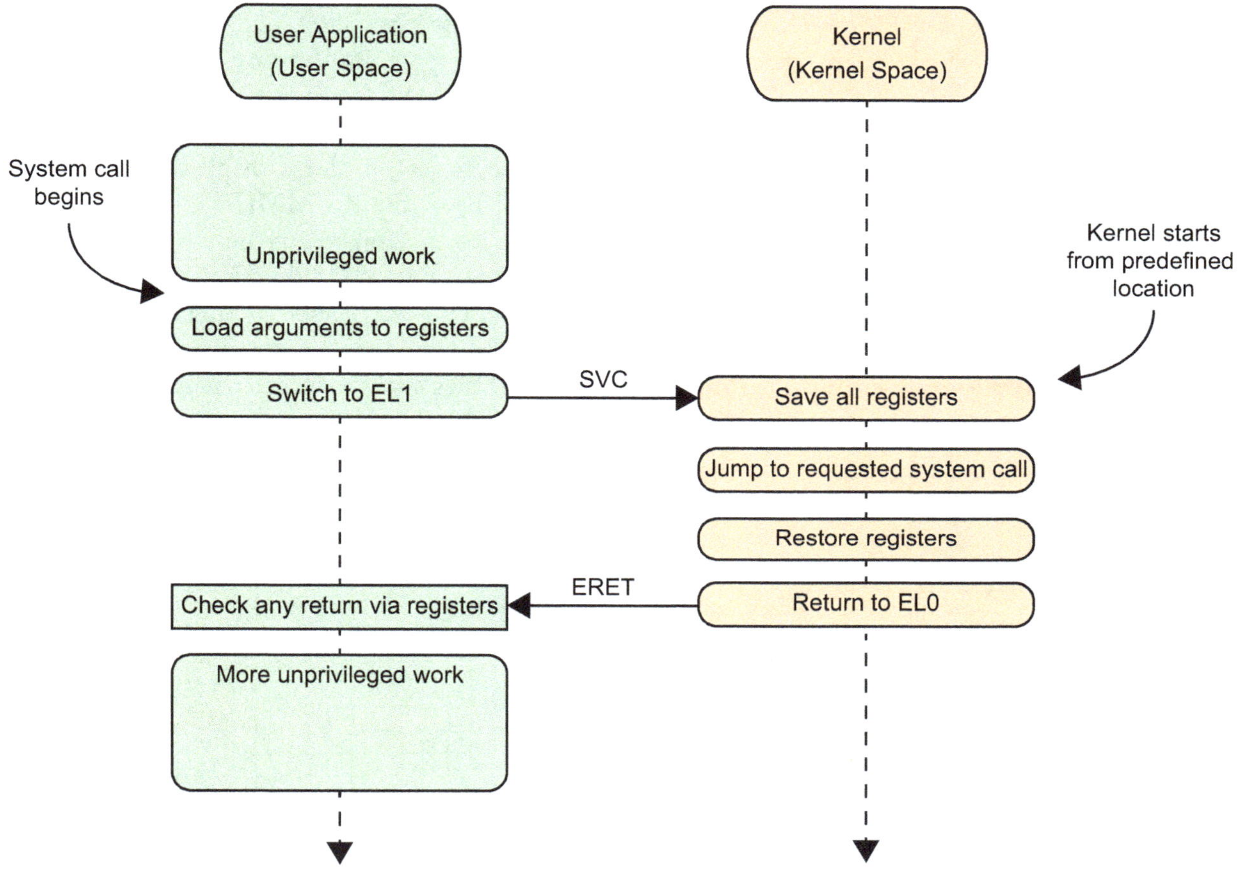

Figure 12.7: Sequence for a Linux system call

12.3.3 Memory Management

To put the Linux memory management in context, first consider the simpler system used in the bare-metal and FreeRTOS stacks. They use a single, flat address space — the software and hardware both work with the same set of addresses. The address space contains memories (off-chip DDR, on-chip memory, etc.) as well as peripheral registers from both the processing systems and programmable logic. By default, there is no protection between regions. Buggy or malicious code from one task can access and overwrite that of another task. These issues become a larger problem when running programs written by untrusted third parties — as is common on top of full operating systems. Also, consider dynamically loading tasks. We must ensure that each task is running from different areas in the address space so one task does not clobber the code and data of another. There can be no hard-coded addresses for storing variables, no absolute addresses for branching, and so on. The flat address space is a simple, intuitive approach but it lacks some neat features which are provided by operating systems such as Linux.

Linux can make use of *virtual memory* [15] (but may also be built without it). This is a scheme which uses virtual address spaces which are mapped into the physical address space. As software executes, it uses virtual addresses which are mapped to physical addresses on the fly by the MMU hardware within each Cortex-A53 core. A simplified version of this system is shown in Figure 12.8. The kernel holds a page table for each process which is the mapping from blocks of virtual memory (pages) to blocks of physical memory (page frames). There is a hardware Translation Lookaside Buffer (TLB) within the MMU which acts much like a cache for the full page table. The TLB enables most load/store instructions to execute using virtual addresses without intervention or any performance penalty. If a mapping is not found in the TLB, the MMU raises an exception and the kernel is responsible for creating the mapping. This can include reading different page table entries, and even fetching pages which have previously been pushed out to backing storage in a swap partition. These activities introduce a substantial performance penalty but they occur very infrequently.

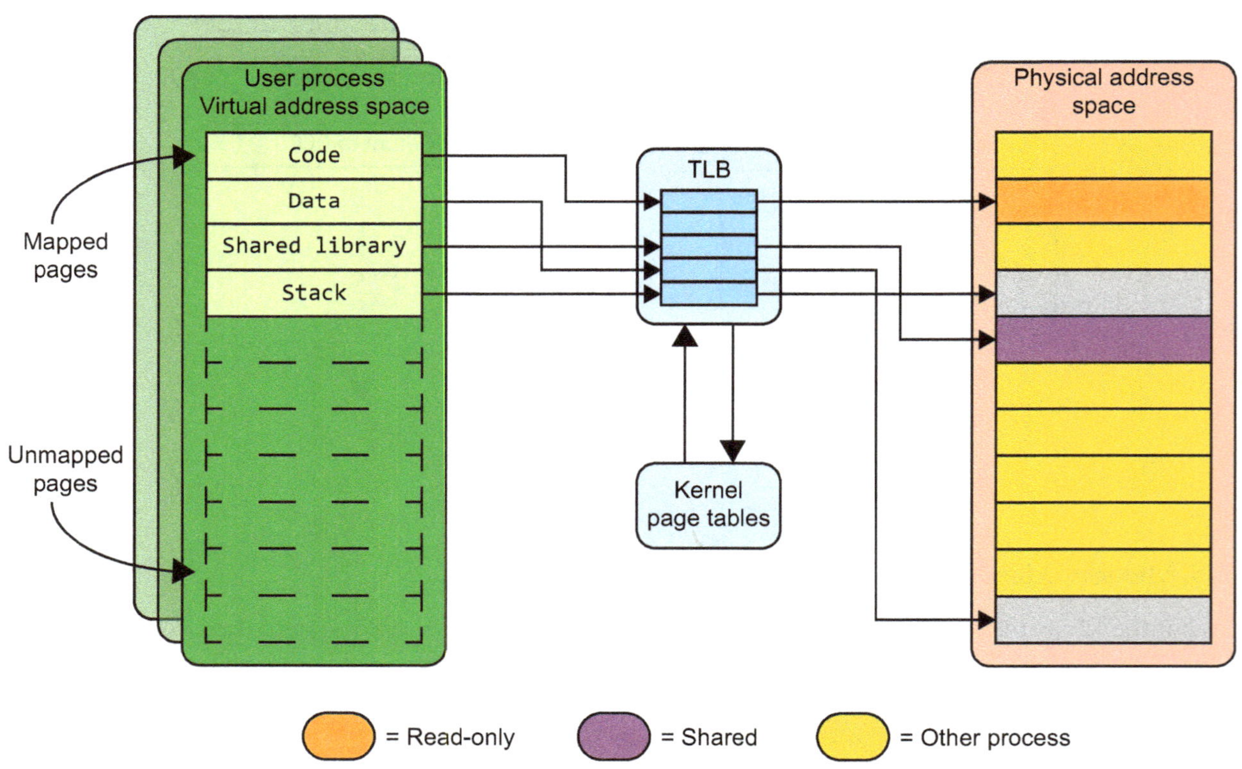

Figure 12.8: Simplified virtual memory translation

This scheme introduces some overhead, so there must be a good reason to do so! In fact, there are several:

- The kernel can maintain **different mappings for each process**. This way there is no way for a process to directly access the private memory of another process — without registering with the kernel.
- We can move the physical location of a page without breaking the process which owns it. This means we can **swap out to backing storage** when RAM is getting full. We can oversell the physical memory available.
- We can still **share frames between processes** when needed. This can be used to let many processes talk to a single instance of a shared library, to minimise duplication when running multiple instances of the same program, or for inter-process communication.

An important performance aspect to consider is the way Linux initially loads pages. Demand paging is used. This means that pages are loaded into physical memory only when first accessed — not when the process is created. This can save huge a amount of RAM, but can cause unpredictable performance if more pages are required later in execution. If timing is key to a program, it may be a good idea to "touch" all of the pages on load. The functions `mlock()` and `mlockall()` can also be used to ensure the pages are never swapped out to backing storage in the future[16].

There is one more difficulty which arises from use of the virtual memory. When communicating with hardware and peripherals, we *need* to use the physical addresses! This is also an issue in the other direction, i.e. when a peripheral or IP core in the PL must access buffers in the processor's virtual address space. We can get around this by creating a mapping from a given physical address to the virtual address space — see [17] for further details. Virtual memory is not often employed in bare-metal systems, making bare-metal drivers free from this concern. So, in order to communicate with any logic implemented in the PL, we will need to navigate the address spaces using Linux device drivers, as discussed in the next section.

12.3.4 Device Drivers

Writing drivers for Linux is seen as somewhat of a black art, with kernel hackers the clandestine magicians keeping the secrets of the code close to their chests; wary of the uninitiated. In reality, the subject is well documented, with many resources available on the development of drivers and even more on the kernel itself. That said, it is a demanding subject with a learning curve which may seem steep to some, it demands good knowledge of the C programming language and at least some basic knowledge of the kernel itself. Because of this, it would be impossible to cover everything in this short section therefore we provide you with a basic overview of some of the more important aspects leaving you to fill in the gaps.

Kernel Space Drivers

Types of Drivers

There are three classes of device drivers available on Linux systems: character (or char), block, and network — each handling data in a different way. Char devices handle streams of bytes, block devices; whole blocks of bytes, and network interfaces; packets. For the most part, the drivers you create will likely fall under the character device category and act very much like a regular file does — you are able to open, close, read and write — and these devices are available as filesystem nodes in the `/dev` directory.

Block devices similarly can be accessed as filesystem nodes and can read and write just like char devices, but use special functionality to handle read and write commands.

A rather over-simplified way of thinking about these two device types is that char devices are like a serial port — one byte at a time goes directly to the hardware, whereas a block device reads and writes in chunks — a hard drive is an example of a block device.

Network devices are a little more involved. The data that is transferred is in packet form and is driven by the network subsystem in the kernel. A network driver does not care about individual connections, it just handles the packets. The interfaces are not mapped into the file system like char or block devices and does not use the read and write calls that these devices use. It uses functions related to packets.

Kernel Modules

Linux provides two ways to install drivers, either by statically building them into the kernel, or as a kernel module which can be dynamically loaded in when needed by the system. When developing drivers it is worthwhile using kernel modules as the need to re-build the kernel after every minor change to your code is both time consuming and unnecessary — statically built drivers are only unavoidable when the driver is needed to help the system boot. Another desirable feature of modules is that they are much faster to maintain and debug. An error in a pre-built driver can be difficult to fix, whereas a module can be unloaded, amended, then loaded back into the kernel all during runtime. No need to reboot your system!

Code Reuse

Thanks to the openness of the Linux kernel and (most of) its drivers, the source code is freely available to view, use and modify by anyone interested in how things work. So when creating your own hardware, it is possible that a driver for a similar device already exists in the kernel, removing the arduous task of writing your own from scratch and giving developers a template to work from. Additionally, the author of Make Linux maintains the Linux Driver Template (LDT) which can be found on his GitHub page here [18].

One thing to note is that if you plan on modifying an existing driver, pay attention to the licensing agreement it falls under. Most of the kernel is GPLv2 which states that any derivative work also falls under the agreement.

User Space Drivers

For some use-cases, writing a complete kernel driver for your hardware may be unnecessary. If all your driver needs is access to memory (and possibly an interrupt) then it is entirely possible to create the driver you need completely in user space. As well as well known APIs such as *libusb* and *i2c-dev*, Linux also provides a few other ways to create drivers while keeping the kernel at arms-length.

Directly Interacting with Memory Through /dev/mem

Linux stores a representation of the main memory of the system in the character device file `/dev/mem`, enabling a user with elevated permissions to read from or write to any location in memory using a program such as *devmem*, or a user mode driver. Similarly `/dev/kmem` allows interaction with kernel virtual memory rather than physical memory.

Programs like *devmem* are excellent tools to quickly poke at a memory to confirm your hardware is functioning correctly but, from a security standpoint, it is not advisable to allow users to interact with memory so freely. This has been addressed from kernel version 2.6.26 by introducing software-based hardware protection at the kernel level. By default, the kernel configuration options *CONFIG_STRICT_DEVMEM* and *CONFIG_DEVKMEM* for most desktop and server distributions are disabled, but for many embedded Linux installations they are not — if you are not building your own kernel it is advisable to check the status of these options on installation before putting your device in the wild.

Xilinx provides an example of what they call a user mode pseudo driver using this method, but in a more formalised way that enables a user to toggle GPIO, available here [19]. In this example `/dev/mem` is opened as a regular file and uses *mmap* to map the device into user space memory. This type of driver is relatively simple to understand and implement, and is very similar to a typical driver you would write for a bare-metal implementation. One downfall of using this type of driver is that they are not able to handle interrupts or any other kernel resources, so there is a limited scope in which they can be used. The example they give is written in C, but could easily be translated into any language that can operate on files and has knowledge of *mmap* (e.g. Python).

Userspace I/O

If the need for a kernel driver is excessive for your project but your hardware requires the use of interrupts, Linux provides the Userspace I/O (UIO) API. The UIO allows users to write drivers that provide access to the memory space of your device and handle interrupts all within userspace, with only the need for a small custom kernel module.

Similarly to the `/dev/mem` approach, UIO allows physical device memory to be mapped to userspace memory but introduces extra security by only allowing the device read and write privileges to memory that belongs to the device. Devices are accessed through device files (e.g. `/dev/uio1`) much like kernel devices, but with the addition of a separate interrupt file (`/dev/uioX`) which can be read when an interrupt occurs.

As mentioned before, the kernel module needed for a UIO driver need only be small, which minimises the amount of kernel code a developer needs to maintain over the lifeteime of the device. And as with most elements of the Linux source code, example modules are available to study and modify to your needs. Tools such as PetaLinux provide a module for you which allows the use of generic interrupts but more complicated drivers may need a more customised kernel module.

Documentation on Userspace I/O is sparse but there is a HOWTO guide available in the kernel documentation which can be accessed here [20], as well as an excellent paper written by the author of the API which expands on the subject in more detail, available here [21].

Additional Resources

This section is only intended to inform the reader of what Linux drivers are and some insight into how they work. As mentioned before, the subject of device drivers is extensive and well beyond the scope of this book but, luckily, there is a myriad of documentation and tutorials out there for developers — if you know where to look.

The first point of call for any developer is the Linux kernel's own website which includes comprehensive documentation on all aspects of the kernel including HOWTOs on drivers, modules and the UIO API. This documentation is well presented in HTML format, easily searchable and is the most up-to-date of any other source [22].

There is also a wealth of materials available for free through various online sources. Bootlin (formerly Free Electrons) offer complete access to all their training material as well as hosting an extremely useful cross-referencing tool, allowing users to explore the Linux source code [23]. The Linux Documentation Project (TLDP) has a host of HOWTOs and free eBooks including guides on kernel modules and device drivers [24]. Finally, the Linux Device Drivers (LDD) book is available in print and as a free eBook from O'Reilly [17]. LDD is renowned as the best all-round resource for Linux device driver development and covers most topics a developer needs. As of writing the current edition (3rd) only covers up to kernel version 2.6 (current stable version is 4.15), which means a lot of the example code does not work on modern installs, although there have been efforts to maintain a somewhat up-to-date repository on Github [25]. Despite its age, LDD is still the go-to resource for many developers.

12.3.5 Process management

A complete system will often be split into logically distinct processes, much like the approach as seen in our discussion of FreeRTOS. If we think of a program as the binary file sitting patiently on the disk, then a process is that program under execution — with all the accompanying state. Linux is responsible for maintaining the supporting structures for each process, managing resource allocations, and scheduling processes to be run on each core.

The state of processes must be stored because the kernel should be able to stop execution of the current process, swap in a new process and have it continue execution as if it was never disrupted. The state is composed of many fields, but can be grouped roughly into the following, non-exhaustive areas[26]:

- **Scheduling state**: Is the process running/runnable, sleeping, sleeping and not interruptible, stopped, or zombie (halted but the process structures are not yet cleaned up)?
- **Identifiers**: Including a process ID along with user and group IDs used to control file access.
- **Family links**: Processes are structured into a tree, with each process storing references to their parent, sibling and child processes.
- **File system**: Each process tracks which files/devices they have open.

- **Virtual memory**: Each process has a distinct page table to map its virtual address space to physical addresses
- **Scheduling information**: Some information is tracked to assist the scheduler when picking the next process to get CPU time.
- **Registers:** The state of the registers must be saved so the process can be resumed from exactly where it left off.

One aspect of Linux process structure that can come as a surprise is its hierarchical organisation. This is partly due to the way we create new processes in Linux — the `fork()` system call. `fork()` creates a new child process by copying the current process. A second function call, `exec()`, or just a branch in the original software usually follows the `fork()` call, making the child process do something different from the parent process. The "fork and exec" approach to process creation leads to the process hierarchy quite naturally. However, the code to do so can look quite unnatural to the uninitiated. The same source code is designed to handle two different process — one process generally takes a "parent" branch of an if statement while the child takes another.

There is a single process, "init", which forms the root of the tree. This is loaded automatically at boot and it does some basic initialisation such as mounting file systems before handing execution off to another, larger initialisation system — a "SysVinit"-style system with the PetaLinux defaults. SysVinit makes it simple to start other processes at boot time, such as network services, consoles, or graphical environments.

We can use the `ps` command on Linux systems (desktop, or Zynq MPSoC) to get summary information about all active processes. Process management is one of the many areas of Linux where the skills are directly transferable between desktop and embedded use. An example embedded system may report a process list similar to Figure 12.9.

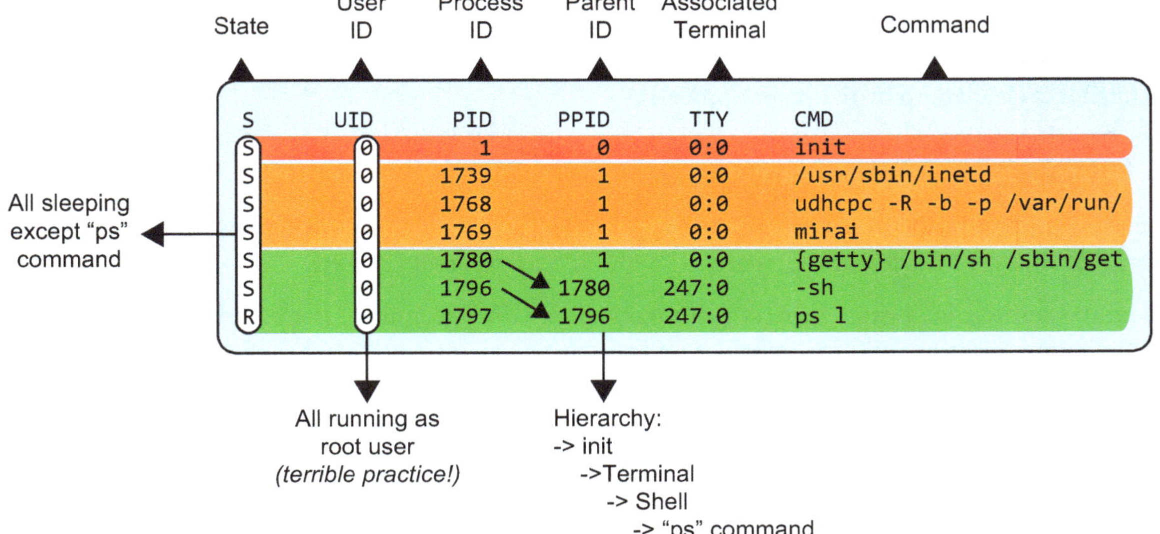

Figure 12.9: Example Linux process list after boot

The last point to note about Linux processes is how they are scheduled. The default scheduling algorithm is the "Completely Fair Scheduler" (CFS) since Linux 2.6.23. This is essentially a weighted fair queuing algorithm, where weights are calculated from a given "niceness" value. The default niceness is 0 but can be altered with system calls or the "nice" command. Higher nice values make a process have a lower priority (more nice), whereas negative nice values have a higher priority (less nice). There is a lot of thought and effort put into developing a good scheduler for a multi-core capable OS such as Linux. Luckily, application developers will likely never need to tinker with the scheduler beyond setting nice values.

To conclude on the dry topic of Linux process management, we provide some levity in the form of a small example of process creation and have a bit of fun with the Linux nomenclature. With just a few lines of C, we can make a *very nice zombie* (in Linux terms, at least). No processes were harmed in the making of this example.

Create child process with fork
Put parent to sleep so child is not cleaned
Make child very nice, then exit (zombify)

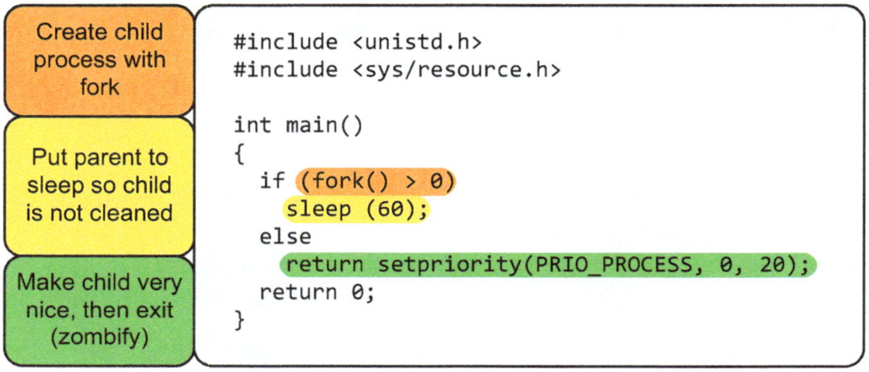

```
#include <unistd.h>
#include <sys/resource.h>

int main()
{
  if (fork() > 0)
    sleep (60);
  else
    return setpriority(PRIO_PROCESS, 0, 20);
  return 0;
}
```

Figure 12.10: Example of creating child processes using fork()

The `fork()` call creates the new child process which is a clone of the parent. It returns 0 to the child process, but returns the new process ID to the parent. The first if statement branch will catch this non-zero return value and only the parent process will be put to sleep. The child process will fall into the else statement instead, setting its *niceness* value very high. When the child process exists, it remains in the system as a *zombie* process until it is cleaned up by the parent process.

12.3.6 Inter-Process Communication

Processes working in isolation can be useful, but processes will often need to communicate with each other. Linux provides many of the features present in FreeRTOS as well as a few extras.

The quintessential Linux Inter-Process Communication (IPC) technique is a *pipe*. Pipes are unidirectional streams for communication, "piping" information from one process to another. From the shell, these are represented by the "|" character between different processes, as seen below. The example takes the output of the `ps` command, which lists all active processes, and passes it via a pipe to the input of "grep" — a command to filter input by a search term.

```
$ ps aux | grep firefox
```

Another set of IPC features comes from the System V IPC interface. These include features familiar to us from FreeRTOS such as semaphores, message queues, and shared memory. For details, we can refer to the extensive Linux Documentation Project at [27].

We could also perform some IPC using network sockets. This provides a channel of communication between processes which *can* be used locally, but also opens the possibility of communicating between different devices.

12.3.7 How can we use it?

For now, keep in mind that Linux is an open, community driven effort. As a result, we have the option of using Xilinx developed tools (which tie in very neatly with the rest of the Xilinx workflow), but also many, many others. There is a vast selection of wonderful open source tools which can be made use of.

To stay within the confines of the Xilinx workflow, PetaLinux provides a simplified method to customise, build and deploy embedded Linux systems onto your device. It uses a set of command line tools and, more recently, a graphical SDK that can installed as a plugin for the Eclipse IDE. It is based on the Yocto Project and many of the functionalities remain the same, so if you have experience working with Yocto, the transition will be a comfortable one. It also supports the use of the Quick EMUlator (QEMU) allowing you to simulate and test your system before deploying to hardware.

The PetaLinux tools allow developers to neatly combine all aspects needed for an embedded Linux system such as the FSBL, U-BOOT, ATF, the Linux kernel, filesystem, libraries and applications. Adding your own software is relatively easy as PetaLinux can automatically build and compile your application source at the same time as the kernel, or just add your pre-built binaries to the appropriate folders. Installation can be even further streamlined by the use of board support packages (BSPs) where pre-built images allow users to quickly boot Linux on a development board.

The Xilinx fork of the Linux kernel is publicly available from their GitHub repository [28]. Nothing necessary for our Linux software development is hidden from us and, if desired, we can build everything from the source code available online ourselves!

Note that building a Linux kernel and file systems is much easier from a Linux desktop than any other operating system. There are ways to work around this for new users though. Linux development can be done in a virtual machine running on top of Windows, for example. Do keep in mind though, a lot of the Linux skills acquired from desktop use are also directly transferable to Linux on the Zynq MPSoC.

12.3.8 Multimedia Software Stack

As a quick exercise to show one of the many ways that Linux can speed up development of a Zynq MPSoC system, we will consider multimedia applications. Considering the stack shown in Figure 12.11, we can make a video processing system without duplicating low-level effort. All the developer needs to consider is developing their graphics application in accordance with the Linux graphics libraries such as OpenGL or OpenVG. There are pre-built drivers and libraries to handle all of the Mali-specific functionality.

Programming directly against the hardware interface of the Mali GPU is a major task. This abstraction saves the developer much time, and thus direct cost, and costs implicit in a device speedily deployed to market.

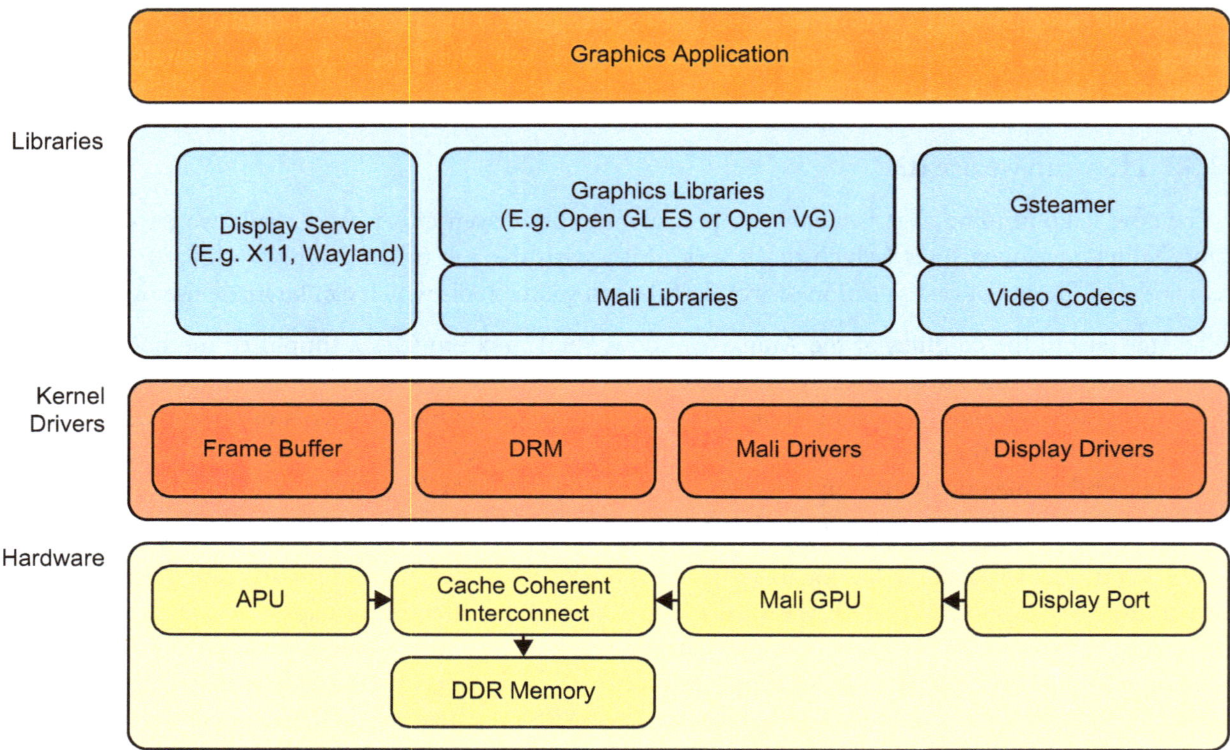

Figure 12.11: Linux multimedia software stack

12.4 Chapter Review

We have introduced three of the main software stacks a developer can exploit on the Zynq MPSoC. These range from bare-metal, in which you have full control but must program everything except the supplied set of drivers, to a full Linux system where we can develop a full video processing system while only programming against a simple graphics library. The reader should remember that there is not a single stack which is "better" than another, it is simply a case of suitability for the task at a hand. We hope that we have given an overview of how and why you may wish to use each software stack.

The software stacks described here are not an exhaustive list. As with the original Zynq-7000 devices there will be some community driven software stacks available eventually. For example, FreeBSD, another Unix based operating system, was ported to the original Zynq[29]. These are always something to consider — we can only discuss the most pertinent ones at the time of writing.

Having discussed the Xilinx supported software stacks already, the reader may be left thinking that their needs straddle two or more of the proposed stacks. In the modern, high-end embedded systems world, there are solutions for this. Chapter 13 provides us with a base for connecting multiple different software stacks. We can use more than one software stack and communicate between them using projects such as the Xen hypervisor and OpenAMP. This can even alleviate the need for custom Linux device drivers when hardware communication is delegated to the R5 processors via OpenAMP.

12.5 References

Note: All online sources last accessed March 2019.

[1] Xilinx, Inc., "UG1137 - Zynq UltraScale+ MPSoC Software Developer Guide", v1.0, November 2015. Available: http://www.xilinx.com/support/documentation/user_guides/ug1137-zynq-ultrascale-mpsoc-swdev.pdf

[2] Xilinx, Inc., "Upstream Newlib + Xilinx branches". Available: https://github.com/Xilinx/newlib

[3] Xilinx, Inc., "UG643 - OS and Libraries Document Collection", v2016.2, June 2016. Available: http://www.xilinx.com/support/documentation/sw_manuals/xilinx2016_2/oslib_rm.pdf

[4] Xilinx, Inc., "Standalone Drivers and Libraries". Available: http://www.wiki.xilinx.com/Standalone+Drivers+and+Libraries

[5] Xilinx, Inc., "Embedded Software Development Git Repository". Available: https://github.com/Xilinx/embeddedsw

[6] FreeRTOS, "FreeRTOS+UDP: A Tiny Embedded UDP/IP Stack Implementation for FreeRTOS". Available: http://www.freertos.org/FreeRTOS-Plus/FreeRTOS_Plus_UDP/FreeRTOS_Plus_UDP.shtml

[7] A. Silberschatz, G. Gagne, P. B. Galvin, *Operating System Concepts*, Chapter 6: "Process Synchronization", 8th edition (ISBN 978-0-470-12872-5). John Wiley & Sons., July 2008.

[8] A. B. Downey, *The Little Book of Semaphores*, 2nd edition, Green Tea Press, June 2016. Available: http://greenteapress.com/semaphores/LittleBookOfSemaphores.pdf

[9] FreeRTOS, "RTOS Task Notifications". Available: http://www.freertos.org/RTOS-task-notifications.html

[10] FreeRTOS, "Software Timers". Available: http://www.freertos.org/RTOS-software-timer.html

[11] E. S. Raymond, "The Cathedral and the Bazaar", Edition 3, August 2002. Available: http://www.catb.org/~esr/writings/cathedral-bazaar/cathedral-bazaar/

[12] FreeRTOS, "RTOS Source Code Download". Available: http://www.freertos.org/a00104.html

[13] FreeRTOS, "Coding Standard and Style Guide".
Available: http://www.freertos.org/FreeRTOS-Coding-Standard-and-Style-Guide.html#NamingConventions

[14] D. Straughan, "Announcing FreeRTOS Kernel Version 10", AWS Open Source Blog, November 2017.
Available: https://aws.amazon.com/blogs/opensource/announcing-freertos-kernel-v10/

[15] A. Ott, "Virtual Memory and Linux", *Embedded Linux Conference*, April 2016.
Available: https://events.linuxfoundation.org/sites/events/files/slides/elc_2016_mem.pdf

[16] M. Kerrisk, "Linux Programmer's Manual: MLOCK".
Available: http://man7.org/linux/man-pages/man2/mlock.2.html

[17] J. Corbet, A. Rubini, G. Kroah-Hartman, *Linux Device Drivers*, Chapter 9, Third Edition.
Available: https://lwn.net/Kernel/LDD3/

[18] C. Shulyupin, "GitHub repo for Linux Driver Template", MakeLinux.
Available: https://github.com/makelinux/ldt

[19] Xilinx Inc., "Linux User Mode Pseudo Driver".
Available: http://www.wiki.xilinx.com/Linux+User+Mode+Pseudo+Driver

[20] H. Koch, "The Userspace I/O HOWTO", Linux Kernel documentation.
Available: https://www.kernel.org/doc/html/v4.12/driver-api/uio-howto.html

[21] H. Koch, "Userspace I/O drivers in a realtime context", Linutronix GmbH.
Available: https://www.osadl.org/fileadmin/dam/rtlws/12/Koch.pdf

[22] The Linux Kernel Organization, "The Linux Kernel Archives".
Available: https://www.kernel.org

[23] Bootlin, "Bootlin — formerly Free Electrons".
Available: https://bootlin.com/

[24] The Linux Documentation Project, "The Linux Documentation Project".
Available: http://www.tldp.org

[25] J. M. Canillas, "GitHub repo for Linux Device Drivers 3 examples updated to work in recent kernels".
Available: https://github.com/martinezjavier/ldd3

[26] D. A. Rusling, "The Linux Kernel", Chapter 9 - Processes, The Linux Documentation Project.
Available: http://www.tldp.org/LDP/tlk/kernel/processes.html

[27] S. Goldt, S. van der Meer, S. Burkett, M. Welsh, "The Linux Programmer's Guide", The Linux Documentation Project.
Available: http://www.tldp.org/LDP/lpg/

[28] Xilinx, Inc., "GitHub Repo for Xilinx linux-xlnx".
Available: https://github.com/Xilinx/linux-xlnx

[29] T. Skibo, "FreeBSD: Zedboard".
Available: https://wiki.freebsd.org/FreeBSD/arm/Zedboard

[30] Xilinx, Inc., "Linux Drivers".
Available: http://www.wiki.xilinx.com/Linux+Drivers

[31] C. Svec, *The Architecture of Open Source Applications, Volume II: Structure, Scale, and a Few More Fearless Hacks*, Chapter 3: "FreeRTOS".
Available: http://www.aosabook.org/en/freertos.html

[32] Arm, Ltd., "ARM Synchronization Primitives Development Article: Implementing a semaphore", 2009.
Available: http://infocenter.arm.com/help/index.jsp?topic=/com.arm.doc.dht0008a/ch01s03s03.html

Chapter 13

Multiprocessor Development

In order to get the most out of the Zynq MPSoC's hardware, we need to design software systems that target the appropriate processors to complete a given task. Developing software that is amenable to multiprocessor systems is a vital skill, not only to exploit all of the Cortex-A53 processors within the APU, but also extending a system to make use of the other available architectures. These include the RPU's Arm Cortex-R5 processors, and even MicroBlaze cores implemented in the PL. Although we only apply multiprocessor development to Zynq MPSoC, properly utilising multiprocessor hardware is becoming a common challenge for many embedded software developers, as SoC devices trend towards offering more and more heterogeneous cores[1].

This chapter introduces the Zynq MPSoC as a heterogeneous multiprocessor system, discusses the motivation for heterogeneousness in embedded systems, and some of the technical nomenclature. Then with a more practical eye, we explain how three different techniques can be applied to the Zynq MPSoC to get the most from the hardware:

1. Writing applications that can make use of all the APU's processors (using Linux).
2. Extending software present in the APU to cooperate with systems running on the RPU and any MicroBlazes (using OpenAMP).
3. Using the APU processors to host diverse software stacks simultaneously (using Xen).

---

1. This is being driven in part by an ever increasing silicon budget, while uniprocessor clock frequency has become limited by the 'power wall' [1] — as well as the push for energy efficiency in mobile devices.

These techniques are cemented with a final example application, which combines all three.

13.1 Introduction to Heterogeneous Processing

The Zynq MPSoC provides multiple sets of processors, each with a different architecture and speciality. The fact that these processors sets are dissimilar to each other (i.e. they are not all just identical cores) means that it is categorised as a 'heterogeneous' system. This is an important topic which is rife with intimidating terminology for simple concepts.

By combining dissimilar processors with different specialities, heterogeneous systems can provide better performance and energy efficiency than an equivalent 'homogeneous' system, that only has one type of processor. This effect is quite intuitive given a simple example. Consider an application such as a smartphone. There will likely be a general purpose processor to support a rich user environment, alongside a different processor type, that has been specifically designed for the radio communication standard's signal processing requirements. The same functionality could, of course, be achieved in a homogeneous fashion by adding more general purpose cores until there is enough raw processing power to handle the radio standard(s). That would be extremely wasteful both in terms of the resources used and also performance per watt though. Heterogeneous processing helps us to avoid this wastefulness.

13.1.1 Processor Sets on Zynq MPSoC

Within the Zynq MPSoC, there are many types of processor. Some are obvious (such as the APU and RPU) while others are more easily overlooked, such as the GPU present on some devices and the possibility for using MicroBlaze soft processors in the PL. None of these share the same instruction set architecture, leaving no doubt that the Zynq MPSoC is a heterogeneous device.

The benefits of heterogeneous processing comes with designing for the unique advantages of each type of processor. We attempt to summarize these advantages in Table 13.1, but keep in mind that there can always be more subtle, per-application considerations too. Clearly, a complex design that demands a Zynq MPSoC device will not be comprised of just one of these "suited applications" — but a set of them. Part of the complexity of designing the software system is identifying these different requirements and splitting tasks between suitable processing blocks. This challenge is briefly summarised in the next section.

13.1.2 Splitting Software Tasks

Software applications that fully utilise the Zynq MPSoC hardware will likely have some architectural complexity — such as the Zynq MPSoC-based Mycroft Mark II open voice assistant [2] and our own software defined-radio example used in "A Hybrid Example for Software Defined Radio Applications" on page 348. Before we dive into how an application can make concurrent use of all the processing blocks available, it is worth touching on how to split a complex application into a set of smaller tasks. The aim here is to make each task have a more manageable complexity and to exploit the advantages of a single processing block.

.

Table 13.1: Summary of Processing Blocks available in Zynq MPSoC

Processing Block	Application Benefit	Suited Application
APU	• Rich OS support for user interaction (e.g. Linux) • Hypervisor based computing that combines many disparate software stacks	• User interface • Network or cloud integration • Business logic layer
RPU	• Real-time software execution (hard, firm, or soft) • Deterministic execution • Low-latency execution of interrupts and memory access	• Safety-critical applications • Standards-compliant software such as radio stack implementations
MicroBlaze(s)	• Low resource usage (for implementation in PL) • Flexibility (can be implemented with or without certain features and extensions)	• Low performance, but deterministic applications • Control blocks for PL accelerators
GPU	• Graphics acceleration (depending on OS support)	• Graphical interfaces • Multimedia applications
PL	• Acceleration of compatible functions	• Highly-parallelisable processing • DSP or other real-time, sample-based applications

As well as splitting a large application by grouping logical functions, there are some less obvious factors to consider. You may actually wish to join two parts of the system which seem disparate in function, but depend heavily on the same data. Noticing an abundance of this type of data dependence can often signal that a part of your algorithm is not well suited to parallelism. You may also wish to further split tasks not only by function, but also by real-time requirements. For example, a communication stack within a larger system will have some lower layers that demand real-time guarantees in order to comply with a standard, while higher layers can be more relaxed.

With an application split into smaller tasks, we can begin to map these tasks onto the processing blocks available in the Zynq MPSoC. Figure 13.1 shows one example decision tree, suggesting how a given task might be assigned to a processing block.

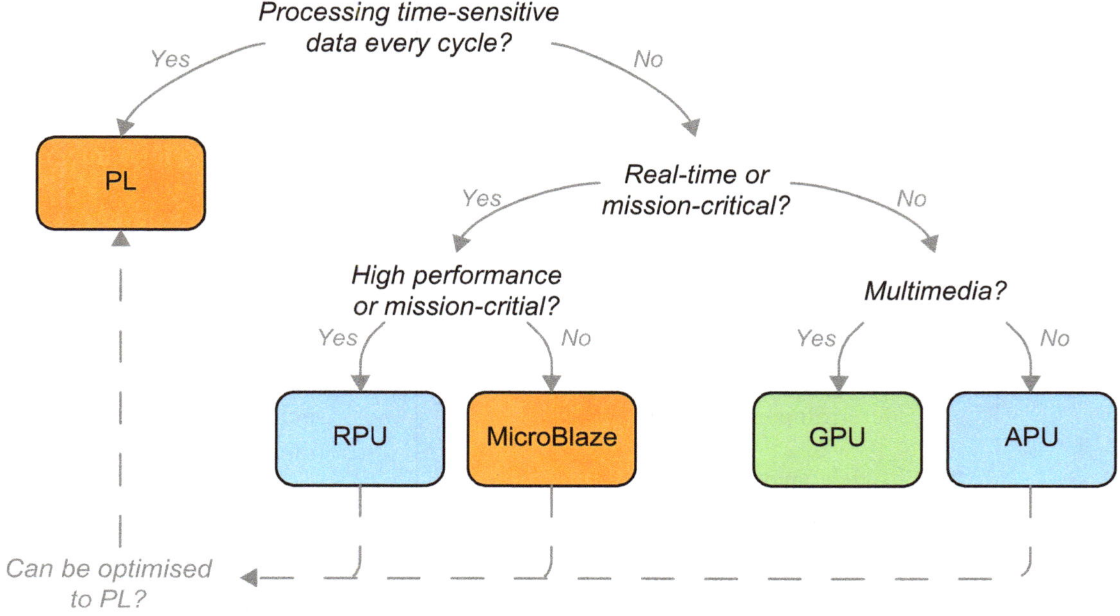

Figure 13.1: Decision tree for assigning tasks to a particular processing block [3]

Note that this is just a rough rule of thumb, especially as the decision tree considers each task independently. Tasks may need to be reshuffled between processing blocks in order to properly balance the load. For example, if the APU is overloaded with tasks but the RPU cores are completely free, it would make sense to move some of this processing to an RPU core, even if it does not have any real-time requirements. Similarly, if the RPU is heavily loaded, we may consider implementing some extra MicroBlaze cores in the PL to handle some of the lower performance real-time tasks.

Of course, power management is also a consideration for most embedded systems. If there is a point (likely in a standby or low-power mode) where all required tasks do fit within a single processing block, load balancing between processing blocks may not be the best use of resources. We could power off the other domains and save energy instead. Refer to Chapter 10 for the details of power management.

13.1.3 Heterogeneous Computing Concepts

The last thing to cover before we discuss some real Zynq MPSoC implementations is the unique (and often intimidating) nomenclature adopted by heterogeneous computing literature. We will briefly introduce these generic concepts now, and then dedicate the rest of this chapter to more Zynq MPSoC specific implementations.

So far in this chapter, we have discussed the Zynq MPSoC in terms of the main processing blocks (RPU, APU, etc.). However, to introduce heterogeneous computing concepts, we must start looking at the system in terms of individual processor cores. Each of these cores are recapped in Figure 13.2 (noting that the MicroBlazes can be configured in many different ways, if used at all).

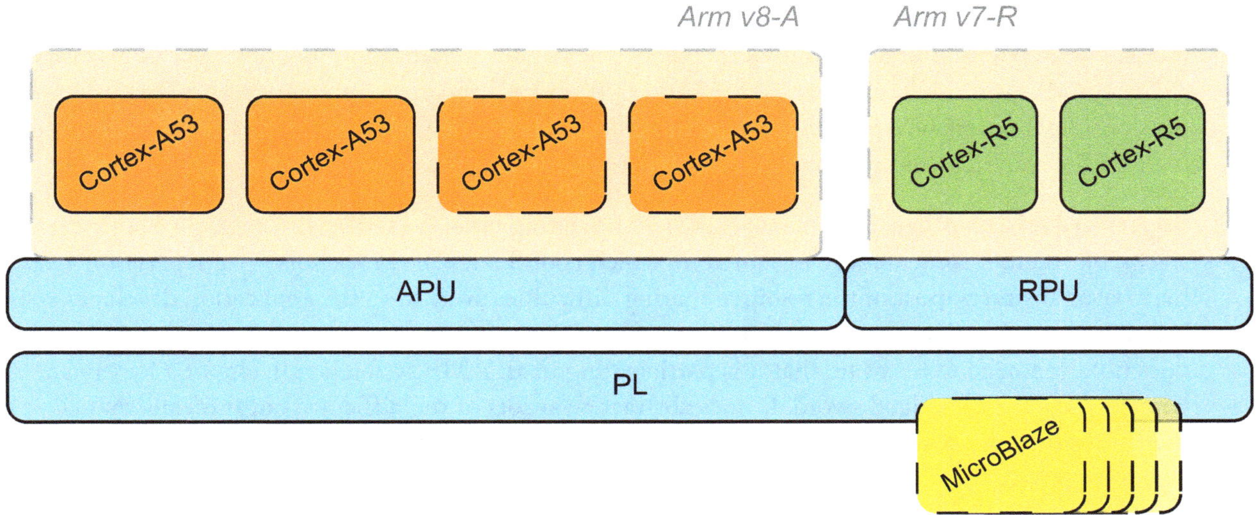

Figure 13.2: Review of the available processing cores

The most important point to consider is how each processor core compares to the others. Comparing any of the Cortex-A53 cores in the APU, any of the Cortex-R5 cores in the RPU, and a MicroBlaze core shows major dissimilarities.

There are a couple of difficulties that arise from this asymmetry, but the rest of this chapter will detail how we overcome them in order to gain the performance and energy efficiency boosts. These difficulties stem from the fact that there is no common instruction set between processing cores[1]:

- We cannot dynamically move execution of a task between dissimilar cores.
- We cannot use a single OS to manage all cores for us.

Due to these difficulties, we will have to adopt a technique called Asymmetric Multiprocessing (AMP) in order to make full use of the Zynq MPSoC hardware. AMP involves explicitly crafting each task to target a particular processor, and will usually require some carefully synchronised communication with other processors.

1. There is the possibility for instruction set compatibility between the Cortex-A53's Armv8-A architecture and Cortex-R5's Armv7-R. This is only when using Cortex-A53 in AArch32 mode and using a subset of the features (i.e. negating the advantages of both architectures!) [4][5]. This is not easily implemented.

There are two types of AMP which we need to distinguish:

1. **Unsupervised AMP**: Where all processors are operating independently, with their own software stacks. Note that there is no central software which coordinates these processors — they are all treated as equals. Because of this, there must be some careful coordination between these cores to protect system resources. For example, while configuring the interrupt controller, each processor must ensure that it does not overwrite the settings used by others (the default initialisation function overwrites every single entry!). As a result, unsupervised AMP can be quite difficult to implement from first principles. Section 13.3.1 covers how we can leverage the OpenAMP framework to make this easier — helping to manage the resources and perform message passing between processors.

2. **Supervised AMP**: Where all processors are still operating independently, with their own software stacks, but there *is* some underlying software which coordinates the processors — a hypervisor. The hypervisor abstracts most of the resource sharing difficulties away from the application developer, making supervised AMP less challenging to implement. The software running on top of the hypervisor does not even need to be aware that it is participating in an AMP system at all. Use of a hypervisor here can be a double-edged sword. It does abstract away lots of the difficulty, but it means that all of the cooperating processors must have the same architecture (otherwise the hypervisor would not be able to supervise them all). Because of this, supervised AMP can be a very useful technique for coordinating all of the Cortex-A53 processors within the APU, but it alone does not allow for AMP between the APU, RPU, and others. Note that supervised AMP is not well suited to implementation in the RPU because the Cortex-R5 processors lack the hardware support for virtualisation that is present on the Cortex-A53s. More discussion of virtualisation is found in Section 13.3.2, along with use of the Xen hypervisor for providing supervised AMP on the cores within the APU.

As touched on for supervised AMP, although the main processing blocks are asymmetric to each other, the APU is composed of multiple 'symmetric' processors. The final concept that we introduce here makes use of these identical, symmetric cores by using a single OS to control all of the processors within the APU. If an OS (such as Linux) has exclusive control of each of the processors, it can schedule and balance tasks between them, completely transparently to the user. This technique is called **Symmetric Multiprocessing** (SMP) and should be familiar to anyone reading a digital copy of this book — your device is most likely using this technique right now! As long as the system is composed of multiple processes or threads, the OS can make use of the multiple processors assigned to it with little extra effort by the application developer. In Section 13.2, we discuss how the APU can support an SMP configuration using Linux.

13.2 Symmetric Multiprocessing with Linux

SMP is a form of multiprocessing where multiple tasks may make use of multiple processors controlled by a single OS. The OS can dynamically assign each workload some CPU time through its scheduling algorithm, and can move tasks between the available processors to perform load balancing. One workload can be assigned to each processor simultaneously, resulting in substantial performance improvements for independent tasks.

All of this scheduling and resource management is performed transparently by the OS, such that the application developer (almost) does not need to think about it! This combination of good multiprocessor utilisation and relative ease of application development makes SMP a good default choice for writing APU software. Note that this not as suitable for use with the RPU, as we will discuss later.

The most popular SMP-capable OS for the Zynq MPSoC is Linux. In Linux, the different tasks come in two forms: processes and threads (where one process can own many threads). Figure 13.3 shows an example SMP configuration of the APU using Linux with multiple processes and threads. This SMP configuration is identical to that of most desktop computers, including the one typesetting this book.

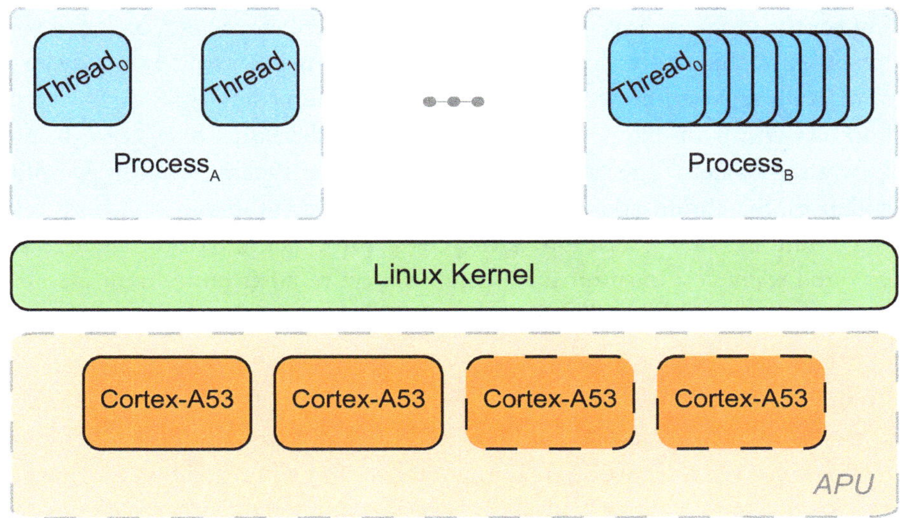

Figure 13.3: Example of SMP facilitated by Linux on the APU

Behind the scenes, the Linux kernel must be able to handle a couple of fundamental features in order to support SMP. Firstly, it must have an algorithm which schedules CPU time for each workload. We introduced Linux scheduling in "Process management" on page 322. Instead of just assigning equal time to each process, it should consider the priority of each process and its 'processor affinity' (i.e. which processor it likes best). Processor affinity is an important element for systems with L1 caches for each processor. If a process is assigned to a different processor each time it is resumed, it will need to waste time refilling that processor's L1 cache. However, if we try to return a process to the same processor (using its processor affinity), any cached data may still be present. As well as scheduling processes, Linux must also be able to pause execution of a process, store any necessary state on the disk, and resume execution on demand. Finally, because SMP promotes multithreading, Linux needs to provide a means of synchronising and communicating between threads. These requirements mean that the complexity of supporting SMP is placed firmly within the OS, and not so much in the user application. So, what is actually required of an application developer in order to exploit SMP?

From an application developer's perspective, consider what enables SMP. SMP is not at all effective if our system has only one, single-threaded process. Therefore, our system must either consist of multiple processes (so the OS can schedule each process simultaneously to a different processor), or at least one multi-threaded process. These techniques can both be used in the same system, and are respectively named with the alliterative tongue twisters, 'process-level parallelism' and a 'parallel processing program' [1].

Although it is likely that a Zynq MPSoC system will consist of multiple processes, it is still useful to consider multithreading for tasks that demand a high-throughput. Multithreading is essentially telling the OS how it can use concurrency within a single process — which can greatly increase that process's performance. Identifying and separating different parts, or instances, of an existing program is the first step towards multithreading. Depending on the program, this can be quite natural and easy to spot, or it can sometimes require quite fundamental changes to the algorithm! Once the program has been split into threads, there is almost always some synchronisation required between them, either when returning final values, or sharing intermediate data. When working with Unix-like OSs, such as Linux, the developer only needs to program against a standard Portable Operating System Interface (POSIX) API to use synchronisation primitives (see [6] and [10] for practical guidance). Synchronisation can allow each thread to stop and wait until other threads have reached a predefined point in their execution. Other uses for synchronisation include ensuring mutual exclusion to protect shared resources. In summary, a Linux system requires either multiple processes (process-level parallelism) or at least one, multithreaded process (a 'parallel processing program') to exploit the performance benefits of SMP.

The developer also has some optional fine control over how Linux performs its SMP scheduling. They may choose to configure the number of processors used, a process's processor affinity, and interrupt affinity. To begin, the number of cores used in a Linux SMP configuration can be dictated by the Linux kernel's *maxcpus* boot argument. This boot argument is most commonly used in conjunction with a hypervisor (see "Supervised AMP with the Xen Hypervisor" on page 344 for details), to reserve some of the APU's cores for a different, independent software stack. A developer may also choose to explicitly set a process's or interrupt's processor affinity in order to manually load balance between processors. This can be done in an effort to provide more deterministic performance for certain processes (with no other processes pinned to the same processor), or to group time-insensitive processes together. Practical information on using these techniques can be found in [7].

As an aside, SMP is not officially supported by the RPU for a couple of reasons. Firstly, this configuration requires an SMP-aware OS, such as Linux. These often (but not always) require an MMU to provide virtual address translation. The RPU does not have an MMU, in order to provide more deterministic execution. More vitally, the RPU does not have hardware-enforced cache coherency between the two Cortex-R5 processors [8], leaving any cache coherence to be software-enforced. This hardware support is expected of most SMP systems. Cache coherency for SMP systems can be implemented in software, but that comes with many restrictions and overheads [9].

13.3 Asymmetric Multiprocessing

Using AMP allows us to expand the borders of our software system, making use of the RPU's Cortex-R5s and any MicroBlaze processors that may have been implemented in the PL. The term AMP implies that we are running independent (and possibly different) software stacks on each processor. There is no single OS in charge of each processor, as there is with SMP configurations. The rest of this section introduces two common use cases for AMP in the Zynq MPSoC. Firstly, we introduce the OpenAMP framework for implementing unsupervised AMP systems, commonly used to offload tasks to the RPU and MicroBlaze processors. Secondly, we consider how the advantages of AMP can also be applied to the APU, using the Xen hypervisor. Before we dive in, let's first consider what are these advantages?

Beyond allowing us to make us of the Zynq MPSoC's heterogeneous processors, AMP also encourages us to split a complex system into smaller, independently developed tasks. This has a couple of nice consequences:

- Ease of development with smaller, independent tasks — especially when dividing work amongst a team.
- We can mix many software stacks. Because there is no single OS controlling all processors, we can opt to use whichever software stack is most suited to each task. Often AMP systems will make use of many different software stacks, including Linux, FreeRTOS, and bare-metal.

AMP also gives us explicit control of which task is executed on which processor. This control is invaluable when developing systems with any hard real-time elements. Such fine control allows a processor to be reserved exclusively for a real-time task, aiding its determinism and minimising its worst-case latency (it will never need to wait to be scheduled CPU time, unlike in SMP configurations).

13.3.1 Unsupervised AMP with OpenAMP

OpenAMP is a framework which aims to enable the development of AMP systems through use of existing, open source components. To appreciate how an OpenAMP system is structured, let's first introduce a simple scenario. Figure 13.4 depicts a situation where one Cortex-A53 processor is delegating a real-time task to one Cortex-R5 processor. Figure 13.4 shows the Cortex-R5 in three stages: in reset, being provisioned by the Cortex-A53, and finally, in execution.

Note that there is some implied hierarchy here — in this example, the Cortex-A53 is acting as a 'master', while the Cortex-R5 is a 'remote' subordinate processor. At run-time, the master decides to offload a task to a remote processor, and sets up the firmware image for that processor in memory. The remote processor is then taken out of reset and begins execution of its delegated task. An important part of this execution is communicating with the master processor — that is absolutely necessary for returning results at the very least!

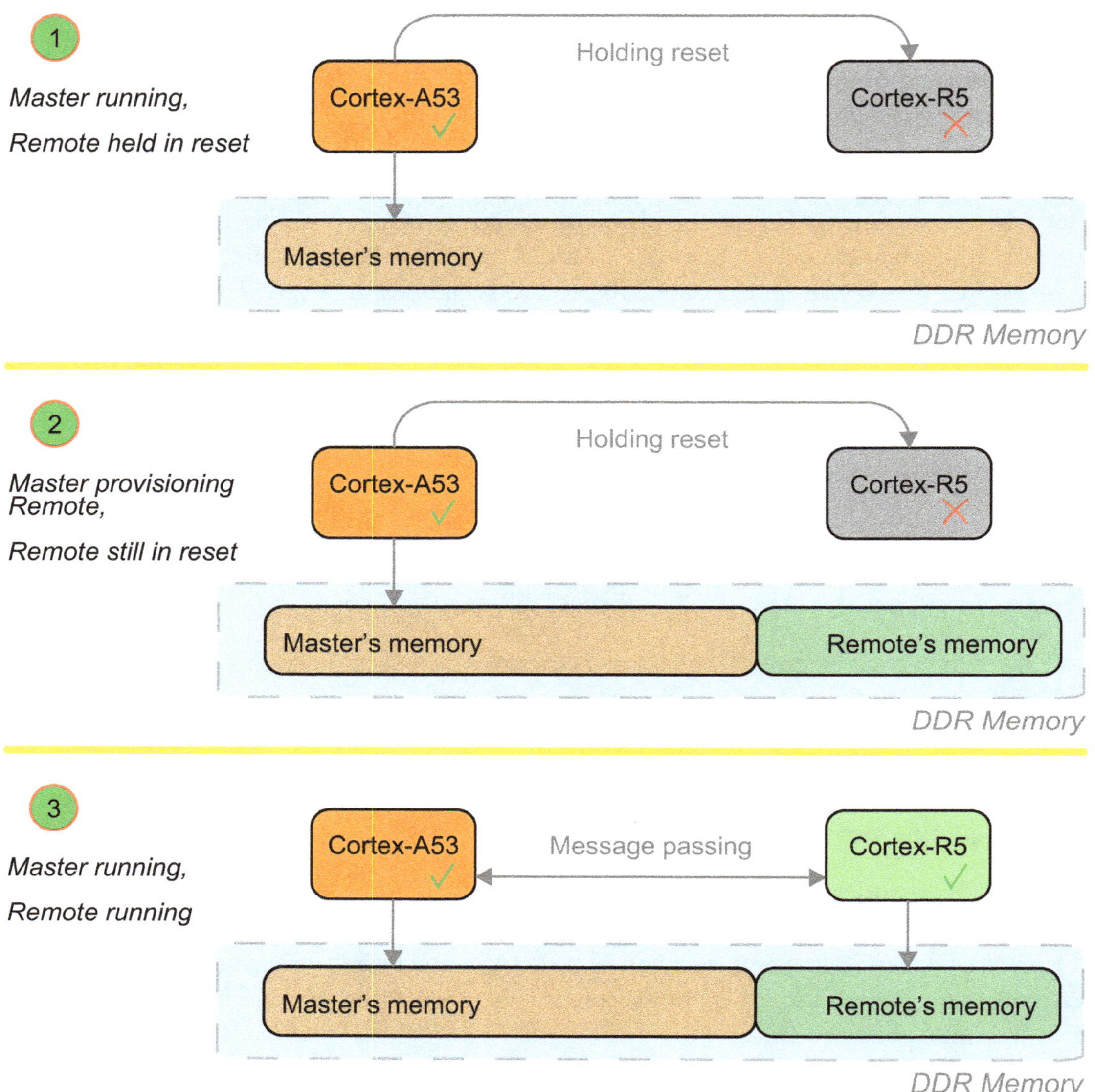

Figure 13.4: Simple OpenAMP scenario

OpenAMP makes implementing this common AMP scenario as easy as possible by providing two sets of features:

1. **Life Cycle Management (LCM)**: allowing a master processor to load, run, and unload a firmware image from a remote processor on demand.

2. **Inter-Processor Communication (IPC)**: allows for message passing between the master processor and the remote processor.

The LCM features are handled by '*remoteproc*', a remote processor framework [11], and the IPC features are handled by '*rpmsg*', a remote processor messaging framework [12]. Both *remoteproc* and *rpmsg* are pre-existing frameworks that are implemented in the Linux kernel. OpenAMP makes use of these existing kernel implementations, but also provides a standalone implementation to support other (non-Linux) software stacks. For example, OpenAMP can support a Linux master communicating with a FreeRTOS or bare-metal remote processor. While this is the most likely configuration, processors running FreeRTOS or bare-metal software stacks may also be used as an OpenAMP master.

In order to get into OpenAMP's technical details, let's consider a more specific AMP scenario, as shown in Figure 13.5. There is a Linux master running on one of the Cortex-A53 processors, communicating with a remote Cortex-R5, running FreeRTOS. Note that there does not have to be a one-to-one relationship between master processors and remote processors, there can be a one-to-many relationship. A single master may choose to spin-up many remote processors — and those remote processors can even be masters to their own remote processors. The example used here was one master and one remote, but "A Hybrid Example for Software Defined Radio Applications" on page 348 concludes with a configuration that uses multiple remote processors.

In Figure 13.5, the remote processor has already been initialised using *remoteproc*. For the rest of the remote process's life, *rpmsg* is responsible for enabling communication between the two processors.

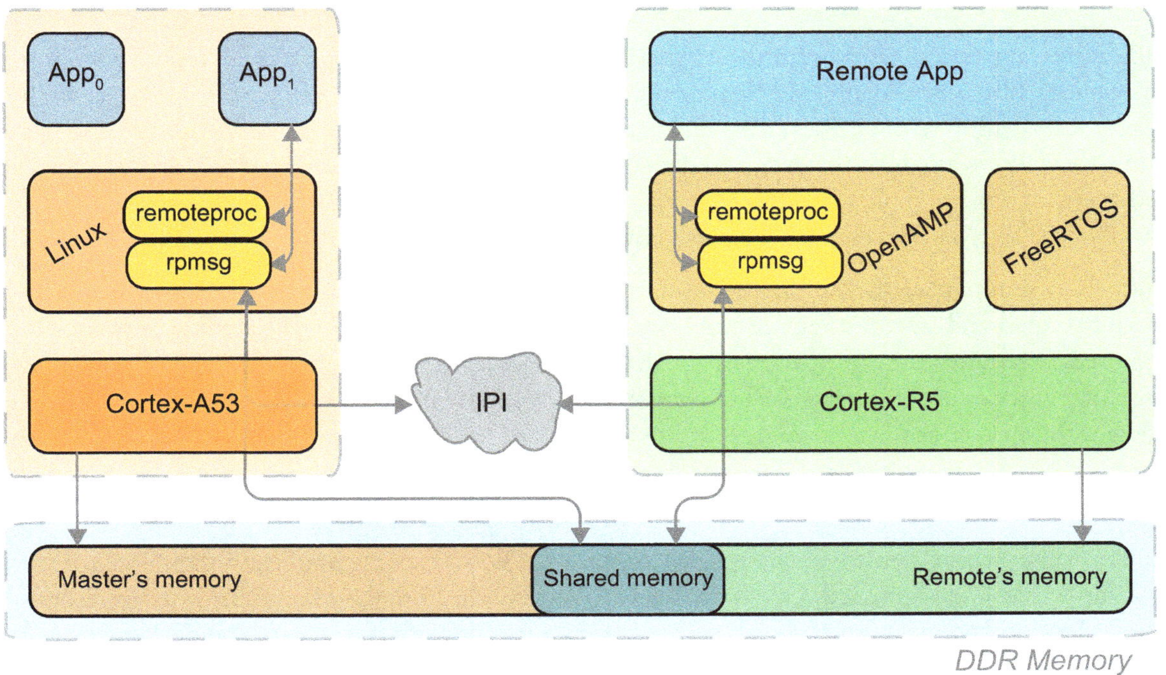

Figure 13.5: An APU-RPU OpenAMP system in more detail

Rpmsg uses two hardware mechanisms to provide inter-processor communication: inter-processor interrupts (IPIs) and shared memory. The IPIs are used to signal the presence of new messages in the shared memory, without the need for any polling in software. Zynq MPSoC has hardware support for IPIs, as detailed in Chapter 13 ("Interrupts") of the Technical Reference Manual [13]. The shared memory is used for storing the actual messages, and is managed by *another* layer of software called 'VirtIO' (we promise not to introduce any more framework names in this section!). VirtIO is used by *rpmsg* as a transport layer, as summarised in [14]. The master processor will have one *rpmsg* channel for each remote processor. Within each channel, the developer can choose to use more than one logical endpoint to separate message types. This is analogous to TCP/IP connections where many port numbers (or 'endpoints' for *rpmsg*) can be used for each individual IP address (or 'channel').

Next, we move on to consider OpenAMP's *remoteproc* component. *Remoteproc's* essential functions all relate to LCM [15], including:

- Loading the remote firmware into memory for execution by the remote processor.
- Starting execution of the remote firmware by releasing the remote processor from reset.
- Establishing communication between the master and remote processor, by initialising an *rpmsg* channel
- Shutting down a remote processor.

With these four functions, we can control the life cycle of a remote OpenAMP application. Some extra information must be embedded in the firmware image, such that *remoteproc* can perform these tasks, however. This extra information defines the system resources needed by the remote processor — including the a CPU identifier, the address at which the code and data of the remote firmware should be placed in memory, and IPI /shared memory information for initialising *rpmsg*. These details are placed in a statically linked 'resource table' within the firmware image, and are parsed by *remoteproc* when preparing the remote processor.

Figure 13.6 summarises the full life cycle of a remote OpenAMP process. The *remoteproc* API is used by the master to load and start the remote firmware. Upon coming out of reset, the remote processor initialises the resources it requires for *rpmsg*, triggering a handshake to establish the *rpmsg* channel. The application is then running and can use the *rpmsg* API to communicate with the master. When the remote processor is to be shut down, the master sends a custom 'shut down' message to the remote, which then performs any application-specific clean up operations. The shut down message is acknowledged, and both sides shut down their *remoteproc* instances.

The last topic to touch on covers how a developer can take an existing application and integrate it with a larger Zynq MPSoC design using OpenAMP. As seen in Figure 13.6, the remote firmware is responsible for some initialisation and clean up tasks, as well as communicating with its master. The developer must ensure they implement:

1. A resource table to describe what resources the remote firmware requires. Examples of this (and much more) can be found in [16].

2. Call `remoteproc_resource_init()` upon starting to initialise *rpmsg*.
3. Define callback functions which describe how to handle *rpmsg* channel creation, receive, and deletion events.
4. Use `rpmsg_send()` to send any application-specific messages back to the master processor.
5. When told to shut down, call `remoteproc_resource_deinit()` to clean up assigned resources.

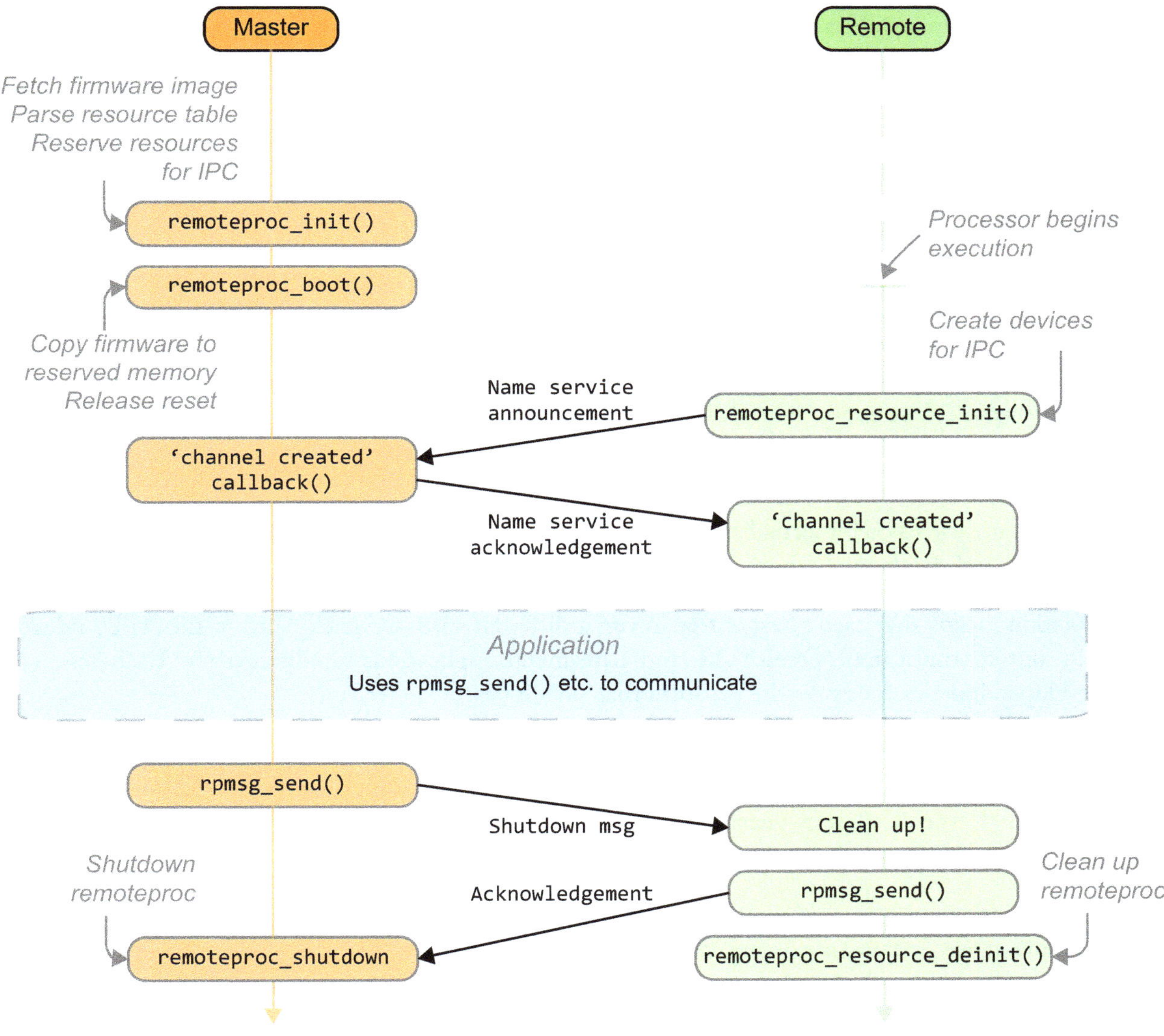

Figure 13.6: The life cycle of an OpenAMP application

Only a very small amount of extra development effort is needed to make an application nicely integrate into an OpenAMP system — especially when using the example programs described in [16] as a starting point. With OpenAMP, we can expand our Zynq MPSoC software to harness not only the APU running Linux, but delegate real-time/hardware-interfacing tasks to the RPU, or even MicroBlaze processors. This is an absolutely vital skill for working with heterogeneous processors, and luckily, OpenAMP makes the implementation relatively painless.

13.3.2 Supervised AMP with the Xen Hypervisor

Software virtualisation has become commonplace in desktop computing. Virtualisation allows a single set of processors to run many different 'guest' operating systems (often called 'Virtual Machines' or VMs). A layer of software called a hypervisor is responsible for managing these VMs. Virtualisation can provide a supervised AMP environment as it can support many different (or asymmetric) software stacks, that are all supervised by the hypervisor. Virtualisation has many different uses in desktops and servers — including keeping servers fully loaded at all times (thus needing fewer processors and saving energy), isolating software for safety or security, and supporting legacy software written for different platforms. In fact, this sentence is even being typed using software running in a VM.

Using virtualisation in embedded systems is a relatively new technique (the Arm support for Xen was first released in 2013!). This has been enabled by devices such as the Zynq MPSoC starting to adopt processors that have hardware support for virtualisation. In the Zynq MPSoC, this hardware support stems from the Cortex-A53's Armv8 architecture and the SMMU. These are vital for running virtualised software at near-native performance. More about this hardware support for virtualisation can been found in "System Virtualisation" on page 155 and "Isolating Software with Virtualisation and Armv8" on page 206. So, now that we *can* use virtualisation in Zynq MPSoC, let's consider *why* we would want to.

There are many use cases for virtualisation in embedded systems, but most of them stem from the isolation between guests, and the dynamic management of guests.

- Isolation means that each guest is free to run a different software stack. Within the APU, a developer may opt to run Linux, FreeRTOS, and bare-metal applications simultaneously. Each task can be developed independently on the platform that suits it best.

- The hypervisor enforces strict isolation such that each guest may only interact with the peripherals that have been explicitly assigned to them. This can help ensure safety-critical parts of a design will remain operational even if another guest crashes. In fact, this sort of decoupling is often required by safety certifications to demonstrate good design practice.

- Decoupling of guests also results in a system that is more easily integrated and maintained. For example, if we update an existing guest, or introduce a new one, we can be confident that it will not functionally impact the existing guests. Likewise, we may test each guest independently and then need only a very simple integration process to orchestrate the full system.

- The hypervisor layer gives us the possibility to detect and respond to faults or crashes in each guest. We can monitor each guest through the hypervisor, spotting faults which would otherwise crash the OS and any monitoring code within. For providing the best possible uptime, guests can be restarted, or redundant fall-back guests can be maintained.

These are just a few examples, but we can start to appreciate why the trend towards virtualisation in embedded systems has begun. There are many different hypervisor solutions, but when working with the Zynq MPSoC, most developers will likely choose the Xen hypervisor. Why Xen in particular?

Xen officially supports the Zynq MPSoC hardware (exploiting Armv8's virtualisation features) and commercial support is available through DornerWorks [18]. Xen is free, in every sense of the word, and uses the GPLv2 license. Further embracing open source software, it has become the de facto virtualisation solution on Linux platforms, and is adopted as part of the Linux Foundation. Clearly, Xen is not going to disappear any time soon. This also means that applications made for Xen on the Zynq MPSoC will be portable to other Xen-capable systems for years to come.

Looking at Xen in particular now, it is a 'layer one' hypervisor, meaning that it sits directly above the hardware. All guest OSs (or 'domains' in Xen nomenclature) then sit above Xen. To offer a control interface to Xen, there is always one privileged domain, called 'dom0'. Dom0 is typically a Linux system which has access to the hardware and can communicate with the Xen kernel via user-friendly tools such as 'xl'. This tool enables a user to manage guest domains through the Linux console — including LCM of guest domains, providing console access to every domain, and controlling per-domain resources (such as number of CPUs, memory, scheduling hints, and much more [19]). Figure 13.7 shows an example of a Xen-based, supervised AMP system running on the APU. This includes a dom0 for control, the Xen kernel, and example guest domains running different software stacks.

Relating this back to the Zynq MPSoC for a moment, how can we interact with the hardware (such as IP implemented in the PL) if the guest domains are so strictly isolated? There are actually two ways of implementing these drivers in a Xen system [17]. The first, and more familiar method, is by setting up a pass-through for a device. A pass-through lets dom0 grant complete access to a peripheral from (usually) a single guest domain. The guest domain can use the peripheral exactly as it would without Xen — i.e. the guest does not even need to know that it is being virtualised. The developer can tell Xen to implement the pass-through using a simple, ASCII configuration file which we will discuss later. Xen then configures the SMMU to pass the guest's memory access through to the peripheral, and configures the GIC to route interrupts from the peripheral to whichever processor the guest is running on. This configuration is shown in Figure 13.8.

The second method of communicating with a hardware peripheral makes use of 'paravirtualisation' (PV) — where guest domains are aware they are being virtualised. With PV, a driver is typically split into two: a front-end driver and a back-end driver. This is visualised in Figure 13.8. The two halves of the driver communicate using shared memory in much the same way that OpenAMP does for IPC.

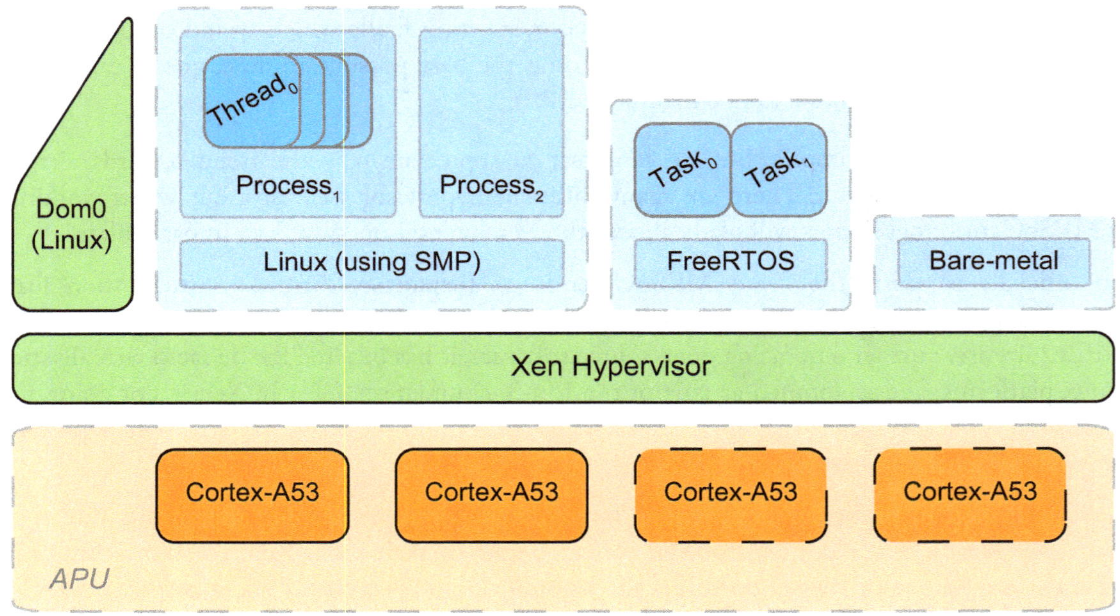

Figure 13.7: Example Xen supervised AMP system running on the APU

While coming at a slight performance impact, there are clear motivations for this structure:

- Multiple guests can safely share the same device, as it is controlled by a single domain. The controlling domain can also implement a scheduling scheme to prevent any one guest from monopolising the device.
- Each guest only needs to implement a generic front-end driver. This introduces a layer of abstraction between the guest domains and the actual hardware implementation, resulting in extremely portable and reusable applications.

Again, the last important topic to touch on is what the developer needs to do in order to harness Xen in their own applications. We face similar challenges as we did with OpenAMP, such as describing the resource requirements of each guest, and communicating between guests when necessary. First of all, the PetaLinux tools can be used to generate a Xen system with a Linux-based dom0 extremely easily. This is, as it stands, the default PetaLinux configuration, with details shown in [20]. For any guest domains in the system we need two things:

1. A firmware image for the guest. For example, a bare-metal elf executable or a Linux image.
2. A plain text configuration file to describe the resource requirements of the guest to Xen.

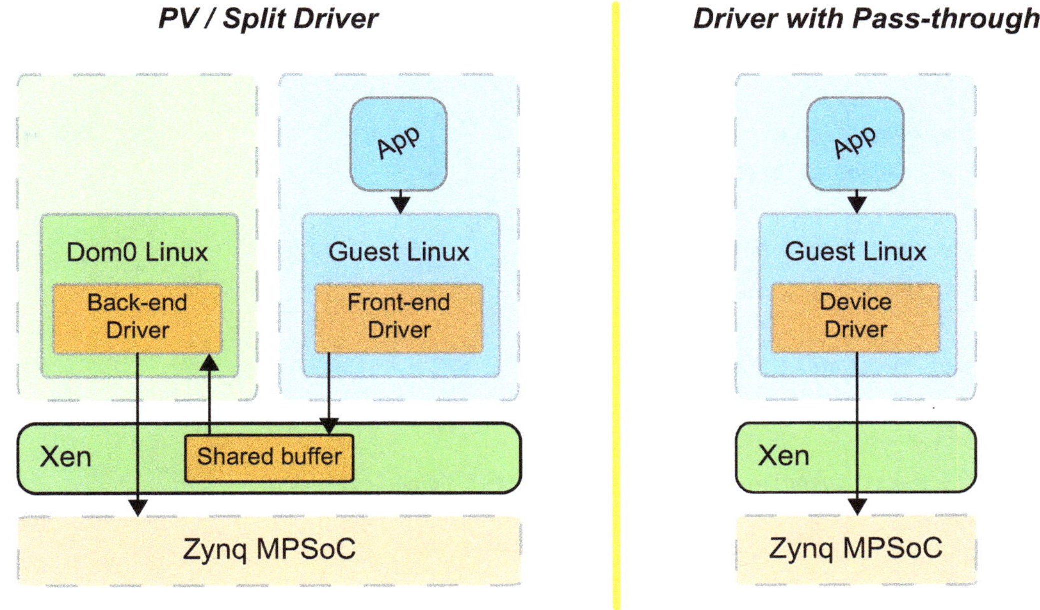

Figure 13.8: Comparison of PV driver and pass-through operation in Xen

For Linux guests, the firmware image can be another PetaLinux project, or even a custom Linux build with the appropriate drivers enabled. Bare-metal images can be created using the SDK, as described in [22], which also demonstrates how to implement a pass-through of a UART device. We provide an example of a plain text configuration file for a bare-metal guest in Figure 13.9. Note that the full list of options available are presented in the manual page [21].

Finally, to create the guest domain described by a configuration file (e.g. bare_guest.cfg), we simply use the 'xl' tool in the dom0 console: `xl create bare_guest.cfg`.

Once we have multiple guests running on Xen, we may want to let them communicate with each other. It is possible to do this with shared memory regions allocated by Xen (as described in [17]). This is a very similar mechanism as seen earlier with OpenAMP. However, it is actually more common for guests, especially Linux guests, to communicate using TCP/IP over a virtual network. This networked configuration can be very simple, needing only one extra line in the guest's configuration file, and using any of the wealth of existing TCP/IP libraries.

We have seen how Xen can be used on the Zynq MPSoC to enable supervised AMP, making the most of the APU's multiple processors with relatively little effort. This can be a very flexible solution that allows mixing of software stacks, promotes decoupled, portable system design, and can use isolation for both safety and security.

```
# Guest's name
  name = "baremetal_guest1"

# Firmware/kernel image to boot
  kernel = "baremetal_app1.bin"

# Allocate 16 MB (It's only wee!)
  memory = 16

# Number of (virtual) CPUs
  vcpus = 1

# Pin this guest to only ever run on the first (physical) CPU
  cpus = [1]

# Pass-through access to UART1 (uses interrupt 54, and base address of 0xff010000)
  irqs = [ 54 ]              # Pass interrupt 54
  iomem = [ "0xff010,1" ]  # Pass 1 physical page (default of 4 KB)
                             # at page number 0xff010. Grants 0xff010000 to 0xff010FFF
```

Figure 13.9: Annotated example Xen guest configuration file

13.4 A Hybrid Example for Software Defined Radio Applications

To conclude the topic of designing software for heterogeneous multiprocessor systems, we examine a software defined radio use case. Borrowing an example from [23], we focus on mobile radios used by services such as the police and ambulances. These are quite complex systems, which must be robust and operate in real-time. Our goal with this example is to demonstrate how a complex system can be split into smaller tasks, suited to the different processing blocks present on the Zynq MPSoC hardware, and to see how these tasks can be easily implemented and integrated using the multiprocessing techniques we have described in this chapter.

There are often *many* tasks in such a design, but they typically fall into three categories: sample-based real-time, real-time, and high-level. These map quite naturally to implementation in the PL, the RPU, and APU respectively. The PL will be responsible for such signal processing tasks as beamforming, real-time FFTs, frequency error correction, and modems. The RPU can handle the next layer which consists of implementing the radio protocol stack on one core, and any relevant security features on the second core (you would hope that police radio communications were encrypted, but let's not get too political!). The deterministic execution of the RPU helps to ensure that we conform properly to the radio standard used. The APU is then responsible for the most high-level applications. This includes decoding and encoding voice data (perhaps with the Cortex-A53's SIMD capabilities), and the user interface running on Linux.

Next, we want to consider how the different software tasks could integrate with each other. The APU applications rely on tasks which ought to be executed on the RPU. This is a clear application for OpenAMP, coordinating between the two processor sets. Furthermore, there are two independent tasks mapped to the APU — the voice codec (likely written in bare-metal), and the user interface (likely making use of Linux drivers and SMP over multiple cores). This scenario is a good fit for the virtualisation capabilities of Xen. This let's us run both software stacks simultaneously, reserving one Cortex-A53 for the bare-metal stack, giving it consistent CPU time. So, this is going to form a hybrid system, using the SMP capabilities of Linux, supervised AMP on the APU using Xen, and unsupervised AMP on the RPU using OpenAMP. A visualisation of this system is shown in Figure 13.10.

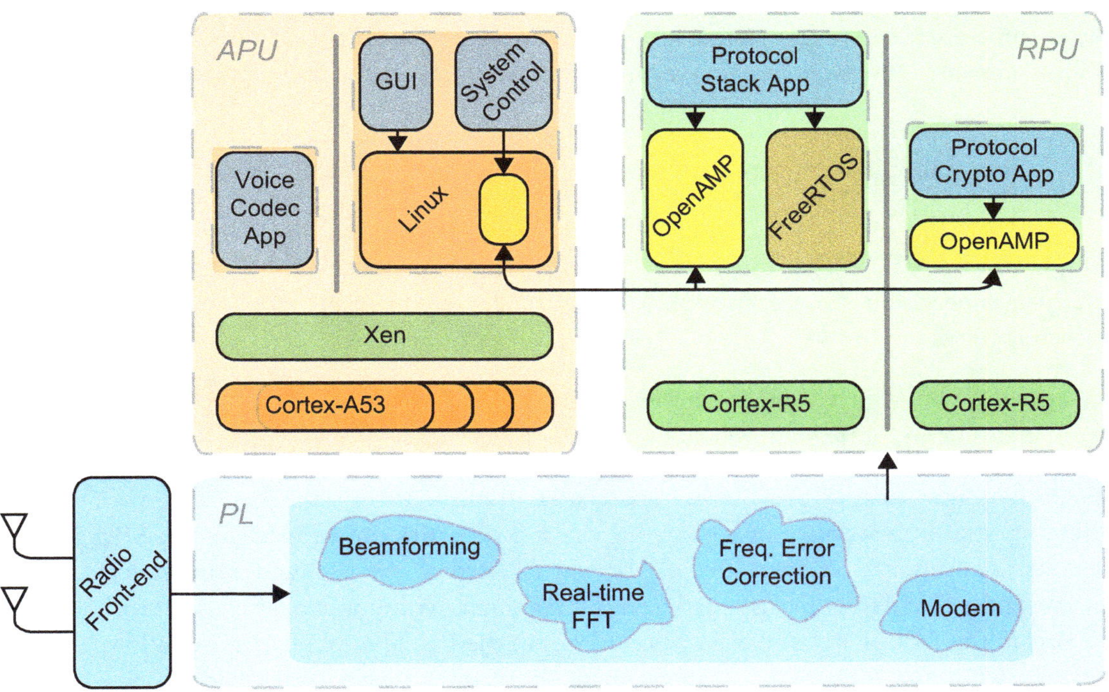

Figure 13.10: An example radio application using SMP, Xen, and OpenAMP simultaneously

As a quick aside, this application can also make use of the power management features of the Zynq MPSoC. These radios spend most of their lives in standby, therefore not needing the GUI and voice codec capabilities. The developer could choose to put the APU and PL in standby, having the RPU wake these up upon detection of a valid message. This power saving can be extremely useful in battery powered embedded devices such as mobile radios. Refer to Chapter 10 for the details of power management.

We have introduced an application which requires use of many processing blocks on the Zynq MPSoC. This application has acted as a case study into how Linux's SMP capabilities, the Xen hypervisor, and OpenAMP can all be used to implement such a system with minimal effort. OpenAMP facilitates using each of the RPU cores as real-time capable coprocessors, while the developer needs only to define a resource table and program to a simple message-passing API. Xen allows us to run two disparate software stacks on the APU cores, and the developer can implement each of them independently. The hypervisor only needs a simple, plain text configuration file describing the resource requirements of both guests. Finally the Linux-based graphical interface can make use of multiple APU cores using SMP with virtually no extra developer effort at all.

13.5 Chapter Summary

This chapter has covered various techniques that may be used to make the most of the Zynq MPSoC's heterogeneous processor sets — a fundamental skill, according to recent trends! After discussing how energy efficiency has motivated the system's heterogeneousness, we covered a few fundamental multiprocessing concepts:

- **SMP**: Where multiple tasks may simultaneously be executed on multiple processors under the control of a single OS.
- **Unsupervised AMP**: Where different software stacks run independently on each processor with no underlying supervision — i.e. each application must take care to ensure it does not clobber the resources of other applications.
- **Supervised AMP**: Where different software stacks run independently on each processor, but there is a layer of coordinating software (such as a hypervisor) that handles the isolation between software stacks.

We then covered how each of these techniques can be implemented on the Zynq MPSoC, using existing software libraries and frameworks. Linux's existing SMP capabilities can be used on the APU, almost transparently to the application developer. OpenAMP can be used to manage and communicate with remote processors, laying the foundation for unsupervised AMP systems with minimal effort. OpenAMP can be used with processors in the APU, RPU, and even MicroBlazes in the PL. The Xen hypervisor allows for supervised AMP systems within the APU, which can facilitate disparate software stacks and various safety/security benefits. Xen exploits the Armv8's hardware virtualisation support for near-native performance.

All three of these options offer simple interfaces, which can be used to orchestrate complex Zynq MPSoC software designs relatively painlessly — as highlighted in our example radio application.

13.6 References

Note: All online sources last accessed March 2019.

Some of the Xilinx documentation referred to below has version-specific URLs. If you are working in a newer version of the tools, check for updates on the Xilinx website, or try adjusting the link according to your version.

[1] D. A. Patterson, J. L. Hennessy, *Computer Organization and Design: The Hardware/Software Interface*, Morgan Kaufmann, 5th edition, 2014.

[2] J. Montgomery, "Mycroft Mark II: The Open Voice Assistant", Kickstarter Project FAQs Page 2018.
Available: https://www.kickstarter.com/projects/aiforeveryone/mycroft-mark-ii-the-open-voice-assistant/faqs

[3] Xilinx Inc., "UG1228 — Zynq UltraScale+ MPSoC Embedded Design Methodology Guide", v1.0, March 2017.
Available:
https://www.xilinx.com/support/documentation/sw_manuals/ug1228-ultrafast-embedded-design-methodology-guide.pdf

[4] Arm, Ltd., "ARMv8-A Foundation Platform: User Guide", Version 9.4, 2012.
Available: https://static.docs.arm.com/dui0677/h/DUI0677H_armv8_a_fp_ug.pdf

[5] Arm, Ltd., "ARM Technical Support Knowledge Articles: Can I link ARMv7-M objects with ARMv7-R or ARMv7-A objects and run the resulting image on an ARMv7-R or ARMv7-A processor?", 2011.
Available: http://infocenter.arm.com/help/index.jsp?topic=/com.arm.doc.faqs/ka16460.html

[6] P. Seebach, "Basic use of pthreads: An introduction to POSIX threads", IBM developerWorks, 2004.
Available: https://www.ibm.com/developerworks/library/l-pthred/index.html

[7] Xilinx Inc., "Xilinx Wiki: Real-Time Linux", SMP Affinity Section.
Available: http://www.wiki.xilinx.com/Real-Time+Linux#Approaches%20to%20Real-Time%20on%20Linux-Native%20Linux-API%20Based%20Solutions-SMP%20Affinity%20(Open%20Source%20Solution)

[8] Arm, Ltd., "Cortex-R5 Technical Reference Manual", vr1p2, September 2011.
Available: http://infocenter.arm.com/help/topic/com.arm.doc.ddi0460d/DDI0460D_cortex_r5_r1p2_trm.pdf

[9] Analog Devices, "uCLinux Wiki: Blackfin SMP 'Like'".
Available: https://blackfin.uclinux.org/doku.php?id=linux-kernel:smp-like

[10] D. Robbins, "POSIX threads explained: A simple and nimble tool for memory sharing", IBM developerWorks, 2000.
Available: https://www.ibm.com/developerworks/library/l-posix1/index.html

[11] B. Swetland, O. Ben-Cohen, Remote Processor (remoteproc) Framework", Linux Kernel Documentation, 2017.
Available: https://github.com/torvalds/linux/blob/master/Documentation/remoteproc.txt

[12] B. Swetland, O. Ben-Cohen, "Remote Processor Messaging (rpmsg) Framework", Linux Kernel Documentation, 2017.
Available: https://github.com/torvalds/linux/blob/master/Documentation/rpmsg.txt

[13] Xilinx Inc., "UG1085 — Zynq UltraScale+ Device Technical Reference Manual", v1.7, December 2017.
Available: https://www.xilinx.com/support/documentation/user_guides/ug1085-zynq-ultrascale-trm.pdf

[14] OpenAMP, "RPMsg Messaging Protocol", OpenAMP's GitHub Wiki pages.
Available: https://github.com/OpenAMP/open-amp/wiki/RPMsg-Messaging-Protocol

[15] OpenAMP, "OpenAMP Life Cycle Management", OpenAMP's GitHub Wiki pages.
Available: https://github.com/OpenAMP/open-amp/wiki/OpenAMP-Life-Cycle-Management

[16] Xilinx Inc., "UG1186 — Libmetal and OpenAMP for Zynq Devices User Guide", v2017.3, January 2016.
Available: https://www.xilinx.com/support/documentation/sw_manuals/xilinx2017_3/ug1186-zynq-openamp-gsg.pdf

[17] DornerWors, "Xen Zynq Distribution: User's Manual", v1.00, October 2017.
Available: http://dornerworks.com/wp-content/uploads/2017/10/Xen-Zynq-Distribution-XZD-Users-Manual2.pdf

[18] DornerWorks Ltd., "Xen Hypervisor on Xilinx Zynq UltraScale+ MPSoC".
Available: http://dornerworks.com/xen/xilinxxen

[19] XenProject, "XL", Xen Wiki, 2016.
Available: https://wiki.xen.org/wiki/XL

[20] Xilinx Inc., "Linux - Xen Dom0", Xilinx Wiki, 2017.
Available: http://www.wiki.xilinx.com/Linux%20-%20Xen%20Dom0

[21] Xen Project, "Xl.cfg Manual Page".
Available: http://xenbits.xen.org/docs/unstable/man/xl.cfg.5.html

[22] Xilinx Inc., "Xen EL1 Baremetal DomU", Xilinx Wiki, 2017.
Available: http://www.wiki.xilinx.com/XEN+EL1+Baremetal+DomU

[23] Xilinx Inc., "WP470 — Unleash the Unparalleled Power and Flexibility of Zynq UltraScale+ MPSoCs", v1.1, June 2016.
Available: https://www.xilinx.com/support/documentation/white_papers/wp470-ultrascale-plus-power-flexibility.pdf

[24] A. Taylor, "Adam Taylor's MicroZed Chronicles, Part 234 (UltraZed Edition 21): OpenAMP—How to use the Zynq UltraScale MPSoC's Arm Cortex-A53 and -R5 processors together in a design", Xcell Daily Blog, January 2018.
Available: https://forums.xilinx.com/t5/Xcell-Daily-Blog/Adam-Taylor-s-MicroZed-Chronicles-Part-234-UltraZed-Edition-21/ba-p/826083

[25] Xilinx Inc., "UG1137 — Zynq UltraScale+ MPSoC Software Developer Guide", v6.0, January 2018.
Available: https://www.xilinx.com/support/documentation/user_guides/ug1137-zynq-ultrascale-mpsoc-swdev.pdf

[26] Xilinx Inc., "WP474 — Enabling Virtualization with Xen Hypervisor on Zynq UltraScale+ MPSoCs", v1.0, March 2016.
Available: http://www.xilinx.com/support/documentation/white_papers/wp474-xen-hypervisor.pdf

[27] A. Raghuraman, Mentor Graphics Corp. "Toward Easier Software Development for Asymmetric Multiprocessing Systems", Xcell Journal 93, Fourth Quarter 2015, p59-65.
Available: https://issuu.com/xcelljournal/docs/xcell_journal_issue_93

[28] OpenAMP, "OpenAMP Overview", OpenAMP's GitHub Wiki pages.
Available: https://github.com/OpenAMP/open-amp/wiki

[29] M. T. Jones, "Virtio: An I/O virtualization framework for Linux", IBM DeveloperWorks, January 2010.
Available: https://www.ibm.com/developerworks/library/l-virtio/

[30] Xilinx Inc., "Building the Xen Hypervisor with PetaLinux 2017.3", 2017.
Available: http://www.wiki.xilinx.com/Building%20the%20Xen%20Hypervisor%20with%20PetaLinux%202017.3

Chapter 14

System Booting

This chapter will cover the Zynq MPSoC's booting process, how developers can influence this process, and some practical guidance. We consider both non-secure boot and secure boot flows, after discussing which types of applications require secure boot features. We step through the 3 main stages of the boot process in detail:

1. The pre-configuration stage, focusing on hardware elements and the PMU.
2. The configuration stage, controlled by the CSU.
3. The post-configuration stage, where user-supplied code takes over.

After becoming familiar with the non-secure boot process, we delve into the authentication and confidentiality provided by secure boot. The final two sections take a more practical perspective, detailing how we can use the Bootgen utility to generate boot images, and configure any security related eFUSEs on the device.

14.1 Introduction to System Booting

Understanding how a Zynq MPSoC system boots is as important to preparing a design for production, as it is for making the development/testing cycle as streamlined as possible. You may actually look at the booting process from different perspectives throughout a project: e.g. quickly deploying a volatile image to hardware from a development PC during prototyping, and then turning an eye to security features to protect your IP before release. The rest of this chapter aims to demonstrate exactly what the Zynq MPSoC hardware does during the boot process, as well as how the developer can work with, and influence it.

One of the most important aspects to consider is whether or not you require a *secure* boot. The secure boot flow introduces some extra configuration steps, so we separate the following sections into a base discussion of non-secure booting, and a complementary discussion of the extra steps in the secure boot flow. In order to help the designer decide on non-secure or secure boot for a given design, we list the main benefits of secure boot in Table 14.1.

Table 14.1: Benefits of using secure boot over non-secure boot [1]

Benefit	Secure Boot Feature
Ensuring that only *your* code will be run on the device	Authentication using the Hardware Root of Trust (HWRoT)
Ensuring that your code is not modified	Authentication via the HWRoT
Protecting your system from being reverse engineered or cloned — protecting sensitive data and intellectual property	Boot image confidentiality
Protecting AES keys from side-channel attacks, which can break confidentiality	DPA countermeasures
Protecting AES keys at rest	Black key storage

If you decide that your design needs the authentication (i.e. only your unmodified design should ever be run) and/or confidentiality (i.e. nobody else should be able to see your design) facilitated by secure boot, be sure to see "Secure Boot Process" on page 364 and "Practical Secure Device Configuration" on page 371. We also highly recommend first reading Chapter 8 for an overview of all Zynq MPSoC security features. If you do not require the secure boot flow, feel free to skip these sections, and instead focus on "Non-Secure Boot Process" on page 356 and "Practical Non-Secure Device Configuration" on page 370.

Before we discuss the boot process in detail, let's stop to consider what a fully configured Zynq MPSoC device looks like before power is applied. Appreciating what non-volatile resources the Zynq MPSoC has available to it upon boot will make it a little easier for us to talk about the boot process. Figure 14.1 depicts such a system, which has been configured for secure boot.

The boot process is initiated by hardware units which we can assume are trustworthy, including a state machine and immutable (i.e. unchangeable), on-chip PMU/CSU ROMs. The other non-volatile memories are the main means that the developer has to influence the first stages of this hard-wired, trusted boot process. These memories include various boot mediums, eFUSEs, and also the BBRAM to a certain extent. The developer can create a boot image, encapsulating user code which will eventually be executed, and store it on one of many boot mediums. The 4 boot mode pins dictate which of the boot mediums is used. There are eFUSE flags that are used to enable optional boot features, and store cryptographic keys or hashes for secure boot. See Section 14.4 and Section 14.5 for details about creating these boot images and configuring security features.

The final piece of background information we must introduce is the purpose of each of the system software images. In our example system in Figure 14.1, we see that the boot image contains many different partitions for different images.

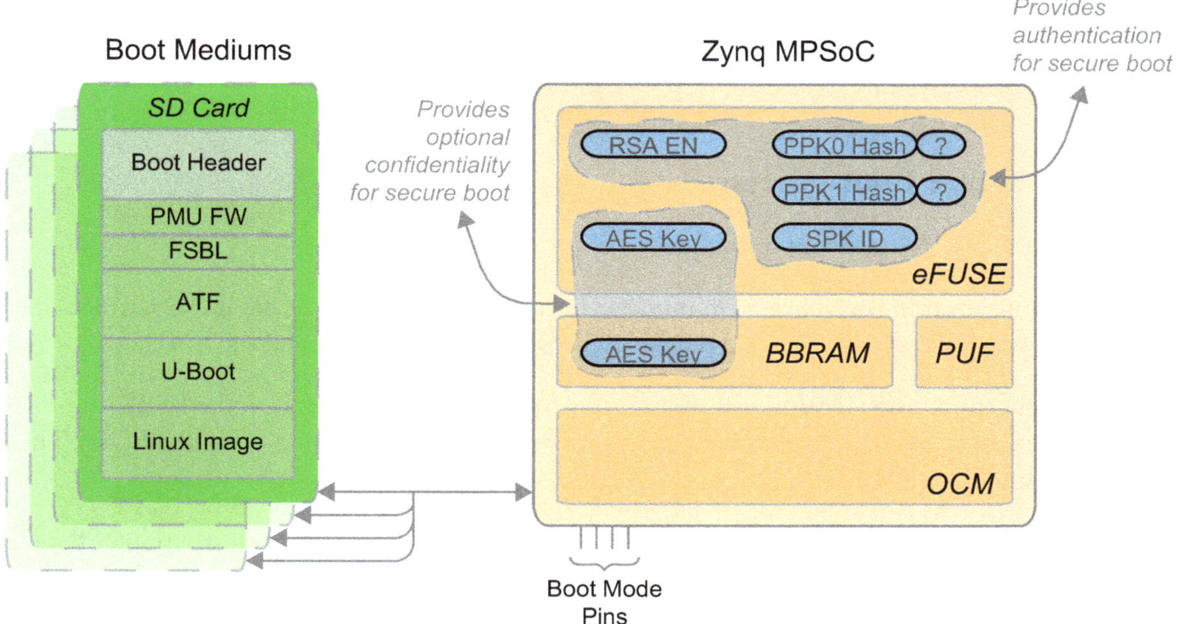

Figure 14.1: Overview of a fully provisioned, bootable system

These include:

- **First Stage Boot Loader** (FSBL): runs in the OCM and is responsible for initialising many parts of the processing system, including clocks, memory, and UART. It also is in charge of loading any subsequent software such as a bare-metal application or second stage boot loader.

- **Platform Management Unit Firmware**: runs in the PMU after boot. This is responsible for power management tasks, such as dynamically shutting down and waking up processors or power domains. The SDK provides this firmware, but developers are free to modify it if required.

- **U-Boot**: is popular Second Stage Boot Loader (SSBL), commonly used to load Linux on the Zynq MPSoC. U-Boot is typically much larger than the FSBL, so runs in DDR memory, but also comes with much more functionality. It initialises the remaining peripherals and supports booting from more complex sources — e.g. over a TCP/IP network or a SATA interface.

- **Arm Trusted Firmware** (ATF): is a small layer of software which runs in EL3 (the most privileged level) of the APU. This facilitates switching between non-secure and secure worlds. The ATF is required by Linux but, similar to U-Boot, this is often provided transparently by Xilinx tools, such as PetaLinux.

Beyond these system software images, the boot image should also contain your custom application(s), be it a simple bare-metal binary, or a more complex system comprised of many partitions. A minimal, bare-metal system may just use an FSBL and the user application. A Linux system, however, requires the FSBL, ATF, and a SSBL, such as U-Boot.

14.2 Non-Secure Boot Process

The boot process is broadly split into three stages:

1. **Pre-configuration stage**: begins to initialise the system to the point where it can start looking for a boot image. This stage is handled by a small hardware state machine and the PMU.

2. **Configuration stage**: interprets the boot image header, which defines how the FSBL and PMU firmware should be loaded, then loads these images before passing execution off to the FSBL. This stage is handled by the CSU.

3. **Post-configuration stage**: encompasses all subsequent functions, from the execution of the FSBL to the execution of the user's application. From a hardware perspective, the device has booted by the start of this stage — i.e. the hard-wired PMU and CSU ROMs have served their purpose, and user code is now in control. The FSBL must still load any user applications however, so the boot process doesn't quite end here from a user perspective!

Taking the common example of loading Linux on the APU, we show how the boot process timeline maps these three main stages to different processing units. Figure 14.2 shows that the pre-configuration stage is handled by a hardware state machine and PMU, the configuration stage is controlled by the CSU alone, and the RPU is responsible for the post-configuration stage. Note that the FSBL may actually be targeted to either the APU or RPU using flags in the boot image header — but we will assume the RPU is used for this example. It is also clear to see that each part of system software does some of its own work and then hands off execution to the next part, forming a chain. Thinking about the boot process as many parts that are chained together becomes very important for secure boot (see Section 14.3.1).

14.2.1 Boot Process Overview

Now we will step through each of the three stages in more detail, discussing exactly what each component shown in Figure 14.2 does, starting with the pre-configuration stage. If you have been reading this book for a little while in one sitting, this might be a good time to fetch a drink of your choice — this topic is quite dry.

Pre-Configuration Stage

The first hardware element to take control on boot — more precisely, a power-on reset (POR) — is a hardware state machine [2]. This performs the bare minimum in order to securely get the PMU up and running.

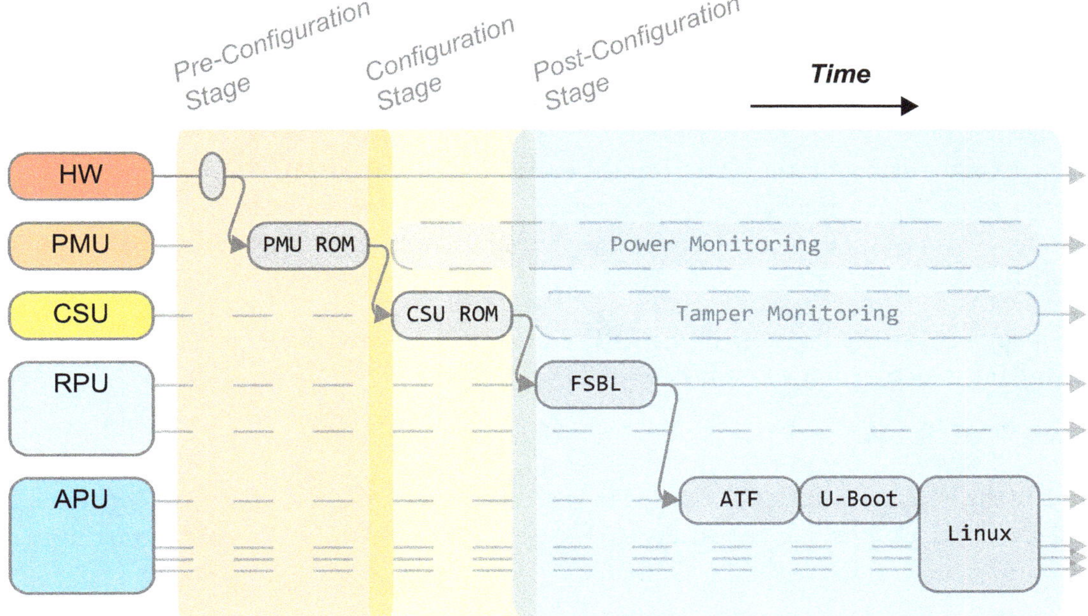

Figure 14.2: One example boot timeline for a Linux application

The steps it performs are:

- **Lockdown of test/JTAG interfaces:** This is done as a security precaution, because the device does not yet know if it should be using secure boot or not! The JTAG interface may be enabled later, either when a non-secure boot flow is detected, or by authenticated software after a successful secure boot (see the CSU's *jtag_sec* register [3]).

- **Zeroise PMU registers**: Zeros are not only written to the PMU registers, but are also read back to ensure each write was successful. Note that whenever we mention zeroisation in this chapter, this read back is also implied.

- ***Optionally* perform the Logic Built-In Self-Test (LBIST)**: This optional step is only performed if the a corresponding eFUSE flag has been blown (*LBIST_EN*). The LBIST is used to pre-emptively check for latent faults in the device's hardware, which is a common requirement of safety-critical applications. It is not enabled by default as the LBIST will have an impact on total boot time. See the "System Test and Debug" chapter in [2] for full details.

- **Perform an integrity check on the PMU ROM**: The PMU ROM contains the boot code that the hardware state machine aims to load and hand-off to the PMU. The ROM is metal-masked (i.e. hard-wired) which precludes any modification. As an extra precaution against any effects of extreme radiation environments (e.g. in space) or sophisticated attacks, the PMU ROM is passed through the hardware SHA-3/384 hashing unit. The output is compared to a stored hash to ensure that the contents are unmodified. Refer back to Chapter 8 for details of the cryptographic blocks, including SHA-3.

- **The PMU reset is released and it begins to execute.**

Next, we turn our attention to the PMU. It is the responsibility of the PMU ROM to initialise and perform checks on the wider system (including clocks, voltage levels, and memories in the CSU, LPD, and FPD) before handing off execution to the CSU. Now that the hardware state machine has handed off control to the PMU, the PMU ROM will:

- ***Optionally* zeroise the registers in the LPD and/or FPD**: Again, these optional steps are enabled by blowing the corresponding eFUSE flags — *LPD_SC* and *FPD_SC* in this case.

- **Zeroise PMU RAM.**

- **Validate the PLL locks**: The PLL is used to generate internal clocks for the booting process. This helps preclude any clock glitching attacks during boot. At this point, the PMU ensures that these clocks are in a stable state.

- **Validate power supply ranges**: The System Monitor is used to check the LPD, auxiliary, and I/O supply voltages. Boot will only continue if all of these voltages are within the normal operating ranges.

- **Zeroise the memories in the CSU, LPD, and FPD.**

- **Perform an integrity check of the CSU ROM**: This uses the SHA-3/384 core to verify the integrity of the metal-masked CSU ROM — just like the hardware state machine did for the PMU ROM.

- **The CSU reset is released and it begins to execute.**

Here, we reach the end of the pre-configuration stage. The core components have been initialised to a known state, and we are ready to start making booting decisions based on the user's design choices in the configuration stage.

Configuration Stage

In the configuration stage, the CSU aims to locate the user supplied boot image, and load the FSBL into the specified location. An optional PMU firmware may also be loaded if specified in the boot image. Note that this is the point at which differences start to appear between non-secure and secure boot — refer to Section 14.3 for secure boot specific details. The CSU ROM performs:

- **Initialise the OCM.**

- **Determine the boot mode**: The boot mode (i.e. where we will look to find a boot image) is determined by four external pins. The state of these pins is captured at the time of a POR, in the *BOOT_-*

MODE_USER register. The various available boot mediums are detailed in "Boot Media Options" on page 360.

- **Search for a valid boot image**: The full search algorithm is detailed in "Boot Image Search" on page 361, and allows for implementing fall-back images.

- **Load the FSBL image partition, and the optional PMU firmware partition if present**: The boot image header defines where the FSBL partition is within the boot image, where it should be placed in memory, and which processor (within the APU or RPU) it targets. The CSU ROM parses this header and loads the FSBL, along with an optional PMU firmware.

- **The target CPU (within APU or RPU) reset is released and it begins to execute**.

Now, the system begins to execute the FSBL in the post-configuration stage. While the FSBL is technically user code, we will still discuss the default behaviour, as almost all applications will opt to use the Xilinx FSBL as supplied through the SDK.

Post-Configuration Stage

Next we cover the default behaviour of the FSBL supplied by Xilinx. This FSBL should be employed without modification, unless the developer has a very good reason. Parts of the FSBL pertain to secure boot, so modifying it recklessly is extremely unwise. To permit some safe customisation, there are compile flags which can exclude unwanted features or enable verbose output, and a small set of function hooks. These hooks allow the FSBL to call user supplied functions at set points in its execution, without the user delving into the core FSBL code. These customisation options are detailed in [4]. Note that the FSBL can be compiled for either the first Cortex-A53 (AArch 32 or AArch64), the first Cortex-R5, or even both Cortex-R5s in lockstep.

The main responsibilities of the FSBL are:

- **Further initialisation**: The PL is powered up and the DDR controller is initialised. The Cortex-A53s' caches, MMU, and stack pointers are initialised, as are the Cortex-R5s' caches, MPU, stack pointers, and TCM. Board-level initialisation is also performed — including the I2C, USB, and PCIe interfaces.

- **Partition loading**: The boot image header is inspected. This can indicate use of a secondary (different) boot medium. The remaining partitions present in the boot image are considered in turn. The header of each partition includes the destination address for the image, the address of the first instruction to execute, the owner (FSBL or U-boot), the destination device, and a pointer to the next partition header. The FSBL uses these attributes to copy any FSBL-owned partition images to the correct target. A bitstream will be sent through the PCAP interface, while any software images will be copied to their destination address — inferring residence on the DDR memory, OCM, TCMs, or PMU RAM.

- **Hand off**: Any processors that are the destination device for a loaded image must be released from reset. Unlike previously, with the PMU ROM and CSU ROM, these images are not stored predictably in a local memory — the user has full control over where each image is loaded. So, before releasing a reset, the program counter of the destination processor is loaded with the execution address, specified in the partition header.

At this point, the user design is running, and is in full control of the initialised Zynq MPSoC. From this point, the process is entirely application dependant. If the system is a simple bare-metal application, nothing else is required. If Linux is used, for example, a SSBL (usually U-Boot) will take charge and subsequently load the Linux image.

This concludes our overview of the non-secure boot process. The remaining subsections expand on this overview by detailing three important aspects that have already been alluded to: where a boot image can be stored, how the CSU ROM finds the boot image, and the format of the boot image.

14.2.2 Boot Media Options

The Zynq MPSoC provides 11 different boot modes, including 6 unique storage mediums. The boot mode is set using four external pins, called *PS_MODE* [5]. The state of these pins is captured at the time of a POR, and stored in the *BOOT_MODE_USER* register. Table 14.2 shows a complete list of the boot modes.

Table 14.2: Full list of available boot modes[2]

Boot Mode		Mode Pins [3:0]	Description
JTAG	PS JTAG	0000	JTAG with dedicated pins, reaching all controllers on the chain.
	PJTAG (MIO #0)	1000	JTAG through MIO, reaching Arm DAP only.
	PJTAG (MIO #1)	1001	JTAG through MIO, reaching Arm DAP only.
SPI	QSPI (24b)	0001	Quad-SPI using 24 bit addressing
	QSPI (32b)	0010	Quad-SPI using 32 bit addressing
SD	SD0 (2.0)	0011	SD card controller for version 2.0
	SD1 (2.0)	0101	SD card controller for version 2.0
	SD1 LS (3.0)	1110	SD card controller for version 3.0, including voltage level-shifters
NAND		0100	NAND flash memory, with 8-bit data bus
eMMC		0110	eMMC version 4.5, using 1.8V
USB0 (2.0)		0111	USB version 2.0

The main caveats to note, concern the USB and various JTAG boot modes [4]. None of them support the 'multi-boot' flow, described in "Boot Image Search" on page 361. Also, as an aside, the JTAG boot modes are not applicable for secure boots. This is because JTAG will remain disabled for a secure boot, after the pre-configuration stage's hardware state machine locks down the test interfaces.

14.2.3 Boot Image Search

Now that we know how to dictate which boot medium the MPSoC should use, let's consider how the CSU ROM searches for a valid image within that boot medium. The way this is implemented actually allows for storing multiple boot images within a single medium, facilitating the use of fallback images for safe updates.

Once the CSU has determined the boot mode, it will start searching for a valid boot image. It does this by starting at address *0x0* of the boot medium (a multi-boot offset of 0), and searching for a magic string present in the boot image header — "XLNX". If this string is not found, or the rest of a detected boot image header is found to be invalid, the multi-boot offset is incremented and the process begins again. Note that the multi-boot offset is aligned to 32 KB blocks — i.e. an offset of 1 translates to an address of *0x8000.* A boot image header can be found as invalid if its in-built checksum fails, or the authentication/decryption fails during a secure boot.

This increment-and-retry process allows for booting a backup image, stored later in the boot medium, in the case of a failed update. For more subtle multi-boot triggers, the FSBL or even user application, can trigger a reboot using a later boot image by writing an offset to the *CSU_MULTI_BOOT* register, and issuing a soft reset with the *RESET_CTRL* register.

It should be noted that, while the multi-boot offset will usually affect the address used, this does not translate well to the SD and eMMC boot modes. These make use of file systems. Instead of searching with raw addresses, we must search through the logical files within the file system. The offset number is concatenated with the filename "BOOT.bin", creating a new filename to search for.

14.2.4 Boot Image Format

Let's quickly look at the format of the boot image used by the Zynq MPSoC. In this section we cover the general structure of the boot image and the purpose of each section, rather than the full, exhaustive details of each field. The details of exact fields is less important for most developers to understand as they are automatically generated for us by the *Bootgen* tool, further discussed in "Generating Boot Images" on page 370. There is even a companion tool, *Bootgen_utility*, which will print the meaning of each field in a given boot image, using human-friendly formatting. Figure 14.3 shows an overview of the boot image structure.

The boot image is essentially a header, followed by a linked-list of partitions. The header provides information about the image as a whole, largely configuring aspects for secure boot, which we will gloss over until "Secure Boot Process" on page 364. There is also some extra information so that the CSU ROM can differentiate between the FSBL and PMU firmware. Each software component is encapsulated in a subsequent partition — the first being the FSBL and optional PMU firmware. There is a partition header that accompanies each software image, specifying what the image is, where it should be loaded in memory, which

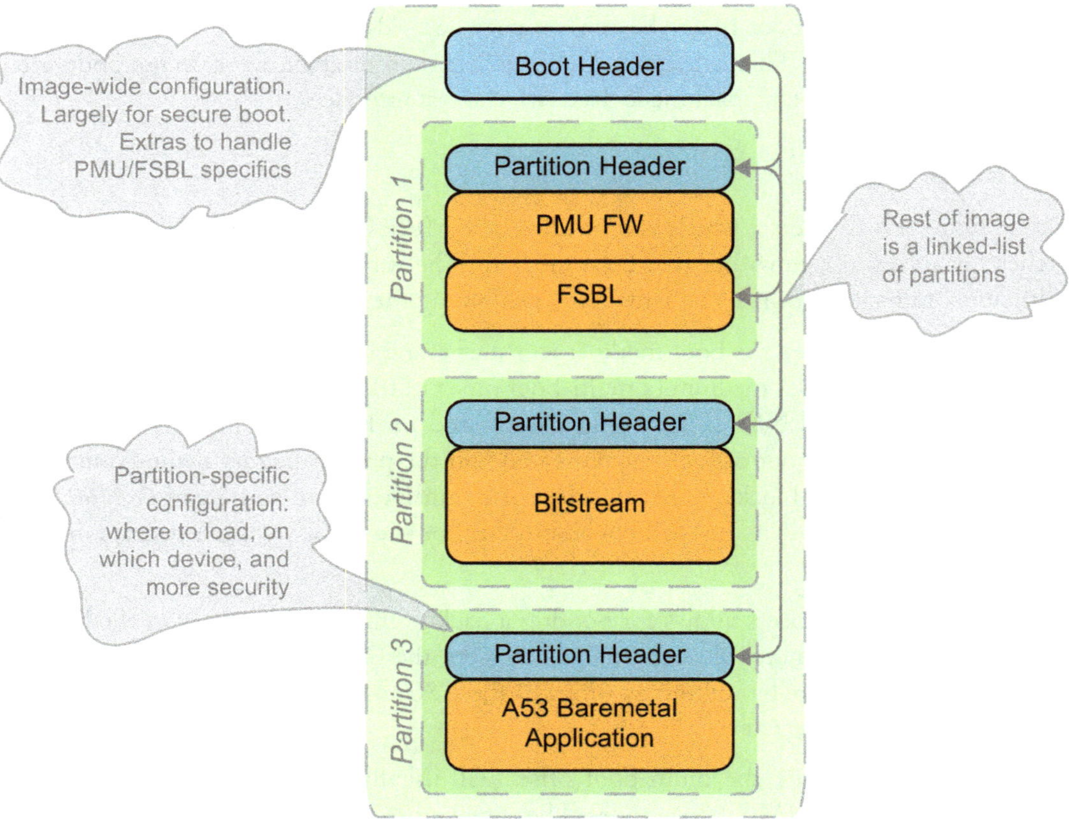

Figure 14.3: Overview of the boot image format

processor it is for, and so on. There are also security flags in the partition headers. Each of the partition headers also stores a pointer to the next partition header, forming the linked-list.

In our example, the boot image is for a simple system comprised of a bare-metal application on a single Corex-A53 processor, and an IP block in the PL. This translates to the three partitions shown in Figure 14.3: the FSBL (required for booting), the PL's bitstream, and the bare-metal application. This configuration is just demonstrative; there can be many other possibilities. More partitions can be added, with images which may target other Cortex-A53 processors, the Cortex-R5s (individually, or in lock-step), or even have no target. Partitions with no target device can be used to easily copy (authenticated/encrypted) data into the address space on boot, including future software stages, such as a Linux image.

The curious reader is referred to [2] for further details of the boot image header fields, and to [4] for further details of the partition headers.

14.2.5 Recap with a Linux/OpenAMP Use-Case

We now summarise the boot process, by returning to a high-level look at the booting timeline for an example application — with a little more complexity this time. Figure 14.4 shows the booting timeline for our example design. This demonstrates that the pre-configuration and configuration stages can be identical for very different designs, as they are controlled by metal-masked ROMs (not user software). Beyond this, the FSBL is used to load the majority of software images (including U-Boot for a Cortex-A53, and an independent application for on Cortex-R5 processor) and the PL configuration, guided by the information stored in the boot image's partition headers. Beyond this, the user software is entirely in control.

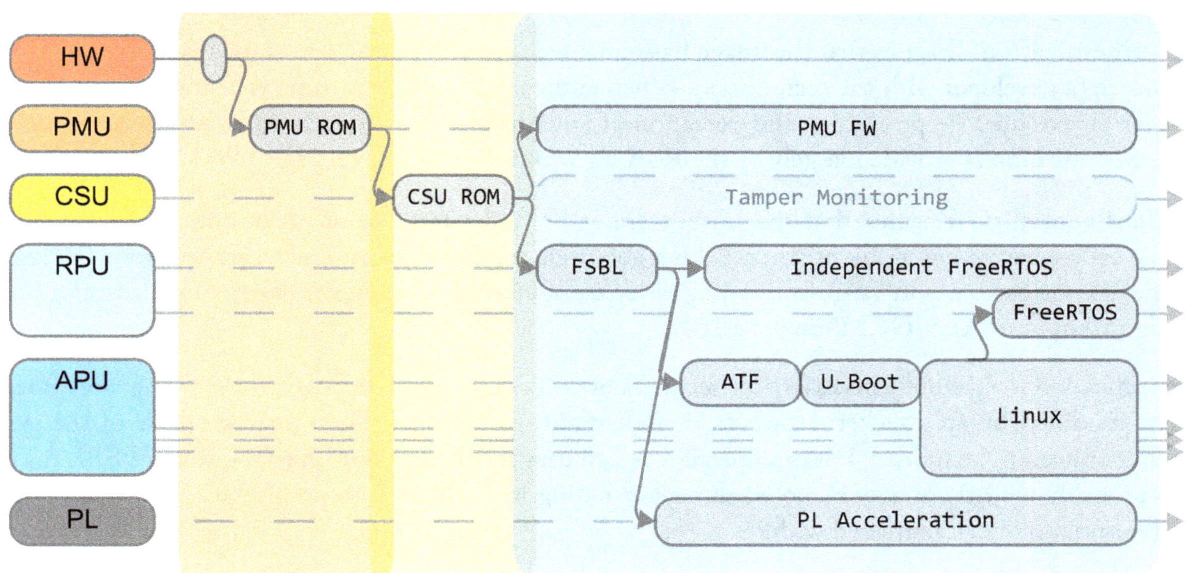

Figure 14.4: Example boot timeline for a complex design

To demonstrate some of the further booting related actions that user software can perform, the system presented in Figure 14.4 makes us of U-Boot and OpenAMP. U-Boot is used as a SSBL in order to load an hand-off to a Linux system on the APU. In practice, most Zynq MPSoC Linux systems will likely make use of U-Boot as provided by the PetaLinux tools, completely transparently to the developer. For those seeking more information on working with U-Boot manually, see [6]. OpenAMP is used on the APU to dynamically boot and shut-down 'remote' processors (the second Cortex-R5 processor in this case) during run-time. This demonstrates that individual processors may still be dynamically booted by a user application throughout run-time, on demand.

14.3 Secure Boot Process

Now we take the basic boot process, as described in Section 14.2, and add in the extra steps that provide authentication and optional confidentiality to the user's design. These extra steps make heavy use of the cryptographic units present in the CSU — so it is highly recommended to read "Information Assurance for Configuration Security" on page 180 (Chapter 8) before continuing with this section.

14.3.1 Secure Boot Fundamentals

To recap, the two most important security properties during boot are:

- **Authentication**: Ensures that the image has remained unmodified and was actually created by a trusted source (a developer with the secret keys). When authentication is used, only your unmodified code will ever be executed — precluding the execution of any unauthorized code from an attacker or customer. Authentication is provided largely by the RSA algorithm (see "RSA" on page 189).

- **Confidentiality:** Ensures that the image cannot be understood by an unauthorised person. This is typically what people think of when talking about encryption. This is used to protect your IP (logic in the PL, or custom software) from being reverse engineered. Confidentiality is provided by the AES algorithm (see "AES-GCM" on page 184).

Authentication is absolutely necessary for secure boot — i.e. a secure system would never boot software that has been modified by an attacker. However, confidentiality is optional. There may be parts of the system which do not need to be hidden. For example, if a design uses the default version of the Xilinx FSBL (as most designs probably will) there may be no advantage in hiding it, as there is no proprietary IP; we just want to ensure that it hasn't been tampered with.

So, how does the Zynq MPSoC ensure authentication and confidentiality throughout the whole secure boot process? This is based on two concepts: the 'chain of trust' and the 'hardware root of trust'. We have already seen how the boot process overtly chains different software components together, where each component loads and hands off to the next. The chain of trust describes the way that each of these software components has been trusted (i.e. has been successfully authenticated) by the previous component, and will ensure that the next component is also trusted. An important property of chaining trust is that it can propagate authentication all the way to the final user application, while allowing for a *flexible* boot process. It accommodates simple, bare-metal systems, as well as more complex systems with SSBLs, ATF, etc.

The chain of trust must have a first element, which has no preceding component to vouch for its trustworthiness. We need to trust this first element completely if we are to have confidence in the secure boot process. This may sound like a big ask, but we base this first element in something that cannot be affected by malicious software — silicon! This immutable, tamper-proof silicon forms the 'hardware root of trust'. Its job is to ensure that the subsequent boot components are authenticated using an on-chip, authorised key. The root of trust achieves this task by utilising the on-chip, metal-masked BootROMs (a collective term for the PMU and CSU ROMs), and the hardware RSA unit present in the CSU. Note that a set of eFUSE bits (*RSA_EN*) tells the hardware root of trust that the secure boot flow must always be used.

Figure 14.5 visualises both the chain of trust, and how the hardware root of trust initiates it. To briefly summarise the fundamentals of the secure boot process:

- The secure boot process starts with a hardware root of trust. We fully trust these hardware components because it cannot be altered by malicious software, and it is protected from hardware attacks by ROM immutability, SHA-3 integrity checks, and the CSU's triplicated processors.
- Each component in the boot process authenticates (i.e. 'trusts') the next component using an on-chip RSA key hash and RSA accelerator. AES decryption can be optionally performed, if the component ought to be confidential.

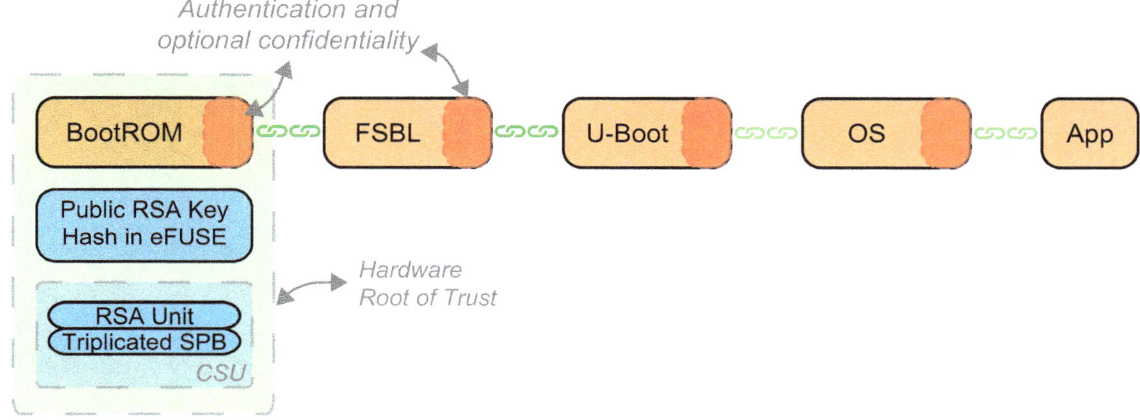

Figure 14.5: Overview of the chain of trust and hardware root of trust

14.3.2 Secure Boot Cryptography

Now, we begin to look at how the Zynq MPSoC actually performs the authentication (and optional decryption) required to maintain a chain of trust for secure boot. Let's first return to an example of the boot timeline — this time, highlighting only the sections specific to secure boot. The parts highlighted in Figure 14.6 are responsible for authentication and decryption.

Authentication and decryption are facilitated by the hardware RSA (Montgomery multiplier), AES-GCM, and SHA-3 blocks in the CSU. These are implemented in silicon, so are immutable, and we consider them to be trusted — the National Institute of Standards and Technology agrees [9][10]! Refer to Chapter 8's "Crypto Blocks" on page 183 for full details of these blocks.

Another important set of features enables the Zynq MPSoC to keep all the sensitive data on-chip during authentication/decryption. On-chip storage is especially important when the device is not physically protected (for example, by a big fence or André the Giant) as external memory could be tampered with. The OCM is generally used to store an image during authentication or decryption. The Zynq MPSoC can also store the keys required of AES and RSA on-chip.

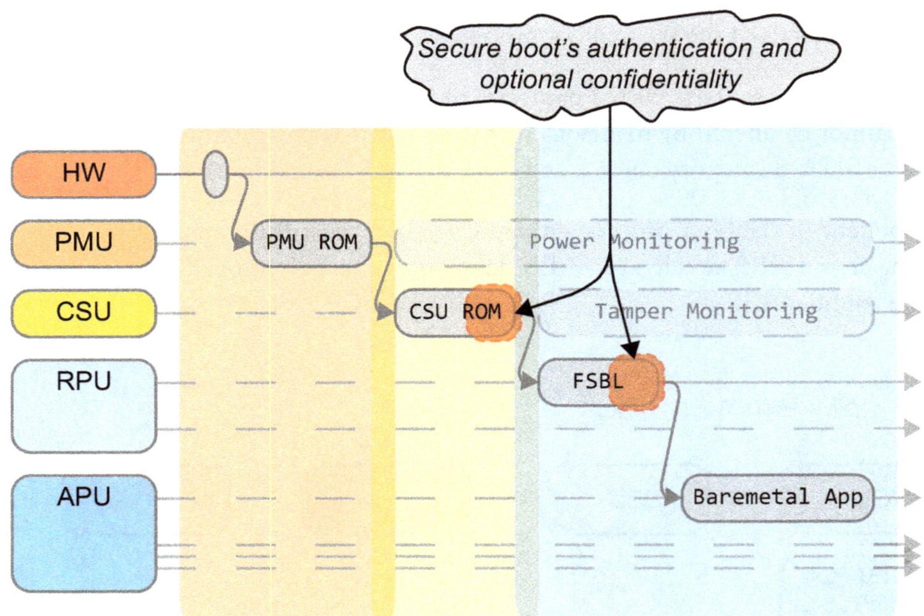

Figure 14.6: Highlighted overview of an example secure boot timeline

The CSU ROM and FSBL both handle these cryptographic features in a similar way, discussed in the following sections.

Image Authentication

The *asymmetric* RSA algorithm is used to facilitate authentication. It is the asymmetry that is particularly important when performing authentication. As mentioned in "RSA" on page 189, RSA uses a private key (known only to the developer) and a complementary public key (which can be shared with anyone). A message which has been 'signed' with the developer's private key can be verified by anyone with the public key. This tells us that the message was definitely created by the developer, as only they know the private key. This process allows a Zynq MPSoC device to determine if a boot image is authentic or not.

For authentication to be possible, the Zynq MPSoC must know the public key of the developer. Note that we are storing a public key here, not a private one. This is another nice property of using RSA for authentication — there is no embedding of any secrets on the device! A total of two Primary Public Keys (PPKs) can be stored using eFUSEs. In reality, RSA keys are very long (4096 bits, in this case) and eFUSE cells are quite large. To make more efficient use of resources, we do not actually store full PPKs, but rather a shorter SHA-3 hash of the PPKs. We can perform integrity checking with this hash, and safely relegate the full PPK to the boot image.

Over the life-time of a device, the developer/company may wish to revoke keys. This is common practice when employees leave a company, for example. The Zynq MPSoC accommodates this in two ways:

1. Providing an eFUSE flag to invalidate each PPK (*PPK0_INVLD* and *PPK1_INVLD*)

2. Using a two-stage authentication process with Secondary Public Key (SPK)

The PPKs should be revoked very sparingly, as the Zynq MPSoC will only be able to use two throughout its entire life. This is alleviated by the use of SPKs. In this scheme, the PPK is only ever used to authenticate a SPK, which is stored in the boot image. The SPK is then used to authenticate the rest of the image. Revocation is facilitated by a 32 bit eFUSE field (*SPK_ID)* that stores the a number to identify the current SPK. This SPK ID is also stored in the boot image, and is authenticated along with the SPK itself. To revoke an SPKs, we blow one extra SPK_ID eFUSE bit, meaning we can have a total of 32 distinct SPKs through the life of a single Zynq MPSoC device. All of these RSA related eFUSEs are summarised in Figure 14.7.

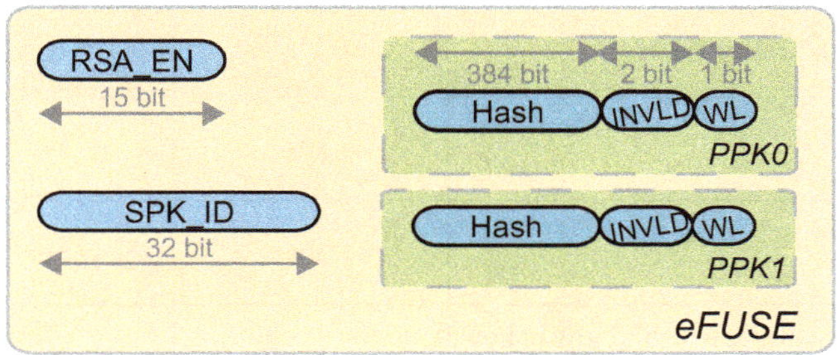

Figure 14.7: Summary of RSA related eFUSEs

The authentication signatures are created by the developer in two steps. First, the image is passed through the SHA-3 block to provide a 384 bit hash. This provides a means of checking the integrity of the image. The hash is then 'encrypted' or 'signed' with the developer's private RSA key. The resulting signature provides both an integrity check and authentication. This signature is included in the boot image for each partition. On boot, the CSU ROM and FSBL will check this signature with the following steps:

- Check if the PPK has been revoked.
- Calculate the hash of the PPK in the boot image, and ensure this matches the hash in eFUSE.
- Use the PPK to authenticate the SPK — by taking a hash of the SPK, SPK ID, etc. and ensuring this matches the SPK's signature after 'decryption' with the PPK.
- Ensuring the SPK ID in the boot image matches the SPK ID in eFUSE.
- Authenticating the partition with the SPK.

Note that there is an interesting caveat when authenticating bitstreams for the PL. These partitions are too big to fit in the OCM. Some more advanced techniques are required to ensure that the bitstream is not modified in external during the authentication process. See "Bitstream Authentication Using External Memory" in [4] for full details.

Image Confidentiality

For secure boot, each partition must be authenticated, and confidentiality is an optional addition for any partitions with contents you would rather keep secret. Confidentiality is provided by the CSU's AES-GCM block (again, see "AES-GCM" on page 184 for full details). This is a symmetric cipher, so unlike RSA, a single key is used for both encryption and decryption of the boot image. This key must be kept as a secret — anyone who knows this key can decrypt the boot image! We must take more care to store the AES key than the public RSA key (which is available for everyone to read in the boot image). Full details of AES key storage can be found in Chapter 8's "Key Management" on page 190. A summary of the key storage options is shown in Table 14.3.

Table 14.3: Summary of AES key storage options

Key State During Storage	eFUSE	BBRAM	Boot Image
Plain text *(or 'red')*	Yes	Yes	No
Encrypted with the PUF KEK *(or 'black')*	Yes	No	Yes
Obfuscated with the Family key *(or 'gray')*	Yes	No	Yes

The AES key may be stored on-chip, in its entirety, in either the BBRAM or eFUSEs. If using eFUSE, the key may be protected further at rest, by storing it in black or gray form (mixing it with a key derived from the PUF, or the hard-coded Family key). If storing the key in either black or gray form, it can also be stored directly in the boot image, rather than eFUSEs.

With a stored, secret AES key, the CSU ROM/FSBL performs decryption simply by passing the image through the CSU's AES-GCM block. One last consideration is about not only protecting the secret key in storage, but also while it is in use. Recent side-channel attacks attempt to recover secret keys by observing physical effects (such as power consumption or EM fields) of the algorithm [11]. In order to protect our AES key against such attacks, the Zynq MPSoC provides some countermeasures:

- We can employ key rolling, as described in "AES-GCM" on page 184. This uses a different key for each block of the image, thus limiting the number of measurements an attacker can make with a single key, under normal operation. The block length can be configured — shorter blocks make attacks more difficult, but also increase the size of the boot image.

- We can employ 'Operational keys'. An Operational key can be encrypted with the main AES key and stored in the boot image. On boot, the main AES key is only used to decrypt the operational key, which is then used to decrypt the rest of the image. This dramatically limits the use of the main AES key, keeping our on-chip secret more safe against attackers.

- All partitions will be authenticated before decryption. This prevents an attacker from getting the device to decrypt their own, maliciously crafted inputs. Chosen inputs can weaken the algorithm, and make side-channel attacks much more effective.

14.3.3 Secure Boot Image Format

The secure boot image format expands on the non-secure boot image format, to facilitate authentication and decryption features. Figure 14.8 shows an example of this format. The main structure (header followed by a linked-list of partitions) remains the same as non-secure images.

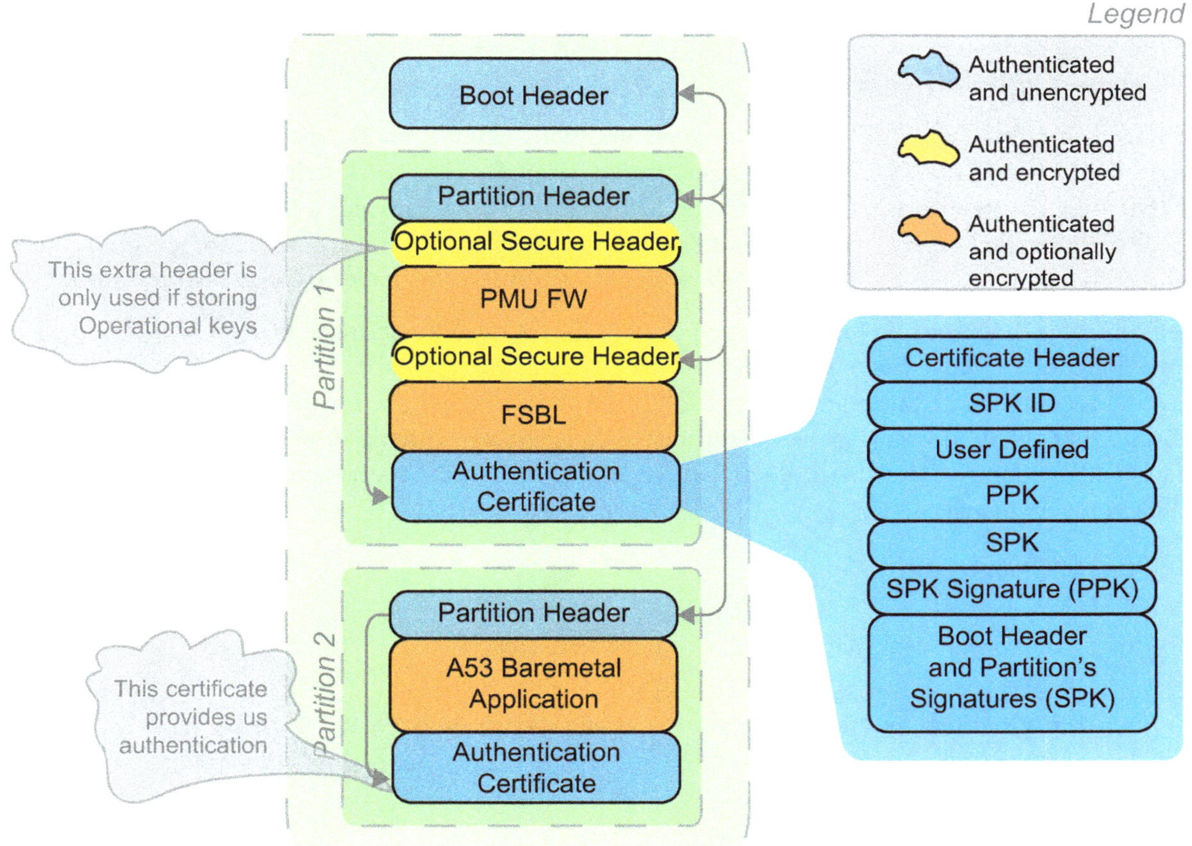

Figure 14.8: Example Secure Boot Image Format with Authentication certificates.

The main addition for secure images is the Authentication Certificate (AC), which follows each partition. This provides the required information for the CSU ROM/FSBL to recover the full PPK, authenticate the SPK, and then authenticate the boot image header and the partition itself (including the AC).

If the design makes use of an Operational key, as specified in the boot image header, we also introduce a 'secure header' before the PMU firmware and FSBL partitions. This secure header contains the Operational key and a corresponding initialisation vector. It is always encrypted with the main AES key, to keep the Operational key confidential.

14.4 Practical Non-Secure Device Configuration

Now that we are familiar with the boot process, we want to look at the practical elements. When not worried about security, we only really need to craft a boot image and place it somewhere that the Zynq MPSoC can access. The following sections will tackle these two steps in turn.

14.4.1 Generating Boot Images

Boot images for the Zynq MPSoC can be generated with Xilinx's standalone Bootgen utility. Bootgen generates a boot image (in BIN or MCS format) from a simple, plain text description called a Boot Image Format (BIF) file. Bootgen is integrated with the Xilinx SDK, which provides a GUI interface for BIF file creation. The Embedded Design Tutorial ([1]) provides a walk-through of using the Bootgen GUI. For the rest of this section, we will use the BIF file without the GUI for the sake of clean, concise descriptions.

To demonstrate the use of Bootgen, let's try to produce the example boot image that we have already introduced in Figure 14.3. The image should contain the PMU firmware, FSBL, a baremetal application for a Cortex-A53 processor, and a bitstream for the PL. Figure 14.9 shows an example BIF file that describes such a boot image. The whole image is given an arbitrary name (`the_ROM_image` in this case) and is enclosed in curly brackets. Each image that we want to include is defined on its own line, and is accompanied by any extra parameters in square brackets.

```
the_ROM_image:
{
    [bootloader,destination_cpu = a53-0] fsbl.elf
    [pmufw_image]                        pmufw.elf
    [destination_device = pl]            design_1_wrapper.bit
    [destination_cpu = a53-0]            my_baremetal_app.elf
}
```

Figure 14.9: Example BIF file

In our example, we make use of parameters to tell Bootgen which part of the Zynq MPSoC each image is targeting, and to identify any images of particular significance (the FSBL and PMU firmware). Note that the PMU firmware does not have an explicit destination device — this can be inferred. The FSBL does have an explicit destination because the user has a choice of target processors. Full specification of all of the BIF file parameters can be found in [4].

Bootgen parses the BIF file and generates a boot image for us. Given a BIF file (`boot.bif`) and an output boot filename (`BOOT.bin`), Bootgen can be invoked using the following command:

```
bootgen -arch zynqmp -image boot.bif -o BOOT.bin
```

Figure 14.10: Example Bootgen command

A full list of Bootgen options can also be found in [4].

14.4.2 Programming the Boot Images

Now that we have a boot image, we need to place it somewhere that the CSU ROM can access. As seen in Table 14.2 on page 360, there are many different boot mediums available. These can be grouped into six groups: JTAG, QSPI, NAND, SD, eMMC, and USB. These require different means of programming:

- JTAG can load an image which is already stored on the developer's PC. This is initiated by the PC, and can be performed easily through the SDK. See [12] for practical usage.
- QSPI and NAND memories must be programmed, for example, using the SDK's Flash writer tool. See [12] for practical usage.
- SD and eMMC cards are a little bit different, as the boot image must be placed within a file system — not just directly in the raw memory. The boot image should be called "BOOT.bin", and placed in a FAT 16/32 file system.
- USB is probably the least common method (it is disabled by default). This makes use of the USB Device Firmware Upgrade (DFU) standard. The *dfu-util* program can be used to prepare a bootable USB medium — see [4] for details.

14.5 Practical Secure Device Configuration

Carefully configuring the Zynq MPSoC is really important for secure boot — misconfiguration can make life very easy for an attacker. The developer needs to consider three different tasks to setup a secure boot system:

1. Crafting a boot image.
2. Device configuration of eFUSEs and BBRAM.
3. Maintaining the chain of trust in user code, beyond the FSBL.

The following sections tackle each of these tasks in turn.

14.5.1 Generating Boot Images

We can use the Bootgen utility to create images for secure boot. This process builds on the basic BIF file format that we introduced in Section 14.4.1. We use the BIF file's security settings (listed in [4]) to describe Armv8 security settings (Exception Level/EL and TrustZone state), the authentication status, and encryption status of each partition.

Let's first look at the new parameters for describing Armv8 settings. A good demonstration of when these settings can be useful is for Linux system on the APU. We need to load the ATF with full privileges (i.e. in EL3 and TrustZone's secure world), and a SSBL with less privileges (EL2 and in the normal world). Figure 14.11 shows an example BIF file which implements this system, with the new parameters in bold.

```
bif_tannen:
{
    [bootloader,destination_cpu = a53-0]                                fsbl.elf
    [pmufw_image]                                                       pmufw.elf
    [destination_cpu = a53-0, exception_level = el-3, trustzone] bl31.elf
    [destination_cpu = a53-0, exception_level = el-2]                   u-boot.elf
}
```

Figure 14.11: BIF file using ELs and TrustZone status for a Linux system

While TrustZone and the Cortex-A53's Exception levels do play an important part in design secure systems, the main thing every secure boot image needs is authentication. To implement this, we add an authentication parameter to every required partition. We must also tell Bootgen how our primary and secondary keys should be configured. To do this, we supply an extra line specifying the PPK on the Zynq MPSoC device to use (PPK0 or PPK1), and what the current SPK ID is (used for key revocation). The keys themselves must also be defined, so we add two lines to link to our different key files — remember we aren't working with the PPK and SPK at this stage, but their secret counterparts, Primary Secret Key (PSK) and Secondary Secret Key (SSK).

The key files for the PSK and SSK can be generated for us by Bootgen, if desired. To do this, we invoke Bootgen with the "`-generate_keys pem`" argument. Also remember that the hashes of the PPK and SPK will need to be programmed into the device (see "Device Configuration for Secure Boot" on page 374). We can request that Bootgen calculate these hashes for us, using the "`-generate_hashes`" argument. Note that we never need to explicitly reference the public keys in the BIF file because the secret key files have all of the information required to recreate their public counterparts.

The last topic to cover for secure boot BIF files is, of course, encryption. Before Bootgen can encrypt a partition, we must first describe where our AES key file is and where the AES key will be found on the device — either in a red form in the BBRAM or eFUSE, or in a black form in the eFUSE or boot header. Note that we will need to program this key into our device later. An example BIF file with an encrypted user application is shown in Figure 14.13.

```
great_scott:
{
    [auth_params] ppk_select=0;spk_id=0x00000000
    [pskfile]     psk.pem
    [sskfile]     ssk.pem

    [bootloader, destination_cpu = a53-0, authentication = rsa] fsbl.elf
    [pmufw_image, authentication = rsa]                          pmufw.elf
    [destination_device = pl, authentication = rsa]              design_1_wrapper.bit
    [destination_cpu = a53-0, authentication = rsa]              my_baremetal_app.elf
}
```

Figure 14.12: BIF file using authentication

```
flux_cap:
{
    [aeskeyfile]        aes_key.nky
    [keysrc_encryption] bbram_red_key

    [bootloader, destination_cpu = a53-0]       fsbl.elf
    [pmufw_image]                               pmufw.elf
    [destination_device = pl, encryption = aes] design_1_wrapper.bit
    [destination_cpu = a53-0, encryption = aes] my_baremetal_app.elf
}
```

Figure 14.13: BIF file using encryption with the BBRAM key

Note that Bootgen can automatically generate AES keys for us too. If we specify a non-existent file as our `[aeskeyfile]` parameter, a key will be generated for us. See [4] for the details of more advanced BIF features, such as black key storage, Operational keys and key rolling.

All of the features shown in our BIF file examples presented in this section can be combined, providing you a complete boot image capable of authentication, confidentiality, and Armv8 security features.

14.5.2 Device Configuration for Secure Boot

Configuring the Zynq MPSoC with on-chip security flags, public RSA keys, and secret AES keys, enables the hardware root of trust that we rely on for secure boot. The Zynq MPSoC is self-programming, not in the sense that this happens automatically(!) but rather, it can program its own eFUSEs and BBRAM. Because of this, most of the device configuration is actually performed with example applications supplied with the SDK, using the XilSKey library. We simply change some constants at the top of the example code, and run the program to blow eFUSES for us. For step-by-step (and blow-by-blow) guide of exactly how to program these values, refer to the excellent documentation in [13].

Instead of repeating the same guidance here, we are going to cover *what* you need to configure in order to properly realise a secure device. First of all, consider enabling the eFUSE flags that enforce authentication of every single boot (*RSA_EN*). This is usually the developer's intent when using secure boot, just remember to check that your RSA keys have been properly programmed before enabling this! This should only really remain disabled if your reason for using secure boot is solely to keep your design confidential — and you have no issue with customers or others running their own software in the future.

Next comes managing the public RSA keys. The eFUSE arrays can store two *hashes* of PPKs. Remember that we can get Bootgen to generate these hashes for us, all we need to do is burn the eFUSEs. Note that each PPK hash has corresponding write lock and invalid flags. After ensuring that the correct hash value has been programmed, always blow the write lock flag. This prevents someone from blowing more of the hash bits in the future, precluding modifications that change our key hash to that of an attacker, or even just bricking our device. More subtly, you may also wish to program both PPKs upfront — an unprogrammed key hash will be zero but still valid!

If the system has images which ought to be kept confidential, you will also need to include a secret AES key. This can be programmed in either the BBRAM (offering key-agility) or in eFUSE. There is also an eFUSE flag to write lock the eFUSE AES key. If you want to keep the key encrypted at rest for added protection (i.e. a 'black' key), you will need to configure the PUF. This process is covered simply in [13]. A black key can be stored in the eFUSE array or in the boot image header.

14.5.3 Maintaining the Chain of Trust Beyond the FSBL

For some parting food for thought on implementing a secure Zynq MPSoC system, consider what the developer's responsibilities are. Xilinx is wholly responsible for properly implementing the BootROM code and all of the surrounding hardware. The developer is responsible for influencing the BootROM's operation through options specified in Xilinx's documentation — mainly boot image headers and eFUSEs, as we have discussed in Section 14.5.1 and Section 14.5.2. However, something we have not discussed, is that the developer is also responsible for any user code they choose to run. This sounds obvious, but it can have a big impact on secure boot.

If user code (typically after the FSBL) introduces any more steps in the booting process, the developer is responsible for properly maintaining the chain of trust. The classic example of this is the use of a SSBL, such as U-Boot, in a Linux system. It is easy as a developer to forget that maintaining the chain of trust between the SSBL and Linux is their responsibility. There can be some really simple issues that arise because of this. Using the default configuration, U-Boot will pause for five seconds, waiting for any user input. If user input is received, an interactive prompt opens and this allows the user to type in any boot commands they desire — including booting an unauthenticated image, completely breaking the chain of trust! This is just a small example to remind developers to think critically about the user software they use after the FSBL. Keep this in mind, and you will be able to confidently deploy Zynq MPSoC systems that boot securely.

14.6 Chapter Summary

Throughout this chapter, we have gained familiarity with the boot process, relevant security concepts, and how developers can make use of these features. We briefly discussed why a design might require secure boot, including the need for authentication in many applications, and the need for confidentiality to protect commercial IP from reverse engineering.

We detailed the boot process's three main stages: pre-configuration, configuration, and post-configuration. After becoming familiar with this process, we extended the process to include a discussion of the optional secure boot features. To conclude, we explored the practical elements of booting, including:

- Using the Bootgen utility to generate boot images from simple, plain text descriptions.
- Programming a boot medium with our boot image.
- Adding security considerations to our boot image, and configuring the eFUSE flags and BBRAM for a secure boot scenario.

The reader is left with an appreciation for how the boot process is structured, how they can influence this process practically, and where to look to find more on each topic.

14.7 References

Note: All online sources last accessed March 2019.

Some of the Xilinx documentation referred to below has version-specific URLs. If you are working in a newer version of the tools, check for updates on the Xilinx website, or try adjusting the link according to your version.

[1] Xilinx Inc., "Zynq UltraScale+ MPSoC: Embedded Design Tutorial", v2017.4, January 2018.
Available: https://www.xilinx.com/support/documentation/sw_manuals/xilinx2017_4/ug1209-embedded-design-tutorial.pdf

[2] Xilinx Inc., "UG1085 — Zynq UltraScale+ Device Technical Reference Manual", v1.7, December 2017.
Available: https://www.xilinx.com/support/documentation/user_guides/ug1085-zynq-ultrascale-trm.pdf

[3] Xilinx Inc., "UG1087 — Zynq UltraScale+ MPSoC Register Reference", v1.3, January 2017.
Available: https://www.xilinx.com/html_docs/registers/ug1087/ug1087-zynq-ultrascale-registers.html

[4] Xilinx Inc., "UG1137 — Zynq UltraScale+ MPSoC Software Developer Guide", v6.0, January 2018.
Available: https://www.xilinx.com/support/documentation/user_guides/ug1137-zynq-ultrascale-mpsoc-swdev.pdf

[5] Xilinx Inc., "UG1075 — Zynq UltraScale+ MPSoC Packaging and Pinouts", v1.4, December 2017.
Available: https://www.xilinx.com/support/documentation/user_guides/ug1075-zynq-ultrascale-pkg-pinout.pdf

[6] DENX Software Engineering, "Das U-Boot — The Universal Boot Loader", Project homepage.
Available: https://www.denx.de/wiki/U-Boot/

[7] Xilinx Inc., "UG1228 — Zynq UltraScale+ MPSoC Embedded Design Methodology Guide", v1.0, March 2017.
Available: https://www.xilinx.com/support/documentation/sw_manuals/ug1228-ultrafast-embedded-design-methodology-guide.pdf

[8] E. Peterson, "XAPP1323 — Developing Tamper-Resistant Designs with Zynq UltraScale+ Devices", Xilinx Inc., v1.0, October 2017.
Available: https://www.xilinx.com/support/documentation/application_notes/xapp1323-zynq-usp-tamper-resistant-designs.pdf

[9] Computer Securtiy Resource Center, "Cryptographic Algorithm Validation Program: AES Validation List", NIST.
Available: https://csrc.nist.gov/projects/cryptographic-algorithm-validation-program/validation/validation-list/aes#4438

[10] Computer Securtiy Resource Center, "Cryptographic Algorithm Validation Program: SHA-3 Validation List", NIST.
Available: https://csrc.nist.gov/projects/cryptographic-algorithm-validation-program/validation/validation-list/sha-3#20

[11] A. Moradi, and T. Schneider, "Improved Side-Channel Analysis Attacks on Xilinx Bitstream Encryption of 5, 6, and 7 Series", Ruhr University Bochum, 2016.
Available: https://eprint.iacr.org/2016/249.pdf

[12] Xilinx Inc., "UG782 — Using Xilinx SDK", v2017.4, December 2017.
Available: https://www.xilinx.com/html_docs/xilinx2017_4/SDK_Doc/index.html

[13] L. Sanders, "XAPP1319 — Programming BBRAM and eFUSEs", Xilinx Inc., v1.0, July 2017.
Available: https://www.xilinx.com/support/documentation/application_notes/xapp1319-zynq-usp-prog-nvm.pdf

System Design with Xilinx SDx Development Environment

Chapter 15

Introduction to System Design with SDx

In this chapter we are going to introduce SDx, a development environment for designing systems to be implemented on Zynq and Zynq MPSoC platforms. SDx is one of Xilinx's development environments aimed at simplifying the development of applications targeted at Xilinx devices by defining the system, including hardware components, using software.

During the course of this chapter we will explore some of the reasons why you might want to design a system using the SDx and how you might approach the task.

Throughout this and the following chapters concerning SDx, we use the term Zynq when referring to both Zynq-7000 and Zynq MPSoC devices. The information presented is based on the 2017.2 version of SDx, and may differ to some degree with other versions.

15.1 Motivation for using SDx

Using traditional development techniques, a system designer targeting a Zynq device needs to provide various different components to produce a working system. Some of these are obvious, such as the application software to run on the PS and the application hardware to run in the PL. However, there are a number of other components necessary in order to create a fully functioning system. The designer must also supply the boot loader(s) required to boot the system, including configuring the PL and loading the application software to run on the PS. In cases where the application depends on an OS, the designer needs to provide this and a boot loader to load it. Furthermore, the designer is obliged to supply the components necessary to interface between the PS and PL, including driver software and data moving hardware to pass data and control signals.

The production of all of the components that comprise a system targeting a Zynq device is a complex process requiring a wide-ranging skill-set. Tools that aim to simplify this process have, therefore, found favour with designers targeting Zynq devices. For example, high-level synthesis tools can be used to synthesise discrete hardware IP blocks from software descriptions. Although, the hardware components produced still need be integrated with the other hardware and software components in a wider system by the developer.

SDx represents a natural extension of the high-level synthesis tools available for hardware development. SDx extends the functionality of the high-level synthesis tools by automating the integration of the synthesised hardware into the system, by producing data moving hardware and OS-specific software drivers. In addition, it also supplies the boot loader(s) required and an OS, if necessary.

The increased level of abstraction provided by SDx can be regarded as advantageous for different reasons, depending on your perspective, and it is upon these reasons that we will expand in the remainder of this section.

Traditionally, a lack of hardware development experience and expertise has presented a hurdle to exploiting the flexibility and low-power performance of FPGA devices when designing a system. SDx seeks to resolve this problem by providing an environment similar to those commonly used to program other platforms, such as CPU, GPU and ASSP, and a SDx compiler allowing the entire system to be defined using C, C++ or OpenCL. A software developer with limited or no FPGA experience is able to take advantage of both the hardware and software resources on the Zynq devices by creating the entire system in software and specifying portions of the system for implementation in hardware. The SDx compiler will generate the necessary hardware and software to enable the system to be used on the targeted platform. Thus, SDx dramatically reduces the prerequisite learning needed to capitalise on not only the PL, but also the closely integrated nature of the PL and PS provided by the Zynq devices.

SDx also provides benefits for experienced hardware designers. One of these benefits is directly derived from the technique of defining systems using software introduced by SDx. By defining the entire system in software, the configuration of the system can be quickly and easily changed, allowing the rapid exploration of design options to find an optimal design. This is likely to increase productivity and accelerate the development of systems targeting Zynq devices. The automation provided by SDx also abstracts the designer from some of the tedious tasks involved in system development. Another benefit for experienced hardware designers is that SDx is underpinned by Xilinx's Vivado suite of design tools. When building a software defined system, SDx uses these tools and generates many of the standard project files. These files are made available within the SDx project, providing a wealth of lower level design information about the system implementation. This means that, although SDx increases productivity by increasing the level of abstraction, the low-level detail of the design is accessible to the designer to help guide the tools to create an optimal system. Familiarity with the Vivado design tools will also prove beneficial when it comes to making use of some of the more advanced features of SDx. We will examine some of these features in later chapters.

So far, we have discussed the benefits of SDx to designers with little hardware development experience as well as to experienced hardware designers. Perhaps the most significant benefit of SDx, however, is to teams composed of both hardware and software developers. SDx provides a common environment for team members to work in, simplifying the collaborative process. The software definition of the overall system provides a clear specification of the interface between software and hardware components for which the low-level functionality is then handled by SDx. This means that both hardware and software developers are, to some extent, abstracted from the interface implementation, which is the area of the system design likely to require the greatest amount of collaboration. Software developers are then able to focus on refining the application code in abstraction from hardware, while hardware engineers are able to concentrate on improving the performance and resource usage of the hardware components. This is likely to increase the productivity of the development team and reduce the time to market for the system.

15.2 About SDx

Now that we have considered why you might want to use SDx, we will introduce how it might be used. In this section, we take a high-level overview of the development environment itself, and the different stages typically involved in designing a system using SDx. We will also take a closer look at the project files and reports generated when running the tool.

15.2.1 The Development Environment

SDx has two interfaces: a Command Line Interface (CLI) and a Graphical User Interface (GUI). We will mainly focus on using the GUI in this book, although we will make use of the CLI from time to time.

The SDx Integrated Development Environment (IDE), shown in Figure 15.1, is based on the Eclipse™ C/C++ Development Tooling (CDT) IDE [1]. Being an Eclipse-based IDE provides the SDx IDE with a familiarity for a broad range of embedded systems and software developers, as well as for those already accustomed to the Xilinx Software Development Kit (SDK) and Vivado HLS tools. For the benefit of those readers with less experience of using an Eclipse-based IDE, we will briefly cover the main concepts underpinning these environments, with particular reference to the SDx IDE itself.

The main window of the development environment is also referred to as a *workbench*, and contains one or more *perspectives*. While there are several perspectives available in the SDx IDE, the majority of work is likely to be carried out in either the SDx perspective, shown in Figure 15.1, or the Debug perspective, shown in Figure 18.3 on page 435. The SDx perspective is the default perspective for the SDx IDE.

A perspective is usually intended to be used to perform a particular task in the development cycle, and comprises a set of *editors* and *views* useful for performing that task. A range of editors are available for editing files of different types. When a file is opened within the development environment, the associated editor is launched in the editor area of the perspective. Views are used to display information and to navigate data associated with the current workspace. Views can be closed and additional views can be opened to customise a perspective and better support the task currently being performed. Figure 15.1 shows an example of the layout of the editor area and views for the SDx perspective. Further information regarding Eclipse-based IDE is available in the Eclipse documentation [2].

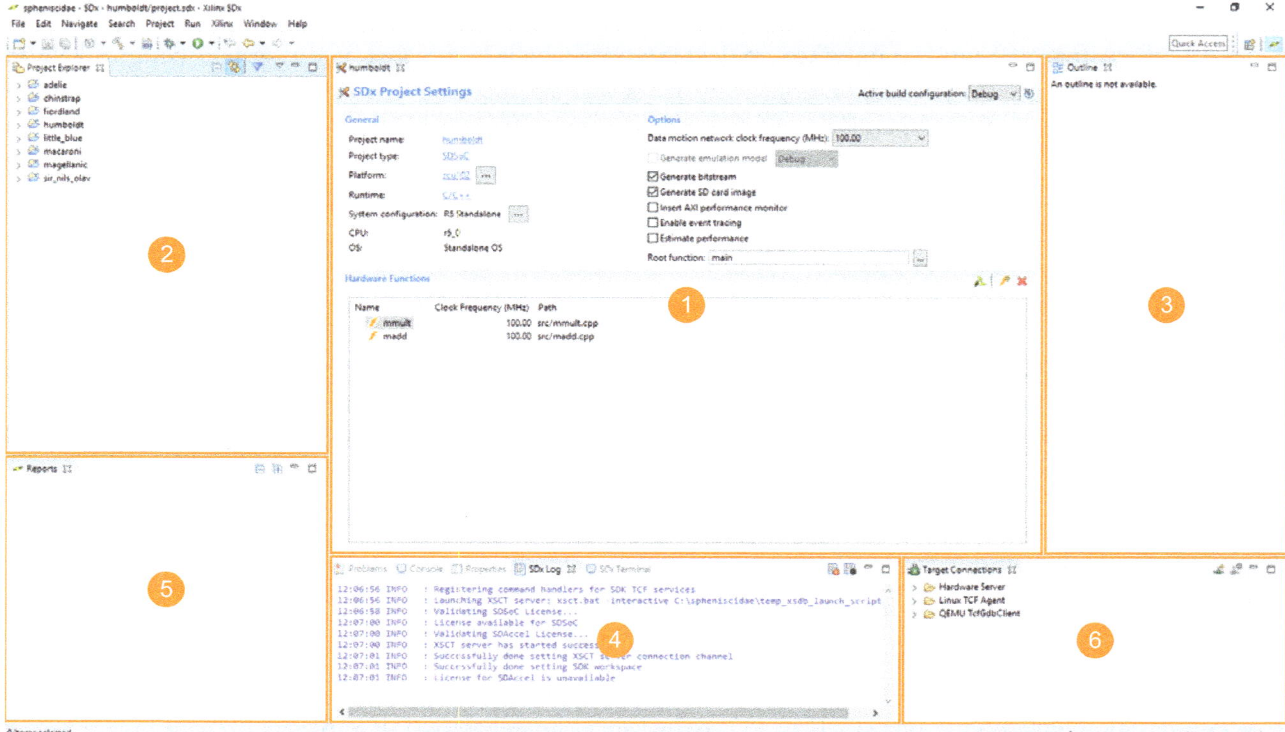

1. Editor Area: used to display files opened within the SDx IDE for which there is an appropriate editor associated

2. Project Explorer View: provides a hierarchical view of the files contained in the SDx projects in the current workspace

3. Outline View: provides an editor specific structural outline of the file open in the editor

4. A stack of five views:
   - Problems View: provides information about errors and warnings encountered during a build
   - Console View: used to display output messages from the SDx IDE
   - Properties View: used to display the properties of the currently selected item
   - SDx Log View: used to display the information logged about activity in the SDx console
   - SDx Terminal View: terminal application to connect to serial ports

5. Reports View: provides links to some of the reports produced by the currently selected SDx project

6. Target Connections View: used to setup and configure connections between the SDx IDE and the target system

Figure 15.1: The default SDx perspective (as in SDx IDE 2017.2)

15.2.2 Introducing the Full-System Optimising Compiler

Having considered the appearance of the SDx IDE, we will now delve a little deeper into its functionality.

The main function of SDx is provided by the full-system optimising compiler. As alluded to in Section 15.1, this tool makes use of some of the Xilinx Vivado suite of design tools, and is perhaps best thought of as a tool-chain itself. The input to the SDx tool-chain is a description of the system written in either C, C++, OpenCL, or a combination of these languages, although our discussion will focus on systems written in C or C++.

A simplified representation of the SDx tool-chain is shown in Figure 15.2, where it can be observed that there are three main sub-tools: ***sdscc*** and ***sds++*** compilers and the ***sds++*** linker. The ***sdscc*** compiler is used to process C language sources into either object code, for execution on the Arm processors, or RTL code, for implementation in the PL. C code marked for software implementation is compiled using a ***gcc*** compiler invoked by the ***sdscc*** compiler. To generate RTL code for C sources, the ***sdscc*** compiler calls the Vivado HLS tool [3]. The process of producing object and RTL code from C++ language sources is similar to the process for C language sources, with the exceptions that the ***sds++*** compiler is used in place of ***sdscc,*** and that a ***g++*** compiler is used to compile the C++ sources to object code. The remaining tools in the SDx tool-chain are invoked by the ***sds++*** linker. A linker is called to link the object code generated by the ***gcc*** and ***g++*** compilers, along with any pre-built libraries, to create an executable file. It is recommended that libraries are built using the same compiler tool-chain and options used by the ***sdscc*** and ***sds++*** tools [4], which, since version 2016.1, have been a Linaro-based GCC compiler tool-chain [5]. A bitstream is generated from the RTL code by invoking the Vivado synthesis and place and route tools [6]. Finally, a boot image is created incorporating the generated executable file and bitstream using the ***bootgen*** utility [7]. This boot image can be loaded onto an SD card and used to boot the Zynq device.

15.2.3 Data Motion Networks

So far, we have described how the C or C++ sources are converted into either hardware or software components, but we have not yet considered how the SDx tool-chain handles the integration of these components to form a coherent system. When the source code is compiled, functions marked for hardware implementation are passed to the Vivado HLS tool to produce RTL code rather than producing any object code. Hence, the SDx tool-chain replaces calls to these functions in the source code with calls to stub functions. These stub functions are automatically generated, and make calls to underlying drivers to handle the transfer of data between the IP block in the PL generated from the function marked for hardware implementation, and the PS memories or other IP blocks.

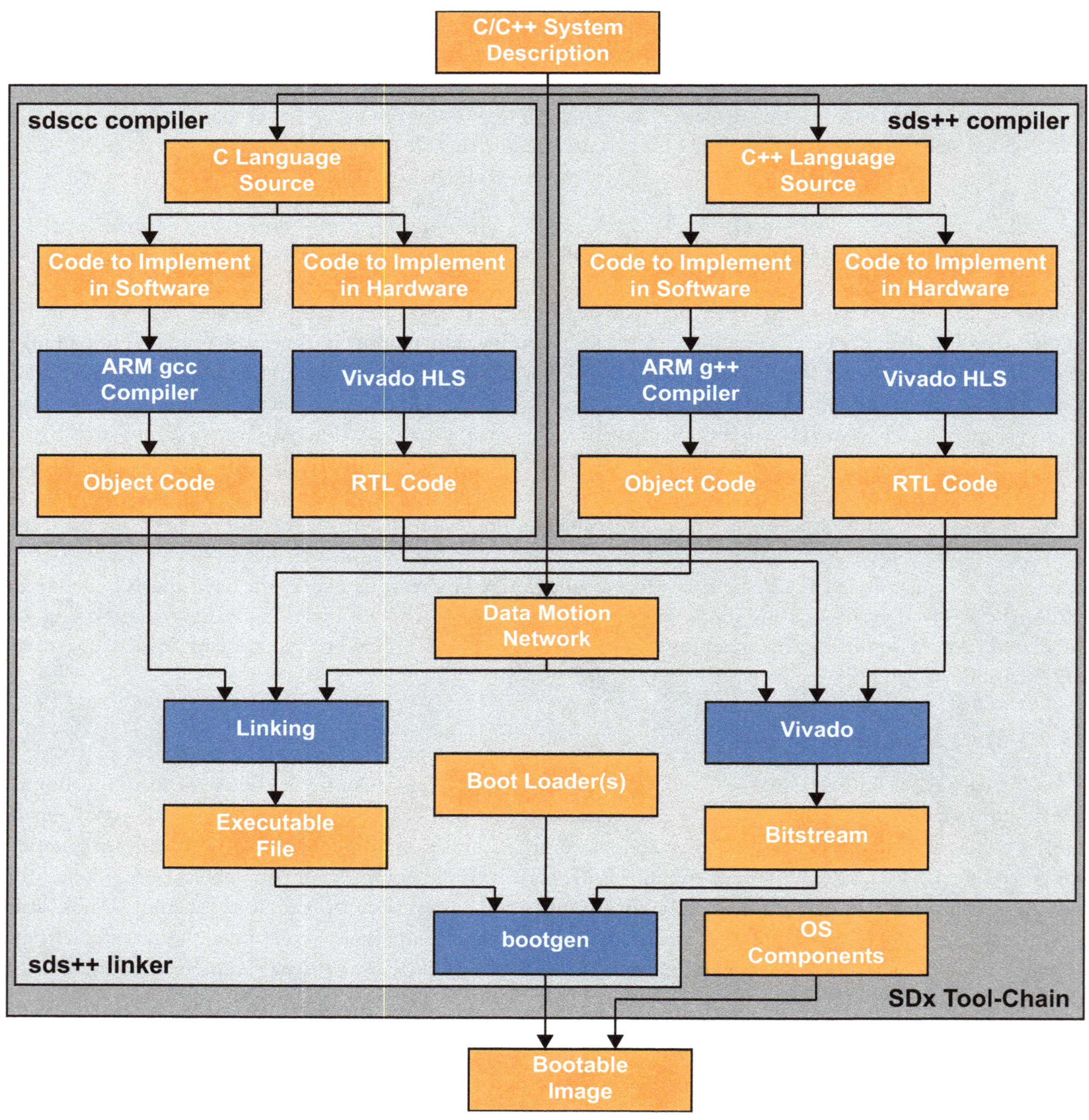

Figure 15.2: SDx compiler tool-chain

SDx creates data motion networks to move data between the created IP blocks and the PS or other IP blocks in the PL. Each data motion network consists of three parts:

- The PS port, or the interface the created IP block is to connect to;
- The data mover;
- The interface on the created IP block.

Unless guided by the designer through pragmas, SDx attempts to infer the optimal data motion network based on an analysis of the source code.

The PS port is the port used to transfer data between the PS and PL. The type of port selected depends on two factors: the amount of data to be transferred, and the requirement for access to caches on the PS. Port selection criteria are also discussed more generally in Section 11.3.7 on page 278.

Data movers are responsible for transferring data between the interface of the created IP block and PS memories or other IP blocks. The inference of a data mover depends on the data structure type used for the function argument, the amount of data to be moved, and whether the memory used to store the data is physically contiguous[1].

For very small amounts of data, no more than 300 bytes, a general purpose port and simple AXI FIFO may inferred as the data transfer time can be less than an AXI Scatter Gather Direct Memory Access (DMA) engine. For larger amounts of data, data may be transferred over a high performance port using a DMA due to the higher data transfer rates it offers compared to general purpose ports. The most efficient data mover in this case is the AXI Simple DMA. This, however, has two limitations: it is best used for transfers of no more than 32MB, and the data must be in physically contiguous memory. If either of these limitations is exceeded then an AXI Scatter Gather DMA will be inferred. The default data movers inferred by SDx are shown in Table 15.1. Only the default data movers have been listed here and others are available. Further information on these is available in [8].

For transfers greater than 300 bytes, data caching arrangements become a factor in the choice of PS port. When data is allocated in cacheable memory, the SDx tool-chain assumes that cache coherency must be maintained for data passed between the PS and PL, and a cache-coherent port is usually inferred. Although, there is an exception to this rule: when the amount of data being transferred is much greater than the size of the cache, which can result in cache thrashing when using a cache-coherent port, a non-cache-coherent port is inferred instead. For cache-coherent ports, cache coherency is maintained by hardware within the PS, providing fast access to data held in cache. Cache coherency when using a non-cache-coherent port, by

1. Physically contiguous memory is a single, continuous area of memory. Obtaining large areas of unused physically contiguous memory on a running system can be difficult and often requires careful memory management. Some OSs implement a virtual memory system to abstract user applications from the task of managing physical memory. The user application uses virtual memory addresses that need to be translated by the OS to physical memory addresses. Contiguous virtual addresses may not necessarily map to contiguous addresses in physical memory.

contrast, is maintained by software. In this case, the stub function generated by the SDx tool-chain is required to perform a cache flush prior to transferring data from the PS to the PL, and a cache invalidation prior to transferring data from the PL to the PS. These additional software instructions incur a performance penalty compared to hardware-enforced cache coherency.

Table 15.1: Default data movers inferred by SDx [8]

Argument	Data Size	Memory Physical Contiguity	Data Mover
scalar	n/a	no	AXI Lite
array	≤300B	no	AXI FIFO
	<32MB	yes	AXI Simple DMA
	>300B	no	AXI Scatter Gather DMA

A non-cache-coherent port is also inferred when the data is allocated in non-cacheable memory. When it is known that data will not be present in cache, cache-coherent ports lose much of their potential performance advantage over non-cache-coherent ports, particularly since the stub function generated by the SDx tool-chain for non-cache-coherent ports does not need to contain code to maintain cache coherency. Moreover, cache-coherent ports tend to share resources for memory access with the processors in the PS, whereas non-cache-coherent ports do not. This means that using a non-cache-coherent port reduces the likelihood of competition for memory resources with the processors, affecting the overall system performance.

The interface on the IP block itself is inferred from the arguments to the function being implemented in hardware. The default interface inferred for different function argument types is shown in Table 15.2. Again, only the default IP interfaces are shown here and there are others available. These will be discussed further in Chapter 16 and the reader is referred to [8] for information as well.

Table 15.2: Default IP interfaces inferred by function argument type [8]

Argument	IP Interface
scalar	register
array	RAM
class/struct	each data member is assigned an interface based on whether it is a scalar or an array

15.2.4 Directing the SDx Tool-Chain

In the course of our discussion of data motion networks, we mentioned the ability of the user to specify alternative components to those inferred by default for the data motion network. The mechanism for doing so is through the specification of pragmas in the C/C++ source code input to the SDx tool-chain. These pragmas serve to direct the SDx tool-chain in optimising the system, and can either directly specify the use of a particular component or provide further information that the compiler is otherwise unable to determine at compile time. All SDx-specific pragmas entered in the source code take the form #pragma SDS <pragma>. Table 15.3 provides a summary of a selection of the SDx-specific pragmas.

Table 15.3: Summary of a selection of SDx-specific pragmas [8]

<table>
<tr><th>Pragma</th><th>Description</th></tr>
<tr><td>data copy(ArrayName[offset:length])</td><td>Specifies that the data is to be explicitly copied between the PS memories and the hardware function. offset and length are optional. offset is currently ignored by the compiler. length specifies the amount of data to be transferred.</td></tr>
<tr><td>data zero_copy(ArrayName[offset:length])</td><td>Specifies that the data will be accessed directly from memory by the hardware function itself. Results in an AXI-Master interface being generated on the hardware accelerator and requires that the data it is accessing is held in physically contiguous memory. offset and length arguments are the same as for the data copy pragma above.</td></tr>
<tr><td>data mem_attribute(ArrayName:cache|contiguity)</td><td>Specifies attributes for the memory allocated to ArrayName. cache can be CACHEABLE or NON_CACHEABLE and instructs the SDx tool-chain whether it needs to maintain cache coherency for this memory. contiguity can be PHYSICAL_CONTIGUOUS or NON_PHYSICAL_CONTIGUOUS and is used by the compiler to infer the optimal data mover.</td></tr>
<tr><td>data access_pattern(ArrayName:pattern)</td><td>Specifies the manner in which array elements are accessed within the hardware function. pattern can either be SEQUENTIAL or RANDOM. When SEQUENTIAL is specified a streaming interface will be generated. When RANDOM is specified a RAM interface will be generated.</td></tr>
<tr><td>data data_mover(ArrayName:DataMover[:id])</td><td>Specifies the data mover to be used for ArrayName. DataMover can be AXIFIFO, AXIDMA_SIMPLE or AXIDMA_SG. id argument is optional, but two hardware functions with the same data mover type and id argument will share a data mover instance.</td></tr>
<tr><td>data dim(ArrayName:dim1 dim2)</td><td>Specifies the size of each dimension of a 2D array. Used exclusively when an AXI 2D DMA data mover is specified.</td></tr>
</table>

Table 15.3: Summary of a selection of SDx-specific pragmas [8]

Pragma	Description
`data sys_port(ArrayName:port)`	Specifies the system port to be used to interface between memory and the hardware function. `port` can be `ACP`, `AFI` or `MIG`. `ACP` specifies a cache-coherent port while `AFI` specifies a non-cache-coherent port. `MIG` specifies a memory interface generator port connecting directly to an off-chip memory from the PL and is only valid for systems with DRAM directly attached to the PL, such as the Xilinx ZC706 platform.
`data buffer_depth(ArrayName:BufferDepth)`	Specifies the depth of FIFO used for FIFO interfaces or the multi-buffer depth used for BRAM interfaces.
`async(ID)`	Specifies that control should be returned to the CPU immediately following the setting up of the hardware function, i.e. it will not wait for the hardware function to return before continuing. This should be partnered with the `wait` pragma below to indicate at what point CPU execution should not continue until the hardware function has returned. `ID` specifies the hardware function being referred to and can be used to infer more than one instance of the same hardware function.
`wait(ID)`	Specifies the point at which CPU execution should wait until the next queued call to the hardware function designated by `ID` has returned. This is the partner pragma to `async` above.
`resource(ID)`	Specifies the hardware function instance that should be used to execute the hardware function call.

With the exception of the last four, all of the pragmas summarised in Table 15.3 should be inserted into the C/C++ source code immediately above the hardware function declaration to which they refer. The last four pragmas in Table 15.3 should be entered into the source code immediately above the specific call to the hardware function to which they refer.

An important point to note when using these pragmas is that they tend to provide information to the SDx tool-chain without enforcing the validity of that information. For example, if we were to use a pragma to specify that an array being passed to a hardware function was allocated in non-cacheable memory, then the SDx tool-chain would not generate the software within the stub function to maintain cache-coherency for that array. If we were then to call that hardware function within our source code and pass it an array allocated in cacheable memory, the required cache flushing operation would not be performed before the data was transferred to hardware. This would mean the data transferred to the IP block may not be valid and our system could produce some very strange results. Care should be exercised when using pragmas to ensure their accuracy.

Further information on the pragmas summarised here, along with some examples of their usage are available in [8] and [9].

15.2.5 SDx API

SDx provides a library, ***sds_lib***, which supplies a number of very useful functions for the embedded systems developer. Table 15.4 summarises some of the available functions. In order to use any of these functions, the library must be included by inserting the line `#include "sds_lib.h"` in the C/C++ source code. It is useful to note that ***sds_lib*** makes use of the `size_t` datatype, which requires the inclusion of ***stdlib.h*** prior to including ***sds_lib***.

Table 15.4: Summary of a selection of functions provided by sds_lib library

Function	Description
`void *sds_alloc(size_t size)`	Allocates a physically contiguous area of memory of `size` bytes and returns a pointer to it.
`void *sds_alloc_non_cacheable(size_t size)`	Allocates a physically contiguous area of memory of `size` bytes, marks it as non-cacheable and returns a pointer to it.
`void sds_free(void *memptr)`	Frees an area of memory allocated using `sds_alloc()`, `sds_alloc_cacheable()` or `sds_alloc_non_cacheable()` functions.
`void sds_wait(unsigned int id)`	Performs a blocking wait for the next queued call to the hardware function identified by `id` to return.
`int sds_try_wait(unsigned int id)`	Polls whether the next queued call to the hardware function identified by `id` has returned. Returns `0` if the hardware function has not returned and `1` if it has.
`void *sds_mmap(void *physical_addr, size_t size, void *virtual_addr)`	Maps an area of physical memory of `size` bytes starting at `physical_addr` to `virtual_addr`.
`void sds_munmap(void *virtual_addr)`	Unmaps an area of physical memory that was previously mapped to `virtual_addr` using the `sds_mmap()` function above.
`unsigned long long sds_clock_counter(void)`	Returns the current value of a free-running counter that increments on each processor clock cycle and wraps to zero.
`void sds_set_counter(unsigned long long val)`	Stops the free-running counter read by the `sds_clock_counter()` function above, loads it with the specified value `val` and starts it again.

15.2.6 SDx Tool-Chain Settings

Within the SDx tool-chain are GCC compilers and some tools from the Xilinx Vivado suite. Under normal circumstances the behaviour of these tools is configurable and, in order to build the optimal system for our specific requirements, it is desirable to retain control over the configuration of these tools despite the simplified interface provided by SDx.

The tools responsible for calling all of the other sub-tools in the SDx tool-chain are ***sdscc*** and ***sds++***, as shown in Figure 15.2 on page 384. It is by passing options to these two tools when they are called, that we exercise control over the sub-tools called by them. From the CLI this is straightforward enough, and full details of the options for these tools are given in [8]. However, when using the SDx IDE, the user is abstracted from direct calls to ***sdscc*** and ***sds++***. Instead the SDx IDE uses *build configurations*, another feature of the Eclipse IDE. A build configuration is a group of tool settings used to build the system. By having multiple build configurations, a system can be built in different ways to suit the purpose of that specific build.

Since version 2016.1, the SDx IDE has provided two build configurations by default: ***Debug*** and ***Release***. The difference between these two build configurations are the settings used for the GCC compilers. The ***Debug*** configuration is intended to be used to build a system for debugging. As such, it sets the `-g` option for the GCC compilers. This causes additional data, which maps between the machine code and source code, to be included in the object files, enabling it to be used by debugging tools. It also compiles the code without applying any optimisations, simplifying the task of mapping between the machine code and source code at the expense of potentially less efficient machine code [10]. The ***Release*** configuration, on the other hand, is aimed at producing the system with the best performance. It therefore omits the additional debugging data from the object files and compiles the code with the highest optimisation settings in an attempt to achieve the best performing machine code at the expense of a potentially larger executable file. The options passed to the ***sdscc*** and ***sds++*** tools by each of these build configurations are shown in Table 15.5. Any options passed to the ***sdscc*** or ***sds++*** tools that are not specifically recognised as being for another sub-tool are assumed to be options for the GCC compilers, and are therefore passed to these compilers when they are called [8]. Further information on the options available for GCC compilers is available from [10] and [11].

Table 15.5: Default build configuration tool settings

Build Configuration	sdscc/sds++ Options
Debug	`-Wall -O0 -g -I”../src” -c -fmessage-length=0 -MT”$@”`
Release	`-Wall -O3 -I”../src” -c -fmessage-length=0 -MT”$@”`

Passing tool settings to the Vivado tools is slightly more complicated than it is for the GCC compilers, and first we must understand how the Vivado tools are called by the SDx tool-chain.

The Vivado HLS tool is called individually for each function marked for implementation in hardware. Options specific to the function being synthesised can be passed on each one of these calls and, as such, do not form part of the more general build configuration we are concerned with here. We therefore leave discussion of directing the Vivado HLS tool to Chapter 16.

Modifying and Creating Build Configurations

Having established how the build configuration controls the settings for the sub-tools within the SDx tool-chain, let us take a look at how we are able to modify the default configurations and create our own.

First, we will examine how to modify existing build configurations. To do this we need to access the C/C++ build settings for the build configuration we want to modify. These can be accessed by selecting *Properties* under the *Project* menu, as shown in Figure 15.3.

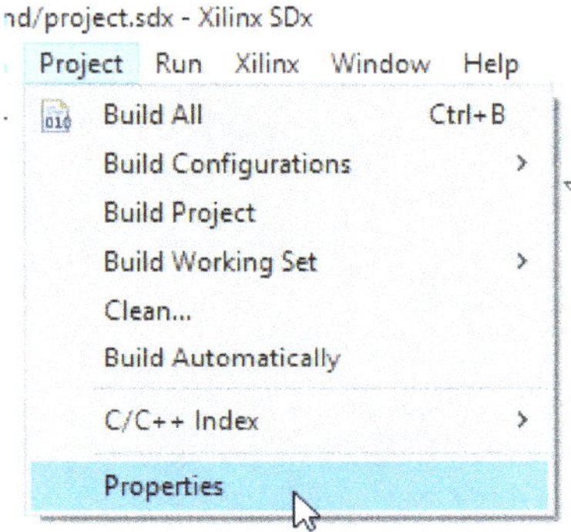

Figure 15.3: Method for accessing C/C++ build settings (as in SDx IDE 2017.2)

This opens the *Properties* dialogue box for the project, shown in Figure 15.4, from where we can set the command line options passed to the ***sdscc*** and ***sds++*** tools from the *C/C++ Build > Settings* menu. Tool options for the three tools making up the SDx tool-chain — ***sdscc*** compiler, ***sds++*** compiler and ***sds++*** linker — can be set individually. A number of the available options can be set using the appropriate tool option categories listed under each tool. The remaining options can be set by entering the command line option under the catch-all *Miscellaneous* category.

Custom build configurations can be created by selecting the *Manage Configurations...* button in the project *Properties* dialogue box, and then selecting *New...*. Settings from existing build configurations can be copied, and indeed copying the settings from one of the default build configurations is likely to be the easiest starting point for creating a custom configuration. A custom build configuration could be used to specify different synthesis or implementation strategies or to link against a specific software library.

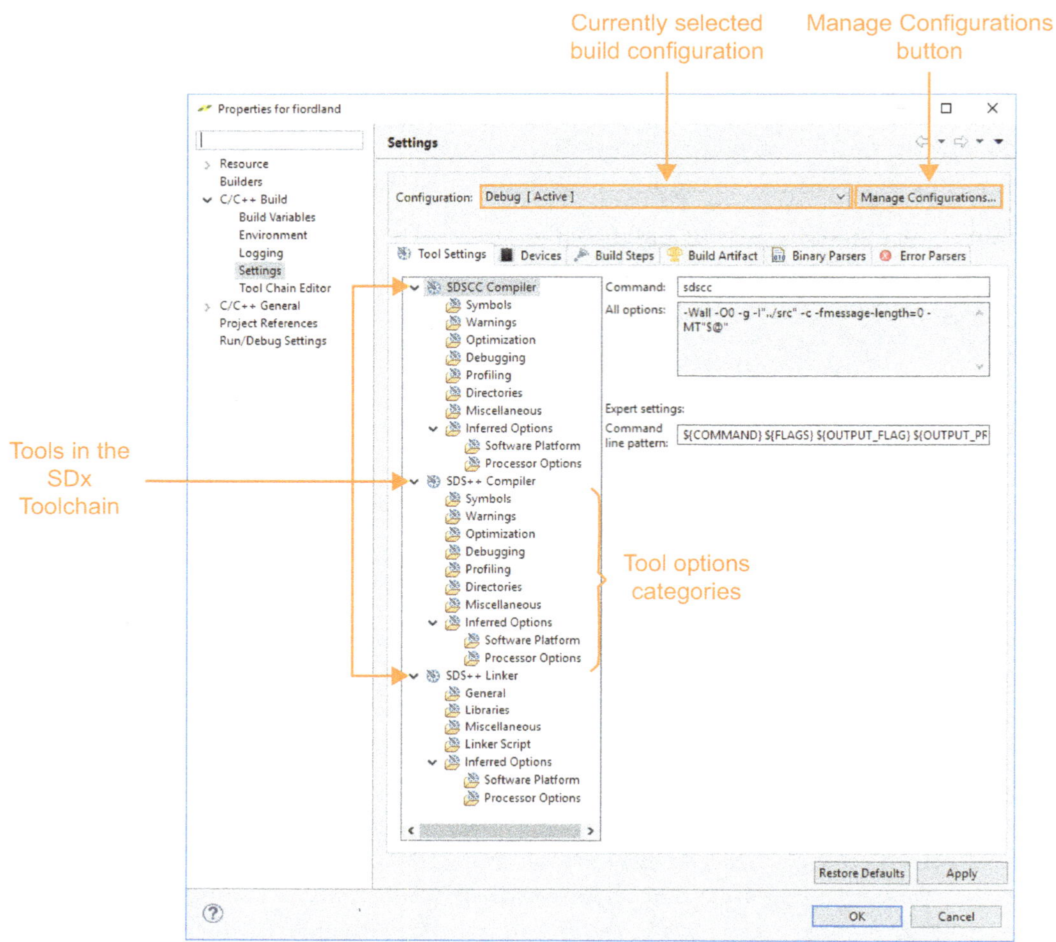

Figure 15.4: Project Properties dialogue box showing C/C++ build settings (as in SDx IDE 2017.2)

15.3 The Design Flow

Having become a little more familiar with SDx in the previous section, in this section we will take a closer look at how we might use it to design a system for a Zynq device. A diagram showing a typical design flow using SDx is shown in Figure 15.5. We will cover each step of this flow in greater detail in subsequent chapters, but for now we will take a brief overview.

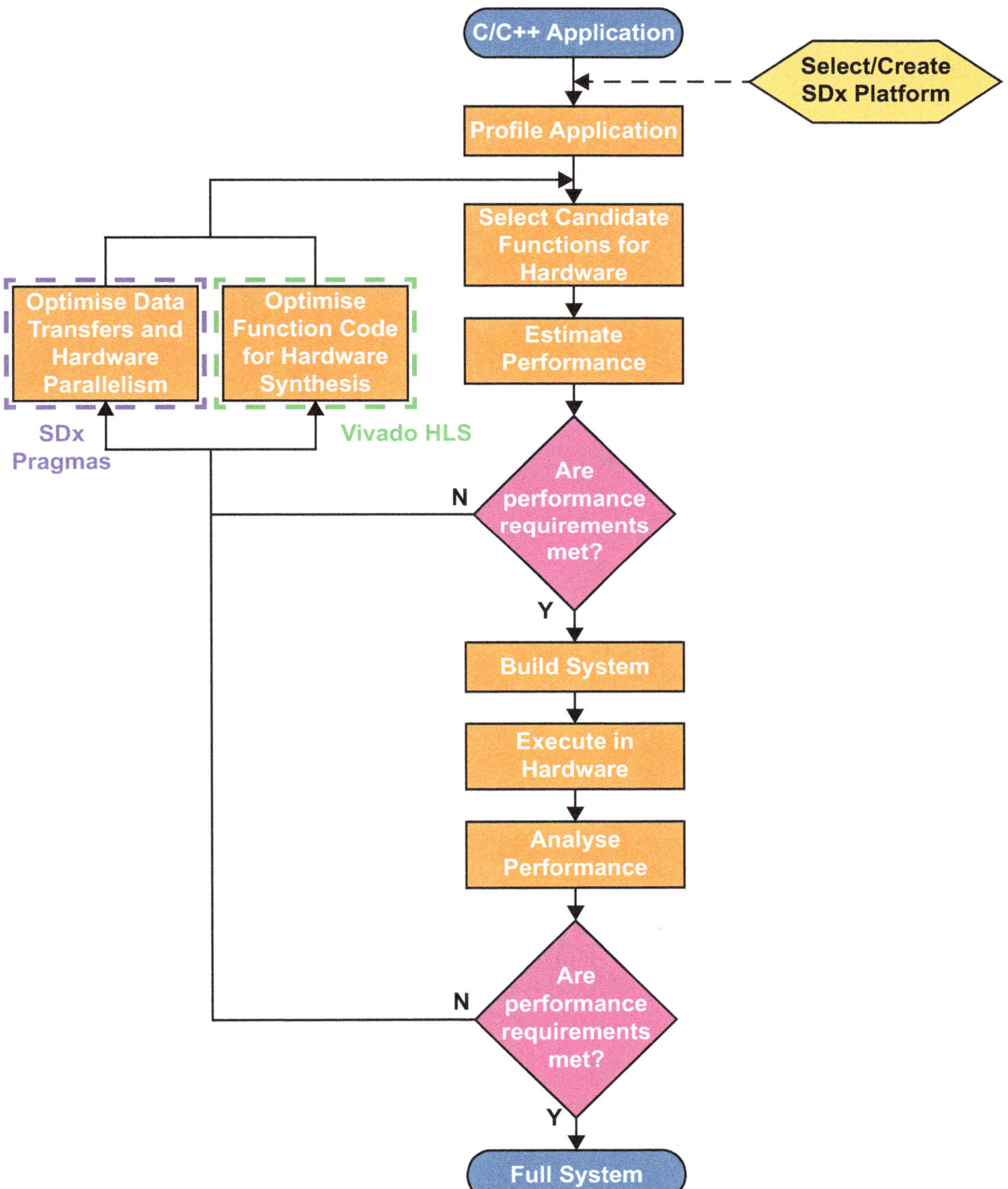

Figure 15.5: Typical design flow using SDx

15.3.1 SDx Platform

The input to the design flow is a C/C++ application that we wish to execute on the Zynq device. This should hopefully not come as a surprise given the foregoing discussion in this chapter. What is perhaps less likely to have been expected is the reference in Figure 15.5 to an SDx Platform. We have omitted any mention of SDx Platforms up until now for the sake of simplification, but now that we are looking at the design process in greater detail, we introduce them here.

An SDx Platform provides a basic system design that acts as the core of the SDx project implementation. It includes both hardware and software elements. On the hardware side, the SDx Platform defines:

- The PS configuration, including interfaces to the PL;
- Interfaces between the PL and external devices;
- Any IP blocks required as part of the SDx Platform.

On the software side, the SDx Platform defines:

- The target operating system(s);
- The boot files required to boot the system;
- Device drivers for the IP included in the platform design;
- Any software libraries to be linked into user applications;
- A root file system, if required.

Every SDx project must target an SDx Platform. The SDx project extends the base platform design by generating hardware coprocessors[1] from the functions marked for implementation in hardware, and interfacing these to the platform hardware via the data motion network. In software, the user application portion of the SDx project is able to make use of the target operating system, device drivers and software libraries provided by the SDx Platform. Conceptually, the SDx project can be thought of as being similar to the software stacks introduced in Chapter 12, with the application running on top of the SDx Platform. A diagrammatic representation of this idea is shown in Figure 15.6.

1. The term coprocessor has been used here to distinguish IP generated as part of an SDx project from IP that may be part of the SDx Platform targeted by the project.

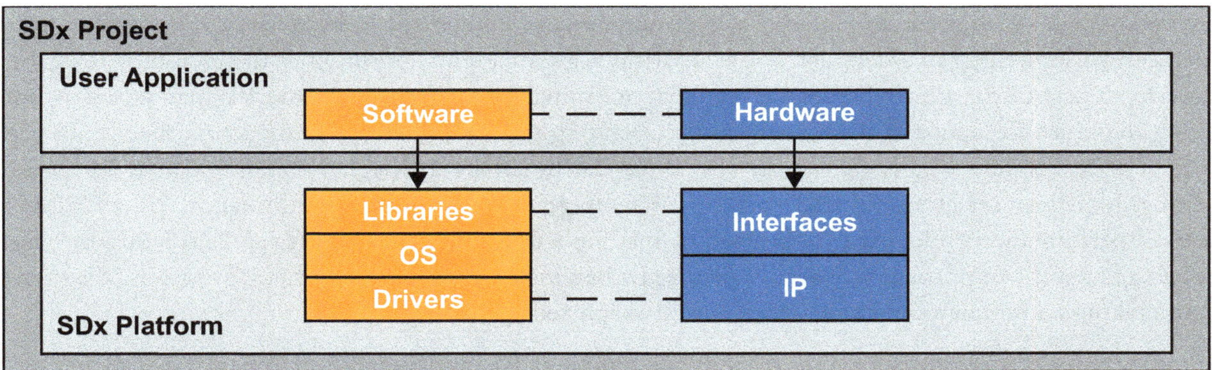

Figure 15.6: Relationship between SDx Platform and project

Several general purpose SDx Platforms supporting a range of development boards are supplied as part of SDx. For example:

- ZCU102;
- ZC702;
- ZC706.

Other SDx Platforms are also available, both for the boards listed above, targeted towards specific application areas, and for alternative boards. A list of the more prevalent SDx Platforms available is maintained in [12]. Using these general purpose platforms is a good way of beginning to develop SDx projects. They can also improve productivity for teams of developers by allowing SDx project development to begin while a custom SDx Platform is developed in parallel. In fact, depending on the specific requirements of the project, it may be found that a general purpose platform is adequate and that there is no need to develop a custom one. This would be the case when the PS configuration defined by the SDx Platform was appropriate for the project and no specific hardware, other than the coprocessors, was required in the PL.

A key point to note at this stage is that any interface to an external device that connects directly to the PL, i.e. one that does not go via the PS, must be implemented in the SDx Platform.

We will go into more detail on SDx Platforms including the steps required to create a custom one in Chapter 19. Further information is also available in [13].

15.3.2 Selecting Initial Candidate Functions for Hardware

At this point in the design flow, we have our C/C++ application and we have chosen an SDx Platform to build our project on. Next, we want to select which functions in our application to target for implementation in hardware.

Considerable performance improvements can be obtained by implementing some functions in hardware. However, being able to accurately predict which functions to implement in hardware to produce an optimal system can be something of a dark art. Some algorithms are inherently unsuited to the architecture of the PL and perform best in the PS. Other algorithms that are suited to acceleration in the PL may not, however, be limiting the execution speed of the system, and accelerating them will have little impact on the overall performance. The availability of PL fabric can also limit the functions a designer can implement in hardware, requiring judicious selection to trade-off resource usage against system performance. In addition, any perceived performance improvement obtained by moving a function between software and hardware must be weighed against the performance overhead incurred when moving data between the PS and PL. We therefore propose taking an iterative approach, which is well supported by SDx.

In order to make an initial selection of candidate functions for implementation in hardware, it is sensible to examine the performance of the C/C++ application running in software. There are a wealth of software profiling tools available and it is a relatively simple task to profile the application running on the host system. However, to gain a more accurate measure of the application's performance, we would ideally like to profile it while it is executing on the Zynq PS. SDx provides utilities for doing just this, the use of which is discussed in greater detail in Chapter 16. From the results of profiling, we are able to identify any computationally intensive hot-spots within the application. Acceleration of these functions is likely to have the greatest impact on overall system performance and, so, these should be considered initially as candidates for hardware acceleration.

15.3.3 Estimating System Performance

Just because a function is a computational bottleneck in the Zynq PS does not mean that it is appropriate for acceleration in the PL. It is also unlikely that the SDx tool-chain and Vivado HLS tools will infer the optimal system design without any further guidance from the designer. We therefore want to test whether our system, when utilising the hardware-accelerated functions, meets our design requirements. The most obvious way to do this would be to build the system, run it and measure the performance, and this is a perfectly valid approach. However, the time taken to build a system that uses customised hardware in the PL is not insignificant. With the iterative approach proposed here, developing an optimal system would be a frustratingly slow process indeed, if we had to perform a full system build each time we made a change to the functions implemented in hardware. Crucially, SDx allows an estimate of system performance to be made without requiring the entire build process to be performed. This can be helpful in providing an indication of the performance of a particular system configuration early in the design process, although it is not always possible to accurately determine the performance at such an early stage. System performance estimation is discussed in further detail in Chapter 16.

15.3.4 Optimising System Performance

In the case of performance estimation results not meeting our requirements, or where there appears to be scope for further performance improvements, we must make some changes to our SDx project in order to improve the performance. There are a few approaches that can be taken to tackle this problem:

- Optimise the software;
- Optimise the code for the functions implemented in hardware;
- Optimise data movement to and from the accelerators;
- Select alternative or additional functions for implementation in hardware.

In a similar manner to the creation of the data motion network, as described in Section 15.2, the Vivado HLS tool attempts to infer an optimal hardware implementation of functions selected for acceleration by analysis of the source code. If the application C/C++ code was not originally written with inference of hardware in mind, then it may be possible to refactor the code in a way that allows the Vivado HLS tool to infer a more efficient hardware architecture. In addition to refactoring the source code, the synthesis behaviour of the Vivado HLS tool can be explicitly directed through the specification of directives. These processes are explained in greater detail in Chapter 16.

Another source of potential performance improvement is in the data movement to and from the hardware accelerators. This is influenced by a number of parameters such as the system port or ports used; whether the memory allocated for the data is cacheable; the type of DMA engine used; the number of instances of a hardware accelerator available; and the use of pipelining when making multiple calls to an accelerator. These parameters can all be configured using the pragmas described in Table 15.3.

The final approach to improving system performance is to reconsider the functions selected for hardware implementation. If an additional function is accelerated in hardware, and the improvement in performance for that function is greater than any data transfer overheads introduced by moving it to the PL, then the overall system performance will be improved. However, if the performance improvement of accelerating a function in hardware is outweighed by the data transfer overhead associated with implementing this function in the PL, then the overall system performance can be improved by moving the execution of this function back to the PS. Additionally, as mentioned earlier, some functions are not appropriate for hardware acceleration. That is, the structure of these algorithms is inherently better suited to the architectures of the processors available in the PS than any custom architecture that could be implemented in the PL. For example, a function that is unavoidably sequential in nature may benefit from the higher processor clock frequencies available in the PS, and the system performance is therefore best served by implementing this function in the PS. Some software constructs cannot be easily mapped to hardware, such as loops whose bounds are not known at compile time, recursive function calls and system calls. Guidelines for coding software for hardware generation are provided in [8].

Once the system optimisations have been applied, the system performance can be estimated again, and this process can be performed iteratively until the user is satisfied with the estimated performance of the system.

15.3.5 Analysing System Performance

Now that we are satisfied with the estimated performance of our system, we want to build it fully and measure the actual performance. This can be done by profiling the application while it is executing on the

Zynq device in a similar way to the profiling used to select the initial candidate functions for hardware acceleration. There is a key difference between these two profiling applications, in that when profiling the application to select candidate functions for hardware acceleration, the *relative* amount of time spent executing each function is sufficient, whereas when measuring the performance of the optimised system we are likely to want to measure *absolute* execution times. Again, the profiling utilities available within SDx are discussed further in Chapter 16.

In the event that the measured performance of our system does not meet the design requirements, we must again seek to optimise the design. The process for doing this is the same as described earlier when optimising the system for performance estimation. This optimisation process is again performed iteratively, measuring the system performance each time, until the designer is satisfied with the measured system performance. At that point we have successfully built our hardware-accelerated system using SDx.

15.4 SDx Project Hierarchy

So far in our brief overview of SDx, we have looked at how it may be used to design a hardware-accelerated system and some of the tools that underpin it. During the course of a build, SDx and many of its underlying tools generate intermediate products and a wealth of information about their progress and status. Much of this data is retained in files within the SDx project hierarchy and these can provide valuable information to the user. It is therefore prudent to examine the structure of an SDx project and identify where we might access this data. We will begin our tour at the top-level of the SDx project hierarchy, journey down into its depths and, like any self-respecting tour guide, highlight the points of interest along the way.

Starting at the very top, the SDx project has a root directory named after the project and situated within the workspace. Figure 15.7 shows the expanded root directory of an example SDx project as it appears in the Project Explorer view of the SDx IDE. The first three sub-directories, namely *Binaries*, *Archives* and *Includes*, are perhaps not quite as they seem. If you were to browse within your file system to the SDx project root directory, you would not find them listed there. This is because they are not actually directories within the file system but are in fact virtual branches. Their purpose is to list the executables, libraries and header files, respectively, associated with the project. The next two directories listed in Figure 15.7 are build directories. These directories are created when the project is built, and they contain data generated during the build

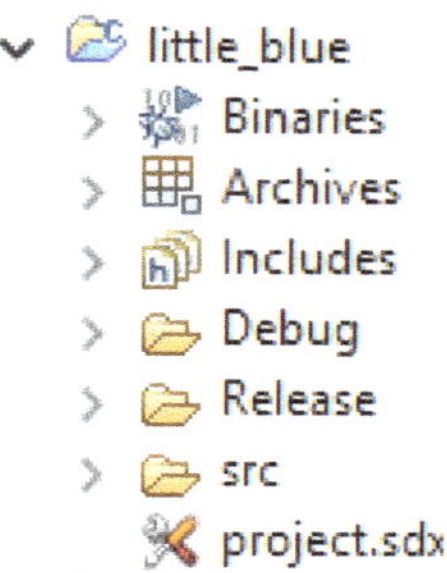

Figure 15.7: Example of an expanded SDx project root directory, as displayed in the Project Explorer view (as in SDx IDE 2017.2)

process. The directory is named after the build configuration used to build the project (we will delve a little deeper into build directories momentarily). The *src* directory simply contains the source files for the project. The final element contained within the root directory is the SDx project file. We have come across this file in passing previously when setting build options, but now we will stop to take a closer look.

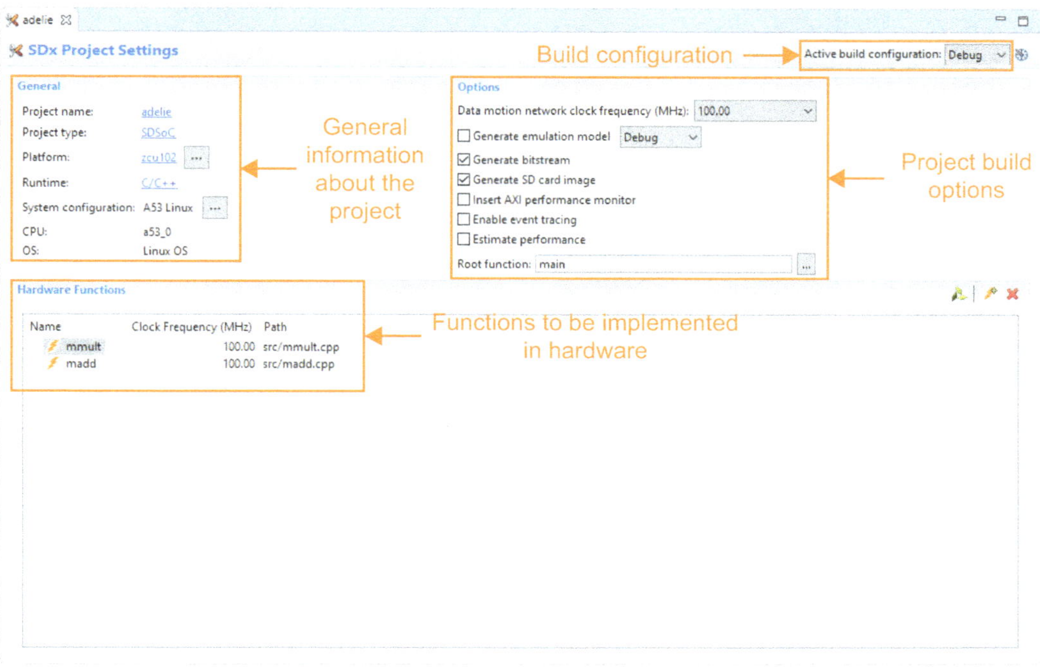

Figure 15.8: Example of an SDx project file (as in SDx IDE 2017.2)

When the SDx project file is opened within the SDx IDE it displays the settings for the project, as shown in Figure 15.8. Most of the settings in this file, with the exception of the *Active build configuration*, are divided into three sections under different headings. General project information is provided under the *General* heading. This includes a link to further information about the SDx Platform being targeted by the SDx Project, an example of which is shown in Figure 15.9. Under the *Hardware Functions* heading the user is able to choose which functions are implemented in hardware and their target clock frequencies. They are also able to launch the Vivado HLS GUI from under this heading. The data motion network clock frequency is specified under the *Options* heading, which also contains a set of build options, and it is possible for the user to specify the project application root function (note that this is not the same as the *perf root* function used for performance estimation).

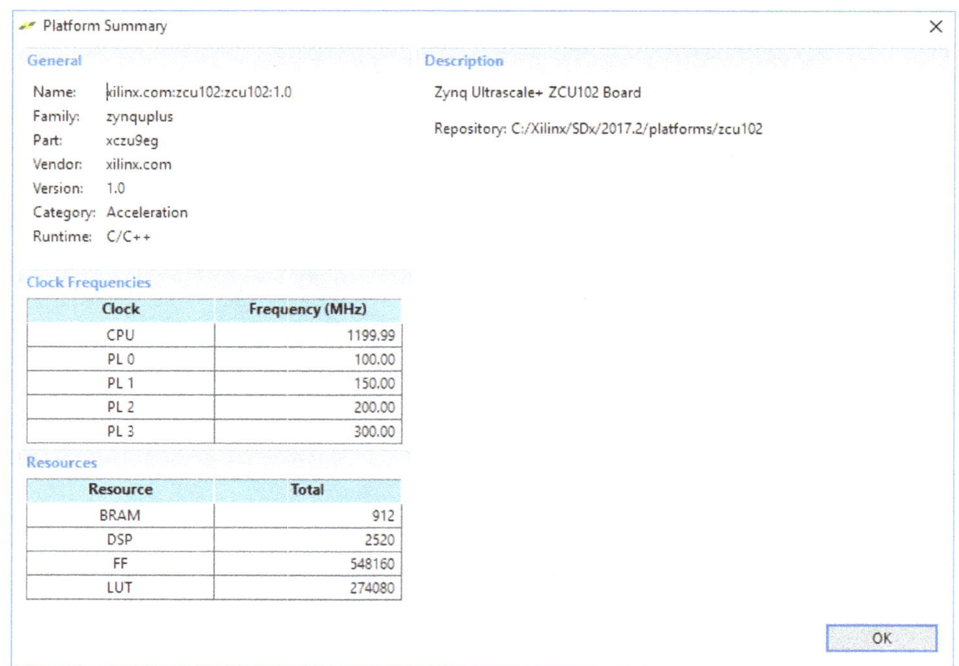

Figure 15.9: Example of the Platform Summary from an SDx project file (as in SDx IDE 2017.2)

Now, as promised earlier, we will continue our descent into the SDx project hierarchy by inspecting the contents of the build directory. An example of an expanded build directory is shown in Figure 15.10. Starting at the bottom of the build directory, there are three makefiles: *makefile*, *objects.mk* and *sources.mk*. These files are automatically generated by SDx and are used to invoke the build process. The *makefile* file is the top-level makefile and includes the other makefiles within the build directory to specify the targets for the build. The build directory also contains a copy of the bitstream (for programming the PL) and application executable (for running on the PS) generated for the project, i.e. the *<project>.elf.bit* and *<project>.elf* files respectively. The *src* directory contains copies of the compiled source files for the project. The precise contents of the *sd_card* directory vary depending on the targeted operating system, but will always contain the boot image, BOOT.BIN, and a text file that should explain the contents of the directory called README.txt. Copying the files contained within the *sd_card* directory to an SD card should allow the board targeted by the project to be booted from the SD card with the system built by the project. The final item to examine in the build directory is the *_sds* directory. This directory contains the reports and intermediate products from the tools within the SDx tool-chain and SDx itself. It therefore warrants a closer look.

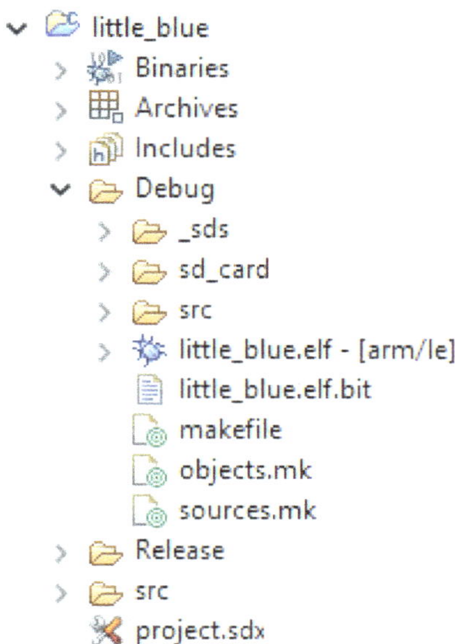

Figure 15.10: Example of an expanded build directory in an SDx project, as displayed in the Project Explorer view (as in SDx IDE 2017.2)

Figure 15.11 shows an example of an expanded *_sds* directory within an SDx project hierarchy. The first thing to mention is the fact that most of the sub-directories in the *_sds* directory appear to be greyed-out, with a line through their icons. This is perfectly normal and is simply the way the SDx IDE indicates that these items have been generated by one of the back-end tools invoked by the SDx tool-chain. The only sub-directories within the *_sds* directory that are generated by SDx itself are the *swstubs* and *trace* directories. It was mentioned in Section 15.2 that the SDx tool-chain replaces calls to hardware-accelerated functions in the application source code with calls to stub functions that handle the hardware setup and data transfers. The *swstubs* directory contains copies of the application source code modified to call these stub functions along with additional software files generated by SDx to support this. The *trace* directory, meanwhile, is used to contain data concerned with performing event tracing on the system, which we will discuss further in Chapter 18.

Continuing down the directory structure, the *iprepo* directory is simply a repository of the IP cores generated by the Vivado HLS tool from the application functions marked for hardware implementation. The *p0* directory contains files and sub-directories related to the Vivado IP Integrator (IPI) project used to create the system bitstream. This includes the actual Vivado IPI project, which can be independently opened and viewed in the Vivado tool, the reports and log-files generated by the Vivado IPI project, and the Tcl scripts used to instantiate the system design in Vivado IPI. The *reports* directory is fairly self-explanatory. It contains reports and log-files documenting the activity of the SDx tool-chain. These include the data motion network report, *data_motion.html*, which reports details of the data motion network inferred by the SDx tool-chain.

The log-files contained in this directory are especially invaluable for debugging problems with the system build. Although many useful messages are printed to the console view of the SDx IDE during system build, the log-files in this directory, in conjunction with the reports contained in the sub-tool specific directories, give a much more comprehensive picture of what is happening during the build process. Vivado HLS project files for the hardware accelerators are contained in the vhls directory. Each of the accelerators is generated using its own Vivado HLS project with the associated project files held within this directory. This directory also includes a log-file for the last run of the Vivado HLS tool and Tcl scripts used to create the hardware accelerators in the Vivado HLS tool.

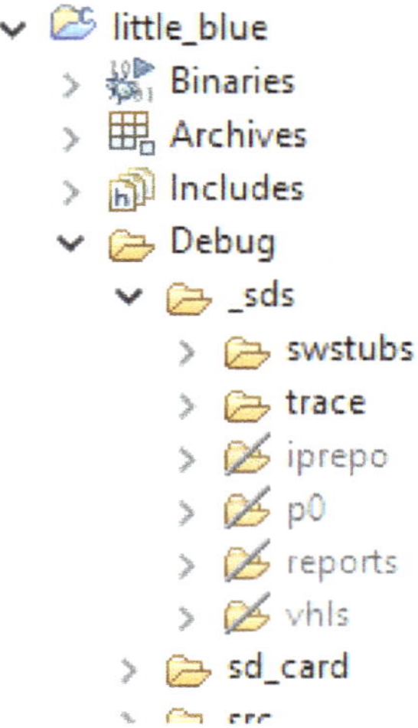

Figure 15.11: Example of an expanded _sds directory within the build directory of an SDx project, as displayed in the Project Explorer view (as in SDx IDE 2017.2)

This concludes our tour of the SDx project hierarchy. We have pointed out the major features of the hierarchy here, but it is always worth investigating the SDx project directories to become familiar with their structure, as there is a huge amount of useful information contained within them.

15.5 Chapter Review

This chapter has provided an introduction to using SDx for designing software-defined systems targeted for implementation on Zynq devices. The main advantages of using SDx over traditional development tools were outlined, and the SDx IDE and its underlying operation were presented. We have also suggested a workflow for using SDx. In subsequent chapters, we will describe the most pertinent features of this workflow in more detail, with the intention that the reader be able to use SDx to develop Zynq-targeted systems.

15.6 References

Note: All online sources last accessed March 2019.

Some of the Xilinx documentation referred to below has version-specific URLs. If you are working in a newer version of the tools, check for updates on the Xilinx website, or try adjusting the link according to your version.

[1] Eclipse Foundation, Inc., "Eclipse CDT" webpage.
Available: https://eclipse.org/cdt/

[2] Eclipse Foundation, Inc., "Help — Eclipse Platform" webpage.
Available: https://help.eclipse.org

[3] Xilinx, Inc., "Vivado Design Suite User Guide: High-Level Synthesis", UG902 (v2017.2), June 2017.
Available: https://www.xilinx.com/support/documentation/sw_manuals/xilinx2017_2/ug902-vivado-high-level-synthesis.pdf

[4] Xilinx, Inc., "SDx Development Environment Release Notes, Installation and Licensing Guide", UG1238 (v2017.2), August 2017.
Available: https://www.xilinx.com/support/documentation/sw_manuals/xilinx2017_2/ug1238-sdx-rnil.pdf

[5] Linaro, "Linaro — Leading collaboration in the ARM Ecosystem" webpage.
Available: http://www.linaro.org/

[6] Xilinx, Inc., "Vivado Design Suite User Guide: Getting Started", UG910 (v2017.2), July 2017.
Available: https://www.xilinx.com/support/documentation/sw_manuals/xilinx2017_2/ug910-vivado-getting-started.pdf

[7] Xilinx, Inc., "Zynq-7000 All Programmable SoC Software Developers Guide", UG821 (v12), September 2015.
Available: http://www.xilinx.com/support/documentation/user_guides/ug821-zynq-7000-swdev.pdf

[8] Xilinx, Inc., "SDSoC Environment User Guide", UG1027 (v2017.2), August 2017.
Available: https://www.xilinx.com/support/documentation/sw_manuals/xilinx2017_2/ug1027-sdsoc-user-guide.pdf

[9] Xilinx, Inc., "SDSoC Environment Tutorial: Introduction", UG1028 (v2017.2), August 2017.
Available: https://www.xilinx.com/support/documentation/sw_manuals/xilinx2017_2/ug1028-sdsoc-intro-tutorial.pdf

[10] B. Gough, *An Introduction to GCC*, Network Theory Ltd. Bristol, 2005.

[11] Free Software Foundation, Inc. "Using the GNU Compiler Collection (GCC): Option Index" webpage.
Available: https://gcc.gnu.org/onlinedocs/gcc/Option-Index.html

[12] Xilinx, Inc. "SDSoC Development Environment: Boards, Kits and Modules" webpage.
Available: http://www.xilinx.com/products/design-tools/software-zone/sdsoc.html#boardskits

[13] Xilinx, Inc., "SDSoC Environment Platform Development Guide", UG1146 (v2017.2), August 2017.
Available: https://www.xilinx.com/support/documentation/sw_manuals/xilinx2017_2/ug1146-sdsoc-platform-development.pdf

Chapter 16

System Profiling and Acceleration with SDx

Having taken a high-level overview of the system design process with SDx in Chapter 15, we now move on to examine in greater detail the steps involved in that process. In this chapter, we are going to consider ways to profile and measure system performance, as well as strategies for accelerating functions in hardware.

16.1 System Profiling

In Chapter 15, we outlined three main tasks where we would wish to estimate, measure or profile system performance. These were:

- Profiling the software-only system to identify candidate functions for hardware acceleration;
- Estimating system performance prior to performing a full system build;
- Measuring the performance of the built system.

We will broadly deal with each of these tasks in turn in this chapter, although there is some unavoidable overlap between them.

16.1.1 Software-Only System Profiling

One of the first tasks when implementing a software application on a Zynq device is to identify the functions to be implemented in hardware. Determining the functions that constitute the largest proportion of the application's execution time in software can aid in the selection of functions for hardware acceleration. We would therefore wish to profile the software application as it executes on the Zynq PS.

There are a number of ways to do this. Here, we will briefly outline some of the options available with the SDx IDE and then go into more detail about our preferred option.

Perhaps the most simple method conceptually is to add calls to a function that inspects the value of a free-running timer to measure the time spent in a function. This is easy to implement, although there is a performance overhead associated with the calls to the timer inspection function and only basic profiling information is obtained. It also requires manual modification of the source code. This method is more suited to the measurement of absolute system performance and we will revisit it in Section 16.1.3.

Another option is to use the GNU profiler, ***gprof***. To use this, the application must be compiled and linked using the `-pg` flag. Details of how to do this are given in Section 15.2.6. The executable created when this flag is specified contains additional instructions to record the time spent in each function. When the executable is run, the profiling data is written to a file called *gmon.out* [1], the contents of which can be visualised in the SDx IDE. The advantages of this method are that it provides quite sophisticated profiling data compared to the previous method. It includes both absolute and relative profiling measures; that is it provides measurements of function execution times in both real time units and as a proportion of the overall execution time. The process is also automated once the flag has been specified to the SDx tool-chain and does not require the source code to be modified. The downside is that it is intrusive since the additional instructions included in the executable file influence the performance of the application.

The final method that we will discuss is our preferred option, the Target Communication Framework (TCF) Profiler. This makes use of the performance monitoring unit in the PS to profile the application as it executes. The advantages of this method are that it does not alter the application executable and does not require any specific options to be set when building the system. A limitation of this method compared with using ***gprof*** is that it only provides relative profiling measurements. This is a fairly minor disadvantage given that, at this point in our design flow, we are only really interested in finding the most computationally intensive functions in the application and relative profiling measurements are perfectly adequate for achieving this. On balance therefore, we feel that the TCF profiler is the best option for our purposes and we will now concentrate on this method.

Target Communication Framework

Before we launch into our exposition of using the TCF profiler, we should perhaps address the small elephant in the room: our surreptitious introduction of the Target Communication Framework or TCF. The TCF is something that we will make use of at various points in our design flow and so it is worthwhile taking a brief aside here to introduce it properly. The TCF is a network protocol that is part of the Eclipse project. It concerns itself with simplifying communication between tools running on the host platform and the devices they are targeting during embedded systems development. It aims to do this by providing a universal framework for tools and targets to communicate regardless of vendor or transport protocol, thus removing the need to setup and maintain individual connections for each tool [2].

From within the SDx IDE, the TCF is largely transparent. The main exception to this being that we need to explicitly establish a connection to the TCF agent on Zynq targets running Linux as the OS. The TCF agents included in the Linux implementations used in the SDx Platforms supplied with SDx all use TCP/IP as the transport protocol. We will therefore focus on establishing a connection using TCP/IP. To do this, we must first connect the Zynq target to a TCP/IP network where it is contactable by the host running SDx. The target must be booted into Linux and assigned an IP address. Within the SDx IDE, we can then configure the *Linux TCF Agent* in the Target Connections view of the SDx perspective with the IP address of the Zynq target. By default, the Linux TCF agents supplied with SDx are configured to use port 1534. Once the connection to the TCF agent is correctly configured, we are ready to use the tools that make use of the TCF with the Zynq target running Linux.

By contrast, Zynq targets running the Standalone or FreeRTOS OSs on the platforms supplied with SDx use the JTAG interface for the TCF connection. Other than connecting the cable between the host system and the target JTAG port, there is usually no manual configuration required.

TCF Profiling

Now that we have been more formally introduced to the TCF, we will resume our discussion of the TCF profiler. First, we must compile and link our software application for execution on the Zynq's PS. We may use any build configuration that sets the `-g` compiler flag to enable the use of debugging tools with the compiled application. The default ***Debug*** build configuration is suitable for this. As we have not yet selected any functions for implementation in hardware, the build time is simply the time taken to compile our application and so should be fairly short. Once the build completes we can set up the TCF connection to our target device. If using the Linux OS, the boot files generated in the *sd_card* sub-directory of the build directory should be used to boot the device into Linux to configure the TCF connection. Right-click the project directory in the Project Explorer view and select *Debug As > Launch on Hardware (SDSoC Debugger)* to launch the application on the target device using the SDSoC debugger. Launching the application using the debugger means that the application will initially launch and then suspend as it enters the main function. We can take this opportunity to switch to the Debug perspective by selecting the *Open Perspective* button and choosing the Debug perspective, as shown in Figure 16.1. The TCF profiler has its own view within the SDx IDE which can be opened from the *Window* menu by selecting *Show View > Other...*, as shown in Figure 16.2. This opens the Show View dialogue box, from whence the *Debug* folder can be expanded and the *TCF Profiler* view can finally be selected, also shown in Figure 16.2. The TCF profiler can be started by clicking the start button in the TCF Profiler view, as shown in Figure 16.3. This opens the Profiler Configuration dialogue box, also shown in Figure 16.3, which allows the behaviour of the profiler to be configured. Table 16.1 summarises the available options.

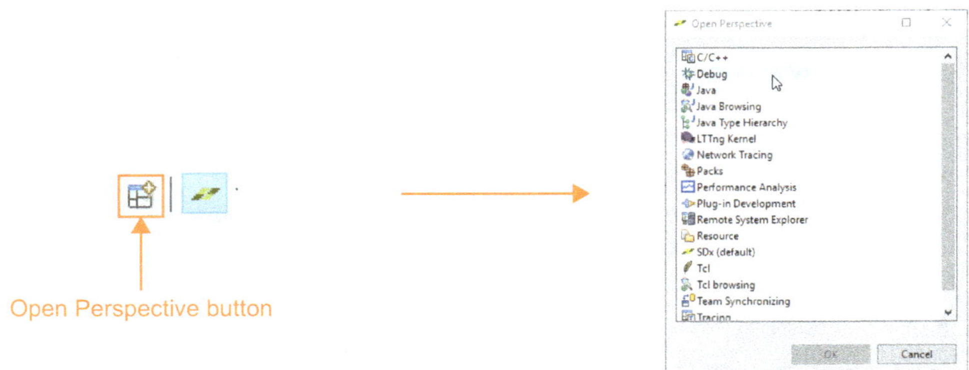

Figure 16.1: Opening the Debug perspective (as in SDx IDE 2017.2)

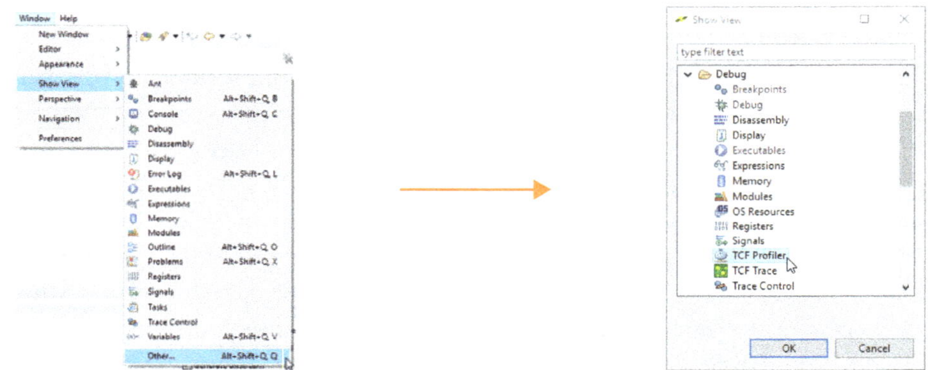

Figure 16.2: Opening the TCF Profiler view (as in SDx IDE 2017.2)

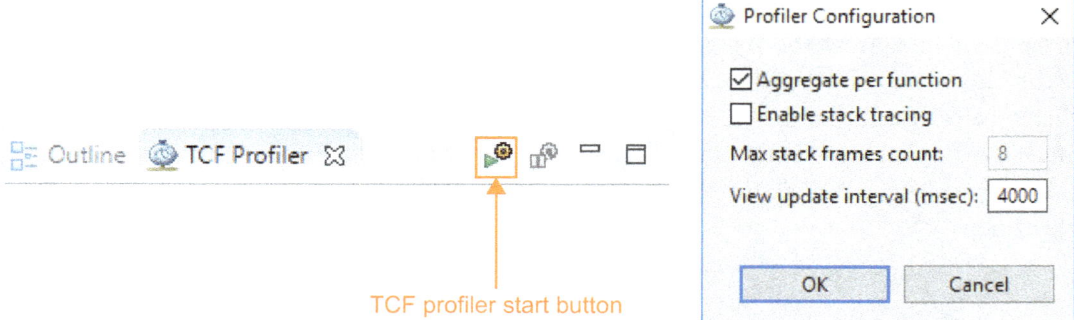

Figure 16.3: Starting and configuring the TCF profiler (as in SDx IDE 2017.2)

Table 16.1: Summary of TCF profiler configuration options

Option	Description
Aggregate per function	Collects calls to the same function together.
Enable stack tracing	Implements thread stack back tracing to enable determination of parent/child relationships between functions.
Max stack frames count	Specifies the number of stack frames to count back when performing stack tracing.
View update interval	Specifies the interval at which the TCF Profiler view is updated with the profiling data (note: this is not the profiler sampling period).

Once the profiler has been configured, the suspended application running on the Zynq device can be resumed by clicking the resume button in the Debug perspective, shown in Figure 16.4. The TCF Profiler view should then begin to display the captured profiling data. Figure 16.5 shows an example of how the profiling data is displayed in the TCF Profiler view. A summary of the profiling data captured is given in Table 16.2.

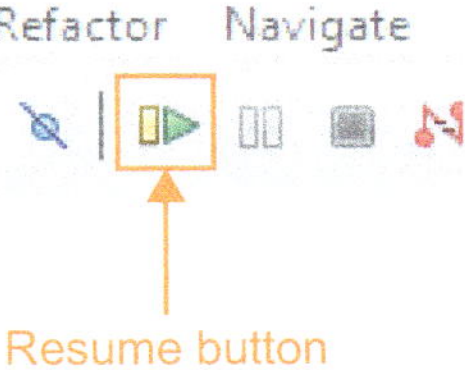

Figure 16.4: Button to resume suspended application (as in SDx IDE 2017.2)

The key parameter to consider when examining the TCF profiler data for compute intensive functions is the *% Exclusive* parameter. This is the proportion of samples captured by the profiler while the application was specifically executing this function, giving an estimate of the proportion of an application's execution time occupied by this function. This differs from the *% Inclusive* parameter since the *% Inclusive* value includes samples captured while the application is executing any functions called by the function as well. The *% Inclusive* parameter can therefore have a high value for functions that call compute intensive functions, but do not themselves perform a lot of computation. The *main* function in the data shown in Figure 16.5 is an example of such a function. Helpfully, the TCF Profiler view also provides the *Called From* and *Child Calls* sections to display the relationship between the selected function and other functions in the application.

Using the *% Exclusive* data from the TCF profiler we are able to identify the most compute intensive functions within our application. If these functions can be accelerated, they are likely to bring the greatest improvement in system performance and are, therefore, good candidates to target for hardware implementation.

Outline | TCF Profiler

Profiler running, 591 samples

Address	% Exclusive	% Inclusive	Function	File	Line
0010d0a8	.000	100	_start		
00100ed4	.000	99.6	main	main.cpp	148
00100b4c	.000	99.6	mmult_test	main.cpp	110
00101130	55.4	55.4	mmult	mmult.cpp	54
0010083c	37.2	37.2	mmult_golden	main.cpp	76
001006dc	2.53	3.38	init_arrays	main.cpp	63
0010094c	1.52	1.52	madd_golden	main.cpp	89
001005e4	1.35	1.35	madd	madd.cpp	44
00112a80	.846	.846	rand		
00100a20	.677	.677	result_check	main.cpp	98
0010d02c	.000	.338	_write		
00187758	.000	.169	_ZSt4endlIcSt11char_traitsIcEERSt13basic_ostreamI...		
0018744c	.000	.169	_ZNSo3putEc		
0011de4c	.000	.169	__swbuf_r		
0010d510	.000	.169	__sflush_r		
00000020	.000	.169			
0011feb0	.000	.169	_write_r		
0012c470	.000	.169	xlnkFreeBuf		
00131ec0	.000	.169	Xil_SetTlbAttributes		
0010d214	.169	.169	XUartPs_SendByte		
00131a38	.169	.169	Xil_DCacheFlush		

Called From

Child Calls

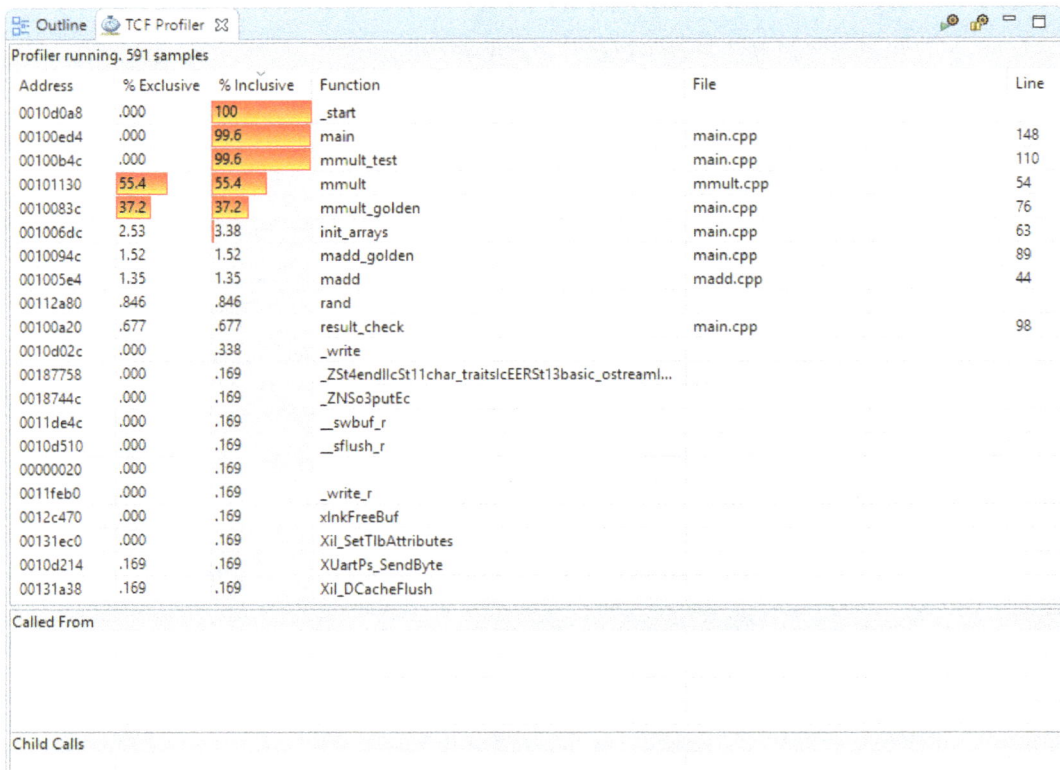

Figure 16.5: Example of TCF Profiler view displaying captured profiling data (as in SDx IDE 2017.2)

Table 16.2: Summary of data captured by TCF profiler

Data	Description
Address	Location of function being sampled in memory.
% Exclusive	Percentage of total samples captured by the profiler where this function was executing..
% Inclusive	Percentage of total samples captured by the profiler where this function was executing, including samples captured when functions called by this function were executing.
Function	Name of function being sampled.
File	File containing the function.
Line	Line number at which the function appears within the file.

16.1.2 Estimating Relative Performance

Before committing to a full system build for a system that includes hardware accelerators, it is prudent to first estimate the performance we are likely to see compared with the software-only application by selecting the *Estimate Performance* option. This may seem like an unnecessary step in the design flow, but by getting a relatively quick estimate of the performance of our system, we can avoid some of the expense of performing full system builds that produce unsatisfactory results. Over the course of a design cycle this is likely to result in an overall time saving.

The performance estimate can be obtained using any build configuration and simply requires that the *Estimate Performance* option is selected in the SDx project file. The build is then invoked in the usual manner and typically only takes a few minutes to complete. This is because only the initial stages of the normal build process are performed for the hardware portion of the system design. The objective of this partial build is to obtain estimates for the latencies of the hardware accelerators synthesised by the Vivado HLS tool and estimates of the time required to transfer data to and from these accelerators. These values are used to estimate the performance of the system when it uses the hardware accelerators. If estimates for the hardware accelerator latencies and data transfer times cannot be determined precisely at compile time, SDx assumes the worst case values, which may be pessimistic when compared to the actual performance of the hardware-accelerated system. The build also compiles the application and generates the boot files necessary to launch a software-only implementation on the Zynq device.

SDx can generate a *Performance Estimation Report*, which includes an estimate of the resources required to implement the hardware accelerators and an estimate of the number of clock cycles latency for the hardware accelerators based on those reported by the Vivado HLS tool. It also allows the performance of a software-only implementation to be measured, as shown in Figure 16.6. This uses the TCF for communication with the target device, hence the host and target systems need to be connected as described on page 406.

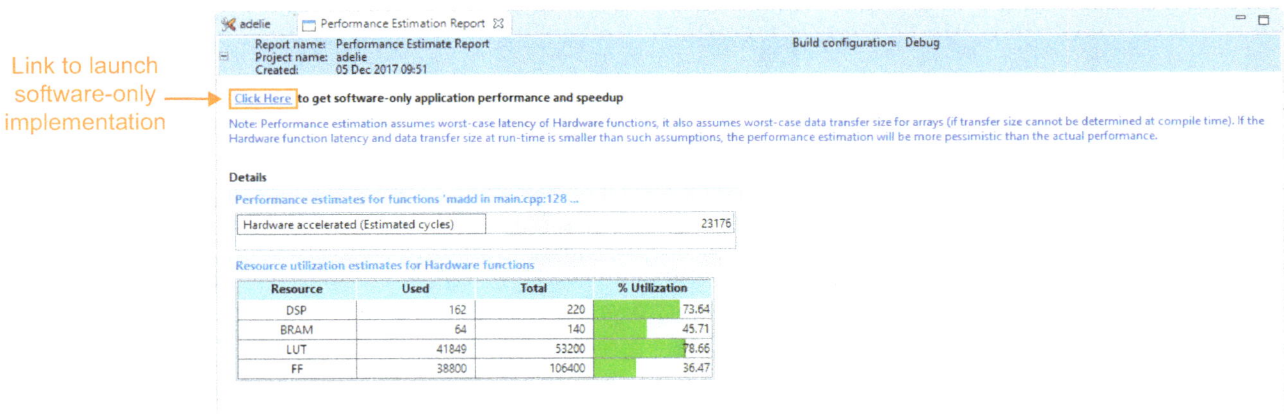

Resource	Used	Total	% Utilization
DSP	162	220	73.64
BRAM	64	140	45.71
LUT	41849	53200	78.66
FF	38800	106400	36.47

Figure 16.6: Example of Performance Estimation Report prior to running software-only implementation (as in SDx IDE 2017.2)

On completion, the *Performance Estimation Report* is updated to reflect the software-only implementation measurements and a comparison is made with the estimated performance of the system using the hardware accelerators. Under *Details*, a comparison is made between the measured software performance and the estimated hardware performance specifically for the functions marked for hardware acceleration. Under *Summary*, the comparison is made for the top-level function.

An example *Performance Estimation Report* is shown in Figure 16.7. It can be seen from the *Details* section of this report that the functions marked for implementation in hardware are likely to execute many times faster than if they were implemented in software. However, it can also be seen from the *Summary* section of the report that the effect on the execution time of the system as a whole is estimated to be much less dramatic. This will be due to a combination of the proportion of the overall system execution time that the accelerated function represents and the overheads of calling functions implemented in hardware.

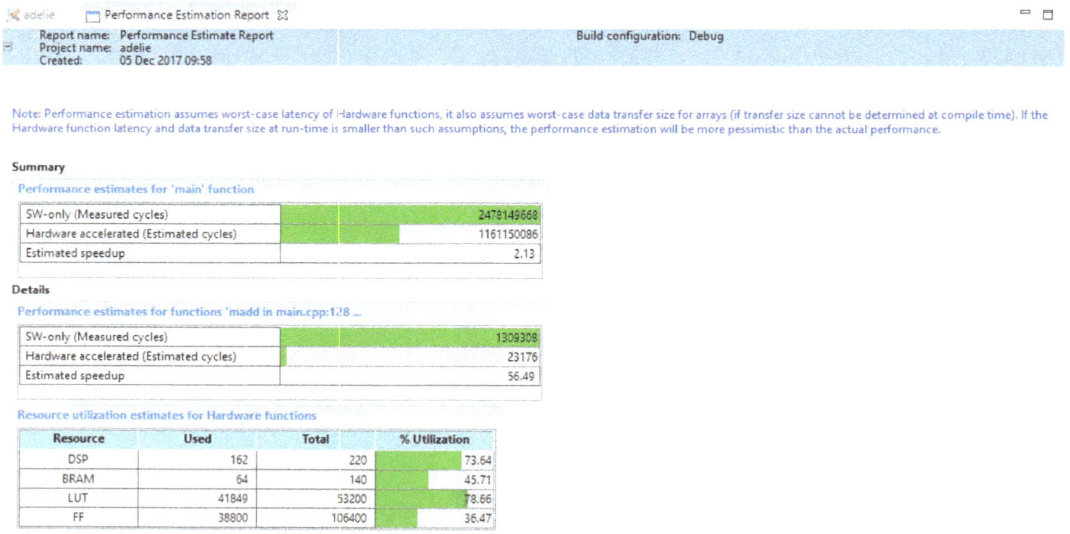

Figure 16.7: Example of Performance Estimation Report after running software-only implementation (as in SDx IDE 2017.2)

The performance estimation flow is a relatively quick way of assessing the performance improvement achieved in a system utilising hardware accelerators and the cost of this improvement in terms of resource usage. Although it may seem like an additional and unnecessary step, it is likely to improve productivity overall by avoiding full system builds of unsatisfactory designs.

16.1.3 Measuring System Performance

Once we have built our system, we want to test whether its performance meets the requirements. While relative measurements were satisfactory in identifying computational hot-spots during system profiling, in assessing whether the performance of the system meets the requirements, absolute measurements are more appropriate. This precludes the use of the TCF profiler, since it provides only relative measurements.

An alternative that we discussed briefly in Section 16.1.1 is ***gprof***. This does provide absolute performance measurements and so is a viable option for this task. The functions targeted for hardware acceleration will not appear directly in the profiler results, however, the stub functions generated by the SDx tool-chain to handle data transfers to and from the hardware accelerator can act as surrogates. There is one note of caution to sound when taking this approach; for any hardware-accelerated function that makes use of the `async()` pragma, described in Table 15.3 on page 387 and [3], the stub function returns once the hardware accelerator and data transfers have been setup and not when the hardware accelerator has finished execution. Accurately determining the execution time of a hardware accelerator in this situation can be problematic. One way to alleviate this problem is by monitoring the performance of the system in greater detail. Methods for doing this are discussed in Chapter 18.

Another way to alleviate this problem in some circumstances, and that we also mentioned in Section 16.1.1, is to inspect the value of a free-running timer at appropriate points during the application's execution. This allows the designer to precisely tailor the points of the application between which they wish to measure the absolute performance. In the case where we want to measure the execution time of a hardware-accelerated function making use of the `async()` pragma, we could obtain an estimate by simply inspecting the timer immediately prior to calling the hardware accelerator stub function and then inspecting it again immediately following our call to either the `wait()` pragma, described in Table 15.3, or one of the API functions `sds_wait()` or `sds_try_wait()`, described in Table 15.4. The reason this is only an estimate of the hardware accelerator execution time is that the hardware accelerator may have completed execution sometime before the application polls it to check, depending on the structure of the application. There are a couple of disadvantages with using this method. Firstly, there is a small performance overhead associated with inspecting the free-running timer, although the other profiling methods presented here also incur performance overheads. The second, and perhaps more significant, disadvantage is that this method requires the application source code to be altered, which none of the other profiling methods we have discussed require. This method does have its advantages too though. Alongside giving the system designer the ability to specify the portions of the application they wish to measure the performance of, it provides a measure of absolute performance as the free-running timer operates at the same frequency as the application processing unit and is relatively simple to implement. This last point is particularly true given that the ***sds_lib*** library provides a function to perform the inspection of the free-running timer: `sds_clock_counter()`.

The methods described here are good for measuring the overall performance of the built system and comparing this to our requirements or the performance of the software-only system. We may also be interested in more detailed analysis of the performance of our system, such as hardware accelerator execution times, data transfer times and CPU workload. As alluded to earlier, methods for performing that level of analysis are discussed in Chapter 18.

16.2 Software Acceleration using Programmable Logic

Now that we have a clearer idea of how to determine which functions in our system to target for hardware acceleration, we can begin to look in greater detail at how we can optimise the performance of our system with functions implemented in the PL. This is the subject that we will deliberate over in this section.

There are two main points of attack when optimising a system with functions implemented in the the PL: the performance of the IP core implementing the function itself and the integration of this IP core into the system. These can, to an extent, be dealt with separately and a two-pronged approach can be adopted; a pitch-fork of optimisation, if you will. As touched upon in Section 15.3.4, code refactoring and pragma specification can be used to direct system synthesis to a more optimal solution. In the case of the IP core, the code structure and pragmas are interpreted by the Vivado HLS tool, while for the integration of the IP core into the system, the interpretation is performed by the SDx compiler. The slight quibble about being able to deal with these two aspects entirely independently is because the interfaces of the IP core will inevitably influence its integration within the system. Generally speaking, the interfaces for the IP core are determined by the SDx compiler rather than the Vivado HLS tool. The exception to this rule is when the SDx compiler fails to generate suitable hardware interface directives, in which case Vivado HLS directives or pragmas can be used to direct synthesis of the appropriate interface types [2]. We will now examine the two tines of our pitch-fork in turn.

16.2.1 IP Core Optimisation

Firstly, we will deal with optimising the IP core itself. Perhaps somewhat counter-intuitively for this chapter, we are going to suggest that you do not use SDx, at least initially, to optimise the IP core. Instead, we would recommend using the Vivado HLS tool directly. The Vivado HLS GUI is also based on the Eclipse IDE and so has a similar appearance to the SDx IDE. The main advantage of using the Vivado HLS tool is that it allows the creation and comparison of multiple solutions in a single project, which greatly aids the efficient and systematic exploration of the available optimisation options.

The Vivado HLS tool infers a hardware architecture based on the C/C++ code. There are, however, limitations to its ability to infer an optimal architecture and the designer may be required to provide some additional assistance. In a similar way to when targeting code for execution on specific processor architectures, the C/C++ code may need to be refactored to enable the Vivado HLS tool to infer a more optimal architecture. The behaviour of the Vivado HLS tool can also be explicitly directed by the use of pragmas inserted directly into the source code or by the specification of directives in a separate Tcl file. In the interests of brevity, we will not enter into an extensive discussion on using the Vivado HLS tool here. Particularly useful reviews on optimising hardware synthesis using Vivado HLS, including coding styles and the use of pragmas and directives, are available in [1] and [2].

When the SDx project is built, a Tcl script is generated for each function marked for hardware implementation, which is used to invoke the Vivado HLS tool to synthesise hardware for the C/C++ function. These scripts are contained within the *_sds/vhls* subdirectory of the project build directory and use the naming convention *<function_name>_run.tcl*. An example of such a script is shown in Figure 16.8. These scripts open a Vivado HLS project for the function of interest and configure it to target the specific Zynq device and clock period determined from the SDx Platform. A directives Tcl file is sourced by these scripts to direct the behaviour of the Vivado HLS tool prior to running hardware synthesis for the target function. By default, the directives file is generated by SDx subsequent to data motion network generation and includes directives to specify appropriate interfaces to connect the IP core to the data motion network. These directives Tcl files are also contained in the *_sds/vhls* subdirectory of the project build directory and use the naming convention *<function_name>.tcl*.

```
open_project mmult
set_top mmult
add_files C:/spheniscidae/chinstrap/src/mmult.cpp -cflags "-IC:/spheniscidae/chinstrap/src -Wall -O0 -g
    -fmessage-length=0 -D __SDSCC__ -I C:/Xilinx/SDx/2017.2/aarch64-linux/include -IC:/spheniscidae/
    chinstrap/src -D __SDSVHLS__ -D __SDSVHLS_SYNTHESIS__ -I C:/spheniscidae/chinstrap/Debug -w"
open_solution "solution" -reset
set_part { xczu9eg-ffvb1156-2-i }
# synthesis directives
create_clock -period 10.000100
set_clock_uncertainty 27.0%
config_interface -m_axi_addr64
config_rtl -reset_level low
source C:/spheniscidae/chinstrap/Debug/_sds/vhls/mmult.tcl
# end synthesis directives
config_rtl -prefix a0_
csynth_design
export_design -ipname mmult -acc
exit
```

Figure 16.8: Example of a Tcl script used to invoke Vivado HLS tool to synthesise hardware for a C/C++ function

As mentioned above, the designer usually has the option to direct the behaviour of the Vivado HLS tool either by entering pragmas into the source code directly or by providing a separate Tcl file containing the directives. When optimising code for use in an SDx project, we would recommend using the former, where possible. In this case, once the C/C++ code has been optimised using the Vivado HLS tool, it can simply be imported into the SDx project as a source file and is ready to use. The disadvantage of this method is that it requires the modification of the source code with the inclusion of the pragmas. In this context, however, this is perhaps not such a significant disadvantage given the likelihood of the code requiring refactoring and the inclusion of SDx-specific pragmas anyway.

It is worth reiterating at this point that care should be taken if manually specifying Vivado HLS pragmas to direct the synthesis of interfaces for any functions marked for hardware implementation in an SDx project. If Vivado HLS interface pragmas are specified, SDx will not generate interface directives for the Vivado HLS tool following data motion network generation. Instead, the Vivado HLS tool will synthesise interfaces based on the user-specified pragmas and it is then the responsibility of the user to confirm the compatibility of the synthesised interfaces with the generated data motion network. The user is still able to exercise influence over the synthesised interfaces without needing to use Vivado HLS-specific pragmas. This can be done using alternative pragmas provided by SDx, which we will discuss further in Section 16.2.2.

It is possible to keep the Vivado HLS directives separate from the source code by specifying a Tcl file containing the necessary directives. An alternative directives file can be specified by right clicking the function marked for hardware implementation under the *src* directory in the Project Explorer and selecting *Vivado HLS > Specify directive TCL file.*

It is important to note that any Vivado HLS directives file manually specified in an SDx project is used as an alternative to the directives file normally generated by SDx. The command to source the automatically generated directives file in the *<function_name>_run.tcl* script is replaced with a command to source the specified directives file. This means that the user is required to direct the synthesis of interfaces for that function and to ensure their compatibility with the generated data motion network.

16.2.2 Optimising IP Core Integration

Secondly, we consider how to optimise the integration of the IP core implementing our function with the rest of our system. The integration of the IP core is centred on the data motion network generated by the SDx compiler and discussed in Section 15.2.3. As mentioned earlier, and similarly to the Vivado HLS tool, the SDx compiler attempts to infer an optimal data motion network through an analysis of the C/C++ code. Again, there are limitations to its ability to infer an optimal data motion network and so, the facility is provided for the user to aid the process by inserting pragmas into the source code to direct its behaviour. The available SDx pragmas are summarised in Table 15.3 on page 387 and we will now briefly outline the effects of some of these pragmas on data motion network generation.

As we discussed in Section 15.2.3, broadly, we are able to consider data motion network generation in three parts:

- IP interface(s);
- Data mover(s);
- System port(s).

The pragmas that we can use to direct data motion network generation specifically direct the generation of these components by providing additional information about the data consumed or produced by the IP core.

In the case of an IP interface being synthesised for an array type function argument, either a RAM interface or streaming interface can be generated. The type of interface to be generated depends on how the array data is accessed by the IP core. A random access interface would, of course, work for any access pattern, however, if the data is accessed in a strictly sequential manner, a streaming interface could be used and requires fewer PL resources. The `data access_pattern` pragma can be used to specify the access pattern for an array argument. It can be specified as `SEQUENTIAL`, in which case a streaming interface is inferred, or as `RANDOM`, in which case a RAM interface is generated. If the access pattern is not specified for an array argument, the SDx compiler assumes that the access pattern is random and infers a RAM interface.

The optimal selection of data mover depends on the amount of data to be transferred and the physical contiguity of the memory allocated for that data, as mentioned in Section 15.2.3. Both of these attributes can be specified using pragmas. The `data mem_attribute` pragma can be used to specify whether the memory allocated for the data is physically contiguous or not and the `data copy` pragma can be used to specify the amount of data to be transferred. Once these two attributes have been explicitly specified, the SDx compiler is able to infer the optimal data mover to generate. There is also a pragma, `data data_mover`, that allows the user to directly specify the type of data mover to be used. However, caution must be exercised when using this pragma as the SDx compiler will generate the specified data mover regardless of its compatibility with the rest of the data motion network and the user is therefore responsible for ensuring that the synthesised system is functionally correct.

The amount of data to be transferred also influences the selection of the system port used to transfer data between the PL and PS, along with whether the memory allocated for the data is cacheable or not. We have described above how the `data copy` pragma can be used to specify the size of data to be transferred. We also mentioned the `data mem_attribute` pragma in the context of specifying the contiguity of the memory allocated for the data, but this pragma can also be used to specify whether the allocated memory is cacheable or not. Similarly to the data mover generation, there is a pragma, `data sys_port`, available to allow the user to directly specify the type of system port to use. This enables the user to dictate whether a cache-coherent or non-cache-coherent port is used. Unlike when using the `data data_mover` pragma, using the `data sys_port` pragma will not influence whether the synthesised system is functional, but it may, obviously, affect the performance of the synthesised system.

There is another pragma to direct data motion network generation that we have not yet mentioned since it specifies the generated IP interface and data mover simultaneously and so does not fit neatly into the foregoing discussion. The pragma to which we refer is the `data zero_copy` pragma, which generates an AXI-Master IP interface. When this pragma is not specified, the data from an array argument is copied to the IP core in the PL by a data mover. When the `data zero_copy` pragma is specified, on the other hand, the data remains within memory until the IP core fetches it, so the IP core effectively shares the allocated area of memory with the PS. This allows the IP core to process data that is too large to copy into the PL and that cannot be streamed to the IP core or to access a large area of memory. There are, however, performance penalties associated with the latency of accessing data in system memory compared to using a local copy stored in the PL. The memory allocated to the data being accessed by the IP core must be physically contiguous.

An important point that warrants repeating here, is that, in general, SDx pragmas provide additional information to the SDx compiler to enable it to infer an optimal system. The SDx compiler will not necessarily verify the accuracy of the specified pragmas, which, could result in the synthesis of a system that does not function as expected. It falls to the user to ensure the information passed to the SDx compiler by the specified pragmas is accurate.

Connecting Two IP Cores

So far we have considered using pragmas and optimisations that could be applicable to any system using a hardware-accelerated function. There are, however, further optimisations available that can be contemplated in cases where the output of one hardware-accelerated function is an input to another.

In this scenario, system performance is best served by avoiding unnecessary data transfers between the PL and system memory. Directly transferring the output data from the first hardware-accelerated function to the input of the second hardware-accelerated function in the PL is therefore usually the optimal strategy. This avoids the need for additional data movers to move the intermediate data between the PL and system memory and the associated performance overhead for setup, execution and clean-up of the data movers. Some concurrency of execution of the two hardware-accelerated functions can also be achieved through pipelining, where the hardware-accelerated functions are connected via streaming type interfaces.

Unlike with the previous optimisations we have considered, the SDx compiler will tend to infer this architecture from the source code without any additional intervention from the user. However, if the data being passed between the two hardware-accelerated functions is accessed between the calls to the two functions, the SDx compiler will not generate the optimised architecture. Instead, it will generate an architecture that transfers the output of the first function to system memory for the programme to access it, and then transfers it back to the PL to be used as input to the second function. Such an architecture has clear disadvantages compared to the optimised architecture. An example of how a seemingly innocuous print statement can radically alter the structure of the generated system is given in Figure 16.9.

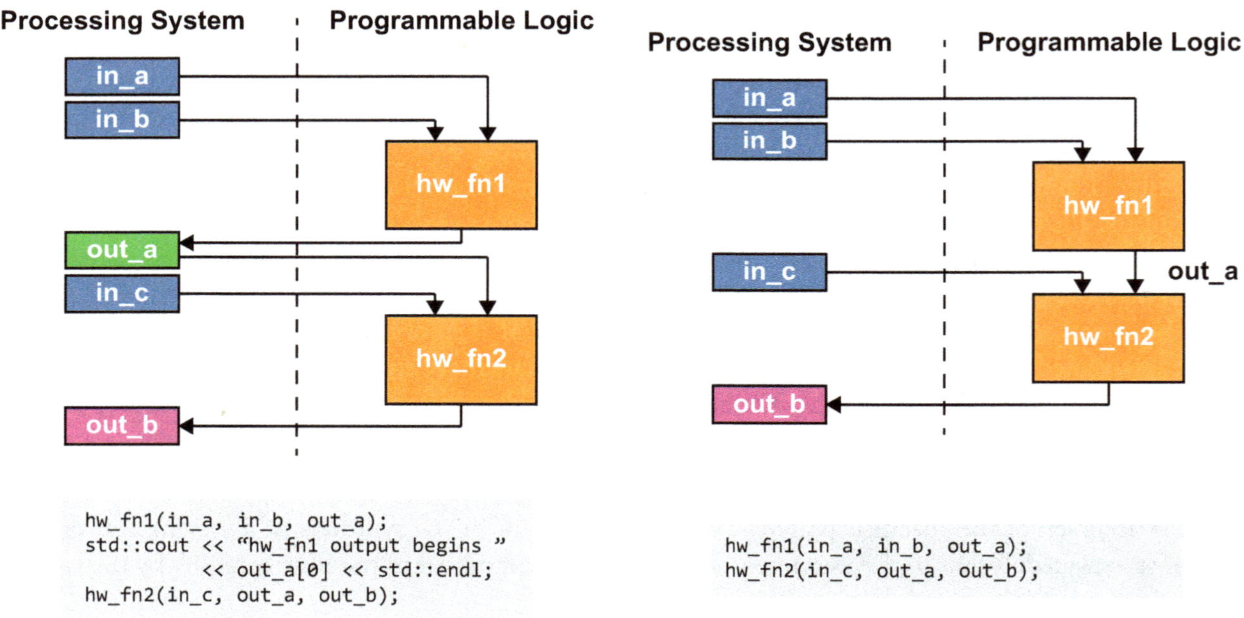

Figure 16.9: Example illustrating inference of hardware-accelerated function connectivity

Specifying Multiple IP Core Instances and Pipelining

Additional optimisations can be applied to systems where there are multiple calls to the same hardware-accelerated function and there is not sufficient time between these calls for the function to complete execution.

In this case, the most obvious way to improve performance is to create multiple instances of the hardware IP core in the PL and to allow these to execute concurrently. We are able to do this using the `resource()` pragma, an example of which is shown in Figure 16.10.

```
// Create the first hardware instance of hw_fn
#pragma SDS resource(1)
 hw_fn(in_a, in_b, out_a);

// Create the second hardware instance of hw_fn
#pragma SDS resource(2)
 hw_fn(in_c, in_d, out_b);
```

Figure 16.10: Example of using the `resource()` *pragma to create multiple hardware instances of a function [3]*

The first `resource()` pragma in Figure 16.10 creates an instance of the `hw_fn()` IP core referenced by the identifier *1*. This first instance of the `hw_fn()` IP core is then setup to execute and control returns to the programme without waiting for the IP core to complete execution. The second `resource()` pragma creates a second instance of the `hw_fn()` IP core referenced by the identifier *2*. This IP core is setup to execute and, again, control returns to the programme without waiting for the IP core to complete execution. This allows both instances of the `hw_fn()` IP core to execute concurrently rather than having one instance executing twice consecutively.

While this is the most obvious way to improve the performance of a system in this case, it may not always be possible to instantiate multiple IP cores for a single function due to limitations in the amount of PL resources available. It is, however, still possible to improve system performance by pipelining multiple calls to a single IP core. This can be particularly effective for IP cores that use RAM type interfaces. Due to the random manner in which they are able to access the data, IP cores using RAM interfaces cannot begin execution until all of the input data has been transferred to a local buffer and output data cannot begin to be transferred until the IP core has completed execution. By transferring input data for the next call to the IP core and the output data from the previous call concurrently with the IP core execution, the amount of time the IP core spends idle while waiting for data to be transferred can be reduced, as illustrated in Figure 16.11.

As is so often the case, the performance benefits achieved through pipelining multiple calls to an IP core come at a cost. In this instance the cost is in the form of the extra memory resources required in the PL to create the additional local buffers to hold multiple sets of data for the IP core.

To pipeline calls to a single IP core we use the `async()` pragma. The `async()` pragma specifies that the call to the hardware function that immediately follows it should setup the transfer of the input data to the IP core

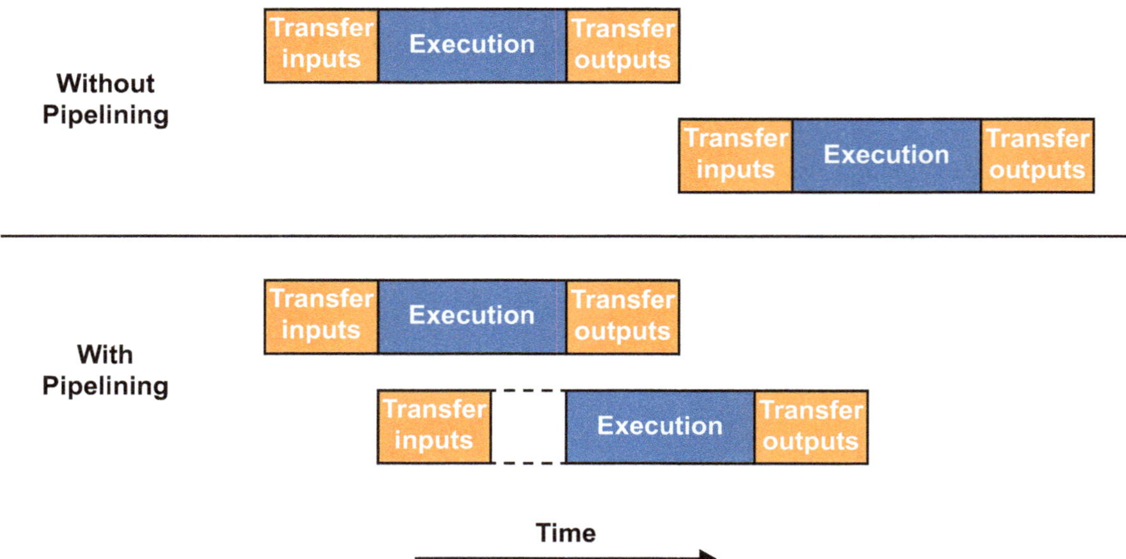

Figure 16.11: Timing diagram illustrating performance benefits from pipelining hardware function calls

and then pass control back to the programme without waiting for the IP core to complete execution. This allows multiple sets of input data to be passed to a single instance of an IP core without waiting for it to finish processing any of the sets of input data, as shown in the first loop in Figure 16.12. The identifier passed to the `async()` pragma as an argument specifies a queue to which the hardware function call request is added. The number of multi-buffers determines the number of sets of input data that can be concurrently passed to the IP core. Once all of the available multi-buffers for an IP core instance have been exhausted, the programme must wait for the IP core to complete processing a set of input data already in the multi-buffers before being able to pass more data to that IP core. This requires that the programme resynchronises with the IP core execution, which is achieved in Figure 16.12 by pairing each `async()` pragma with a corresponding `wait()` pragma. The programme pauses at the `wait()` pragma until the next hardware function call request on the queue specified by the argument passed to the `wait()` pragma completes execution. This signals that a set of input data buffers are free and more input data can be transferred to the IP core, if necessary. Although we have used `wait()` pragmas here, the ***sds_lib*** library provides functions that act as alternatives to the `wait()` pragmas; namely the `sds_wait()` and `sds_try_wait()` functions summarised in Table 15.4 on page 389. While the `sds_wait()` function's behaviour is similar to the `wait()` pragma, the `sds_try_wait()` function allows the status of the IP core to be polled without blocking the programme from executing other instructions in the meantime. A slight advantage of using the `wait()` pragma over using the ***sds_lib*** functions is that if the source code needs to be compiled at any point with a compiler other than ***sdscc*** or ***sds++***, it can be since the alternative compiler will simply not interpret the `wait()` pragma.

```
 // The first calls to the hardware function can simply be made using the async() pragma until the
 // multi-buffers are all used
 for(int i = 0; i < NUM_BUFS; i++) {
#pragma SDS async(1)
     hw_fn(in_a[i], in_b[i], out_a[i]);
 }

 // Once the multi-buffers are all used we must wait for one to become available again before making
 // another call to the hardware function
 for(int i = NUM_BUFS; i < NUM_CALLS; i++) {
#pragma SDS wait(1)
#pragma SDS async(1)
     hw_fn(in_a[i], in_b[i], out_a[i]);
 }

 // Synchronise the programme with the hardware function by waiting for the execution of the final
 // calls to complete
 for(int i = 0; i < NUM_BUFS; i++) {
#pragma SDS wait(1)
 }
```

Figure 16.12: Example of pipelining multiple calls to a hardware-accelerated function [3]

16.3 Chapter Review

In this chapter we have taken a closer look at techniques and strategies for profiling and optimising the performance of an SDx system. We discussed techniques for profiling a software application in order to identify computationally intensive functions and to estimate the benefit we may derive from accelerating these functions in hardware. A simple method for measuring the performance of a built system was also presented. We took a more detailed look at how to optimise the performance of systems that make use of hardware-accelerated functions in the PL, including optimising the performance of the IP cores using the Vivado HLS tool and the integration of these IP cores into the system using the SDx compiler. Further information and examples of these strategies are also available in [3], [4] and [5].

16.4 References

Note: All online sources last accessed March 2019.

Some of the Xilinx documentation referred to below has version-specific URLs. If you are working in a newer version of the tools, check for updates on the Xilinx website, or try adjusting the link according to your version.

[1] B. Gough, *An Introduction to GCC*, Network Theory Ltd. Bristol, 2005.

[2] The Eclipse Foundation, "TCF — Eclipsepedia" webpage, 6th December 2017.
Available: http://wiki.eclipse.org/TCF

[3] Xilinx, Inc., "SDSoC Environment User Guide", UG1027 (v2017.2), August 2017.
Available: https://www.xilinx.com/support/documentation/sw_manuals/xilinx2017_2/ug1027-sdsoc-user-guide.pdf

[4] Xilinx, Inc., "Vivado Design Suite User Guide: High-Level Synthesis", UG902 (v2017.2), June 2017.
Available: https://www.xilinx.com/support/documentation/sw_manuals/xilinx2017_2/ug902-vivado-high-level-synthesis.pdf

[5] Xilinx, Inc., "SDSoC Environment Tutorial: Introduction", UG1028 (v2017.2), August 2017.
Available: https://www.xilinx.com/support/documentation/sw_manuals/xilinx2017_2/ug1028-sdsoc-intro-tutorial.pdf

Chapter 17

Reusing Existing IP in SDx

One of the main aims of SDx is to increase productivity for designers targeting Zynq devices. It is therefore important that designers are able to reuse IP cores developed outside of SDx to avoid replicating the effort expended on previously developed functionality. In this chapter we present the steps required to integrate pre-existing IP cores within an SDx project.

17.1 Creating a C-Callable Library

In order to reuse a pre-existing IP core in our SDx project, we need to be able to call the IP core and pass data to and from it within the C/C++ source code that describes our project. To achieve this, we must create a C-callable software library containing functions to allow the IP core to be referred to within the SDx project source code. It must also provide the information required by the SDx compiler to configure the IP core and integrate it within the system. This may seem like a complex task, however, the SDx development environment greatly simplifies it by providing the ***sdslib*** utility to create the C-callable library from a set of inputs, as illustrated in Figure 17.1, and we will discuss this process here.

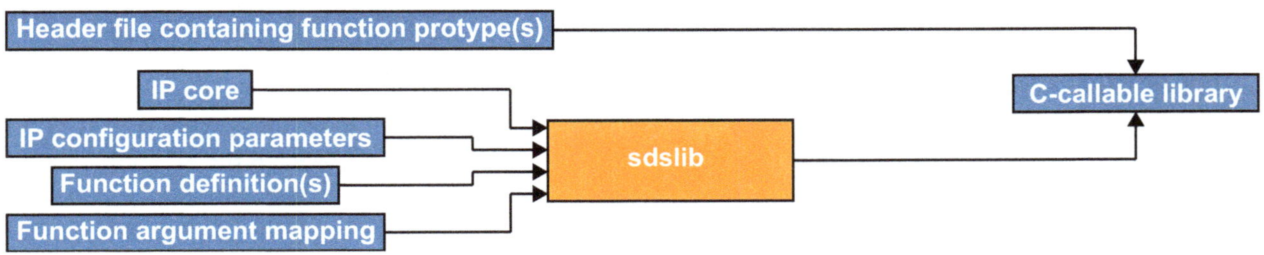

*Figure 17.1: Creating a C-callable library using **sdslib***

17.1.1 IP Core Requirements

We will firstly discuss the IP core itself, as this is perhaps the most obvious input to the ***sdslib*** utility. While, ideally, we would be able to use any IP core without any further modification in our SDx project, practicality dictates that the IP core must conform to some common standards to allow the SDx compiler to generate the necessary software and hardware to interface with it. SDx requires the IP core to use only AXI4, AXI4-Lite or AXI4-Stream interfaces. It also requires that the IP core has a control register at address offset *0x0* [1], so that the software stubs generated by the SDx compiler are able to correctly control the IP core once it is integrated within the system.

There are two protocols that can be supported by the control register: *none* and *axilite*. The *none* protocol should be used for IP cores that run continually from power-on and require no additional synchronisation through the control register. When the *none* protocol is used, the control register must be tied to the constant value *0x6* [1], which effectively marks the IP core as always ready to execute. The *axilite* protocol, on the other hand, provides additional synchronisation of the IP core through the control register. The *axilite* protocol requires that the control register uses the layout shown in Table 17.1. This is the same configuration as IP generated by Vivado HLS [1].

Table 17.1: Control register layout for IP cores using the axilite protocol [1] [2]

Bits	Signal
0	Start — initiates operation of the IP core.
1	Done — indicates whether the IP core has completed operation.
2	Idle — indicates whether the IP core is idle.
3	Ready — indicates whether the IP core is ready to accept new inputs.
7	Auto restart — indicates and sets whether the IP core automatically restarts operation on completion.
others	reserved

The majority of data movers that can be instantiated by the SDx compiler only support stream data supplied in packets [3]. Generally, it is therefore necessary that any AXI4-Stream interfaces on the IP core support the *tlast* side-band signal to ensure compatibility with the data motion network generated by the SDx compiler. Similarly, all AXI4 interfaces on the IP core must also support the *tready* side-band signal.

Another requirement for the IP core is that it must be packaged to conform with the IP-XACT standard [4]. This ensures that the files defining the IP core are in the correct structure to be interpreted by the Vivado tool when it is invoked by the SDx compiler. The Vivado tools can be used to package the IP core appropriately according to the process described in [5].

17.1.2 IP Configuration Parameters

Closely related to the IP core is the IP configuration parameters file. This is an XML file used to configure parameters for IP cores that are customisable at the time of synthesis. An example showing the IP configuration parameters file for customising a Complex Multiplier IP logiCORE [6] is given in Figure 17.2 and demonstrates the structure of such a file. The parameter names used to customise the IP core should match the names of the generics or parameters in the IP core and their possible values can be obtained from the IP core documentation. Alternatively, the parameters can be obtained from the Tcl Console following appropriate customisation of the IP core using the Vivado Design Suite [5].

```
<?xml version="1.0" encoding="UTF-8"?>
<xd:component xmlns:xd="http://www.xilinx.com/xd" xd:name="cmpy">
    <xd:parameter xd:name="APortWidth" xd:value="8"/>
    <xd:parameter xd:name="HasATLAST" xd:value="true"/>
    <xd:parameter xd:name="HasATUSER" xd:value="false"/>
    <xd:parameter xd:name="BPortWidth" xd:value="8"/>
    <xd:parameter xd:name="HasBTLAST" xd:value="true"/>
    <xd:parameter xd:name="HasBTUSER" xd:value="false"/>
    <xd:parameter xd:name="MultType" xd:value="Use_Mults"/>
    <xd:parameter xd:name="OptimizeGoal" xd:value="Resources"/>
    <xd:parameter xd:name="FlowControl" xd:value="Blocking"/>
    <xd:parameter xd:name="RoundMode" xd:value="Truncate"/>
    <xd:parameter xd:name="OutputWidth" xd:value="17"/>
    <xd:parameter xd:name="OutTLASTBehv" xd:value="Pass_A_TLAST"/>
    <xd:parameter xd:name="MinimumLatency" xd:value="9"/>
</xd:component>
```

Figure 17.2: Example of an IP configuration parameters file

The IP configuration parameters file is a mandatory input for the ***sdslib*** utility, even if the IP core being used does not have any parameters that can be configured at the time of synthesis. In this situation, the IP configuration parameters file should simply not list any parameters. An example of such a file is shown in Figure 17.3.

```
<?xml version="1.0" encoding="UTF-8"?>
<xd:component xmlns:xd="http://www.xilinx.com/xd" xd:name="arraycopy_axis"/>
```

Figure 17.3: Example of an IP configuration parameters file for an IP core with no configurable parameters

17.1.3 Function Definition

The inputs to the ***sdslib*** utility that we have discussed so far have dealt with defining the hardware of the IP core. We also need to define a function or set of functions to allow the IP core to be called from the source code of an SDx project. Functions intended to control the IP core, or to transfer data to or from it, are defined in C or C++ source files referred to as function definition files. The SDx compiler will replace the bodies of these functions with calls to stub functions to handle the hardware interaction in a similar fashion as it does with calls to hardware-accelerated functions. The bodies of these functions have no influence over the system created by the SDx compiler. They can, therefore, contain any code the designer likes, such as emulating the function of the IP core for simulation purposes, or indeed be left empty. The SDx compiler is able to discern the appropriate stub functions to use based on the function argument mapping, which we will discuss momentarily. An example of a function definition file is shown in Figure 17.4.

```
#define N_DATA 8

 // Function to execute a complex multiplication function on the A and B inputs and produce the P
 // output
 void cmpy( short A[N_DATA], short B[N_DATA], long long P[N_DATA] ) {
     // SDx will replace the function definition with a stub function to control the IP and transfer
     // data between system memory and hardware.
 }
```

Figure 17.4: Example of a function definition file for a Complex Multiplier IP core

17.1.4 Function Argument Mapping

The final input to the ***sdslib*** utility is the function argument mapping file. This file is used to describe how the software functions defined in the C-callable library interface with the IP core. Specifically, it is used to describe which IP core port each function argument maps to.

The file is in an XML format with a *function mapping element* for each function. *Argument elements* are children of the *function mapping element* and set a number of attributes to describe how that argument maps onto the IP core ports. A selection of the attributes set within function mapping and argument elements are given in Table 17.2. The *argument elements* must use the same name for the argument as used in the function definition and must appear in exactly the same order.

Table 17.2: Selection of function argument attributes set within function argument map

Attribute	XML Attribute	Description
Function name	`fcnName`	The name of the function being mapped.
Component reference	`componentRef`	The name of the IP core being mapped to as it appears in the IP-XACT Vendor-Library-Name-Version (VLNV) identifier.
C argument name	`name`	The function argument being mapped as it appears in the function definition.
Argument direction	`direction`	This can be either `in` or `out`, depending on whether the argument is an input or output respectively.
Bus interface reference	`busInterfaceRef`	The name of the port on the IP core the argument is being mapped to.
Port interface type	`portInterfaceType`	The interface type of the port on the IP core the argument is being mapped to. This can be `aximm`, `axis` or `axilite`

Table 17.2: Selection of function argument attributes set within function argument map

Attribute	XML Attribute	Description
Address offset	`offset`	Hexadecimal address offset for arguments mapping onto memory mapped ports.
Data width	`dataWidth`	The number of bits per data value for the argument data type.
Array size	`arraySize`	The number of elements in an array argument.

In addition to the *argument elements* within each *function mapping element*, elements to provide estimates of the resource utilisation and latency for the IP core can also optionally be provided. An example of a function mapping file for the function defined in Figure 17.4 is given in Figure 17.5.

```
<?xml version="1.0" encoding="UTF-8"?>
<xd:repository xmlns:xd="http://www.xilinx.com/xd">
  <xd:fcnMap xd:fcnName="cmpy" xd:componentRef="cmpy">
    <xd:arg xd:name="A"
            xd:direction="in"
            xd:portInterfaceType="axis"
            xd:dataWidth="16"
            xd:busInterfaceRef="S_AXIS_A"
            xd:arraySize="8"/>
    <xd:arg xd:name="B"
            xd:direction="in"
            xd:portInterfaceType="axis"
            xd:dataWidth="16"
            xd:busInterfaceRef="S_AXIS_B"
            xd:arraySize="8"/>
    <xd:arg xd:name="P"
            xd:direction="out"
            xd:portInterfaceType="axis"
            xd:dataWidth="64"
            xd:busInterfaceRef="M_AXIS_DOUT"
            xd:arraySize="8"/>
  </xd:fcnMap>
</xd:repository>
```

Figure 17.5: Example of a function mapping file for the function defined in Figure 17.4

17.1.5 The *sdslib* Utility

Now that we have all of our inputs, we are ready to generate our static C-callable library. To do this, we will use the ***sdslib*** utility. The ***sdslib*** utility is command line based and is called from the SDx Terminal. Once in our working directory, the ***sdslib*** utility can be invoked using the `sdslib` command and it is passed details of the input files as arguments. The arguments required by the utility are described in Table 17.3. There are also a number of options that can be specified and these are outlined in Table 17.4. An example of the invocation of ***sdslib*** to create a library for the function definitions in Figure 17.4 is given in Figure 17.6.

*Table 17.3: **sdslib** utility arguments*

Argument	Description
`-lib <libname>`	The name of the C-callable library.
`<function file>`	Specifies the name of a function to be added to the library and the file it is defined in. More than one function/file pair can be specified at a time.
`-vlnv <v>:<l>:<n>:<v>`	The IP-XACT VLNV identifier for the IP core to be used.
`-ip-map <file>`	Specifies the file to be used as the function argument mapping file.
`-ip-params <file>`	Specifies the file to be used as the IP configuration parameters file.

*Table 17.4: **sdslib** utility options*

Option	Description
`-pfunc`	Specify that IP core is a platform function.
`-ip-repo <path>`	Specify an additional IP repository search path.
`-target-os <name>`	Specify the target operating system. Can be `linux` (default) or `standalone`.

```
sdslib -lib libcmpy.a -target-os linux \
    cmpy cmpy_stub.cpp \
    -vlnv xilinx.com:ip:cmpy:6.0 \
    -ip-map cmpy.fcnmap.xml \
    -ip-params cmpy.params.xml
```

*Figure 17.6: Example of **sdslib** utility invocation*

Once the ***sdslib*** utility has successfully executed it produces the C-callable library file used to link into SDx projects targeting the pre-existing IP core. An interesting point to note is that, if the ***sdslib*** utility is invoked using the name of a library that already exists within the working directory, the utility will not overwrite the library, but will simply append the new functions to it.

17.1.6 Library Header File

There is one more file that is needed to enable use of the C-callable library created by the ***sdslib*** utility: the library header file. This file requires very little explanation, as it is simply a C/C++ header file containing prototypes for the functions defined in the library. Including this header file in an application provides the interface to the C-callable library functions. The only point to note with this file is that it can be used to direct the choice of data mover and system port the IP core is connected to by specifying SDx pragmas in the usual way within it.

```
#ifndef CMPY_H_
#define CMPY_H_

#define N_DATA 8

 void cmpy( short A[N_DATA], short B[N_DATA], int P[N_DATA] );

#endif
```

Figure 17.7: Example of a header file for a C-callable library

17.2 Using a C-Callable Library

Using a C-callable library to target a pre-existing IP core is straightforward once the library has been created. All that is required is to specify the location of the library to the SDx tool-chain, and to call the library functions at the appropriate points in the source code.

In order to specify the location of the library to the SDx tool-chain there are a number of steps to perform:

- Include the library header file at an appropriate point in the project source code;
- Specify the directory containing the library header file to the compiler;
- Specify the library file and its location to the linker.

The first of these steps is largely self-explanatory.

The other steps can be achieved by changing the build configuration settings, as described in Section 15.2.6. The directory containing the library header file should be specified using the `-I` option to the ***sdscc*** compiler, if the source code is C, or the ***sds++*** compiler, if the source code is C++. The name of the library file should be specified to the ***sds++*** linker using the `-l` option and the directory where it is contained should also be specified to the ***sds++*** linker, but using the `-L` option.

When an SDx project targeting a pre-existing IP core using a C-callable library is built, the SDx compiler automatically instantiates the pre-existing IP core in the PL of the Zynq device and integrates it with the rest of the system, as specified by the SDx project.

17.3 Chapter Review

In this chapter we have discussed the process of reusing existing IP within SDx projects by creating a C-callable library to represent it. The steps necessary to create a C-callable library using the ***sdslib*** utility were introduced, as were those needed to instantiate the IP core within an SDx project using the library. Simple examples have been provided to help illustrate the process. Further reading is also available on this subject in [1].

17.4 References

Note: All online sources last accessed March 2019.

Some of the Xilinx documentation referred to below has version-specific URLs. If you are working in a newer version of the tools, check for updates on the Xilinx website, or try adjusting the link according to your version.

[1] Xilinx, Inc., "SDSoC Environment User Guide", UG1027 (v2017.2), August 2017.
Available: https://www.xilinx.com/support/documentation/sw_manuals/xilinx2017_2/ug1027-sdsoc-user-guide.pdf

[2] Xilinx, Inc., "Vivado Design Suite User Guide: High-Level Synthesis", UG902 (v2017.2), June 2017.
Available: https://www.xilinx.com/support/documentation/sw_manuals/xilinx2017_2/ug902-vivado-high-level-synthesis.pdf

[3] Xilinx, Inc., "SDSoC Environment Platform Development Guide", UG1146 (v2017.2), August 2017.
Available: https://www.xilinx.com/support/documentation/sw_manuals/xilinx2017_2/ug1146-sdsoc-platform-development.pdf

[4] IEEE, "1685-2014 — IEEE Standard for IP-XACT, Standard Structure for Packaging, Integrating, and Reusing IP within Tool Flows", 2014.
DOI: 10.1109/IEEESTD.2014.6898803

[5] Xilinx, Inc., "Vivado Design Suite User Guide: Designing with IP", UG896 (v2017.2), June 2017.
Available: https://www.xilinx.com/support/documentation/sw_manuals/xilinx2017_2/ug896-vivado-ip.pdf

[6] Xilinx, Inc., "Complex Multiplier v6.0 LogiCORE IP Product Guide", PG104, November 2015.
Available: https://www.xilinx.com/support/documentation/ip_documentation/cmpy/v6_0/pg104-cmpy.pdf

Chapter 18

Debugging and Performance Monitoring with SDx

Systems that effectively exploit the architecture of Zynq devices are, by their very nature, complex systems consisting of closely integrated hardware and software components. The prospect of attempting to debug or analyse the performance of such a system can be a rather daunting one. The SDx development environment once again comes to the rescue, however, by providing a number of features to assist the user in performing these tasks. In this chapter, we will review some of the available features and how they can be used to investigate system behaviour.

Debugging is sometimes divided into two categories based on the context of the debugging. These categories are functional debugging, where there is an issue that is causing unexpected behaviour in the system, and performance debugging, where the system is behaving as expected but performance is unsatisfactory (in other words, an increase in the clock speed would cause the system to fail due to it not meeting the timing constraints). Although each of the features we discuss here may prove useful in both debugging scenarios, we have categorised them based on the situations in which they are likely to be of most use. We will firstly discuss the utilities most helpful for functional debugging of both software and hardware before moving on to discuss the performance monitoring features best suited to performance debugging.

18.1 System Emulation

The system emulation flow feature of SDx can be valuable for debugging complex hardware and software systems targeting Zynq devices. This flow allows both the hardware and software components of the system to be simulated in concert with one another. The PS is emulated using QEMU [1], enabling the compiled software components to be executed on this emulated system. Meanwhile, the hardware components are simulated using Vivado Logic Simulator. The QEMU emulated PS and the Vivado Logic Simulator simulated PL are interfaced to pass the appropriate signals between them, allowing the entire system to be simulated.

System emulation can be carried out prior to generating a bitstream for the hardware components of the system. Bitstream generation constitutes a significant proportion of the full system build time, so being able to test and verify the functionality of the system before committing to a time-consuming system build is a considerable advantage. In addition, the use of the Vivado Logic Simulator to simulate the PL design can allow any of the hardware signals to be probed.

The first step in the system emulation flow is essentially as simple as selecting the *Generate emulation model* option in SDx Project Settings and building the project. This generates the files needed to simulate the system using QEMU and Vivado Logic Simulator. When emulating the system it is not necessary to generate a bitstream. The emulation model can be generated in one of two modes: *Debug* or *Optimized*, with the mode being selectable from a drop-down menu in SDx Project Settings. The mode selection determines the type of model used to simulate the PL. When the *Debug* mode is selected, a PL simulation model is generated that is capable of capturing time-resolved data for the signals in simulated hardware that can be visualised as waveforms within the Vivado Logic Simulator. When the *Optimized* mode is selected, the simulated hardware signal data is not recorded and cannot be viewed as waveforms, however, system emulation does execute faster.

Once the emulation model has been generated, system emulation can be performed. The emulator is started by selecting the *Start/Stop Emulator* option under the *Xilinx* menu, shown in Figure 18.1. This opens the Start/Stop Emulator dialogue box, also shown in Figure 18.1. In the Start/Stop Emulator dialogue box, the project to be emulated should be selected, along with the configuration appropriate to the mode of the generated emulation model. There is an option to *Show Waveform (Programmable Logic only)*, which is applicable only when using a *Debug* mode emulation model. When this option is selected the Vivado Logic Simulator GUI will launch when the emulator is started, to allow the simulated PL signals to be visualised. The emulator is started by clicking the *Start* button in the dialogue box, which should be confirmed by messages in the console view of the SDx IDE.

If the *Show Waveform* option was selected when starting the simulator, the Vivado Logic Simulator GUI will launch and look similar to Figure 18.2. By default no PL signals are selected for visualisation. It is at this point, prior to launching our application on the emulated system, that we can select the PL signals we wish to monitor during application execution. To add signals to the wave window for monitoring, the appropriate entity from the hardware design hierarchy should be selected in the Scopes pane of the Vivado Logic Simulator GUI. The signals belonging to this entity are then shown in the Object pane. The desired signals can then be added to the wave window by right-clicking them and selecting the *Add to Wave Window* option.

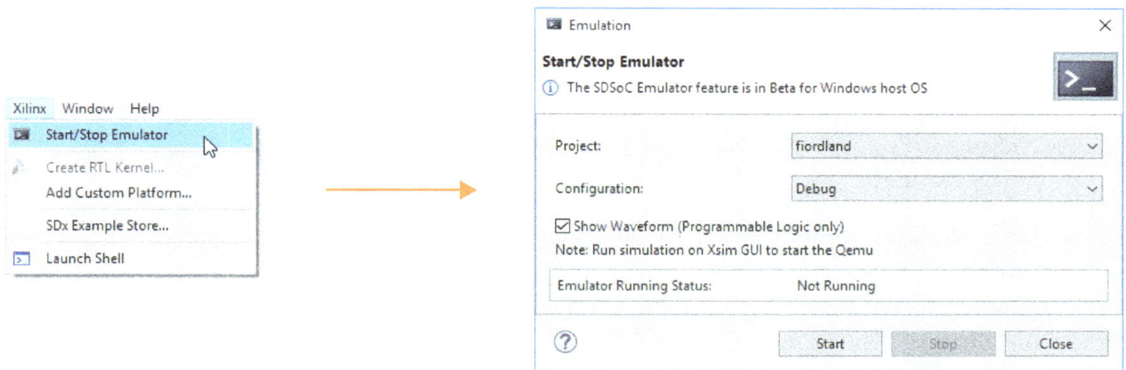

Figure 18.1: The location of the Start/Stop Emulator option and the Start/Stop Emulator dialogue box (as in SDx IDE 2017.2)

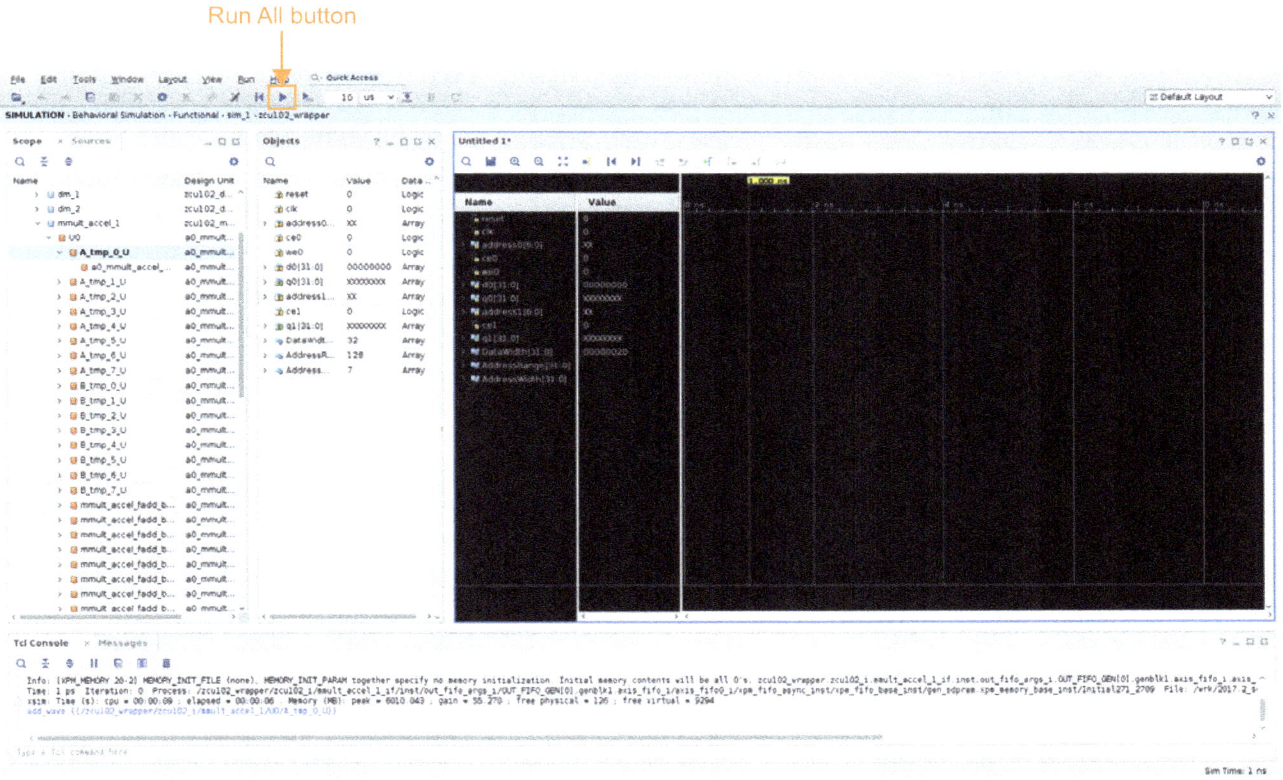

Figure 18.2: Example of the Vivado Logic Simulator GUI showing the location of the Run All command button

Once all of the signals to be monitored have been added to the wave window, the *Run All* button should be selected from the toolbar of the Vivado Logic Simulator GUI, as shown in Figure 18.2. This starts the Vivado Logic Simulator capturing and displaying signal data, ready for when the application is executed.

The application can be launched on the emulated system by right-clicking the project in the Project Explorer view of the SDx IDE and selecting the *Run As > Launch on Emulator (SDSoC Debugger)* option. The standard output from the emulator is directed to the Emulation Console in the SDx IDE. If the PL signals were being monitored using the Vivado Logic Simulator then these will be visible in the wave window of the Vivado Logic Simulator GUI.

It is also possible to launch the application on the emulated system using the *Debug As > Launch on Emulator (SDSoC Debugger)* option when the project is right-clicked in the Project Explorer view. This enables software debugging to be carried out while the application executes on the emulator, but requires that an appropriate build configuration was used to build the application. More details on this are provided in Chapter 18.2 on software debugging.

An important point to note when observing PL signals and using breakpoints on an emulated system is that the breakpoints pertain to the software only, and so activity may be observed on some PL signals when the application is paused by a breakpoint. This behaviour is also consistent with what would be observed if performing software debugging on a physical device.

As of SDx 2017.2, the system emulation flow is fully supported on Linux hosts, while a beta version is available on Windows hosts.

18.2 Software Debugging

The main software debugging utility provided by the SDx IDE is Eclipse-based and called System Debugger. It is conventional in both appearance and functionality, lending it a sense of familiarity for any user with software debugging experience. System Debugger is a GUI-based debugger with a dedicated perspective within the SDx IDE.

In order to be used with the debugger, the application executable needs to contain debug symbol information. This information is included in the executable by using an option at compile time. The `-g` option will accomplish this with the ***sdscc*** or ***sds++*** compilers. The ***Debug*** build configuration also includes this option by default.

The process of launching the application on the target platform ready for debugging is exactly the same as was described for launching the software-only system profiler in Section 16.1.1, up to the point of switching to the Debug perspective.

Once the application has been launched on the target platform and the user has switched to the Debug perspective, they will be presented with a scene similar to the one shown in Figure 18.3. At this point, the application is suspended at the entry to the *main* function. The Debug perspective provides the user with convenient access to a range of conventional debugging features. These include the ability to set breakpoints and watchpoints, inspect the contents of registers, variables and memory, and to step the execution of the application. Using these features, the user is able to thoroughly investigate the execution of the software portion of the system as it executes on the Zynq device.

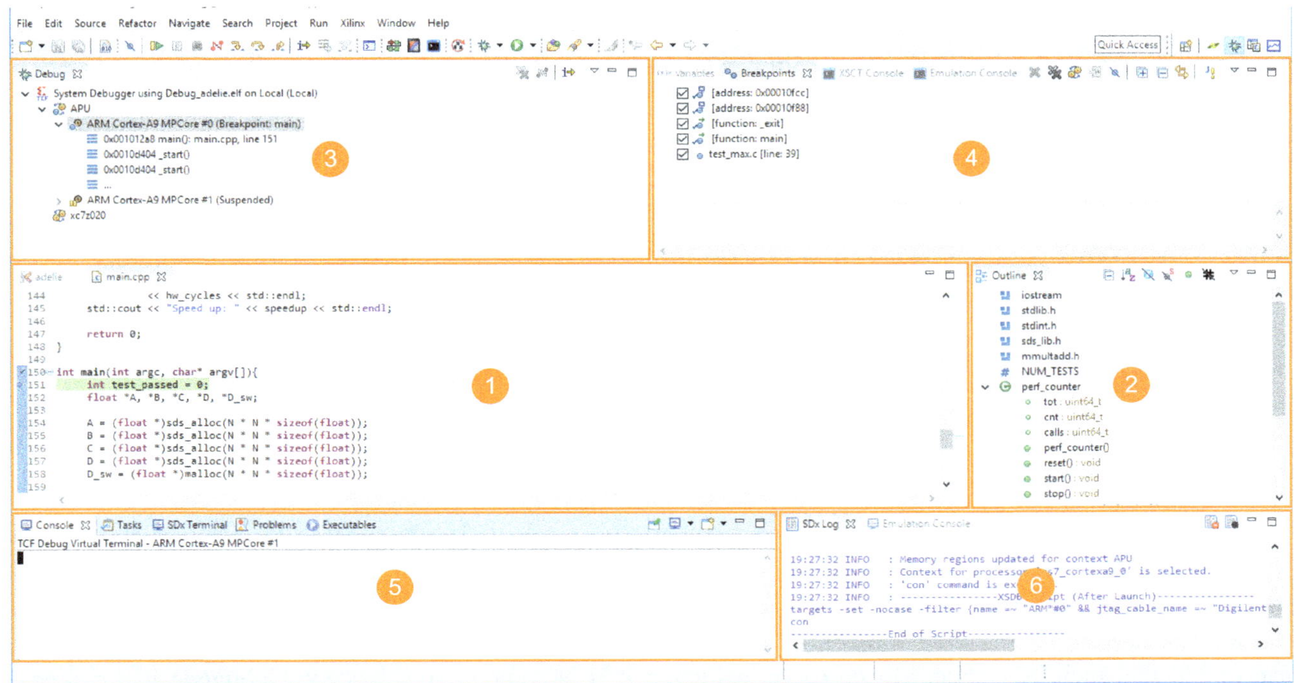

1 Editor Area: used to display files opened within the SDx IDE

2 Outline View: provides an editor specific structural outline of the file open in the editor

3 Debug View: provides status information for current debug sessions

4 A stack of views including:
- Variables View: used to display the value of variables during debugging
- Breakpoints View: lists any breakpoints set

5 A stack of five views:
- Console View: used to display output messages from the SDx IDE
- Tasks View: used to list tasks associated with the project
- SDx Terminal View: terminal application to connect to serial ports
- Problems View: provides information about errors and warnings encountered during a build
- Executables View: lists executable files in the current workspace

6 A stack of two views:
- SDx Log View: used to display the information logged about activity in the SDx console
- Emulation Console View: used to display the standard output from the emulator

Figure 18.3: The default Debug perspective (as in SDx IDE 2017.2)

18.3 Hardware Debugging

Hardware debugging is a little more complicated than software debugging and the flow proposed here is not directly supported by SDx. It requires that we are able to probe signals within the PL itself. In order to do this, debug IP cores must be instantiated in the PL to interface between these signals and the user. Here, we have concentrated on using the Virtual Input/Output (VIO) [2] and System Integrated Logic Analyser (System ILA) [3] LogiCORE IP cores — other debugging hardware could be used in a similar manner.

There are a number of steps involved in adapting an SDx project to incorporate debugging IP cores. Firstly, the debugging IP cores need to be included in the hardware portion of the system design and the bitstream generated from this. A bootable image that includes the bitstream needs to be created to boot the Zynq device. Finally, the PL needs to be programmed to allow the debugging IP cores to be initialised, before the system begins executing. We will now outline each of these steps in more detail.

Incorporating Debugging Hardware in a System Design

To incorporate the debugging IP cores, the system design must first be created by performing a build. Synthesis, implementation and bitstream generation should be postponed until after the debugging IP cores have been included. The build process can therefore be expedited by deselecting the *Generate bitstream* option in the SDx Project Settings. A build configuration that includes debug symbol information in the application executable file, such as ***Debug***, can be useful as, when it is launched using System Debugger, it provides a breakpoint at the entry of the *main* function. This pauses the application prior to execution, giving an opportunity to initialise the debugging IP cores.

Once the build has completed, the Vivado IPI project created by the SDx compiler for the hardware portion of the system can be accessed from within the project build directory. After opening this project within Vivado, the debugging hardware can be added to the design. Instructions for adding the VIO and System ILA LogiCORE IP can be found in [2], [3], [4] and [5]. A bitstream for the hardware design that includes the debugging hardware can then be generated using Vivado.

Creating a Bootable Image

The next step is to create an image that includes the generated bitstream from which the Zynq device can be booted. The process for doing this varies depending on the OS being targeted. Here, we will outline the steps necessary to boot the Zynq device with the newly generated bitstream for debugging Standalone- and Linux-based designs.

With Standalone-based designs targeting the platforms supplied with SDx, there is no real need to create an image, as such, since the Zynq device can be booted over the JTAG interface. All that is necessary is make sure that the application executable and bitstream files are in the appropriate directory and adhere to naming conventions. Following the build process, the application executable file should already be located in the *<workspace>/<project>/<build_configuration>/* directory with the appropriate naming convention of *<project>.elf*. The bitstream file generated from the Vivado project can be copied into this directory and renamed to follow the convention *<project>.elf.bit*. We are then ready to launch our project on the Zynq device over the JTAG connection.

For designs that target the Linux OS, a bootable image is needed to be able to boot the Zynq device from an SD card. Here, we will cover the steps necessary to create a basic boot image, based on the SDx Platforms supplied with SDx, to be used for hardware debugging purposes. Creating boot images for Zynq devices is discussed in more detail in Chapter 14 and Chapter 19.

SDx once again provides a tool to simplify this operation, which is the ***bootgen*** utility we have previously mentioned as being used by the SDx compiler to create boot images. The ***bootgen*** utility is driven by a Boot Image Format (BIF) file, which is a text file that lists the partitions to be created in the boot image along with the configuration options and source files for the partitions. The partitions contained in the boot image differ, depending on whether a Zynq-7000 or Zynq MPSoC is being targeted.

For the Zynq-7000, three partitions are specified: a First Stage Boot Loader (FSBL), the bitstream and a Second Stage Boot Loader (SSBL). The FSBL and SSBL should be available from the targeted SDx platform. Here, we have assumed that the SSBL is based on Das U-Boot [8], as is the case for the platforms supplied with SDx. We have also assumed the file path and naming conventions used for the platforms supplied with SDx. Figure 18.4 shows an example BIF file with the appropriate partitions and settings to create a boot image for the Zynq-7000.

```
image: {
     [bootloader] <platform_path>/sw/linux/boot/fsbl.elf
     <path_to_bitstream>
     <platform_path>/sw/linux/boot/u-boot.elf
}
```

Figure 18.4: Example BIF file for creating a Linux boot image for the Zynq-7000

Boot images for Zynq MPSoC devices are a little more complex and, consequently, so are the BIF files used to create them. As well as having additional partitions compared to Zynq-7000 devices, there are also additional configuration settings, chiefly required to specify which processor should execute the partition and how it should be configured for doing so. In addition to the FSBL, bitstream and SSBL partitions used in the Zynq-7000 boot image, a Linux boot image for Zynq MPSoC also requires a platform management unit (PMU) firmware partition and an Arm Trusted Firmware (ATF) partition. The source files for both of these partitions should be available from the targeted SDx Platform. Figure 18.5 shows an example BIF file with the appropriate partitions and settings to create a boot image for the Zynq MPSoC. Again, we have assumed that the SSBL is based on Das U-Boot, and the file path and naming conventions used for the platforms supplied with SDx.

Once the appropriate BIF file has been created, it can be used to drive the ***bootgen*** utility to create the boot image necessary to boot the Zynq device. The ***bootgen*** utility can be invoked from the CLI of SDx. The command shown in Figure 18.6 invokes the ***bootgen*** utility using *bootimage.bif* as the BIF file to direct the creation of the boot image, which is named *BOOT.bin*. The boot image used for booting Zynq devices from

```
image: {
   [fsbl_config] a53_x64
   [bootloader] <platform_path>/sw/a53_linux/boot/fsbl.elf
   [pmufw_image] <platform_path>/sw/a53_linux/boot/pmufw.elf
   [destination_device=pl] <path_to_bitstream>
   [destination_cpu=a53-0, exception_level=el-3, trustzone] <platform_path>/sw/a53_linux/boot/bl31.elf
   [destination_cpu=a53-0, exception_level=el-2] <platform_path>/sw/a53_linux/boot/u-boot.elf
}
```

Figure 18.5: Example BIF file for creating a Linux boot image for the Zynq MPSoC

```
bootgen -image bootimage.bif -o BOOT.bin
```

Figure 18.6: Example command to invoke the bootgen utility

SD card should always be named *BOOT.bin*. Further information on using the ***bootgen*** utility to create boot images is available in [6] and [7].

The final step is to copy the files necessary to boot the Zynq device to an SD card. Three files are required to achieve this. The first file is the boot image we have just created: *BOOT.bin*. The second file is the application executable created by the SDx compiler during the build process, and the final file is the Linux image that the SSBL in the boot image will load. An appropriate Linux image for the platforms supplied with SDx can be found within the platform's directory structure. The SD card is then ready to be used to boot the Zynq device and run our design, including the debugging IP cores.

Debugging the System

Once the appropriate boot image has been created, we are ready to launch our system on the Zynq device. If a build configuration that includes debug symbol information in the application executable file was used, then this can be done using System Debugger. First, a TCF connection between the host running the SDx IDE and the target Zynq device must be established. The process for doing this is described in Section 16.1.1 for both Standalone- and Linux-based systems. The application can then be launched on the target device by right-clicking the project in the Project Explorer view and selecting *Debug As > Launch on Hardware (SDSoC Debugger)*. This programs the Zynq device with the bitstream and pauses the application execution as it enters the *main* function.

At this stage, there is an opportunity to initialise the debugging IP cores in the PL, ready to probe the appropriate signals on resumption of the application. To do this for the VIO or System ILA IP cores, a connection to the debugging hardware in the PL of the Zynq device must be established. This can be done using a JTAG connection from within the Vivado Design Suite using the Hardware Manager feature. Details of connecting to, configuring, and interacting with VIO and System ILA IP cores using the Hardware Manager are available in [2], [3] and [4].

After the debugging IP cores have been initialised, the paused application can be resumed from the SDx IDE debug perspective. The software portion of the system can now be debugged in the conventional way while the hardware portion of the system is debugged simultaneously.

The hardware and software debugging techniques we have discussed so far are perhaps most applicable to low-level, functional debugging. We may also be interested in examining the performance of our system in greater detail, in order to improve system performance or alleviate performance bottlenecks. With this in mind, we will now move on to consider some of the utilities available within SDx that are suited to this purpose.

18.4 Performance Monitoring

It is often critical to be able to accurately characterise the behaviour of the system and to identify areas where performance can be improved. Happily, there are a number of features available within SDx to aid the user in performing just this task:

- Event tracing;
- AXI performance monitoring;
- PS performance monitoring.

18.4.1 Event Tracing

Perhaps the most comprehensive performance monitoring feature available in SDx for analysing systems running on a Zynq device is event tracing. This feature is available for any system with at least one function implemented in hardware, and records the sequence and duration of software, hardware and data transfer events. The events are recorded using a combination of additional hardware implemented in the PL, and instrumentation of the software stub function used to interface to IP cores in hardware. A single, global timer implemented in the PL is used to timestamp both hardware and software events, with the resulting timestamped data being collected in a buffer in the PL. This buffer is accessed by the host running SDx over a JTAG interface to record and visualise the trace data.

With event tracing enabled, the instrumentation of the software stub function results in calls to the `sds_trace` function immediately before and after any software tracing event. The software events traced include:

- Hardware function setup and initiation;
- Data transfer setup;
- Hardware/software synchronisation barriers.

Calls to the `sds_trace` function write to the PL over an AXI4-Lite interface to obtain a timestamp from the global timer and to record this in the buffer in the PL.

Hardware trace events are recorded using a combination of hardware monitor IP cores:

- Accelerator Monitor IP core — timestamp the start and stop of functions compiled to hardware by the SDx compiler and that use the *ap_start* and *ap_done* signals in their control interface;
- AXI4-Stream Monitor IP core — timestamp the start and stop of data transfers across AXI4-Stream interfaces based on the handshake and *TLAST* signals;
- Integration IP core — combines the event signals from up to 63 individual monitor IP cores and records them in a buffer.

More detailed information on the implementation of the event tracing feature on Zynq devices is available from [9].

Implementing the additional software and hardware required for event tracing in SDx is very simple. Selecting the *Enable event tracing* option in the SDx Project Settings before building the system will cause the SDx compiler to include the additional instrumentation code in the software stub functions and implement the additional hardware in the PL.

Launching the application for event tracing on the Zynq device must be done using the SDx IDE. The Zynq target must therefore be connected to the host running the SDx IDE in the manner described in Section 16.1.1. For Zynq targets running the Linux OS, an additional JTAG connection should be made between the host and the target to allow the host to access the trace data buffer. For Zynq targets running Standalone or FreeRTOS, the host will simply use the JTAG interface used for the TCF connection to access this buffer.

The application is launched on the Zynq device by right-clicking the project in the Project Explorer view and selecting *Run As > Trace Application (SDSoC Debugger)*. It is also possible to launch a trace application using the *Debug As* option, although care must be taken in the placement of breakpoints for debugging, to avoid affecting the timing of trace events. Once the application launches, the host begins retrieving trace data from the buffer on the Zynq device in real-time in order to prevent the buffer, which has a capacity for 1024 events, from overflowing. The trace data, however, is only displayed to the user once the application has exited successfully.

The trace data is displayed graphically to the user as a timeline of events, similar to that shown in Figure 18.7. The size of the block represents the duration of the event and its position indicates the time at which it occurred. The events are also colour-coded, with software events, hardware function events and data transfer events being represented by different coloured blocks. This visualisation of the event data provides a fast, simple and intuitive way for the designer to analyse how the system is performing calls to functions targeted for implementation in hardware. The trace data is also archived within the project build directory. The designer is able to view archived trace data by right-clicking on the archive in the Project Explorer view and selecting the *Import and Open AXI Trace* option.

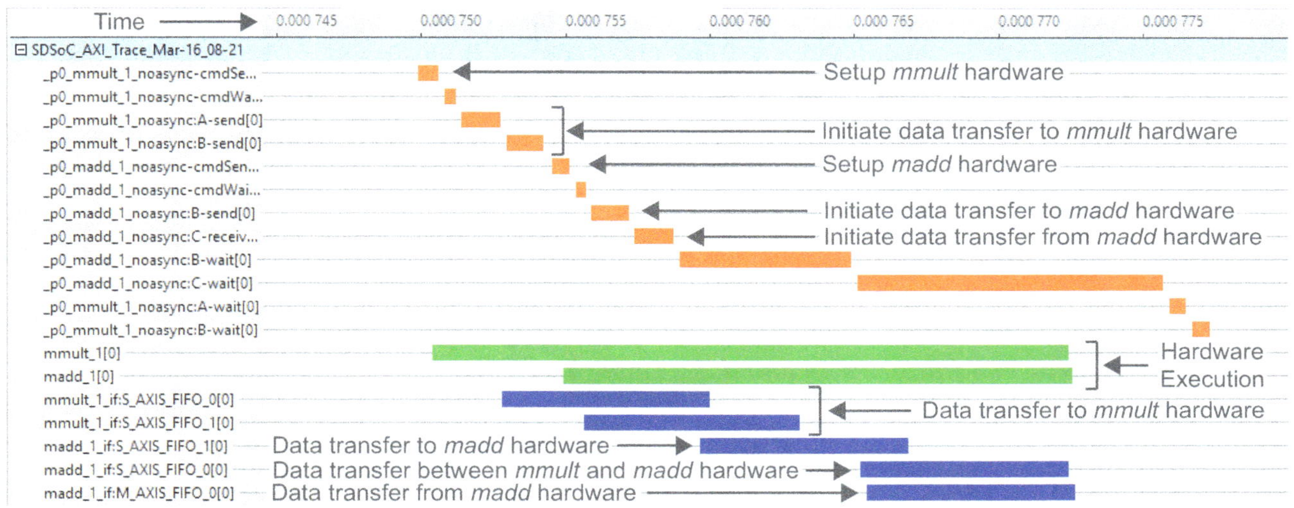

Figure 18.7: Example of a portion of trace visualisation (as in SDx IDE 2017.2)

18.4.2 AXI Performance Monitor

The rate at which data is moved between system memory and the PL can often have a significant impact on the performance of a system. The system designer may therefore wish to examine in greater detail the performance of the interfaces between the PL and PS. To assist in this task, SDx provides the AXI Performance Monitor feature. When this option is selected, the SDx compiler includes an AXI Performance Monitor (APM) LogiCORE IP block [10] in the Vivado project to record the performance of the interfaces between the PS and PL. The data recorded by this IP block can then be accessed and visualised in the SDx IDE.

To include the APM IP block in the system design, the *Insert AXI performance monitor* option should be selected in the SDx Project Settings before building the system. Similarly to building a system with debugging hardware, a build configuration that includes debug symbol information in the application executable file, such as the ***Debug*** build configuration, can be useful. This allows the application to be launched on hardware using System Debugger, and provides an opportunity to setup and initiate the APM IP block during the breakpoint at the entry of the *main* function.

We will once again connect between the host running the SDx IDE and the Zynq target using the TCF, as described in Section 16.1.1. For Zynq targets running the Linux OS, a connection between the host and target using a JTAG interface also needs to be made, in a similar fashion as described for event tracing. This is done in order for the host to access the data captured by the APM IP block.

The application is launched using System Debugger in the customary manner, namely by right-clicking the project in the Project Explorer view and selecting *Debug As > Launch on Hardware (SDSoC Debugger)*. This launches the application on the Zynq device but pauses the execution as it enters the *main* function.

While the application is paused, there is an opportunity to setup and initiate the APM IP block. First, the Performance Analysis perspective must be opened, if it is not open already. We have thus far made extensive use of the SDx and Debug perspectives within the SDx IDE and the Performance Analysis perspective is simply another available perspective. It is configured to suit the task of analysing performance with the inclusion of views for managing performance sessions and visualising the results obtained from performance monitoring. The Performance Analysis perspective is opened by selecting the *Open Perspective* button, as shown in Figure 16.1 on page 408 and choosing the Performance Analysis perspective.

At this point, the method for setting up and initiating the APM IP blocks differs depending on the OS being used on the Zynq target. We will therefore discuss them separately.

APM IP Block Setup and Initiation on Standalone and FreeRTOS

Setup and initiation of the APM IP block on Zynq targets running the Standalone or FreeRTOS OS is the much simpler of the two methods. From the Performance Analysis perspective, the appropriate PS core must be selected in the Debug view. This is the core currently paused due to the breakpoint in our debugging session, as shown in Figure 18.8. In the Performance Session Manager view, the System Debugger session pertaining to our application can be selected and started using the start button, also shown in Figure 18.8, to begin monitoring the performance of the interfaces.

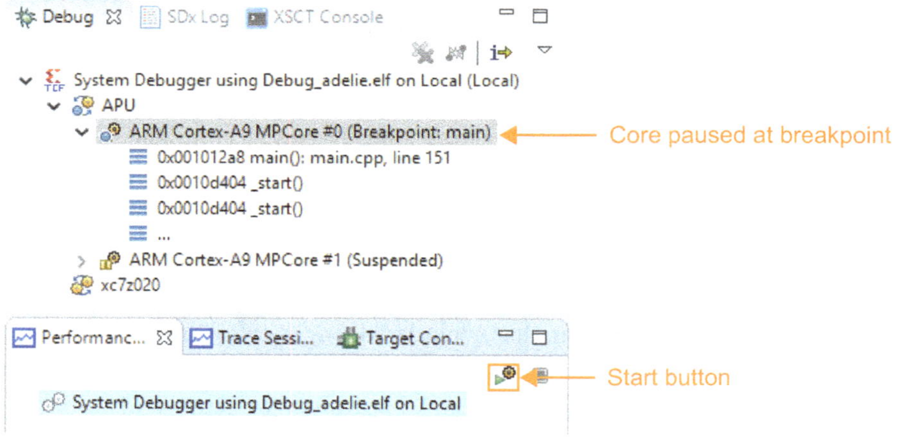

Figure 18.8: A portion of the Performance Analysis perspective showing the Debug and Performance Session Manager views (as in SDx IDE 2017.2)

APM IP Block Setup and Initiation on Linux

Setup and initiation of the APM IP block on Zynq targets running the Linux OS has an additional step, needed to establish a second TCF connection using the JTAG interface (in addition to the first TCF connection we have already established using the TCP/IP interface). This is required to enable the host to retrieve the performance monitoring data from the APM IP block.

To establish the second TCF connection a new run configuration must be created by selecting the *Run Configurations...* option from the *Run* menu. This opens the Run Configurations dialogue box. We wish to create a new *Xilinx C/C++ application (System Debugger)* run configuration, and so this option should be chosen. Under the Target Setup tab of this new run configuration, the Debug Type should be set to *Attach to running target*, as the target we are attempting to connect to is already running, and the Connection should be *Local* to connect using the JTAG interface. Clicking *Run* will establish the second TCF connection to the Zynq target, as should be visible within the Debug view of the Performance Analysis perspective.

Now that the second TCF connection using the JTAG interface has been established, the APM IP block needs to be setup and initialised to monitor the interfaces. The newly created run configuration should now be available for selection as a session in the Performance Session Manager view of the Performance Analysis perspective.

Starting this session opens the Performance Analysis Input dialogue box, where the *Enable APM Counters* option can be selected to indicate that we wish to obtain information from an APM IP block. The option to edit the APM Hardware Information can also be selected to specify the details of the APM IP block we wish to obtain performance data from. While it is possible to enter the APM IP block details manually, it is perhaps easiest to obtain them automatically from the hardware definition file produced by the IP Integrator project created during the system build. To do this, the location of the hardware definition file, which can be found within the project build directory, should be given.

Once these details have been loaded successfully, we are ready to begin monitoring the performance of our interfaces.

Capturing and Analysing AXI Performance Data

Now that the interfaces are being monitored, the paused application can be resumed. Once the application has exited, the performance session can be stopped from the Performance Session Manager view and the results will be displayed in the Performance Analysis perspective in the APM Performance Graphs view and APM Performance Counters view. Figure 18.9 shows an example of the results displayed in the APM Performance Graphs view.

AXI performance monitoring provides values for the following for each interface being monitored:

- Read and write transactions per ms;
- Minimum, maximum and average read and write latencies;
- Read and write throughput in MB/s.

It also provides graphs of these values plotted against the time over which monitoring took place, as illustrated in Figure 18.9. This information can be invaluable for analysing the flow of data between system memory and the PL, enabling the system designer to identify and address any performance bottlenecks. For example, a system streaming video data between system memory and the PL may begin to slow down due to traffic on another AXI interface. This would manifest itself in the APM Performance Graphs as a reduction in

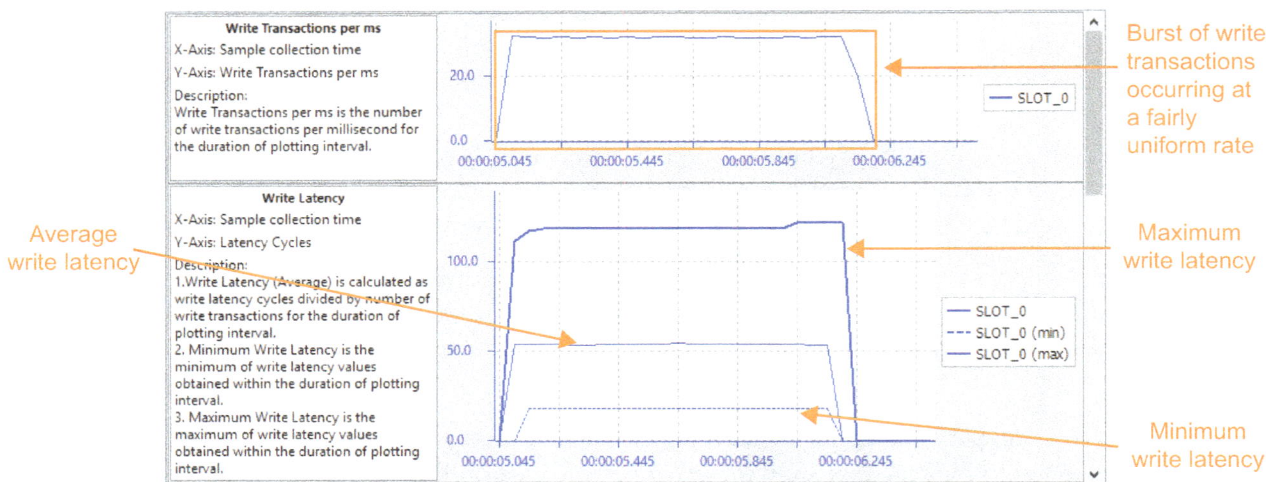

Figure 18.9: Example of the APM Performance Graphs view showing the results from AXI performance monitoring (as in SDx IDE 2017.2)

throughput on the interface used for streaming the video data, coinciding with an increase in transactions on the other AXI interface.

18.4.3 Processing System Performance Monitoring

A serendipitous side-effect of conducting AXI performance monitoring is that the SDx IDE automatically collects PS performance information at the same time. The PS performance information is gathered from the Arm performance monitor unit on the Zynq device across the JTAG interface, and is displayed in the PS Performance Graphs view and PS Performance Counters view of the Performance Analysis perspective. The information gathered includes:

- CPU utilisation — percentage of clock cycles for which the CPU was active;
- Instructions per cycle — average number of instructions executed per clock cycle when the CPU was active;
- L1 data cache miss rate — percentage of L1 cache accesses that missed;
- L1 data cache access rate — number of L1 cache accesses per millisecond;
- Read and write stall cycles per instruction — average number of clock cycles per instruction the CPU was stalled waiting for a read from or write to memory.

These values are also plotted against time, an example of which is shown in Figure 18.10. This information allows the system designer to analyse the utilisation of the PS within their system more closely, and adapt the design to make the best use of the PS and PL.

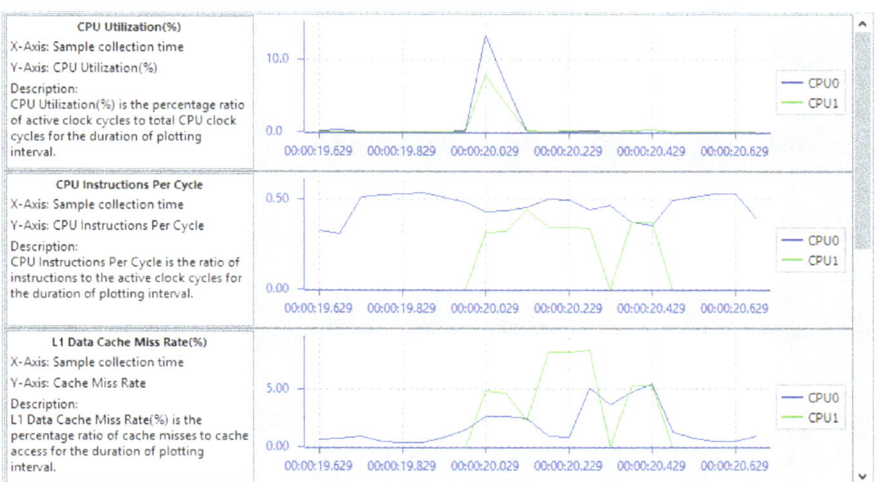

Figure 18.10: Example of the PS Performance Graphs view showing the results from performance monitoring (as in SDx IDE 2017.2)

The PS performance monitoring data can also be obtained without the need to conduct AXI performance monitoring. In fact, it is an almost identical procedure to the one described for conducting AXI performance monitoring.

The procedure differs in that the step to select the *Insert AXI performance monitor* option in the SDx Project Settings before building is omitted. In the case of Zynq targets running the Linux OS, the step to select the *Enable APM Counters* option in the Performance Analysis Input dialogue box before beginning recording performance data is also omitted.

Once recording has completed, there will be no information contained in the APM Performance Graphs view or APM Performance Counters view, but the PS performance data will be visible in the PS Performance Graphs view and PS Performance Counters view, as when conducting AXI performance monitoring.

18.5 Chapter Review

This chapter has provided an introduction to many of the features provided by the SDx for debugging and monitoring the performance of systems targeted for implementation on Zynq devices. These features were described in both the context of functional debugging of a system and performance characterisation. We have presented system emulation for the purposes of functional debugging, along with techniques for debugging the software and hardware portions of the system independently. Methods were also discussed for monitoring the performance of the software and hardware portions of system, as well as the interfaces between software and hardware. Further reading on these topics is available in [4] and [9].

18.6 References

Note: All online sources last accessed March 2019.

Some of the Xilinx documentation referred to below has version-specific URLs. If you are working in a newer version of the tools, check for updates on the Xilinx website, or try adjusting the link according to your version.

[1] QEMU, "QEMU the FAST! processor emulator" webpage.
Available: http://www.qemu-project.org/

[2] Xilinx Inc., "Virtual Input/Output v3.0 LogiCORE IP Product Guide", PG159, October 2017.
Available: https://www.xilinx.com/support/documentation/ip_documentation/vio/v3_0/pg159-vio.pdf

[3] Xilinx Inc., "System Integrated Logic Analyzer v1.0 LogiCORE IP Product Guide", PG261, June 2017.
Available: https://www.xilinx.com/support/documentation/ip_documentation/system_ila/v1_0/pg261-system-ila.pdf

[4] Xilinx Inc., "Vivado Design Suite User Guide: Programming and Debugging", UG908 (v2017.2), June 2017.
Available:
https://www.xilinx.com/support/documentation/sw_manuals/xilinx2017_2/ug908-vivado-programming-debugging.pdf

[5] Xilinx Inc., "Vivado Design Suite User Guide: Designing IP Subsystems Using IP Integrator", UG994 (v2017.2), June 2017.
Available: https://www.xilinx.com/support/documentation/sw_manuals/xilinx2017_2/ug994-vivado-ip-subsystems.pdf

[6] Xilinx Inc., "Zynq-7000 All Programmable SoC Software Developers Guide", UG821 (v12.0), September 2015.
Available: https://www.xilinx.com/support/documentation/user_guides/ug821-zynq-7000-swdev.pdf

[7] Xilinx Inc., "Zynq Ultrascale+ MPSoC Software Developer Guide", UG1137 (v6.0), January 2018.
Available: https://www.xilinx.com/support/documentation/user_guides/ug1137-zynq-ultrascale-mpsoc-swdev.pdf

[8] DENX Software Engineering, "WebHome < U-Boot < DENX" webpage, 20th April 2017.
Available: http://www.denx.de/wiki/U-Boot/WebHome

[9] Xilinx Inc., "SDSoC Environment User Guide", UG1027 (v.2017.2), August 2017.
Available: http://www.xilinx.com/support/documentation/sw_manuals/xilinx2017_2/ug1027-sdsoc-user-guide.pdf

[10] Xilinx Inc., "AXI Performance Monitor v5.0 LogiCORE IP Product Guide", PG037, October 2017.
Available: https://www.xilinx.com/support/documentation/ip_documentation/axi_perf_mon/v5_0/pg037_axi_perf_mon.pdf

Chapter 19

Custom SDx Platforms

SDx Platforms provide the base hardware design and software components that are augmented by application-specific hardware and software to create an SDx project. The composition of an SDx Platform can, therefore, be fundamental in determining the capabilities of SDx projects targeting it.

In this chapter, we will discuss the process of creating a custom SDx Platform. We will examine why we may wish to create a custom SDx Platform, the key components required and how they can be generated. We will also briefly outline procedures for testing and using a custom SDx Platform.

19.1 Why might we want to develop a custom SDx Platform?

Every SDx project must target an SDx Platform. While there are a limited range of SDx Platforms supplied with SDx and there are others available from third parties that will fulfil the requirements of many system designs, some system designs will require a custom SDx Platform to be developed. The most obvious reason that a custom SDx Platform would need to be developed is when targeting a board not supported by existing platforms. However, a system designer may also need to develop a custom SDx Platform to cater for a specific application by:

- Adding drivers or changing the OS;
- Changing PS settings;
- Adding IP blocks to the PL or changing interfaces;
- Providing a completed PL design as a starting point for software development.

Development of a custom SDx Platform is much more akin to the traditional hardware development techniques than the software defined flow employed by SDx. This is because the SDx Platform provides the low-level hardware design and software components that enable the abstraction exploited by the software defined flow. When developing a custom SDx Platform it can be helpful to refer to the SDx Platforms supplied with SDx, which can be found in the */platforms/* and */samples/platforms/* subdirectories of the SDx installation directory.

19.2 SDx Platform Structure and Components

To begin our description of the SDx Platform creation process we will examine the structure and the components that make up an SDx Platform. This will give a clearer idea of the elements that need to be generated in order to create one.

An SDx Platform is made up of many constituent parts. An illustration of the principal parts and structure of a typical SDx Platform is shown in Figure 19.1. One of the main points to note from the directory structure shown in Figure 19.1 is that the majority of the components making up the platform are contained in either the *hw* or *sw* directories. These directories represent the hardware and software portions of the platform respectively. It can be useful to consider these two elements somewhat separately from one another and, indeed, when creating an SDx Platform, the hardware portion tends to be developed first, followed by the software portion.

The hardware portion of an SDx Platform is based on a Vivado Design Suite project. It defines the base hardware design that can be extended by the SDx compiler when building projects targeting the platform. The base hardware design defines the configuration of the PS, the interfaces and clock signals available between the PL and PS, any interfaces directly between the PL and external devices and any additional IP blocks to be included as part of the platform. We go into further detail about the hardware component in Section 19.3.

The software portion of an SDx Platform, at a minimum, defines the OSs supported by the platform and provides the files necessary to boot the Zynq target using these OSs. In addition to these there are a number of optional elements that can be included in the software portion of an SDx Platform. As such, the contents and structure of the *sw* directory of a platform, as shown in Figure 19.1, are subject to a much greater degree of variation than the *hw* directory. Optional elements of the software portion of a platform include software libraries that can be included in applications targeting the platform, a pre-built version of the base hardware design to expedite building projects that do not include any additional hardware and files to support the SDx emulation flow with the platform. A closer examination of the software component is made in Section 19.4.

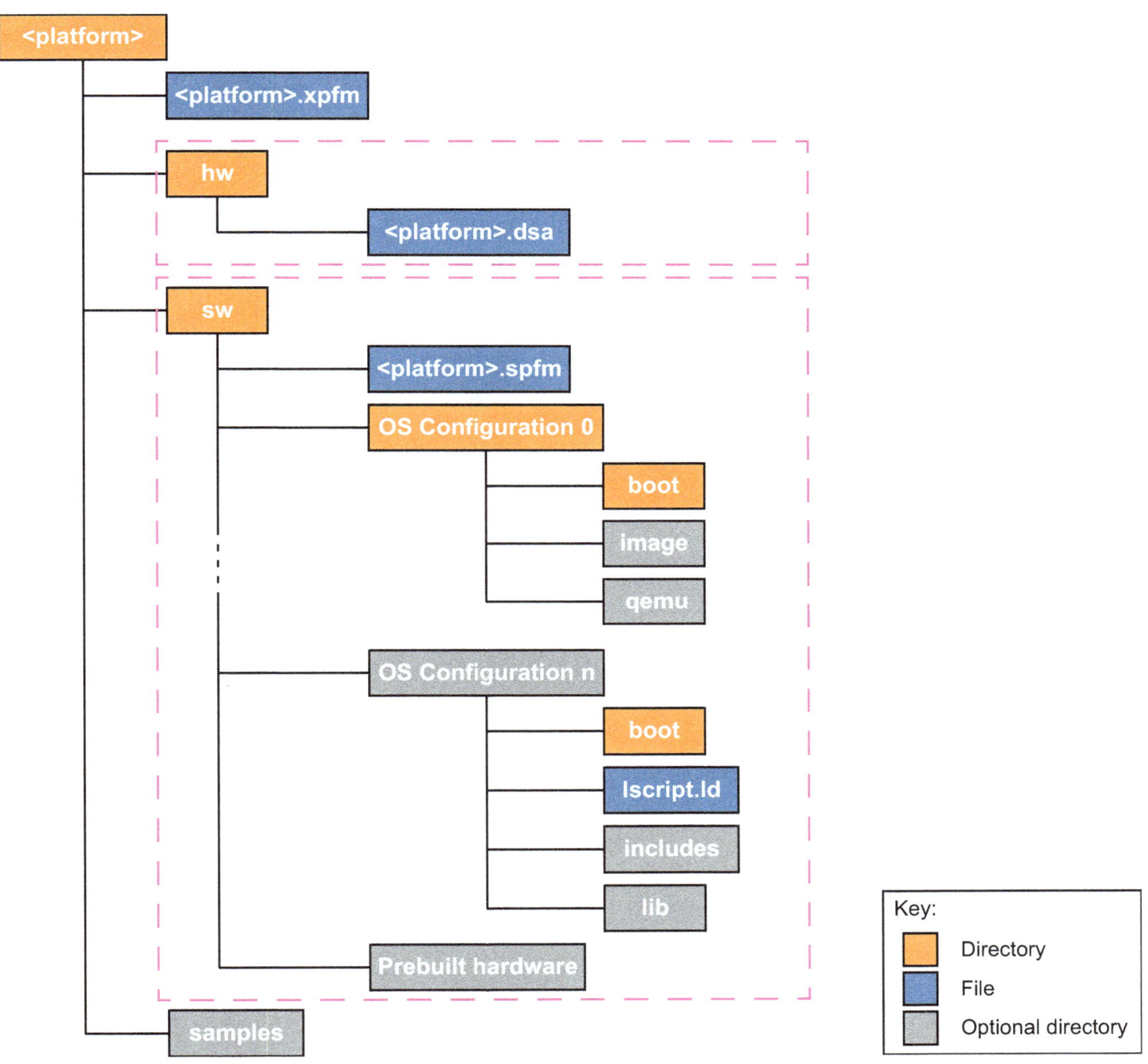

Figure 19.1: Illustrative example of principal components and structure of an SDx Platform

19.2.1 Metadata Files

There are three metadata files present in all SDx Platforms: a hardware metadata file, a software metadata file and a top-level platform metadata file. The purpose of these files is to convey the necessary information about the hardware and software components of the platform to the SDx compiler to enable a system targeting the platform to be built. Once all of the hardware and software components of the platform are available, SDx's ***sdspfm*** utility can be used to generate the metadata files and this is discussed further in Section 19.6.

The hardware and software metadata files are discussed further in Section 19.3 and Section 19.4, respectively. The top-level platform metadata file is found in the top-level directory of the platform and is named *<platform>.xpfm*, as shown in Figure 19.1. An illustrative example of a top-level platform metadata file produced by the ***sdspfm*** utility is shown in Figure 19.2. This file contains the location of the platform Device Support Archive (DSA) and software metadata files and a description of the platform. The DSA file contains the Vivado Design Suite project for the platform hardware design, as well as the hardware metadata file.

```
<?xml version="1.0" encoding="UTF-8"?>
<sdx:platform sdx:vendor="strath.ac.uk"
              sdx:library="sdx"
              sdx:name="zavodovski"
              sdx:version="1.0"
              xmlns:sdx="http://www.xilinx.com/sdx">

    <sdx:description>Platform to demonstrate custom platform generation</sdx:description>

    <sdx:hardwarePlatforms>
        <sdx:hardwarePlatform sdx:path=”hw” sdx:name=”zavodovski.dsa”/>
    </sdx:hardwarePlatforms>

    <sdx:softwarePlatforms>
        <sdx:softwarePlatform sdx path=”sw” sdx:name=”zavodovski.spfm”/>
    </sdx:softwarePlatforms>

</sdx:platform>
```

Figure 19.2: Example of top-level platform metadata file

19.2.2 Sample Applications

The final part of the directory structure shown in Figure 19.1 is the optional *samples* directory. This directory contains example applications to be used with the platform. When located within this directory, along with the appropriate metadata, the example applications are made available as templates when an SDx project targeting this platform is created within the SDx IDE. A more detailed explanation of how to include sample applications in a custom SDx Platform is given in Section 19.5.

19.3 SDx Platform Hardware Component

Typically, the first step in the creation of a custom SDx Platform is to generate the hardware component. This step can be further subdivided into two main tasks: development of the platform hardware design in a Vivado Design Suite project and the generation of the platform hardware metadata file.

19.3.1 Developing the Platform Hardware Design

In essence, the platform hardware design is simply a hardware design targeting the appropriate Zynq device or board. More information on creating such a design can be found in [1], [2] and [3]. In order to make the platform hardware design compatible with SDx, however, there are several criteria that the design must meet.

The design must be contained in a Vivado Design Suite project whose name matches that of the platform being developed and includes an IP Integrator block diagram. The IP Integrator block diagram must contain a PS IP block that is used to define the PS configuration. Any additional hardware required as part of the platform, such as hardware required to interface the PL to external devices, must also be included. Any platform hardware that is not part of the standard Vivado IP catalogue needs to be included locally to the Vivado design project using the procedure described in [1].

In order for the SDx compiler to be able to add data movers and hardware function blocks when building SDx projects, the SDx Platform base hardware design must make interfaces available to the SDx compiler. At the very minimum, one general purpose AXI master port and one AXI slave port must be made available on the PS. The general purpose AXI master port is required to enable software executing on the PS to control IP added by the SDx compiler, while the AXI slave port is required to allow IP added by the SDx compiler to access system memory.

Of course, it is likely that the platform designer will wish to make more interfaces than the bare minimum available to the SDx compiler. The following types of interfaces are compatible with the SDx compiler:

- AXI4 or AXI4-Stream;
- Clock;
- Reset;
- Interrupt.

Although these types of interfaces are compatible for use by the SDx compiler, there are still some caveats associated with using them. AXI4-Stream interfaces must include the *TLAST* and *TKEEP* sideband signals to be compatible with the data movers instantiated by the SDx compiler. Clock signals must be connected via *Processor System Reset* IP blocks to provide synchronised reset signals, as shown in Figure 19.3. Interrupt signals must be collected together and connected to the interrupt ports of the PS IP block via a *Concat* IP block, also shown in Figure 19.3. Other interfaces can be used internally in the base hardware design, but SDx (or any IP cores generated by SDx) will only connect to the interfaces mentioned above.

Figure 19.3: Example of a base hardware design IP Integrator block diagram showing the use of Processor System Reset and Concat IP blocks

Where platform hardware is connected to the AXI ports of the PS, the designer may want to have the option to share this AXI port with hardware generated by SDx projects targeting the platform. To do this, an *AXI Interconnect* IP block must be connected to the AXI port in question. The platform hardware can be connected to the AXI port through the *AXI Interconnect* IP block, while unused AXI ports on the *AXI Interconnect* IP block can be made available to the SDx compiler to allow other hardware to be connected to the PS AXI port as well. The ports on the *AXI Interconnect* IP block with the least significant indices should be used to connect to any IP cores in the base hardware design, leaving the port with the highest index unused and available for SDx. A similar scenario exists for the interrupt signals connected to the interrupt ports of the PS through *Concat* IP blocks. Interrupt signals for platform hardware should be connected to the least significant bits of the interrupt ports.

Once complete, the platform hardware block design should be validated. The output products of the IP it contains can then be generated and a top-level RTL wrapper for the design created. This allows a hardware definition file describing the platform hardware design to be exported from the Vivado Design Suite project for use in generating the platform software components. Details on carrying out these tasks can be found in [3].

To enable its long-term use, it is likely that a platform designer would wish to maintain their SDx Platform through multiple releases of SDx. Each release of SDx includes and targets the latest version of the Vivado Design Suite tools. With each release of the Vivado Design Suite tools, many of the IP blocks from the standard Vivado IP catalogue are upgraded. If any of these IP blocks form part of the platform base hardware design then the new version of the Vivado Design Suite tools will require that they are upgraded in the platform Vivado Design Suite project. This is a simple process described in [3]. If the IP blocks within the SDx Platform base hardware design are not adequately upgraded prior to a build within SDx being attempted, *IP Locked* errors may occur when the Vivado Design Suite tools are invoked by the SDx compiler.

19.3.2 Platform Hardware Metadata File

The SDx compiler needs to be able to determine the interfaces and signals made available to it by the platform base hardware design for connecting additional hardware generated during a system build. This information is conveyed to the SDx compiler by the platform hardware metadata file, which forms part of the DSA file shown in Figure 19.1.

The platform hardware metadata file can be generated automatically by the ***sdspfm*** utility based on a Tcl script that defines the interfaces and signals available to the SDx compiler in the base hardware design. This Tcl script is manually written and can be considered to have up to six steps:

- Specify the name of the platform hardware metadata file with an identifier and description for the platform;
- Specify the platform clock signals available to the SDx compiler;
- Specify the AXI4 and AXI4-Stream interfaces available to the SDx compiler;
- Specify platform input/output (I/O) devices that will be used as Linux I/O platform devices;
- Specify the platform interrupt signals available to the SDx compiler;
- Write the platform hardware metadata file.

The specification of platform I/O devices and interrupt signals are optional steps depending on the content of the platform hardware design and the intended use of the platform.

We will describe the steps necessary to create a platform hardware metadata file and the Tcl commands used to perform them using an example Tcl script. The script has been divided into sections here to aid the discussion of each step.

The process of generating the platform hardware metadata file begins by creating a new platform file object, which is achieved using the `sdsoc::create_pfm` command, as shown in Figure 19.4. This command is passed an argument specifying the name of the platform hardware metadata file using the naming convention *<platform>.hpfm*. The command returns a handle to the platform file object, which can be used to refer to the platform file subsequently in the script. In Figure 19.4, the handle to the platform file object is assigned to the variable `pfm`.

The name and description fields of the platform file object are set using the `sdsoc::pfm_name` and `sdsoc::pfm_description` commands respectively and these are largely self-explanatory. The only point to note with regard to the name field is that it should be given in the vendor, library, name, version format.

Specification of the clock signals available to the SDx compiler follows the setting of the name and description fields in Figure 19.5. This is performed using the `sdsoc::pfm_clock` command, which has six arguments. The first argument is the handle to the platform file object we are setting the values in. The second and third arguments are the names of the clock port and of the IP block instance that the clock port

```
# create new platform file object, specifying the name of the platform hardware metadata file and
# assign the handle to this object to a variable
set pfm [sdsoc::create_pfm zavodovski.hpfm]

# set the name and description fields of the platform file object
sdsoc::pfm_name $pfm "strath.ac.uk" "xd" "zavodovski" "1.0"
sdsoc::pfm_description $pfm "Platform to demonstrate custom platform creation"
```

Figure 19.4: Example of Tcl commands used to specify the name of a platform hardware metadata file with an identifier and description for the platform

belongs to respectively. The fourth argument assigns a unique identifier to the clock signal, which can be any non-negative integer. The fifth argument designates whether the clock signal is the default clock signal for the platform, of which there must be exactly one in every SDx Platform. The final argument is the instance name of the *Processor System Reset* IP block associated with that clock signal. These entries must have corresponding instances in the design as the tool will check for them.

```
# specify clock signals for use by the SDx compiler
sdsoc::pfm_clock $pfm pl_clk0 zynq_ultra_ps_e_0 0 true proc_sys_reset_0
sdsoc::pfm_clock $pfm pl_clk1 zynq_ultra_ps_e_0 1 false proc_sys_reset_1
sdsoc::pfm_clock $pfm pl_clk2 zynq_ultra_ps_e_0 2 false proc_sys_reset_2
sdsoc::pfm_clock $pfm pl_clk3 zynq_ultra_ps_e_0 3 false proc_sys_reset_3
```

Figure 19.5: Example of Tcl commands used to specify the platform clock signals available to the SDx compiler

AXI4 and AXI4-Stream interfaces available for use by the SDx compiler are designated using the `sdsoc::pfm_axi_port` and `sdsoc::pfm_axis_port` commands respectively, as shown in Figure 19.6. Both of these commands take the same four arguments. The first three arguments are the handle to the platform file object, the name of the port and the name of the IP block instance that the port belongs to respectively. The final argument describes the type of port being specified. For specifying AXI4-Stream ports the final argument can either be `M_AXIS` or `S_AXIS`, depending on whether the port is a master or slave. When specifying AXI4 ports there is a wider range of possible values for the final argument, which are summarised in Table 19.1.

Table 19.1: AXI4 port type argument values

Argument Value	Port Type
`M_AXI_GP`	General purpose master
`S_AXI_HP`	High performance slave
`S_AXI_ACP`	Accelerator coherent slave
`MIG`	Slave connected to Memory Interface Generator memory controller

```
# specify AXI4 ports for use by the SDx compiler
sdsoc::pfm_axi_port $pfm M_AXI_HPM0_FPD zynq_ultra_ps_e_0 M_AXI_GP
sdsoc::pfm_axi_port $pfm M_AXI_HPM1_FPD zynq_ultra_ps_e_0 M_AXI_GP
sdsoc::pfm_axi_port $pfm M_AXI_HPM0_LPD zynq_ultra_ps_e_0 M_AXI_GP
sdsoc::pfm_axi_port $pfm S_AXI_HP1_FPD  zynq_ultra_ps_e_0 S_AXI_HP
sdsoc::pfm_axi_port $pfm S_AXI_HP2_FPD  zynq_ultra_ps_e_0 S_AXI_HP
sdsoc::pfm_axi_port $pfm S_AXI_HP3_FPD  zynq_ultra_ps_e_0 S_AXI_HP


# specify AXI4 ports on platform AXI Interconnects for use by the SDx compiler
for {set i 1} {$i < 16} {incr i} {
    sdsoc::pfm_axi_port $pfm S[format %02d $i]_AXI axi_interconnect_0 S_AXI_HP
}


# specify AXI4-Stream ports for use by the SDx compiler
sdsoc::pfm_axis_port $pfm M_AXIS stream_fifo M_AXIS
```

Figure 19.6: Example of Tcl commands used to specify the AXI4 and AXI4-Stream interfaces available to the SDx compiler

When specifying the ports available for use by the SDx compiler on platform *AXI Interconnect* IP blocks, it can be beneficial to specify as many free ports up to the maximum allowed on a single *AXI Interconnect* IP block for that type of port. This avoids the SDx compiler unnecessarily connecting additional *AXI Interconnect* IP blocks to the original in a cascade due to a lack of available AXI ports on the platform *AXI Interconnect* IP block. The specification of multiple ports available to the SDx compiler on a single *AXI Interconnect* IP block can be done programmatically within the Tcl script, as shown in Figure 19.6. Table 19.2 shows the maximum number of ports permissible on a single *AXI Interconnect* IP block by port type.

Table 19.2: Maximum number of ports permissible on a single AXI Interconnect IP block [4]

Port Type	Maximum Number of Ports
General purpose master	64
Accelerator coherent slave	8
High performance slave/Memory Interface Generator slave	16

Any platform I/O devices that are to be used as Linux I/O platform devices should be declared using the `sdsoc::pfm_iodev` command, as shown in Figure 19.7. This is necessary so that other Linux I/O platform devices can be correctly configured in code generated by the SDx compiler. The first argument of the `sdsoc::pfm_iodev` command is the handle to the platform file object. The second argument is the name of the I/O port with the third being the instance name of the I/O device IP block. The final argument of the `sdsoc::pfm_iodev` command is the type of the device. For example, if the driver for the device uses the Linux user-space I/O (UIO) framework it would be `uio`, whereas if it uses a Linux kernel driver it would be `kio`. The Linux UIO framework will be discussed a little further under "Linux Kernel" on page 462.

```
# specify I/O devices for use by the SDx compiler
sdsoc::pfm_iodev $pfm S_AXI axi_gpio_0 uio
```

Figure 19.7: Example of Tcl commands used to specify platform I/O devices that will be used as Linux I/O platform devices

The final fields of the platform file object to be assigned values in Figure 19.8 are populated by specifying the interrupt signals available to the SDx compiler. This is accomplished using the `sdsoc::pfm_irq` command, which takes three arguments, these being the handle to the platform file object, the name of the interrupt signal being specified and the instance name of the *Concat* IP block the signal belongs to, respectively. Similarly to specifying available ports on an *AXI Interconnect* IP block, multiple interrupt signals can be specified programmatically, as shown in Figure 19.8.

Once all the fields of the platform file object have been populated, the platform hardware metadata file is ready to be generated, which is achieved by invoking the `sdsoc::generate_hw_pfm` command, as shown in Figure 19.8. The handle to the platform file object is passed as an argument to the command, which generates the platform hardware metadata file with the name specified when creating the platform file object.

```
# specify interrupt signals for use by the SDx compiler
for {set i 1} {$i < 8} {incr i} {
   sdsoc::pfm_irq $pfm In$i xlconcat_0
}


# generate platform hardware metadata file
sdsoc::generate_hw_pfm $pfm
```

Figure 19.8: Example of Tcl commands used to specify the platform interrupt signals available to the SDx compiler and write the platform hardware metadata file

When using the ***sdspfm*** utility, generation of the platform hardware metadata file is abstracted from the platform creator as the utility produces the DSA file, which incorporates both the platform hardware metadata file and the base hardware Vivado Design Suite project.

A more detailed description of the platform hardware metadata file creation Tcl API is available in [4].

19.4 SDx Platform Software Component

The software component of an SDx Platform contains all of the software elements required by the SDx compiler to build boot images for each of the OSs supported by the platform. Optionally, it can also include pre-built hardware files, to speed up the system build process when there is no hardware implemented in the system other than the platform base hardware design, and files required to support the SDx emulation flow.

19.4.1 Boot Files

We will begin by discussing the software elements required by the SDx compiler to create boot images. One of the objectives of the SDx compiler when building an SDx project is to create a set of files that can be used to:

- Boot the target Zynq device;
- Configure the PS and PL;
- Load the target OS and/or the application executable.

These files constitute the boot image. While the SDx project is responsible for supplying the application-specific executable and bitstream, the SDx Platform supplies the remainder of the components used to build a boot image. Exactly what these components are depends on the Zynq device and OS being targeted. A summary of the boot image components that need to be supplied by the SDx Platform are given in Table 19.3 arranged by the target Zynq device and OS.

We will divide our discussion into two, based on the target OS. Firstly, we will describe the boot files supplied by the SDx Platform to create boot images for systems targeting the Standalone and FreeRTOS OSs and then follow this by describing the boot files required for systems targeting the Linux OS. Some of the software elements that we will discuss, notably the First Stage Boot Loader (FSBL) and readme file templates, are common to more than one OS and so there will be some overlap between our descriptions of the boot files for the different OSs. In the case of the Zynq MPSoC, there are two processing units, the APU and RPU, that can be used to boot the device. A separate set of boot files is required for each processing unit that the SDx Platform supports to boot the device.

Table 19.3: Summary of boot image components supplied by the SDx Platform arranged by target Zynq device and OS

Zynq Device	OS	Boot Image Components
Zynq-7000	Standalone/FreeRTOS	First stage boot loader
		Linker script
		Readme file template
	Linux	First stage boot loader
		Second stage boot loader
		Linux kernel
		Device tree
		Root file system
		Readme file template
Zynq MPSoC	Standalone	First stage boot loader
		Platform management unit firmware
		Linker script
		Readme file template
	Linux	First stage boot loader
		Platform management unit firmware
		Arm trusted firmware
		Second stage boot loader
		Linux kernel
		Device tree
		Root file system
		Readme file template

First Stage Boot Loader

The FSBL is the first part of the boot image to be loaded into memory and executed when the Zynq device boots. It is responsible for initialising the PS, configuring the PL and then loading and handing off to either the Second Stage Boot Loader (SSBL) or the application executable. The initialisation of the PS carried out by the FSBL is based on the configuration defined in the platform hardware design and contained in the hardware design file exported from the platform hardware Vivado Design Suite project. The content of the FSBL is not influenced by the target OS and so only one FSBL need be built for each platform processing unit, which can then be used to boot all of the OSs supported by that platform processing unit.

The FSBL can be built from a template as an application project within SDx. The FSBL project should be based on the Standalone OS to make use of low-level software providing access to basic PS features. A target hardware platform must be created based on the hardware design file exported from our platform hardware project and targeting the PS processor that the FSBL will be executed on. The hardware platform provides the code to initialise the PS based on the configuration defined in the platform hardware project. The Zynq FSBL or Zynq MP FSBL application templates should be selected for the Zynq-7000 and Zynq MPSoC respectively.

The executable file produced from building the FSBL project is the FSBL file for inclusion in the SDx Platform. Further information on building an FSBL can be found in [5].

Linker Script

For SDx projects targeting either the Standalone or FreeRTOS OSs, a linker script is required. This script specifies the locations in memory to be used by the application when it is executed. The linker script for SDx projects is supplied by the SDx Platform. The platform designer can write the linker script manually, although it is perhaps more efficient to use an automatically generated script, at least as a basis.

Automatic generation of a linker script is as simple as creating a new application project targeting the OS, processor and hardware platform on which applications targeting this SDx Platform are intended to be executed. The hardware platform generated when creating the FSBL can be reused for this purpose. Following creation of the application project, the linker script can be found in the directory containing the source files for the application project. The platform designer may wish to make modifications to this script, such as the size of the stack or heap, before including the script in the SDx Platform.

Readme File Template

When the option to generate an SD card image is selected in the SDx Project Settings, the SD card image produced by the SDx compiler includes a text file called *README.txt*. Typically, this text file contains information pertaining to the system built by the SDx project and instructions on its use. The text file is generated from a template text file supplied by the SDx Platform. An example of one of these templates is shown in Figure 19.9. A pleasing feature of these templates is the use of place-holder strings, such as *<platform>* and *<elf>* in Figure 19.9, that are automatically replaced in the README.txt file included in the SD card image with the appropriate values obtained from the project being built.

```
-= SD card boot image =-

Platform: <platform>
Application: <elf>

ZCU102 R5 Application

1. Copy the contents of this directory to an SD card
2. Set boot mode to SD
     DIP switch SW6:
          Mode0 (#1) OFF
          Mode1 (#2) ON
          Mode2 (#3) OFF
          Mode3 (#4) ON
3. Insert SD card and turn board on
```

Figure 19.9: Example of a readme file template from the ZCU102 SDx Platform

Platform Management Unit Firmware

For SDx Platforms targeting Zynq MPSoC devices, platform management unit (PMU) firmware is also required to create boot images. The PMU firmware is loaded into PMU RAM during system boot and implements the PMU-side of the API used to manage the power consumed by the Zynq MPSoC. This is a simplified description of the PMU firmware, but is sufficient for our needs here. Further details of the PMU firmware and the power management API are available in Chapter 10, [6] and [7].

The PMU firmware is a user-defined program and it is possible that the platform designer would wish to supply a custom implementation for use with their platform. We will assume here, however, that the template PMU firmware program supplied by SDx is to be used. The process for generating the template PMU firmware is very similar to the one used for the FSBL. It is created using a new application project targeting the hardware platform created when building the FSBL, the processor on the PMU and the Standalone OS. A new board support package should be created and the *ZynqMP PMU Firmware* template application should be selected.

The executable file created from building the PMU firmware application project is the file to be included in the SDx Platform. The same PMU firmware executable file can be used when either the APU or RPU is used to boot the Zynq MPSoC.

Linux Boot Files

A number of the boot files used to support the Linux OS are generated in the same way as they are to support the Standalone or FreeRTOS OSs. In fact, if they have already been generated to support the Standalone or FreeRTOS OSs, then they can simply be reused to support the Linux OS as well.

These files are the:

- FSBL;
- PMU firmware;
- Readme file template.

The remainder of the boot files necessary to support the Linux OS are specific to this task and must be created afresh.

An entire book could be written about embedded Linux development. Indeed, there are many fine volumes on the subject. We do not, therefore, intend to provide a comprehensive review of the tools and options available for creating Linux boot files. Instead, we intend to give the reader a general overview of the components required, highlight specific points pertinent to SDx Platform development for Zynq devices and suggest useful support and resources for the task.

Arm Trusted Firmware

The first Linux-specific component we will discuss is the Arm Trusted Firmware (ATF). The ATF is required to support the Linux OS running on Zynq MPSoC devices. When the Linux OS is used, implementation of the power management API requires more privileged access to processing resources than is typically available to user applications. The ATF is therefore used to implement the power management API at this higher level of privilege and to provide an interface to user applications using Secure Monitor Calls [6]. Although the ATF is an extensive piece of firmware, only the runtime services portion is necessary to implement the power management API. The runtime services portion of the ATF is initialised by the Stage 3-1 Boot Loader (BL31) of the ATF. This is an executable file and is loaded by the FSBL prior to loading the second stage Boot Loader [8]. BL31 can be built for the Zynq MPSoC platform from the sources and instructions supplied by Xilinx [9][10], although a ready-made BL31 file is also available as part of the ZCU102 SDx Platform supplied with SDx, which can be copied into new custom platforms.

Second Stage Boot Loader

The next component that we will discuss is the SSBL, which is necessary to support the Linux OS for every Zynq device. The purpose of the SSBL is to load the Linux kernel into memory and pass control to it. There are a wealth of second stage boot loaders available that could be used to boot Linux on a Zynq device, however, we will focus our discussion here on Das U-Boot [11] as it is perhaps the best supported SSBL for Zynq devices. Xilinx makes available the sources to build the Das U-Boot executable file required to boot Linux, including configuration files used to configure the build for some popular Zynq development boards [12]. When developing an SDx Platform for a Zynq board that is not yet supported with a Das U-Boot build configuration, these files can serve as templates for developing appropriate custom build configuration files. Instructions for building the Das U-Boot SSBL executable file can be found in [13].

Linux Kernel

As we mentioned in our above discussion of the SSBL, the Linux kernel is loaded into memory by the SSBL. The Linux kernel is, of course, freely available [14] and can be configured and compiled to run on Zynq devices. The amount of effort required is likely to be significant, however, and the prospect of tackling this task may seem more than a little terrifying to the inexperienced Linux developer. Happily, this undertaking can be substantially ameliorated by making use of the Xilinx fork of the Linux kernel [15]. This fork not only augments the mainline Linux kernel with Zynq-specific drivers, but also includes files for configuring the kernel with default settings suited to the Zynq devices that can be used as a starting point for building a custom Linux kernel. Instructions for building the Linux kernel are available in [16].

When creating a Linux kernel for use in an SDx platform, there are a couple of configuration settings that may need to be adjusted to make the kernel compatible with, and improve the functionality of, SDx projects targeting the Linux OS. The first of these is to increase the default memory area size for the DMA Contiguous Memory Allocator (CMA) driver. This determines the amount of memory that is specifically managed by the kernel to allow large areas of physically contiguous memory to be allocated when required [17]. The recommended default memory area size is 256MiB for the Zynq-7000 and 1024MiB for the Zynq MPSoC [4], although, the amount to specify will also depend on how much memory the system has. The second setting that may need to be adjusted is the inclusion of the Xilinx DMA engine driver. Details of making both of these adjustments are given in [4].

While on the subject of Linux kernel configuration, there is another family of device drivers that we will specifically draw attention to, namely the Linux User space I/O (UIO) framework drivers [18]. Linux device driver development and maintenance is notoriously challenging, particularly when the hardware the driver is to interface with does not fit well within any of the kernel subsystems, as is often the case with custom hardware developed in the PL. As was discussed in Chapter 12, device drivers are implemented in the kernel space of the Linux software stack, which is where much of the difficulty in their development and maintenance stems from. The Linux UIO framework is intended to offer a solution to this problem by providing device drivers that perform a minimal and highly generic set of driver tasks within the kernel space. This allows the vast majority of the device-specific driver behaviour to be written in user space, where most users have more development experience. When an SDx Platform that contains custom hardware in the PL is being developed, the Linux UIO framework is often the logical choice to implement the Linux driver for the custom hardware. In order to support this, the Linux kernel used in the SDx Platform must be configured to include the appropriate Linux UIO driver module. The user space portion of the custom hardware device driver can be incorporated into the application, just like any other piece of code, including as a library in the SDx Platform, which we will discuss further in Section 19.4.4.

Device Tree

The purpose of the device tree is to specify the hardware present in the system to the Linux kernel. The information passed about each device can include:

- The address(es) of the device,

- The driver that should be used to control the device,
- The interrupt signal(s) used by the device [19], [20].

A binary representation of the device tree, called a binary large object or blob, is passed to the kernel at boot time by being loaded to a specific address in memory. It is this device tree blob that is included in the SDx Platform boot files. The device tree blob is produced by compiling a device tree structure file using a device tree compiler. The simplest way to obtain a device tree compiler is as a pre-built package on the host OS. Alternatively, a device tree compiler can be built from source files provided in both the mainline and Xilinx fork of the Linux kernel using the instructions found at [21] and [22] respectively.

Device tree structure files are simply text files describing the device tree data structure. These source files can be created manually and, if this path is chosen, [19] and [20] are invaluable sources of information for accomplishing this task. However, it is also possible to have SDx automatically generate the device tree structure files using the hardware design file exported from the platform hardware project and it is this process that we will now describe.

Device tree generation is not supported by SDx by default and so we need to obtain some additional source files for the device tree generator and add these as a repository in SDx. Xilinx makes these source files available to be cloned from [23]. To add these files as a repository and to generate the device tree structure files we will make use of the Hardware Software Interface (HSI) environment [24]. The HSI environment is invoked from within an SDx Terminal using the `hsi` command. The HSI commands used to generate our device tree structure files are shown in Figure 19.10.

```
% Open the platform hardware design
open_hw_design <platform_hardware_design_file>

% Add the cloned device tree generator repository
set_repo_path {<cloned_device_tree_generator_repository>}

% Create a new software design module targeting the device tree generator and appropriate CPU
create_sw_design device-tree -os device_tree -proc <target_cpu>

% Generate the device tree source files
generate_target -dir dts_files
```

Figure 19.10: HSI commands to generate device tree source files [22][24]

The commands shown in Figure 19.10 are largely self-explanatory, however, we include some comments here to assist in elucidating their usage and meaning. The command to add the cloned device generator repository shows braces around the cloned device tree generator repository path. These are actually only necessary on hosts using a Windows OS. The value of the `target_cpu` argument in the command to create the software design module can be `ps7_cortexa9_0` or `psu_cortexa53_0` depending on whether the Zynq-7000 or Zynq MPSoC device is being targeted.

After running the commands shown in Figure 19.10 a directory called *dts_files* is created that contains the device tree structure files for our platform hardware design. The platform developer is free to edit these device tree structure files prior to compiling them into the device tree blob.

A particularly important edit that must be made to the top-level device tree structure file in order to make it compatible for use with SDx, is to append an additional device tree fragment to the end of the file to represent the Xilinx AXI DMA engines that may be added to the system by the SDx compiler. The necessary fragment is shown in Figure 19.11 [4].

```
/{
    xlnk {
        compatible = "xlnx,xlnk-1.0";
    };
};
```

Figure 19.11: Device tree to be appended to the device tree source files in order to use SDx with a booted Linux system [4]

It is also recommended to edit the *bootargs* value in the top-level device tree structure file to include the *quiet* option [4]. This option suppresses output messages from the kernel during boot.

Root File System

The root file system is mounted by the kernel during the boot process and usually contains the first application loaded by the kernel; the *init* application. Root file systems can also contain kernel modules, other applications and the libraries to support them. The precise content and structure of a root file system can be extremely varied, particularly for embedded Linux systems where these factors are predominantly driven by the intended application of the system. Guidelines for the construction of root file systems are provided by the Linux Foundation's Filesystem Hierarchy Standard [25], which is relatively short and easy to read.

Embedded Linux systems, such as those likely to target Zynq devices, often aim to minimise the size of the root file system, particularly if the root file system is mounted in system memory rather than on an external device. One project that has found particular popularity in such scenarios is BusyBox [26], which provides replacements for many of the standard utilities found in Linux systems in a much smaller, single executable file. The BusyBox project has, in fact, been used in many of the root file system images made available by Xilinx. Some examples of the Xilinx root file systems, along with instructions on how to build and modify a root file system can be found at [27].

In the event that the size of the root file system is less of an issue than having a wealth of features, then the Linaro project [28] has a selection of root file systems available that have been pre-built for some of the Arm architectures used in the Zynq devices. While the use of these pre-built images is convenient, they are likely to be significantly less efficient than a root file system built specifically for the embedded Linux system and may be unsuitable for mounting in system memory.

An important consideration when building a root file system for an SDx Platform is the inclusion of a TCF Agent. In Chapter 16 we described how the TCF is used to implement many of SDx performance monitoring and debugging features on target systems running the Linux OS. To operate, the TCF requires that the target implements a TCF Agent. A TCF Agent therefore needs to be included in the Linux root file system to enable many of the performance monitoring and debugging features that we described in Chapter 16 and Chapter 18. Instructions for building a TCF Agent can be found in [29].

There is quite a range of possible boot components required in an SDx Platform and we have covered the generation of each of them here. This may seem a little daunting at first, but by using the guide to the necessary components in Table 19.3 and the information provided here, we hope the prospect of producing the boot files for an SDx Platform seems a little less intimidating.

19.4.2 PetaLinux Tools

The task of generating the list of boot files we have just described for implementing the Linux OS in an SDx Platform is not trivial, particularly for those developers with little experience of Linux. Serendipitously, Xilinx provides a set of tools called the PetaLinux Tools that can be used to generate all of these boot files, including the PMU firmware and ATF required for Zynq MPSoC devices. The PetaLinux Tools are freely available [30] and are almost certainly the simplest and quickest way of generating the boot files necessary to support a basic Linux implementation in an SDx Platform. Instructions for using the PetaLinux Tools generally can be found in [31], along with instructions for generating the Linux boot files for an SDx Platform specifically given in [4].

19.4.3 Boot Image Format Files

The boot files supplied by the SDx Platform are used by the SDx compiler to build boot image files for projects targeting the SDx Platform. In our introduction to the SDx compiler in Chapter 15, we mentioned that the ***bootgen*** utility is used by the compiler to create a boot image file incorporating the executable file and bitstream generated from the application in our SDx project. The ***bootgen*** utility creates a single boot image file by building the necessary boot header, appending tables describing the partitions contained in the boot file and processing the input data files, such as the application executable and bitstream, into partitions within the boot image file. The ***bootgen*** utility is driven by boot image format (BIF) files, which are text files describing the partitions to be contained in the boot image file and the input data files that constitute the partitions.

The SDx Platform needs to contain BIF files to direct the ***bootgen*** utility in creating a boot image file that incorporates the project-specific executable file and bitstream along with the appropriate boot files from the SDx Platform. A BIF file should be created for each OS and processor combination supported by the SDx Platform. Examples of basic BIF files for creating boot image files for the Standalone OS on a Zynq-7000 and the Linux OS booting on the APU of a Zynq MPSoC are shown in Figure 19.12 and Figure 19.13 respectively. Other examples of BIF files can be found within the SDx Platforms supplied with SDx and further information on using the ***bootgen*** utility is available from [32] and [33].

```
boot_image: {
    [bootloader] <standalone/boot/fsbl.elf>
    <bitstream>
    <elf>
}
```

Figure 19.12: Example of a BIF file supporting the Standalone OS on a Zynq-7000

```
boot_image: {
    [fsbl_config] a53_x64
    [bootloader] <a53_linux/boot/fsbl.elf>
    [pmufw_image] <a53_linux/boot/pmufw.elf>
    [destination_device=pl] <bitstream>
    [destination_cpu=a53-0, exception_level=el-3, trustzone] <a53_linux/boot/bl31.elf>
    [destination_cpu=a53-0, exception_level=el-2] <a53_linux/boot/u-boot.elf>
}
```

Figure 19.13: Example of a BIF file supporting the Linux OS to boot on the APU of a Zynq MPSoC

19.4.4 Libraries

It may be desirable to include additional platform-specific functionality implemented in libraries within the SDx Platform. Applications targeting the platform that include the appropriate header files are automatically linked with the platform libraries when the application is built, thus saving the application developer the trouble of manually modifying the application build configuration settings to link against these libraries.

Perhaps the most obvious example of platform-specific functionality that could be included as a library within an SDx Platform is software to interact with custom hardware implemented in the PL as part of the platform. In such a situation it would be sensible to include software to control the platform custom IP as a library within the SDx Platform. The included library could be one generated to control existing IP using the ***sdslib*** utility described in Chapter 17. It may contain manually written driver implementations for the Standalone or FreeRTOS OSs or it may contain the user space portion of a driver for a device using the UIO framework on the Linux OS. Ultimately, the platform developer can include any library they wish within the SDx Platform. All that is required is that the library is compiled into a static library suitable for execution on the intended processor and the appropriate header files are available. There are sample SDx Platforms included with SDx and documented in [4] that demonstrate the inclusion of libraries within the platform to control platform IP.

19.5 SDx Platform Sample Applications

Sample applications can be included as part of an SDx Platform and are particularly useful for demonstrating the use of platform-specific features or for providing application templates to the platform user. The sample applications are presented to the user when they create an SDx project targeting the SDx Platform.

In addition to the source code for each sample application, the SDx Platform must also provide an XML file that passes pertinent information about the sample applications to SDx. This XML file must be named *template.xml* and reside, along with the application source files, in a directory called *samples* within the root directory of the platform, as shown in Figure 19.1 on page 449.

An example of a *template.xml* file is shown in Figure 19.14. The structure of this file is relatively simple with each sample application being described by an individual *template* sub-element of a single *manifest* element. There are three attributes belonging to each *template* element. The *location* attribute is used to specify the path to a directory containing the source code for the sample application. The path to this directory is given relative to the location of the *template.xml*. When an SDx project using the sample application is created, the contents of the directory designated using the *location* attribute are copied into the *src* directory of the SDx project ready for use. The *name* and *description* attributes provide a name and description of the sample application that will be displayed to SDx Platform user used by the SDx IDE.

Each *template* element can contain a number of sub-elements that are used to define features of the sample application represented by the enclosing *template* element. The first of these sub-elements shown in Figure 19.14 is the *supports* element. This is used to specify which OSs the sample application will be available for selection with. The structure of the *supports* element is, perhaps, a little convoluted with the use of *and* and *or* sub-elements to define a boolean function that is used to test whether the target OS selected by the SDx project creator is appropriate for the sample application. The first application shown in Figure 19.14 will be available for selection when the target OS is either Linux or Standalone, while the second application will only be available for selection when the target OS is FreeRTOS.

The *accelerator* sub-element of a *template* element can be used to mark functions from the sample application for implementation in hardware by default. The *name* and *location* attributes of an accelerator element are mandatory. These define the name of the function to be implemented in hardware and the path to the source file containing that function respectively. The path to the source file is given relative to the path designated by the *location* attribute of the enclosing *template* element. Optionally, the *clkid* attribute of the *accelerator* element can be used to assign a specific platform clock for use with the function marked for hardware implementation instead of using the default platform clock. In cases where the function to be implemented in hardware calls code that resides in a separate source file, SDx is normally able to infer the path to the other source file. However, in some circumstances, it may be necessary to specify the path to the other source file. This can be achieved by including an *hlsfiles* sub-element within the accelerator element, as shown in Figure 19.14. The path to the other source file is given in the *hlsfiles* sub-element and should, again, be given relative to the path designated by the *location* attribute of the enclosing *template* element.

```
<?xml version="1.0" encoding="UTF-8"?>
<manifest:Manifest xmi:version="2.0"
                   xmlns:xmi="http://www.omg.org/XMI"
                   xmlns:manifest="http://www.xilinx.com/manifest">
    <template location="application_1" name="First application" description="Sample application">
        <supports>
            <and>
                <or>
                    <os name="Linux"/>
                    <os name="Standalone"/>
                </or>
            </and>
        </supports>
        <accelerator name="hw_fn" location="app_1.cpp"/>
        <system dmclkid=”2”/>
        <includePaths>
            <path location=”include”/>
        </includePaths>
        <libraryPaths>
            <path location=”lib”/>
        </libraryPaths>
        <libraries>
            <lib name=”andersonian”/>
        </libraries>
        <exclude>
            <directory name=”exclude”/>
            <file name=”excludeme.txt”/>
        </exclude>
        <compiler inferredOptions=”-D SAMPLEAPP”/>
        <linker inferredOptions=”-static”/>
    </template>
    <template location="application_2" name="Second application" description="Sample application">
        <supports>
            <and>
                <or>
                    <os name="FreeRTOS"/>
                </or>
            </and>
        </supports>
        <accelerator name="hw_fn" location="app_2.c" clkid=”3”>
            <hlsfiles name=”sub_fn.c”/>
        </accelerator>
    </template>
</manifest:Manifest>
```

Sample Applications
OS Support
Function marked for hardware by default
Default data mover clock

Figure 19.14: Example of a template.xml file

In a similar vein to assigning a specific platform clock for use with a function marked for hardware implementation, it is also possible to specify a platform clock to be used by the data motion network. This can be achieved by inserting a *system* sub-element into the *template* element. There is one attribute belonging to a *system* sub-element, *dmclkid*, which is used to identify the platform clock for use by the data motion network in the sample application. An example of a *system* sub-element is shown in Figure 19.14.

Most of the remaining sub-elements of a template element that we have not yet discussed are used to stipulate settings for building the sample application. The *includePaths*, *libraryPaths* and *libraries* sub-elements can be used to detail libraries that the sample application should be linked against. Examples of the use of these sub-elements are given in Figure 19.14. The paths specified within the *includePaths* and *libraryPaths* elements should be given relative to the path designated by the *location* attribute of the enclosing *template* element. It is also possible to set application-specific compiler and linker options by using the compiler and linker elements, examples of which are also shown in Figure 19.14.

There may be occasions when the platform creator wishes to include files or directories within the directory containing a sample application that are not necessary to build that application. In such circumstances, copying these files into an SDx project increases the size of the project unnecessarily. It is, however, possible to designate files and directories within the sample application directory that should not copied when an SDx project for the sample application is created. This can be achieved using the *exclude* sub-element. There are two types of sub-element that an *exclude* element can contain: *directory* and *file*. These sub-elements are used to identify directories and files that should not be copied into an SDx project using the sample application respectively. An example of the use of an exclude element is shown in Figure 19.14.

This is a relatively brief, yet broad-ranging, overview of the process of including sample applications within a custom SDx Platform. While we have provided the salient details of the structure and elements contained in a *template.xml* file, the reader may be interested in more detailed information on this subject, which can be found in [4]. Further examples of *template.xml* files can also be found in some of the platforms supplied with SDx.

19.6 The sdspfm Utility

It is possible to manually generate the metadata files and directory structure for an SDx Platform, however, a more convenient solution is to use the ***sdspfm*** utility. The content of the SDx Platform still needs to be generated manually, as outlined in the foregoing, but the ***sdspfm*** utility will populate the platform directory structure it creates with these components and produce the appropriate metadata files. Since version 2017.4, the ***sdspfm*** utility has been superseded by the Platform SDx project type. Further information on using the Platform SDx project type can be found in [34].

The ***sdspfm*** utility is invoked from the SDx Terminal and can be used either from the command line or through a GUI. We will focus our attention on using the GUI and the command to invoke this is shown in Figure 19.15.

```
sdspfm -gui
```

Figure 19.15: SDx Terminal command to invoke sdspfm utility GUI

Within the ***sdspfm*** GUI the platform creator is able to specify the SDx Platform name, which must match the name of the Vivado Design Suite project containing the base hardware design, and, optionally, a vendor, version and description. The platform creator is also able to designate the location where the platform will be created.

The path to the Vivado Design Suite project containing the base hardware design should be provided along with the Tcl script created to generate the platform hardware metadata file and, once these have been specified, the platform creator can select the name of the appropriate IP Integrator block design for the base hardware project. This information is used by the ***sdspfm*** utility to produce the DSA file for the platform.

For SDx Platforms that include sample applications, the path to the directory containing the application source code and *template.xml* file should be indicated for inclusion in the appropriate field.

Based on the contents of the platform base hardware design, the ***sdspfm*** utility provides a list of processor types that the platform creator can designate support for in the platform. Only one instance of each processor type need be selected as it will support SDx projects on any of the cores of that processor type.

For each processor type added, the platform creator can indicate the list of OSs that the SDx Platform supports. This is done by selecting the appropriate OS type and providing an ID, which will also be used as the name of the directory in the SDx Platform to contain the files supporting the OS on that processor type. A more descriptive name can be specified as well. Regardless of the type of OS selected, the location of the appropriate BIF file, readme file template and directory containing the boot files must be given. For Linux OSs, the path to the directory containing the Linux image components must also be identified, while, for Standalone and FreeRTOS OSs, the location of the linker script is required.

The contents of the boot files directory will vary depending on the Zynq device and OS being supported by the platform, however it should contain an appropriate subset from:

- First stage boot loader;
- PMU firmware;
- Arm Trusted Firmware;
- Second stage boot loader.

For Linux OSs, the directory containing the image components should contain the kernel image, device tree blob and ramdisk image, if being used. These may be present as individual components or as a Flattened Image Tree (FIT) boot image.

Libraries to be included in the SDx Platform should be identified under the OS and processor combination with which they are intended to be used. These are added by designating the location of a directory containing the appropriate header files and the path to the static library file.

Once all of the information about the SDx Platform components has been entered into the GUI, the SDx Platform can be generated. The progress of the generation process is reported in the SDx Terminal used to invoke the ***sdspfm*** utility.

The SDx Platform produced by the ***sdspfm*** utility is still able to be manually augmented by the platform creator by adding components to the platform directory structure, such as pre-built hardware, and editing the appropriate metadata files. Further information on using the ***sdspfm*** utility is available from [4].

19.7 Pre-Built Hardware

Generally, the most time-consuming task of an SDx project build is generating the bitstream for the hardware portion of the project. In cases where the SDx project is purely based in software, the bitstream to be generated will consist solely of the base SDx Platform hardware design. In these cases, the time taken for the build process could be significantly shortened by using a pre-built version of the platform hardware rather than creating a new one. A pre-built version of the platform base hardware design can optionally be included in the SDx Platform for just this purpose. When an SDx project targeting the platform is built without any functions marked for implementation in hardware, the SDx compiler will opt to use the pre-built hardware included with the platform rather than generating a new version, thus reducing the amount of time required to build the project.

Usually, the files to describe the pre-built hardware are stored in a directory such as */sw/prebuilt* within the SDx Platform directory structure. The simplest way to generate the files required to describe the pre-built hardware is to build an SDx project targeting the platform with an application without any functions marked for implementation in hardware. Clearly, this requires that the platform is functional before it is augmented with the addition of the pre-built hardware.

The files necessary for defining the pre-built hardware can then be copied from within the project build directory to the SDx Platform directory structure. The location of these files in projects built using SDx 2017.2 and the names they should be given in the SDx Platform are provided in Table 19.4.

Table 19.4: Files required to describe pre-built hardware (as in SDx 2017.2)

Project Build File	File Name in SDx Platform
<workspace>/<project>/<build_directory>/_sds/p0/vpl/system.bit	*bitstream.bit*
<workspace>/<project>/<build_directory>/_sds/p0/vpl/system.hdf	*<platform>.hdf*
<workspace>/<project>/<build_directory>/_sds/.llvm/apsys_0.xml	*apsys_0.xml*
<workspace>/<project>/<build_directory>/_sds/.llvm/partitions.xml	*partitions.xml*
<workspace>/<project>/<build_directory>/_sds/swstubs/portinfo.c	*portinfo.c*
<workspace>/<project>/<build_directory>/_sds/swstubs/portinfo.h	*portinfo.h*

In order to indicate to the SDx compiler that a pre-built version of the platform base hardware design is available, the platform creator must edit the software metadata file, *<platform>.spfm*, found within the SDx Platform directory structure. Figure 19.16 shows an example of a platform software metadata file with the `sdx:prebuilt` elements added within each configuration element specifying the path to the directory containing the pre-built hardware data, relative to the platform software metadata file.

```
<?xml version="1.0" encoding="UTF-8"?>
<sdx:platform sdx:vendor="strath.ac.uk" sdx:library="sdx" sdx:name="zavodovski" sdx:version="1.0"
              sdx:schemaVersion="1.0" xmlns:sdx="http://www.xilinx.com/sdx">
    <sdx:description>Basic Platform to demonstrate custom platform creation</sdx:description>

    <sdx:systemConfigurations sdx:defaultConfiguration="linux">
        <sdx:configuration sdx:name="linux" sdx:displayName="Linux on Cortex-A53"
                           sdx:defaultProcessorGroup="a53" sdx:runtimes="cpp">
            <sdx:description>Linux running on Zynq UltraScale+ Cortex-A53</sdx:description>
            <sdx:prebuilt sdx:data="prebuilt" />
            <sdx:bootImages sdx:default="standard">
                <sdx:image sdx:name="standard" sdx:bif="a53_linux/boot/linux.bif"
                           sdx:imageData="a53_linux/image" sdx:mountPath="/mnt"
                           sdx:readme="a53_linux/boot/generic.readme" />
            </sdx:bootImages>
            <sdx:processorGroup sdx:name="a53" sdx:displayName="Cortex-A53"
                                sdx:cpuInstance="psu_cortexa53_0" sdx:cpuType="cortex-a53">
                <sdx:os sdx:name="linux" sdx:displayName="Linux" sdx:includePaths="a53_linux/inc/inc"
                        sdx:libraryPaths="a53_linux/lib"
                        sdx:libraryNames="andersonian:leighton:playfair" />
            </sdx:processorGroup>
        </sdx:configuration>

        <sdx:configuration sdx:name="standalone_r5" sdx:displayName="Standalone OS on Cortex-R5"
                           sdx:defaultProcessorGroup="r5_0" sdx:runtimes="cpp">
            <sdx:description>Standalone OS running on Zynq UltraScale+ Cortex-R5</sdx:description>
            <sdx:prebuilt sdx:data="prebuilt" />
            <sdx:bootImages sdx:default="standard">
                <sdx:image sdx:name="standard"
                           sdx:bif="r5_standalone/boot/standalone.bif"
                           sdx:readme="r5_standalone/boot/generic.readme" />
            </sdx:bootImages>
            <sdx:processorGroup sdx:name="r5_0" sdx:displayName="R5_0" sdx:cpuInstance="psu_cortexr5_0"
                                sdx:cpuType="cortex-r5">
                <sdx:os sdx:name="standalone" sdx:displayName="Standalone"
                        sdx:includePaths="r5_standalone/inc/inc" sdx:libraryPaths="r5_standalone/lib"
                        sdx:libraryNames="andersonian:leighton:playfair"
                        sdx:ldscript="armr5-none/lscript.ld" />
            </sdx:processorGroup>
        </sdx:configuration>

    </sdx:systemConfigurations>

</sdx:platform>
```


Figure 19.16: Example of a platform software metadata file

19.8 Testing and Using a Custom SDx Platform

Once a custom SDx Platform has been created, SDx provides some basic tests that can be used to check the validity of the platform and its compatibility with SDx. The validity test only checks that the metadata files for the platform are valid and can be read by SDx, but is a useful first test for a new custom SDx Platform nonetheless. The test is performed from the directory containing the platform using the SDx Terminal. The test is invoked by calling the ***sdscc*** compiler with the `-sds-pf-list` command line option. This causes SDx to print a list of all of the available SDx Platforms, including those it finds in the current directory. If the platform metadata files are valid, then the custom SDx Platform will appear in this list with the description from the top-level metadata file and a list of the system configurations as defined in the software metadata file. An example of this is shown in Figure 19.17. This test can be further extended by calling the ***sdscc*** compiler with the `-sds-pf-info <path_to_platform>/<platform>` command line option. In addition to the information about the specified platform that was printed when listing all of the available SDx Platforms, the device targeted by the platform and the frequencies of the platform clocks are also printed in this test.

```
<directory_containing_custom_platform>>sdscc -sds-pf-list
Platform:    zavodovski (<directory_containing_custom_platform>/zavodovski)
Description: Platform to demonstrate custom platform creation
Available system configurations:
  linux (Linux running on Zynq UltraScale+ Cortex-A53)
  standalone_r5 (Standalone OS running on Zynq UltraScale+ Cortex-R5)

Platform:    microzed (C:/Xilinx/SDx/2017.2/platforms/microzed)
Description: Basic platform targeting the MicroZedBoard, which includes 1 GB of DDR3, 128 Mb QSPI Flash
   and a MicroSD card interface. More information at http://www.zedboard.org/products/microzed
Available system configurations:
  linux (linux Linux OS on a9_0)
  standalone (standalone Standalone OS on a9_0)
  freertos (freertos FreeRTOS on a9_0)
  ocl (ocl Linux OS on a9_0)
    .
    .
    .
```

Figure 19.17: Example output from sdscc -sds-pf-list command

After checking that the custom SDx Platform is valid, a suite of further tests to confirm the suitability of the custom platform for use with SDx is provided. These tests include checking that at least 64MiB of physically contiguous memory can be allocated, the full range of data movers that can be instantiated in the PL by the SDx compiler work with the appropriate PS interfaces and that the frequency of the clock signals available in the PL are within 1% of their expected values. The source code for this suite of tests can be found in the */samples/platforms/Conformance* subdirectory of the SDx install directory. To build an SDx project to run this suite of tests on the custom SDx platform, the *Conformance* subdirectory should be copied from the SDx install directory to a suitable workspace from where the `make` command shown in Figure 19.18 can be used in the SDx terminal. The resulting project files can be used to run the tests on hardware, with status messages from the tests being printed to standard output.

```
make OS=<LINUX or STANDALONE> PLATFORM=<path_to_platform>/<platform> PLATFORM_TYPE=<ZYNQ or MPSOC>
```

Figure 19.18: SDx Terminal command to build a suite of tests for a custom SDx Platform

Once the validity and compatibility of a custom SDx Platform has been determined, the final step in testing the platform is to create and build SDx projects that target the platform. SDx projects targeting a custom platform are created in the same way as those targeting a standard platform. When it is time to specify the target platform in the SDx IDE, the option to add a custom platform should be selected and the location of the custom SDx Platform identified. In all other respects, the creation of an SDx project targeting a custom platform is the same as for one targeting a standard platform.

19.9 Chapter Review

In this chapter we have discussed custom SDx Platforms. In addition to considering some of the reasons that a custom SDx Platform may be necessary or desirable, we have described the process of creating and using a custom platform. The use of an SDx Platform as the basis for a project is what enables much of the abstraction that is achieved by using SDx to develop systems targeting Zynq devices. Consequently, the process of creating the components that constitute an SDx Platform is much more akin to the traditional development techniques used in developing Zynq-based systems.

19.10 References

Note: All online sources last accessed March 2019.

Some of the Xilinx documentation referred to below has version-specific URLs. If you are working in a newer version of the tools, check for updates on the Xilinx website, or try adjusting the link according to your version.

[1] Xilinx, Inc., "Vivado Design Suite User Guide: Embedded Processor Hardware Design", UG898 (v2017.2), June 2017

Available: https://www.xilinx.com/support/documentation/sw_manuals/xilinx2017_2/ug898-vivado-embedded-design.pdf

[2] Xilinx, Inc., "Vivado Design Suite User Guide: Designing with IP", UG896 (v2017.2), June 2017

Available: https://www.xilinx.com/support/documentation/sw_manuals/xilinx2017_2/ug896-vivado-ip.pdf

[3] Xilinx, Inc., "Vivado Design Suite User Guide: Designing IP Subsystems Using IP Integrator", UG994 (v2017.2), June 2017

Available: https://www.xilinx.com/support/documentation/sw_manuals/xilinx2017_2/ug994-vivado-ip-subsystems.pdf

[4] Xilinx, Inc., "SDSoC Environment Platform Development Guide", UG1146 (v2017.2), August 2017

Available: https://www.xilinx.com/support/documentation/sw_manuals/xilinx2017_2/ug1146-sdsoc-platform-development.pdf

[5] Xilinx, Inc., "FSBL" Xilinx wiki page

Available: http://www.wiki.xilinx.com/FSBL

[6] Xilinx, Inc., "Zynq Power management Framework User Guide For Zynq Ultrascale+ MPSoC Devices", UG1199 (v2.0), November 2016

Available: https://www.xilinx.com/support/documentation/user_guides/ug1199-zynq-power-management.pdf

[7] Xilinx, Inc., "Zynq Ultrscale+ MPSoC Technical Reference Manual", UG1085 (v1.7), December 2017

Available: https://www.xilinx.com/support/documentation/user_guides/ug1085-zynq-ultrascale-trm.pdf

[8] Arm, Ltd., Arm Trusted Firmware Design

Available: https://github.com/Xilinx/arm-trusted-firmware/tree/master/docs/firmware-design.md

[9] Xilinx, Inc., "GitHub - Xilinx/arm-trusted-firmware: ARM Trusted Firmware" Git repository

Available: https://github.com/Xilinx/arm-trusted-firmware

[10] Xilinx, Inc., "Build ARM Trusted Firmware (ATF)" Xilinx wiki page

Available: http://www.wiki.xilinx.com/Build+Arm+Trusted+Firmware+(ATF)

[11] DENX Software Engineering, "WebHome < U-Boot < DENX" webpage, 20th April 2017

Available: http://www.denx.de/wiki/U-Boot/WebHome

[12] Xilinx, Inc., "GitHub - Xilinx/u-boot-xlnx: The official Xilinx u-boot repository" Git repository

Available: https://github.com/Xilinx/u-boot-xlnx

[13] Xilinx, Inc., "Build U-Boot" Xilinx wiki page

Available: http://www.wiki.xilinx.com/Build+U-Boot

[14] Linux Kernel Organization, Inc., The Linux Kernel Archives webpage

Available: https://www.kernel.org/

[15] Xilinx, Inc., "GitHub - Xilinx/linux-xlnx: The official Linux kernel from Xilinx" Git repository

Available: https://github.com/Xilinx/linux-xlnx

[16] Xilinx, Inc. "Build kernel" Xilinx wiki page

Available: http://www.wiki.xilinx.com/Build+kernel

[17] J. Corbet, "A reworked contiguous memory allocator", 14th June 2011

Available: https://lwn.net/Articles/447405/

[18] H.J. Koch, "The Userspace I/O HOWTO", 22nd May 2008

Available: http://www.osadl.org/UIO.uio0.0.html

[19] eLinux.org, "Embedded Linux Wiki Device Tree Usage - eLinux.org" webpage

Available: http://elinux.org/Device_Tree_Usage

[20] eLinux.org, "Embedded Linux Wiki Device Tree Reference - eLinux.org" webpage

Available: http://elinux.org/Device_Tree_Reference

[21] Xilinx, Inc., "Build Device Tree Compiler (dtc)" Xilinx wiki page

Available: http://www.wiki.xilinx.com/Build+Device+Tree+Compiler+(dtc)

[22] Xilinx, Inc., "Build Device Tree Blob" Xilinx wiki page

Available: http://www.wiki.xilinx.com/Build+Device+Tree+Blob

[23] Xilinx, Inc., "GitHub - Xilinx/device-tree-xlnx: Linux device tree generator for the Xilinx SDK (Vivado > 2014.1)" Git repository.

Available: https://github.com/Xilinx/device-tree-xlnx

[24] Xilinx, Inc., "Generating Basic Software Platforms: Reference Guide", UG1138 (v2017.2), June 2017

Available: https://www.xilinx.com/support/documentation/sw_manuals/xilinx2017_2/ug1138-generating-basic-software-platforms.pdf

[25] The Linux Foundation, "Filesystem Hierarchy Standard v3.0", 19th March 2015

Available: http://refspecs.linuxfoundation.org/FHS_3.0/fhs-3.0.pdf

[26] E. Andersen, BusyBox webpage

Available: https://busybox.net/

[27] Xilinx, Inc., "Build and Modify a Rootfs" Xilinx wiki page

Available: http://www.wiki.xilinx.com/Build+and+Modify+a+Rootfs

[28] Linaro, "Linaro — Leading collaboration in the ARM Ecosystem" webpage

Available: https://www.linaro.org/

[29] Wind River Systems, Inc., "Target Communication Framework: Getting Started" webpage

Available: http://git.eclipse.org/c/tcf/org.eclipse.tcf.git/plain/docs/TCF%20Getting%20Started.html

[30] Xilinx, Inc., "Downloads" webpage

Available: https://www.xilinx.com/support/download/index.html/content/xilinx/en/downloadNav/embedded-design-tools.html

[31] Xilinx, Inc., "PetaLinux Tools Documentation: Workflow Tutorial", UG1156 (v2017.2), June 2017

Available: https://www.xilinx.com/support/documentation/sw_manuals/xilinx2017_2/ug1156-petalinux-tools-workflow-tutorial.pdf

[32] Xilinx, Inc., "Zynq-7000 All Programmable SoC Software Developers Guide", UG821 (v12.0), September 2015

Available: https://www.xilinx.com/support/documentation/user_guides/ug821-zynq-7000-swdev.pdf

[33] Xilinx Inc., "Zynq Ultrascale+ MPSoC Software Developer Guide", UG1137 (v6.0), January 2018

Available: https://www.xilinx.com/support/documentation/user_guides/ug1137-zynq-ultrascale-mpsoc-swdev.pdf

[34] Xilinx Inc., "SDSoC Environment Platform Development Guide", UG1146 (v2017.4), January 2018

Available: https://www.xilinx.com/support/documentation/sw_manuals/xilinx2017_4/ug1146-sdsoc-platform-development.pdf

PYNQ, and Machine Learning Applications

Chapter 20

Deep Learning

Sarunas Kalade, University of Strathclyde

Deep Learning (DL) methodologies have seen a great deal of success in many application domains in the past few years, and interest in the area continues to grow. Applications such as classification, analysis and captioning of images, and language translation in the cloud, are just a few examples of where DL has been incredibly effective.

As the field grows, so do the needs of the end users — the amount of data that must be processed, the size of the neural network models, and in some applications latency has also become a concern — requiring significant investment in data centre infrastructure. The Zynq MPSoC is a technology that can potentially benefit a great deal of AI applications, both in the cloud and in edge computing cases, in accelerating the current state-of-the-art neural network models in a low latency and power efficient manner.

This chapter will introduce, at a high level, the fundamental concepts of current DL, some of its main application domains, and a selection of the primary tools and frameworks available in the industry.

20.1 Machine Learning is Everywhere (or will be!)

In this section, we will clear up any potential ambiguity around DL, with reference to the related terms Machine Learning and Artificial Intelligence. The expanding selection of DL applications will be introduced, and we will establish how far it has come in the past few decades, in terms of developing support.

20.1.1 Machine Learning versus Deep Learning

The terms Deep Learning (DL), Machine Learning (ML), and Artificial Intelligence (AI) are often used interchangeably when breakthroughs in various fields are made and covered in the media — but in fact, each is a subset of the next, as illustrated in Figure 20.1.

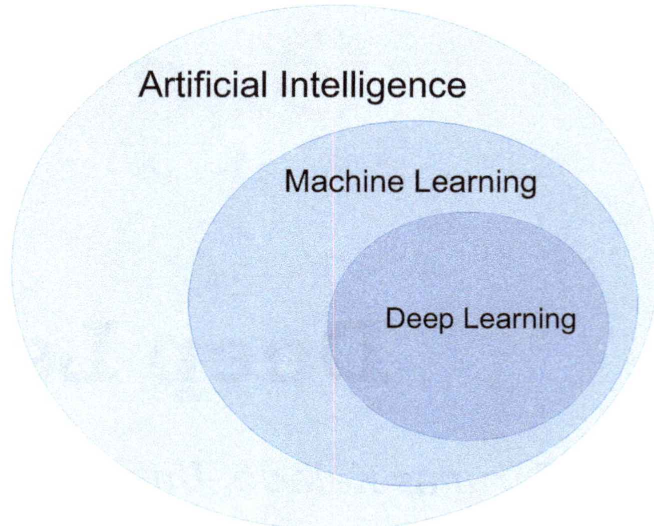

Figure 20.1: Artificial Intelligence, Machine Learning, and Deep Learning

AI is the science of developing intelligent systems, with the ultimate goal of achieving artificial general intelligence (AGI) - the ability of a machine to learn and perform tasks to the same level of competence as a human. Machine Learning is a subfield of AI, which allows programs to learn to make decisions based on experiences, previous observations, or data. Finally, Deep Learning is the subfield of machine learning that harnesses raw data in order to accomplish incredibly complex computing tasks.

Typical ML algorithms will usually be trained on data or features that human experts have deemed important in order to achieve a certain task. For example, to determine whether a patient is at risk of a cardio-vascular disease, doctors can employ a machine learning system that takes in features such as resting blood pressure, maximum achieved heart rate, and other measurements that are relevant to the task [1]. A significant amount of human effort goes into engineering appropriate features, and cleaning and pre-processing the data, so that a machine learning algorithm can achieve sufficient competence.

Where DL has found its stride, and is currently flourishing, is performing tasks for which humans would find it very difficult to express features algorithmically. For example, how would you begin to write a program to recognise that a cat is present in an image? Or a program to recognize multiple different objects in a video? Deep Neural Networks (DNNs) have become so successful because they do not require humans to carefully craft inputs tailored to specific applications; instead they operate on raw pixel inputs, audio samples, or other supplied data. From there, they learn to extract the necessary features, such as edges or basic shapes in images, and from these abstract features, they are capable of making decisions.

The downside is that, generally, a DL approach requires much more data and computational power than popular ML models with pre-determined input features. This is however becoming less of a problem in the age of information, where image data (for example) is abundant, and the power of available processing platforms continues to increase.

20.1.2 Machine Learning Overview

There are three main machine learning paradigms: supervised learning, unsupervised learning and reinforcement learning. In this section, we will introduce all three at a high level. The remainder of the chapter will then focus primarily on supervised learning, as that is currently the most intensively researched and applied methodology in various industries (by far).

Supervised Learning

In supervised learning, we normally deal with labelled data. Labelled data means that there is a dataset of inputs, as well as the desired outputs. The problem becomes developing a model that can perform a function $y = f(x, w)$, where x represents the inputs, y represents the outputs, and w are the parameters that the model must learn in order to best fit the training set. Figure 20.2 shows an example classification boundary for a two-class classification dataset, as learned by a function $f(\ )$.

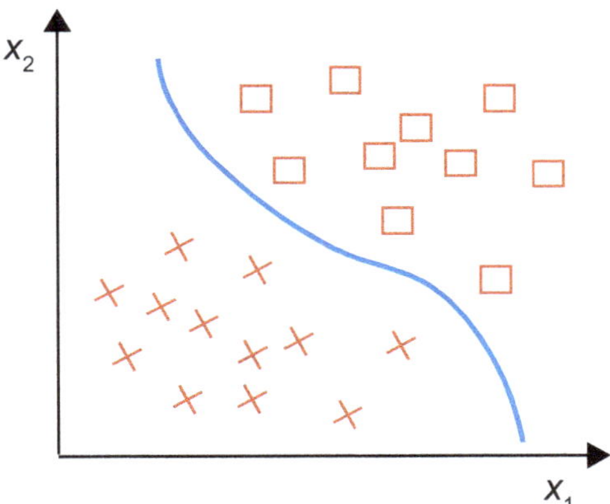

Figure 20.2: Classification of two classes

Examples of supervised learning include image classification, language translation, stock price prediction, and medical diagnosis, and there are many more.

Unsupervised Learning

In unsupervised learning, the data is unlabelled. We apply unsupervised learning methods when we want to discover some structure in the data, find irregularities, reduce dimensionality, or extract some features that can be used in a supervised learning task.

Another common example of unsupervised learning is that of an autoencoder — a form of neural network that is trained in a fully unsupervised fashion. It uses an encoder-decoder structure to encode an image into a latent feature space, and then uses the decoder to try and reconstruct the image at the output. In this case we do not care about what is in the image, just that the autoencoder is capable of reproducing it from a set of features. This network can then be used as a part of other DNNs.

Most commonly, unsupervised learning is used for clustering and discovering distinct categories in data. An example of clustering is shown in Figure 20.3, where three individual classes are identified by an unsupervised clustering algorithm.

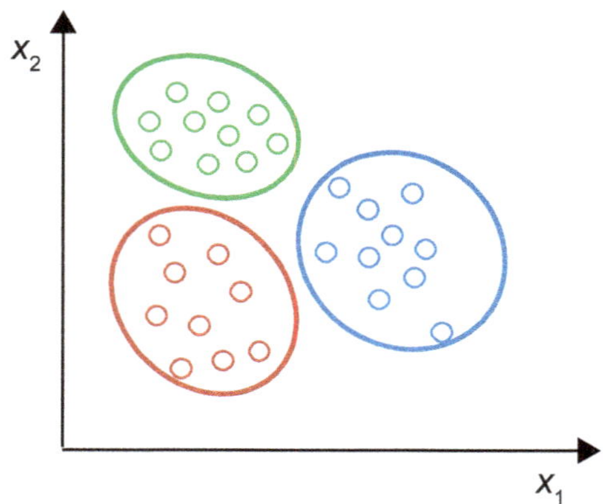

Figure 20.3: An example of clustering with three classes

Reinforcement Learning

In reinforcement learning, an agent acts on the environment and learns from its mistakes, as illustrated in Figure 20.4. It is an intuitive approach to learning, showing very good recent results in games, and it represents DL taking a step closer to AGI.

One of the most popular recent achievements in AI research has been AlphaGo — the DNN developed by Deepmind that was able to beat the top Go player multiple times in a head-to-head match [2]. This was achieved with a combination of reinforcement learning and a DNN architecture. While achievements in gaming have been making headlines, these techniques can be readily applied to other applications such as autonomous vehicles.

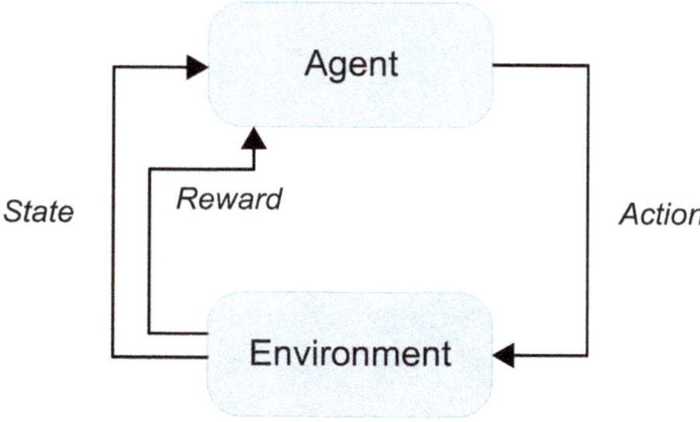

Figure 20.4: Basic Reinforcement Learning Flow

20.1.3 Established Applications

ML is prominent in a great deal of applications such as search engines, military systems, finance, medicine, driverless cars, and so many more. By harnessing DL, two very large domain spaces have impacted a number of different industries in various ways: Computer Vision (CV) and Natural Language Processing (NLP).

Computer Vision (CV)

Computer Vision is a major inter-disciplinary research field that aims to establish human-like vision capabilities in machines. It is a very broad topic that includes image processing, object classification and recognition, image understanding, objected tracking and various other operations. In essence it is the science of extracting knowledge from image data by using computer programs. An example CV application of image detection is illustrated in Figure 20.5.

Deep Learning has arguably made a comeback due to its success in the Computer Vision field back in 2012 when Alex Krizhevsky et. al. [3] won the ImageNet Large Scale Visual Recognition Challenge (ILSVRC) of that year — an annual competition where research teams test their classification algorithms on a large dataset of images [4]. A common training dataset is freely available to anyone, and it can be used to train the competitors' models on various images, with a portion of the dataset kept secret by the organizers of the competition. This suppressed data is later used to test the competing models. The model that makes the most correct predictions on the test images wins the competition. That year, the winners used a deep Convolutional Neural Network (named 'AlexNet'), and drastically outperformed all other contestants who were using feature engineering and classical machine learning techniques.

Since the inception of AlexNet, every winner of the competition in subsequent years has used a deep neural network, and the winning architectures have been influential in both academia and industry. The winning model architectures, as well as the code, are published and freely available to anyone. Examples of freely available models are the original and various subsequent versions of AlexNet, VGG Net (which did not win,

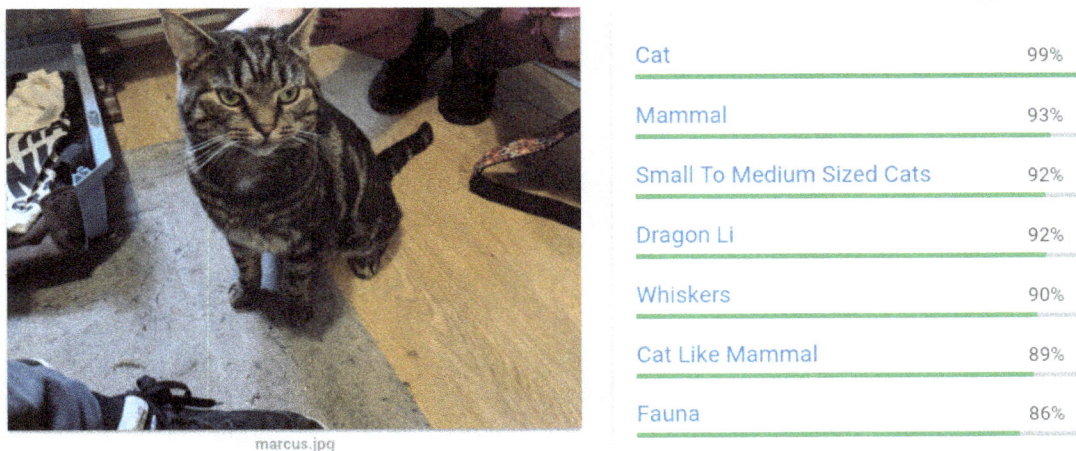

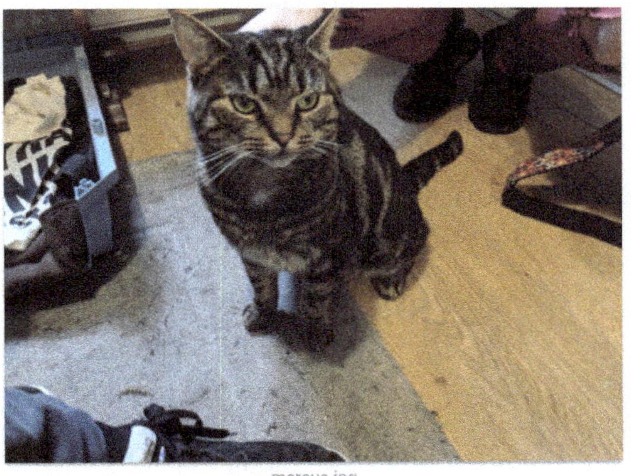

Figure 20.5: An example of image detection

but used an intuitive architecture and is still widely used today), GoogLeNet, and Microsoft's ResNet. Different versions of these models exist: for example, Resnet-50 or ResNet-101 correspond to the same architecture, but they contain 50 and 101 layers respectively [5]. Other published architectures are often included just because they were published and presented exceptional results at a given task, such as the Yolo (You Only Look Once) Net [6].

The openness of this approach has made iterative research and quick deployment of pre-trained neural nets in various different applications extremely accessible. Having reference models and datasets available such as MNIST, CIFAR-10, ImageNet, and many others, makes it a very straightforward for *anyone* interested to get started experimenting, not just developers in companies that can afford massive HPC clusters! It is incredibly valuable to have shared datasets and reference models publicly available. This promotes code sharing and has greatly contributed to the growth of DL over the years.

Presently, most prevalent state-of-the-art computer vision systems — from identifying cats in facebook pictures, to assisting driverless cars by recognising pedestrians and traffic signs — are using some form of deep convolutional neural network.

Natural Language Processing (NLP)

As smart home assistants and voice-activated technology is becoming increasingly popular, so is the need for good human speech recognition. Devices such as the Amazon Alexa use DL in the cloud to convert the speech data (i.e. what a user has said) into readable English language, and then use Natural Language Understanding (NLU) to make sense of the spoken sentence.

In the context of smart helpers and assistive technologies, realistic, pleasant-sounding voices are also needed — text-to-speech systems are used to generate human-sounding audio from typed sentences. For example, Google's speech synthesis model (called WaveNet) uses advanced DL techniques, such as dilated convolutions, to produce the human-sounding voices that you can hear from Google Assistant.

Language translation is another application where DL has been particularly beneficial. Services such as Google Translate use Sequence-to-Sequence models to transform words, or sequences of words, from one language to another, as shown in Figure 20.6 [7]. This type of neural network is implemented using two separate Recurrent Neural Networks (covered later in Section 20.2.4) — one trained as an encoder, which condenses the input sentence into what is termed a 'thought vector', and another, called the decoder, which converts the 'thought vector' into the desired language.

Along with computer vision, NLP remains one of the most important major fields where DL is thriving, and driving research forward.

20.1.4 Emerging Applications

Due to the successes and progress made possible by DL in CV, NLP and various other applications, DL research floodgates have opened in various other fields. Researchers are experimenting with many different architectures in trying to apply DL methodologies to the data from their field. Here, we will cover just a few examples of early DL applications.

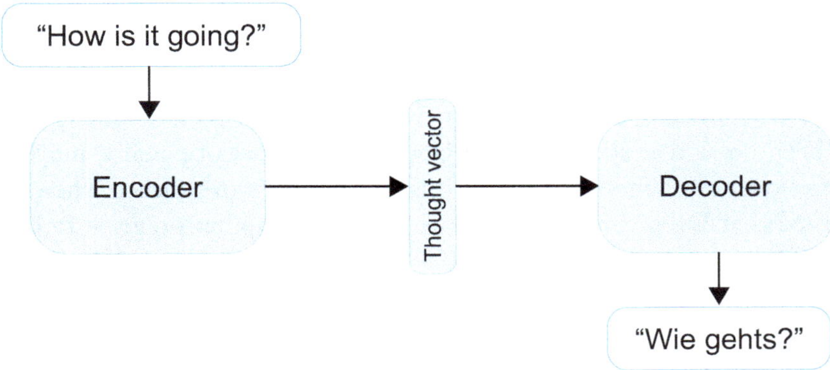

Figure 20.6: Text translation using a sequence-to-sequence model

Autonomous Vehicles

Self-driving cars are widely consideration to be the future of transportation, promising to improve the efficiency of travel and the transportation of goods, and more importantly to save many of the lives lost due to traffic accidents caused by human drivers. To make this vision a reality, a plethora of technologies are being researched and developed. Lately, DL has become an integral part of the many subsystems that comprise a driverless vehicle. Due to the number of sensors, and the volume of raw data that the vehicle must process, it makes sense to take advantage of data-driven ML techniques.

It has been established for some time that DNNs are able to recognise traffic signs, pedestrians and obstructions [8]. Some of the most exciting developments in DL that have enabled highly advanced vision systems on vehicles is semantic segmentation [9], which not only recognises targets in the camera lens of an AI-powered vehicle, but also extracts the precise pixel-wise location of the pedestrians, obstructions, signs, and other objects of interest. This is accomplished using CNNs that are trained in a fully convolutional way (meaning that only convolutional and pooling layers are used, rather than fully connected layers as seen in many classification tasks).

When exactly autonomous vehicles will be the norm is still widely speculative, but it is fair to say that DL will play a significant role in the future of autonomous driving, in one way or the other. Right now, the DNNs used can be effective as assistive technology [10], however in the future we may see driverless cars that are operated entirely by neural networks.

AI in Games

A long standing goal in AI research has been to achieve (and exceed!) human competence in various challenging domains. While it may not seem apparent, AI breakthroughs in games often result in great breakthroughs in the research field as a whole, and even signify milestones on the way towards AGI. One of the first significant breakthroughs was when an IBM supercomputer called 'Deep Blue' beat Garry Kasparov (world chess champion at the time) in a chess match of six games [11]. This match was the first time that a chess world champion was defeated by a computer program under tournament conditions.

Chess has a relatively small search space of available moves, when compared to another famous strategy board game, *Go*. While algorithms exist that routinely beat the best chess players in the world, until quite recently there have been no similar successes with Go, as the number of possible moves on a 19x19 grid far exceed the possibilities on an 8x8 chessboard. This however has been overturned when AlphaGo, a computer program based on DNNs at DeepMind [12], became the first computer program to beat a professional Go player (Lee Sedol, in October 2015). Part of AlphaGo was developed using DNNs, which were trained in a supervised learning scheme on data collected from expert player matches. Once it reached a sufficient level of proficiency in the game, it was trained using self-play and reinforcement learning, by playing matches against instances of itself.

While AI finally beating a human expert in the game of Go has been a very significant achievement, and definitely one for the history books, Go, just like chess, is a turn-based game with complete information — meaning that the state of the board is known to all parties, and technically the best moves could be calculated (although with current technology, this would be impractical). What we want from artificial intelligence is the ability to perform well in real world situations, and in the real world, action is more messy and fluid.

Keeping this in mind, the next big milestone will be in solving one of the complex real time strategy games that are being played professionally, like *Dota 2*. OpenAI developed a 1v1 bot capable of beating top professional Dota 2 players at 1v1 matches under tournament rules in August 2017 [13]. They have then further improved on the idea and created OpenAI Five, a team of bots that were pitched against some of the best players in the world during the biggest tournament of the game [14]. While the AI team did not win against the top teams in the world, it has shown that reinforcement learning is an effective means of teaching agents cooperation and long-term planning. Just like AlphaGo, the OpenAI Five team was trained using a mix of supervised learning and reinforcement learning, or self-play. Bots like AlphaGo and OpenAI Five are currently the most prominent driving forces in supervised learning research, and in return will have great implications for various other industries.

Wireless Communications

DL is slowly seeping into most fields of research, and wireless communications is no exception. There has been a surge of research into different machine learning applications in various sections of the field. DL has been demonstrated as very effective for performing tasks like channel equalisation (recovering signals that have been distorted when passing through a radio channel, e.g. due to bouncing off of buildings and other obstructions) [15]. These problems have been addressed using analytical models since the 1960's, and now with the advent of DL, it seems there is scope to apply drastically different approaches in the pursuit of advancing communications technology.

There are also exciting opportunities to apply DL techniques in various areas of new mobile networks, such as 5G [16]. The demand for data is higher than ever, and meeting these demands with sufficiently high Quality of Service (QoS) is extremely challenging. DL techniques could be highly beneficial, for example, in high level prediction of user demand, and organising mobile network resources in such a way that would guarantee the required QoS to all users. DL techniques could also be used to improve the physical layer performance of the radios used in the base stations and phones themselves.

Some researchers are even beginning to model and train an entire radio physical layer (for transmission and reception) composed entirely out of neural networks, by using an autoencoder approach [17], [18]. An autoencoder is usually two neural networks connected in such a way that one tries to recreate the original input, after having been passed through some bottleneck. In this case the bottleneck can be a model of a wireless channel, and the receiver network must learn how to deal with these simulated channel impairments, such as you would see distorting a signal by passing through walls or trees [17].

20.1.5 Platforms for Machine Learning

The concept of Artificial Neural Networks (ANNs) is not new — the first instance of a neural network being modelled was recorded in 1943, when a neurophysiologist named Warren McCulloch, and a young mathematician, Walter Pitts, wrote a paper on how neurons might work. Backpropagation, discussed later in Section 20.3.2, is the fundamental learning algorithm used to train every popular AI-based application we use in our daily lives and was derived mathematically as early as 1960. In 1982, it was applied by Paul J. Werbos to train neural networks as we do today.

Training such neural networks may have been challenging, or even impossible, at the time of these early developments. However, with recent (early 2010's) advancements in computing power, and access to unprecedented amounts of data, deep learning models have been shown to produce incredible results. The obtainable results are only getting better by throwing more computation and training data at the problem. Much of the recent computation gains have been achieved with modern Graphical Processing Units (GPUs), which are now widely used in High Performance Computing (HPC) clusters to accelerate deep learning training.

20.2 Common Neural Network Architectures

This section will cover the main neural network archetypes that are most commonly applied in the field of Deep Learning for various applications. We begin with the fundamental computational building block of many networks — the neuron.

20.2.1 The Neuron

The basic neuron (otherwise known as the perceptron) is an integral building block of numerous DNN architectures, and is shown in Figure 20.7. Historically, the term 'neuron' originates from the fact that early models were derived by trying to simulate the human brain (the reason why Artificial Neural Networks are given their name). Much like the neurons in the brain make connections by using synapses, the weights of a computational neuron are often referred to as synaptic weights, and early researchers tried to approximate those complexities by creating neural networks composed of artificial neurons.

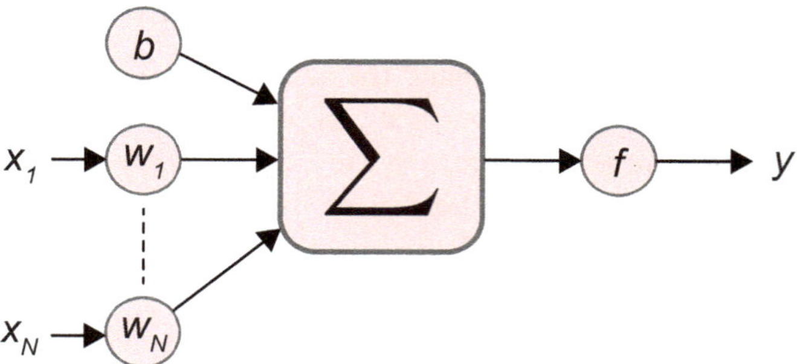

Figure 20.7: The neuron

The output of a single neuron is defined in Eq. (20.1),

$$y = f\left(\sum_{i=1}^{N} x_i w_i + b\right) \tag{20.1}$$

where x are the inputs to the neuron, w are the weights associated with each input, b is the bias added to the neuron, and y is the output. The simplest linear neuron will simply sum all of the input values multiplied by corresponding weights to calculate the output.

The bias is an important addition, as it shifts the decision boundary of the neuron. This is useful, for example, when processing an image: it guarantees that the neuron triggers an activation only after a strong enough correlation with the weights has been detected. Alternatively, a learned high bias will reduce the sensitivity of a neuron and cause it to activate more frequently.

The activation function f is included to introduce non-linearity into the neuron — this allows for highly complex, deep neural architectures that are capable of fitting non-linear data structures. Some common activation function choices often seen in the literature are listed in Table 20.1, and shown in Figure 20.8.

Table 20.1: Common activation functions

Sigmoid	**Tanh**	**ReLU**
$\frac{1}{1+e^{-x}}$	$\frac{2}{1+e^{-2x}} - 1$	$max(x, 0)$

There are two general qualities that an activation function is preferred to possess: it should to be non-linear; and it must be differentiable, or at least have a computational approximation of its gradient (as is the case with ReLU (Rectified Linear Unit), where its gradient is undefined at input values of 0). Activation functions are

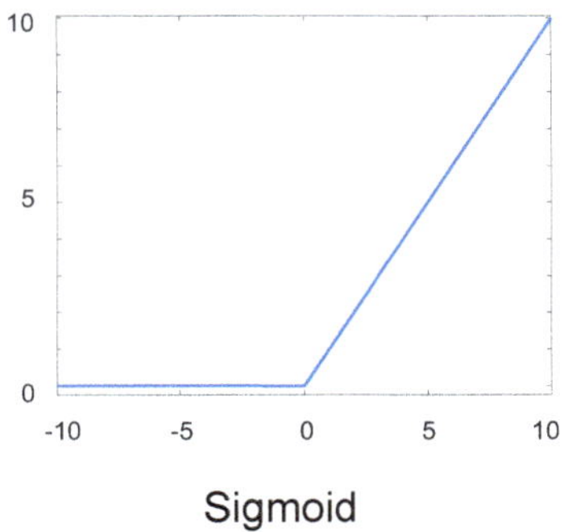

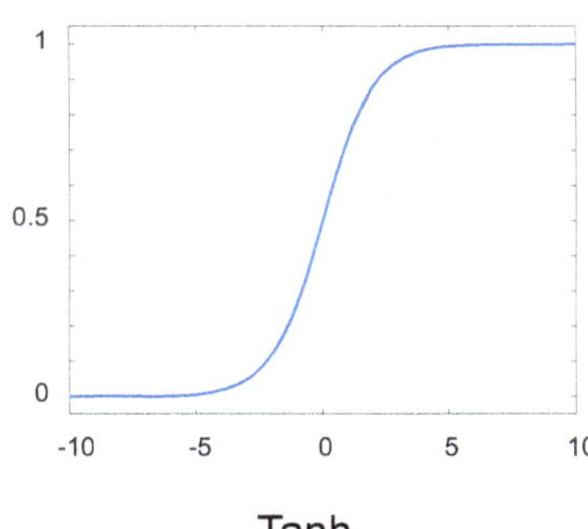

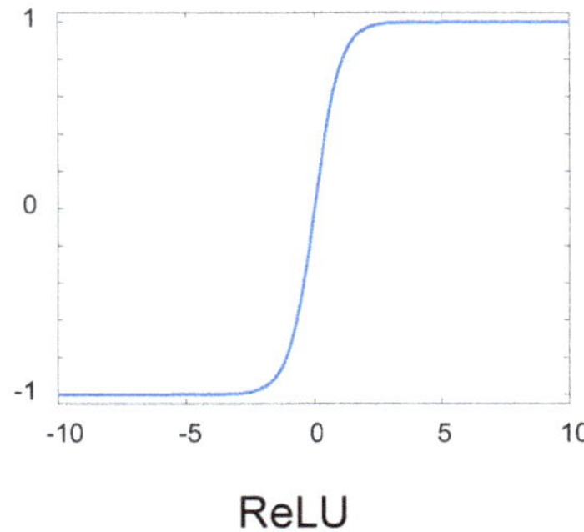

Figure 20.8: Common activation functions

still an active research topic, and it is highly likely that in the future, opinion regarding good choices of activation functions for a specific application may change.

Originally, the sigmoid activation function, which squeezes the neuron output value $a = f(x)$ to a range of $a \in [0, 1]$ was a very popular choice, with a simple-to-compute gradient for backpropagation (discussed in Section 20.3.2). It is convenient for small neural networks, however deep NNs were hard to train because very small or very large values would saturate the neuron output and stagnate the learning process. *Tanh* has similar properties to the sigmoid, however it is capable of outputting both negative and positive values, which provides more range for individual neurons to work with.

Presently the most dominant choice for most applications, but especially computer vision, is the ReLU (Rectified Linear Unit). The implementation of ReLU is very simple — just a *max*() function that passes an input if it is above zero, and outputs a zero if it is negative. ReLU was first shown to work well on image data as part of the AlexNet implementation for ImageNet, and reported an increase in training convergence speed by a factor of 6! Since then, it has become the most popular choice of activation function.

While Sigmoid and Tanh functions are not used as much in computer vision, they can still be found in very successful DNN architectures used for NLP tasks. For example, LSTM (Long Short Term Memory) networks use Sigmoids and Tanh functions at their core.

20.2.2 Multilayer Perceptron

The Multilayer Perceptron (MLP) is among the simplest and most popular examples of a feedforward neural network (meaning that the connections between the nodes in the network propagate information in one direction and do not form loops, as would be the case with a recurrent neural network). Neural networks are primarily composed of layers — the layered structure is what allows DNNs to have the multiple levels of abstraction required to learn how to solve complex, non-linear tasks. Generally speaking, what makes a Deep Neural Network 'deep' is that the number of layers is quite high. The very basic MLP network consists of at least three layers: the input, hidden and output layers. An example of a simple MLP with four layers (including two hidden layers) is shown in Figure 20.9.

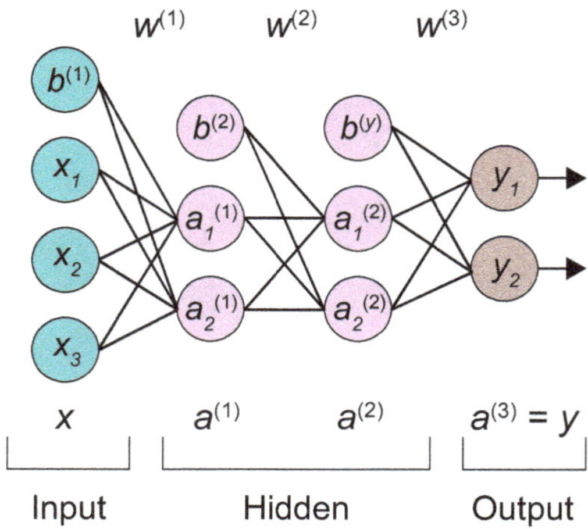

Figure 20.9: Multi Layer Perceptron

MLPs are fully connected ANNs, which means that each node (or neuron) is connected to each node of the following layer. Because of this structure, it is convenient to represent the neural network layers as matrices of weights. This also allows the use of linear algebra to quickly compute the outputs of each layer.

The general form of computing the outputs of each MLP layer, l, can be represented by Eq. (20.2). Note that the convention for describing neural networks is to indicate the layer number l as a superscript to the weights and activations (as opposed to raising to the power of l),

$$a^{(l)} = \sigma(w^{(l)}a^{(l-1)} + b^{(l)}) \tag{20.2}$$

where $a^{(l-1)}$ are the activations of the previous layer (or in the case of input layer, simply the input vector), and where $w^{(l)}$ and $b^{(l)}$ are the weights and biases associated with the current layer, l. The summation is then passed through some activation function, σ, such as the sigmoid from Table 20.1. Different layers can have different activation functions.

For classification problems, the final layer will almost certainly include a softmax activation function — an activation function that is applied to the outputs of the entire layer, squeezing each value between the range of 0 and 1, and making sure that all outputs sum up to exactly 1. It essentially returns a probability distribution of possible classes for a given output. This allows us to more intuitively interpret results output by a classifier DNN.

The example shown in Figure 20.9 is only a basic version of the MLP, with an input layer, two hidden layers and an output. Technically this model could be considered a DNN as the number of hidden layers is larger than one. However, no matter how many layers are added (even a DNN with $L = 100$ hidden layers!) the same function as in Eq. (20.2) will be used; it is just a series of matrix multiplications, additions and activation functions.

20.2.3 Convolutional Neural Networks (CNNs)

Convolutional Neural Networks (CNNs) are a type of specialised neural network that use filters and the convolution operation, rather than matrix multiplications as seen in MLPs. CNNs are strongly preferred to MLPs in computer vision because they scale much better to full images.

Most implementations using CNNs do not employ only convolutional layers (though there is a subset of such architectures called Fully Convolutional Neural Networks, primarily used for image segmentation). The most common arrangement is to use CNN layers as building blocks for the first part of the Deep Learning model pipeline for feature extraction. Then, after a sufficient number of CNN layers, all resulting activations are passed onto a familiar MLP network, which will contain the final output layer for performing tasks like classification, as shown in Figure 20.10.

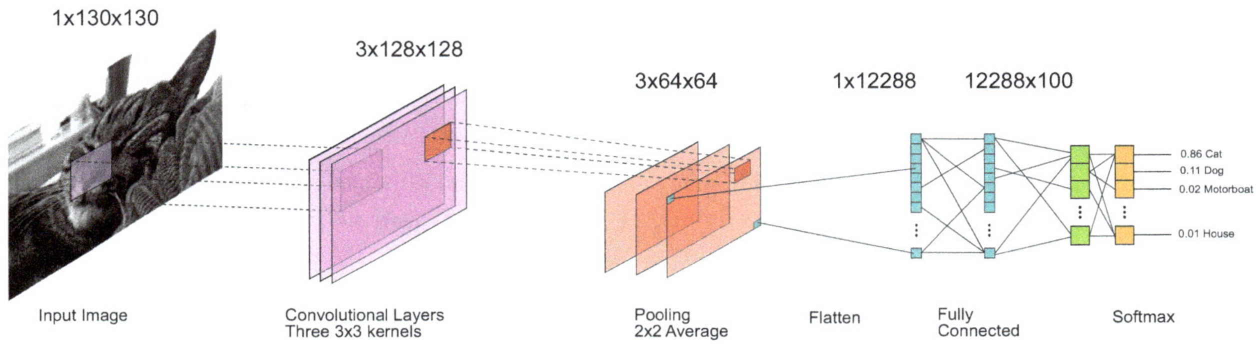

Figure 20.10: Image classification using a CNN

Next, we will go on to consider the individual building blocks of a CNN in more detail, namely the convolutional layers, pooling layers and flattening. MLPs are also involved, but have already been covered Section 20.2.2.

Convolutional Layers

Convolutional layers are special locally-connected layers, meaning that they activate on small receptive fields of the input image, rather than globally or fully connected layers (like the MLP), where each neuron of the previous layer is connected to each neuron of the following layer. This allows a very useful feature of CNNs called parameters sharing, where locally-connected neurons can be applied to multiple sub-fields of the input, learning the weights on a small receptive field, but being continuously applied to the entire input.

A convolutional layer is composed of a number of kernels (or filters) that are convolved, or more precisely, correlated with the input. Kernels have associated parameters in the form of $w \times h \times d$, where w is the input image width, h is the height, and d is the dimensionality or depth [19]. For example, a typical colour image could be $128 \times 128 \times 3$ (3 in this case representing three channels for the RGB values of each pixel). A single kernel of size $3 \times 3 \times 1$ would produce a feature map of size $126 \times 126 \times 1$. A simple convolutional layer, consisting of 1 kernel, is shown in Figure 20.11.

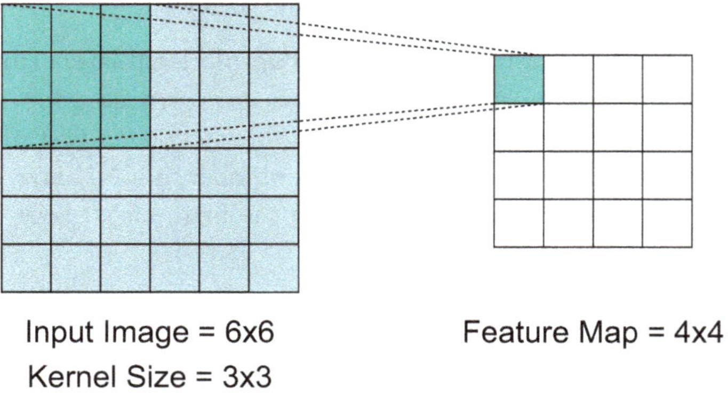

Figure 20.11: Example of a convolutional layer

The outputs of these convolutions are called feature maps, which can either be passed through more convolutional layers, downsampled using pooling layers, or flattened and used as inputs to a fully connected network, which would typically perform the final classification.

Pooling Layers

It is common practice to add pooling layers in between convolutions, in order to reduce the number of parameters and computation required by the model. The smaller amount of parameters also helps to reduce overfitting, as the network cannot as easily memorize training examples.

There are many possible ways to implement a pooling layer — amongst the most popular are the MAX or AVG operations, i.e. simply taking the largest value in a region, or the average of that region. An example of average pooling is shown in Figure 20.12. These implementations are also very computationally efficient and do not add any parameters to the model.

Flattening Layer

While not exclusive to CNNs, the flattening layer is most commonly used as the transitioning layer between the CNN layers and final, fully-connected output layers. Convolutional layers are often associated with the task of extracting features from an image, rather than making direct inference. To make meaningful decisions based on these features we need to employ fully connected layers (an MLP). The process of flattening reduces an N-dimensional input into a single column vector, as shown in Figure 20.13. The flattening layer can also be referred to as the input layer to the fully-connected network. This is very commonly done in CNNs, because in most CNN implementations the outputs of convolutional layers need to be connected to fully-connected layers to facilitate classification.

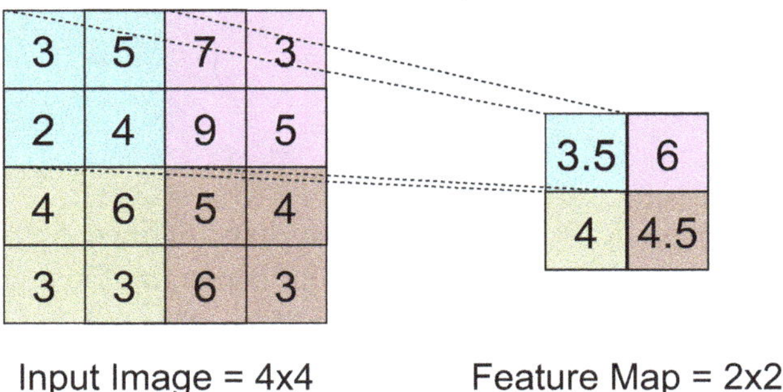

Figure 20.12: An example of average pooling

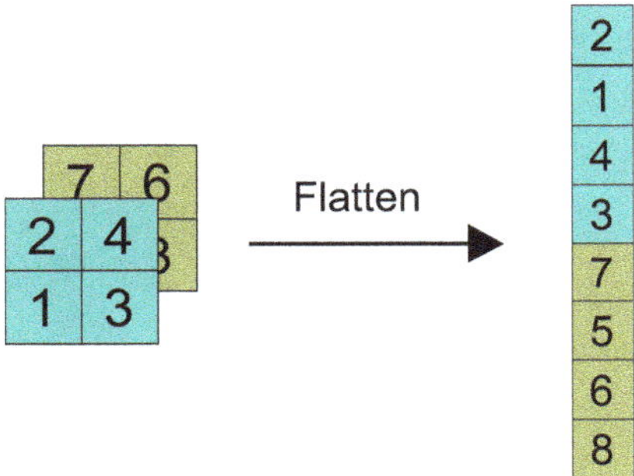

Figure 20.13: An example of the flattening process

20.2.4 Recurrent Neural Networks (RNNs)

Just as CNNs are a specialized architecture type of ANN for image data, RNNs are best suited for time-series and sequential data, such as audio or text. These networks perform best on data where context and input order matter, because they are capable of modelling time-dependencies. RNNs are seeing prominent use in NLP, speech recognition, language translation, and even video captioning, among many other applications!

There are many different ways of implementing RNNs, and different sub-types depending on the cell you choose to use. Here, we will introduce the basic RNN cell, and give a brief overview of some of the more popular variants to be deployed in real life applications.

Basic Cell

What differentiates RNNs from CNNs and MLPs is the presence of a hidden state, due to the recurrent nature of this architecture. Rather than just outputting an output value for each time step, the RNN updates an internal hidden state representation, therefore acting as additional memory for storing context about the input sequence.

A simplified RNN can be implemented with just a couple of equations [20]:

$$h_t = \tanh(W_h h_{(t-1)} + W_x x_{(t)}) \quad (20.3)$$

$$y_t = W_y h_{(t)} \quad (20.4)$$

The difference between these equations and the ones we have defined for solving MLPs earlier, is the additional weights w_h and activations h_t representing the hidden state. These activations are then fed back into the network by the recurrent connections, allowing the hidden state to be updated.

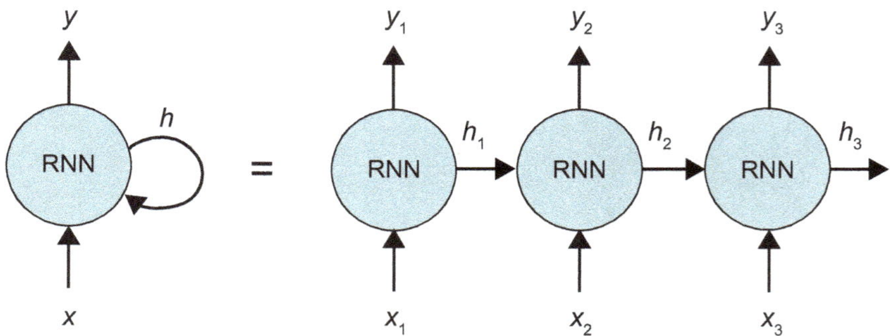

Figure 20.14: A recurrent neural network (two equivalent representations)

While simple to conceptualize and implement, these 'basic' RNN cells are hardly used in practice, due to issues with scaling to longer sequences. Basic RNN cells are difficult to train on longer sequences, and are notoriously vulnerable to a training problem called *vanishing gradient* — a problem in machine learning whereby DNN training on a long sequence or exceptionally deep model becomes stagnant. This occurs as, due to the nature of the backpropagation algorithm (discussed in Section 20.3.2), the applied weight changes can become negligibly small, halting the training process.

Long Short-Term Memory Cell

In order to solve the vanishing gradient problem and allow RNNs to be trained on very long sequences, the LSTM (Long Short-Term Memory) cell has been introduced. In addition to having a hidden state, h_t, LSTM cells also carry a cell state, c_t, which is manipulated using gate functions. A diagram of an LSTM is shown in Figure 20.15.

Instead of blindly computing and adding to the hidden state, as with the basic RNN cell, the LSTM gated approach allows the network, at each time step, to choose what data to keep, and what to pass into the cell state.

The forget gate, f_t, chooses which inputs are unimportant to the overall goal of the LSTM — for example, it could discard background noise during silent periods of an audio example containing speech. The input gate i_t is what controls the insertions of new information into the cell state, while the output gate, o_t, ultimately controls what goes into the hidden state of the LSTM cell. These are all implemented using sigmoid

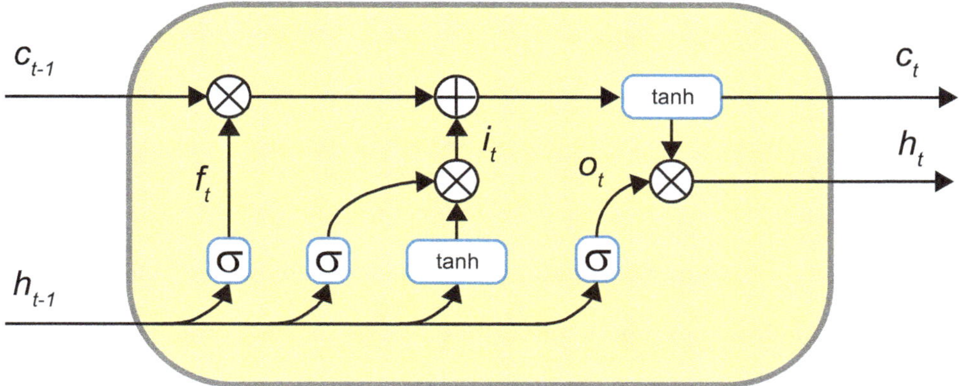

Figure 20.15: An LSTM cell

activation functions, because it is one of the few functions that limits the output in the range of between 0 and 1. Intuitively, this allows the cell to weigh how much of its input it wants to forget — in this case a 0 at the forget gate will deem the input at that time step redundant or unnecessary, and the LSTM cell will not waste any memory on it. The relevant mathematical derivations and implementation details can be found in the original LSTM paper, [21].

Most DNN libraries do have support for LSTMs and abstract all of these functions. They can essentially be used as regular RNNs, however, due to the mechanics of the gating functions, LSTMs offer much better performance on long sequences.

LSTMs are currently the most popular form of implemented RNN types, however there have been many variants proposed. The most popular cell variant is the GRU (Gated Recurrent Unit) cell, [22]. It takes a similar gating approach in order to prevent vanishing gradients, however it has no cell state and only operates on the hidden state, which makes it much more lightweight, requiring less computation.

There are no hard rules regarding which cell type is the best to use overall. In some cases, even the basic cell can outperform advanced LSTMs and GRUs due to its simplicity. Meanwhile, LSTMs and GRUs have been shown to outperform one another on similar tasks. While LSTMs are by far the most popular, it is often worth experimenting with different architectures.

20.3 Training Neural Networks

The goal of training a neural network is adapting its weights so that it 'fits' the training data. This is achieved through a number of steps.

In the beginning of its life a DNN will have randomly initialized weights. It will then go through the following steps (illustrated in Figure 20.16), each of which are expanded upon in the subsequent subsections:

- Training data gets passed into the DNN and a prediction is computed.
- A loss metric is evaluated based on how far the prediction is from the desired value stored in the training label.
- Based on the loss and gradients computed in the forward pass, the backpropagation algorithm returns a vector of all partial derivatives associated with the network parameters.
- A training step is taken according to some update rule — this could be a stochastic gradient descent step to change the weights, so that next time a similar input would produce a lower error value.

Regardless of the DNN architecture choice, the training process will be fundamentally the same — at least, as long as backpropagation is used.

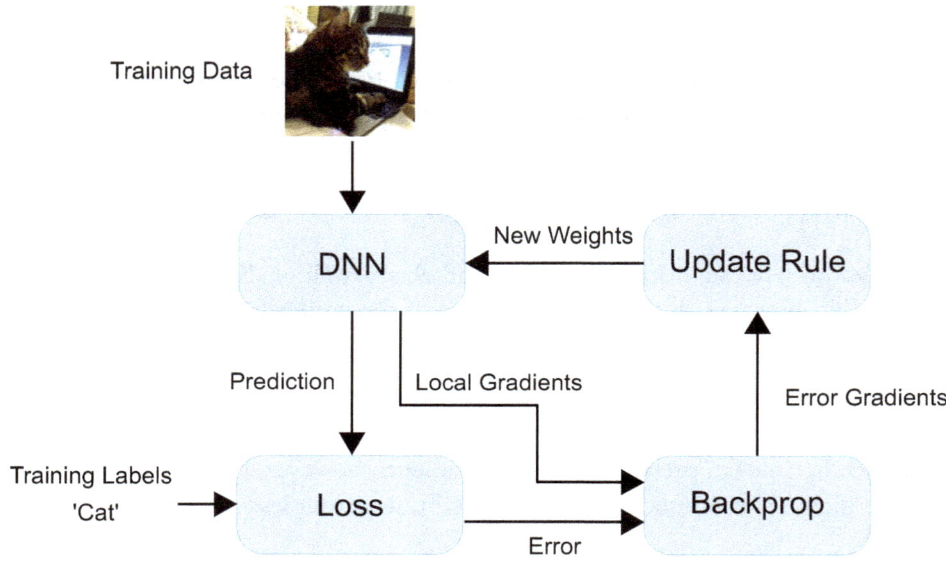

Figure 20.16: Gradient-based neural network training

20.3.1 Loss Functions

The loss function (sometimes referred to as the cost function) is a metric of how well a DNN fits the training data. During training, the desire is to minimise this. There are various loss functions available, and the best choice often depends on the task that the neural network needs to perform. Some of the more popular loss functions are Mean Squared Errors (MSE), or Categorical Cross Entropy (CCE), which are respectively defined for a single example as

$$L_{MSE} = (y - \hat{y})^2 \tag{20.5}$$

and

$$L_{CCE} = y\log_2\hat{y} \tag{20.6}$$

where y is the known labelled value and $\hat{y}$ is the prediction from the DNN.

Log Loss, or Categorical Cross-Entropy loss, is by far the most popular loss function in computer vision, as well as general classification tasks. In contrast, MSE is better suited to (and often used in) regression problems, where the output can be more continuous rather than discrete classes.

While loss functions are a part of a the more general field of optimisation, and not strictly a part of the DL field, the prominence of DNNs has spiked a great deal of interest in loss function design and optimisation techniques targeting NNs specifically. Just like activation functions, training and selection (or engineering) of appropriate loss functions is an active research area in optimisation.

20.3.2 Backpropagation

The loss given by the loss function will determine how drastically the network should adjust itself — if it made a prediction very far from the ground truth, it will have to nudge its weights more drastically than if it had made the right decision. How does the neural network determine which weights need nudging, in order to make the correct decision? This is where backpropagation comes in.

The backpropagation algorithm is the basis of the training procedure of all DNNs [23]. In the context of ANNs, 'learning' simply means that given a training example (e.g. a labelled picture), the neural network will adjust each of its weights in such a way that, the next time it sees the same or a similar training example, it will be able to more confidently predict the correct label.

The main goal of backpropagation is to compute the error gradients of the network, which is basically a set of values that characterises how much each parameter of the DNN contributed to the computed loss. The process is split into two parts: the forward pass and the backward pass, as illustrated in Figure 20.17.

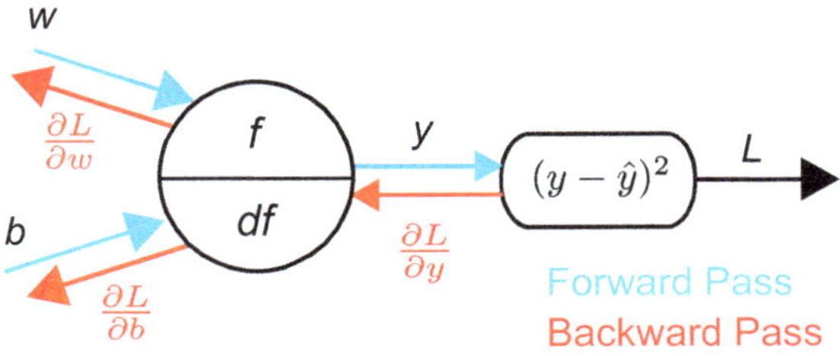

Figure 20.17: Backpropagation for a single perceptron

The forward pass is what the neural network would do once deployed, which is also referred to as *inference*. Given a training example, the inputs of this training example are passed into the neural network and an output is produced. For example, the input could be the pixels of an image, while the output is one of a number of possible classes. During inference, this would produce a target class, and the network's job would be done.

However, in backpropagation the additional step of computing the training loss is performed, as shown in Figure 20.17. Once the training loss is obtained, the error signal can be backpropagated through the DNN.

On the backward pass, the error signal from the loss is propagated through the net backwards, allowing the computation of the partial derivatives of all parameters with respect to the loss value. This will instruct the optimisation algorithm how much each weight should be changed.

Forward Pass

Let's assume we are training a single perceptron using MSE loss, as shown in Figure 20.17. It has a single weight w, bias b and activation function f, which outputs a value y, with the ground truth label being $\hat{y}$ (for simplicity we are ignoring the input x as that is not parametarisable unit). The forward pass simply involves multiplying the inputs and passing them through the activation function $y = f(wx+b)$. The loss can then be computed as

$$L = (y - \hat{y})^2 . \tag{20.7}$$

Additionally, the local gradients or activation function gradients can also be computed on the forward pass, and stored for later computation of the error gradients. Now this value can be used to determine the partial derivatives of the computational components.

Backward Pass

The main goal of the backward pass is to compute the partial derivatives of the tunable parameters, which in this case are

$$\frac{\partial L}{\partial w} = \frac{\partial L}{\partial y}\frac{\partial y}{\partial w} \tag{20.8}$$

and

$$\frac{\partial L}{\partial b} = \frac{\partial L}{\partial y}\frac{\partial y}{\partial b} . \tag{20.9}$$

This can be done by computing the local gradients of the forward pass, and applying the chain rule with the backpropagated gradients.

Backpropagation is essentially the chain rule from calculus applied to ANNs. By computing the partial derivatives with respect to every single weight, it is possible to determine which weight has contributed the most to the error signal given by the loss output at the output of the neural network.

Once the gradients are known, we can use a stochastic optimisation method to determine how much to nudge the weights in the direction of the gradients.

20.3.3 Stochastic Gradient Descent (SGD)

Almost all training of deep neural network models is done using SGD-derived methods. It is an iterative statistical approach to finding optimum solutions (see Figure 20.18). In this case, SGD is used to update the weights of the neural network to reach an optimum fit to the training data.

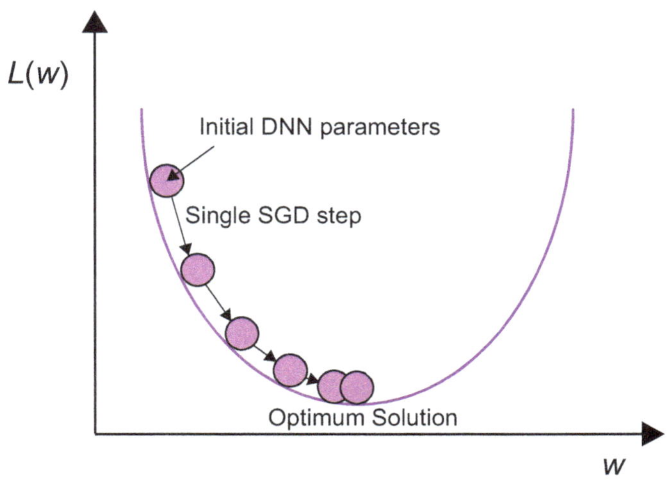

Figure 20.18: Finding the optimum solution by gradient descent

Normally one would take an entire training set in a single batch, perform backpropagation on the batch to compute the full gradient, and then take a gradient descent step towards that direction. In DL, however, the datasets are so incredibly large that this method turns out to be highly inefficient. Instead, the accepted practice is to split the dataset into mini-batches, which could be hundreds or thousands of pictures, and then perform backpropagation and take a gradient descent step, one batch at a time. The training period required to pass all mini-batches is called an epoch.

Additionally to the training set, it is often useful to include a validation dataset in the training procedure — this is data that was either collected separately to the training data, or it is a portion of training data that has been held back for testing purposes. Now, instead of performing a forward pass on just the training data, the network will make a prediction on both training and validation datasets. The second forward pass measures how the network performs with data that it has not been specifically trained on, or in other words, how well it generalises to new examples. This is very useful to do, because if unchecked and given a big enough model, a neural network could have the capacity to memorise a training set — this is called overfitting (to the training set).

It is useful to monitor both losses over time in epochs, and to make sure that the validation loss decreases together with the training loss (shown in Figure 20.19). If the two start to diverge, it means that the network is overfitting, and that would indicate a good stopping point for the training procedure.

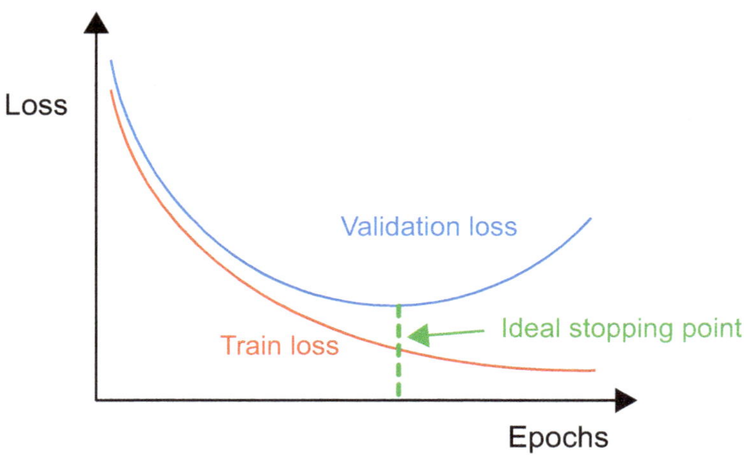

Figure 20.19: Typical training/validation losses over time

20.3.4 Regularisation

While a validation set and early stopping of the training help combat overfitting, in practice, it is not the only method applied. Regularisation is a method of preventing overfitting by keeping the weights relatively small and avoiding too much biasing. There are many techniques to regularisation, but here we will cover just a couple of popular ones, in order to introduce the idea.

Weight Decay

Weight decay (or L2 Regularisation) is one of the most common methods of regularisation, and works by simply adding a new term to the loss function, thus producing a new loss function,

$$Loss_{reg} = Loss + \frac{\lambda}{2m} \times \sum w^2 . \tag{20.10}$$

This introduces a new hyper-parameter, λ, which controls how much penalisation is being applied to the neural network. Like most hyper-parameters, there is no hard rule for the value to set it to. Most DL libraries will apply a default value that should work well on most problems, but the best value will often need to be determined empirically.

It is preferable to keep the weights of a DNN small and encourage it to use all available weights. If some weights become too large and contribute too much to the output, it might prevent the network from utilising other weights. By penalising the network for having large weights, the customized cost function forces it to learn to use small weight values, and hopefully encourages it to use all of the synapses available. This is why it is often referred to as weight decay.

Dropout

Dropout is a very simple, yet elegant way of regularising a DNN, and it was introduced in [24]. Dropout works by randomly disabling or 'dropping' some proportion of neurons in the network at each training step, as illustrated in Figure 20.20.

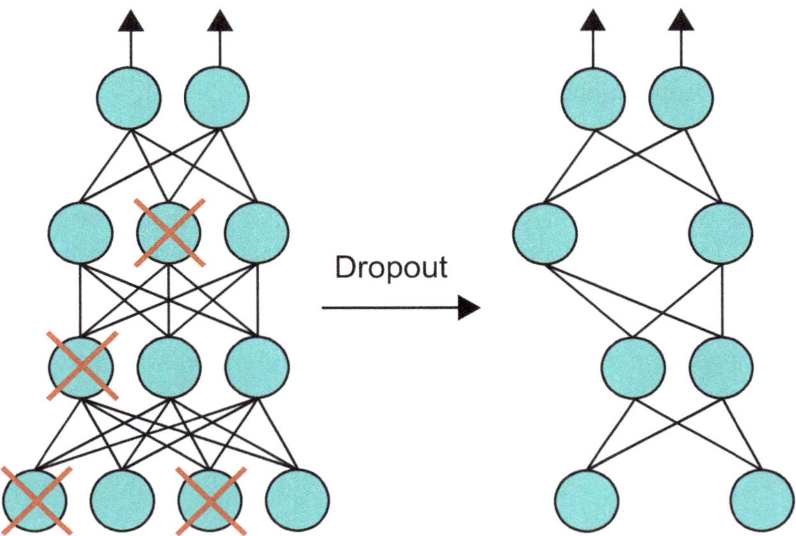

Figure 20.20: Dropout, resulting in a thinned neural network

Intuitively this works, because it prevents the network from learning a single feature and developing an over-reliance, or bias associated with it. Also, according to the original paper, neurons in the later layers have a tendency of co-adapting and fixing the mistakes of previous neurons — these co-adaptations can cause it to memorise patterns in a training dataset, causing overfitting. By using dropout we prevent these co-adaptations from forming, and reduce the possibility of overfitting as a result.

Dropout can also be considered as a form of ensemble modelling, because it causes the DNN to split into multiple thinned networks which, after training, are automatically 'ensembled' into a single network. Generally ensemble models, which are composed of multiple networks, and whose results get averaged together, perform better than single DNNs, and dropout approximates this effect.

20.4 Deep Learning Tools and Frameworks

Training neural network models generally requires a lot of computational power and parallel computing. Fortunately, since the early 2010's, computational power has become much more accessible not only to small research groups, but also individuals in the forms of graphics cards or Graphical Processing Units (GPUs), which have become very powerful thanks to the video gaming industry. Writing highly parallel and optimised code for GPUs is not an easy task, however multiple programming frameworks exist now to ease this task, such as Nvidia's proprietary CUDA platform, and open source alternatives such as OpenCL.

Presently the market is dominated by CUDA-enabled cards, and consequently this is the primary coding platform for Deep Learning applications. However, most researchers do not define their DL models in CUDA, as that is still considered a fairly low level interface, with many development challenges. Instead, many higher level frameworks have emerged, which abstract most of the interaction with the GPU, and provide high level APIs in languages such as Python, allowing developers to quickly prototype models, while also taking advantage of the computational power of the GPU.

Some of the most popular frameworks at the moment are Tensorflow [25] and PyTorch [26], developed and maintained by Google and Facebook respectively. Both primarily allow developers to write high level code in Python, which later gets translated into CUDA code that allows models to be trained and deployed on GPUs. Apache MXNet [27] is another popular solution for developers who want to deploy their DNNs for commercial applications. Caffe [28] is one of the older frameworks, and many researchers still prefer it for quick prototyping. Caffe features a plethora of pre-trained models to experiment with and, even though it is not built in a high level language like Python, it has numerous interfaces with various language support to perform experiments. Several other frameworks are also available, such as Microsoft's Cognitive Toolkit (CNTK), Torch, DeepLearning4J, Keras, and Caffe2, as reviewed in [29].

20.5 The Need for Acceleration

Much of the computation in modern DL applications is being done by GPUs, especially for training. However, platforms such as GPUs and CPUs are sub-optimal for fast inference, and can also suffer from poor power efficiency. Most state-of-the-art deployed neural network models are also trained using standard 32-bit precision floating point parameters, which has been shown to be an unnecessarily high precision, leading to models with significantly heightened memory requirements. This is an issue for deployments both in data centres, and on the edge.

20.5.1 Cloud and Edge Applications

Energy costs, and the data transmission overheads of offloading to the cloud, are important factors for data centre deployment of machine learning algorithms. In particular, transmission times may render the datacentre option unsuitable for low-latency, real-time or safety-critical application requirements, such as augmented reality, drone control or autonomous driving. Therefore, machine learning applications may require to run on the edge, even in battery powered systems. Among such applications, speech recognition and computer-vision based problems (such as object detection and image recognition) are very important in several markets, like automotive.

Also, in the context of cloud computing itself, operators face ever-increasing throughput requirements to process astronomical scales of data, bringing additional challenges in energy efficiency to minimise operating expenses. While cloud service latency is less critical compared to embedded scenarios, it still translates directly into customer experience for interactive applications. For instance, Jouppi et al. [30] quote a maximum response time limit of 7 ms for an interactive user experience in the Google cloud-based services.

Both Google [31] and Facebook [32] report high variety of machine learning tasks in their datacentres, with different neural networks topologies and architectures deployed according to the task, spanning CNNs, MLPs, and LSTMs.

20.5.2 Compression Techniques

Several techniques for reducing the heightened memory requirements of modern DNNs are actively being researched. Recently, research has been focused on decreasing the memory and compute footprint of DNNs by mean of techniques like parameter quantisation (which will be extensively covered in the next chapter). Other common techniques are:

- **Network distillation** (also known as knowledge distillation) aims at 'distilling' a large and complex model composed of a large DNN, or an ensemble of neural networks, into a much smaller network, with the smaller network approximating the original function learnt by the complex 'cumbersome' model [33]. This method allows usage of a large amounts of computational power in order to train a model having large capacity and memory requirements, which are much more lax than in deployment. However the smaller model, taught in a teacher/student kind of way, will be able to meet much stricter memory footprint requirements.

- **Pruning** is based on the idea that, among the many network parameters, some are redundant and contribute very little to the output [34], [35]. Thus, pruning seeks to remove parameters from the network, reducing the on-chip memory, memory bandwidth and computational requirements, whilst minimising or even avoiding any loss in accuracy. In general, pruning consists of avoiding or skipping the multiplication with weight parameters close to 0, since their effect is minimal on the overall result of the accumulation. Fine-grained pruning methods, such weights pruning, results in a fully sparse neural network, having serious impacts on the regularity of access patterns, with implication on the actual achievable speed-up. By contrast, coarse-grained pruning methods, such as filter pruning, result in a structured sparse representation that maintains the regular data access patterns of the original network, making it easier to translate the compute savings into an actual throughput increase.

20.6 Chapter Summary

Recognising the potential of the Zynq MPSoC in deep learning applications, this chapter has introduced some of the main areas of DL that could benefit from this hardware. Basic DL concepts, primary architecture types, and training procedures have been covered. We have also introduced some of the currently popular platforms and tools used in the development and deployment of deep artificial neural networks.

20.7 References

Note: All online sources last accessed February 2019.

[1] A. Hazra, S. K. Mandal, A.Gupta, A. Mukherjee, and A. Mukherjee, "Heart Disease Diagnosis and Prediction Using Machine Learning and Data Mining Techniques: A Review", *Advances in Computational Sciences and Technology*, 2017.
DOI: 10. 2137-2159.

[2] The Google DeepMind Challenge Match, March 2016. Deepmind.
Available: https://deepmind.com/research/alphago/alphago-korea/

[3] Krizhevsky, Alex et. al., "ImageNet Classification with Deep Convolutional Neural Networks", *Proceedings of the 25th International Conference on Neural Information Processing Systems*, Volume 1, 2012, pp. 1097-1105.
DOI: 10.1145/3065386

[4] ImageNet, *Large Scale Visual Recognition Challenge (ILSVRC)* webpage.
Available: http://image-net.org/challenges/LSVRC/

[5] K. He, X. Zhang, S. Ren, and J. Sun, "Deep Residual Learning for Image Recognition", *Proceedings of the IEEE Conference on Computer Vision and Pattern Recognition*, Las Vegas, USA, June 2016.
Available: https://arxiv.org/abs/1512.03385 (arXiv pre-print)

[6] J. Redmon, S. Divvala, R. Girshick, A. Farhadi, "You Only Look Once: Unified, Real-Time Object Detection", *Proceedings of the IEEE Conference on Computer Vision and Pattern Recognition (CVPR)*, Las Vegas, USA, June 2016.
Available: https://arxiv.org/abs/1506.02640 (arXiv pre-print)

[7] I. Sutskever, O.Vinyals, and V. Le Quoc, "Sequence to Sequence Learning with Neural Networks", *Proceedings of the 28th Neural Information Processing Conference*, Montreal, Canada, December 2014.
Available: https://arxiv.org/abs/1409.3215 (arXiv pre-print)

[8] G. von Zitzewitz, "Survey of neural networks in autonomous driving", 2017.
Available: https://www.researchgate.net/publication/324476862_Survey_of_neural_networks_in_autonomous_driving

[9] J. Long, E. Shelhamer, and T. Darrell, "Fully Convolutional Networks for Semantic Segmentation", *Proceedings of the IEEE Conference on Computer Vision and Pattern Recognition*, Boston, USA, June 2015.
Available: https://arxiv.org/abs/1411.4038 (arXiv pre-print)

[10] L. Fridman et. al., "MIT Autonomous Vehicle Technology Study", 2017.
Available: https://hcai.mit.edu/avt/

[11] D. Silver, et al., "Mastering the game of Go with deep neural networks and tree search", *Nature* 529, pp. 484-489, January 2016.
DOI: 10.1038/nature16961

[12] D. Goodman and R. Keene, *Man Versus Machine: Kasparov Versus Deep Blue*, H3, 1997.

[13] OpenAI, *Dota 2* webpage.
Available: https://blog.openai.com/dota-2/

[14] OpenAI, *OpenAI Five* webpage.
Available: https://blog.openai.com/openai-five/

[15] K. Burse, R. N. Yadav and S. C. Shrivastava, "Channel Equalization Using Neural Networks: A Review", *IEEE Transactions on Systems, Man, and Cybernetics, Part C (Applications and Reviews)*, vol. 40, no. 3, pp. 352-357, May 2010.
DOI: 10.1109/TSMCC.2009.2038279

[16] C. Zhang, P. Patras, and H. Haddadi, "Deep Learning in Mobile and Wireless Networking: A Survey", *arXiv Computing Research Repository*, 2018.
Available: https://arxiv.org/abs/1803.04311

[17] T. O'Shea and J. Hoydis, "An Introduction to Deep Learning for the Physical Layer", *IEEE Transactions on Cognitive Communications and Networking*, vol. 3, no. 4, pp. 563-575, Dec. 2017.
DOI: 10.1109/TCCN.2017.2758370

[18] S. Dörner, S. Cammerer, J. Hoydis and S. ten Brink, "Deep Learning-Based Communication Over the Air", *arXiv Computing Research Repository*, 2017.
Available: https://arxiv.org/abs/1707.03384 (arXiv pre-print)

[19] A. Karpathy, L. Fei-Fei, and J. Johnson, *Convolutional Neural Networks for Visual Recognition*, Stanford University CS231n class webpage.
Available: http://cs231n.github.io/convolutional-networks/

[20] L. Fei-Fei, J. Johnson, and S. Yeung, "Lecture 10: Recurrent Neural Networks", Stanford University CS231n class lecture notes, May 2017.
Available: http://cs231n.stanford.edu/slides/2017/cs231n_2017_lecture10.pdf

[21] S. Hochreiter and J. Schmidhuber, "Long Short-Term Memory", *Neural Computation* 9, No. 8, November 1997, pp. 1735-1780.

[22] J. Chung, C. Gulcehre, K. Cho, and Y. Bengio, "Empirical Evaluation of Gated Recurrent Neural Networks on Sequence Modeling", *NIPS Deep Learning and Representation Learning Workshop*, 2014.
Available: https://arxiv.org/abs/1412.3555 (arXiv pre-print)

[23] David E. Rumelhart, Geoffrey E. Hinton & Ronald J. Williams, "Learning representations by back-propagating errors", *Nature* 323, pp. 533-536, October 1986.
DOI: 10.1038/323533a0

[24] N. Srivastava, G. Hinton, A. Krizhevsky, I. Sutskever, and R. Salakhutdinov, "Dropout: A Simple Way to Prevent Neural Networks from Overfitting", *Journal of Machine Learning Research*, Vol. 15, pp. 1929-1958, June 2014.
Available: http://www.jmlr.org/papers/volume15/srivastava14a/srivastava14a.pdf

[25] *Tensorflow* website.
Available: https://www.tensorflow.org/

[26] *PyTorch* website.
Available: https://pytorch.org/

[27] *Apache MXNet* website.
Available: https://mxnet.apache.org/

[28] *Caffe* website.
Available: http://caffe.berkeleyvision.org/

[29] W. G. Hatcher and W. Yu, "A Survey of Deep Learning: Platforms, Applications and Emerging Research Trends", *IEEE Access*, vol. 6, pp. 24411-24432, April 2018.
DOI: : 10.1109/ACCESS.2018.2830661

[30] N. P. Jouppi, et al, "In-Datacenter Performance Analysis of a Tensor Processing Unit", *Proceedings of the 44th International Symposium on Computer Architecture (ISCA)*, Toronto, Canada, June 2017.
Available: https://arxiv.org/abs/1704.04760 (arXiv pre-print)

[31] Google, Inc., "Cloud TPU" webpage.
Available: https://cloud.google.com/tpu/

[32] K. Hazelwood, et al, "Applied Machine Learning at Facebook: A Datacenter Infrastructure Perspective", *Proceedings of the IEEE International Symposium on High Performance Computer Architecture (HPCA)*, Vienna, Austria, February 2018.
Available: https://research.fb.com/wp-content/uploads/2017/12/hpca-2018-facebook.pdf (pre-print)

[33] G. Hinton, O. Vinyals, and J. Dean, "Distilling the Knowledge in a Neural Network", *arXiv Computing Research Repository*, 2015.
Available: https://arxiv.org/abs/1503.02531 (arXiv pre-print)

[34] S. Anwar, K. Hwang, and W. Sung, "Structured Pruning of Deep Convolutional Neural Networks", *arXiv Computing Research Repository*, 2015.
Available: https://arxiv.org/abs/1512.08571 (arXiv pre-print)

[35] P. Molchanov, S. Tyree, T. Karras, T. Aila, and J. Kautz, "Pruning Convolutional Neural Networks for Resource Efficient Inference", *arXiv Computing Research Repository*, 2016.
Available: https://arxiv.org/abs/1611.06440 (arXiv pre-print)

Chapter 21

Reduced Precision Neural Networks

Giulio Gambardella, Thomas Preusser, Yaman Umuroglu, and Michaela Blott, Xilinx

Artificial intelligence and deep learning represent incredibly exciting and powerful machine learning-based techniques used to solve many real world problems. Neural networks, which are inspired by the human brain, are now the predominantly used vision processing algorithms, exceeding humans in accuracy in multiple applications. They are capable of modelling and processing nonlinear relationships between inputs and outputs, and they are characterized by containing adaptive weights along paths between neurons, whose parameters are tuned during training time. Once all parameters are learned to perform a given task with a desired accuracy, neural networks can be deployed.

21.1 Reduced Precision Neural Networks

Weights, or network parameters, in neural networks are traditionally implemented with 32-bit float data types, which can represent values between 10^{-38} to 10^{+38}. Nonetheless, it has been demonstrated that there is much inherent redundancy within the neural network, and that the extended dynamic range of numbers that can be represented is not fully exploited. Therefore, alternative data representations could provide equivalent performance in regard to accuracy, while providing a more efficient representation of the neural network.

Neural networks are typically amenable to direct quantization down to 8 bits, or to even fewer bits using what is referred to as trained quantization, yielding Quantized Neural Networks (QNNs). The quantization scheme may be uniform or non-uniform, and different quantization schemes may be used for different parts of the network. A conceptual diagram of a quantised neural network is shown in Figure 21.1. Using fewer bits requires much less computation and memory, but may result in a slight degradation of accuracy.

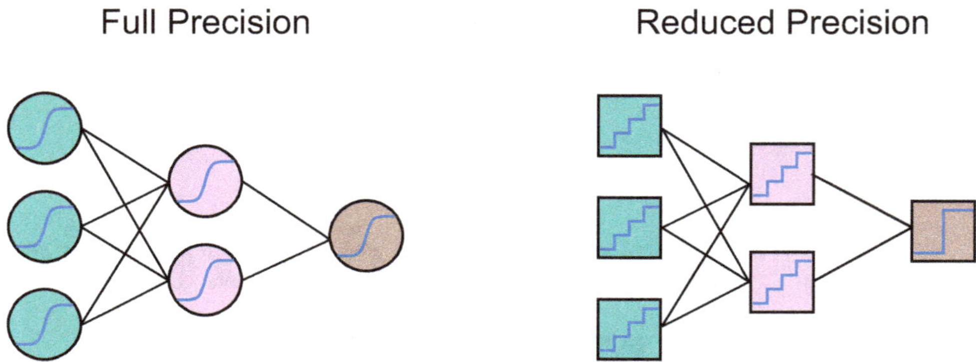

Figure 21.1: Reduced precision in neural networks

QNNs, on smaller image classification benchmarks such as MNIST, SVHN and CIFAR-10, have been demonstrated to achieve nearly state-of-the-art accuracy [15]. Kim and Smaragdis [16] consider full binarization (where weights, inputs and outputs are binarized) with a predetermined portion of the synapses having zero weight, and all other synapses with a weight of one. They report 98.7% accuracy with fully-connected networks on the MNIST dataset, and observe that only XNOR and bitcount operations are necessary for computing with such neural networks. XNOR-Net by Rastegari et al. [11] applies convolutional Binarized Neural Networks on the ImageNet dataset with topologies inspired by AlexNet, ResNet and GoogLeNet, reporting top-1 accuracies of up to 51.2% for full binarization and 65.5% for partial binarization (where only part of the components are binarized).

For the more challenging ImageNet benchmark, there is a noticeable accuracy drop when using QNNs compared to their floating-point equivalents. For example, DoReFa-Net by Zhou et al. [17] explores reduced precision during the forward pass as well as the backward pass. Their results include configurations with partial (with only binarized weights) and full binarization, including best-case top-1 accuracies of 43% for full and 53% for partial binarization. Furthermore, there is significant evidence that increasing network layer size can compensate for this drop in accuracy [24].

21.1.1 What is 'Reduced Precision'?

Intuitively you might think that the usage of reduced precision parameters (instead of full 32-bit floating point representations) would impact on the accuracy of the neural network, but it has been meanwhile demonstrated for numerous popular networks, that if the training is aware of the adoption of quantized weights and activations, neural networks can maintain a very reasonable level of accuracy.

The quantization of neural networks down to 8-bit integers has been meanwhile widely adopted and is well supported by designated and optimized software libraries, such as gemmlowp [10] by Google and the Arm Compute Library [14].

QNNs can have independent data representation of inputs, weights and output activations, as well as different bitwidths in different layers of the same networks. In the extreme case, where we quantize parameters down to a single bit precision, we refer to the networks as Binarized Neural Networks (BNNs). This is a very favourable case, since the Multiply and Accumulate (MAC) arithmetic can be simplified to XNOR and pop-count operations [11].

21.1.2 Motivation

The key advantages of quantized arithmetic in inference are threefold:

- The quantized weights and activations have a significantly lower memory footprint, and as a result the working set of some QNNs may entirely fit into on-chip memory and avoid memory access bottlenecks;
- Replacing floating-point with fixed-point representations and reducing bit precision inherently reduces processing power by orders of magnitude;
- The hardware resource cost of quantized operators is significantly smaller than that of floating-point ones. Therefore, many more arithmetic operations can be implemented and occupied in parallel as neural networks are incredibly parallel, thereby significant performance scaling can be achieved.

21.1.3 The Impact on Scaling

Multiple techniques, like quantization, can be applied to neural networks to decrease the memory footprint and the hardware cost of the operations. Roofline models [13] can be used to showcase the potential of these methodologies, providing an intuitive visual representation of the performance estimation for a system. The model shows theoretical peak performance over a spectrum of applications. Each application is represented by its arithmetic intensity (the number of operations performed for every byte read or written from external memory). The application peak performances can be either limited by the memory bandwidth (on the left side of the graph) or by the peak compute (right side of the graph).

Figure 21.2 shows a roofline model [13] for a Xilinx ZU19 device (in the legend, the first number is the weight precision and the second is the activation bitwidth; T stands for ternary). For each precision, it is possible to calculate the peak performances of the device, given the overall available resources and the hardware cost of a single operation (intended as half the cost of a Multiply and Accumulate) at the specified precision.

It can be noticed how the peak performance, reported as trillion of operations per second, increases by an order of magnitude when reducing the bitwidths of the operands. For example, the graphs illustrate that the peak compute performance increases by a factor of 82 when going from half-precision floating-point to reduced-precision fixed point.

The roofline also shows how the accompanying reduction in model size enables an increase of Arithmetic Intensity, thanks to the storage of network parameters in on-chip memory. This is illustrated by the arithmetic

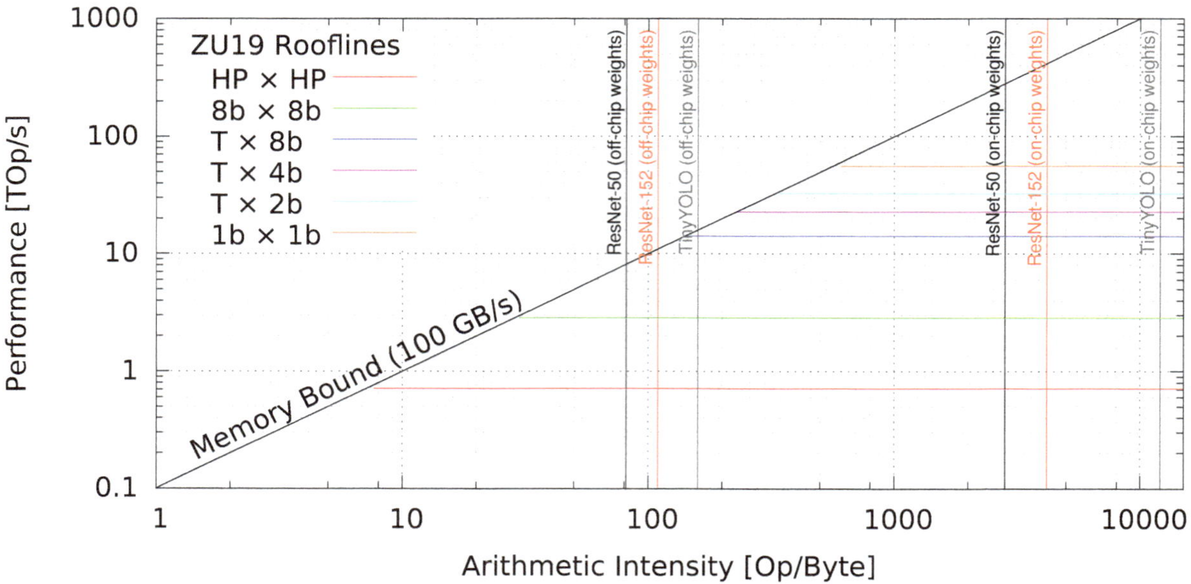

Figure 21.2: ZU19 Rooflines across a range of precisions - 400MHz, 90% DSP and 70% LUTs utilisation

intensity of three different neural network topologies, specifically ResNet-50, ResNet-152 and Tiny-Yolo, which are shown for the two cases of half-precision floating-point (requiring access from the external memory) and 1-bit operation (using on-chip memory resources). The arithmetic intensity for the 1-bit variants is greatly increased and, as a result, the implementation is no longer memory bound. In summary, reducing the precision brings an increased arithmetic intensity which eliminates memory bottlenecks, combined with a higher theoretical peak compute, enabling significant performance scaling within the same device.

The minimal reduction in accuracy combined with significant performance gains and power savings provides highly interesting design trade-offs. Figure 21.3 illustrates the results of one such design space exploration. In this example, different quantization schemes are used to produce networks of varying compute cost. The x-axis numerically expresses roughly how many FPGA LUTs and DSPs would be used for the entire computation, while the y-axis represents accuracy. The red line is the Pareto frontier, containing the design points which are best-in-class for compute cost and accuracy. In this case, we observe that a lower-precision deep network (ResNet-50 with 2-bit weights, 8-bit activations) outperforms a higher-precision shallower network (ResNet-18 with 8-bit weights, 8-bit activations), both in terms of lower compute cost and lower error rate.

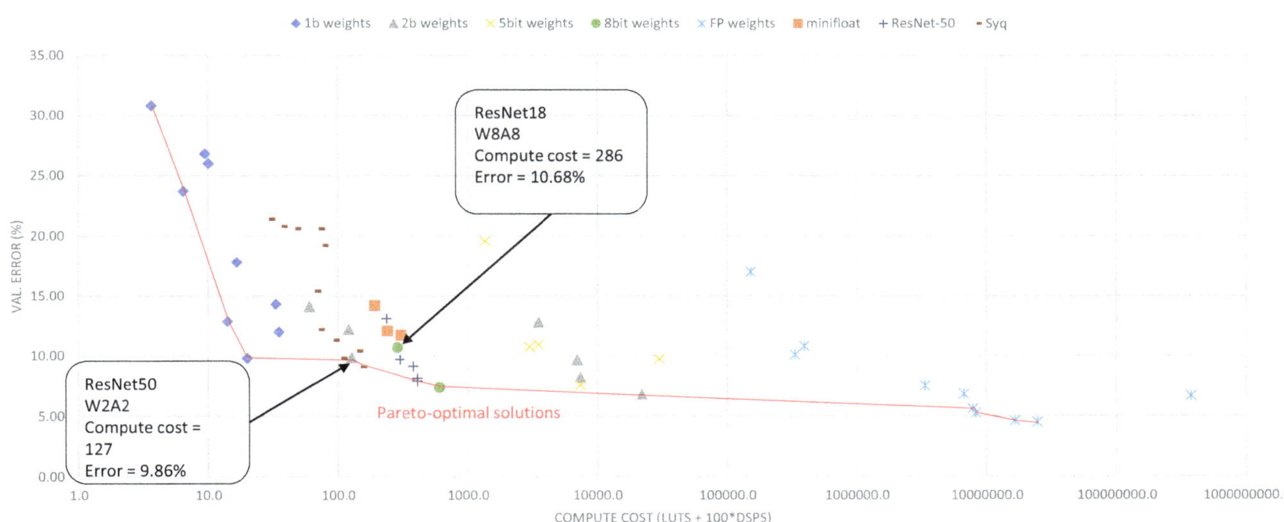

Figure 21.3: Pareto graph of top Accuracy vs Hardware cost

21.1.4 Neural Network Implementation Options

Modern CNNs feature enormous memory bandwidth and high computational needs, challenging existing hardware platforms to meet throughput, latency and power requirements. The high parallelism provided by GPUs have been the first choice for machine learning training tasks in the cloud. Additionally, GPU vendors have developed less power-hungry SoCs for deployment in the edge, tailored for inference tasks [12].

The usage of CPUs for machine learning tasks is common for inference tasks, especially in non-computer-vision tasks, while a mixture of GPUs and CPUs are typically used for training [5].

Driven by the hard computational demands and high volumes, ASICs have also been designed specifically for machine learning tasks. Most prominently among them, Google's Tensor Processing Unit (TPU) has been used to power several of their major products, including Translate, Photos, Search, Assistant, and Gmail [6] in the cloud. Similarly, Edge TPUs [7] have been also designed for inference tasks in the IoT Edge.

FPGAs, thanks to their flexibility and programmability, are ideal for exploiting low-precision inference engines leveraging custom precisions to achieve the required numerical accuracy for a given application. Additionally, according to the size and features of the problem, different architectures can be implemented in the same device according to the applicable requirements.

21.2 Introducing FINN-R

One of the main challenges for machine learning engineers is: given a set of design constraints and a specific machine learning task, how can the best possible trade-off within the vast design space be identified? This answer to this question entails two parts: one is concerned with deriving the most implementation-friendly neural network achieving the target accuracy; and the second part looks at the hardware implementation itself. Deriving systematic accuracy results is a time-intensive process given typical neural network training times. Even the potential computational benefits cannot be easily understood, because hardware implementations are time intensive, the deployment space is complex, architectural choices are myriad and the prediction of power, performance and latency is complex. To find an optimal implementation given a set of design constraints requires a framework that provides insights and estimates given a set of design choices, and automates the customization of the hardware implementation.

FINN-R is trying to address this problem by expanding upon the initial FINN tool [8] with support for more architectural choices as well as mixed and variable precisions beyond binary.

FINN supports feed-forward dataflow implementations for fully binarized neural networks. FINN-R uses a quantization-aware intermediate representation to enable QNN-specific optimizations and has a modular frontend / transform / backend structure for flexibility and increased support, as will be discussed in the following sections.

21.2.1 Origins of FINN-R

The first version of the tool, namely FINN [8], was a framework developed to build scalable and fast BNN inference accelerators for Xilinx FPGA devices. In the original version, the only precision supported was binary for both input and weights, resorting to XNOR and bitcount (a.k.a. Hamming weight, popcount or the number of 1's in the word) operations to perform multiply and accumulate. Additionally, the only supported hardware architecture was feed-forward dataflow (DF), meaning that all the layers in the accelerated neural network are concurrently executed in hardware, each of them having a compute element and communication via FIFOs.

21.2.2 Reduced Precision Optimisations

FINN-R enhanced the features of the original tool by providing support for arbitrary precision. The highly templated HLS library provides the ideal backend for the hardware deployment using Vivado HLS, while internally, a flexible quantization-aware intermediate representation has been developed to provide support within the framework.

21.2.3 Architectural Optimisations

FINN-R provides scalability to bigger networks on small FPGA devices by offering more architectural choices beyond the original feed-forward dataflow.

The Multilayer Offload (MO) architecture is beneficial when the minimal footprint of the fully unrolled DF architecture exceeds the target device capabilities, or when the fragmentation overhead is unattractive. The MO architecture is based on a subset of layers of the neural network being implemented in the PL, and executing the complete neural network by means of multiple calls of the hardware itself.

21.2.4 FINN-R Design Flow

FINN-R provides insights, and estimates and automates the customization of the hardware implementation of a QNN accelerator in programmable logic.

The tool is to be used interactively to explore a given CNN in terms of high-level concepts of the target platform, architecture, and precisions to achieve specific design goals and satisfy given constraints. Given the choices, the tool then customizes the hardware accelerator, either DF or MO style, to meet the constraints. It currently supports resource footprint and throughput constraints.

The key functionality of the tool for the MO is the generation of the runtime schedule that sequences the computation onto the hardware engines, and for DF, the calculation of the folding factors that generate a balanced dataflow whereby the whole architecture gets incrementally unfolded until the design targets are met.

As shown in Figure 21.4, FINN-R has a modular structure inspired by LLVM compiler infrastructure, with frontends, passes and backends, and a quantization-aware intermediate representation (IR) of QNNs. The frontend is responsible for interfacing with a selection of training frameworks such as Caffe, DarkNet and Tensorflow, and translating trained QNNs into the IR.

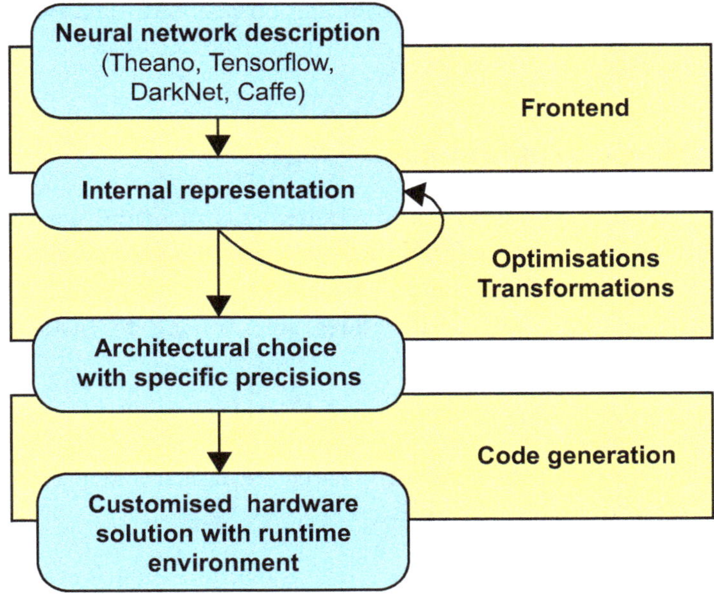

Figure 21.4: FINN-R Framework Overview

The IR is used as the basis of the performance estimation tool. FINN-R supports several transformations that help generate more efficient representations. Finally, the backend contains a code generator that creates executable inference accelerators for a selection of platforms, including PYNQ-Z1, PYNQ-Z2, Ultra96, and AWS F1.

21.2.5 Working with FINN-R

Within FINN, many end-to-end example designs have been released, providing demonstrations that start from training a quantized neural network, all the way down to an accelerated design running on hardware. The examples span a range of datasets and network topologies, relying on the open source templated Vivado HLS library of streaming components: FINN comes with an HLS hardware library that implements convolutional, fully-connected, pooling and LSTM layer types as streaming components. The library uses C++ templates to support a wide range of precisions and layer types.

A set of pre-made accelerators on real hardware are available on multiple boards, namely:

- **BNN-PYNQ** [1] shows examples of image classification accelerators using feed-forward dataflow at multiple precisions;
- **QNN-MO-PYNQ** [2] shows examples of image classification accelerators using multi-layer offload at multiple precisions;
- **LSTM-PYNQ** [3] implements optical character recognition with LSTMs.

The Python tool-flow supporting multiple Xilinx devices is also publicly available [4], transforming a trained, quantized Caffe network to an accelerator running on real hardware. All of the material is available and their features summarized at [23].

21.3 Neural Networks on Programmable Logic

As mentioned earlier, computation and memory requirements in neural networks can be very large: for instance, image classification with a popular DNN such as ResNet-50 requires 7.7 billion operations for every single input image. On the upside, DNNs are highly parallel in nature, which can be exploited. As such, numerous forms of customized hardware architectures are evolving to make the deployment of these algorithms a reality.

The inference computation for a DNN contains multiple levels of parallelism, that can be summarized as:

1. Coarse-grain topology parallelism between consecutive layers, and parallel branches such as those found in GoogLeNet or in DNN ensembles;
2. Neuron and synapse parallelism inside a layer, such as multiple input/output feature map channels and pixels in convolutional layers;

3. Bit-level parallelism inside arithmetic, when individual bits of weights and activations are viewed separately.

A great deal of work on mapping neural networks to hardware exists for both FPGAs and ASICs, exploiting those levels of parallelism. The reader may refer to the work by Misra and Saha [18] for a comprehensive survey.

In general, recent works targeting FPGA hardware implementation of neural networks can be roughly divided into three main categories, based on their basic architecture:

1. A single processing engine, usually in the form of a systolic array, which processes each layer sequentially;

2. A streaming architecture, consisting of one processing engine per network layer where inputs are streamed through the dataflow architecture and all layers are computed in parallel;

3. A vector processor with instructions specific to accelerating the primitive operations of convolutions.

The main advantage of overlay architectures, like (1) and (3), is that a single hardware design can be used for a varied set of networks just by changing the code, or the scheduling of the network on the systolic array/vector processor. Streaming architectures are usually customized for the target network to be accelerated.

21.3.1 A Neuron in Programmable Logic

The basic structure of a neuron is composed of an X input, connected to the neuron with a synaptic connection. The basic function of a neuron is defined in Section 20.2.1 on page 489; it is a non linear function applied to the weighted sum of the inputs.

When a neuron is implemented in programmable logic, it consists of at least one Multiply and Accumulate (MAC) engine, performing the multiplications of the inputs and their relative weights, as well as a block performing the neuron activation. Common activation functions are:

- **Rectified Linear Units (ReLU):** function that takes the maximum of the accumulation value and 0;
- **Linear:** or identity, the function returns the value of the accumulation;
- **Sigmoid:** or logistic function is an S-shaped function between 0 and 1 following the formula $\frac{1}{1+e^{-x}}$;
- **Hyperbolic tangent function:** is an S-shaped function between -1 and +1 following the *tanh* formula.

Figure 21.5 shows the internal architecture of a Processing Element (PE) in FINN-R [9] implementing a single neuron. For neurons working on reduced precision weights (W-bits wide) and producing reduced precision output activations (A-bits wide), it is possible to perform ad-hoc optimizations. The neuron activation, output quantization and any other linear transformations performed on the accumulation value can be simplified by performing comparisons of the neuron accumulator value and a set of thresholds. Those thresholds, evaluated offline for each channel, have the same width as the accumulator (T-bits wide) for easy comparison at the end of the accumulation phase. If the output activation is A-bit wide, 2A-1 thresholds need to be stored in the threshold memory.

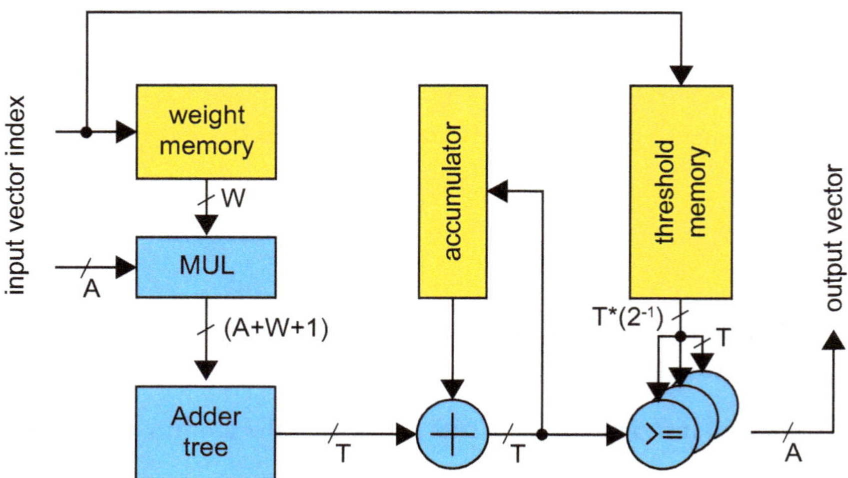

Figure 21.5: Processing element architecture

21.3.2 Creating Layers

Layers in a neural network are basically formed by a set of neurons, connected in different ways. Fully connected layers connect all the inputs of one layer to every neuron of the following. Mathematically speaking, the input vector of the layer is multiplied by a matrix of weights, resulting in an output vector, as explained in (21.1).

$$\begin{pmatrix} W_{1-1} & W_{1-2} & W_{1-3} \\ W_{2-1} & W_{2-2} & W_{2-3} \\ W_{3-1} & W_{3-2} & W_{3-3} \end{pmatrix} \begin{pmatrix} I_1 \\ I_2 \\ I_3 \end{pmatrix} + \begin{pmatrix} b_1 \\ b_2 \\ b_3 \end{pmatrix} = \begin{pmatrix} O_1 \\ O_2 \\ O_3 \end{pmatrix} \tag{21.1}$$

The hardware implementation of a fully connected layer resembles the hardware implementation of a matrix-vector multiplication, as shown in Figure 21.6, where multiple neurons process the input image in parallel as implemented in FINN [8].

Similarly, convolutional layers can be simplified as matrix-matrix multiplications. The simplification is usually performed by taking the input feature (for the first layer, an image) and applying the im2col [19] technique, which rearranges and potentially duplicates portions of the input feature map, generating a so-called lowered image, in order to express the compute as a matrix-matrix multiplication, as shown in Figure 21.7.

In FPGAs the im2col can be implemented through smart addressing schemes, in the so-called Sliding Window Unit (SWU), which avoids explicit input duplication.

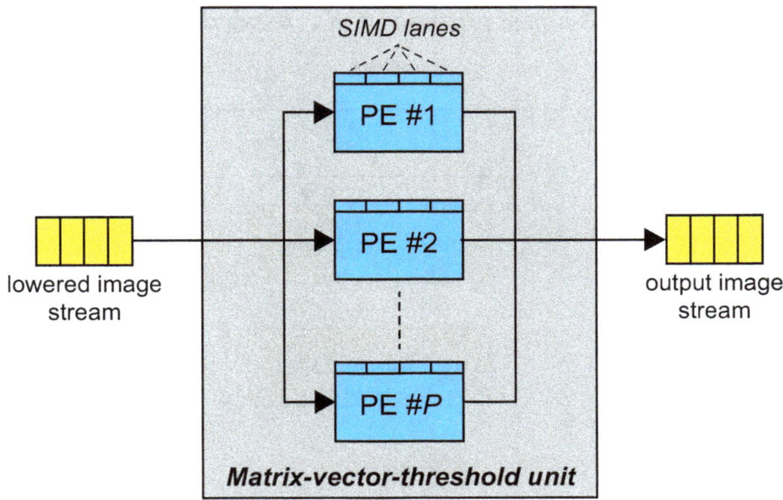

Figure 21.6: Matrix-vector threshold unit

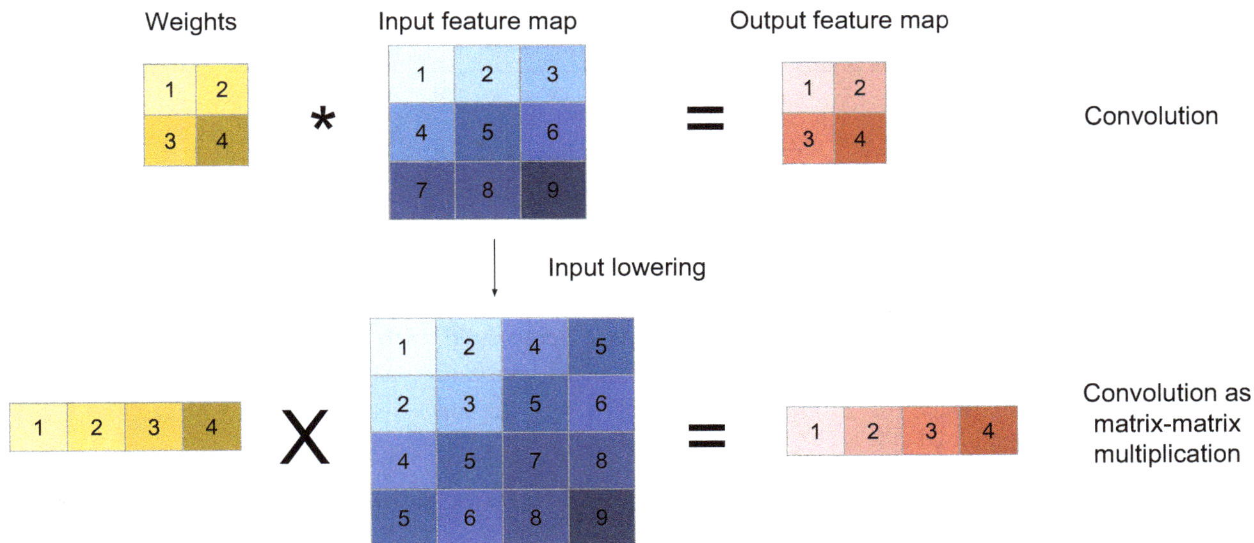

Figure 21.7: Lowering convolution

Figure 21.8 visually shows the SWU operation for a convolutional layer. The input feature map, assumed to be squared with $N \times N$ pixels and C channels, arrives at the input as a raster scan, meaning all channels of a pixel, one line after another from left to right. The SWU leverages multi-line buffering and programmable address logic, to generate the lowered image providing a $K \times K$ window of the input data for the following convolution.

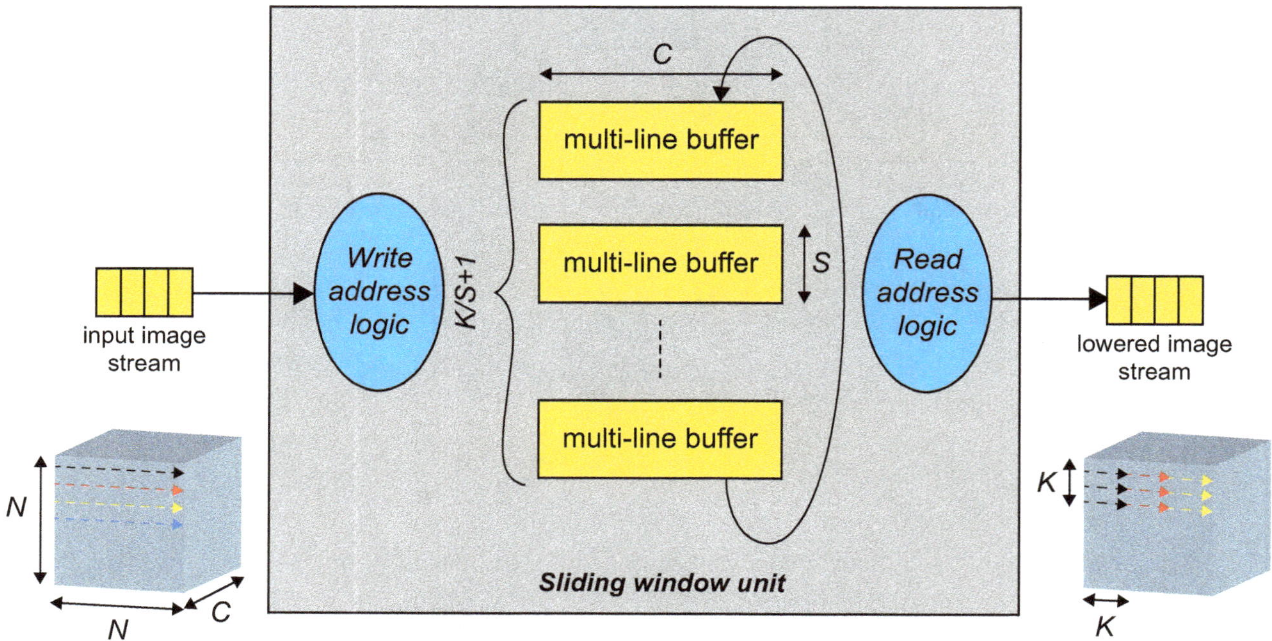

Figure 21.8: Sliding window unit

21.3.3 Design Methods and Tools

The clear majority of machine learning engineers are acquainted with a very high-level programming language. Modern machine learning frameworks, like TensorFlow, Caffe, mxnet, Torch and PyTorch are built on top of Python/C++/Lua, while FPGA design usually relies on hardware description languages at a much lower abstraction level, like VHDL or Verilog.

Recently, FPGA manufacturers have provided tools enabling faster hardware development, such as Vivado HLS, which enables the description of hardware in C/C++. New tools such as SDAccel and SDSoC provide an even easier design experience for datacentre FPGA, Zynq, and Zynq MPSoC devices facilitating the integration of the hardware implementation with the host environment.

Providing a framework to connect the machine learning tools to the hardware deployment is pivotal. The Xilinx ML Suite enables developers to optimize and deploy accelerated ML inference engines. It provides support for many common machine learning frameworks such as Caffe, MxNet and Tensorflow, as well as

Python and RESTful APIs, using xfDNN compiler/optimizer and quantizer (to 8-bit) for deployment on multiple platforms [20]. Similarly, FINN-R [9] automates the deployment of QNNs by using an open source Vivado HLS library of layers.

21.4 FINN-R Case Study: An Inference Accelerator for Zynq MPSoC

Object detection, which aims to classify and localize individual objects within images, is one of the most popular machine learning tasks, especially in the automotive market. A popular reference dataset for this challenge is the Pascal Visual Object Classes (Pascal VOC), and one of the most state-of-the-art networks to accomplish the object detection task is Yolo [21] and its smaller version, Tiny-Yolo.

Performing object detection with a popular neural network such as Tiny-Yolo requires close to 7 billion floating-point operations to process a single frame. This creates significant challenges for acceleration on an embedded device with real-time performance.

Zynq MPSoC platforms are truly heterogeneous, and offer more than just an FPGA accelerator. As already addressed, these platforms include a powerful Arm multicore processor, which not only allows applications to be run from a convenient OS environment, but also offers additional compute power through thread-based concurrency. In the context of intense linear algebra, the NEON vector extension available in these processors is also of utmost interest, as it enables parallel SIMD operations, e.g. in four single-precision floating-point lanes, or in eight 16-bit integer lanes.

Figure 21.9 shows the network topology called Tincy-Yolo, which has been modified from the original "Tiny" implementation. The input and output layers of the network have proven sensitive to quantization, thus were kept at higher precision (8-bit), while the remaining parts of the neural network, accounting for over 97% of the overall operations, where heavily quantized. The arithmetic was simplified enormously by using binary weights (-1, +1) and 3-bit feature map pixels, while keeping the detection accuracy within 3% of the full precision network, by successful retraining.

Low bitwidth quantization is a natural fit for programmable logic implementation, as it can truly exploit the reduced cost for low bitwidth operations, thereby enabling high parallelism for the acceleration of the core part of the neural network.

The remaining parts of the neural network, the first and last non-quantized layers, needed acceleration to meet the real-time requirements. An obvious way to increase the number of arithmetic operations per cycle is vectorization, as offered by the NEON extension of the platform processor. Using 128-bit registers, equivalent parallel computations can be performed in four 32-bit lanes and up to sixteen 8-bit lanes. Using the developed offload mechanism integrated in Darknet and already used for the communication between PS and PL, we thus implemented a custom layer with an im2col [19] implementation that quantized the image data while arranging the multiplicand matrix and a matrix multiplication performed through the gemmlowp library [10].

We applied some neural network surgery, like channel pruning on the last 2 layers and dismissal of the first MaxPool layer, while increasing the stride of the first convolutional layer. Combined, all of these optimiza-

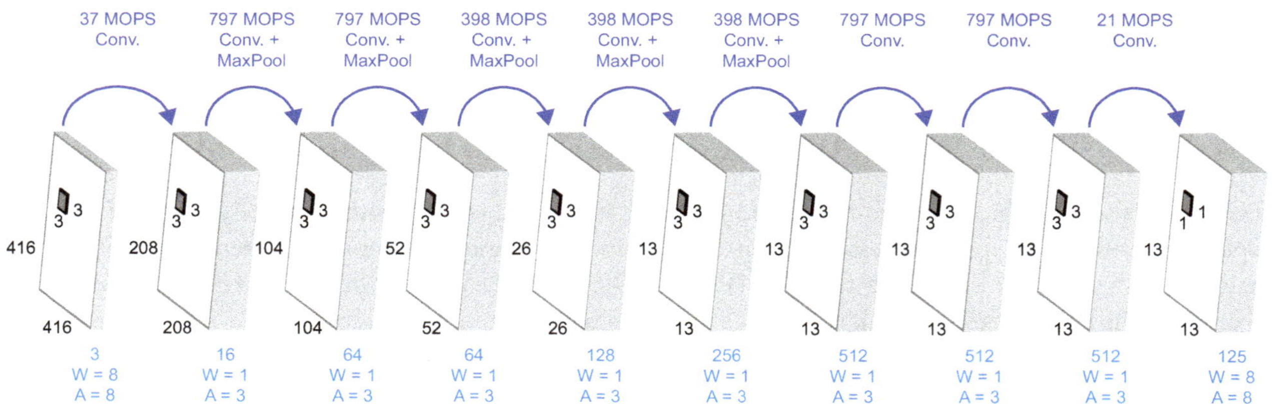

Figure 21.9: Tincy-Yolo network topology

tions led to a 160X speed-up of the inference execution, also thanks to the multi-threaded execution enabled by Zynq MPSoC. The demo achieved almost a threefold speed-up, resulting in a frame rate of 16 fps on a small ZU3 device (on the Ultra96 board). This allows live video to be played in a way that it is perceived as smooth in practice [22].

21.5 Chapter Summary

This chapter has introduced the concept of reduced precision in the context of machine learning, especially neural networks. Programmable logic enables the possibility of scaling performance by orders of magnitude with respect to standard floating-point neural networks, creating an interesting design trade-off in exchange for a small reduction in accuracy, as shown in the Pareto graph in Figure 21.3. Especially in conjunction with FPGAs, a broad spectrum of architectures for inference engines can be explored, whereby the minimal precision to achieve precisely the numerical accuracy required for an application can be exploited.

We provide examples of tooling in our open-source framework FINN-R [9], which assists designers in finding interesting trade-offs to implement machine learning tasks, and should be of interest to all those who are keen to enable machine learning tasks on Xilinx Zynq MPSoC devices. All examples are cross-linked through our landing page at:

https://www.pynq.io/ML .

21.6 References

Note: All online sources last accessed March 2019.

[1] Xilinx BNN-PYNQ (Binarized Neural Network on PYNQ) GitHub repository.
Available: https://github.com/Xilinx/BNN-PYNQ

[2] Xilinx QNN-MO-PYNQ (Quantised Neural Network on PYNQ using a Multi-Layer Offload architecture) GitHub repository.
Available: https://github.com/Xilinx/QNN-MO-PYNQ

[3] Xilinx LSTM-PYNQ (Quantised LSTM on PYNQ) GitHub repository.
Available: https://github.com/Xilinx/LSTM-PYNQ

[4] Xilinx FINN GitHub repository.
Available: https://github.com/Xilinx/FINN

[5] K. Hazelwood et al, "Applied Machine Learning at Facebook: A Datacenter Infrastructure Perspective", *Proceedings of the 24th IEEE International Symposium on High Performance Computer Architecture*, Vienna, Austria, February 2018.
Available: https://research.fb.com/wp-content/uploads/2017/12/hpca-2018-facebook.pdf (preprint)

[6] Google, Inc., "Cloud TPU: Train and run machine learning models faster than ever before" webpage.
Available: https://cloud.google.com/tpu/

[7] Google, Inc., "Edge TPU: Google's purpose-built ASIC designed to run inference at the edge" webpage.
Available: https://cloud.google.com/edge-tpu/

[8] Y. Umuroglu et al, "FINN: A Framework for Fast, Scalable Binarized Neural Network Inference", *Proceedings of the 25th International Symposium on Field Programmable Gate Arrays (FPGA)*, Monterey, California, USA, February 2017.
Available: https://arxiv.org/abs/1612.07119 (arXiv preprint)

[9] M. Blott et al, "FINN-R: An End-to-End Deep-Learning Framework for Fast Exploration of Quantized Neural Networks", 2018.
Available: https://arxiv.org/abs/1809.04570 (arXiv preprint)

[10] Google gemmlowp (A Small Self-Contained Low-Precision GEMM Library) GitHub repository.
Available: https://github.com/google/gemmlowp

[11] M. Rastegari, V. Ordonez, J. Redmon, and A. Farhadi, "XNOR-Net: ImageNet Classification Using Binary Convolutional Neural Networks", *Proceedings of the European Conference on Computer Vision*, Amsterdam, Netherlands, October 2016.
Available: http://ai2-website.s3.amazonaws.com/publications/XNOR-Net.pdf (preprint)

[12] NVIDIA, Inc., "NVIDIA JETSON SYSTEMS" webpage.
Available: https://www.nvidia.com/en-us/autonomous-machines/embedded-systems-dev-kits-modules/

[13] Roofline: An Insightful Visual Performance Model for Floating-Point Programs and Multicore Architectures
Available: https://www2.eecs.berkeley.edu/Pubs/TechRpts/2008/EECS-2008-134.pdf

[14] Arm, Ltd., "Compute Library: A software library for computer vision and machine learning" webpage.
Available: https://developer.arm.com/technologies/compute-library

[15] M. Courbariaux et al, "Binarized Neural Networks: Training Neural Networks with Weights and Activations Constrained to +1 or -1", March 2016.

Available: https://arxiv.org/pdf/1602.02830.pdf

[16] M. Kim and P. Smaragdis, "Bitwise Neural Networks", *Proceedings of the 31st International Conference on Machine Learning (Workshop on Resource-Efficient Machine Learning)*, Lille, France, July 2015.

Available: https://arxiv.org/pdf/1601.06071.pdf

[17] S. Zhou et al, "DoReFa-Net: Training Low Bitwidth Convolutional Neural Networks with Low Bitwidth Gradients", June 2016.

Available: https://arxiv.org/pdf/1606.06160v1.pdf

[18] J. Misra and I. Saha, "Artificial neural networks in hardware: A survey of two decades of progress", *Neurocomputing*, Vol. 74, Issues 1 - 3, December 2010, pp. 239 - 255.

DOI: 10.1016/j.neucom.2010.03.021

[19] MathWorks, "im2col: Rearrange image blocks into columns" webpage.

Available: https://www.mathworks.com/help/images/ref/im2col.html

[20] Xilinx, Inc. "AI Inference Acceleration" webpage.

Available: https://www.xilinx.com/applications/megatrends/machine-learning.html

[21] J. Redmon, S. Divvala, R. Girshick and A. Farhadi, "You Only Look Once: Unified, Real-Time Object Detection" *Proceedings of the 29th IEEE Conference on Computer Vision and Pattern Recognition (CVPR)*, Las Vegas, Nevada, USA, June 2016.

Available: https://arxiv.org/abs/1506.02640 (arXiv preprint)

[22] T. Preusser, G. Gambardella, N. Fraser and M. Blott, "Inference of Quantized Neural Networks on Heterogeneous All-Programmable Devices", *Proceedings of the Design Automation & Test in Europe Conference & Exhibition (DATE)*, Dresden, Germany, March 2018.

Available: https://arxiv.org/pdf/1806.08085.pdf (pre-print)

[23] PYNQ, "Machine Learning on Xilinx FPGAs with FINN" webpage.

Available: http://www.pynq.io/ml

[24] N. Fraser et al, "Scaling Binarized Neural Networks on Reconfigurable Logic", *Proceedings of the 2017 HiPEAC Conference*, Stockholm, Sweden, January 2017.

Available: https://arxiv.org/pdf/1701.03400.pdf (arXiv preprint)

Chapter 22

PYNQ

with Patrick Lysaght and Cathal McCabe, Xilinx

PYNQ is an open source framework for creating and using applications with Zynq and Zynq MPSoC devices. The name 'PYNQ' is derived from 'Python productivity for Zynq'; and, as the name suggests, PYNQ uses the Python programming language to simplify the process of creating applications with Zynq devices. This has two important consequences: it improves the productivity of designers already working with Zynq; and it reduces the barrier-to-entry for new users, especially those with limited experience of hardware design.

This chapter will explain the key concepts of PYNQ and provide practical guidance on how to start creating your own designs, while also featuring some illustrative examples from the open source community.

22.1 Introducing PYNQ

PYNQ (pronounced "pink") is a framework integrating both software and hardware components, and it is designed to facilitate faster, more productive development using Zynq devices, including Zynq MPSoC. Over the next few pages, we define what is meant by a 'framework' in this context, and also explain what PYNQ is (as well as clarifying what it isn't!). We also discuss why Python was adopted for PYNQ, and outline the goals of the PYNQ initiative.

22.1.1 PYNQ as a Framework

Throughout this chapter, we refer to PYNQ as a framework — why?

In a general sense, a framework is defined as a structure that underpins a system or concept. More specifically the Wikipedia entry for 'Software framework' [22], provides a good overview.

From Wikipedia:

"In computer programming, a software framework is an abstraction in which software providing generic functionality can be selectively changed by additional user-written code, thus providing application-specific software. A software framework provides a standard way to build and deploy applications. A software framework is a universal, reusable software environment that provides particular functionality as part of a larger software platform to facilitate development of software applications, products and solutions. Software frameworks may include support programs, compilers, code libraries, tool sets, and application programming interfaces (APIs) that bring together all the different components to enable development of a project or system."

PYNQ fits Wikipedia's description well, as it provides a set of integrated software and hardware components that designers can use as the basis for their Zynq-based systems. There is no necessity to use all of the available components, and there is much freedom to build upon them, customise, and innovate. As such, the PYNQ framework represents a firm, established starting point that can be used in its original form, adapted, and extended as desired.

As will be highlighted in the remainder of this chapter, the PYNQ framework complements the mainstream Xilinx tools, rather than replacing them — in particular, PYNQ accelerates the software/ hardware co-design aspects of Zynq-based embedded systems, and makes the task of interfacing between the PS and PL easier.

22.1.2 What is PYNQ?

As discussed earlier in the book, Zynq-based embedded systems can be conceived as a stack of layers, extending from a hardware system design at the bottom; through an OS layer, which includes low-level software drivers and APIs for interfacing with hardware; to software applications running on top of the OS. Depending on the target design, the creation of a complete system can involve significant design effort, spanning a range of different skills, expertise, and design tools.

PYNQ addresses the complexity of co-design by providing a pre-configured software stack, augmented by libraries of hardware and software components that can be selectively reused, depending on the target application. Figure 22.1 illustrates the general concept of the PYNQ framework, and shows how the various PYNQ layers relate to the those of a typical Zynq-based embedded system.

Upper Layer (Applications)

At the top of the PYNQ stack, user interaction is facilitated by one or more *Jupyter Notebooks*. Jupyter is an open-source project that emerged from academia, with its roots in data science, and a key aim of the project was to further 'reproducible science' [11]. Briefly, Project Jupyter enables scientific results to be presented in a manner that enables readers to reproduce and validate the claims of the authors, with reasonable effort.

Jupyter allows users to create interactive documents, known as 'notebooks', which are served via a standard web browser. These notebooks can contain a variety of different content, including live executable code and visualisations, as well as textual, graphical, and mathematical documentation, and they are organised in cells

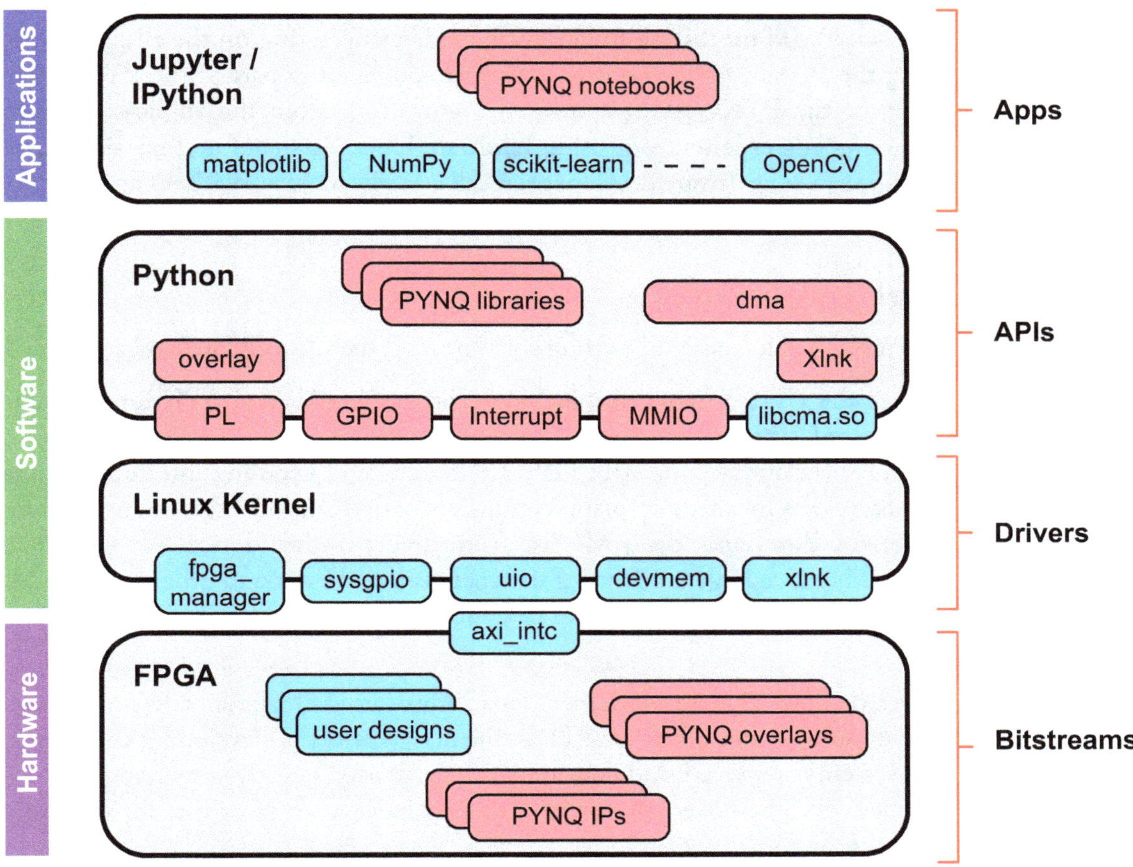

Figure 22.1: The PYNQ Framework

that may be executed individually. Originally known as *IPython* (Interactive Python) notebooks and featuring only Python programming, Jupyter Notebooks now support a variety of programming languages [20]. More recently, *Jupyter Labs* has extended the concept to a complete web-based Integrated Development Environment (IDE). Jupyter Notebooks are an integral part of Jupyter Labs.

A distinguishing feature of PYNQ is that its Jupyter Notebooks are hosted on the Zynq's Arm processor (i.e. an embedded device), whereas the Jupyter project was originally conceived for desktop and server computing. The notebooks reside on a webserver on the Zynq PS, and the user accesses them from a standard web browser over a network connection. We will see examples of Jupyter Notebooks a little later, in Section 22.2.

Within a PYNQ Jupyter notebook, the developer creates their own custom functionality by writing their own Python code, and selectively reusing third-party code from the many open source Python libraries that are available. They can also add documentation and visualisation content to help others understand and use the design.

In addition to developing software-based functionality to run on the Arm processor in the PS, Python code within the notebook can also offload processing to hardware modules operating on the PL. Interaction with hardware is achieved using the Python APIs and drivers that are provided as part of the PYNQ framework (passing through all layers of the PYNQ stack shown in Figure 22.1, from the application layer to the hardware layer). The programmer's experience of using hardware blocks is therefore very similar to calling functions from a software library — a software developer can call a hardware block without any need to understand the internals of the hardware design.

Middle Layers (Software)

The mid-layers of the PYNQ stack consist of Python software, and the OS and low level software drivers.

In the upper middle layer, the PYNQ framework includes Python libraries and APIs for interacting with various elements of Zynq-based systems. For instance, there are Python APIs for downloading overlays (bitstream files) to the PL, for communicating with GPIO resources, and handling interrupts generating in the PL. There are also Python APIs for memory mapped transfers (MMIO), and DMA transfers. One of the most significant advantages of Zynq and Zynq MPSoC compared to other devices (or combinations of devices) is the ability to quickly move large amounts of data between processor and programmable logic, and vice versa. The use of PYNQ enables these transfers to be controlled in a very straightforward fashion using Python code.

In the lower middle layer, the PYNQ framework includes a Linux-based OS, bootloaders to initiate system start-up, and a web server to host Jupyter notebooks. Thus, the design effort of developing common software elements of an embedded system is removed, and new users can get started quickly with Zynq — this is a key benefit of the PYNQ framework. The lower middle layer also includes a set of drivers for interacting with elements of the Zynq hardware system.

Lower Layer (Hardware)

The bottom layer of the stack represents a hardware system design, which would normally be created in Vivado using IP Integrator and associated design tools, and then generated to a bitstream (*.bit) file. The bitstream file is transferred onto the memory card inserted into the target board. The process of programming the hardware system onto the PL can then be initiated directly from within a Jupyter notebook (running on the PS), using a single line of code:

```
my_overlay = Overlay("/path/to/your/overlay/file/bitstream.bit")
```

In PYNQ, hardware system designs are often referred to as *overlays*. They can be used in a manner analogous to software libraries, wherein a hardware system is developed for a particular application domain, but with an aspect of generality that facilitates enhanced sharing and reuse. Details of hardware designs can be abstracted and their functionality exposed in Python via an API, which enables a very software-centric style of using PYNQ. As outlined in Section 22.4, one of the goals of PYNQ is to enable designers without hardware expertise to develop applications in software, based on pre-existing overlays. It also helps hardware engineers to create designs that can be evaluated and used by software engineers.

Although overlays are often generalised designs, a more traditional hardware/software co-design approach could also be taken, wherein highly customised hardware is developed for a specific use case. Here, several advantages of the PYNQ framework can be leveraged, including the availability of a ready-made software stack, the ease of interfacing with elements of the developed hardware design, and the potential to adapt and extend the software programming environment. The set of available PYNQ IPs can also be freely reused — this includes interfacing blocks for DMA, audio, video, and I2C, and components from *logictools* (an overlay described later, in Section 22.3.3).

Taking all of these resuable components into consideration, we can define the term *hardware libraries*. This is an umbrella term, and it refers to the set of IPs and overlays that are available as part of the PYNQ framework for flexible reuse. Hardware libraries may be considered analogous to software libraries. Of course, additional hardware libraries can also be created by developers for their own use, or for sharing with others.

22.1.3 Why Python?

Python is an interpreted language, meaning that it offers intuitive, high level programming at the expense of slower code execution compared to a compiled language such as C or C++. The use of an interpreted language aids interactivity, because changes and additions to the code can be run and evaluated immediately. In PYNQ, the Python interpreter is run on Zynq, which makes cross-compilation unnecessary and therefore greatly simplifies the overall design process. The PYNQ framework also includes GCC (the GNU Compiler Collection) natively on the target, allowing Python to be extended with C and C++.

Execution speed is usually not a significant factor in notebooks for PYNQ, but in cases where it is, performance can be boosted by incorporating optimised C/C++ functions and libraries that are called by the primary Python code (for this reason, PYNQ adopts an implementation of CPython). Much more important is Python's status as a 'productivity language', meaning that useful code can be developed quickly and reliably.

Further, Python is really popular! Many readers of this book will already be familiar with the language, whether via study, work, or hobbies. There has been a huge increase in the adoption of Python in recent years, such that, as of 2018, it is now the most popular programming language *of all*, according to IEEE Spectrum's annual survey (interestingly, Python was also listed as an embedded language for the first time) [2]. Students and recent graduates are often particularly well exposed to Python, given that it is commonly taught in introductory software classes at university and college, as well as in schools. Therefore, PYNQ offers programming for Zynq-based embedded systems in a language that many potential users are already comfortable with.

Another key advantage of Python is that a rich set of resources can be leveraged from the open-source community, including packages such as NumPy, Matplotlib, OpenCV, and many others. In fact, the selection of available software libraries is constantly expanding and evolving, which provides great scope for designers to reuse code in their projects, rather than having to spend time developing the desired functionality from scratch — this is another factor that greatly enhances designer productivity.

Bearing in mind these major attributes, the use of Python is central to the 'productivity' concept of PYNQ.

22.1.4 What PYNQ Is Not!

Having described what PYNQ is, at a high level, it is also useful to clarify what PYNQ *is not*, and thereby to clear up some misconceptions that may arise.

PYNQ is Not... a board

As mentioned in earlier sections, PYNQ is a software framework. It was first released along with a specially supported PYNQ-Z1 development board, featuring a pink-coloured PCB (more recently, a PYNQ-Z2 has been released, which is also pink). There are two common misconceptions relating to these boards.

The first is that the term 'PYNQ' refers to these pink-coloured boards (e.g. "I've just bought a PYNQ!"). Rather, PYNQ-Z1 or PYNQ-Z2 are physical development platforms that are supported by PYNQ, where PYNQ refers to the *software framework* that can be run on them.

A second common mistake is to assume that the PYNQ software framework can *only* be deployed on a pink-coloured board. While the PYNQ-Z1 and PYNQ-Z2 are supported by the PYNQ project, so too are other development boards, including the ZCU104 and Ultra96 Zynq MPSoC boards. The primary form of support for development boards is the release of pre-tested PYNQ images that run upon them (more on images in Section 22.3.1), and Figure 22.2 shows a selection of boards for which PYNQ images are available. If a prepared image is not available for your target board, it is straightforward to create one.

PYNQ is Not... for synthesising circuits from Python

As PYNQ involves Python coding for Zynq SoC platforms, it is sometimes inferred that Python coding is used for hardware description and verification, i.e. to generate circuits for implementation in the PL section. This is not the case — in PYNQ, Python is used for programming on the PS only, which includes interaction with hardware via the PS-PL interfaces. Generation of HDL code from Python descriptions is not supported, but instead, designers have several other options for hardware development, including the use of pre-verified IP Integrator blocks, custom design using the High Level Synthesis (HLS) and System Generator tools, conventional VHDL / Verilog coding... or indeed any other method they currently use.

Third party tools for Python-based HDL generation and verification do exist, and readers interested in that area may wish to investigate the free open-source tool, *myHDL* [4].

PYNQ is Not... just for students and hobbyists

With the availability of low-cost development boards, the adoption of an open-source model, and extensive resources provided for teaching and projects, it might be tempting to conclude that PYNQ is targeted only at students and hobbyists.

While it is true that these communities are well supported, PYNQ is also suited for research, and for professional developers, who can leverage PYNQ's extensibility to cater for their own custom boards and requirements. The potential gains in productivity attainable using the PYNQ framework, its open source nature, and the availability of third-party software libraries are all good reasons to consider PYNQ for

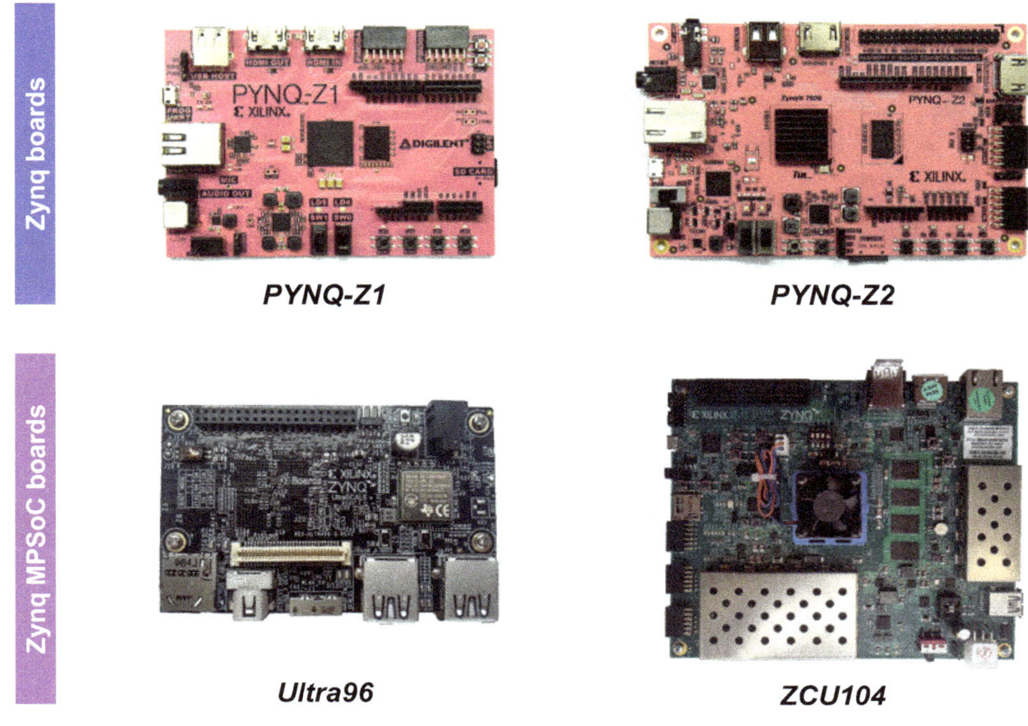

Figure 22.2: A selection of development boards for which a PYNQ image is available

commercial projects, particularly for prototyping and developing demonstrators. To illustrate this point, a case study is provided in Chapter 23 demonstrating the successful use of PYNQ for prototyping a commercial satellite imaging system. A good example of PYNQ-enabled research is the acceleration framework discussed in [6].

22.2 An Example PYNQ Application

To demonstrate the Jupyter notebook interface in PYNQ, two example notebooks will be presented over the next two pages. The notebooks implement the same functionality: in each case, an image file is reduced from its original dimensions to a new, smaller size, where the *x* and *y* axes are both reduced by a user-specified integer factor. The distinction is that the first notebook (Figure 22.3) is a software implementation executing exclusively on the Arm processors. In contrast, the second design (Figure 22.4) also leverages the PL to accelerate the image re-sizing operation. Please note that some steps, such as importing data and libraries, are omitted for ease of presentation.

library declarations omitted for brevity
also present in HW accel. notebook (omitted for brevity)
software resizing
resizer_PS
Load image from the SD card and create an image object
In [2]: image_path = "./images/paris.jpg"
original_image = Image.open(image_path)
original_image.load()
Read image from file and save to a NumPy array.
Create a numpy array of the pixels
In [3]: input_array = np.array(original_image)
Display the image to be resized
In [4]: input_image = Image.fromarray(input_array)
display(input_image)
Display the input image.
Original image!
Original image size
In [5]: old_width, old_height = original_image.size
print("Image size: {}x{} pixels.".format(old_width, old_height))
Image size: 640x360 pixels.
Resizing
Setting image resize dimensions
In [6]: resize_factor = 2
new_width, new_height = int(old_width/resize_factor), int(old_height/resize_factor)
Determine old and new image dimensions.
In [7]: resized_image = original_image.resize((new_width, new_height))
Display resized image
In [8]: output_array = np.array(resized_image)
result = Image.fromarray(output_array)
display(result)
Display the resized image.
Resized image!
Resized image size
In [9]: width, height = resized_image.size
print("Resized image size: {}x{} pixels.".format(width, height))
Resized image size: 320x180 pixels.
Print out new image size.

Figure 22.3: The Software-Only Image Resizer Notebook

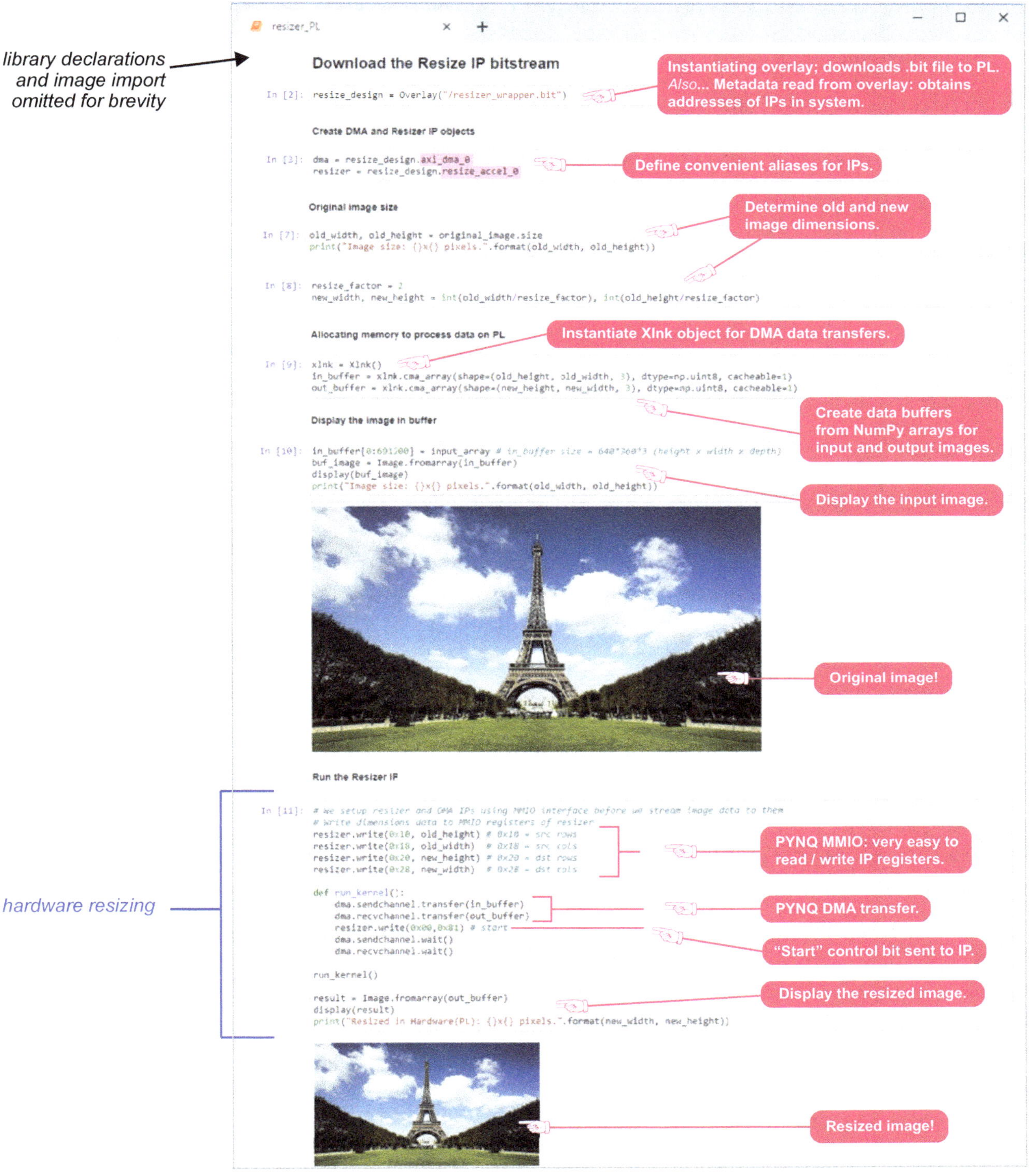

library declarations and image import omitted for brevity
resizer_PL
Download the Resize IP bitstream
In [2]: resize_design = Overlay("/resizer_wrapper.bit")
Instantiating overlay; downloads .bit file to PL. Also... Metadata read from overlay: obtains addresses of IPs in system.
Create DMA and Resizer IP objects
In [3]: dma = resize_design.axi_dma_0
resizer = resize_design.resize_accel_0
Define convenient aliases for IPs.
Original image size
In [7]: old_width, old_height = original_image.size
print("Image size: {}x{} pixels.".format(old_width, old_height))
Determine old and new image dimensions.
In [8]: resize_factor = 2
new_width, new_height = int(old_width/resize_factor), int(old_height/resize_factor)
Allocating memory to process data on PL
Instantiate Xlnk object for DMA data transfers.
In [9]: xlnk = Xlnk()
in_buffer = xlnk.cma_array(shape=(old_height, old_width, 3), dtype=np.uint8, cacheable=1)
out_buffer = xlnk.cma_array(shape=(new_height, new_width, 3), dtype=np.uint8, cacheable=1)
Create data buffers from NumPy arrays for input and output images.
Display the image in buffer
In [10]: in_buffer[0:691200] = input_array
buf_image = Image.fromarray(in_buffer)
display(buf_image)
print("Image size: {}x{} pixels.".format(old_width, old_height))
Display the input image.
Original image!
Run the Resizer IP
In [11]: # we setup resizer and DMA IPs using MMIO interface before we stream image data to them
# write dimensions data to MMIO registers of resizer
resizer.write(0x10, old_height)
resizer.write(0x18, old_width)
resizer.write(0x20, new_height)
resizer.write(0x28, new_width)
PYNQ MMIO: very easy to read / write IP registers.
def run_kernel():
dma.sendchannel.transfer(in_buffer)
dma.recvchannel.transfer(out_buffer)
resizer.write(0x00,0x81) # start
dma.sendchannel.wait()
dma.recvchannel.wait()
PYNQ DMA transfer.
"Start" control bit sent to IP.
run_kernel()
result = Image.fromarray(out_buffer)
display(result)
print("Resized in Hardware(PL): {}x{} pixels.".format(new_width, new_height))
Display the resized image.
hardware resizing
Resized image!

Figure 22.4: The Hardware Accelerated Image Resizer Notebook

Some of the main PYNQ features demonstrated by this example are:

- The use of Jupyter Notebooks for programming and user interaction.
- The simple method of downloading a bitstream onto the board, to program the PL.
- The interpretation of overlay *metadata*, including details of IPs, by parsing a file generated by Vivado and included as part of the overlay file set (this aspect will be discussed further in Section 22.3.5).
- The use of Python APIs to easily transfer data between the PS and PL.

This example was developed by Xilinx[1], and is available from the *Community* pages of the PYNQ website,

www.pynq.io/community.html

under the heading 'PYNQ Hello World'. If you have access to a supported development board, you can install the example and try it out for yourself.

22.2.1 Python Notebook (Software Only)

The software-only example is shown first, in Figure 22.3 on page 532. In this case, all functionality is implemented in Python (specifically, the resizing operation is performed by a function from the Python Imaging Library (PIL)), and all processing takes place on the PS part of the Zynq device. The notebook includes documentation along with the Python code to explain the steps involved, and the images before and after resizing are displayed in the notebook.

It is worth mentioning that a Jupyter Notebook can be saved with results embedded, which makes it easy to share outcomes with others. Recipients can also easily rerun the notebook and reproduce the results.

22.2.2 PYNQ Notebook (Hardware Accelerated)

The second, software-hardware notebook is shown in Figure 22.4 on page 533. It is similar to the first, but integrates a hardware IP to accelerate resizing of the image, which is the most computationally demanding task within the notebook. The IP is a Xilinx implementation of the OpenCV resizing function. Notably, the hardware block for image resizing can be called in the same way as software function.

The hardware system underlying this notebook, and associated software interaction, is reviewed in Section 22.2.3.

22.2.3 System Architecture (Hardware Accelerated Version)

The second notebook (Figure 22.4 on page 533), which features PL-based processing for the resizing operation, is underpinned by the hardware system design shown in Figure 22.5. Although this is a simple application, the design makes use of several important features of PYNQ.

1. The example is featured here under the BSD 3-Clause License; copy available on page 583.

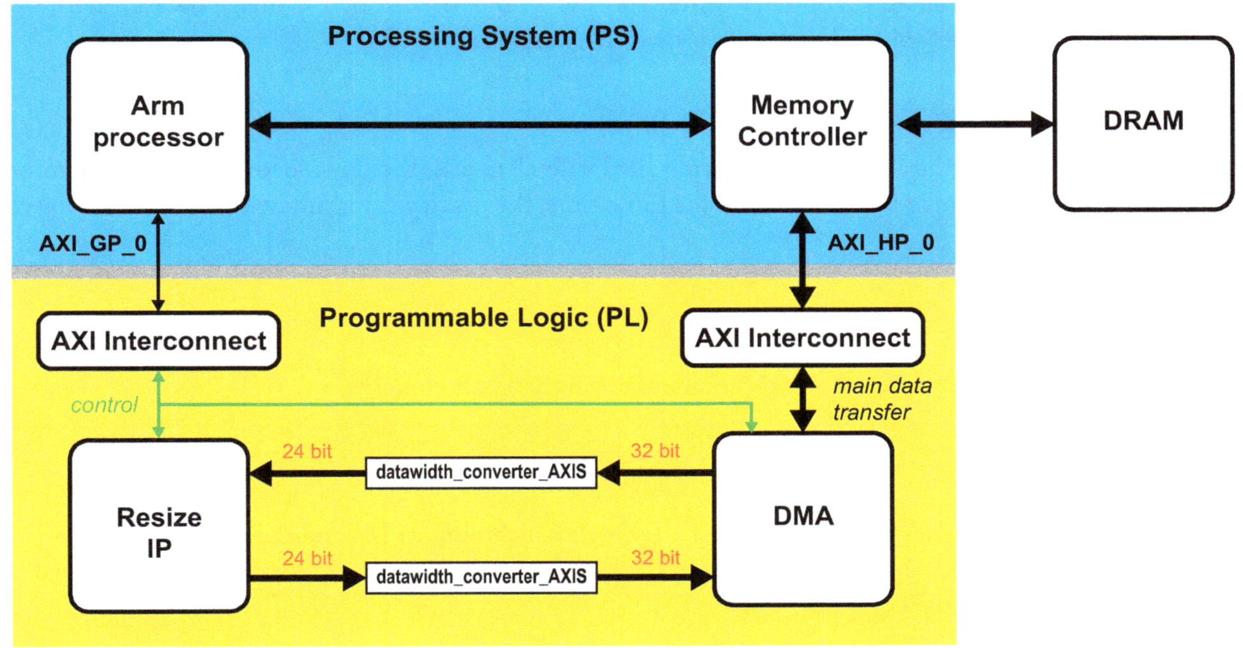

Figure 22.5: Hardware system for the 'Image Resizer' PYNQ design

Operation

The original image is read from a file on the memory card, and stored within the notebook as a NumPy array. The dimensions of the output image are determined, and an appropriately sized NumPy array is created. Within the Python code in the notebook, contiguous memory is allocated in DRAM, where the input and output image arrays are buffered.

The DMA and Resizer both exist as IP blocks in the PL, with the DMA sitting between the PS memory and Resizer in the data path. When the overlay for the resizer design is instantiated in the Jupyter Notebook, the metatdata in the design files is parsed by the 'Overlay' class of the PYNQ framework. This process exposes the IP blocks contained within the overlay for interaction via Python (different names can be assigned if desired; in this case 'resizer' and 'dma' have been used).

The Resizer IP is first configured with the dimensions of the input and output images, by writing the widths and heights of the input and output images to the appropriate registers, as shown in the highlighted section of the notebook in Figure 22.4. In hardware, this takes place over AXI-Lite, via the general purpose AXI port AXI_GP_0, and it is enabled in PYNQ by the Python 'MMIO' class. A DMA transfer via the AXI high performance port AXI_HP_0 is then initiated (similarly, this uses the PYNQ 'DMA' class), and a register command is written to the Resizer IP over AXI-Lite to start operation (again using 'MMIO'). Once complete, the data corresponding to the resized image can then be read from the output buffer and displayed in the notebook.

Conversion between standard 32 bit data words, and the 24 bits needed by the image data format (8-bits each for red, green and blue) takes place between the DMA and Resizer IP.

Enabling PYNQ Features

In this notebook, as for any notebook featuring hardware elements, the downloading of a bitstream onto the PL is initiated by a simple line of code. At the same time, the overlay is instantiated as a Python object in the notebook. In this case:

```
resize_design = Overlay("/resizer_wrapper.bit")
```

It is then possible to query the overlay object to establish the IPs it contains,

```
resize_design?
```

and this generates an inline report with details of the overlay, including its IPs and their associated drivers. In this case, the PYNQ 'DMA' driver is associated with the DMA IP block, and a generic 'Default IP' driver is associated with the Resizer IP.

```
IP Blocks
----------
axi_dma_0               : pynq.lib.dma.DMA
resize_accel_0          : pynq.overlay.DefaultIP
```

PYNQ therefore provides an easy way to program hardware, and automatically exposes IPs that can be interacted with programmatically using Python.

The image resizer design features two different interface types between PS and PL: DMA and AXI-Lite.

Primary data transfer (i.e. of image data) is achieved using a DMA, which is supported in PYNQ by a combination of features. Firstly, the facilities of the NumPy open-source library are exploited: image data is stored as a NumPy array, and this is the data format supported by the PYNQ Xlnk class for contiguous memory allocation in PS memory. This represents the first stage of data transfer to the Resizer IP. Transfer between PS memory and the DMA engine is then handled using PYNQ's DMA class.

Control data is passed between the Arm processor and the DMA and Resizer IPs over AXI-Lite, which makes use of the MMIO API, another part of the PYNQ framework.

Referring back to Figure 22.1, it can be confirmed that this example features components from all layers of the PYNQ framework.

- In the ***Apps*** layer, a Jupyter notebook has been developed, exploiting the Python Image Library (PIL), as well as the NumPy library, and IPython for displaying images.

- In the ***APIs*** layer, the PYNQ Overlay, MMIO, Xlnk and DMA APIs are all used.
- Beneath that, the ***Linux OS and Drivers*** layer operates transparently to the user, but exercises kernel drivers including 'fpga-manager' for programming the PL, and 'xlnk' for memory allocation.
- Finally, the ***Bitstreams*** layer features a prepared PYNQ overlay for the image resizer.

22.2.4 Key Observations

Comparing the two notebooks (software-only, versus software and hardware) it can be observed that the two are actually quite similar — the hardware resizer is called from Python in the same way as a software function. The effort of developing a PYNQ application that exploits the features of the Zynq / Zynq MPSoC architecture (for instance, to interface with peripherals, or use a hardware accelerator) is comparable to that of creating a software-only Jupyter notebook. Any peripherals based in the PL can be exposed to the Jupyter notebook using standard interface types. Leveraging the PYNQ framework makes it is as simple to invoke a hardware-based processor as it is to call an equivalent software function.

In this simple image resizer example, the effect of moving the image resizing function into PL permitted a speed-up by roughly a factor of three, from to 24.6ms to 7.6ms, based on the PYNQ-Z2 board. It may be possible to achieve far greater degrees of hardware acceleration for other functions, in all cases, benefiting from the user-friendly Jupyter interface.

22.3 Components of a PYNQ System

Having considered an example PYNQ application, we will now review the components of PYNQ systems in more general terms, and explain the steps involved from a designer's perspective. Opportunities for more advanced users to customise PYNQ will also be highlighted where appropriate.

22.3.1 PYNQ Image

The PYNQ image is the practical embodiment of the software layers that comprise the PYNQ framework as presented in Figure 22.1, and provides the basis of any PYNQ system design. As shown in Figure 22.6, the PYNQ image boots the board and runs on the PS section of the target Zynq or Zynq MPSoC chip, meaning that the Jupyter user interface and all notebooks execute on the embedded device.

The applications layer of the image includes Jupyter and IPython, along with a set of useful open-source Python libraries, and example PYNQ notebooks for getting started. As described earlier, APIs for interacting with overlays and hardware are included in the second layer, and then the Linux OS and drivers in the third. A Jupyter webserver is included as part of the OS for hosting the Jupyter interface over a network connection. Depending on the board, a small set of example overlays are provided in the hardware layer, including the *base* and *logictools* overlays (to be discussed in Section 22.3.3).

PYNQ images for supported boards can be downloaded from the PYNQ website, extracted, and directly written to a memory card (e.g. an SD or MicroSD card, depending on the target board). The memory card is then placed into its slot on the development board and, with the jumpers correctly set, the board will then boot from the image.

The user connects to Jupyter interface running on the board over a network connection; an Ethernet link direct from a host computer is shown in Figure 22.6, but more generally the connection can also be made wirelessly (via WiFi), and/or over a network. The only software requirement for interacting with PYNQ running on the development board is a standard web browser such as Chrome, FireFox, or Safari [14].

22.3.2 File Handling

After booting the image, users can transfer files over the network connection to the PYNQ system running on the development board, normally via the Jupyter interface running in a web browser, or by connecting to

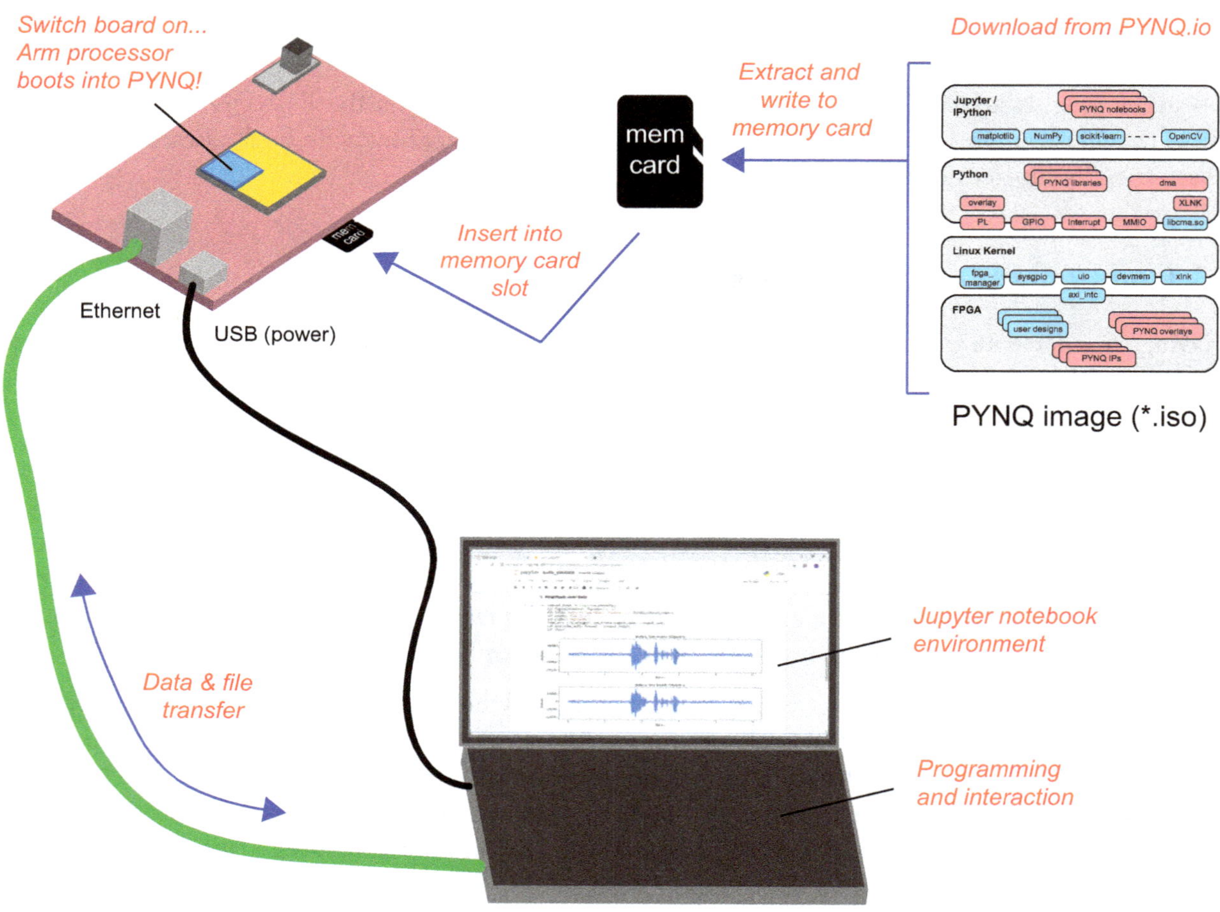

Figure 22.6: PYNQ image, booting, and interaction

the board as a network drive. There is no requirement to modify, rebuild and transfer the image, in order to add user files.

22.3.3 Overlays

An overlay is a complete hardware system that will be programmed onto the PL, and it represents part of the hardware layer of the PYNQ framework. Developers may use an existing overlay (available via the PYNQ project and community) either directly or with modifications, or develop their own custom overlay. The uniqueness of overlays (as opposed to regular bitstreams) is the ability to interact with these hardware designs from Python code, running in a Jupyter notebook on the PS.

PYNQ.io Overlays

The PYNQ project makes example overlays available. These are board-specific, and include the following:

- **Base overlay** — The *base* overlay includes logic to interface with all external peripheral interfaces on supported development boards. Depending on the target board, this includes GPIO (buttons, switches, LEDs), HDMI in and out ports, audio ports, Pmod interfaces, and headers for Raspberry Pi, and Arduino. A Trace Buffer is also included — this component allows signals passing to and from the external pins to be captured for analysis and debugging purposes. Figure 22.7 shows a block diagram of the PYNQ-Z2 base overlay and connected interfaces.

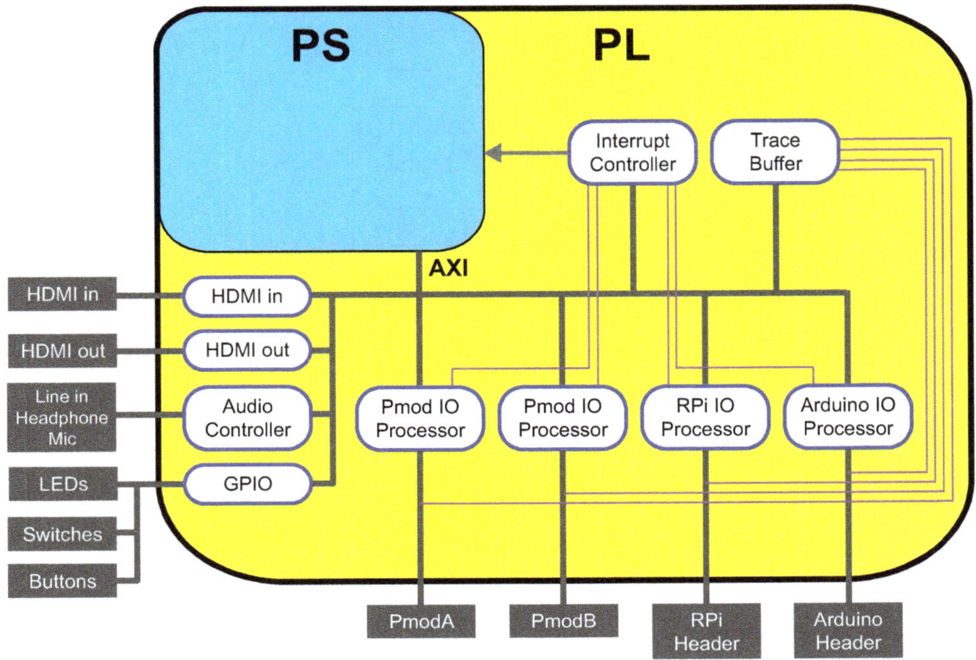

Figure 22.7: Interface functionality implemented by the base overlay (for PYNQ-Z2)

- **Logictools overlay** — *logictools* comprises functional blocks for testing external digital logic circuits. It enables Boolean logic functions, Finite State Machines (FSMs) and internal and external digital signal patterns to be generated and output to pins on the board, with functionality entirely specified in Python. A trace analyser is also included for hardware introspection, in particular to capture and analyse signals on input pins. The screenshots shown in Figure 22.8 provide a flavour of the functionality enabled by the logictools overlay.

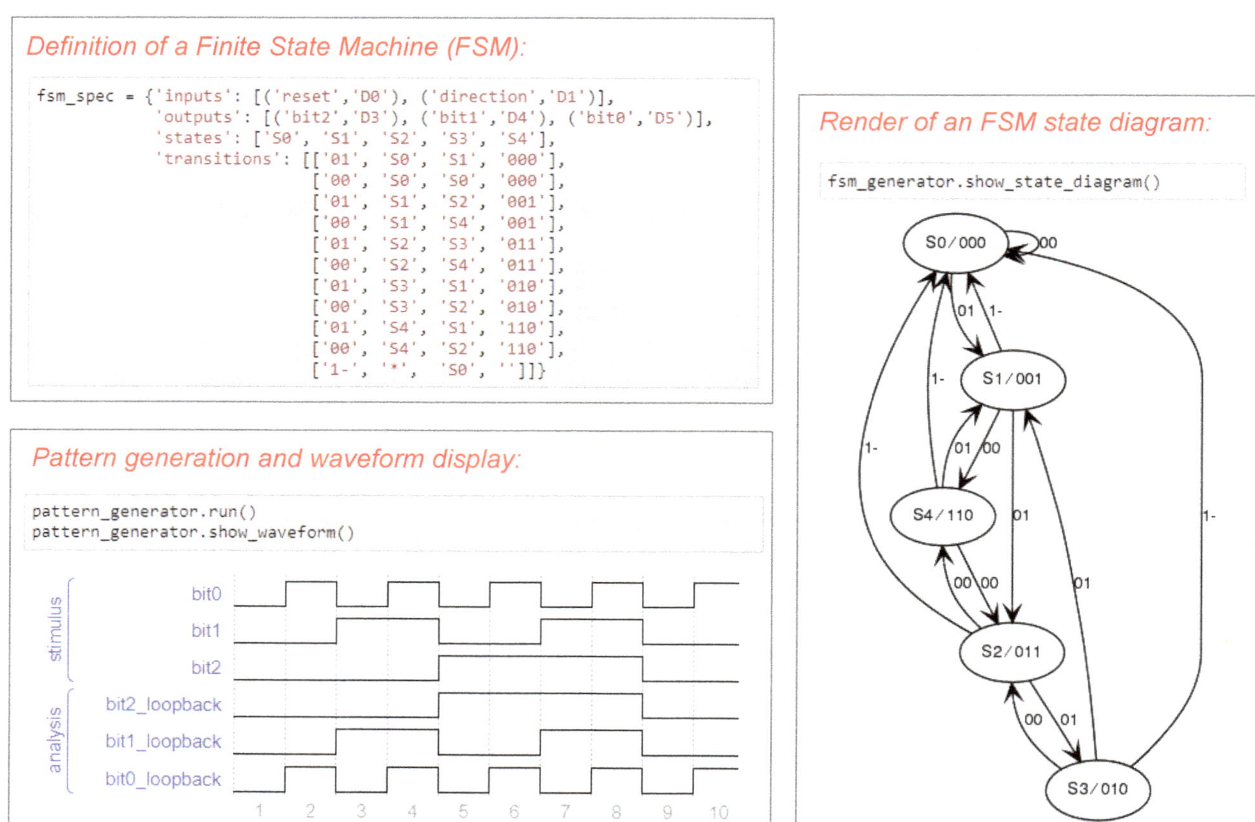

Figure 22.8: Example functionality of the logictools overlay (in Jupyter notebooks)

These two overlays are designed to be used 'out-of-the-box' in conjunction with the example PYNQ notebooks that accompany them. Additionally, the Vivado projects and all source code are available for download on an open source basis, which permits experimentation, modification and reuse in whole or in part.

Community Projects

A number of community projects are featured on the PYNQ website (www.pynq.io), covering topics such as signal, image, and video processing, machine learning, communications, robot and motor control, and many others. The sources files for community projects can normally be obtained from their owners' GitHub repositories, and are naturally self-documenting by virtue of the accompanying Jupyter notebooks. Most projects additionally provide high-level 'Readme.md' files in markdown format, which makes it easy to install and run examples, and make use of the PIP installation method, which is described later in Section 22.3.5.

Creating a Custom Overlay

Standard PYNQ.io overlays, and community overlays, may provide a good starting point for creating a new design, either by modifying the original, or by serving as useful references.

Alternatively, some developers may wish to design their own, completely custom overlay. This can be achieved using established Zynq hardware design methods, making use of the PYNQ libraries [12], and standard interface types that are supported by PYNQ. Supported interfacing between the PS and PL includes AXI-Lite via the general purpose AXI (GP) ports, and streaming interfaces via the high performance AXI (HP) ports, as well as interrupts and GPIO. The PYNQ libraries also include IPs for use with the audio and video interfaces, amongst others, and facilities for incorporating MicroBlaze subsystems in the PL.

To create a custom overlay, designers can used any of the conventional tools for hardware design, including Vivado (in particular its *IP Integrator* tool), SDx, etc.. If custom functionality is required, bespoke IPs can be created using methods such as traditional HDL coding, block-based design with System Generator or similar tools, or high level synthesis using Vivado HLS.

Overlay Files

To use a pre-existing overlay without modification, users must simply copy the bitstream file (*.bit) and Hardware Handoff (*.hwh[1]) file to the development board. The bitstream file is used to program the PL, while the HWH file contains information about the hardware system design, including all of its constituent settings, IPs, and attributes. This file is automatically parsed by the PYNQ Overlay class to infer details (metadata) of the hardware design in software, and thus permit interaction with elements of the overlay from Python. The bitstream and HWH files must have corresponding names, e.g. 'my_overlay.bit' and 'my_overlay.hwh'.

When developing a custom overlay, the bitstream file is created by running through the standard Vivado design flow. The HWH file is generated when exporting a Vivado hardware design for subsequent software development.

1. Prior to PYNQ v2.4, Tcl (*.tcl) was used to read Overlay metadata. It may still be useful to include a Tcl file in the project, as it can be used to regenerate the IP Integrator block diagram for the hardware system.

22.3.4 Jupyter Notebooks

Example Jupyter notebooks for the Resizer application were featured in Section 22.2, and this illustrated some of their capabilities. In particular, we saw that the notebook content is organised into cells that can be executed individually, with the results of execution (e.g. formatted text, images etc.) being displayed underneath the corresponding cell. Documentation content, such as text-based explanations, mathematical expressions, and diagrams, can also be included as cells within the notebook.

Interactive Features

One of the most compelling aspects of Jupyter notebooks is the interactivity that they offer. Designers can edit code within the browser and integrate live plotting and visualisation into their notebooks, as well as interactive widgets such as slider bars and buttons. Animations, audio and video output, and web integration (such as displaying a web page or YouTube video inside a notebook cell) are all possible too. This makes it very straightforward to alter an aspect of the code, re-execute, and observe any changes in the generated results.

With PYNQ, there is a further level of interactivity possible, because the Jupyter notebook includes aspects of interactivity with design components on the PL. This may include hardware acceleration of algorithm execution, interaction with external board peripherals, configuration of the parameters of a hardware processing element, or even capturing 'live' signals from within the PL, to be displayed in the notebook. In this last example, the Zynq device could even act as a self-contained logic analyser.

Figure 22.9 provides some further examples of interactive content in Jupyter notebooks (as far as the static pages of a book will allow!).

Documentation

Jupyter notebooks can include rich multimedia (basically, anything possible in a browser is supported!), enabling the creation of documentation that extends well beyond the traditional commenting of code. Video and audio players, and interactive visualisations are three notable examples. Headings can be added to segregate different sections of a notebook, and documentation cells may be included to provide more detailed information.

Text-based documentation, including features such as tables, web links, and bulleted lists, are defined using the popular Markdown format [3], while equations can be edited using LaTeX commands and the notebook renders them within the browser using MathJax [9]. Diagrams or other images can be imported using standard formats including PNG, JPG, and TIF. Pre-existing documentation, such as web pages or PDF documents, can be included in a notebook within interactive frames.

Notebooks are excellent resources for development and demonstration, and to act as user guides; it is also possible to extract the code and convert it to another file format if desired.

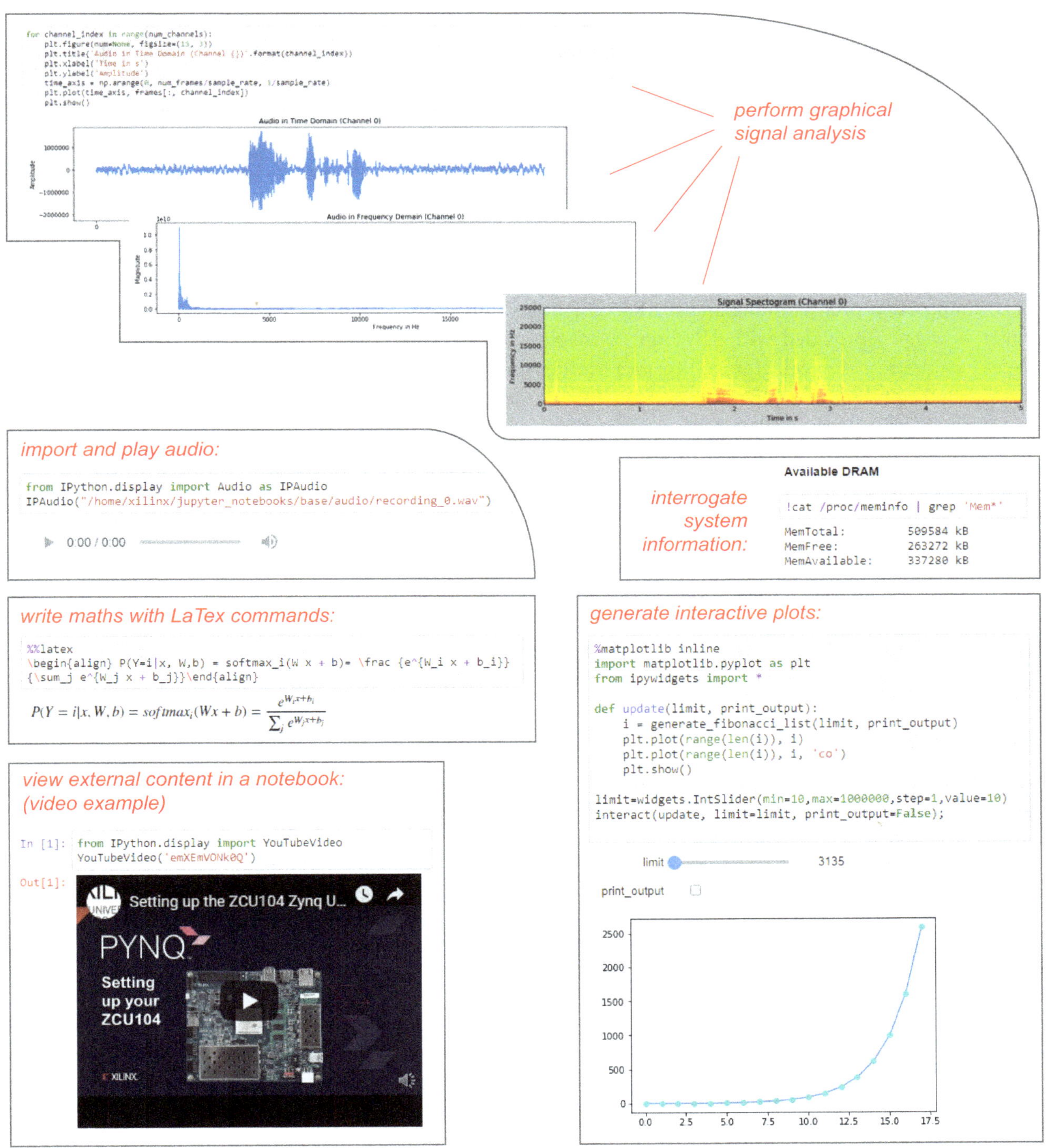

perform graphical signal analysis
Audio in Time Domain (Channel 0)
Audio in Frequency Domain (Channel 0)
Signal Spectogram (Channel 0)
import and play audio:
from IPython.display import Audio as IPAudio
IPAudio("/home/xilinx/jupyter_notebooks/base/audio/recording_0.wav")
0:00 / 0:00
interrogate system information:
Available DRAM
!cat /proc/meminfo | grep 'Mem*'
MemTotal: 509584 kB
MemFree: 263272 kB
MemAvailable: 337280 kB
write maths with LaTex commands:
%%latex
\begin{align} P(Y=i|x, W,b) = softmax_i(W x + b)= \frac {e^{W_i x + b_i}} {\sum_j e^{W_j x + b_j}}\end{align}
generate interactive plots:
%matplotlib inline
import matplotlib.pyplot as plt
from ipywidgets import *
def update(limit, print_output):
i = generate_fibonacci_list(limit, print_output)
plt.plot(range(len(i)), i)
plt.plot(range(len(i)), i, 'co')
plt.show()
limit=widgets.IntSlider(min=10,max=1000000,step=1,value=10)
interact(update, limit=limit, print_output=False);
limit 3135
print_output
view external content in a notebook: (video example)
In [1]: from IPython.display import YouTubeVideo
YouTubeVideo('emXEmVONk0Q')
Out[1]:
Setting up the ZCU104 Zynq U...
PYNQ
Setting up your ZCU104
XILINX

Figure 22.9: Selected Jupyter notebook features (as demonstrated by example notebooks provided with PYNQ)

Off-Chip Development

While is it easy and convenient to work with Jupyter notebooks running on the Zynq device, there may be occasions when it is not possible to access a development board, and it would be useful to have another mechanism for notebook authoring. Fortunately, aspects of Jupyter notebook authoring, including documentation and non-hardware-dependent code, can be progressed independently of the PYNQ target board by running the Anaconda data science distribution [1] on a host PC, for example.

Magics

The primary programming language used by PYNQ notebooks is Python, but it is also possible to include cells running other programming and markup languages. This is accomplished using the magic functions (also known as *magics*) provided with IPython. A magic is a command that causes the rest of a line (for 'line magics') or the rest of a cell (for 'cell magics') to be interpreted in another defined way. For instance, typing `%%js` at the beginning of a cell will cause the rest of the cell to be executed as a block of JavaScript code; or `%%latex` will prompt the cell to be interpreted and rendered as LaTeX. IPython magics also provide very convenient access to the underlying Linux shell. This makes the IPython prompt a very powerful interface to all of a system's resources.

22.3.5 Files and Installation

In this section, we describe the set of files that are necessary to deploy a PYNQ system (and their purposes), as well as outlining optional files. A larger set of files is required for the development of the project, and for sharing with others if desired. The installation of shared PYNQ projects can be enhanced through the use of a common directory structure, and support for the PIP installation method.

Deployed PYNQ Files

PYNQ projects can vary in complexity, but in all cases, the following files should be present on the board in order to successfully run a system:

- A bitstream file (*.bit) — for programming the hardware overlay onto the PL.
- A hardware handoff file (*.hwh[1]) — can be parsed to obtain metadata (system information).
- A Jupyter notebook (*.ipynb) — the interactive notebook for the project.

Secondary files such as C libraries, images or other data files (whether accessed programmatically, or used as part of the documentation) may also be required. More advanced projects might include multiple notebooks or hardware overlays, or even partial bitstreams in designs employing partial reconfiguration (i.e. the re-programming of part of the PL, while the rest continues to operate).

1. Tcl (*.tcl) files were used for design metadata up to and including PYNQ v2.3.

Project Files and Sharing

When developing a PYNQ project, IP blocks will be included as part of the PL hardware system design (whether selected from the Vivado library, or custom-developed). The entire hardware system can be regenerated from the combination of the Tcl file and IPs, if desired. It is therefore useful to provide the IP files as part of a shared project — this makes it easier for designers to learn from, adapt, or reuse components.

Shared projects are most reusable when good documentation is provided, and developers sharing PYNQ projects are encouraged to use GitHub for disseminating their work, and to include a ReadMe.md file (i.e. a markdown file rendered in the GitHub interface) for providing guidance to others. Especially for more complex projects, it is also useful to include a 'makefile', which will automate the rebuilding of the project from source.

PIP Installation

PIP is a Python command for installing and managing packages. In this context, the PIP command can be exploited to greatly simplify the installation of a PYNQ project onto a Zynq device. The use of PIP in Pynq is enabled by the concept of 'hybrid packages', i.e. extending the concept of a Python package to include both software and hardware files. A PYNQ hybrid package includes the Python, bitstream and design metadata files that were outlined earlier in this section (under the heading '*Deployed PYNQ Files*').

To give an example, the Hello World (image resizer) project featured in Section 22.2 can be installed onto a target PYNQ board simply by opening a terminal within the Jupyter notebook GUI (hosted on the board), and then typing the command:

```
sudo pip3 install --upgrade git+https://github.com/xilinx/pynq-helloworld.git
```

This command will cause the project files to be obtained from the PYNQ Hello World GitHub repository, transferred onto the board, and installed. The PIP method has the advantage of being quick and convenient (for instance, it ensures that all of the files are installed to the correct locations), and it also provides more advanced functionality, like the scripting of source code compilation. Guidance is available for developers wishing to share their projects, in terms of best practice and the steps necessary to make their work PIP-installable [16].

22.3.6 Advanced User Customisation

The PYNQ website provides images for selected boards, but you may wish to generate a different one for your target board. In this case, the source code can be downloaded from the PYNQ GitHub repository, then adapted and regenerated for the new board. This is a more advanced task, but permits PYNQ to be used on any Zynq, Zynq MPSoC or Zynq RFSoC board. Guidance is available on the PYNQ website [13], and a good third party tutorial is also available online [21].

The functionality enabled by PYNQ notebooks can be extended by installing additional Python software libraries. A subset of libraries are included in the images available via the PYNQ website (including those shown in the upper layer of Figure 22.1); these are popular libraries that should be of greatest use to the widest variety of users. If other libraries are desired, they can be installed to the base image. Alternatively, a modified version of the PYNQ image can installed, thereby extending the base image.

Indeed, as the project is an open-source framework, developers are encouraged to extend and adapt PYNQ as fits the needs of their projects. We go on to discuss the PYNQ project in further detail in the next section.

22.4 The PYNQ Project

The PYNQ project originated as a collaboration between teams from Xilinx Research Labs and the Xilinx University Program (XUP). The project now benefits from contributions from many other collaborators, both within Xilinx and externally. All of these collaborators share a common excitement about the tremendous capabilities of programmable systems-on-chip, such as Zynq and Zynq MPSoC devices. At the same time, they recognise the need to make the benefits of the technology more accessible to a broader community by simplifying the process of creating applications on programmable systems-on-chip. It is inevitable that, as more system functions are integrated into a single chip, the process of designing with that device becomes more complicated. The challenge of taming this complexity is what motivates the PYNQ project. Much of the inspiration for solving the problem comes from ideas from the Free Open Source Software (FOSS) community.

This section briefly outlines the original goals of the PYNQ project, before reviewing some of the influences that shaped its system-level architecture and design philosophy.

22.4.1 PYNQ Goals

The three main goals of the PYNQ project can be summarised as:

1. To simplify the process of creating new applications on Zynq devices for people who understand hardware and software design.

2. To make it possible for people who understand software, but have limited knowledge of hardware design, to create applications which use both the processor and programmable logic subsystems of Zynq devices.

   To achieve this second goal requires the use of overlays. Overlays, as we have seen already, are designs that are tailored for different application domains. However, they are created so that they can be reused across many different applications, within their respective domains. This is achieved by equipping them with software APIs that programmers can invoke to customise their functionality.

3. To simplify the process of creating these overlay designs.

These goals are defined with the various different sets of expertise involved in SoC design creation in mind, and recognising that relatively few people possess the full set of necessary skills. PYNQ is a framework that better enables collaboration, design reuse, and therefore productivity.

22.4.2 PYNQ Influences

A number of factors have influenced the PYNQ project in approaching its goals. These influences are outlined in the remainder of this section.

Adopting Open Source Ideas

The PYNQ project embraces the open source model in two important ways. First, it makes extensive use of existing open source ideas, tools, libraries and methodologies, especially from the software community. Second, the project contributes the PYNQ framework and all of its components to the community as open source material. The importance of the FOSS model is that it fosters openness and maximises the opportunities for third-parties to extend and improve the PYNQ project.

Moving to Higher Levels of Design Abstraction

An outstanding example of the successful application of design abstraction is provided by the Linux kernel. Only a few hundred very talented engineers develop and maintain the Linux kernel. Yet, thanks to the POSIX abstraction layer, millions of Linux users and application developers benefit from the work of the kernel developers.

Similarly, by enabling work to be done at higher levels of design abstraction, PYNQ makes designing applications simpler and more productive. A software programmer, for example, should not have to understand how an overlay design is crafted: she should only have to load and instantiate an overlay, before invoking its functionality by calling into its Python API.

A comparison is drawn between these two models in Figure 22.10.

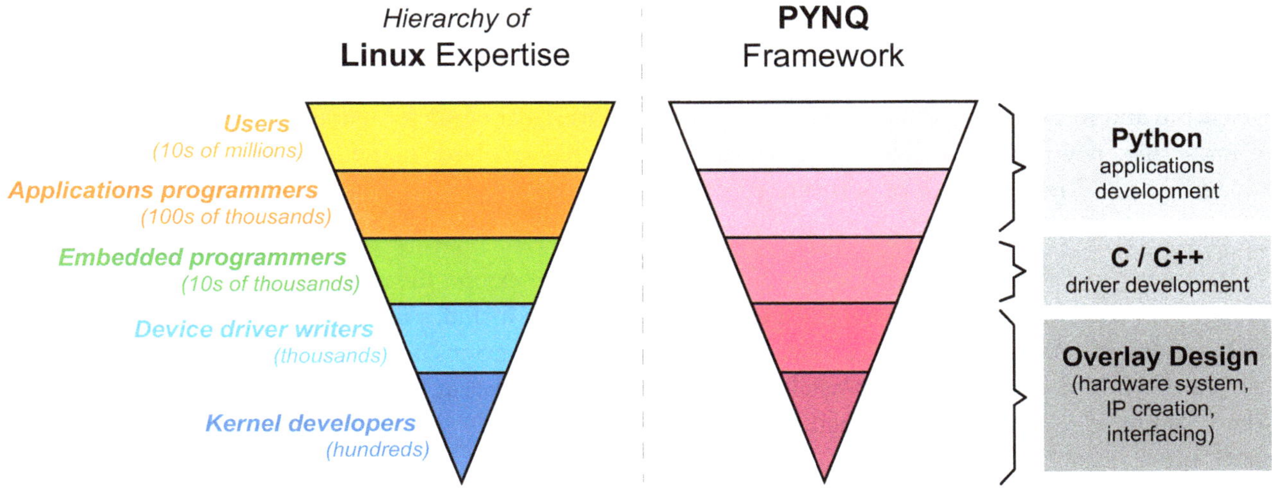

Figure 22.10: Layers of SoC design skills (left - Linux; right - the PYNQ framework)

Extensive Reuse and Re-Targeting of Good Ideas

Project Jupyter introduced a set of very powerful abstractions to the Data Science community, of which Jupyter Notebooks are often the most visible example. PYNQ reuses these innovations in the context of programmable systems on chip. The project team deliberately re-purposed these ideas in the embedded context because the case for reproducible embedded engineering is as compelling as the need for reproducible science.

Of course, the fact that Machine Learning techniques are being increasingly incorporated into embedded systems makes the relationship between Jupyter notebooks and embedded systems even more natural. The rich set of Python and JavaScript libraries for *interactive* data visualisation, available within the Jupyter lab ecosystem, is another example of an opportunity for systemic reuse.

Embracing Ease-of-Use and Productivity as First-Class Design Objectives

PYNQ embraces IPython and Jupyter Notebooks because they are leading examples of a bigger evolution toward productivity languages and environments in the broader software community.

It is possible for valuable technologies to remain inaccessible, if they are too awkward and difficult to master. For instance, for many designers, configuring cross-compilation environments for embedded targets is simply outside their capabilities, even if they have applications that would benefit from programmable systems on chip. By supplying a desktop Linux image, with a full set of native programming tools, PYNQ lowers the barrier-to-entry for Zynq designs.

In general, the PYNQ architecture assumes that performance-critical functions will be assigned to the programmable logic, while the processor system will be used for communication and control functions. Thanks to the use of the CPython kernel in PYNQ, developers can also augment critical sections of their Python code with more performant C or C++ code, when necessary.

Making Hardware-Software Co-Design More Like Software Programming

Reusing and re-purposing established paradigms is a proven technique for reducing the cognitive challenge in mastering new material. The software community has evolved many excellent solutions for the management and maintenance of software ecosystems. The PYNQ project aims to learn from and reuse as many of these ideas as possible.

One example of this approach is the use of *hybrid* packages to extend the already very successful concept of Python packages. In addition to software files, a hybrid package includes files which describe programmable hardware designs and their associated metadata. With this simple extension, PYNQ can now package and re-distribute hybrid designs, using well-established Python techniques, such as PIP installing a hybrid package from GitHub. This synergistic approach extends existing best practice and emphasises commonality wherever practical.

Having explained PYNQ and the philosophy behind it, we will now close the chapter by reviewing some of the resources available to those interesting in exploring further.

22.5 Resources

To start using PYNQ, a few different types of resources are required, including a suitable development board, some simple additional hardware components, and files and resources that can be freely downloaded from the web. This section outlines the requirements, from a practical viewpoint.

22.5.1 How Do I Get PYNQ?

PYNQ is an open-source project with all software resources available online at:

www.pynq.io

This includes pre-compiled images for supported boards, guidance, documentation, community examples, and useful links.

In order to work with PYNQ, a Zynq, Zynq MPSoC or Zynq RFSoC development board will be required. For first-time users may find it easiest to select a board for which a prepared and tested PYNQ image is available. At the time of writing, this includes:

- **PYNQ-Z1** — low-cost Zynq development board from Digilent;
- **PYNQ-Z2** — low-cost Zynq development board from TUL;
- **ZCU104** — professional Zynq MPSoC development board from Xilinx;
- **Ultra96** — community-based Zynq MPSoC development board from Avnet.
- **ZCU111** — professional Zynq RFSoC development board from Xilinx.

The first four of these boards were pictured in Figure 22.2 on page 531. The ZCU111 is a specialist board featuring an RFSoC device, which is likely to be of interest to researchers and professional designers working in the radio communications domain.

The PYNQ webpages provide guidance on the additional hardware and software requirements for programming and interacting with development boards, which are modest (e.g. SD or microSD card, Ethernet and USB cables).

22.5.2 Documentation

Reference documentation is accessible via the PYNQ website, and is presented and rendered online using 'Read the Docs' [17]. There is an option to automatically generate a PDF of the documentation, if desired. All documentation is open source, and available along with source code on the PYNQ GitHub repository (discussed in the following section). The documentation files include a step-by-step guide for getting started, as well as more advanced reference material and code examples.

22.5.3 Files and Source Code

For a quick start with a supported board, the PYNQ image (discussed in Section 22.3.1) can be downloaded and installed directly onto the board. It includes example overlays and Jupyter notebooks that demonstrate the key features of PYNQ, and hence it enables a complete "out-of-the-box" experience within about half an hour!

If you would like to deploy PYNQ on a board without a pre-packaged image, the first step would be to refer to the documentation and prepare an image for the target board, before proceeding.

For more detailed investigation and development work, the full set of source code for PYNQ is available from the project's GitHub repository:

https://github.com/Xilinx/PYNQ

The repository provides access to all of the component parts of PYNQ, enabling reuse in user designs. The source code can also be used to re-build the PYNQ image, if desired; alternatively, it serves as the basis for generating a specialised version of PYNQ to target an unsupported board, or with modified software components (for instance to include a customised selection of Python libraries).

Source code for third party PYNQ projects is also normally hosted on GitHub.

22.5.4 Hardware

In addition to the PYNQ development board and core hardware components for programming and connectivity, users may wish to obtain further hardware peripherals depending on their intended projects. PYNQ boards, especially the PYNQ-Z1 and PYNQ-Z2, provide extensive interfacing opportunities via their Pmod, Arduino, and Raspberry Pi ports. Seeed's Grove peripherals [19] can also be attached to the Pmod ports via an adapter [5]. Audio and video sources and outputs can be connected to the board via standard cables, and USB peripherals such as webcams can also be integrated.

22.5.5 Community Resources

Several different types of community resources are available for PYNQ, and these are briefly highlighted in this section. Many of the resources mentioned here are curated on a dedicated *Community* page on the PYNQ website; other third party PYNQ projects and resources of course exist independently (one good example being *Fabric GSM Determination Using Ultra96 + PYNQ* [10]).

Reference Projects and Tutorials

Various reference projects are available via the PYNQ website's *Community* pages. This includes contributions from Xilinx (notably the *Machine Learning with FINN* project, see also Section 21.2 on page 514), academic institutions around the world, and others interested in PYNQ.

Additionally, a collection of tutorials have been prepared by various members of the community to guide others through the development of example applications using PYNQ and associated tools. One excellent example is the "PYNQ HLS Tutorial", contributed by Dustin Richmond [18]. This tutorial demonstrates how to successfully develop IPs using Vivado HLS, and integrate them into a PYNQ system using different types of PS-PL interfaces including streaming and memory-mapped.

Academic Resources

Aided by the availability of workshop material on PYNQ, the PYNQ framework has already been adopted in university teaching programs, one interesting example being a project class focused on video processing with dynamic partial reconfiguration [8]. An advantage of running student projects with PYNQ is that students can gain confidence quickly from early results, before venturing into the details. The availability of open-source projects and reference materials provides a firm basis for independent investigation and development. Students at all levels (undergraduate through to PhD) can also take the opportunity to enter a design contest, such as the annual *OpenHW* contests run in Europe and Asia by Xilinx University Program.

PYNQ has also featured in a number of research projects to date, as evidenced by a growing body of published work (for the latest results, search for "PYNQ" on Google Scholar or IEEE Explore!).

22.6 Chapter Summary

This chapter has introduced the PYNQ framework, and explained the motivation for this Python-based development methodology as an enhancer of productivity for Zynq system design. The various constituent components of the PYNQ framework were explained, including the PYNQ image, the concept of hardware overlays, and the use of Jupyter notebooks as an interactive user interface.

An example was presented, showing a 'real-world' image resizer application, and comparing software and hardware-accelerated notebooks. The comparison demonstrated that a PYNQ notebook can be of similar complexity to a software equivalent, with the calling of the main processing function called in the same way, regardless of its implementation in software or hardware. Some of the notable features of a PYNQ notebook were also highlighted via this example.

Later in the chapter, the PYNQ project was discussed, including the goals of the project, and the factors that have been influential in its development. Practical issues were also covered, such as how to get started with PYNQ, and access the relevant files, documentation and hardware. The various types of available resources and support for PYNQ were also outlined.

22.7 References

Note: All online sources last accessed March 2019.

[1] *Anaconda* webpages:
Available: https://www.anaconda.com/

[2] S. Cass, "The 2018 Top Programming Languages", IEEE Spectrum, July 2018.
Available: https://spectrum.ieee.org/at-work/innovation/the-2018-top-programming-languages

[3] Daring Fireball, *Markdown* webpages:
Available: https://daringfireball.net/projects/markdown/

[4] J. Decaluwe, "MyHDL: From Python to Silicon" webpage.
Available: http://myhdl.org

[5] Digilent, Inc., "PYNQ Grove System Add-On Board" product webpage.
Available: https://store.digilentinc.com/pynq-grove-system-add-on-board/

[6] GitHub website.
Available: https://github.com/

[7] J. Gray, "An 80-core GRVI Phalanx Overlay on PYNQ-Z1: Pynq as a High Productivity Platform For FPGA Design and Exploration", PYNQ Workshop, *25th IEEE Symposium on Field-Programmable Custom Computing Machines*, Napa, CA, USA, May 2017.
Available: http://fpga.org/wp-content/uploads/2017/09/GRVI-Phalanx-on-Pynq-FCCM2017-Pynq-Workshop.pdf

[8] B. Hutchings and M. Wirthlin, "Rapid Implementation of a Partially Reconfigurable Video System with PYNQ", *Proceedings of the 27th International Conference on Field Programmable Logic (FPL)*, Ghent, Belgium, September 2017, pp. 1 - 8.
DOI: 10.23919/FPL.2017.8056845

[9] *MathJax* webpages.
Available: https://www.mathjax.org/

[10] S. Peiris and P. Liyanage, "Fabric GSM Determination Using Ultra96 + PYNQ" *Hackster.io* webpage.
Available: https://www.hackster.io/yasaspeiris/fabric-gsm-determination-using-ultra96-pynq-6c5499

[11] *Project Jupyter* webpages.
Available: https://jupyter.org/

[12] PYNQ, "PYNQ Libraries" documentation page.
Available: https://pynq.readthedocs.io/en/latest/pynq_libraries.html#

[13] PYNQ, "PYNQ SD Card" documentation page.
Available: https://pynq.readthedocs.io/en/latest/pynq_sd_card.html

[14] PYNQ, "The Jupyter Notebook" documentation page ("Browser Compatibility" section).
Available: https://jupyter-notebook.readthedocs.io/en/latest/notebook.html#browser-compatibility

[15] Python Package Library (PyPI) website.
Available: https://pypi.org/

[16] Python Packaging *ReadTheDocs* webpages, "How to Package Your Python Code".
Available: https://python-packaging.readthedocs.io/en/latest/index.html

[17] *Read the Docs* webpages.
Available: https://readthedocs.org/

[18] D. Richmond, *PYNQ-HLS Tutorial*, GitHub repository.
Available: https://github.com/drichmond/PYNQ-HLS

[19] Seeed, "Grove System" webpage.
Available: http://wiki.seeedstudio.com/Grove_System/

[20] H. Shen, "Interactive Notebooks: Sharing the Code", *Nature*, Vol. 515, 6th November 2014, pp. 151 - 152.
DOI: 10.1038/515151a

[21] A. Taylor, "Making Cora PYNQ" *Hackster.io*.webpage.
Available: https://www.hackster.io/adam-taylor/making-cora-pynq-582854

[22] Wikipedia, "Software framework" definition.
Available: https://en.wikipedia.org/wiki/Software_framework

Chapter 23

Case Study: Prototyping CNNs on Zynq for Space Applications, Using PYNQ

Phil Karagiannakis, Murray Ireland, and Steve Greenland,
Craft Prospect Ltd.

Editor's Note: This chapter has been contributed by Craft Prospect, Ltd.[1]*, a space engineering practice that delivers mission enabling products and develops novel mission applications. The company architects and delivers 'smarter satellites', targeting products at the global small satellite market. The case study demonstrates how Craft Prospect have successfully used the PYNQ framework (from Chapter 22), along with the Binarised Neural Network (BNN) overlay developed by Xilinx Research Labs (see Chapter 21), in their prototyping of an imaging system for small satellites.*

In the last decade, miniaturised low-cost satellites such as CubeSats have enabled a rapid growth in space missions and applications. Growing in maturity, many missions now target commercial applications requiring tens or hundreds of satellites each. Although launch and satellite costs have decreased over this period, operational costs have not followed this trend.

Craft Prospect aims to address this opportunity by repackaging commoditised machine learning, big data, and autonomous vehicles technologies for space applications [4]. By analogy to the self-driving car, this will deliver a more autonomous and responsive self-flying satellite. Our Forwards Looking Imager (FLI) subsystem [3] uses AI to identify features in the upcoming environment (e.g. clouds, agriculture, natural disasters), enabling an increase in real-time operational efficiency of up to 30%. The following case study

1. First Floor, Tontine Building, 20 Trongate, Glasgow, Scotland, UK. G1 5NA. hello@craftprospect.com

outlines the algorithm design considerations for the FLI and highlights the rapid prototyping made possible by adopting the PYNQ development framework.

23.1 Design Concept

The FLI CubeSat payload provides analysis of the upcoming environment of the satellite to allow real-time decision making. This can take several forms: it may be the prioritisation of sensing targets from a list, repositioning and slew manoeuvres to capture targets of opportunity, or reassignment of targeting tasks among a constellation.

The cloud detection case represents a near-term opportunity for responsive Earth observation imaging. Real-time detection of clouds using forwards looking, wide-field-of-view imagers allows a second near-Nadir[1] -pointing payload to be targeted at gaps in the cloud, such as for high resolution Earth imaging, or to enable optical laser communications. In all avoidance scenarios, a rapid reacquisition of the payload is necessary, such that the overhead is significantly less than the active operations, i.e. less than 10%. In the Earth imaging case, in subsequent passes re-tasking the satellite or satellites within the same network can fill the gaps, provided that information on previous acquisitions can be efficiently exchanged.

A preliminary specification for the FLI was defined as illustrated in Figure 23.1. The FLI sightline is pitched at some angle from the vertical, providing a view ahead of the satellite. This gives a short amount of time for image capture, processing, decision and action time (collectively referred to as response time) before the captured terrain passes under the satellite. At a design altitude of 400km, the relationship between pitch angle and response time is roughly linear up to around 45°, where the available response time is almost a minute. Our example case uses a reference pitch of 15° from the vertical, providing roughly 15 seconds of response time while avoiding feature distortion associated with large angles of incidence. With a field of view of 4° and 1600 x 800 resolution, the FLI can capture an area of approximately 32.3 x 16.7 km. The FLI is matched with a corresponding narrow field imager, providing 1m resolution at nadir. The narrow field imager, or the CubeSat itself, may then be pointed based on the FLI's output to maximise its utility, as in the case of pointing around identified clouds.

23.2 Cloud Detection with Neural Networks

We now detail how the algorithm for cloud detection was initially selected, and discuss its development.

23.2.1 Algorithm Selection

In the realm of feature identification in static images, cloud detection is relatively simple due to their familiar shapes and their ubiquity in the Earth's atmosphere. A number of solutions exist for identifying and locating clouds in satellite imagery, such as *Fmask* [2] and *TextureCam* [7], and many fall under the umbrella

1. Nadir: the point on the earth directly under the satellite.

of machine learning. However, such methods are ill-suited to deployment on a CubeSat payload due to computational overheads and some inflexible requirements, as summarised in Table 1.

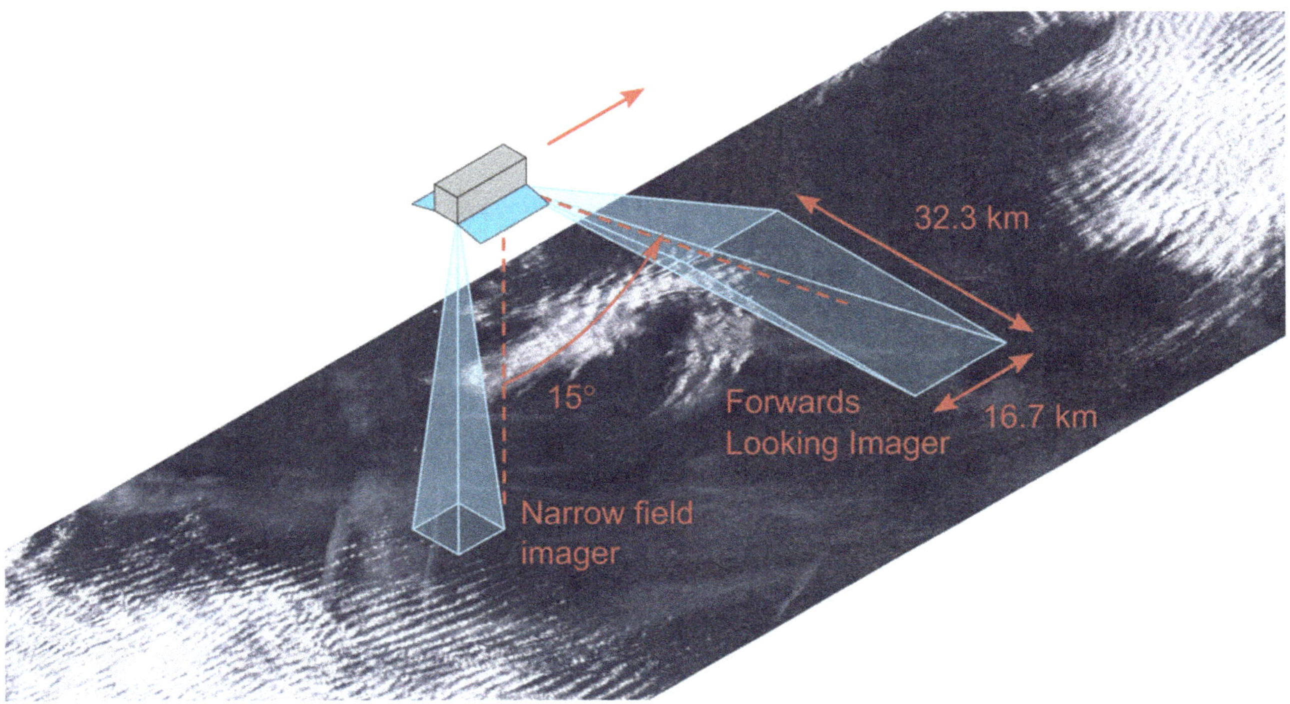

Figure 23.1: Geometry of CubeSat with FLI payload
The FLI captures images of the terrain ahead of the satellite and classifies these images with a neural network. The knowledge of where upcoming clouds are located allows the narrow field imager to be pointed around the clouds to ensure unobscured imagery of the ground.

Table 23.1: Candidate algorithms for cloud detection

Algorithm	Class	Advantages	Disadvantages
Fmask [2]	Rule-based	Well known and understood; Deterministic.	Requires specific wavelength inputs including infrared; Requires a catalogue of georectified reference images, which is not available in real time; Computationally expensive.
TextureCam [7]	Random forest	Well known and understood; Rapid inference.	Adequate pre-processing of data set might be challenging on-board; Requires good feature engineering.
CNN	Neural network	Rapid inference.	Requires training data set; Training is computationally expensive.

Neural networks have the benefit of accommodating a wide selection of spectral bands and requiring little information beyond a large number of training images. A network trained on clouds would therefore require only several thousand examples of cloud varieties and little else. An additional benefit is that, while the training is time-consuming and computationally expensive, this step may be performed on the ground, while the onboard activities are limited to inference: this is fast and lightweight in comparison to training.

23.2.2 Algorithm Development

Having selected neural networks as the basis for our cloud detection algorithm, we trained a variant of the award-winning VGG19 network [6] on around 40 thousand labelled satellite images. Each image covered an area of roughly a square kilometre, and was described by one or more labels. Exactly one label for each image describes the state of the cloud cover: either cloudy, partly cloudy, hazy or clear. Training on these images and the cloud cover label ultimately results in a network which can take an image (or batch of images) of approximately the same ground area and return the predicted cloud cover.

An example of the algorithm in action is provided in Figures 23.2 and 23.3. Using our CubeSat simulation, written in Python, we can generate a typical image capture of the FLI. As per the specifications defined previously, the FLI is looking ahead of the satellite at an angle of 15° from the vertical. With a field of view of 4°, the camera covers an area of approximately 32.3km x 16.7km.

The image from Figure 23.2 is split into a 34 x 17 segment grid, and each segment is classified by the neural network, returning the results shown in Figure 23.3. Here, the segments identified as cloudy are red, and those identified as partly cloudy are yellow.

Taking into consideration the satellite's orbital velocity and the projected area of the FLI camera, a capture must be taken and classified every two seconds to ensure unbroken coverage of the scene below the satellite. In other words, our neural network must be able to classify 289 images per second, at least. This requirement strongly affected our choice of hardware, as there are few platforms that can classify images as this rate, while adhering to the CubeSat paradigm of low-cost and power.

23.3 Hardware Selection

With the CubeSat-imposed power and performance requirements, we quickly selected an FPGA as the intended device for our neural network. The goal was to accelerate our neural network on the FPGA by leveraging its parallel processing abilities.

23.3.1 Selecting the FPGA

A primary requirement when choosing the FPGA to deploy our neural network in space was flight heritage — referring to use of the device on previous satellites. This aspect is important, as we want to de-risk whatever uncertainties there might be in putting our payload in space, and ensure that the FPGA would not be one of them. Additionally, we needed a development board with adequate connections and performance for prototyping and debugging.

Figure 23.2: Simulated FLI image capture

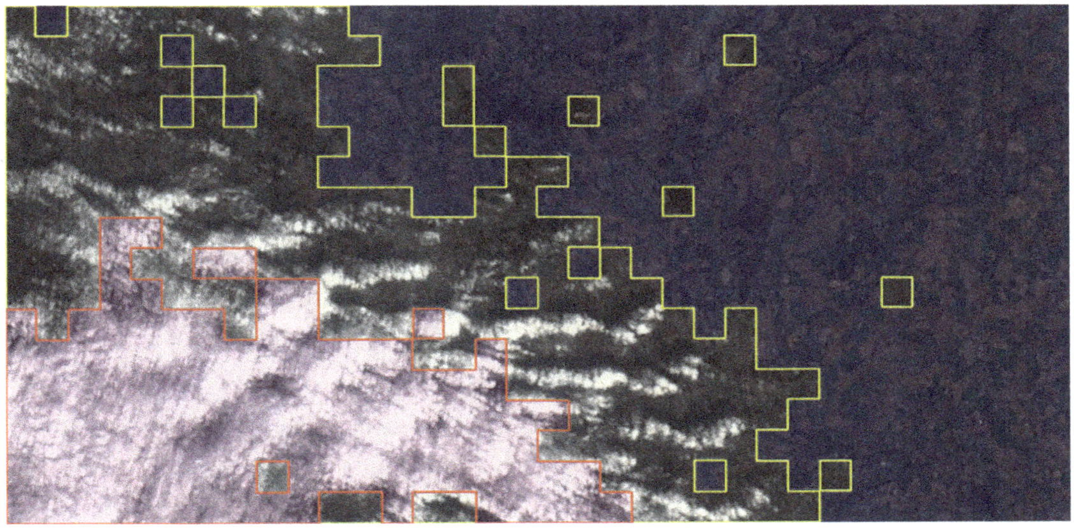

Figure 23.3: Classified image capture with 34x17 grid (red — cloudy, yellow — partly cloudy)

The Xilinx Zynq SoC comes in various sizes, with enough resources for our development, and has an established space heritage. It also has an active community and the potential to match our Python-driven simulation environment. We settled on the Z7020 chip on a MicroZed developer board, and began investigating the acceleration of Python-based neural networks using HLS.

It was only after beginning this work, and generating our first results, that we discovered a platform that not only satisfied our hardware requirements, but also ticked a lot of our software compatibility boxes too.

23.3.2 Discovering PYNQ

We purchased a couple of PYNQ-Z1 boards in order to develop Python-based networks and simulation code in parallel to the MicroZed work. Almost immediately, the benefits of the PYNQ framework became clear. The serial and Ethernet connections made interfacing with a host computer straightforward, while the PYNQ image's pre-installed Python distribution meant near-instant compatibility with our CubeSat simulation, and easy transference of algorithms onto the PYNQ's storage to run on the Arm processor. The supplied overlays on the image let us access the board's LEDs, switches and GPIO headers with ease.

Most crucially, however, was the discovery of PYNQ's community projects, which included an implementation of an FPGA-based neural network developed by Xilinx Research Labs [5]. This overlay enabled Binarised Neural Networks (BNNs), a simplified type of CNN, to be run on the programmable logic of the PYNQ-Z1.

23.4 Cloud Detection with PYNQ

We now describe how a prototype for the cloud detection system was prototyped using the PYNQ framework, using the PYNQ-Z1 development board, and summarise the key results that were obtained.

23.4.1 Developing a Prototype with PYNQ

As discussed in Chapter 21, a BNN is essentially a variant of a CNN, where the weights and activations have been binarised to ±1. This produces a much simpler network and reduces the size of the neural network implementation. The reduction in precision has the potential to limit the network's ability to classify some more complex features, but for straightforward phenomena such as clouds, it worked perfectly. We modified the provided training algorithms to use our own training set and generated a set of binary weights for the BNN's custom convolutional neural network (called CNV) which could identify cloud, partial cloud and clear sky in images covering around a square kilometre. Using this network was then simply a case of calling the required Python package and loading our generated weights.

We integrated PYNQ into our simulation testbench, using our simulation to generate realistic captures for the FLI, and sending these to the PYNQ-Z1 over Ethernet. The BNN on the PYNQ-Z1 would then segment and classify the captures, returning the results as a list of the labels for each segment via the serial connection. The simulated satellite could then use the return labels to geolocate cloudless terrain beneath, allowing a high-resolution camera to point around clouds and ensure a high frequency of clear, usable imagery. Figure 23.4 shows a still photograph from a recent demonstration of the FLI prototype.

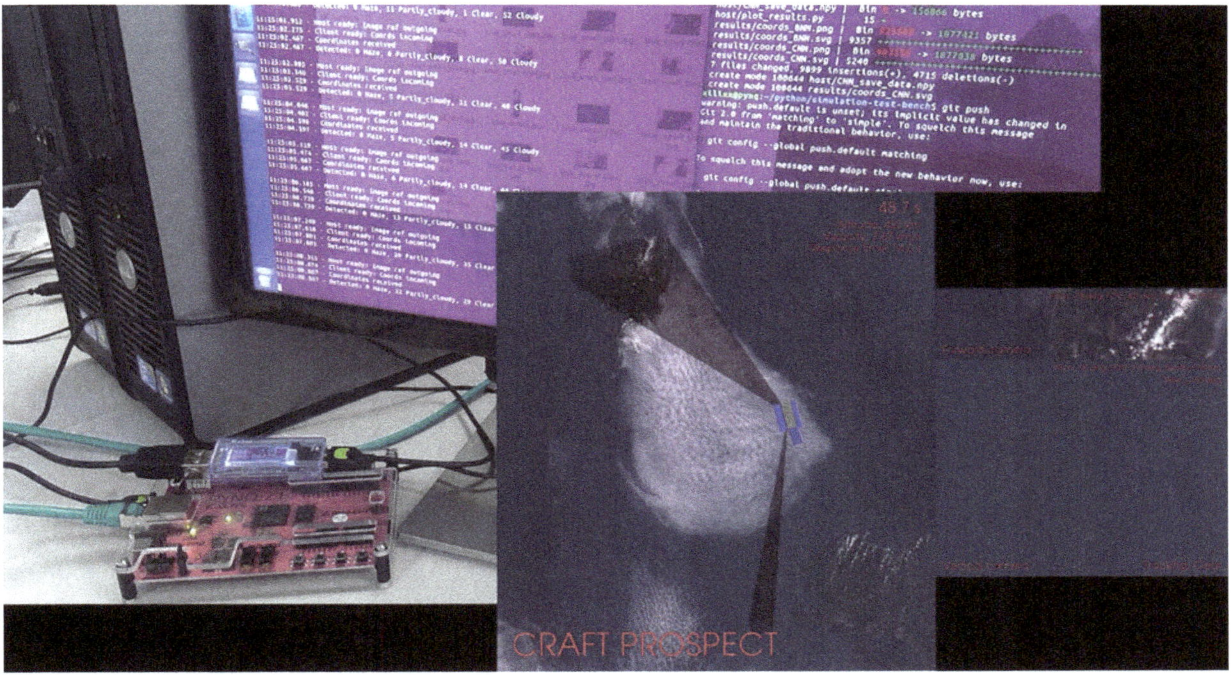

Figure 23.4: Still from a video showing a trained BNN running on the PYNQ, allowing real-time classification of clouds and responsive pointing of a high-resolution camera at cloudless terrain (full video available online [1]).

23.4.2 Results

The classification rate of the PYNQ-Z1 means that it is more than capable of meeting the requirements for real-time processing of captured images at a sufficiently high resolution. Table 2 compares the classification rates and accuracy of the cloud detection BNN, for both the PYNQ-Z1's PL, and in software running on the PS, with the original VGG-19 network.

Table 23.2: Metrics for cloud-detection neural networks

Property	CNV (PL)	CNV (PS)	VGG-19 (i5)
Classification time per image	0.003s	0.83s	0.057s
Image classification rate	337.7 im/s	1.2 im/s	17.5 im/s
Accuracy	98%	98%	99%
Average power	2.02W	1.85W	
Power increase over resting	0.23W	0.06W	

From these results, it is clear that the BNN as run on the PL is around 20 times faster than the VGG-19 network running on a mid-range laptop, and almost 300 times faster than the BNN when running in software on the PYNQ-Z1's processor. Additionally, the BNN achieves this increase in speed with a negligible drop in accuracy, one which may be negated through further training and post-processing of results.

23.5 Reflections and Ruminations

This work has outlined the design and prototyping of the Forwards Looking Imager embedded with a deep learning algorithm for cloud detection. The work has shown that a cost and power effective system may be implemented using emerging deep learning concepts to enable more responsive onboard operations for Earth observation in real time.

The rapid progress of the FLI from design concept through prototyping and into early engineering models was made possible through the PYNQ framework. The PYNQ framework allows those unfamiliar with hardware acceleration to work with the higher level programming language, Python, while accessing programmable logic overlays to perform the machine learning acceleration. The quick turnaround in demonstrating the proof-of-concept on real hardware has helped to gain industry traction, and paved the way for our future development.

Utilising such a heterogeneous platform as the Zynq SoC allows the inference network to be implemented within the programmable logic of the device, providing a significant increase in performance while keeping power consumption low. The low cost to implement and run this advanced detection capability demonstrates that such a system may be accommodated within a CubeSat platform for immediate application, and in support of more complex operations of high value missions or payloads. Following space ruggedisation, a maiden FLI-enabled mission is planned for 2019.

23.6 References

Note: All online sources last accessed March 2019.

[1] Craft Prospect, Ltd., "FLI Prototype in Simulation Loop" webpage and video.
Available: https://craftprospect.com/fli-prototype-in-simulation-loop/

[2] Global Environment Remote Sensing Lab, *Fmask* GitHub repository.
Available: https://github.com/gersl/fmask

[3] S. Greenland, et al, "Design & Prototyping of a Miniaturised Forwards Looking Imager using Deep Learning for Responsive Onboard Operations", *European Space Agency Symposium on Small Satellite Systems and Services*, Sorrento, Italy, May 2018.
Available: https://strathprints.strath.ac.uk/63922/ (pre-print)

[4] S. Greenland, et al, "System Framework for Autonomous Data Processing Onboard Next Generation of Nanosatellite", *15th Reinventing Space Conference*, Glasgow, UK, October 2017.
Available: https://strathprints.strath.ac.uk/62228/ (pre-print)

[5] PYNQ, "Machine Learning on Xilinx FPGAs with FINN" webpage.
Available: http://www.pynq.io/ml

[6] K. Simonyan and A. Zisserman, "Very Deep Convolutional Networks for Large-Scale Image Recognition", September 2014.
Available: https://arxiv.org/abs/1409.1556v6 (arXiv pre-print)

[7] D. R. Thompson, *TextureCam* GitHub repository.
Available: https://github.com/davidraythompson/texturecam

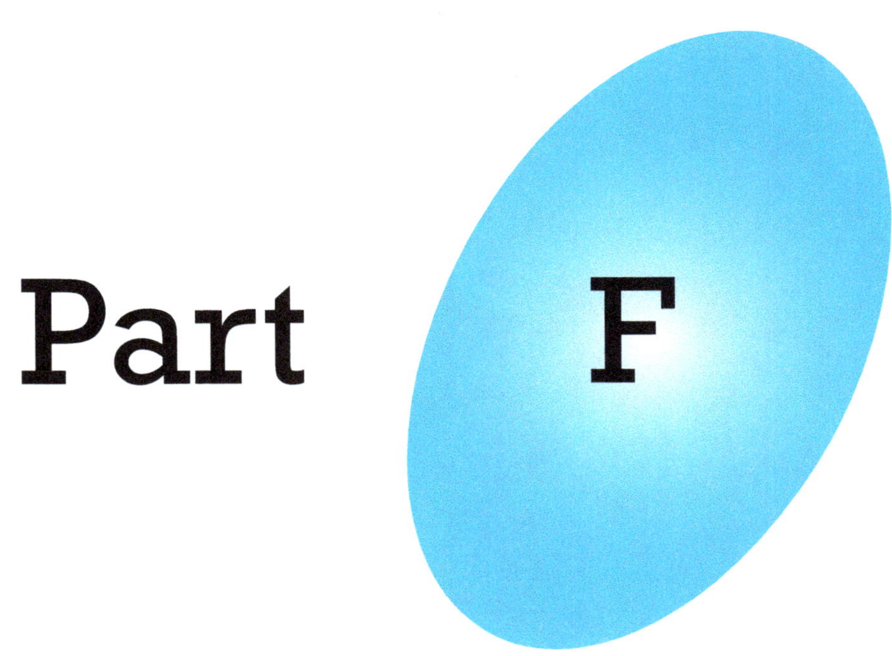

Part F

And Finally…

Chapter 24

Academic Case Studies

The variety of processing architectures and interfaces available on the Zynq MPSoC allow it to be the processing power behind many embedded applications. However, it is not always clear how applications can be partitioned across the device and which processing elements are most suitable for performing a task. In light of this, and given the relative maturity of the Zynq SoC device compared to the Zynq MPSoC, we will explore a number of academic case studies in this chapter associated with both devices. The intention of this chapter is to provide inspiration and practical insights into the selection of applications for implementation on a Zynq device, and how best to leverage the available processing architectures.

Since its release in 2011, the Zynq SoC has been employed in a myriad of disparate application areas. It is impractical to attempt to discuss all of these areas here, so a few case studies from prominent application areas have been selected. It is hoped that the general principles examined in these case studies for targeting applications to a Zynq device, may provide useful insight into the considerations necessary when designing your own Zynq-based applications and systems.

24.1 Embedded Computer Vision

Computer vision applications and algorithms have been popular with the Zynq SoC. This has predominantly been due to the low power consumption of the device coupled with the availability of the FPGA fabric, providing the opportunity to accelerate portions of the algorithm.

Sample based operations that are applied to individual pixel values, such as convolutional filters, can be well suited to implementation in the PL. These operations are suitable due to the parallel processing capabilities of FPGAs and their ability to support deep processing pipelines. In contrast, computationally complex operations, such as classification, are inherently sequential in nature. Therefore, these operations are more suited to implementation in the PS with its higher clock rate.

Case 1: Hardware Acceleration of Feature Detection and Description Algorithms on Low-Power Embedded Platforms [1]

Ulusel and colleagues [1] used a Zynq 7020 device to implement image feature detection and description algorithms that are fundamental to many computer vision applications. The performance of this implementation on the Zynq 7020 was compared to implementations on an embedded CPU-based platform and an embedded GPU-accelerated platform. The platform used for both the CPU and GPU-accelerated implementations was an NVIDIA Jetson TK1 development kit, sporting an NVIDIA Tegra K1 SoC that comprises of an Arm Cortex-A15 quad-core CPU and an NVIDIA Kepler GPU on a single die.

The CPU and GPU-accelerated platforms used OpenCV library implementations of feature detection and description algorithms [2]. The Zynq implementation, meanwhile, used the PL to perform the feature detection and description. Moreover, these two operations were pipelined in the FPGA fabric. The upstream operations determined when downstream hardware was switched off (when not required) to reduce power consumption.

For feature detection alone, the Zynq implementation had superior performance to both the CPU and GPU-accelerated implementations in respect of execution time and energy consumption. Compared to the CPU implementation, the Zynq system executed nearly four times faster and consumed less than a tenth of the energy [1]. The GPU-accelerated implementation compared a little better than the CPU implementation, with an execution time around twice as long as the Zynq and consuming just under four times the amount of energy [1].

When feature description and detection was included, the relative runtime performance and energy consumption of the Zynq system further improved compared to both the CPU and GPU-accelerated implementations. Depending upon the feature description algorithm used, the execution time for the Zynq implementation was around 15 to 16 times lower than for the CPU implementation, and used approximately 43 times less energy [1]. Again, GPU-acceleration compared better than the CPU. The execution time for GPU-acceleration was only between two and a half to four times greater than the Zynq implementation and consumed between seven and ten times as much energy [1].

The Zynq system derived its advantage over the others primarily from pipelining the algorithm in the PL. Memory access could be reduced using Zynq by propagating image data through the PL to each stage of the algorithm, rather than retrieving it from memory each time [1]. Furthermore, multiple stages of the algorithm could be executed in parallel by using different resources within the PL to implement the different stages and by propagating the data between stages in the PL [1].

It may seem strange to suggest that the Zynq derived its performance advantage over the GPU-accelerated implementation through greater parallelism, since GPUs are well known for their highly parallel architecture. However, the parallelism provided by a GPU tends to be in performing the same instruction with multiple inputs in parallel. The application investigated in this work derived greater benefit from multiple instructions being performed in parallel with multiple inputs [1], which is better suited to the FPGA contained in the PL of the Zynq.

Case 2: Real-Time Implementation of Contextual Image Processing Operations for 4K Video Stream in Zynq UltraScale+ MPSoC [11]

Kowalczyk and colleagues demonstrate the Zynq MPSoC's ability to apply contextual image processing algorithms to 4K video data, in particular 3840 x 2160 pixel resolution at 60 frames per second (fps). Video resolutions such as High Definition (HD) and Full High Definition (FHD), which are 1280 x 720 pixels and 1920 x 1080 pixels respectively, use serial-pipeline architectures. These designs typically process one pixel per clock cycle, as shown for a local filter in Figure 24.1. Assuming the above video resolutions operate at 60 fps, the pixel clock is 74.25 MHz for HD video and 148.5 MHz for FHD video. Video processing architectures can easily meet timing constraints for these clock frequencies on most FPGA devices.

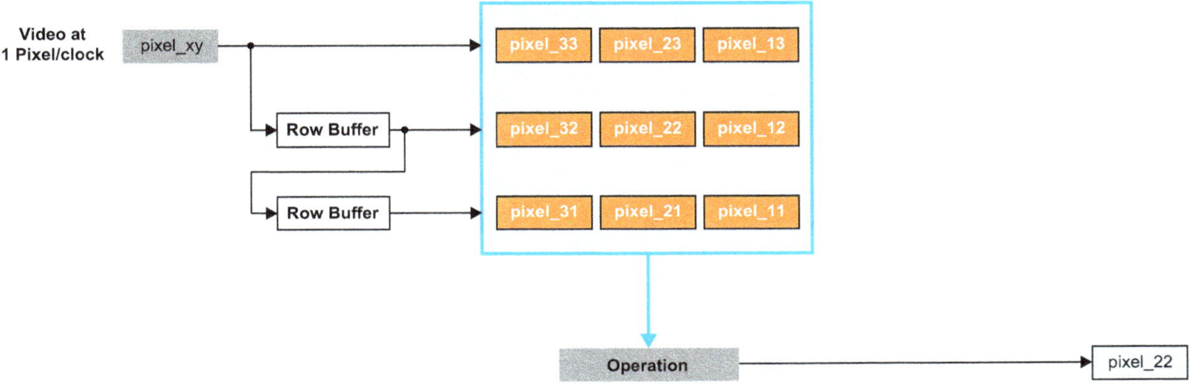

Figure 24.1: A serial-pipeline architecture for a 3x3 local filter window that requires one pixel per clock

4K resolutions, also known as Ultra High Definition (UHD), approximately require a 600 MHz pixel clock to achieve a frame rate of 60 fps. It is difficult to design an architecture that can meet timing constraints of this magnitude, even in modern FPGA devices. To reduce the pixel clock and maintain or improve the overall data rate of UHD video streams, 2 or 4 pixels per clock (ppc) is typically required. This allows architectures to be designed with higher parallelism and use a lower pixel clock. These are known as parallel-pipeline architectures and are an emerging and interesting area of research for 4K video processing on FPGAs. An example of a parallel-pipelined local filter using 2 ppc is illustrated in Figure 24.2 on page 570.

Kowalczyk and colleagues address the problem of designing an FPGA architecture for applying image operations to UHD resolutions at 60 fps. This work details and discusses video processing architectures for an averaging filter, Gaussian filter, Sobel and Canny edge detection operators, median filter and morphological erosion and dilation [13]. The ZCU102 development board [15], host to the XCZU9EG-2FFVB1156 Zynq MPSoC device, was selected to implement the contextual based image processing architectures for 4K video. This development board is suitable as it has two High Definition Multimedia Interface (HDMI) ports connected to the PL of the Zynq MPSoC device. The HDMI ports allow 4K video data to propagate to the hardware accelerators implemented in the FPGA logic fabric and then output the video to a monitor.

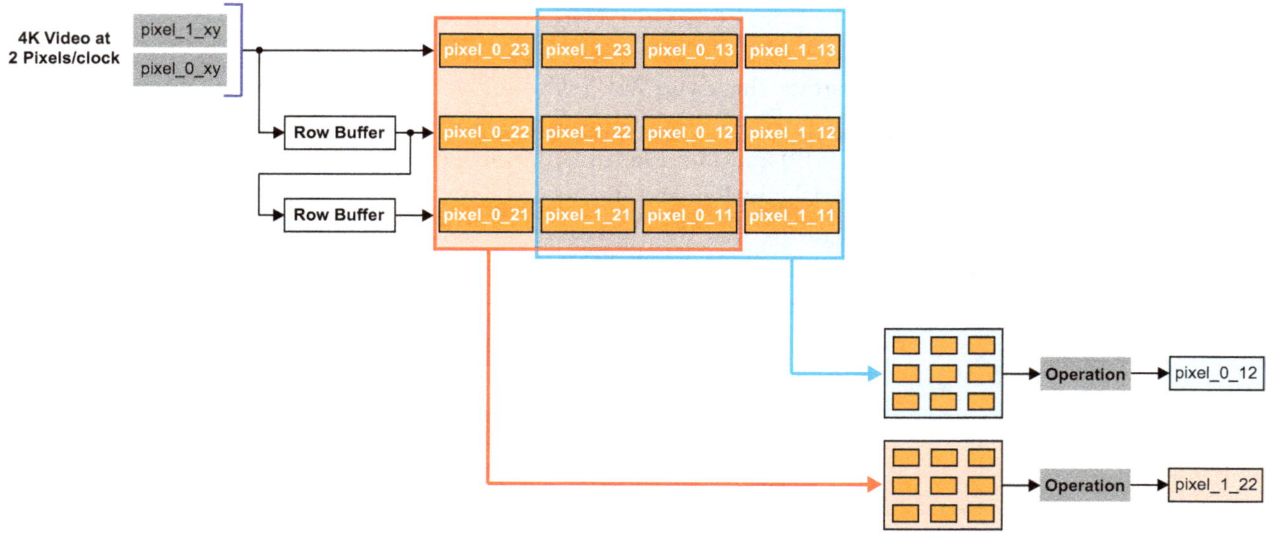

Figure 24.2: A parallel-pipeline architecture for a 3x3 local filter window that requires two pixels per clock [11]

4K video processing architectures were developed using Verilog Hardware Description Language (HDL) and Vivado HLS [18]. Kowalczyk and colleagues implemented each of the aforementioned image processing operations using Verilog HDL resulting in efficient FPGA resource consumption. The Vivado HLS implementation required a custom HLS function to implement the 4 ppc interface. This was achieved using a four channel interface with a width of 8-bits to read and write data. Upon implementing Sobel and median filtering, the Sobel architecture consumed more FPGA resources than the corresponding Verilog HDL implementation and the median filter consumed less.

The authors used the Vivado xPower [12] tool to approximate the power consumption of the Zynq MPSoC when using 2 ppc and 4 ppc architectures. This investigation was performed on the Sobel and median filter modules. The 2 ppc Sobel architecture was estimated to consume 1.269 W for the PL alone. The equivalent 4 ppc architecture was estimated to consume 1.211 W. The 2 ppc median architecture consumed 0.641 W and the 4 ppc architecture consumed 0.535 W.

The difference in power consumption between 2 ppc and 4 ppc architectures arise from the FPGA resource consumption and the speed of the pixel clock. In general, a higher number of pixels per clock will achieve a lower pixel clock required by the associated architecture. The resource consumption also increases as the FPGA architecture is operating on more pixels in parallel. Conversely, decreasing the pixels per clock will result in a higher pixel clock and reduced resource consumption. All of these factors impact the power consumption directly. It can be seen that using 4 ppc for the Sobel module decreases the dynamic power consumption by approximately 4.79% in comparison to the 2 ppc equivalent. Similarly, using 4 ppc for the median module decreases the dynamic power consumption by approximately 16.54% in comparison to the 2 ppc architecture.

In conclusion, the dynamic power consumption required by each architecture could be reduced using parallel-pipelined video processing architectures. Although more FPGA resources were used to implement 4 ppc video processing architectures, the slower pixel clock rate resulted in lower dynamic power consumption. The ZCU102 development board and Vivado tools proved to be very useful in applying and analysing 4K video architectures.

24.1.1 Advanced Driver Assistance Systems

An application area of embedded computer vision algorithms that has seen particularly strong growth in recent years is ADAS. Such has been the interest and popularity of the Zynq in this area that it warrants specific mention within this chapter.

Improvements in the camera technology used in vision-based ADAS has made it increasingly difficult to process the acquired image data in real-time using more traditional hardware architectures, such as microprocessors [3]–[5]. This has fuelled interest in developing hardware that is designed specifically to perform real-time computer vision algorithms for ADAS. Furthermore, combined with the low power and small size requirements of ADAS applications [3]–[5], both FPGA-based and application specific integrated circuit (ASIC) solutions have been proposed.

The competitiveness of the automotive industry gives advantages to FPGA-based systems compared to their ASIC counterparts, as they tend to have much lower upfront costs, a shorter time to market and can be easily updated over their lifetime [5], [6]. The combination of the tightly integrated PS and PL provides further advantage to Zynq-based systems. The computer vision component of ADAS does not tend to exist in isolation from other systems within the vehicle. To make an effective system, the computer vision component must integrate with the other system components. While this is possible to achieve with the PL alone, it is a task that can often be performed much more simply using a microprocessor such as those available in the PS of Zynq devices [5].

Case 3: Traffic Light and Sign Recognition for Advanced Driver Assistance [3], [4]

Borrmann and colleagues [3], [4] used a Zynq 7020 device to design a traffic light and sign recognition system for use in ADAS. They found that it was not possible to meet the real-time processing requirements of their system using only software on an embedded processor that also met their size, power and cost constraints. Therefore, it was necessary to implement hardware accelerators for their application [3], [4]. A Zynq device was chosen as it was a single chip solution that offered an embedded CPU, tightly integrated with FPGA fabric for implementing hardware accelerators. Furthermore, it also provides CAN, Gigabit Ethernet and DRAM controllers for integrating their design with other systems [3]. The Zynq 7020 device was specifically selected based on cost considerations [3].

The traffic light and traffic sign recognition algorithms were implemented using a similar processing pipeline. This incorporated a multi-scale segmentation stage followed by feature extraction and classification stages and a final object tracking stage. The output of the object tracking stage was also fed back to the segmentation stage of the next iteration of the algorithm to improve its robustness [3], [4].

In their implementation, one core of the Arm Cortex-A9 CPU was used to implement traffic light recognition, while the other core implemented traffic sign recognition [4]. For traffic light recognition, hardware accelerators were used to perform the multi-scale segmentation and the feature vector comparison during the classification stage [3], [4]. For traffic sign recognition, hardware accelerators were used only for the feature vector comparison in the classification stage [4]. Four instances of the hardware accelerator used for feature vector comparison were implemented to improve runtime performance, and a strategy controller was implemented in the PL using a Xilinx MicroBlaze [10] soft-core CPU. The MicroBlaze processor coordinates access to the shared hardware accelerator instances, and handles synchronisation between hardware and software [4].

When either core of the PS CPU completes execution of a stage in its algorithm, it configures an interrupt request memory structure in On-Chip Memory (OCM) with status flags and parameters representing the algorithm state, and asserts an interrupt to the MicroBlaze strategy controller in the PL. The strategy controller accesses the interrupt request memory structure in OCM over a General-Purpose AXI (GP-AXI) interface between the PS and PL. It uses a dedicated timer implemented in the PL to measure the execution time for that stage of the algorithm and for the algorithm as a whole [4]. The measured execution time was used by the strategy controller to arbitrate requests for access to the hardware accelerators in the PL that are shared between both the traffic light and sign recognition algorithms.

The hardware accelerators are configured and controlled by the strategy controller using the GP-AXI bus, while image data is transferred between the accelerators and DDR memory using a high-performance AXI (HP-AXI) interface [4]. Since the OCM is heavily used for synchronisation between the algorithms executing in the PS and the hardware in PL, it is marked as non-cacheable [4]. This was done to maintain the coherency of the memory structures that are accessed by both the strategy controller and the PS CPU cores.

In addition to the parts of the system used to execute the traffic light and sign recognition algorithms, Borrmann and colleagues also implemented components to interface their system with external devices. This included an interface in the PL to connect the type of cameras commonly used in automotive applications. As an alternative source of input image data, the system was also designed to receive pre-recorded video data over an Ethernet connection. The Gigabit Ethernet controller in the PS of the Zynq was used for this purpose, and was controlled by one core of the PS CPU [4].

To demonstrate the functionality of the system, an HDMI interface was implemented in the PL. This fetched video data from DDR memory using an HP-AXI bus, and drove an HDMI controller external to the Zynq device [4]. The final additional component implemented in their system was a general purpose input/output (GPIO) interface used to drive LEDs external to the Zynq device and controlled using the GP-AXI bus [4]. A simplified architecture of the system is shown in Figure 24.3.

Borrmann and colleagues demonstrated that their system was able to meet the real-time processing requirements for automotive applications that they were unable to meet using software alone [3], [4]. They showed this to be the case when using either pre-recorded video or video from an in-vehicle camera. Moreover, they established that their system was able to meet the real-time processing requirements while also conforming to power usage constraints (consuming 5.07–5.14W depending on the input video source and content [4]).

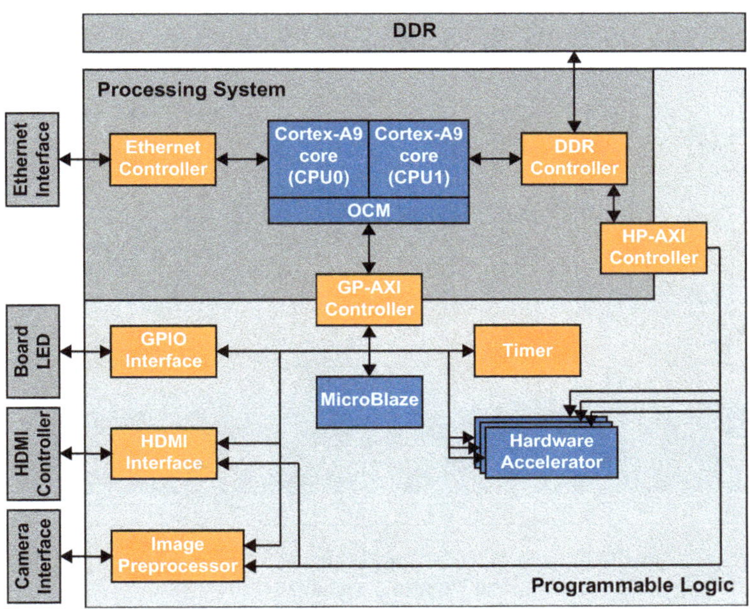

Figure 24.3: Simplified architecture of traffic light and sign recognition system implemented by Borrmann and colleagues [4]

24.2 Spaceflight Systems

Traditionally, spaceflight systems require radiation hardened components and processors to withstand the harsh environment of space and the influence of single-event effects (SEEs). SEEs affect computing equipment due to an energetic particle influencing the operation of electronic components. There are a small selection of electronic components today that are radiation hardened and capable of sustained operation on satellites and spacecraft. Due to high-demand for improved computing performance in space applications, off-the-shelf components such as the Zynq MPSoC are investigated for there energy-efficiency, cost, and performance.

Many off-the-shelf components are susceptible to the harsh environments of space and will be affected by SEEs. The problem to be addressed is the design and development of reliable computing systems for space applications using off-the-shelf components, such as the Zynq MPSoC.

Case 4: Towards Resilient Spaceflight Systems with Virtualization [14]

Sabogal and George propose the Virtualized Space Applications (ViSA) software framework to withstand the influence of SEEs. The framework uses the Xen hypervisor [17] for virtualisation support. The research team develop and test their system on the ZCU102 development board [15] and UltraZed-EG System on Module (SOM) [16]. The ViSA framework exploits the virtualisation capabilities of the Zynq MPSoC's Cortex-A53 application processor to create a voting system in Userspace. This is illustrated in the software system shown in Figure 24.4.

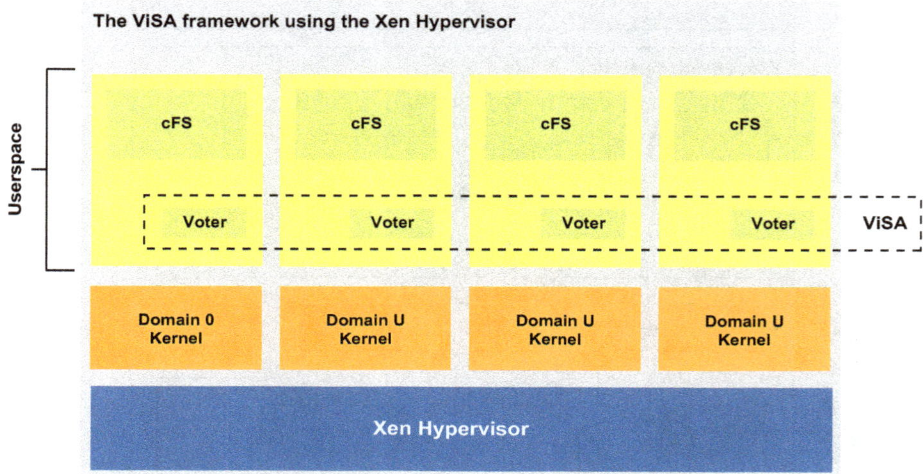

Figure 24.4: A reduced diagram of the ViSA software framework using the Xen hypervisor (a full diagram can be found in the author's original work [14])

In the Userspace of each Virtual CPU (VCPU) is an instance of the NASA core Flight Software (cFS) consisting of a series of scripts, utilities, and applications for flight tasks. An instance of the ViSA software is also included. ViSA is an intuitive framework that deploys mechanisms for redundancy throughout the software system (and much more [14]). A request is broadcast to each Userspace cFS instance and the result of this request (a vote) is returned to the ViSA instance local to that Userspace. The vote across all Userspace cFS instances determine the result of the requested task. The virtual voting system creates a redundant network for performing tasks and processing applications on a single-chip.

The primary aim of Sabogal and George was to investigate a fault tolerant technique for space applications. The research team carried-out several experiments to test the ViSA framework as detailed in [14]. Their findings concluded that ViSA was capable of masking errors and determining unreliable results caused by corruption in data during system operation.

24.3 Machine Learning

Machine learning is an area that has seen a resurgence in popularity in recent years. This interest has extended to implementing machine learning algorithms on embedded platforms to create intelligent systems. It is in these applications that the Zynq has shown itself to be well suited.

Machine learning algorithms are generally computationally intensive. Embedded machine learning platforms need to be capable of performing these computations efficiently to meet the processing and power consumption requirements of the system. The PL of the Zynq lends itself well to this purpose, where hardware customised to the algorithm can provide computational performance per Watt that cannot be achieved on other general purpose processing architectures.

Case 5: A Dynamically Reconfigurable High Density Myoelectric Prosthesis Controller [7]

Boschmann and colleagues implemented a system to provide autonomous control of a multi-movement prosthetic hand, based on electrical signals generated by muscle movements in an amputee's residual limb [7]. They used machine learning in the form of a support vector machine (SVM) to classify the detected signals into movements of the prosthetic using a Zynq 7020 device [7].

The electrical signals generated by muscle movements were detected using an array of 192 high density electromyographic (HD EMG) sensors, arranged in 8 rows and 24 columns [7]. For this application, the HD EMG signals need to be processed in real-time to produce control signals for the prosthetic. These control signals should be produced less than 125ms after the corresponding user generated input signals [7]. An embedded solution was required, as the short latency prevented the use of an external device to which data could be transmitted, processed and control signals transmitted back to the prosthetic [7]. The group found that the embedded microprocessors currently used in prosthetics were unable to meet the real-time processing requirements of this application. Therefore, a Zynq device was chosen to accelerate computationally intensive parts of the control algorithm using the PL [7]. Furthermore, the system was ultimately intended to be installed within the prosthetic and powered by batteries. The modest power consumption of FPGA devices and the small physical size of the Zynq device make it suitable for this application [7].

In their work presented in [7], Boschmann and colleagues used a separate system to acquire the HD EMG signals. These signals were transmitted to the Zynq device using a TCP/IP connection. The control signals for the prosthetic were generated on the Zynq device and transferred to another system for transmission to the prosthetic using a *Bluetooth*® device[7].

The system implemented on the Zynq was composed of a number of processing stages. The main part of the system had data acquisition, feature extraction and classification stages [7]. The data acquisition stage received the HD EMG data from the TCP/IP interface and performed some preprocessing for the feature extraction stage. After features had been extracted from the data, they were input to the classification stage, which used a linear SVM to classify the input features into one of six hand movement classes. These classes determine the appropriate output from the controller [7]. The data acquisition stage was implemented using software running on the PS. The feature extraction and classification stages were implemented as hardware using the PL, as they were the most computationally expensive parts of the algorithm [7].

Algorithms were also implemented to train the SVM used for classification. These algorithms were implemented in software, as the SVM training was deemed to be less time critical [7].

The system developed by Boschmann and colleagues used a free and open source runtime environment called ReconOS [7]–[9] to assist with communication and synchronisation between portions of the system in hardware and software. This runtime environment extends the standard software multi-threading paradigm to include threads that execute in hardware as well. The programming model for ReconOS partitions the application into separate threads, each of which can either be implemented in software or hardware. A standardised interface is used for each thread thus making the thread's implementation, either in software or hardware, transparent to the rest of the system [8]. Extending this concept a little further, the ReconOS runtime environment can facilitate inter-thread communication and synchronisation using the standard multithreading features of the host OS, regardless of whether the threads are implemented in software or hardware [7], [8]. A major advantage of this approach is the ease in which threads can be moved between hardware and software to explore the range of available system configurations [8].

In this case, Boschmann and colleagues used the ReconOS runtime environment on top of a Linux OS on the PS's APU [7]. Each hardware thread in an application, using the ReconOS runtime environment, has a similar structure that enables it to initiate and respond to system calls to and from the host OS kernel. Figure 24.5 shows this structure.

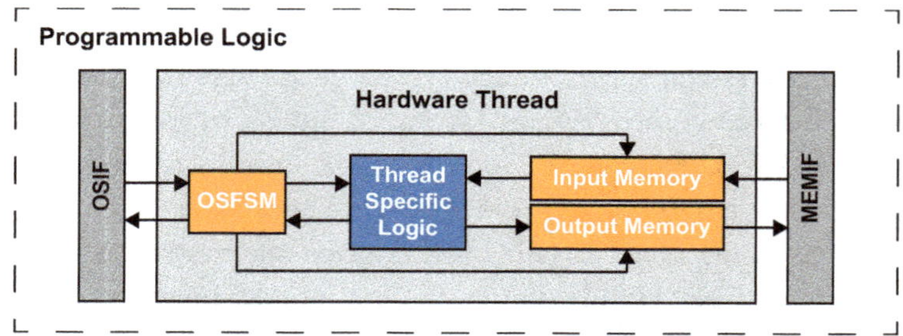

Figure 24.5: Structure of a hardware thread in the ReconOS runtime environment [7], [8]

A key component in the operation of a hardware thread is the OS synchronisation finite state machine (OSFSM). The OSFSM controls the operation of the hardware thread by initiating system calls to the OS kernel and controlling the flow of data to and from the hardware thread [8]. In order to synchronise with other threads, the OSFSM initiates and responds to system calls to and from the OS kernel via the OS interface (OSIF) and a lightweight software thread running on the PS's APU called a delegate thread. A delegate thread is created by the ReconOS runtime environment for each hardware thread. The delegate thread makes system calls to the OS kernel on behalf of the hardware thread when requested by the OSFSM, and communicates responses back to the OSFSM, via the OSIF [8]. Data consumed and produced by the hardware thread is buffered locally using Block RAM resources, as shown in Figure 24.5. The flow of data between these local buffers and system memory is also controlled by the OSFSM to enforce the coherency of system memory shared between threads. Data flow to and from each hardware thread is performed via the memory interface (MEMIF). The MEMIF provides access to system memory through the ReconOS memory subsystem located in the PL, as shown in Figure 24.6. The ReconOS memory subsystem serves to arbitrate system memory requests from hardware threads and supports host OS virtual memory addressing through the provision of an MMU [7], [8].

Boschmann and colleagues implemented the feature extraction and SVM classification operations as separate hardware threads in the PL of the Zynq device [7], as shown in Figure 24.6. As we are chiefly concerned with machine learning in this section, we will focus our discussion on the implementation of the SVM classifier. The specific logic in the SVM classifier hardware thread performed a matrix multiplication between the input feature vector extracted from the HD EMG signals and the SVM training matrix [7]. The input feature vector was buffered locally to the hardware thread as single precision floating point values using

Block RAM resources, as was the vector resulting from the matrix multiplication. There were insufficient Block RAM resources available to store the entire training matrix locally to the hardware thread. Only the row of the matrix being used for the multiplication at any one time was held in local memory. Training matrix values were stored as 16-bit fixed point values with two values stored per Block RAM word [7]. The data types used to represent the input feature vector, output vector and training matrix were chosen to maintain the most accurate classification results, whilst also reducing the PL resource usage and calculation time [7].

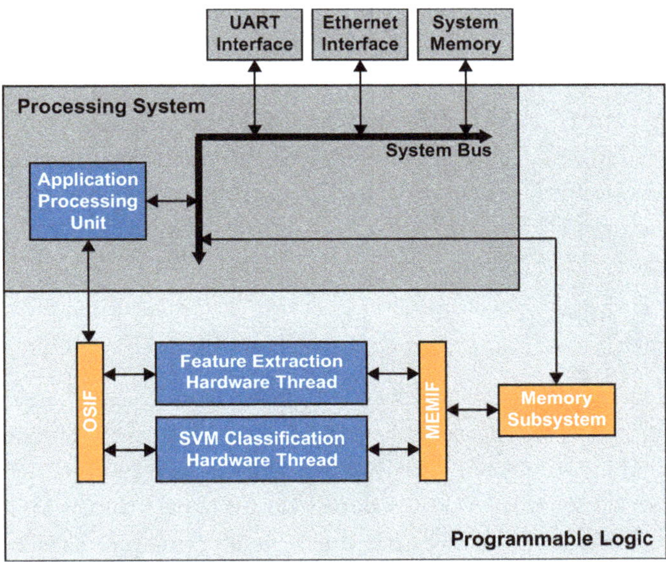

Figure 24.6: Simplified architecture of myoelectric prosthesis controller implemented by Boschmann and colleagues [7]

The system designed by Boschmann and colleagues employed an innovative approach, to deal with classifying noisy HD EMG signals, whilst also minimising system latency. Accurate classification of HD EMG signals in the presence of noise is a major challenge [7]. The robustness of noisy signal classification achieved with the SVM could be improved by using more complex features. However, this increased the time required to compute the classification. It also increased the PL resources used by the classification hardware thread to accommodate the larger feature vectors and training matrix [7]. The solution proposed by Boschmann and colleagues was to dynamically reconfigure the PL with hardware threads that used the most appropriate feature complexity, based on the noise detected in the HD EMG signals. A software thread was executed in the PS periodically to estimate how much noise was present on the input HD EMG signals. This estimate was used to index a look-up table to find the most appropriate feature complexity to use, based on balancing classification accuracy and processing time [7].

Five pre-compiled bitstreams were available to configure the PL, each with hardware thread implementations using different feature vector complexities. The most appropriate bitstream, as determined by estimating the noise on the HD EMG signals, was used to reconfigure the PL, if it had not already configured with this bitstream [7]. The dynamic reconfiguration scheme introduced a performance overhead. The software thread that was used to estimate noise on the HD EMG signals took around a second to execute. The PL reconfiguration took around half a second [7]. However, on their test data, which had varying degrees of noise, the system using dynamic reconfiguration was able to achieve similar classification accuracies and provide a lower average controller delay compared to a static system using the highest feature vector complexity [7].

Compared to a purely software version of their system executed on the PS of the Zynq device, the hardware accelerated system proposed by Boschmann and colleagues was able to perform signal classification in around half the time [7]. This performance improvement was crucial for the real-time requirements of this application, i.e. a delay of no more than 125ms between the user generating the input signal and the system producing the appropriate control signal. The software-only implementation was unable to meet these requirements, with a controller delay of around 135–165ms, depending on the complexity of the feature vectors. The hardware accelerated implementation had a controller delay of less than 120ms regardless of the complexity of the feature vectors [7].

24.4 Chapter Summary

This chapter has provided a flavour of some of the areas in which the Zynq-7000 SoC device has been used. We have specifically discussed applications in embedded computer vision and machine learning. Details of designs created for these areas have been given with particular reference to the partitioning of the system between the PS and PL, and the schemes used to coordinate these parts of the system. While this chapter has focused on the Zynq-7000 device, we believe that the general principles presented can also be applied to designs targeting the Zynq MPSoC device.

24.5 References

Note: All online sources last accessed March 2019.

Some of the Xilinx documentation referred to below has version-specific URLs. If you are working in a newer version of the tools, check for updates on the Xilinx website, or try adjusting the link according to your version.

[1] O. Ulusel, C. Picardo, C.B. Harris, S. Reda and R.I. Bahar, "Hardware acceleration of feature detection and description algorithms on low-power embedded platforms," *Proceedings of the 26th International Conference on Field Programmable Logic and Applications (FPL)*, Lausanne, Switzerland, September 2016, pp. 1–9.
DOI: 10.1109/FPL.2016.7577310

[2] OpenCV Team, "OpenCV library" webpage
Available: http://opencv.org/

[3] J.M. Borrmann, F. Haxel, D. Nienhüser, A. Viehl, J.M. Zöllner, O. Bringmann and W. Rosenstiel, "STELLaR — a Case-Study on SysTEmaticaLLy Embedding a Traffic Light Recognition," *Proceedings of the 17th IEEE International Conference on Intelligent Transportation Systems (ITSC)*, Qingdao, China, October 2014, pp. 1258–1265.

DOI: 10.1109/ITSC.2014.6957860

[4] J.M. Borrmann, F. Haxel, A. Viehl, O. Bringmann and W. Rosenstiel, "Safe and Efficient Runtime Resource Management in Heterogeneous Systems for Automated Driving," *Proceedings of the 18th IEEE International Conference on Intelligent Transportation Systems (ITSC)*, Las Palmas, Spain, September 2015, pp. 353–360.

DOI: 10.1109/ITSC.2015.67

[5] G. Velez, A. Cortés, M. Nieto, I. Vélez and O. Otaegui, "A reconfigurable embedded vision system for advanced driver assistance," *Journal of Real-Time Image Processing*, Volume 10, Issue 4, December 2015, pp. 725–739.

DOI: 10.1007/s11554-014-0412-3

[6] S.M. Trimberger, "Three Ages of FPGAs: A Retrospective on the First Thirty Years of FPGA Technology," *Proceedings of the IEEE*, Volume 103, Issue 3, March 2015, pp. 315–331.

DOI: 10.1109/JPROC.2015.2392104

[7] A. Boschmann, G. Thombansen, L. Witschen, A. Wiens and M. Platzner, "A Zynq-based dynamically reconfigurable high density myoelectric prosthesis controller," *Proceedings of Design, Automation and Test in Europe Conference and Exhibition (DATE)*, Lausanne, Switzerland, March 2017, pp. 1002–1007.

DOI: 10.23919/DATE.2017.7927137

[8] A. Agne, M. Happe, E. Lübbers, B. Plattner, M. Platzner and C. Plessl, "ReconOS: An Operating System Approach for Reconfigurable Computing," *IEEE Micro*, Volume 34, Issue 1, January-February 2014, pp. 60–71.

DOI: 10.1109/MM.2013.110

[9] University of Paderborn, "ReconOS — Operating System for Reconfigurable Computing" webpage.

Available: http://www.reconos.de/

[10] ilinx, Inc., "MicroBlaze Processor Reference Guide", UG984, (v.2016.2), June 2016.

Available: http://www.xilinx.com/support/documentation/sw_manuals/xilinx2016_2/ug984-vivado-microblaze-ref.pdf

[11] M. Kowalczyk, D. Przewlocka and T. Krvjak, "Real-Time Implementation of Contextual Image Processing Operations for 4K Video Stream in Zynq UltraScale+ MPSoC," *2018 Conference on Design and Architectures for Signal and Image Processing (DASIP)*, Porto, Portugal, 2018, pp. 37-42.

DOI: 10.1109/DASIP.2018.8597105

[12] Xilinx, Inc, "Xilinx Power Estimator User Guide", UG440, (v2018.3), December 2018.

Available: https://www.xilinx.com/support/documentation/sw_manuals/xilinx2018_3/ug440-xilinx-power-estimator.pdf

[13] R. Gonzalez R. Woods, *Digital Image Processing*, Pearson Education 2011.

[14] D. Sabogal and A. D. George, "Towards resilient spaceflight systems with virtualization," *2018 IEEE Aerospace Conference*, Big Sky, MT, 2018, pp. 1-8.

DOI: 10.1109/AERO.2018.8396689

[15] Xilinx, Inc., "Xilinx Zynq UltraScale+ MPSoC ZCU102 Evaluation Kit" webpage.

Available: https://www.xilinx.com/products/boards-and-kits/ek-u1-zcu102-g.html

[16] Avnet, "UltraZed System-on-Module" webpage

Available: http://www.zedboard.org/product/ultrazed

[17] DornerWorks Ltd., "Xen Hypervisor on Xilinx Zynq UltraScale+ MPSoC"
Available: https://dornerworks.com/xen/xilinxxen

[18] Xilinx Inc., "Vivado Design Suite User Guide", UG902 (v2017.1), Apr. 2017.
Available: https://www.xilinx.com/support/documentation/sw_manuals/xilinx2017_1/ug902-vivado-high-level-synthesis.pdf

Postscript

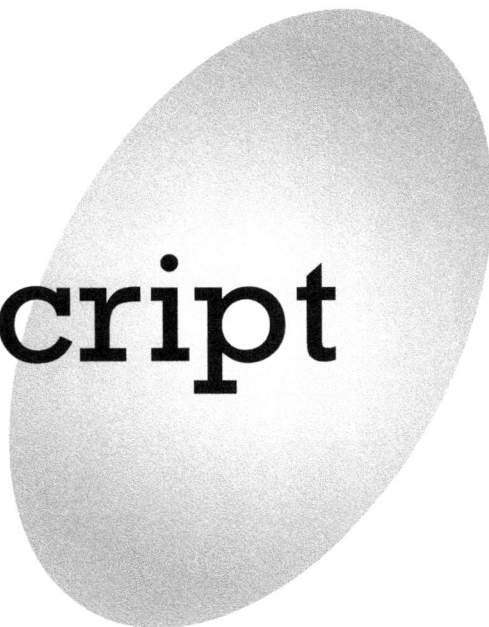

Appendix: BSD 3-Clause Licence

The licence shown below relates to the 'Hello World' example presented in Chapter 22.

List of Acronyms

123

5G	5th Generation
5G NR	5th Generation New Radio

A

ABI	Application Binary Interface
AC	Authentication Certificate
ACE	AXI Coherency Extension
ACP	Accelerator Coherency Port
ADALM	Analog Devices Active Learning Module
ADAS	Advanced Driver Assistance System
ADC	Analogue to Digital Converter
ADS	Automated Driving System
AES	Advanced Encryption Standard
AFI	AXI FIFO Interface
AGI	Artificial General Intelligence

AHB	Advanced High-Performance Bus
AI	Artificial Intelligence
AMBA	Advanced Microcontroller Bus Architecture
AMD	Advanced Micro Devices, Inc.
AMP	Asymmetric Multi Processing
AMS	Analog Mixed Signal
ANN	Artificial Neural Network
ANSI	American National Standards Institute
APB	Advanced Peripheral Bus
API	Application Programming Interface
APM	AXI Performance Monitor
APU	Application Processing Unit
ARC	Applied Reconfigurable Computing
ARM	Advanced RISC Machines
ASIC	Application-Specific Integrated Circuit
ASIL	Automotive Safety Integrity Level
ATB	Advanced Trace Bus
ATF	Arm Trusted Firmware
ATP	Authorized Training Providers
AVC	Advanced Video Coding
AWS	Amazon Web Service
AXI	Advanced eXtensible Interface

B

BBRAM	Battery-Backed RAM
BIF	Boot Image Format
BIST	Built-In Self-Test

BNN	Binary Neural Network
BOM	Bill Of Materials
BPD	Battery Power Domain
BPU	Battery Power Unit
BRAM	Block RAM
BSP	Board Support Package

C

CAN	Controller Area Network
CBRS	Citizen Band Radio Service
CCE	Categorical Cross Entropy
CCF	Common Causes of Failure
CCI	Cache Coherent Interconnect
CDT	C/C++ Development Tooling
CES	Consumer Electronics Show
CFI	Common Flash Memory Interface
CFS	Completely Fair Scheduler
CIB	Crypto Interface Block
CLB	Configurable Logic Block
CLI	Command Line Interface
CMA	Contiguous Memory Allocator
CMOS	Complementary Metal Oxide Semiconductor
CMT	Clock Management Tiles
CNN	Convolutional Neural Network
CNTK	CogNitive ToolKit
CPLD	Complex Programmable Logic Devices
CPRI	Common Packet Radio Interface

CPSR	Current Program State Register
CPU	Central Processing Unit
CRC	Cyclic Redundancy Check
CSU	Configuration Security Unit
CTI	Cross Trigger Interface
CTM	Cross Trigger Matrix
CV	Computer Vision

D

DAC	Digital to Analogue Converter
DAP	Debug Access Port
DCU	Data-Cache Unit
DDC	Digital Down-Converter
DDR	Double Data Rate
DDRC	DDR Controller
DF	DataFlow
DFU	Device Firmware Upgrade
DHCP	Dynamic Host Configuration Protocol
DIP	Dual In-line Package
DIS	Digital Image Stabilisation
DL	Deep Learning
DMA	Direct Memory Access
DMAC	DMA Controller
DNN	Deep Neural Network
DOI	Digital Object Identifier
DP	Display Port
DPA	Differential Power Analysis

DPD	Digital Pre-Distortion
DPU	Data-Processing Unit
DRAM	Dynamic Random Access Memory
DRC	Design Rule Check
DRE	Data Realignment Engine
DRM	Digital Rights Management
DSA	Device Support Archive
DSA	Dynamic Spectrum Access
DSP	Digital Signal Processing
DUC	Digital Up-Converter

E

ECC	Error Correction / Correcting Code *or* Error Checking and Correction
EDA	Electronic Design Automation
EL	Exception Level
ELR	Exception Link Registers
EM	ElectroMagnetic
EMEA	Europe, the Middle East and Africa
EMG	ElectroMyoGraphic
EMIO	Extended MIO
ENOB	Effective Number of Bits
ETF	Embedded Trace FIFO
ETR	Embedded Trace Router

F

FAT	File Allocation Table
FCC	Federal Communications Commission

FEC	Forward Error Correction
FET	Field Effect Transistor
FF	Flip-Flop
FFT	Fast Fourier Transform
FHD	Full High Definition
FIFO	First In First Out
FIQ	Fast Interrupt reQuest
FIR	Finite Impulse Response
FIT	Failures In Time
FLI	Forwards Looking Imager
FMC	FPGA Mezzanine Card (see FPGA)
FOSS	Free Open Source Software
FP	Full Power
FPD	Full Power Domain
FPGA	Field Programmable Gate Array
FPU	Floating Point Unit
FSBL	First Stage Boot Loader
FSM	Finite State Machine
FW	FirmWare

G

GB	GigaByte
GCC	GNU Compiler Collection
GCM	Galois/Counter Mode
GEM	Gigabit Ethernet MAC
GIC	Generic Interrupt Controller
GigE	GigaBit Ethernet

GNU	GNU's Not Unix (recursive acronym)
GP	General Purpose *or* Geometry Processor
GPI	General Purpose Input
GPIO	General Purpose Input Output
GPL	General Public License
GPO	General Purpose Output
GPP	General Purpose Processor
GPS	Global Positioning System
GPU	Graphics Processing Unit
GRU	Gated Recurrent Unit
GSM	Grams per Square Meter
GT	Gigabit Transceiver (inc. GTH, GTR, GTX, GTY variants)
GUI	Graphical User Interface

H

HD	High Density
HD EMG	High Density Electromyographic
HDL	Hardware Description Language
HDMI	High Definition Multimedia Interface
HDR	High Dynamic Range
HEVC	High-Efficiency Video Coding
HLS	High Level Synthesis
HP	High Performance
HPC	High Performance Coherency *or* High Performance Computing
HPM	High Performance Master
HR	High Range
HSI	Hardware Software Interface

HTML	HyperText Markup Language
HVC	HyperVisor Call
HW	HardWare
HWRoT	HardWare Root of Trust

I

I2C	Inter-Integrated Circuit
IC	Integrated Circuit
ID	IDentity
IDE	Integrated Development Environment
IDF	Isolation Design Flow
IEC	International Electrotechnical Commission
IEEE	Institute of Electrical and Electronics Engineers
IF	Intermediate Frequency
IFU	Instruction Fetch Unit
IIoT	Industrial Internet of Things
ILA	Integrated Logic Analyser
ILSVRC	ImageNet Large Scale Visual Recognition Challenge
IMU	Inertial Measurement Unit
I/O *or* IO	Input / Output
IOB	Input / Output Block
IOP	Input / Output Peripherals
IoT	Internet of Things
IOU	Input / Output Unit
IP	Intellectual Property
IPA	Intermediate Physical Address
IPC	Inter-Process Communication

I/Q	In-Phase / Quadrature
IPI	IP Integrator
IR	Intermediate Representation
IRQ	Interrupt ReQuest
ISM	Instrumentation, Scientific and Medical
ISO	International Organization for Standardization
ISR	Interrupt Service Routine
IV	Initialisation Vector

J

JPG	Joint Photographic Experts Group
JTAG	Joint Test Action Group

K

KB	KiloByte
KEK	Key-Encryption Key

L

LBIST	Logic Built-In Self-Test
LCM	Life Cycle Management
LDPC	Low-Density Parity Check
LDT	Linux Driver Template
LED	Light Emitting Diode
LFM	Latent Fault Metric
LiDAR	Light Detection And Ranging
LLPP	Low Latency Peripheral Port
LP	Low Power
LPD	Low Power Domain

LPS	Low Power Subsystem
LSTM	Long Short-Term Memory
LSU	Load Store Unit
LSVRC	Large Scale Visual Recognition Challenge
LTE	Long Term Evolution
LUT	LookUp Table

M

MAC	Media Access Control
MB	MegaByte
MBISR	Memory Built-In Self Repair
MBIST	Memory Built-In Self Test
MDM	MicroBlaze Debug Module
MEC	Mobile Edge Computing
MEMIF	Memory Interface
MIC	Message Integrity Code
MIG	Memory Interface Generator
MIMO	Multiple Input Multiple Output
MIO	Multiplexed Input/Output
MIT	Massachusetts Institute of Technology
ML	Machine Learning
MLP	Multilayer Perceptron
MMU	Memory Management Unit
MNIST	Modified National Institute of Standards and Technology
MO	Multilayer Offload
MP	Multi-Processor *or* Multi-Processing
MPE	Media Processing Engine

MPSoC	Multi-Processor System on Chip
MPU	Memory Protection Unit
MSE	Mean Squared Error
MUX	Multiplexer

N

NCO	Numerically Controlled Oscillator
NIC	Network InterConnect
NIST	National Institute of Standards and Technology
NLP	Natural Language Processing
NLU	Natural Language Understanding
NN	Neural Network
NRE	Non Recurring Engineering
NSD	Noise Spectral Density

O

OCM	On-Chip Memory
OEM	Original Equipment Manufacturers
Ofcom	Office of Communications
ONFI	Open NAND Flash Interface
Open AI	Open Artificial Intelligence
OpenAMP	Open Asymmetric MultiProcessing
OpenCL	Open Computing Language
OpenCV	Open Computer Vision
OpenGL	Open Graphics Library
OpenHW	Open HardWare
OpenRTOS	Open Real-Time Operating System

OpenVG	Open Vector Graphics
OS	Operating System
OSFSM	Operating System Synchronisation Finite State Machine
OSIF	Operating System Interface
OTG	On-The-Go

P

PBR	PMU Boot ROM
PA	Physical Address
PC	Program Counter (or Personal Computer!)
PCAP	Processor Configuration Access Port
PCB	Printed Circuit Board
PCIe	Peripheral Component Interconnect Express (alternatively known as PCI Express)
PD	Power Domain
PE	Processing Element
PFU	PreFetch Unit
PIL	Python Imaging Library
PIP	Python Package Installer
PIT	Programmable Interval Timers
PJTAG	Parallel JTAG
PL	Programmable Logic
PLB	Polygon List Builder
PLL	Phase-Locked Loop
PLPD	Programmable Logic Power Domain
PM	Power Management
Pmod	Peripheral Module
PMU	Platform Management Unit

PNG	Portable Network Graphics
PoR	Power-On Reset
POSIX	Portable Operating System Interface
PP	Pixel Processors
PPI	Private Peripheral Interrupts
PPK	Primary Public Key
PS	Processing System
PSIO	PS Input Output
PSJTAG	Processing System JTAG
PSK	Primary Secret Key
PSTATE	Processor State
PU	Processing Unit
PUF	Physical Unclonable Function
PV	ParaVirtualisation
PYNQ	Python Productivity for Zynq

Q

QEMU	Quick EMUlator
QM	Quality Managed
QNN	Quantized Neural Network
QoS	Quality of Service
QSPI	Queued Serial Peripheral Interface

R

RADAR	RAdio Detection And Ranging
RAM	Random Access Memory
RAN	Radio Access Network

RAW Reconfigurable Architectures Workshop

ReLU Rectified Linear Unit

RF Radio Frequency

RF-ADC Radio Frequency Analogue to Digital Converter

RF-DAC Radio Frequency Digital to Analogue Converter

RFSoC Radio Frequency System on Chip

RFE Return From Exception

RGMII Reduced Gigabit Media Independent Interface

RISC Reduced Instruction Set Computer

RNN Reduced Neural Network

ROM Read-Only Memory

RPU Real-time Processing Unit

RSA Rivest–Shamir–Adleman

RTC Real-Time Clock

RTL Register-Transfer Level

RTOS Real-Time Operating System

Rx Receiver

S

SATA Serial Advanced Technology Attachment

SCA Side-Channel Analysis

SCSI Small Computer System Interface

SCU Snoop Control Unit

SD Secure Digital

SD-FEC Soft Decision Forward Error Correction

SDK Software Development Kit

SDR Software Defined Radio

SDSoC	Software-Defined System-on-Chip
SEC-DED	Single-Error Correction, Double-Error Detection
SEE	Single Event Effect
SEL	Single Event Latch-up
SEM	Soft Error Mitigation
SEU	Signle Event Upset
SFDR	Spurious Free Dynamic Range
SFP	Small Form-factor Pluggable
SGD	Stochastic Gradient Descent
SGI	Software Generated Interrupt
SGMII	Serial Gigabit Media Independent Interface
SHA	Secure Hash Algorithm
SIL	Safety Integrity Level
SIMD	Single Instruction, Multiple Data
SIO	Serial Input Output
SIOU	Serial Input Output Unit
SLCR	System-Level Configuration Register
SMA	SubMiniature version A
SMC	Secure Monitor Call
SMMU	System Memory Management Unit
SMP	Symmetric Multi Processing
SoC	System on Chip
SOM	System on Module
SP	Stack Pointer
SPB	Secure Processor Block
SPFM	Single Point Fault Metric

SPI	Serial Peripheral Interface
SPK	Secondary Public Key
SRAM	Static RAM
SRS	Save Return State
SRST	System Reset
SSBL	Second Stage Boot Loader
SSK	Secondary Secret Key
SSS	Secure Stream Switch
STL	Safety Test Library
SVHN	Street View House Numbers
SVM	Support Vector Machine
SWDT	System WatchDog Timer
SWU	Sliding Window Unit
SYSMON	Xilinx System Monitor

T

TAP	Test Access Port
TCB	Task Control Block
TCF	Target Communication Framework
Tcl	Tool Command Language
TCM	Tightly Coupled Memory
TCP	Transmission Control Protocol
TEE	Trusted Execution Environment
TIF	Tagged Image Format
TLB	Translation Look-aside Buffers
TLDP	The Linux Documentation Project
TMC	Trace Memory Controller

TMR	Triple Mode Redundancy
TPIU	Trace-Port Interface Unit
TPU	Tensor Processing Unit
TRD	Targeted Reference Design
TRM	Technical Reference Manual
TTC	Triple-Timer Counter
TV	Television
TVWS	TV White Space
Tx	Transmitter

U

UART	Universal Asynchronous Receiver/Transmitter
UAV	Unmanned Aerial Vehicle
UDP	User Datagram Protocol
UHD	Ultra High Definition
UHF	Ultra High Frequency
UI	Unique Identifier
UIO	Userspace I/O
URL	Uniform Resource Locator
USB	Universal Serial Bus
USRP	Universal Software Radio Peripheral

V

V2I	Vehicle to Infrastructure
V2V	Vehicle to Vehicle
VA	Virtual Address
VCPU	Virtual CPU

VCU	Video Codec Unit
VDMA	Video DMA
VFP	Vector Floating Point
VGG	Visual Geometry Group
VHDL	VHSIC Hardware Description Language (see VHSIC)
VHSIC	Very High Speed Integrated Circuit
VIO	Virtual Input/Output
ViSA	Virtualized Space Applications
VIV	Vivado Isolation Verifier
VLNV	Vendor-Library-Name-Version
VM	Virtual Machine
VPU	Video Processing Unit

W

WFE	Wait for Exception
WFI	Wait for Interrupt
WL	Write Lock

X

XDC	Xilinx Design Constraints
XML	Extensible Markup Language
XMPU	Xilinx Memory Protection Unit
XPPU	Xilinx Peripheral Protection Unit
XUP	Xilinx University Program

Y

YOLO	You Only Look Once

Z

ZynqMP	Zynq Multi-Processor

Index

D

E

N

O

P

Q

R

S

www.ingramcontent.com/pod-product-compliance
Lightning Source LLC
Chambersburg PA
CBHW051747200326
41597CB00025B/4476

9780992978754